SETON HALL UNIVERSITY
QL807 .L4
Comparative histology :
MAIN
3 3073 00257018 0

D1519704

COMPARATIVE HISTOLOGY

COMPARATIVE HISTOLOGY

An introduction to the microscopic structure of animals

LUCY D. LEAKE

Department of Biological Sciences, Portsmouth Polytechnic Portsmouth, England

1975

ACADEMIC PRESS

LONDON NEW YORK SAN FRANCISCO

A subsidiary of Harcourt Brace Jovanovich, Publishers

ACADEMIC PRESS INC. (LONDON) LTD.
24/28 Oval Road,
London NW1

United States Edition published by
ACADEMIC PRESS INC.
111 Fifth Avenue
New York, New York 10003

Library of Congress Catalog Card Number: 73 15687
ISBN: 0 12 441050 2

Printed in Great Britain by
T. and A. Constable Ltd.
Hopetoun Street, Edinburgh

Preface

In the study of biological sciences, as in most other subjects, there are trends both in research and training. The present-day trend is to concentrate on the experimental approach, and to reduce the amount of time spent on morphological and systematic studies. It often happens that a tendency is carried too far, and there is a danger of having students who may recognize the importance of morphological studies, but who are unable to identify the tissues or animals on which they are working.

The morphological and experimental approaches are, of course, two aspects of the same subject. One cannot fully understand function without knowing about structure. Biochemists are becoming aware that the full secret of the organization of living cells cannot be appreciated until the precise arrangement of various chemicals within the cell is known. Similarly, the way in which tissues operate rests on the organization of their components, and the functioning of animals depends upon the nature and arrangement of their constituent tissues. This book aims to offer a simple guide to the microscopical structure of standard "type" animals in relation to function.

The animals chosen are locally occurring or convenient examples of tripoblastic metazoans. They, or quite closely related species, are often used in experimental studies. The animal kingdom has been considered as a whole, although more space has been allocated to the Phylum Chordata than to any other phylum. This is due partly to the considerable evolutionary diversity found among chordates, and partly to the importance accorded the members of this phylum as experimental animals. Embryological stages of animals have, in general, been mentioned only briefly, but two key larval forms have been dealt with in more detail. The cyclostome ammocoete larva is important because of its intermediate position between the invertebrates and vertebrates, while the amphibian tadpole is included because it represents the stage between aquatic and terrestrial vertebrates.

The light microscopical study of the structure of animal tissues is receiving new impetus with the increasing use of histochemical staining techniques. The electron microscope has contributed a great deal towards an understanding of cellular organization, but electron microscopy should always be based on light microscope studies. The problem with all microscopical studies is one of recognition and interpretation. Whereas the shape and position of a particular cell or tissue may be obvious to an expert, it often appears as an heterogenous mass to the uninitiated. This book gives photographs of histological sections, and also provides simple diagrams so that students can identify the various cells, tissues and organs, and their relation to one another. The diagrams have been simplified purposely so that the salient structures illustrating a particular function can be emphasized. Table I (p. xi) gives a list of all the features illustrated and the figures that show them. It is most important that students draw from their own slides and not from the diagrams given in this book. It is assumed that the student will be familiar with the histological appearance of the basic cell types, and will have access to dissection guides.

January 1975

LUCY D. LEAKE

Contents

Abbreviations

ANS = autonomic nervous system
Ant = anterior
BM = basement membrane
CNS = central nervous system
csf = cerebrospinal fluid
CT = connective tissue
D = dorsal
HLS = horizontal longitudinal section (see Planes of Section)
L = left
LS = longitudinal section
Post = posterior
R = right
rbc = red blood corpuscle
TS = transverse section (see Planes of Section)
V = ventral
VLS = vertical longitudinal section (see Planes of Section)

Planes of Section

Planes of histological sections are labelled with reference to the animal's normal position when moving slowly:

viz.:

VLS (vertical longitudinal section)

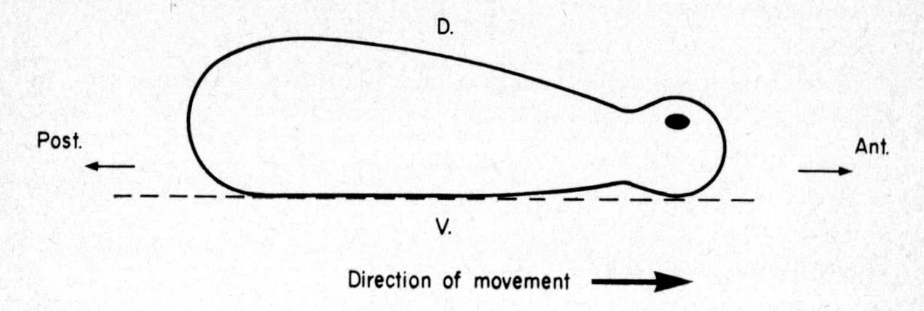

HLS (horizontal longitudinal section)

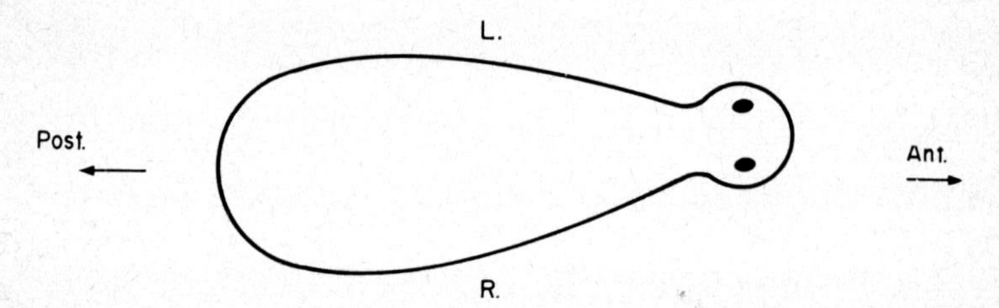

TS (transverse section)

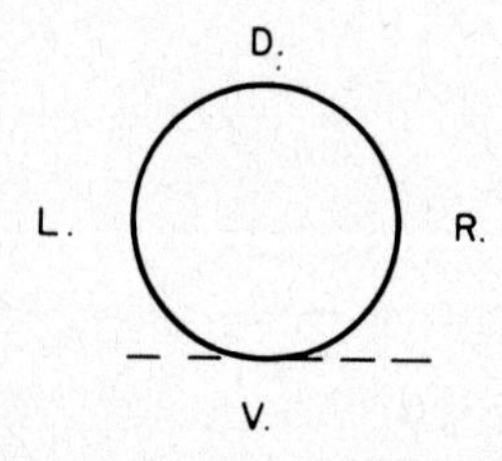

TABLE I

Distribution of figures comparing different features of triploblastic animals

	Platyhelminths	Acanthocephalans	Nematodes	Annelids	Onychophorans	Arthropods	Molluscs	Echinoderms	Hemichordates	Chordates: Protochordates	Chordates: Cyclostomes	Chordates: Elasmobranchs	Chordates: Teleosts	Chordates: Amphibians	Chordates: Reptiles	Chordates: Birds	Chordates: Mammals
Body cavity	—	Pseudo-coelom	Pseudo-coelom	Coelom	Haemo-coel	Haemo-coel	Haemo-coel	Coelom	Coelom	Atrium and coelom	Coelom	Coelom	Coelom	Coelom	Coelom	Coelom	Coelom
		2·1	3·1	4·1 4·2 4·18 4·19 4·25 4·26	5·1 5·2 5·16	5·1 6·19 6·44	7·1 7·22 7·35 7·36	8·1 8·17 8·25 8·26	9·1	10·1 10·2 10·8 10·16 10·17	11·2	12·2	13·1	14·1	15·1	16·1 16·16	17·1
Integument skin	1·2 1·15 1·16 1·18 1·33	2·2– 2·4	3·2 3·8	4·3– 4·5 4·27	5·3	6·2 6·20 6·27 6·45	7·3 7·22 7·36 7·37	8·2 8·3 8·17 8·27	9·2	10·3 10·18	11:3	12·3	13·2	14·2	15·2 15·25	16·2 16·16	17·2 17·3
Muscles	1·2 1·15 1·16 1·18 1·34 1·35	2·3 2·5	3·1– 3·3	4·2 4·3 4·19 4·25 4·27 4·29	5·3	6·2 6·20 6·45	7·1 7·22 7·36 7·37	8·9 8·25 8·27	9·1 9·2	10·3 10·16 10·18	11·2 11·25	12·1– 12·3 12·6	13·1 13·2	14·19 14·21 14·23	15·8 15·17 15·21 15·30	16·2 16·16	17·2 17·15 17·18
Skeletal material						6·2 6·7 6·20 6·45	7·38	8·1 8·3 8·4 8·18 8·19	9·1	10·16 10·19 10·21 10·22	11·1 11·2 11·4 11·5 11·8 11·11 11·16	12·1 12·2 12·4– 12·6	13·1 13·19	14·3– 14·6	15·3 15·4 15·12 15·16	16·1 16·9– 16·11 16·16 16·20	17·1 17·5

										Chordates							
	Platyhelminths	Acanthocephalans	Nematodes	Annelids	Onychophorans	Arthropods	Molluscs	Echinoderms	Hemichordates	Protochordates	Cyclostomes	Elasmobranchs	Teleosts	Amphibians	Reptiles	Birds	Mammals
Locomotory appendages				4·2 4·5 4·19 4·20	5·2 5·2	6·19	7·1 7·22 7·35 7·36	8·1 8·8 8·9 8·19 8·25		10·15		12·6					
Nervous system	1·1 1·2 1·4 1·17 1·35	2·6 2·7	3·1 3·3 3·8	4·1 4·6– 4·8 4·18 4·19 4·21 4·22 4·28	5·1 5·2 5·5 5·16	6·3 6·4 6·19 6·21– 6·25 6·44 6·46	7·4 7·5 7·22 7·23 7·35 7·39– 7·41	8·1 8·2 8·4 8·8 8·19 8·25 8·27	9·2 9·3	10·4 10·5 10·15 10·19 10·20	11·1 11·5– 11·14	12·1 12·2 12·5– 12·8	13·3– 13·6 13·19	14·5– 14·7	15·1 15·4– 15·7	16·1 16·4– 16·6 16·8– 16·10	17·1 17·6 17·7
Sense organs	1·5 1·16 1·18			4·18 4·20 4·22	5·3 5·4 5·6– 5·8	6·1 6·5– 6·7 6·26 6·47	7·6 7·7 7·35 7·42– 7·44	8·19 8·27		10·15 10·20	11·1 11·6 11·8 11·14– 11·17	12·9– 12·11	13·1 13·7– 13·11 13·21	14·1 14·8 14·9 14·11– 14·13	15·1 15·4 15·6 15·92– 15·13	16·7– 16·11	17·8– 17·10
Endocrine organs						6·21 6·25 6·28					11·18 11·19	12·1 12·12 12·13	13·12– 13·15	14·6 14·14– 14·16 14·24	15·6 15·14 15·15	16·4 16·5 16·12– 16·14 16·17	17·1 17·11 17·12
Other glands	1·3 1·34			4·4	5·1 5·11	6·44 6·50 6·51	7·2 7·21			10·4– 10·6	11·17		13·16	14·10	15·8 15·13 15·25	16·3	17·10 17·29
Respiratory system					5·3	6·1 6·8 6·25 6·27 6·48	7·1 7·3 7·24– 7·26 7·34 7·35 7·45	8·1 8·3 8·25 8·32 8·33	9·1 9·4 9·5	10·1 10·9 10·24	11·1 11·2 11·20	12·1 12·14 12·15	13·1 13·16– 13·18	14·1 14·17 14·18	15·1 15·16	16·15– 16·17	17·1 17·13

Circulatory system				4·1 4·2 4·8 4·19 4·25 4·26	5·2	6·1 6·28 6·44 6·49	7·8 7·31 7·36 7·46	(tubular coelomic systems) 8·1 8·4–8·10 8·20 8·21 8·25 8·28	9·1 9·6 9·8	10·1 10·7 10·26	11·21	12·1 12·16–12·18	13·1 13·19 13·20 13·23 13·24	14·19	15·1 15·17 15·18	16·16 16·18 16·19	17·1 17·14 17·28
Attachment organs/jaws/teeth	1·16 1·17 1·19 1·31	2·1 2·5		4·18 4·23	5·8	6·19 6·31 6·44	7·1 7·10 7·35 7·47	8·17				12·3	13·21	14·1	15·19	16·20	17·1 17·17
Digestive tract	1·1 1·6 1·7 1·14 1·19 1·20 1·22		3·4–3·6	4·1 4·2 4·9–4·12 4·18 4·19 4·23 4·25 4·26 4·29–4·31	5·1 5·2 5·8–5·10 5·16	6·1 6·9–6·15 6·19 6·29–6·34 6·44 6·50	7·1 7·9–7·14 7·22 7·27–7·31 7·35 7·48–7·54	8·1 8·11 –8·14 8·17 8·22 8·23 8·25 8·29–8·31	9·1 9·6–9·8	10·1 10·2 10·8–10·12 10·15–10·17 10·21–10·27	11·1 11·2 11·22–11·26	12·2 12·19–12·25	13·1 13·21–13·23	14·1 14·20–14·24	15·1 15·19–15·27	16·1 16·17 16·20–16·23	17·1 17·4 17·15–17·22
Excretory system	1·8 1·32 1·36		3·7 3·8	4·1 4·2 4·11–4·13 4·26 4·32 4·33	5·2 5·12	6·16 6·35 6·36	7·1 7·8 7·15 7 32 7·55	8·1 8·3	9·6 9·8 9·9	10·12 10·13 10·28 10·29	11·27 11·28	12·19 12·26–12·28	13·1 13·19 13·24	14·5 14·25	15·28–15·28	16·1 16·24 16 25	17·23
Reproductive system	1·1 1·9–1·14 1·21–1·30 1·32 1·37–1·39	2·1 2·8–2·13	3·1 3·9–3·14	4·1 4·14–4·18 4·24 4·26 4·33–4·36	5·2 5·13–5·16	6·17 6·18 6·37–6·44	7·16–7·21 7·33 7·34 7·56–7·59	8·15 8·16 8·24 8·34	9·10	10·13 10·14 10·16 10·30	11·28	12·29–12·32	13·1 13·25 13·26	14·26–14·28	15·29 15·31 15·32	16·25	17·24–17·28

Acknowledgements

Firstly, my thanks go to Professor G. A. Kerkut, without whom this book would not have been started or completed.

On the technical side I would like to thank, in particular, Miss M. Milne for the many histological sections that she has prepared, and also Miss A. Clinton, Mrs A. L. Evans, Mrs A. Herbert, Miss N. Newton, R. J. Timms and R. J. Webb for their help in preparing sections. Thanks for photographic assistance go especially to Mr B. D. Hamilton, Miss S. Hopkins and Mrs S. Martin, and also to Mr J. L. Mauger, Miss V. Nind, Mr I. M. Strong and Mrs B. Taylor.

I am very grateful to Portsmouth Polytechnic for the facilities provided, and to various of my colleagues there who have checked through different parts of the text: Drs G. H. Charles, C. Herbert, T. Jenkins, A. L. Martin, A. M. Pilgrim, R. C. Reay, F. R. Stranack, S. C. Turner and Mr I. D. Tunks. They are in no way responsible for any faults that remain. My thanks too to Dr H. Fox for interpreting Fig. 14.5.

Finally, I must thank my husband, Dr M. A. Carter, for his encouragement, forbearance, helpful discussion and assistance at all stages in the preparation of this book.

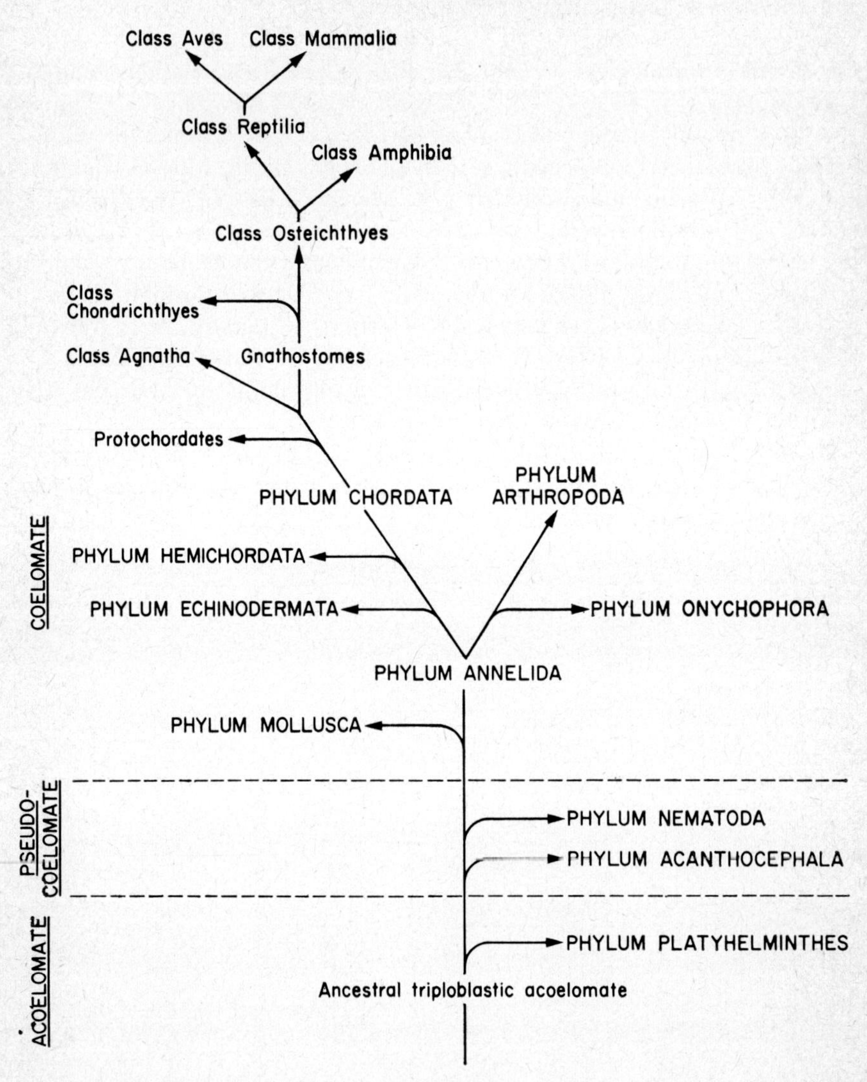

FIG. i. A simplified phylogenetic tree to show the possible evolutionary relationships between the phyla and classes of animals included in this book.

Introduction

Animals live in many different environments. Some simple forms are very successful within restricted habitats, but are vulnerable to even quite small environmental fluctuations. Their high reproductive rate and their ability to produce relatively inert but resistant forms under adverse conditions are important factors in their survival.

As well as these simple forms, there are a variety of more complex animals. In general, their increasing size and complexity are paralleled by a progressive independence of the environment, so that they are able to exploit a greater range of environmental conditions. Such animals carry out the same basic life processes as the simpler ones, i.e.: (i) appreciation of environmental changes and responses towards them; (ii) movement from one situation to another (to seek new food, protection, or other members of the same species); (iii) intake of raw materials (e.g. oxygen, food, water) and their subsequent transport round the body and treatment to yield energy; (iv) removal of waste products; and (v) reproduction.

However, in the more complex animals there is specialization and differentiation of individual structures concerned with particular functions. With such division of labour, each structure can be adapted optimally for a given function, and each function can be controlled independently. Co-ordination of all these different activities will allow a greater degree of flexibility and integration of the whole animal. For life to continue, animals must survive and reproduce viable offspring. Different environments impose varying restrictions on life within them (e.g. freshwater→hydration, terrestrial→dehydration and abrasion). The animals that continue to survive in any of these conditions do so because they possess particular structures that enable them to adapt to the conditions.

There is a considerable body of evidence which suggests that the more complex animals have evolved from simpler ones as a result of interaction between the environment and the genetic variations found in living organisms. Figure i is a phylogenetic tree showing a possible pattern of evolution. It has been simplified to include only the phyla described in this book.

Importance of a body cavity

Unicellular organisms and many diploblastic animals are relatively small and live in a fairly constant environment. There is some differentiation of function among their tissues, but most of their basic requirements are satisfied by diffusion from and into their surrounding medium.

One major step in the evolution of larger animals was the interposition between the two body layers (**ectoderm** and **endoderm**) of another layer of cells (**mesoderm**). The mesoderm is often differentiated to form distinct organs responsible for different functions. In the simplest triploblastic animals (Fig. ii), such as the Platyhelminthes, this mesoderm forms a spongy middle layer (**mesenchyme**) on which discrete ectodermal muscles can act to effect body movements, thus forming a simple **hydrostatic skeleton.** However, in larger animals where the mesoderm increases in thickness and differentiates into organs, its cells require more oxygen and nutrients. A system is developed to provide these factors, in the form of an internal body cavity filled with circulating fluid. Such a cavity appears first of all as a **pseudocoelom** between the mesoderm and endoderm (Fig. iii), in the Nematoda, Acanthocephala and five other minor phyla.

Most other animals from the annelids up to the chordates have a true **coelom,** or body cavity formed within the mesoderm (Fig. iv). There is still controversy as to exactly how it is formed, but a coelom has probably arisen more than once during evolutionary history. The coelom is surrounded by mesoderm, which thus forms both an outer lining to the gut, largely as (splanchnic) muscles, and an inner lining to the body wall which differentiates into (somatic) muscles and visceral organs. These organs are suspended within the coelom by mesodermal **mesenteries**, which may disappear secondarily. The coelom is lined with a single-layered epithelium (**peritoneum**), and contains coelomic fluid which is a transport medium for nutrients, respiratory gases and metabolites.

An important function of the coelom is that it allows for independent movement of internal organs relative to one another. Freedom of movement of a gut allows larger food to be handled and permits certain regions of the gut to expand or move more than others. Movements are assisted by mesodermal muscles and the gut is buffered in a fluid medium. Similarly, free movement of a pulsatile heart can also take place.

The coelom serves as a basis for a **hydrostatic skeleton**, with muscles working against back pressure of the incompressible coelomic fluid. A hydrostatic skeleton allows more extensive movements to occur than is possible when an animal relies on simple muscle sheets or locomotor aids such as cilia and flagella. Further, muscle pressure exerted at one point can be transmitted through the fluid to bring about distant effects.

Another use of the coelom is the storage of gonad products. In certain animals it is thought that the coelom arises from enlarged gonadial cavities, and in some of these (e.g. polychaete worms) gametes are released directly into the coelom where they develop and eventually are extruded through ducts or by rupture of the body wall. In others, a small portion of the

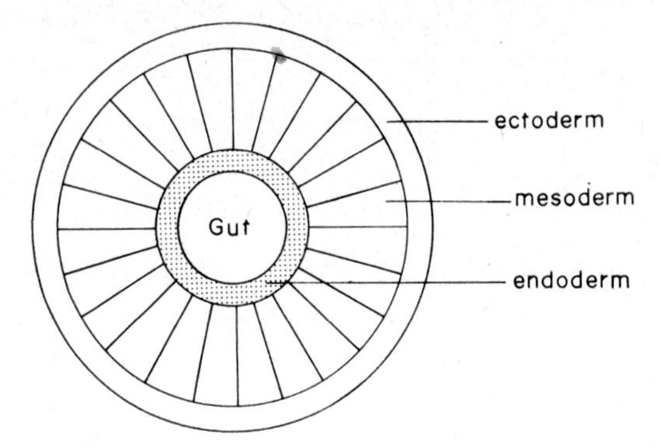

FIG. ii. Schematic TS of an acoelomate triploblastic animal showing the arrangement of its body layers.

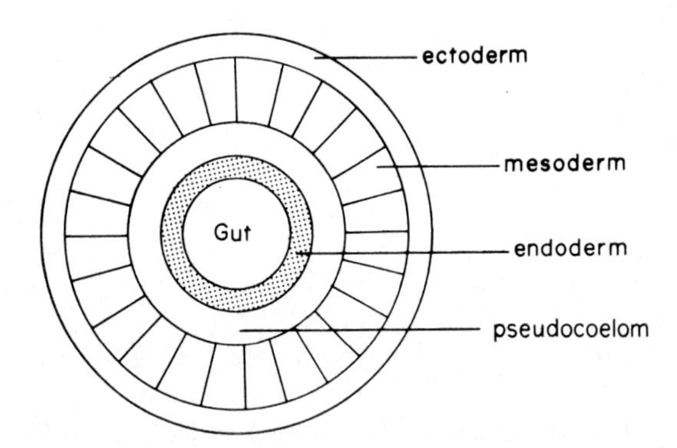

FIG. iii. Schematic TS of a pseudocoelomate triploblastic animal showing the arrangement of its body layers.

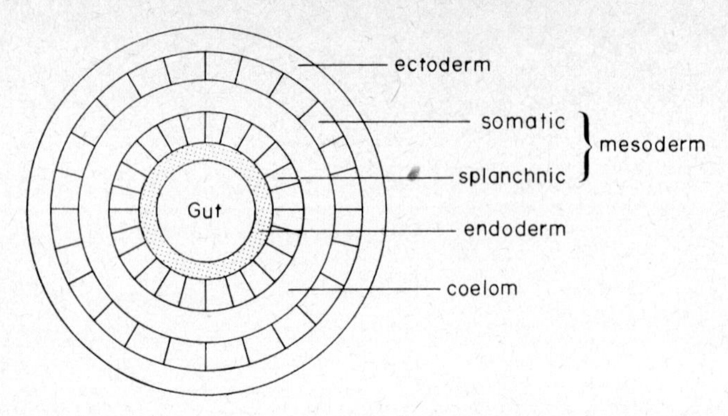

FIG. iv. Schematic TS of an acoelomate triploblastic animal showing the arrangement of its body layers.

coelom (**gonocoel**) is set aside for the accommodation of gametes (e.g. the seminal vesicles of oligochaetes or the testes of leeches).

The coelom connects with the outside by ducts which are basically of two types. The mesodermally-derived **coelomoducts** are concerned primarily with transport of gonad products, although they acquire additional functions. There are also ectodermally-derived **nephridioducts** used mainly for excretion. The two ducts can be completely separate or there may be various degrees of association between them. In the higher vertebrates coelomic ducts are lost and new special tubes develop for the transportation of gametes and excretory products.

Coelomic cavities arise in each segment in segmented animals and are separated by **septae** which may be secondarily broken down (as in polychaetes), resulting in one cavity throughout the animal. When this happens the efficiency of the hydrostatic skeleton is reduced and such animals have to rely on accessory locomotor structures such as parapodia. The single cavity is sometimes reduced due to invasion by mesenchyme (as in leeches), or partially replaced by a blood-filled cavity (**haemocoel**). Such a structure is found in onychophorans, arthropods and molluscs and is probably formed by breakdown of blood-vessels. In more complex animals the coelom becomes partly or completely divided into a connected series of compartments (e.g. the pericardium, renocoel and gonocoel of molluscs and protochordates; the tubular systems of echinoderms). In vertebrates the coelom is divided into an anterior **pericardium** enclosing the heart, and a posterior abdominal cavity. This may be marked off by a **transverse septum**, as in fish, amphibia and reptiles, or by a pleuroperitoneal membrane (**diaphragm**) which in mammals also encloses the lungs. In birds the area between pericardial membrane and transverse septum is split up by a pair of horizontal longitudinal septae into cavities which are taken over by lungs and large air-sacs.

Movement

Movement in simple animals is achieved by the use of cilia or flagella, or by a hydrostatic skeleton and interacting muscles (assisted by suckers, chaetae, parapodia, etc). Ciliary and flagellar movement is efficient only for small organisms; and a hydrostatic skeleton is wasteful of energy, depending as it does on contraction of all the body wall musculature and often including lateral undulations. It also depends for efficiency on a segmental arrangement of coelom and body muscles.

More efficient locomotion can be achieved with the use of a rigid **skeleton** composed of some hard material. Insertion of discrete muscles onto such a structure, associated with the development of jointed limbs, allows for more precise movements and greater leverage. This achieves more extensive movements for the same expenditure of energy. A hard external skeleton (**exoskeleton**), such as is found in the arthropods, protects against injury and also assists body support and reduces desiccation on land. However, it has some major disadvantages, chief of which is that it restricts growth. It, therefore, has to be shed and renewed at intervals during the growing period, leaving the temporarily soft-bodied animal in an extremely vulnerable position.

An important evolutionary step was the development of an **internal skeleton** around which the body could grow. Such a structure is seen in the cephalopods where the typical molluscan shell has become reduced and internal, thereby conferring increased mobility on this group. In the simpler chordates, which are relatively sluggish, there is a light internal skeleton in the form of a notochord; while in the more active gnathostomes (vertebrates with jaws) the notochord is reinforced or replaced by cartilage. Contractions of muscles against an incompressible central axis cause movements of the body which are used in locomotion. In most modern vertebrates the cartilaginous skeleton is replaced by a much-reduced, harder bony skeleton. Part of this forms a jointed rod (**vertebral column**) which permits great mobility. In birds, as an adaptation to flight, the skeleton is even further reduced by the creation of cavities in the bones which become filled with air-sacs.

Locomotion is assisted in many animals by **limbs**. In lower invertebrates (such as annelids and onychophorans) these limbs are segmental and alike. In those animals with a hard exoskeleton, however, the limbs are jointed and are often differentiated to serve a variety of functions. Those limbs performing similar functions become grouped together, thereby increasing efficiency of operation. Finally, in the more advanced arthropods, the segments bearing these groups of limbs have become split (**tagmosis**) into groups or regions (**tagmata**); e.g. the **head** bears appendages for feeding and collecting sensory information, the **thorax** carries locomotory appendages and the **abdomen** bears limbs modified for respiratory or reproductive purposes. In the vertebrates, the number of limbs is reduced and each may be adapted to serve a variety of functions, thus giving greater economy and plasticity.

Nervous organization

The fact that animals are mobile allows them to move towards favourable environmental conditions (e.g. food, light, oxygen). They are provided with a cellular system capable of sensing these factors and conveying the relevant information about them to the locomotor muscles. In very simple animals there are single cells (**independent effectors**) which perform this function and which have a sensory surface and a motile portion. In slightly more advanced forms **sensory** and **motor** functions are separated, and finally there is the interposition of a system of conducting cells (**interneurones**) between input and output components. In many invertebrates these interneurones are arranged into a diffuse sub-epidermal **nerve net** containing scattered cell bodies, which serves to conduct information and instructions over the whole body. Gradually differentiation of the nerve net occurs into functionally distinct parts (e.g. the longitudinal nerve cords, motor and sensory-association nets of echinoderms). Aggregation of the nerve cell bodies occurs to form **ganglia**, and later a central nervous system (**CNS**). Increasing localization of the major sensory components at the animal's "head" end, associated with movement in one direction, is paralleled by increasing cephalization, or location of ganglia in the same anterior region to form a **brain**. In metamerically segmented invertebrates the ganglia are segmental and control local (i.e. segmental) reflex responses. Such ganglia are linked by longitudinal nerve cords, allowing co-ordination of these local responses. In many higher invertebrates ganglia are detailed to control specific functions in different parts of the body (e.g. in arthropods the heart is controlled by a cardiac ganglion), and, indeed, they show a capacity for action that can be independent of the brain. However, in the more advanced invertebrates and in the vertebrates there is increasing dominance of the cephalic end of the CNS over the rest of the organism, and increasing differentiation of function within the brain. In the larger chordate brains a central fluid-filled canal provides the layers of nerve cells with nutrients and oxygen.

Nerve nets are normally located sub-epidermally, but increasing complexity of the nervous svstem is accompanied by its movement inwards from the peripheral position. Ultimately, in the vertebrates, the CNS is surrounded completely by hard skeletal elements which protect it against damage.

Nerve fibres conduct impulses at a rate that is proportional to diameter. Therefore, fast motor responses are achieved in many invertebrates and some aquatic vertebrates by the development of **giant nerve cells** with fibre diameters of up to 1mm which can conduct impulses at rates up to

60m/s. Giant fibres appear to have arisen many times during evolutionary history, and may be formed from single cells (e.g. polychaete giant fibre, vertebrate Mauthner cell) or by the fusion of many neurones (e.g. squid giant fibre, cyclostome Muller's cell). In higher vertebrates fast conduction is achieved by a different mechanism. The nerve is **myelinated**, i.e. surrounded by an insulative sheath of fatty material (myelin) which is interrupted at intervals by nodes. Impulses pass down myelinated nerves by a series of jumps from node to node (saltatory conduction), which allows for even faster conduction, at rates up to 120m/s.

Sensory receptor systems, which in lower invertebrates are generally simply modified nerve cells, become aggregated together in higher animals to form organized **sense organs**. Each sense organ responds to a single modality (e.g. the eye is stimulated by light), and communicates with the brain by special nerve tracts. In the vertebrates, specific association areas in the brain are linked with individual sense organs. In general, any animal's supply of sense organs depends on its habitat and mode of life (i.e. a nocturnal terrestrial predator will have different requirements from a sessile particulate-feeder living on the sea floor).

Nearly all nerves secrete chemicals from their endings. In most cases these chemicals are used in the transmission of impulses from a nerve to an adjacent nerve or effector organ (e.g. muscle or gland). Some nerves, however, are specialized to produce considerable quantities of chemical (**neurosecretion**) which acts as a **hormone** in controlling various functions of the body distant from the site of secretion. In invertebrates nearly all hormones are neurosecretions, and these are normally released by a "glandular" structure (**neurohaemal organ**) which consists of swollen neurosecretory nerve endings and associated blood vessels or sinuses. In vertebrates, although neurohaemal organs exist, the majority of hormones are produced in glandular tissues specialized for that purpose (**endocrine organs**).

Respiration

All animals deal with molecular oxygen at the cellular level in much the same way, but the processes by which they ensure a plentiful supply of oxygen to the tissues vary enormously. Very small animals obtain their oxygen by diffusion through their external surface. They have a large surface area : volume ratio and are therefore able to obtain sufficient for their needs. As surface area : volume ratio decreases with increasing size, larger animals often cannot obtain enough oxygen and therefore various features have developed which enable their tissues to receive an adequate supply.

Some animals live in fast-moving water where they are constantly in contact with freshly oxygenated medium, but others have to circulate the respiratory medium (using cilia, body movements or **respiratory pumps**) so that oxygen supplies are renewed continuously.

Larger animals need an increased surface area for respiratory exchange. One solution to this problem was the development of **gills**, thin-walled structures formed as invaginations or evaginations of the permeable integument. Finely divided gills of this type occur in nearly all aquatic animals. Often the rest of the integument shows restricted permeability to water, so that control of water and salt balance can be effected across the gills. Terrestrial animals encounter the problems of water loss, and lack of support for a finely divided respiratory surface. In such animals a new range of respiratory structures arises in the form of intucked, non-collapsible tubes (**trachea**) and folded internal respiratory surfaces (e.g. **lungs**, air-bladders, lung-books). Respiratory surfaces have thin, heavily-vascularized walls so that there is a minimum distance for oxygen exchange between respiratory medium and blood.

Circulation

Under normal conditions of pressure and temperature there is a low limit to the amount of oxygen which can be carried in physical solution (6·4 ml O_2/l of water at 10°C). Specialized **respiratory pigments** have evolved which are adapted for transporting higher concentrations of oxygen. Such pigments occur in most coelomate animals, although they have evolved the greatest degree of efficiency in the higher vertebrates. In all but a few invertebrates the pigments are dissolved in colloidal solution in the **blood**, while in the vertebrates the pigment molecules (**haemoglobins**) are smaller and are carried in special circulating cells in the blood (red blood cells or **erythrocytes**).

In smaller invertebrates the blood circulates round the body in a system of open sinuses between the cells. This is known as an **open circulatory system**. It is a low-pressure system, but circulation of blood in a particular direction is ensured by passing it through a pump. The pump may take the form of pulsatile sinuses or blood-vessels as in annelids, or a contractile "heart" as in molluscs or arthropods. In the vertebrates there is a completely **closed circulatory system** where the blood is confined to narrow channels (arteries, veins and capillaries) and does not come into direct contact with the tissues. A closed circulatory system is a high-pressure system because of the hydrodynamic resistance of the narrow channels, and therefore the circulating pump (**heart**) in vertebrates has to be more powerful.

In the relatively simple vertebrates blood is pumped from the heart to the respiratory surface, thence to the tissues and back to the heart in a **single circulation**. In the most advanced vertebrates (birds and mammals) a complete **double circulation** develops. Thus blood is pumped from the heart to the lungs to be oxygenated, and back to the heart for another boost before it goes to the brain and other regions of the body. The heart becomes completely divided to form separate but linked pumps. This ensures separation of oxygenated and deoxygenated blood so that maximally deoxygenated blood is passed close to the fresh oxygen supply, and fully oxygenated blood is removed rapidly to those areas where it is required. The demand for oxygen is particularly high in mammals and birds because they maintain a constant high body temperature to ensure optimal enzyme activity, and have, therefore, an exceptionally high metabolic rate.

Digestion

Intake of nutrient occurs in small animals by diffusion or active transfer across the body surface. In larger animals greater quantities of food are needed to satisfy their energy requirements, and so a special digestive system developed. The simplest guts (**enterons**) are mere bags with a single opening for intake of food and discharge of waste products. They are lined with a single layer of epithelium which is both secretory and absorptive, and intracellular digestion takes place.

In more complex invertebrates the gut forms an open-ended tube allowing one-way traffic of food; in at the **mouth** and out at the **anus**. As larger food particles are ingested differentiation of function occurs within the gut so that special areas are set aside for grinding (trituration) storage, digestion and absorption. Structures are developed for capturing and macerating the food outside the digestive tract (e.g. complex ciliary tracts, tentacles, suckers, jaws, teeth, radulae, biting and sucking mouthparts). Digestion is at least partly extracellular with quantities of enzymes being poured out onto food masses. In higher invertebrates and the vertebrates the gut lining is differentiated into regions secreting divers enzymes, so that an "enzyme chain" is established for the sequential breakdown of large molecules. Accessory **digestive glands** (e.g. liver and pancreas) also assist digestion.

Water balance and excretion

The processes of water balance and excretion are intimately linked. The chief end products of metabolism are water and carbon dioxide (which are easily disposed of), and ammonia which is highly toxic and must be removed immediately from the body. In small aquatic animals this will diffuse out into the surrounding medium very quickly, but larger animals have disposal problems. Many have solved these by converting ammonia to a less toxic compound (urea) which, however, is still soluble and therefore involves some water loss when it is excreted. In terrestrial animals the problem of dehydration becomes acute, and many terrestrial animals excrete an insoluble nitrogenous product (uric acid) which needs little or no water for its elimination.

Hydration and dehydration problems of larger animals are partly solved by the acquisition of an impermeable integument or one through which water movements are restricted (by scales or a mucus covering), and partly by the development of specialized tissues or organs for controlling salt and water balance. Most animals have a series of tubules which filter varying amounts of water off from the body fluids and return it to the outside. These **excretory organs** are derived either from nephridioducts or from coelomoducts.

Nephridioducts give rise to the excretory organs (**nephridia**) of many invertebrates. **Protonephridia,** such as occur in acoelomates, pseudocoelomates and *Amphioxus*, end internally by blind tubules containing tufts of cilia (flame cells) or flagella (solenocytes). **Metanephridia**, as found in the annelids, open into the body cavity by a ciliated funnel (**nephrostome**).

The excretory organs of onychophorans, arthropods, molluscs and vertebrates are derived from coelomoducts. They all consist of a coelomic fragment (which filters off fluid from the body), an excretory canal (which is the site of ion or water control) and a storage bladder.

All excretory organs show characteristics correlated with the medium in which the animal lives. Thus a marine animal which has body fluids virtually isotonic with seawater, will have a simple filtering system to remove only excess water taken in with the food; while a marine animal with its body fluids hypotonic to seawater will need an elaborate mechanism for getting rid of salts. A freshwater animal, on the other hand, has a complicated excretory organ which filters off a great deal of excess water that may be taken in by osmosis, and will have mechanisms for retaining salts that would tend to diffuse out. Many terrestrial animals have very efficient excretory organs that reabsorb as much water as is required and control salt balance as well. Most excretory organs work on the filtration principle, but there are some exceptions such as the secretory Malpighian tubules of insects.

Reproduction

All living things have the ability to reproduce their kind. Many simple organisms usually do this asexually, i.e. by budding and fission, but all animals can reproduce sexually. **Sexual reproduction** involves the fusion of **gametes** of two different sexes. The exceptions are some platyhelminthes, annelids and protochordates which can divide by fission or autotomy followed by regeneration, and certain insects which have parthenogenetic stages in their life history. Male and female gametes may be produced by one (**hermaphrodite**) individual or by two individuals of the opposite sex (**dioecious**). All animals have structural and behavioural mechanisms for ensuring the meeting of gametes of opposite sexes.

In aquatic animals generally, large numbers of male and female gametes are shed together and the eggs are fertilized outside the body. Internal fertilization does occur in some species, particularly those which live in fast-moving water. Aquatic animals are usually egg-laying (**oviparous**). **Eggs** and **larvae** have various mechanisms for buoyancy and motility which maintain them in media rich in oxygen and nutrients, and which allow for their dispersion.

Eggs of terrestrial animals face problems of water loss and of accumulation of waste products. Two solutions have been developed by truly terrestrial animals. The first of these is the **cleidoic egg** of oviparous amniotes (e.g. reptiles and birds), or its equivalent in other groups (e.g. insects), which provides the embryo with a controlled microenvironment. The second is the acquisition of **viviparity** (birth of live young), which involves mechanisms for internal fertilization, and continuing care of the young both inside the body and during their early independent existence.

Ovo-viviparous species give birth to live young, but here the **embryos** merely develop inside the mother's body as a protective measure, and derive little or no nutrient from the mother.

Most vertebrate embryos are provided with an extra-embryonic membrane (**yolk sac**) which is continuous with the gut and encloses yolk on which the developing embryo lives. The young of most higher vertebrates (Fig. v) develop within a liquid-filled sac (**amniotic cavity**) formed by a second extra-embryonic membrane (**amnion**) in which they are protected from dehydration and mechanical disturbance, and can develop independently of water. A third membrane (**allantois**) grows out from the embryonic gut to surround the embryo, and comes to lie very close to

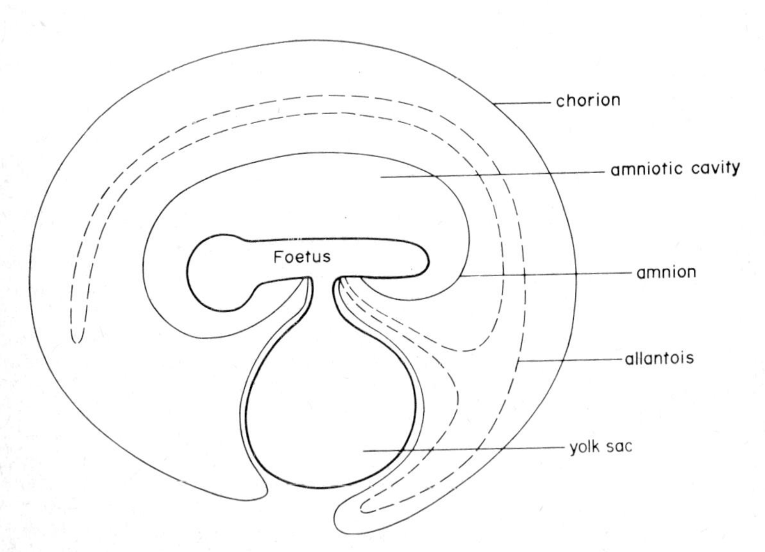

FIG. v. Diagrammatic arrangement of the embryo and extraembryonic membranes of an amniote vertebrate (based on the bird).

yet another membrane (**chorion**); thus forming the chorio-allantoic membrane. This complex membrane is highly vascularized. It is closely applied to the egg-shell in oviparous terrestrial vertebrates and to the uterus of viviparous species, and forms an organ for respiration and exchange of metabolites. In viviparous species it gives rise to a **placenta**, and assumes the function of providing for nutrient exchange, in which case the yolk sac is lost. Placentas can be formed of any one of the extra-embryonic membranes. In certain lower animals which exhibit ovo-viviparity, the developing embryos may receive some nourishment from the mother, but a true placenta is not formed.

Development of metazoan animals from the fertilized ovum normally follows one of two basic patterns. One group of animals (including the platyhelminths, annelids, molluscs and arthropods) shows **determinate cleavage**, i.e. the ultimate destination of the cells is determined at an early stage of development. In this group cleavage is generally spiral, and the adult mouth is derived from the embryonic blastopore: they are thus called **protostomes.** The other group (including the echinoderms, hemichordates and chordates) shows radial cleavage, and the adult anus develops from or near the blastopore while a new mouth appears some distance from it. The members of this group are therefore known as **deuterostomes**. They show **indeterminate cleavage**, i.e. the fate of the embryonic cells remains unfixed until a late stage of development.

Chapter 1

Phylum Platyhelminthes

Platyhelminths (flat-worms) are unsegmented, bilaterally symmetrical triploblastic animals with no coelom or haemocoel. They are usually dorsoventrally flattened and have a distinct head at the anterior end. Flatworms can be either free-living or parasitic, and much of their structure reflects their mode of life.

An enteron (gut) is absent from many parasitic species, while when present it takes the form of a blind-ended sac with only one opening. Mechanisms for attachment range from sticky secretions and simple adhesive patches to hooks, spines, teeth and complex suckers. There are no respiratory or circulatory systems as diffusion of oxygen can take place rapidly across the flat body surface. The nervous system follows the elementary pattern of a nerve net, thickened in places to form longitudinal nerve cords. There is generally a protonephridial excretory system, although this is lost in some forms. Flat-worms have a hydrostatic skeleton, and the body is filled with mesenchyme cells (parenchyma) forming a solid mass sometimes with intercellular spaces. Platyhelminths are usually hermaphrodite, but cross-fertilization is characteristic. Complicated gonads develop from mesenchyme cells. These are numerous in primitive forms but become reduced in number in more advanced members. An unusual feature of the females is that ovaries and vitelline glands are separate. The terminal parts of the genital tracts are of ectodermal origin. Internal fertilization is facilitated by the development of complex copulatory structures. Young are sometimes born live, but usually the life cycle involves one or more larval stages.

Class Turbellaria

Turbellarians form a large group of flat-worms living freely in aquatic or moist terrestrial conditions. The flattened and unsegmented body is covered by an epidermis bearing tracts of cilia. These are particularly well-developed on the ventral surface, which is used for locomotion. The epidermis characteristically possesses rhabdites, which are curved rods secreted by gland cells and which possibly have a protective function. Attachment organs are rarely present, although they are found in a few species that lead commensal lives. A sac-like gut is present in all turbellarians except the Acoela, and classification within the class is based on gut morphology. As a group they have simple life cycles.

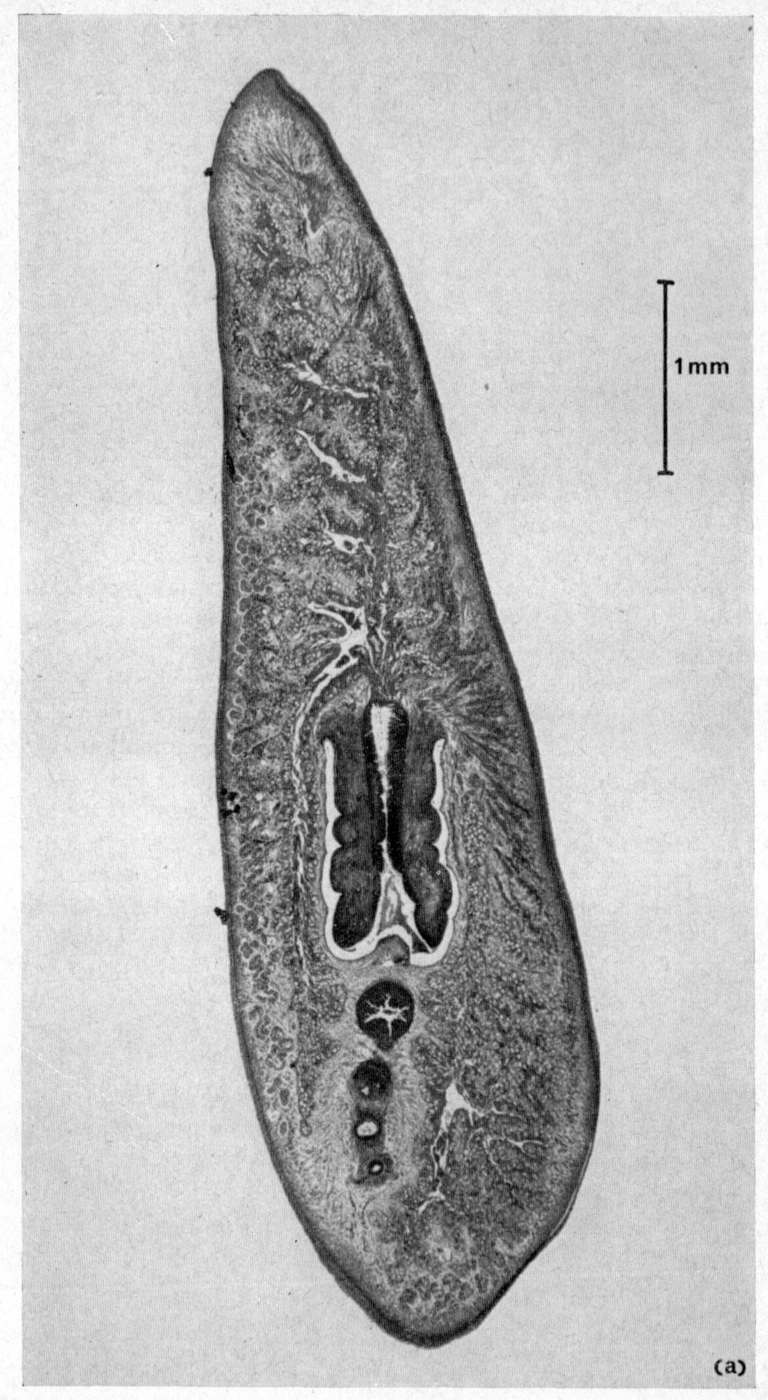

FIG. 1.1 (a) Oblique HLS and (b) explanatory drawing of the planarian *Dugesia lugubris*.

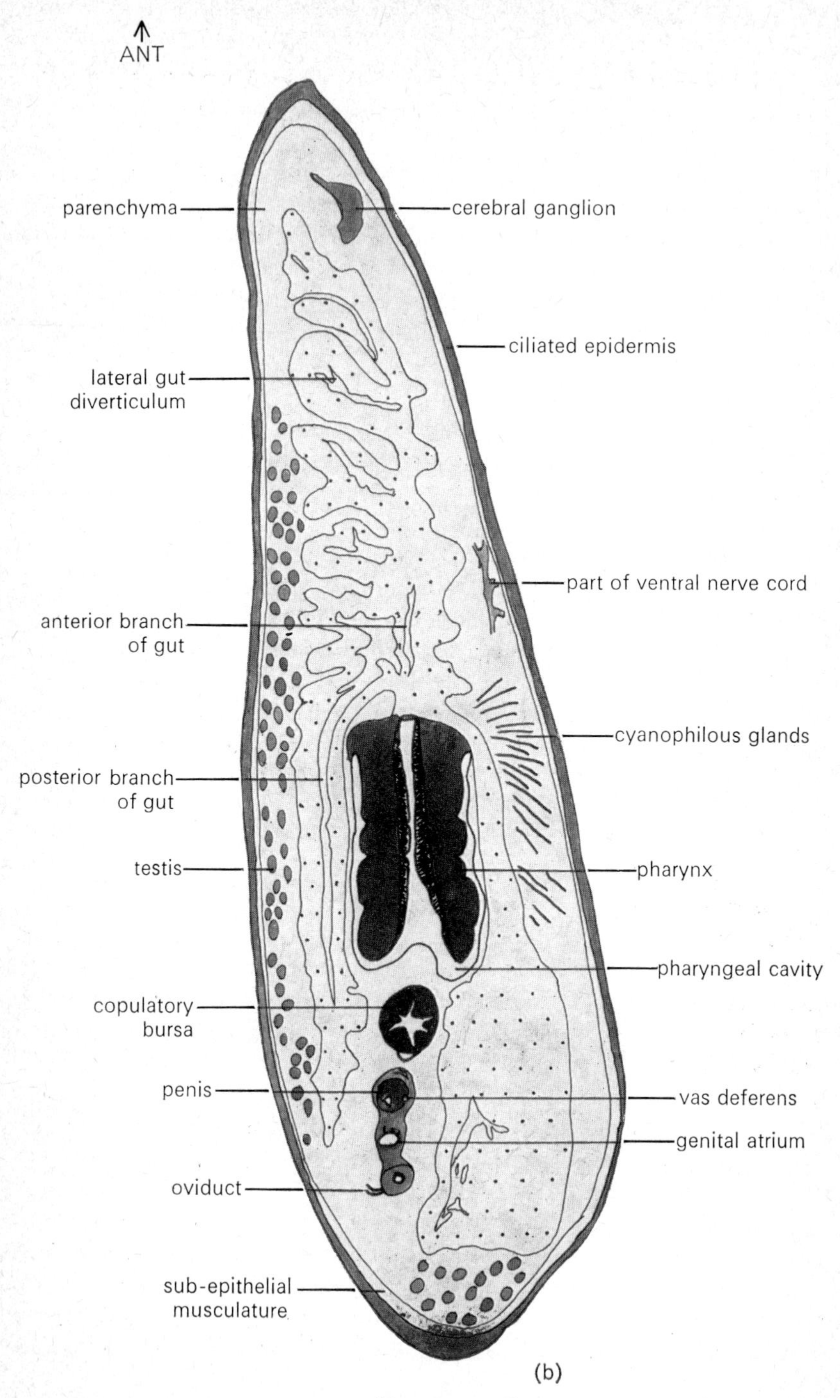
ANT
parenchyma
cerebral ganglion
ciliated epidermis
lateral gut diverticulum
part of ventral nerve cord
anterior branch of gut
cyanophilous glands
posterior branch of gut
testis
pharynx
pharyngeal cavity
copulatory bursa
penis
vas deferens
genital atrium
oviduct
sub-epithelial musculature
(b)

FIG. 1.1

Example: *Dugesia* (*lugubris*)

D. lugubris is a freshwater turbellarian belonging to the order Tricladida (planaria) which possess a three-branched gut; two branches directed backwards and one branch anteriorly. There is a characteristic plicate pharynx; that is, a free muscular cylinder projecting backwards into the pharyngeal cavity when at rest. These features are shown in Fig. 1.1, which is an oblique HLS of *Dugesia lugubris*, a black planarian with a triangular-shaped head and two large eyes.

Turbellarians are covered by a distinctive epithelium possessing rhabdites, large oval glands cells and cilia, but no cuticle. The cilia in freshwater species such as *Dugesia* occur only on the ventral surface. This can be seen in Fig. 1.2 which shows portions of the dorsal and ventral body walls. The epidermis overlies layers of circular, diagonal and longitudinal muscle. Below the muscle one can see the parenchyma traversed by muscle fibres and enclosing irregular intercellular spaces filled with fluid and wandering amoeboid cells, which may be considered to form a primitive type of vascular system. Also in the parenchyma are pigment cells and gland cells which may be sunken epidermal cells and which open by long ducts onto the surface. These gland cells are of four different kinds: (1) rhabdite-forming cells occurring chiefly on the dorsal surface; (2) cyanophilous glands producing slime and mucus for lubrication; (3) eosinophilous glands secreting sticky material for adhesion, capturing prey, sticking eggs to the substrate and other functions; (4) marginal adhesive gland cells, also eosinophilous and used for adhesion. Some of these glands can be seen in Fig. 1.2, while their distribution is shown in Fig. 1.3.

Nervous system

The nervous system in triclads is in the form of a complex network. There is a fine sub-epidermal nerve net, but the main part of the nervous system has sunk into the mesenchyme below the sub-epidermal musculature (sub-muscular nerve net). This network is thickened into longitudinal cords which have lateral branches and are interconnected by transverse commissures. Although in more primitive turbellarians there are dorsal, ventral and lateral cords, in the triclads only the ventral ones remain [Fig. 1.4(a)]. Fig. 1.4(b) is a longitudinal section of one ventral nerve cord showing how at the point of entry of the lateral nerves, the nerve cord appears to be thickened. Anteriorly two cerebral ganglia fuse to form a bilobed "brain" [Fig. 1.4(c)] part of which is visible in Fig. 1.1. There is a rich nervous supply to the pharynx, adhesive and copulatory organs. Nervous connections occur between the two nerve nets. Neurones are scattered randomly throughout the nervous system.

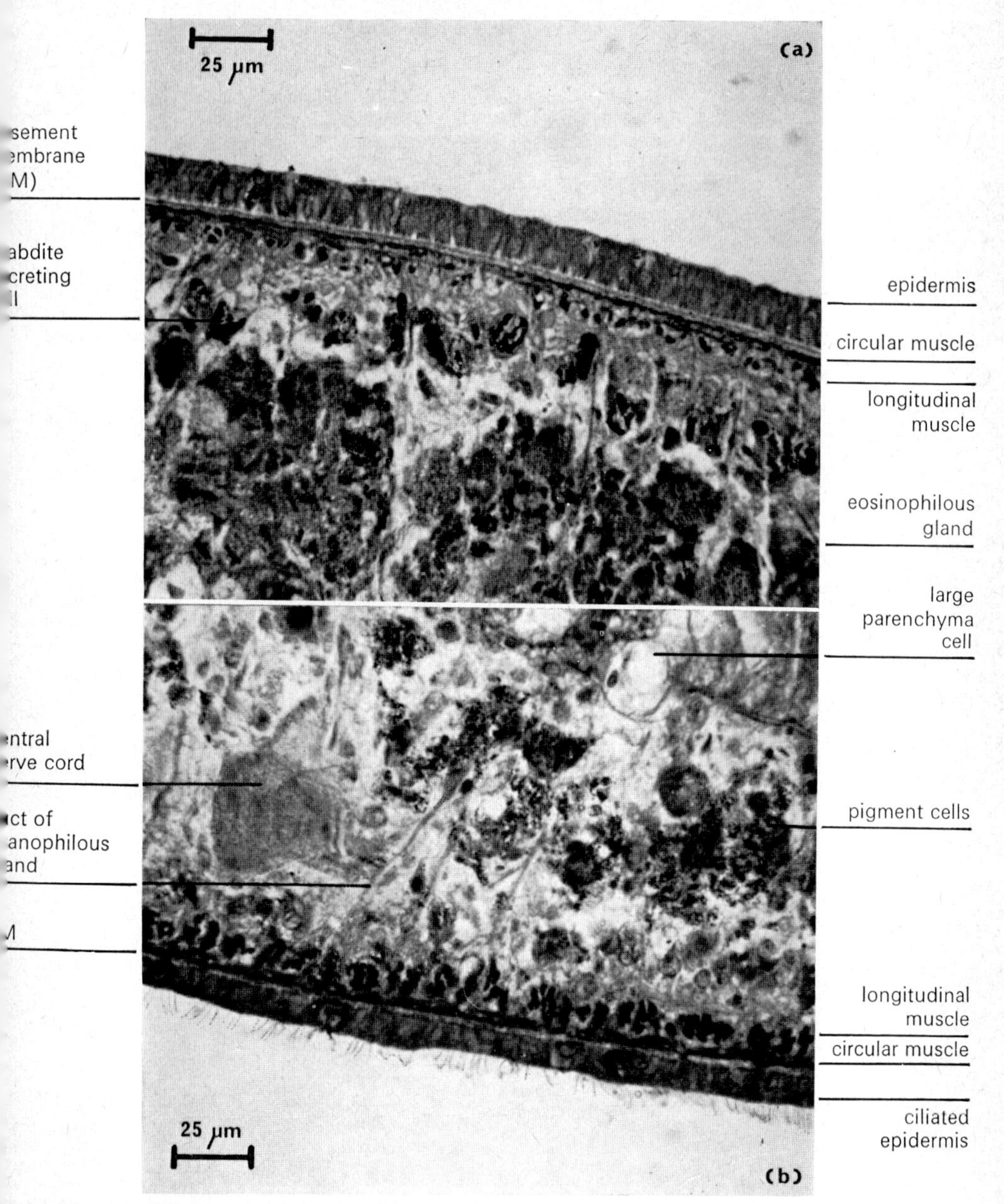

FIG. 1.2. TS through (a) the dorsal body wall and (b) the ventral body wall of *D. lugubris*. Note the dorsal position of the rhabdite-secreting cells, the thicker ventral body wall muscles, and the ventrally placed cilia used in locomotion.

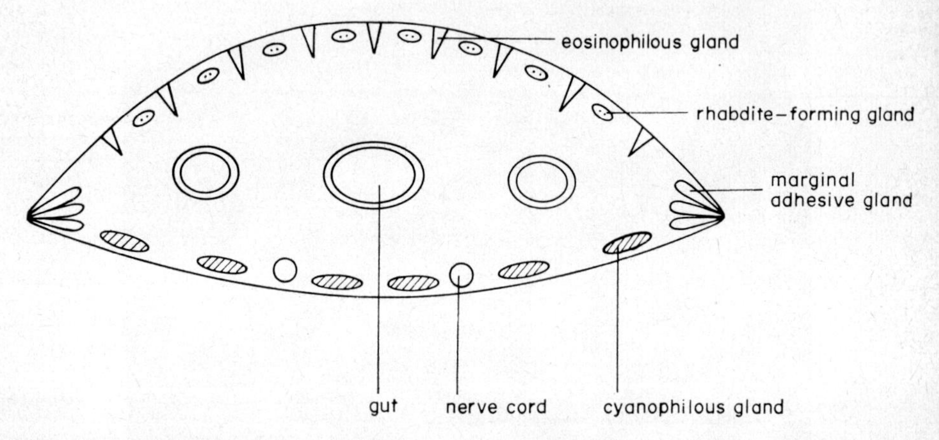

FIG. 1.3. Plan of a TS through a freshwater triclad turbellarian showing the distribution of gland cells.

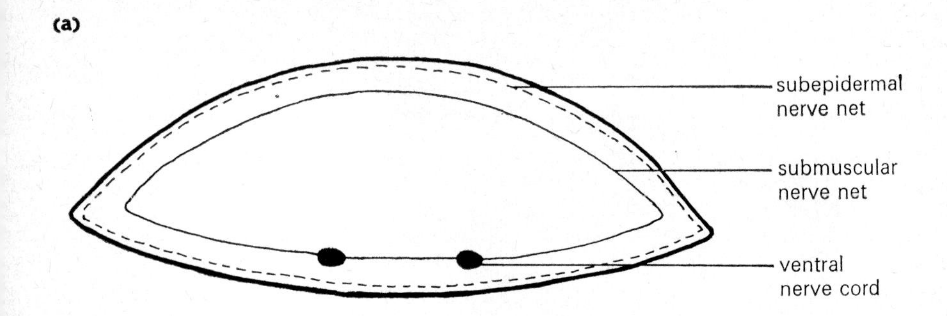

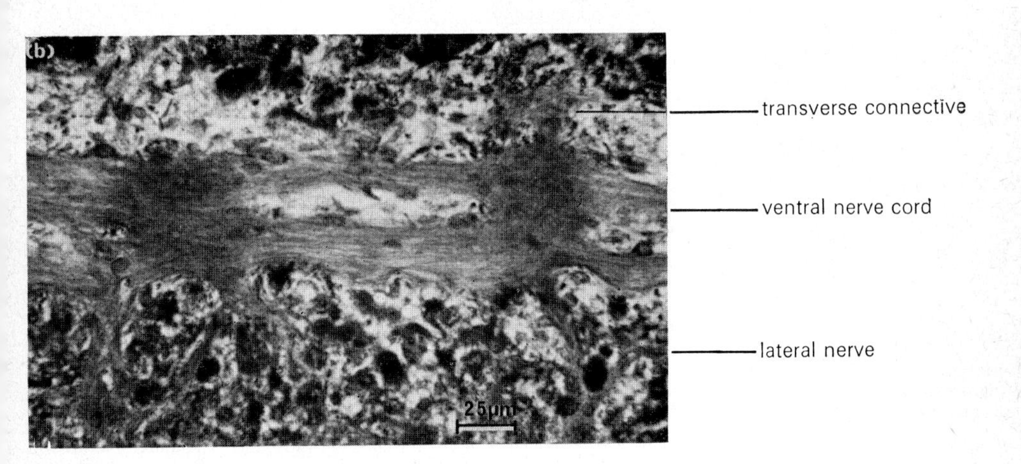

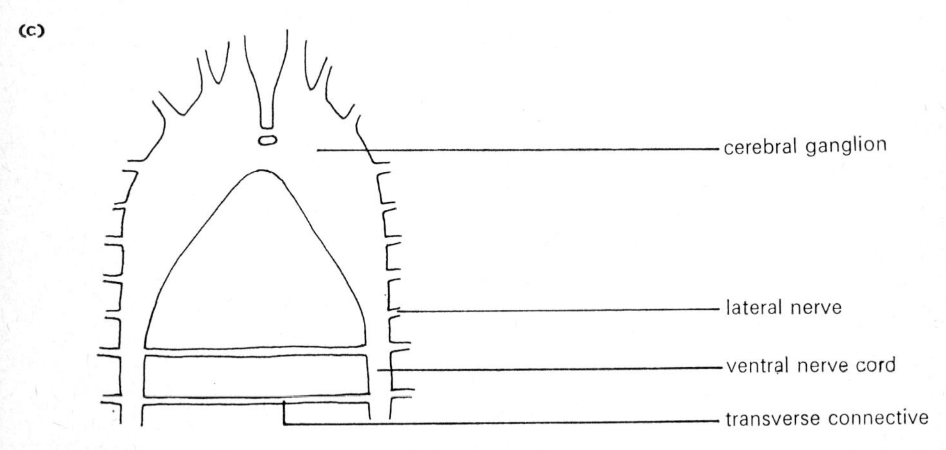

FIG. 1.4 (a) Plan of a TS through a freshwater triclad turbellarian showing the position of the nervous system; (b) LS through part of the ventral nerve cord of *D. lugubris*; (c) plan of the anterior end of the nervous system of *Dugesia*.

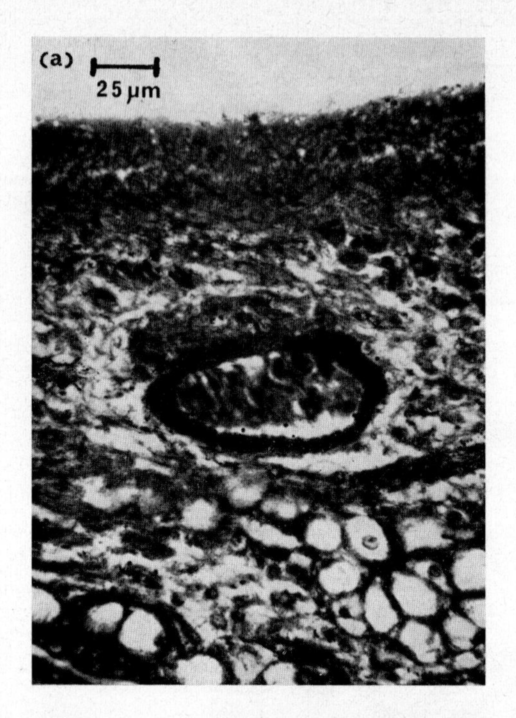

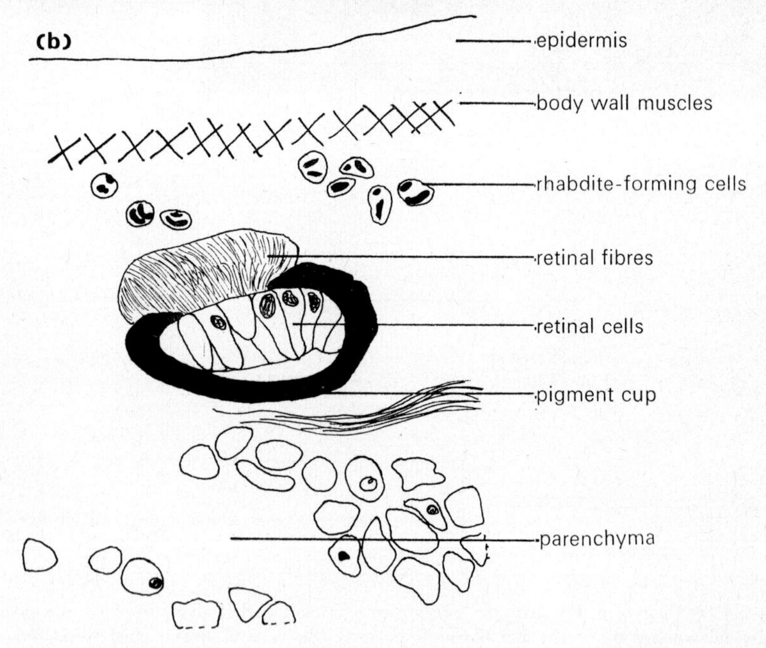

FIG. 1.5 (a) Section and (b) drawing of the eye of *D. lugubris*.

Sense organs

Turbellarians are well supplied with sense organs. These are mainly tactile or chemoreceptors taking the form of ciliated pits and grooves from which sensory bristles project above the body surface. A feature of this group is the possession of eyes, which in some land planarians can be quite complex. *D. lugubris* has two large eyes which can probably only distinguish light from dark: there are no refractive structures present such as would be associated with image formation. The eyes are simple ocelli or cups of pigment cells sunk into the parenchyma and lined with receptor cells (Fig. 1.5).

Digestion

Dugesia is carnivorous, preying on molluscs. The plicate pharynx is provided with powerful muscles and is eversible [Fig. 1.6(a)]. It can be protruded through the ventral mouth into the host's body, and sucks up the soft tissues. The pharynx wall [Fig. 1.6(b)] is complicated in structure, possessing numerous gland cells which produce proteolytic enzymes as well as sticky and slimy secretions. An unusual feature is that the ectodermal nuclei have sunk through the basement membrane and lie between the muscle layers and the parenchyma, remaining attached to the epithelium by long protoplasmic strands.

Food is passed through the pharynx into the enteron, a blind-ended sac with three branches, all possessing lateral diverticula. This is the site of both digestion and absorption. The enteron wall is composed of a single epithelial layer of large columnar phagocytic cells interspersed with narrow granular gland cells (Fig. 1.7). The gland cells secrete enzymes to break down food into fragments which are then ingested by the phagocytic cells and digested intracellularly. There is no anus, so waste material is expelled through the mouth.

Excretion

All turbellarians, except the Acoela, have a protonephridial excretory system. Interstitial fluid is filtered into minute flame cells and driven by a tuft of long cilia into an elaborate network of branched tubules. It is voided to the outside through numerous nephridiopores (Fig. 1.8). The components of the excretory system are very small and not easy to see in microscope sections.

Reproduction

Turbellarians can reproduce asexually (by transverse fission followed by regeneration) or sexually. *Dugesia* is an hermaphrodite: a plan of the reproductive system is shown in Fig. 1.9. There is mutual exchange of

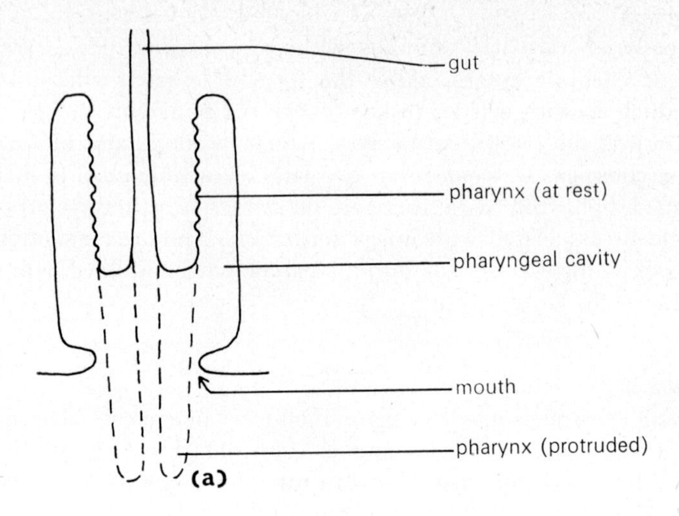

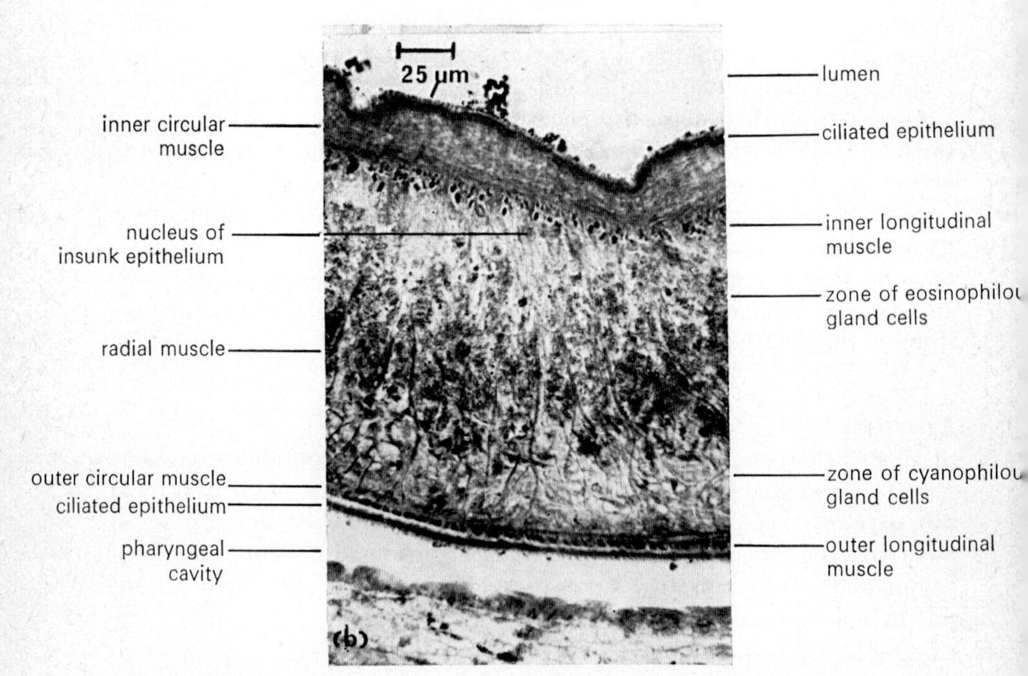

FIG. 1.6 (a) Diagram of the plicate pharynx of a triclad, in the resting and protruded positions; (b) TS through the pharynx wall of *D. lugubris*.

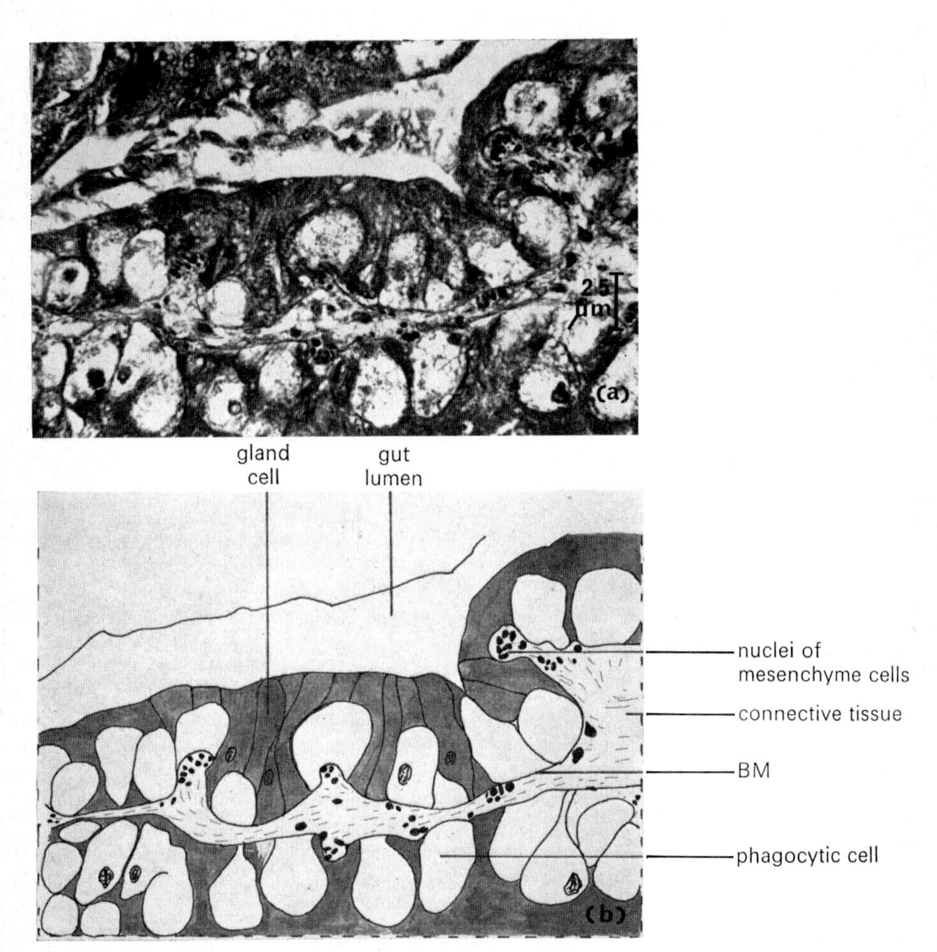

FIG. 1.7 (a) LS and (b) drawing of part of the enteron (gut) wall of *D. lugubris*. The single layer of epithelium rests on a basement membrane (BM).

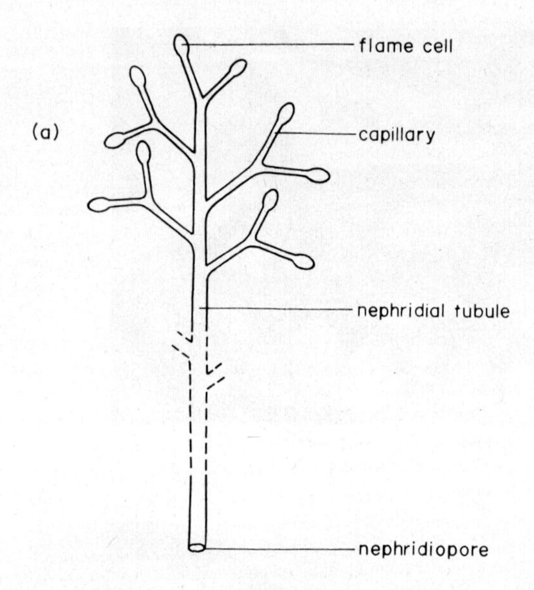

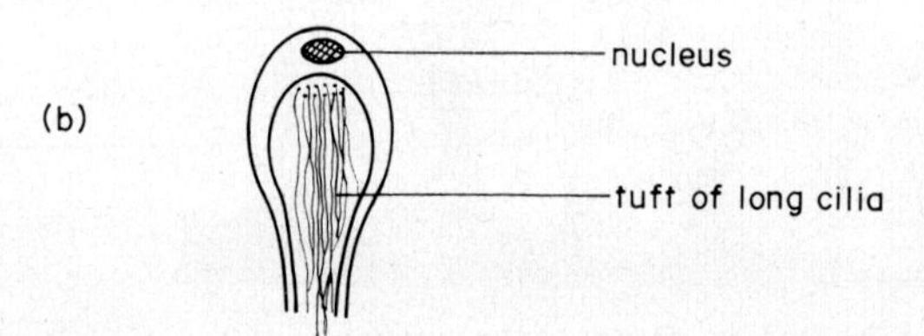

FIG. 1.8 (a) Plan of a generalized planarian protonephridial excretory system; (b) diagram of a flame cell from a planarian protonephridium.

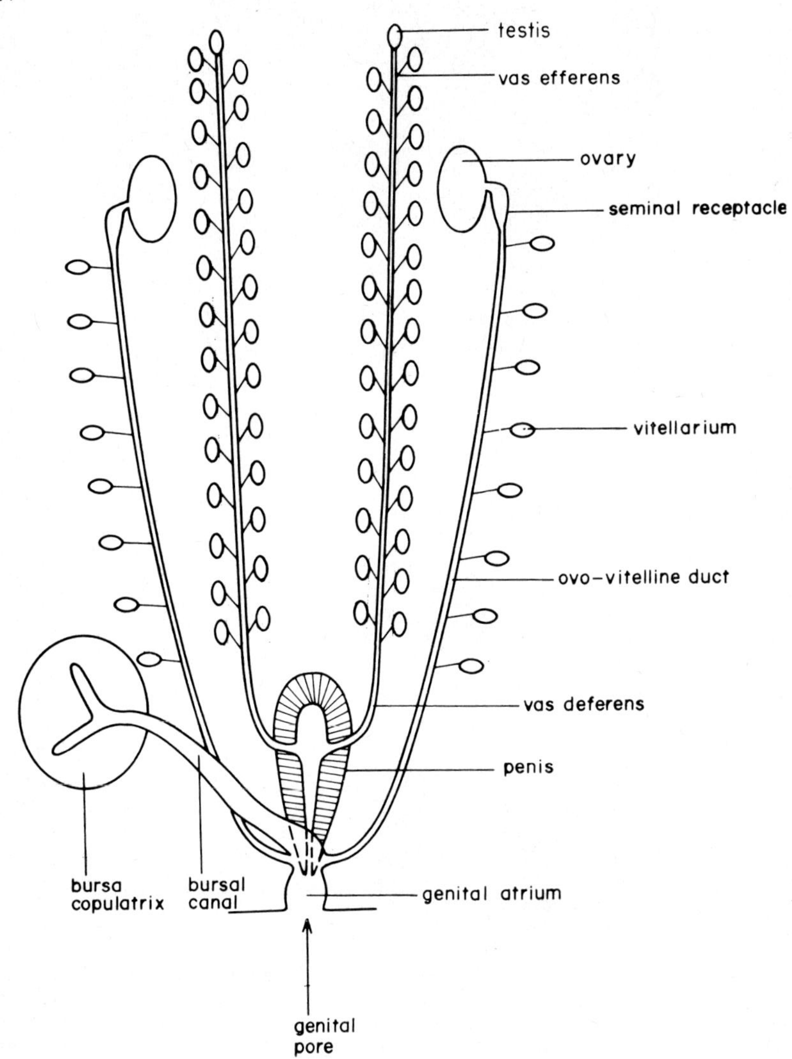

FIG. 1.9. Plan of the hermaphrodite reproductive system of *Dugesia*, as viewed from the dorsal side. The bursa copulatrix has been displaced to the left.

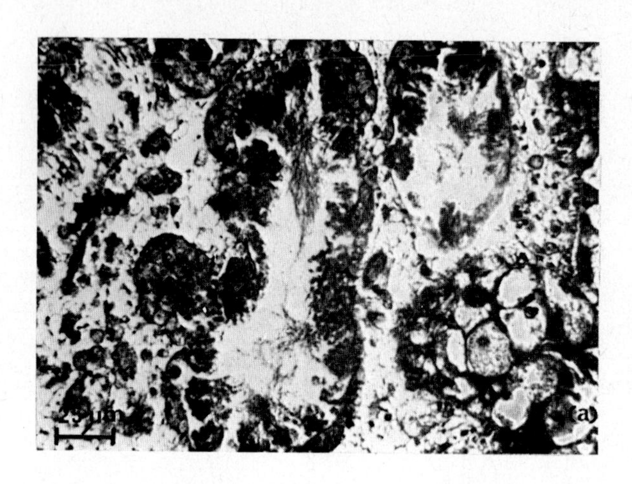

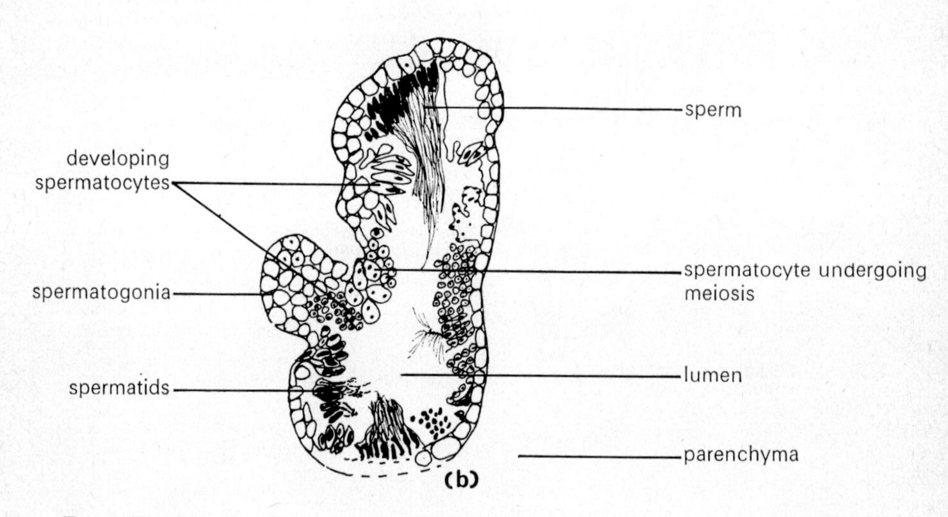

FIG. 1.10 (a) Section through part of the testis of *D. lugubris* showing two follicles containing germinal epithelium undergoing spermatogenesis; (b) drawing of one of the testis follicles seen in (a).

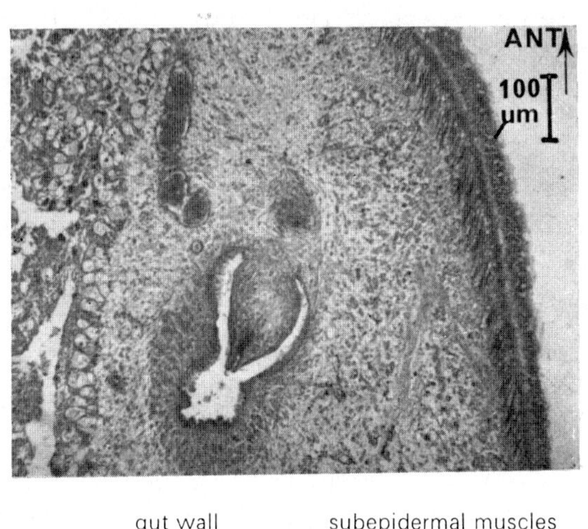

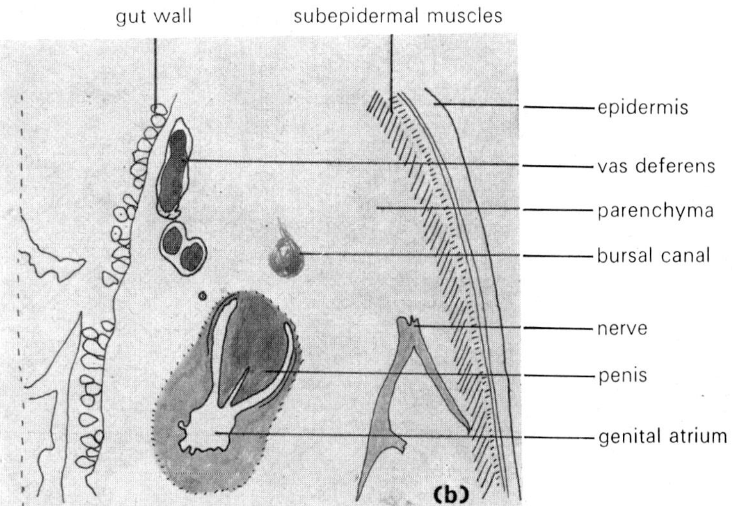

FIG. 1.11 (a) Section and (b) drawing of part of the male genital apparatus of *D. lugubris*.

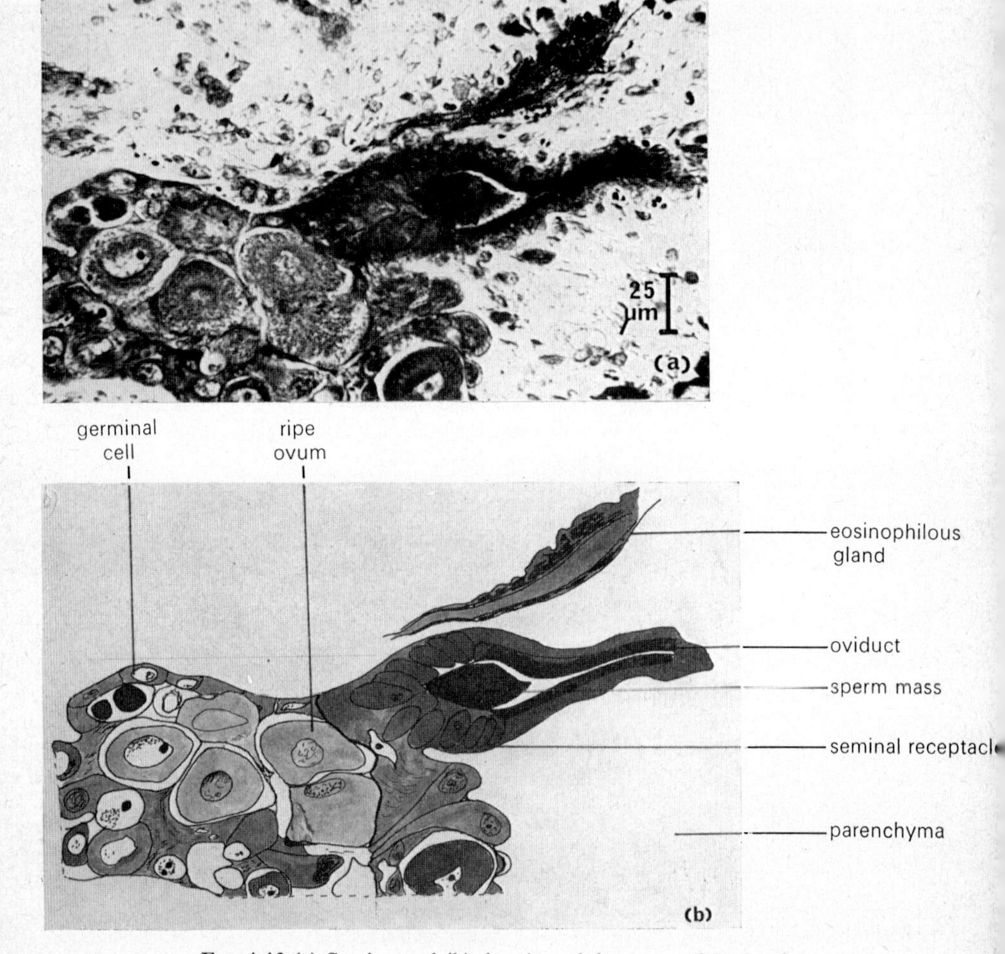

FIG. 1.12 (a) Section and (b) drawing of the ovary of *D. lugubris*.

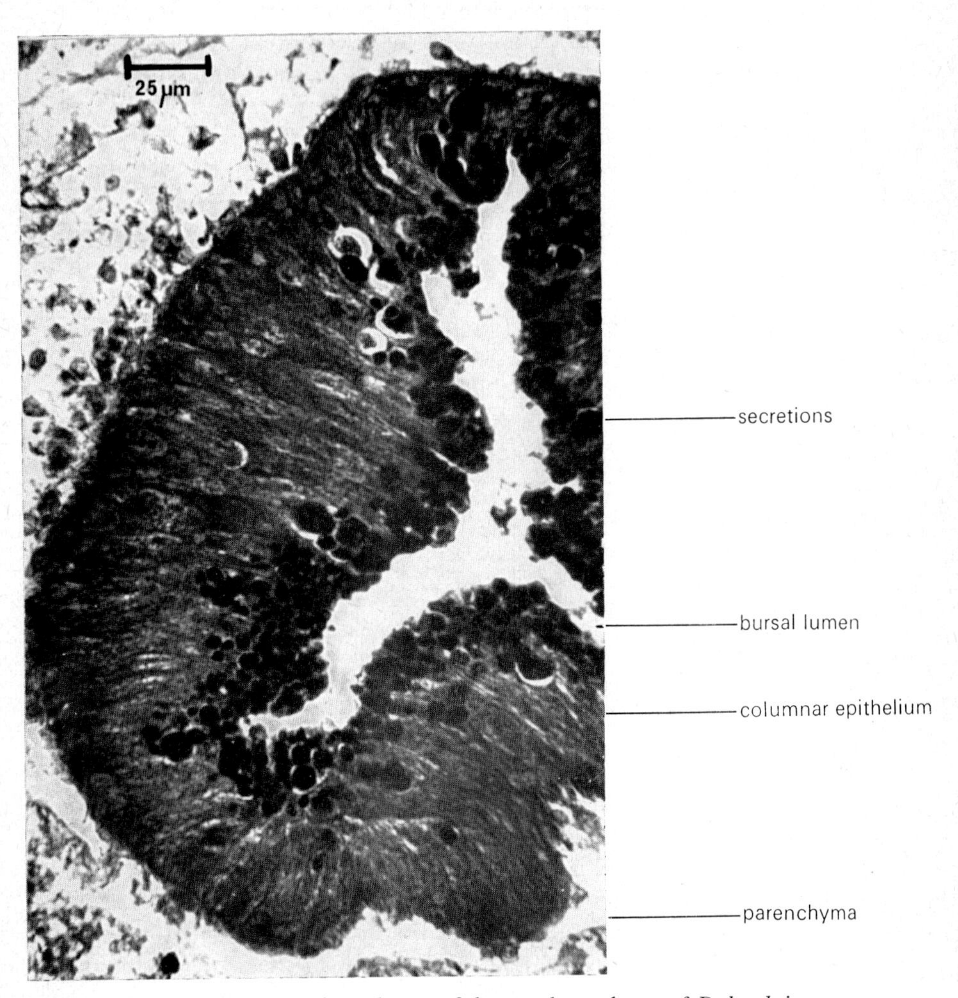

FIG. 1.13. Section through part of the copulatory bursa of *D. lugubris*.

sperm by copulating planarians, but self-fertilization is prevented by the arrangement of the copulatory apparatus.

Sperm are produced in numerous small follicular testes (Fig. 1.10) connected by short vasa efferentia to a pair of large vasa deferentia which eventually unite in a muscular penis. The penis normally projects into the common genital atrium (Fig. 1.11), but can be extruded by muscle action and inserted into the genital atrium of another individual, completely blocking off the oviduct entrance. Sperm pass up the bursal canal and are stored for short periods in the copulatory bursa of the second individual. It seems probable that sperm are only released during ejaculation.

D. lugubris possesses only a single pair of small ovaries. Each is composed of a germinal part producing ovocytes, and a series of follicles that increase in size towards the oviduct (Fig. 1.12). Fertilization occurs in the paired oviducts. Yolk cells (vitellaria), which are spatially separate from the ovary, discharge yolk and shell material into the oviducts (which are therefore termed ovo-vitelline ducts). The yolk becomes attached to the egg and enclosed in a common shell. In *Dugesia*, the oviducts open into a bursal canal and thence into the common genital atrium. The copulatory bursa lies dorsally and is lined with a tall columnar epithelium similar in structure to the intestinal epithelium (Fig. 1.13). It is capable of intracellular digestion and removes excess sperm and other secretions.

In *Dugesia*, when the cocoons containing eggs and yolk cells are laid, they are attached to the substrate by an adhesive secretion. Young worms emerge in most turbellarian species, but a larval stage occurs in polyclads.

Class Trematoda

The trematodes or flukes resemble the turbellarians in shape. They are all parasitic and possess one or more suckers for attachment to their host; one surrounding the mouth and another on the ventral surface. A broad classification of the flukes is based on the structure of these suckers and on the number of hosts required for a life cycle. Thus the Monogenea, which are chiefly ectoparasites, make use of only one host and possess two stiffened suckers; while the endoparasitic Digenea have more complicated life cycles requiring two or more hosts (which are components of the same food chain), but their attachment organs are much simpler and often reduced in number. When there are two hosts in the life cycle the larval stages usually live in molluscs while the adults parasitize vertebrates. Internally the trematodes show the same general body form as the turbellarians in that the body is undivided and filled with parenchyma. Trematodes are covered externally by a tegument which may bear spines but is unciliated in the adult.

Example: *Fasciola* (*hepatica*)

The liver fluke is a leaf-shaped digenetic trematode with two suckers. Its larval stages are found in the freshwater snail *Lymnaea* while the adults live in the bile ducts of sheep and cattle. Figure 1.14 shows a transverse section of *Fasciola hepatica*.

The body wall is covered by a characteristic cytoplasmic (non-cuticular) tegument bearing spines (Fig. 1.15). It is absorptive and is a site for the exchange of respiratory gases and nutrient. Beneath the basement membrane are layers of circular, longitudinal and diagonal muscle, while dorso-ventrally running muscle fibres are seen especially in the lateral regions of the body.

The suckers of *Fasciola* are unstiffened and muscular, with fibres running round the rim and centre of the sucker and many radiating fibres in between (Fig. 1.16). They are provided with spines, sensory papillae and tactile sensory endings (see also Fig. 1.18).

Nervous system

The nervous system is similar to that of the Turbellaria. Paired cerebral ganglia are joined by a commissure (Fig. 1.17) and the sub-muscular nerve net is thickened into longitudinal cords with transverse connections. Three pairs of nerves (dorsal, lateral and ventral) run anteriorly from the cerebral ganglia and three pairs run posteriorly, of which the ventral nerves are especially well developed.

Sense organs

Simple ocelli occur in most Monogenea but they are present only in the larvae of the Digenea. Adult flukes have few sense organs and these consist mainly of tactile sensory endings in the tegument and suckers. They are bulb-shaped nerve endings with sensory bristles projecting above the body surface (Fig. 1.18).

Digestion

As an adult, *Fasciola* sucks up and absorbs the nutrient-rich medium in which it lives. The mouth is encircled by an oral sucker and a small muscular pharynx lined with tegument which pumps food towards the enteron. The enteron divides into two main branches each of which bears short lateral caecae (Fig. 1.19, see also Fig. 1.22). The intestinal wall is lined with an epithelium that is both secretory and absorptive. Its internal surface area is enlarged by fine protoplasmic processes which project into the gut lumen (Fig. 1.20). The height of the epithelial cells varies in relation to their activity. Thus when food is present the cells are short, while in the absence of food the cells are tall and columnar.

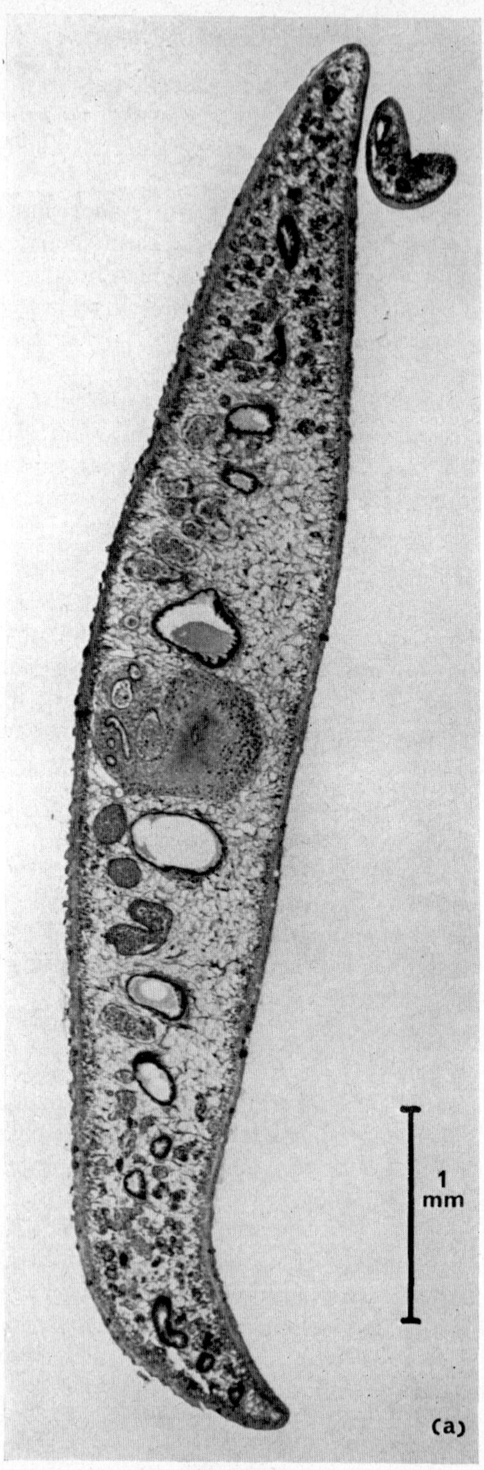

FIG. 1.14

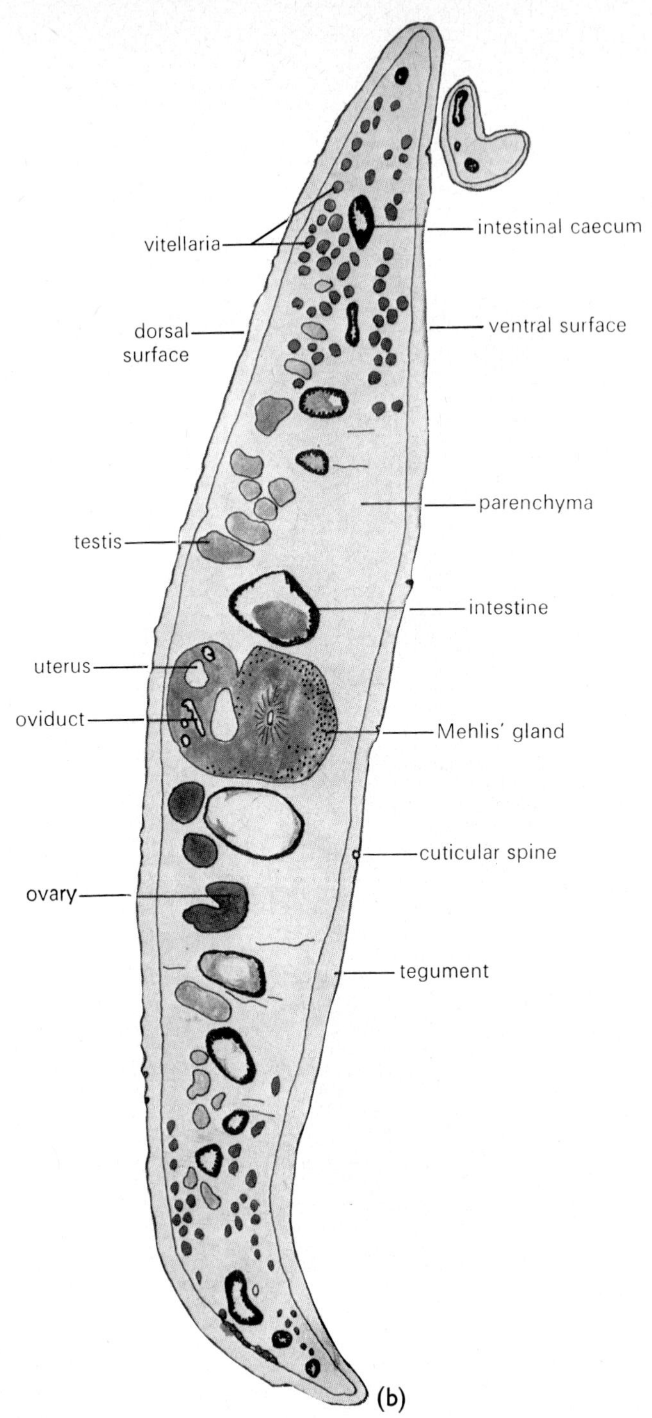

FIG. 1.14 (a) TS and (b) explanatory drawing of the trematode *Fasciola hepatica* in the region of the female genital system.

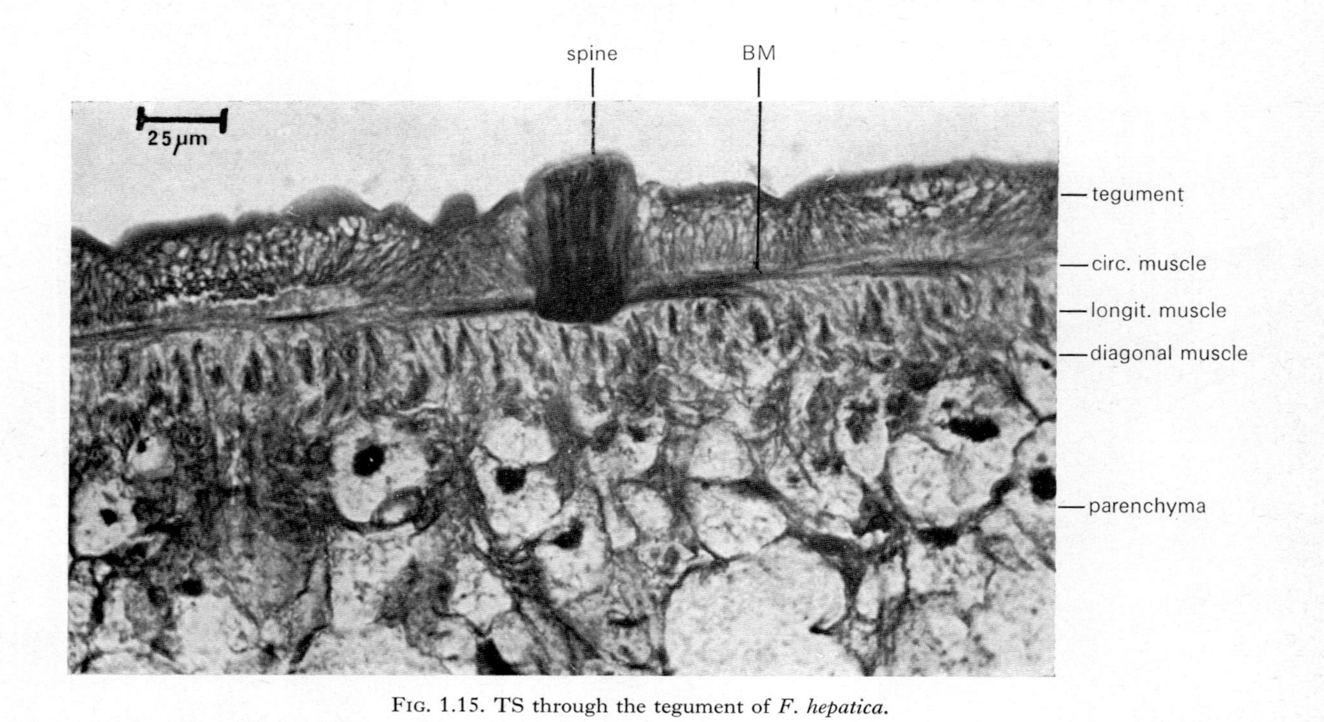

FIG. 1.15. TS through the tegument of *F. hepatica*.

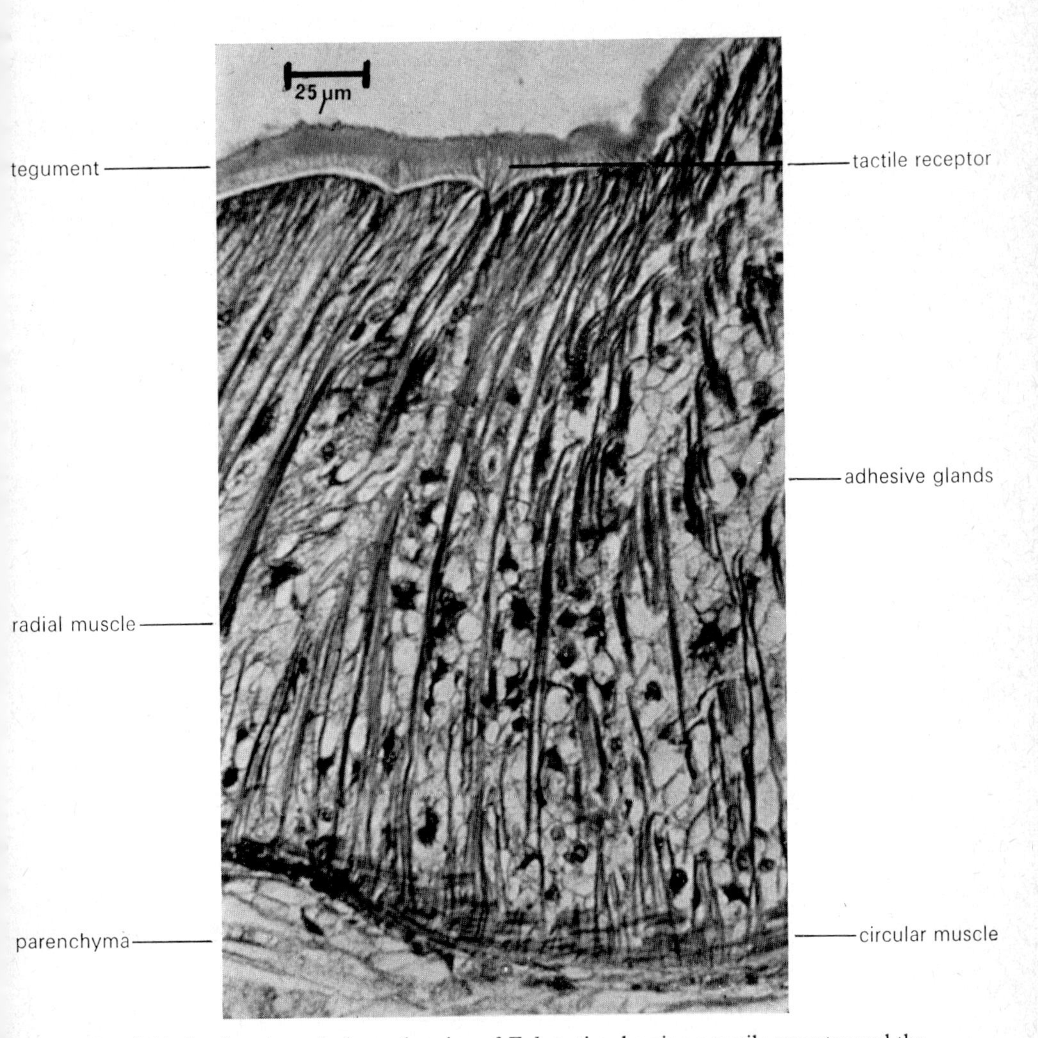

FIG. 1.16. Section through the oral sucker of *F. hepatica* showing a tactile receptor and the well-developed radial muscles.

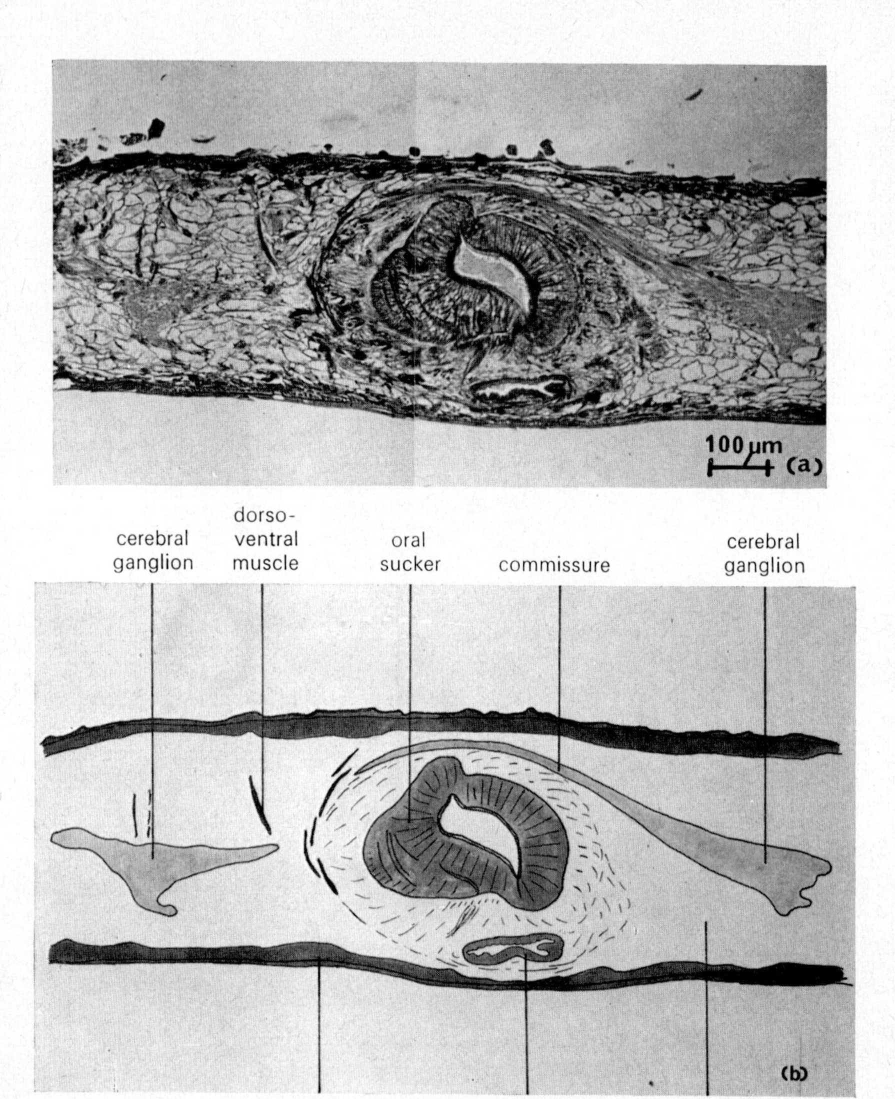

FIG. 1.17 (a) TS and (b) explanatory drawing of *F. hepatica* in the region of the oral sucker, showing the two cerebral ganglia and linking commissure.

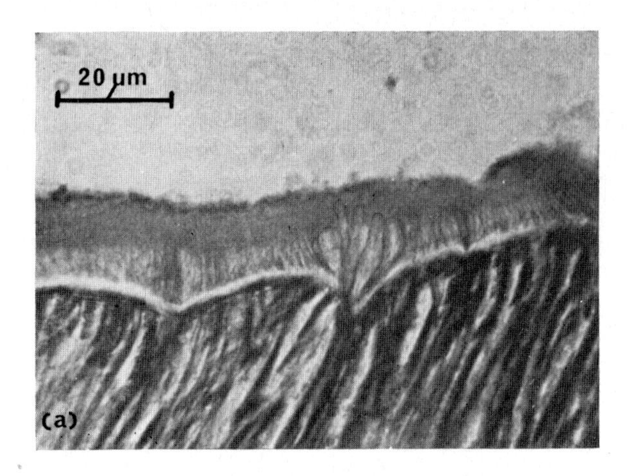

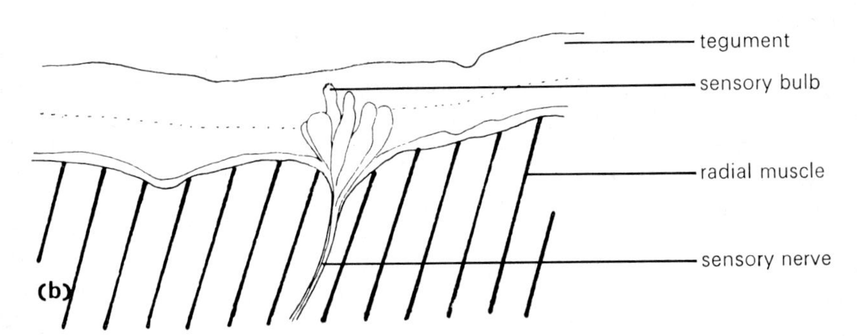

FIG. 1.18 (a) Detail and (b) drawing of a tactile sense organ on the oral sucker of *F. hepatica*.

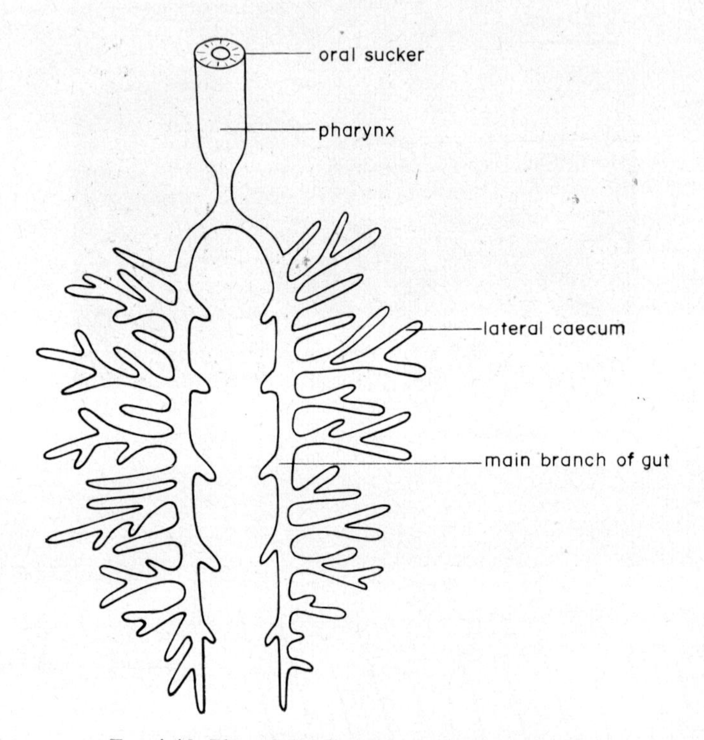

FIG. 1.19. Plan of the digestive tract of *F. hepatica*.

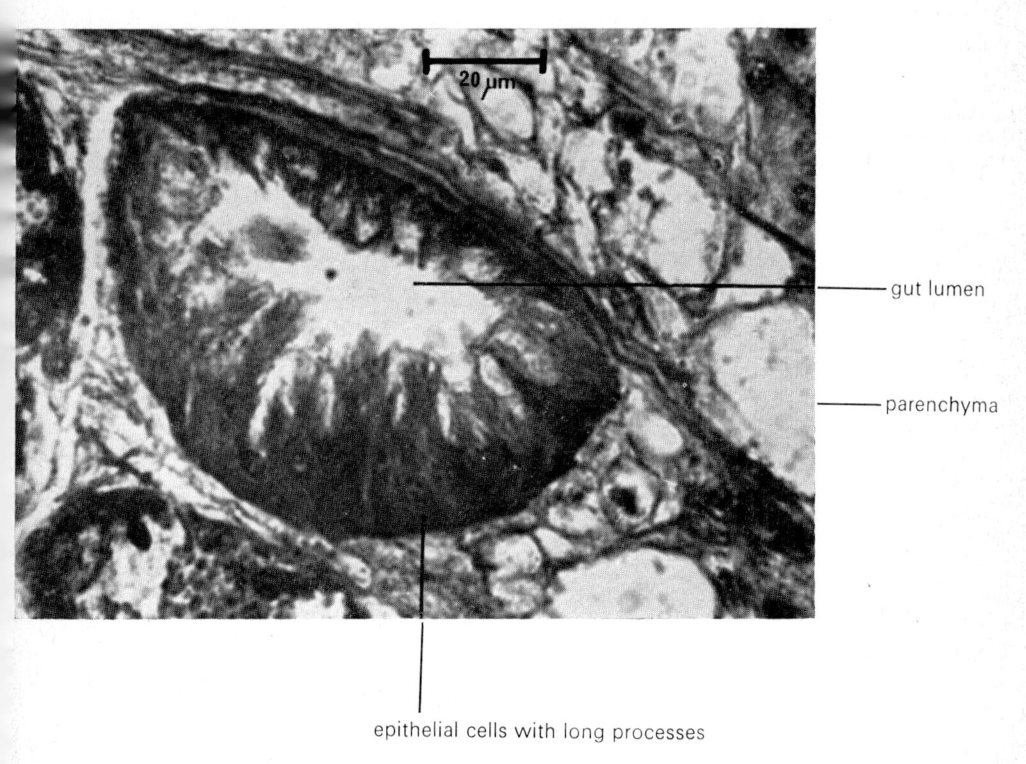

FIG. 1.20. TS through a small branch of the gut of *F. hepatica*.

Excretion

The excretory system of trematodes is very similar to that of other flatworms, and is equally difficult to see in histological sections. Flame cells lead into branch ducts and thence into two main lateral excretory canals. These fuse to form a main excretory duct which opens to the exterior by a posterior pore. The flame bulb arrangement in trematodes is used for taxonomic classification, especially in larval forms.

Reproduction

Trematodes are generally hermaphrodite although dioecious species do occur. A plan of the hermaphrodite system in *F. hepatica* is shown in Fig. 1.21. The gonads, normally simple in trematodes, are very branched in this species (Fig. 1.22).

Paired testes lie behind one another in the posterior part of the body and are packed with clusters of sperm-forming cells (Fig. 1.23). The thread-like sperm are passed down long coiled vasa deferentia into a seminal vesicle for temporary storage (Fig. 1.24). A twisted ejaculatory duct leads from the seminal vesicle to the penis or cirrus. The cirrus is lined with tegument bearing spines and can be everted by muscles during copulation (Fig. 1.25). It is enclosed in a cirrus sac which also surrounds extensive prostate glands that add their secretions to the sperm (see also Fig. 1.24).

The female tract consists of a single ovary (Fig. 1.26) connected to a short oviduct into which opens a muscular canal leading to the dorsal body surface. This is Laurer's canal, homologous to the vagina of monogeneans and cestodes (see Fig. 1.39). Voluminous paired vitelline glands secrete both yolk and shell into the oviduct via a common duct (Fig. 1.27). In many trematodes the oviduct is enlarged at this point to provide a chamber for egg formation (ootype), but in *Fasciola* the ootype is absent (Fig. 1.28). This area is surrounded by secretory cells (Mehlis' gland) the function of which is obscure (Fig. 1.29). They possibly secrete hormones, but are not, as was once suggested, involved with shell formation. Fertilization occurs in this region, and the eggs are provided with yolk and shells before passing on to the enlarged ovo-vitelline duct or uterus where the shells harden and where large numbers of eggs are stored together with vitelline cells (Fig. 1.30).

Male and female ducts converge in a common genital atrium before opening to the outside via a genital pore. During copulation the cirrus is normally inserted into the atrium of another individual, and sperm are then stored in the proximal part of the uterus (sometimes called the seminal receptacle). However, the cirrus can be inserted into the Laurer's canal of the second individual and fertilization may be effected directly in some species. Self-fertilization is possible, but cross-fertilization normally occurs.

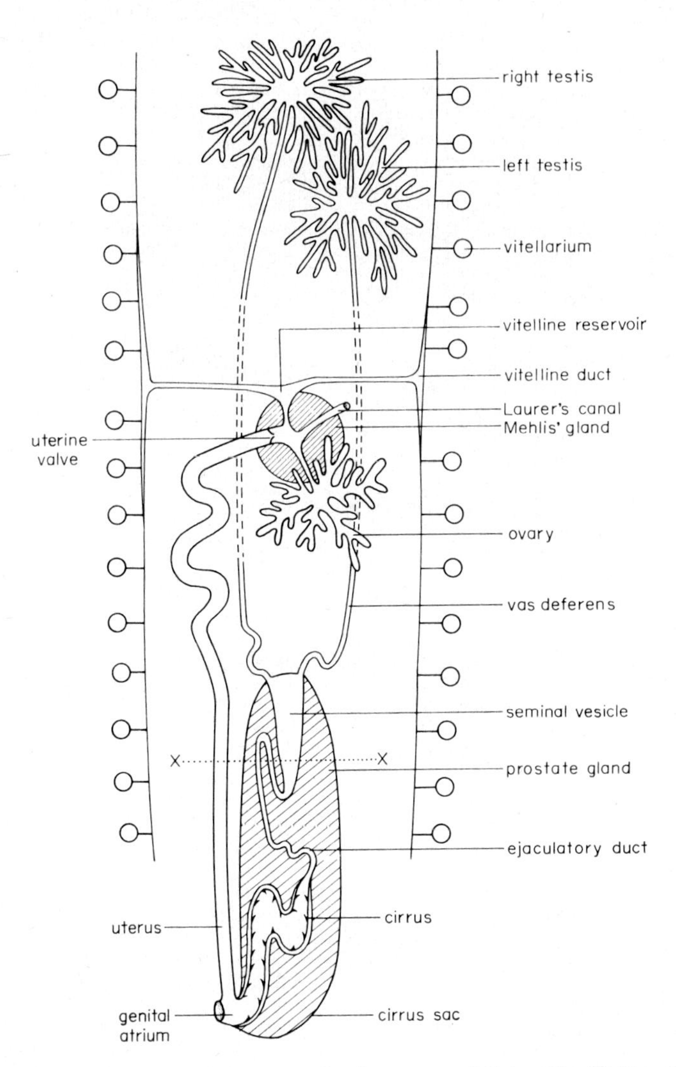

FIG. 1.21. Plan of the hermaphrodite reproductive system of *F. hepatica* (X-X is the section shown in Fig. 1.24).

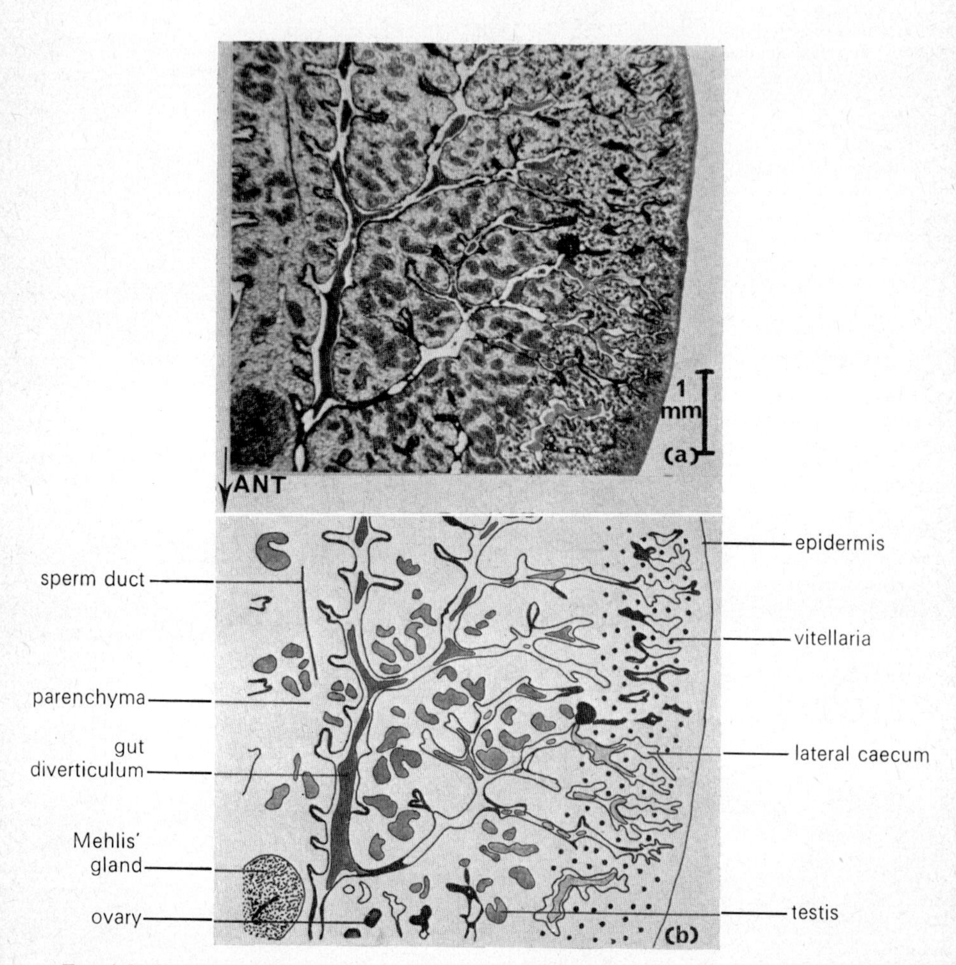

FIG. 1.22 (a) HLS and (b) drawing of part of *F. hepatica* showing the relationships of the branching gut, testes, ovary and vitallaria.

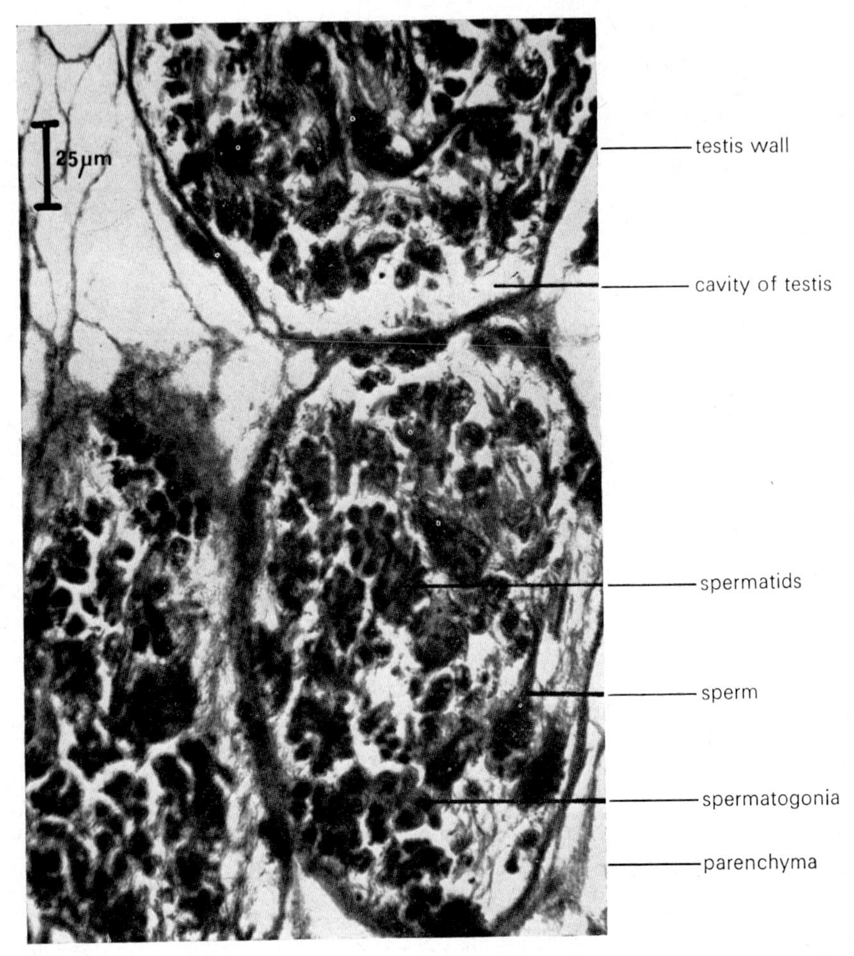

FIG. 1.23. Section through part of the branching testis of *F. hepatica* showing spermatogenesis.

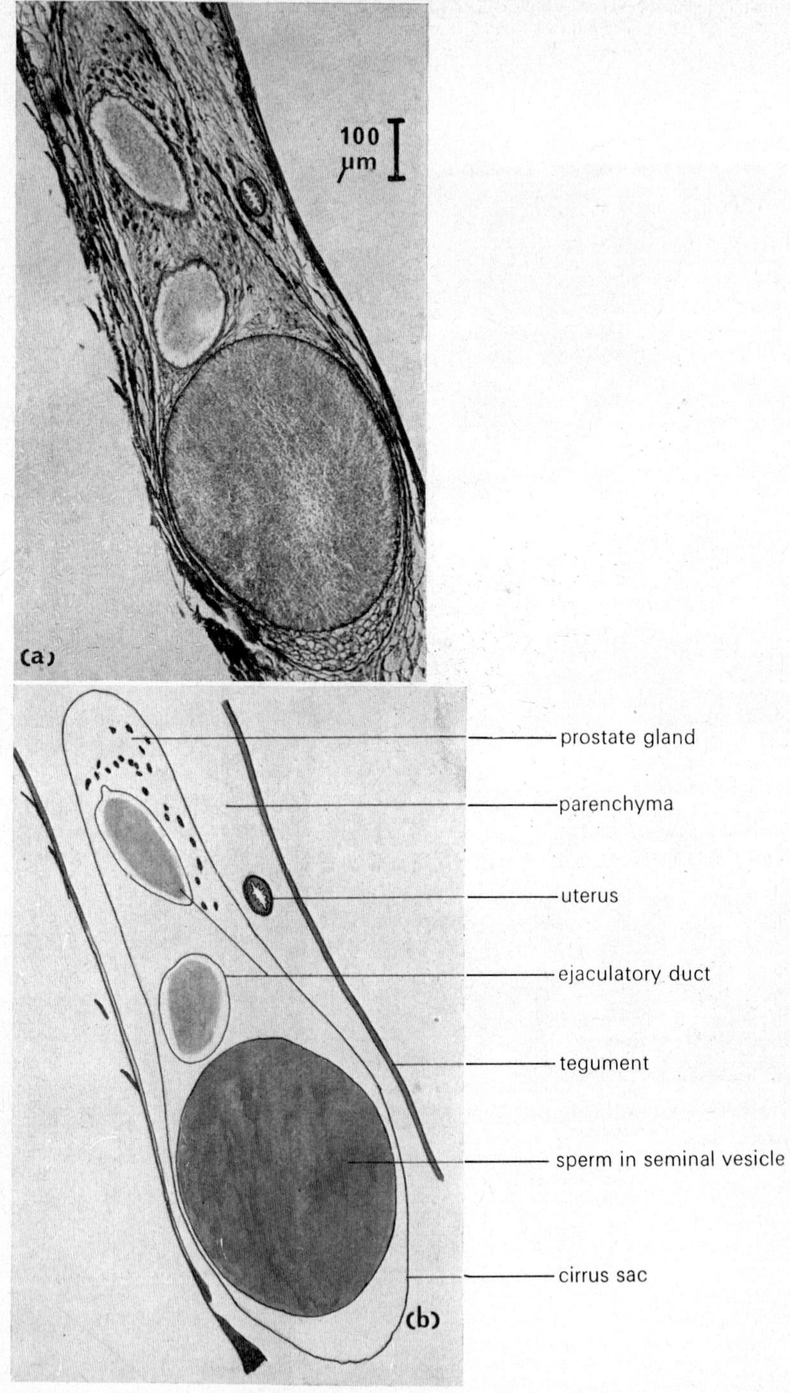

FIG. 1.24 (a) TS and (b) explanatory drawing of part of *F. hepatica* showing the cirrus sac at the level of the seminal vesicle (see X-X in Fig. 1.21).

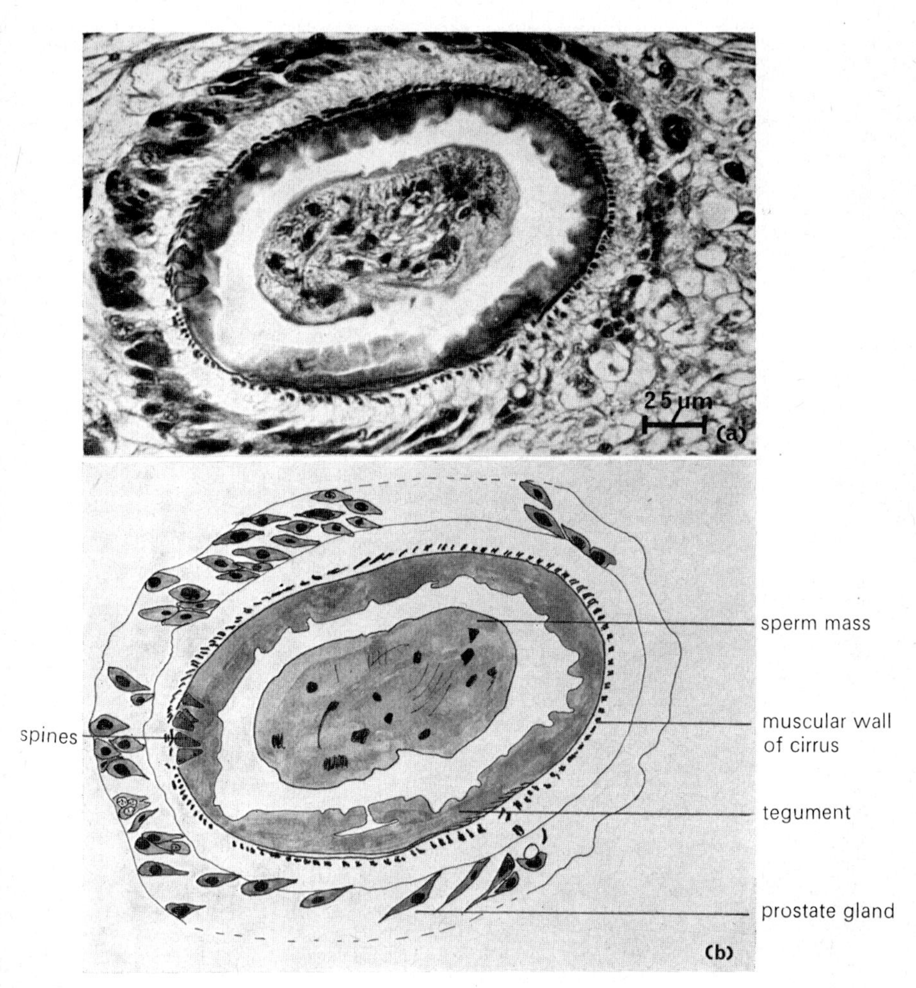

FIG. 1.25 (a) TS and (b) drawing of the cirrus of *F. hepatica* in its resting (inverted) position.

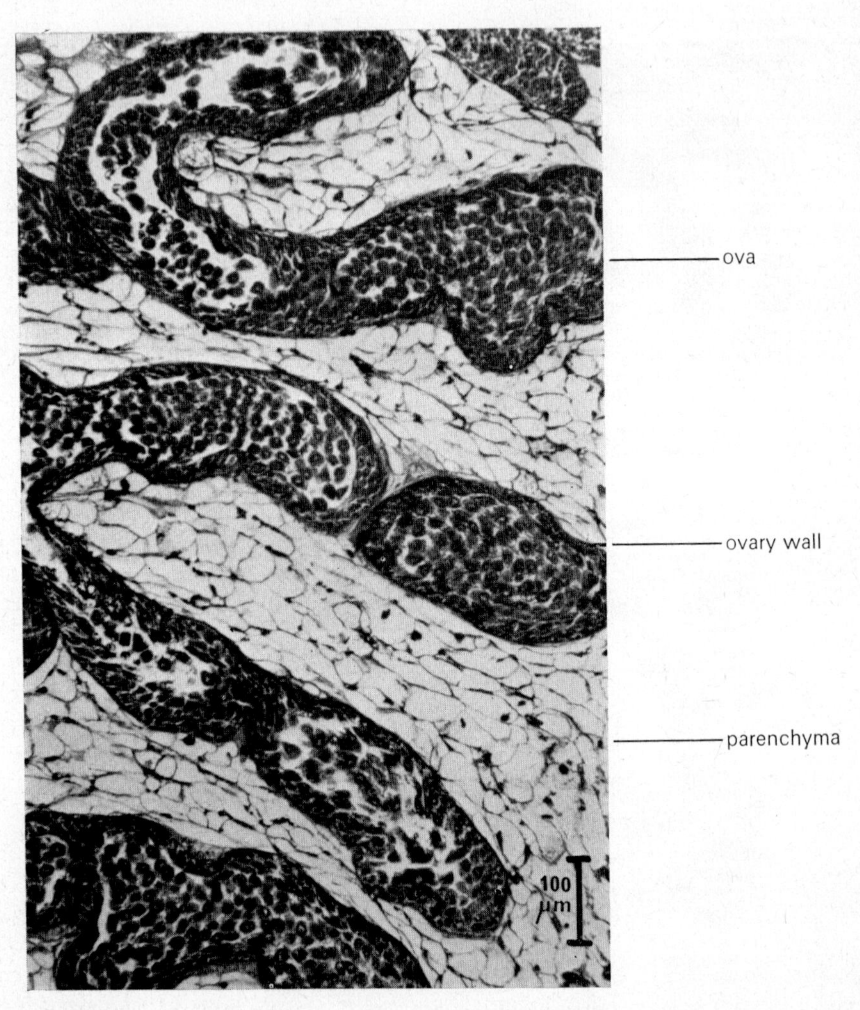

FIG. 1.26. Section through part of the branched ovary of *F. hepatica*.

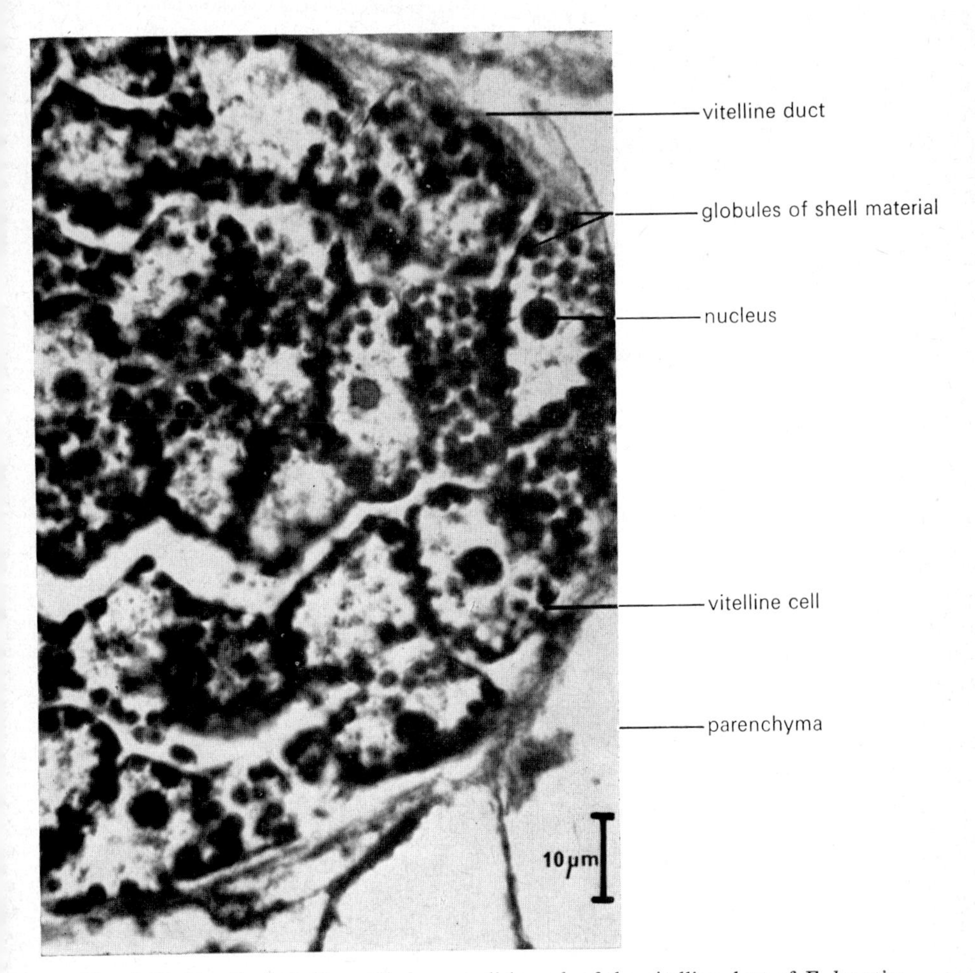

FIG. 1.27. Detail of vitelline cells in a small branch of the vitelline duct of *F. hepatica*.

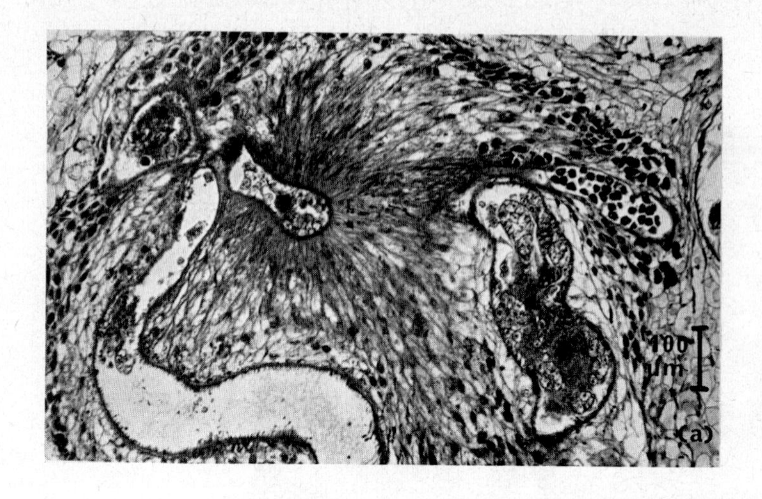

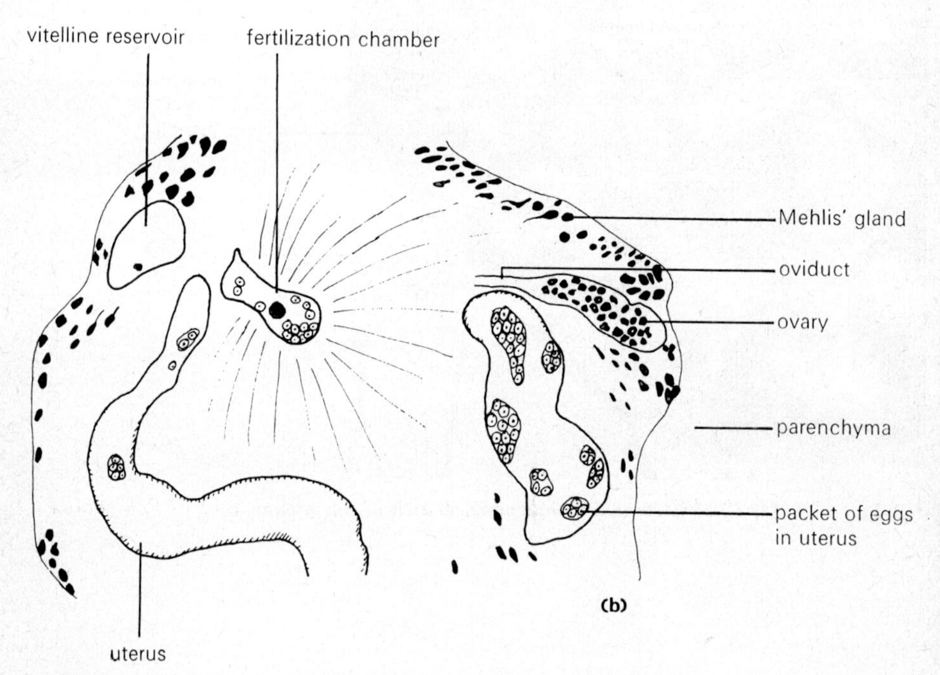

FIG. 1.28 (a) Section and (b) explanatory drawing of part of the female reproductive system of *F. hepatica*.

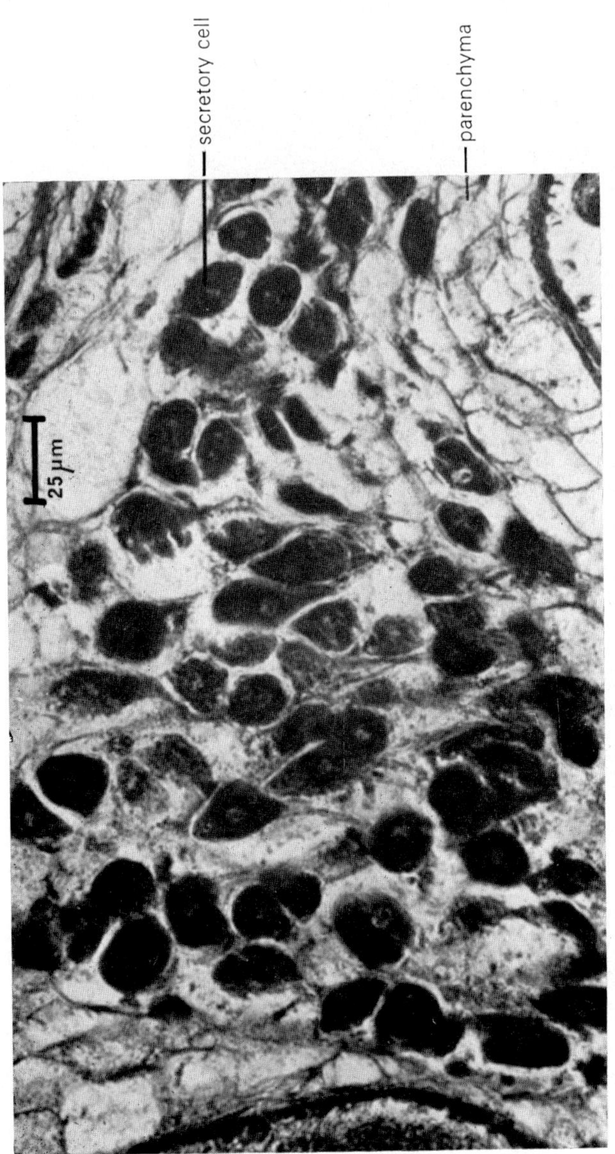

FIG. 1.29. Secretory cells of the Mehlis' gland in *F. hepatica*

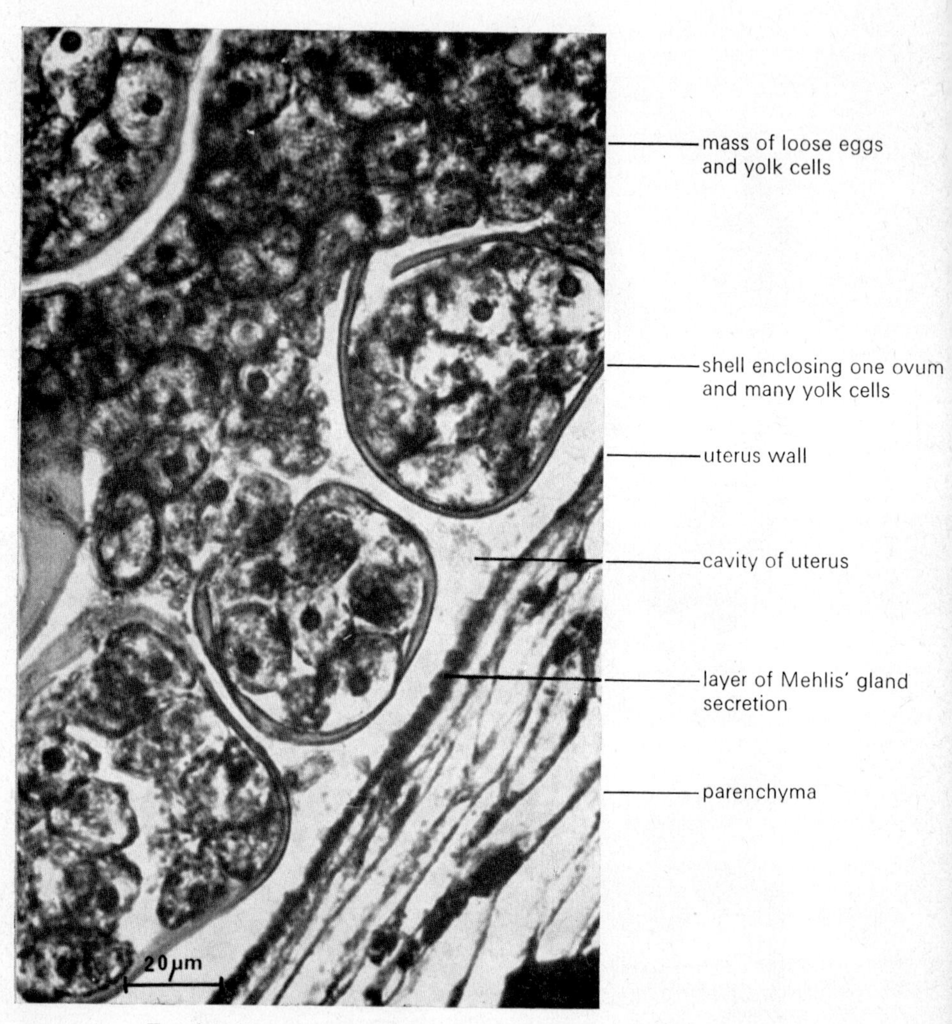

FIG. 1.30. Fertilized eggs and yolk cells in the uterus of *F. hepatica.*

The life cycle of trematodes generally includes several larval stages. In *Fasciola* the first larval stage is ciliated and free-swimming, but dies if it does not find an invertebrate host within a few hours. The two later larval stages are found parasitizing molluscs, while an encysted stage is passed on to the vertebrate host.

Class Cestoda

The cestodes or tapeworms are endoparasitic, and require two or more hosts for a complete life cycle. Larvae can be found in a variety of intermediate hosts, but adult cestodes normally live in the digestive tracts of vertebrates. They are well adapted for life in an anaerobic, nutrient-rich habitat and are without mouth, intestine and respiratory structures.

Tapeworms are usually ribbon-like in shape with an anterior head (scolex) provided with suckers and sometimes hooks for attachment to the host. The scolex is joined by a short neck to a string of flattened proglottids which form a strobila. A complete set of reproductive organs, and very little else, is found in each proglottid. In many species proglottids detach from the main strobila as they mature, and are thus a specialized means of reproduction. A small group of cestodes (Monozoa) resemble other platyhelminthes in that they lack a scolex and are not divided.

Example: *Moniezia* (*expansa*)

Moniezia is a eucestode (which has a 6-hooked larva), and occurs commonly in the intestine of sheep. It belongs to the order Cyclophyllidae, which are highly adapted to terrestrial hosts, and which possess a solid mesenchymatous scolex (Fig. 1.31) bearing four true suckers (acetabula) similar in structure to those of the digenetic trematodes (see Fig. 1.16). *Moniezia* consists of a long strobila of short wide proglottids (Fig. 1.32), which arise due to transverse constrictions of the neck region. The most recently formed proglottid is therefore nearest the scolex. The more distal proglottids are shed into the host's gut as they mature, and are passed out of the animal with the faeces, as bags packed with eggs.

The body wall is covered by a thick tegument: a metabolically active cytoplasmic layer. This is the site of exchange of all nutrients and metabolites since there is no gut. The tegument surface is thrown up into elongated villi (microtriches) which increase the surface area for absorption (Fig. 1.33). Underlying the tegument is a thin layer of circular and longitudinal sub-tegumentary muscles. Beneath this lies a sub-tegumentary layer composed mainly of gland cells interspersed with large cells which have protoplasmic extensions up into the tegument and are thought to secrete it. The parenchyma contains calcareous or lime cells and tracts of longitudinal muscle which run from the suckers in the scolex down through the strobila from one proglottid to another and finally atrophy in

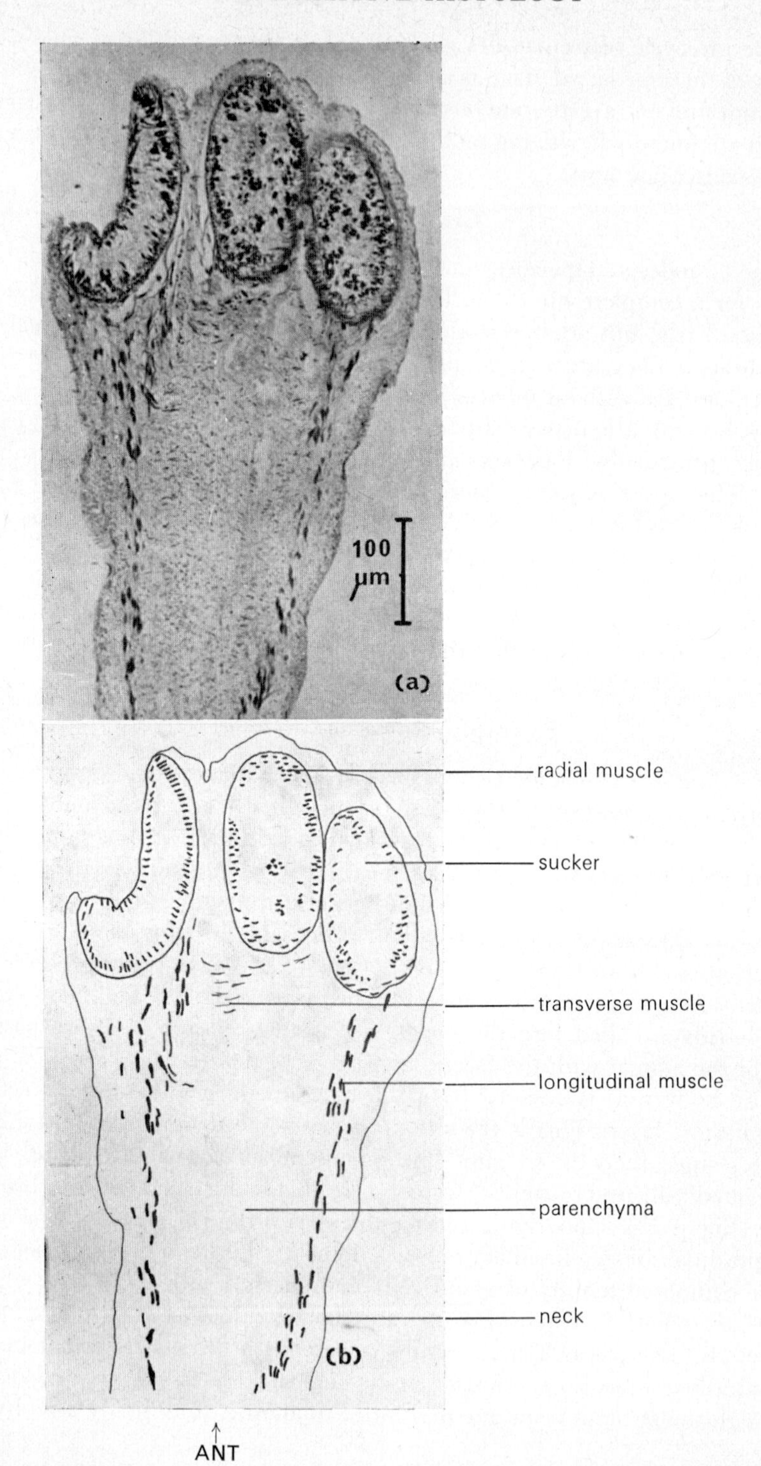

FIG. 1.31 (a) VLS and (b) drawing of the anterior end (scolex) of the tapeworm *Moniezia expansa*.

the ripe proglottids. A few transverse muscle fibres are found, particularly between suckers in the scolex. In developing proglottids the parenchyma is much more dense and heavily nucleated than in the ripe ones, and contains groups of cells from which the gonads develop.

Moniezia is unique among Cyclophyllidae in possessing interproglottidal glands along the posterior edge of each proglottid [Fig. 1.34]. Each gland consists of a cluster of gland cells around a central sac. Their function is as yet unknown.

Nervous system

The nerve net of cestodes has sunk right into the parenchyma to a position just beneath the longitudinal muscle bands (see Fig. 1.31). Paired cerebral ganglia in the scolex are joined by transverse commissures to a ring commissure. Paired dorsal and ventral nerves and a pair of large ganglionated lateral nerve trunks run the length of the animal from this ring. At least one nerve ring connects these longitudinal nerves in each proglottid. The nerve cell bodies are mainly located in the commissures. Accessory lateral cords are found in some species. Figure 1.35 shows a plan of the nervous system in transverse section in both anterior and posterior regions of the body.

Sense organs

Cestodes do not possess any special sense organs but the body surface and suckers are well supplied with sensory nerve endings (such as in Fig. 1.18) and simple stretch receptors.

Excretion

The excretory system of cestodes is very similar to that of turbellarians. Clusters of minute flame-cells are connected to paired dorsal and ventral excretory canals lying in the parenchyma. These canals run from the scolex, where they form a plexus, down the length of the worm. As the mature proglottids are shed the dorsal canals die out and the fractured ends of the larger ventral canals act as excretory pores. The ventral canals are connected by transverse canals in the posterior part of each proglottid (Fig. 1.36, see also Fig. 1.32).

Reproduction

With very few exceptions the cestodes are hermaphrodite, and are generally protandrous (i.e. male gametes ripen before female). In *Moniezia* there are two complete sets of male and female reproductive organs in each mature proglottid (Fig. 1.37), but many other cestodes possess only one set of each. In most species the male and female ducts open into a common genital atrium which opens to the outside by a genital pore.

In the male system numerous small round testes lie scattered in the

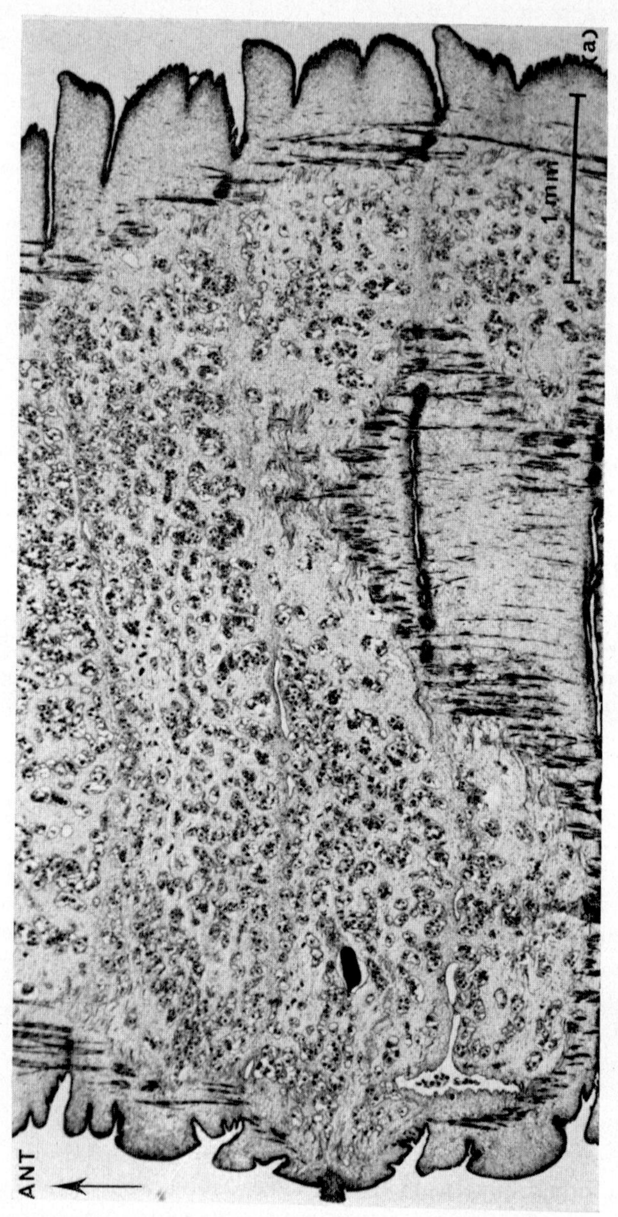

FIG. 1.32

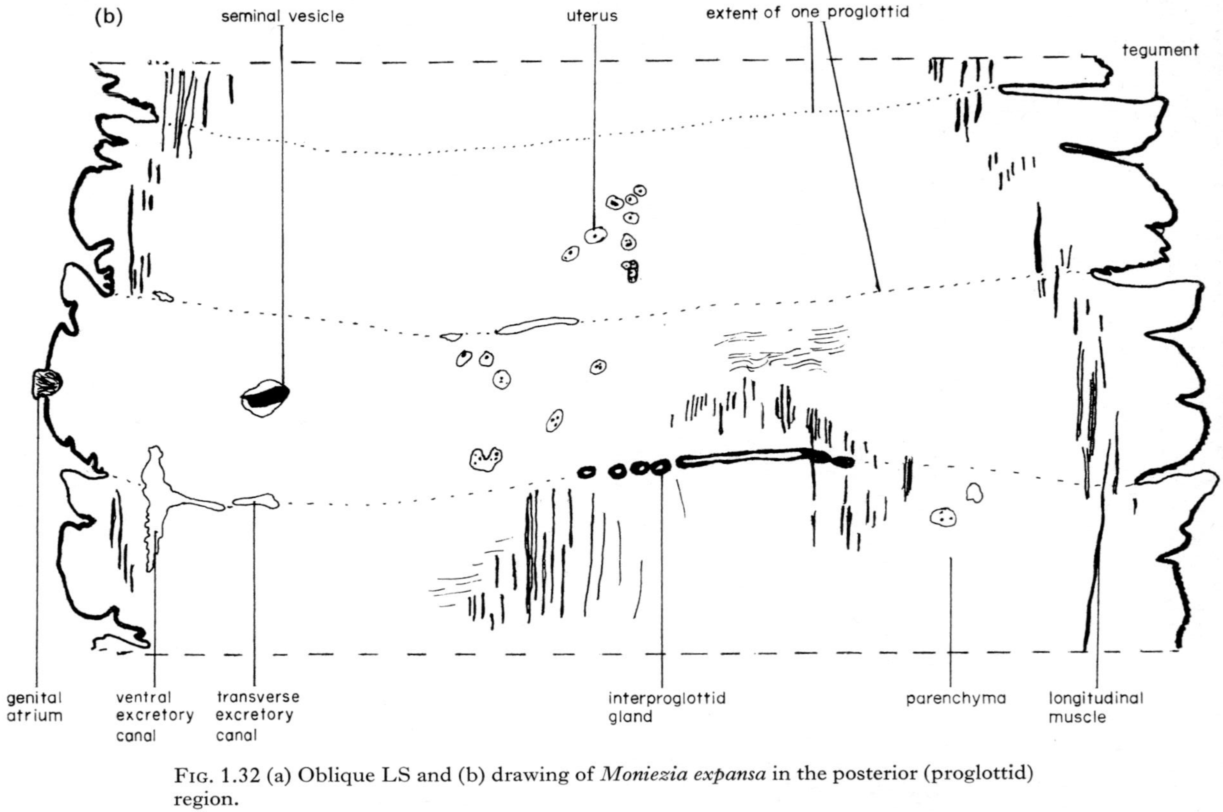

FIG. 1.32 (a) Oblique LS and (b) drawing of *Moniezia expansa* in the posterior (proglottid) region.

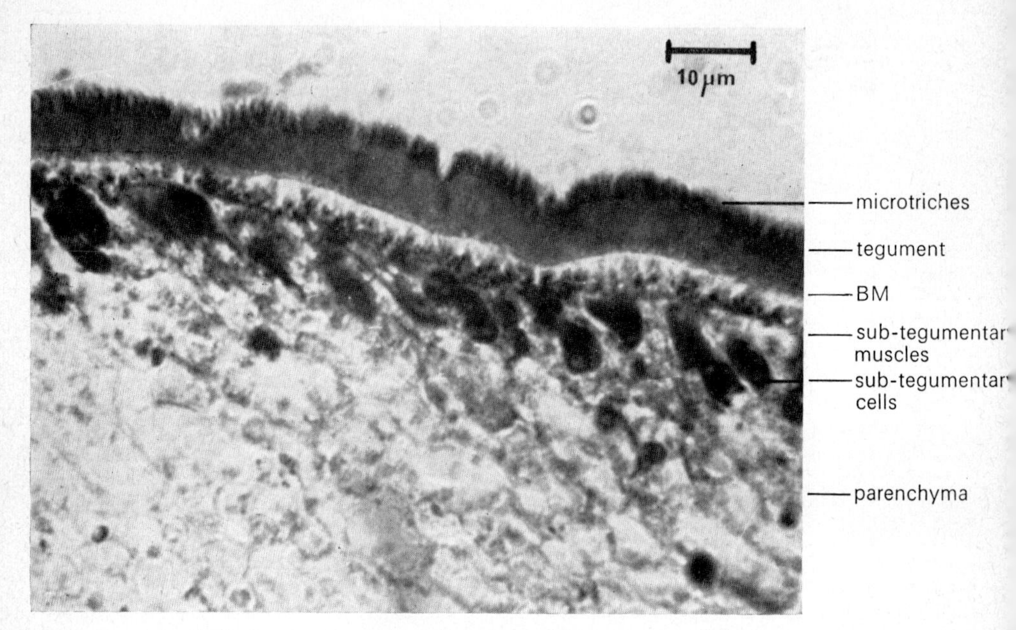

FIG. 1.33. Section through the tegument of *M. expansa*. Note the microtriches.

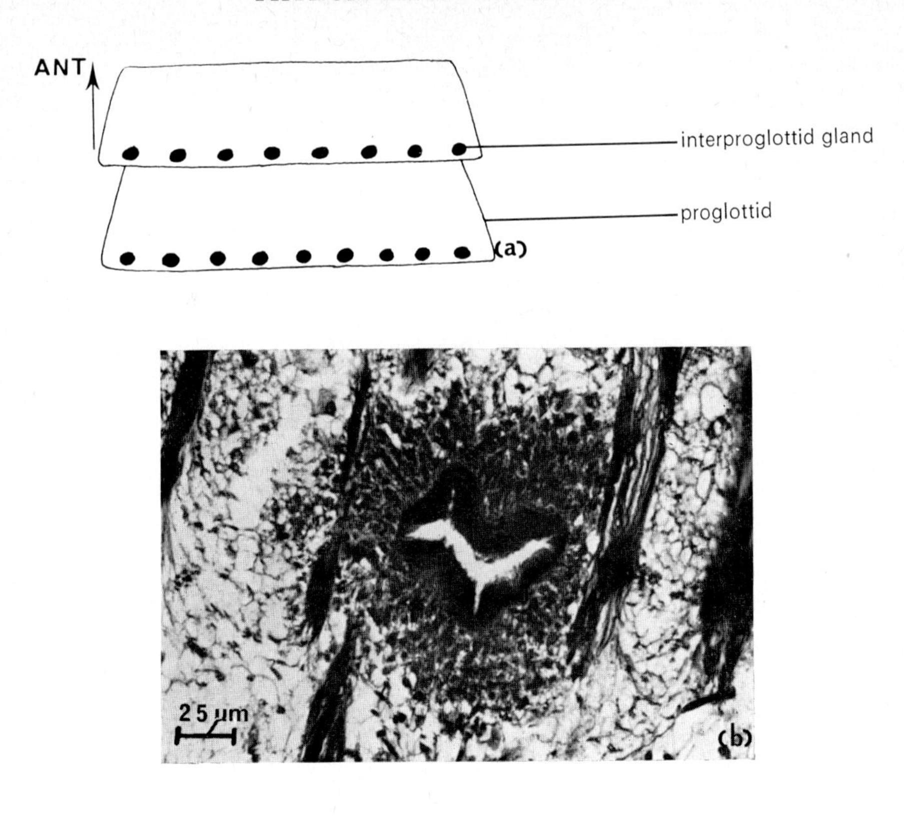

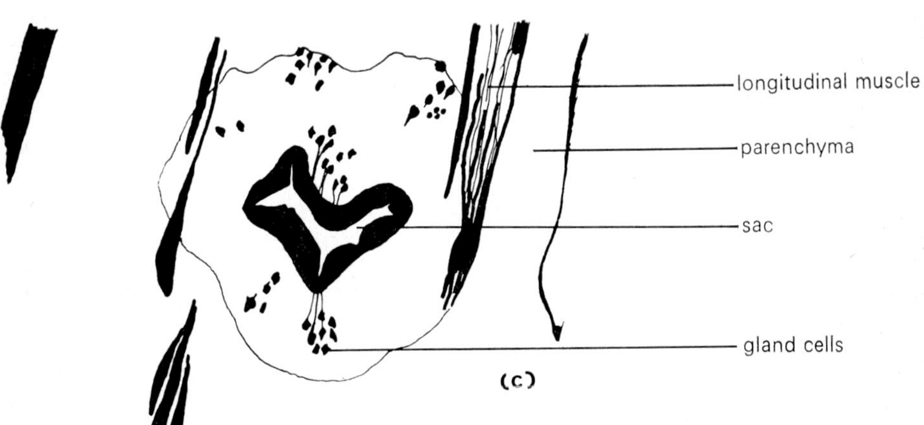

FIG. 1.34 (a) Diagram of two proglottids of *Moniezia* showing the position of the interproglottid glands; (b) detail and (c) explanatory drawing of an interproglottid gland of *M. expansa*.

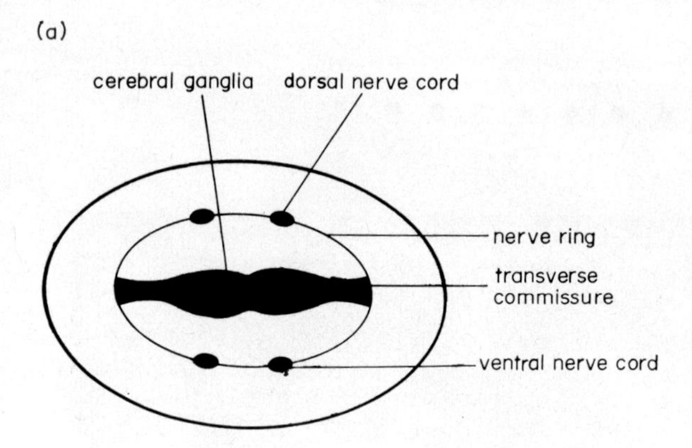

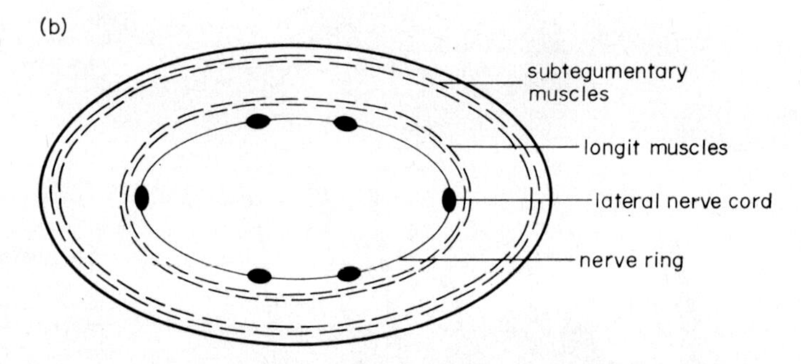

FIG. 1.35. Plan of a TS of *M. expansa* in (a) the scolex region and (b) the proglottid region, showing the arrangement and position of the nervous system.

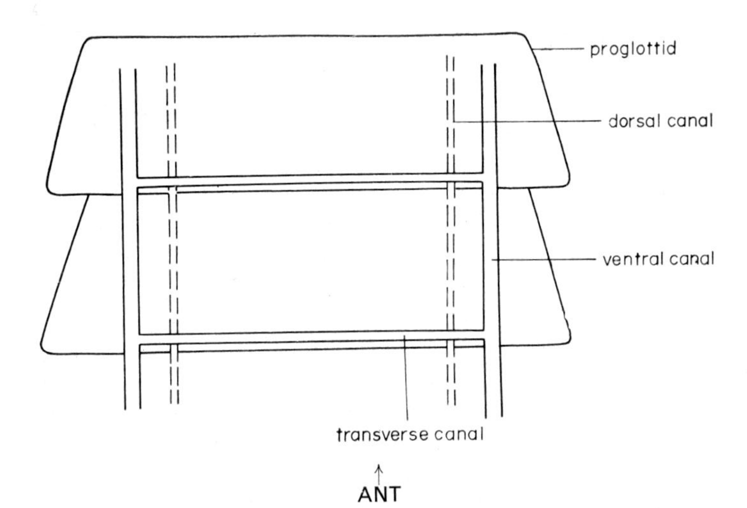

FIG. 1.36. Plan of the excretory canals in the proglottid region of *M. expansa* (viewed from the ventral side).

dorsal parenchyma. Fine sperm ductules lead from them into a large coiled sperm duct which opens into a muscular cirrus sac on each side of the proglottid. The cirrus is eversible (Fig. 1.38), and in some species is lined with bristles, hooks or spines.

The female system is like that of the digenetic trematodes although there are several major differences. The ovary is similar but the vitelline gland, which is smaller and more compact, is chiefly concerned with yolk production. It lies behind the ovary and opens by a short duct into a slightly enlarged ootype where fertilization occurs. The eggs develop in a uterus, which is blind-ended in the Cyclophyllidae [Fig. 1.39(a)]. The uterus in *Moniezia* has many lateral diverticula, thus forming a reticulate sac that fills the entire ripe proglottid (see Fig. 1.32). It begins to develop only after the gonads are mature, and has no direct connection with the genital atrium, unlike the trematodes. The oviduct is linked to the atrium by a long vagina (homologous to the Laurer's canal of the Digenea) which has its middle portion enlarged to form a seminal receptacle for the storage of sperm [Fig. 1.39(b)]. Also opening into the ootype is a small "shell-gland", similar in structure to the Mehlis' gland of trematodes. It also plays no part in shell formation, but secretes mucopolysaccharides which are possibly hormonal in nature.

Embryos are released by disintegration of the shed proglottid, and development is by a hooked larva which attaches to the tissues of the intermediate host.

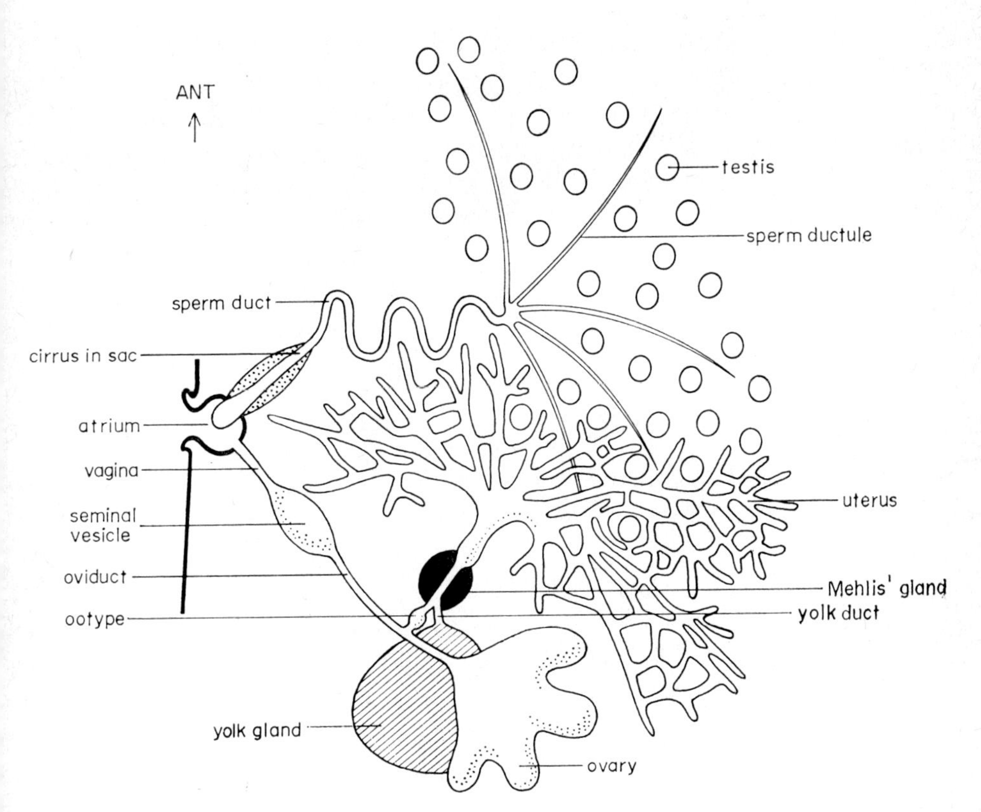

FIG. 1.37. Plan of one hermaphrodite reproductive system of a proglottid of *M. expansa*. Each proglottid bears two such sets of reproductive organs.

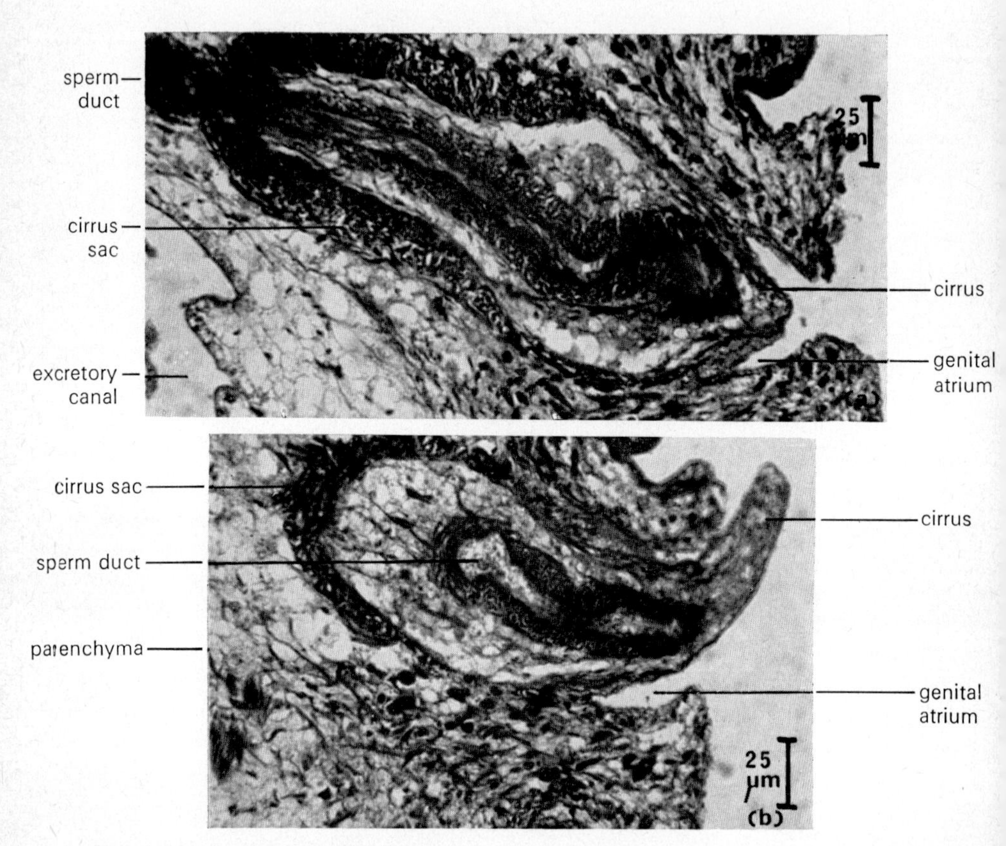

FIG. 1.38. LS of a cirrus of *M. expansa* in (a) the resting position in the genital atrium; and (b) the semi-everted position.

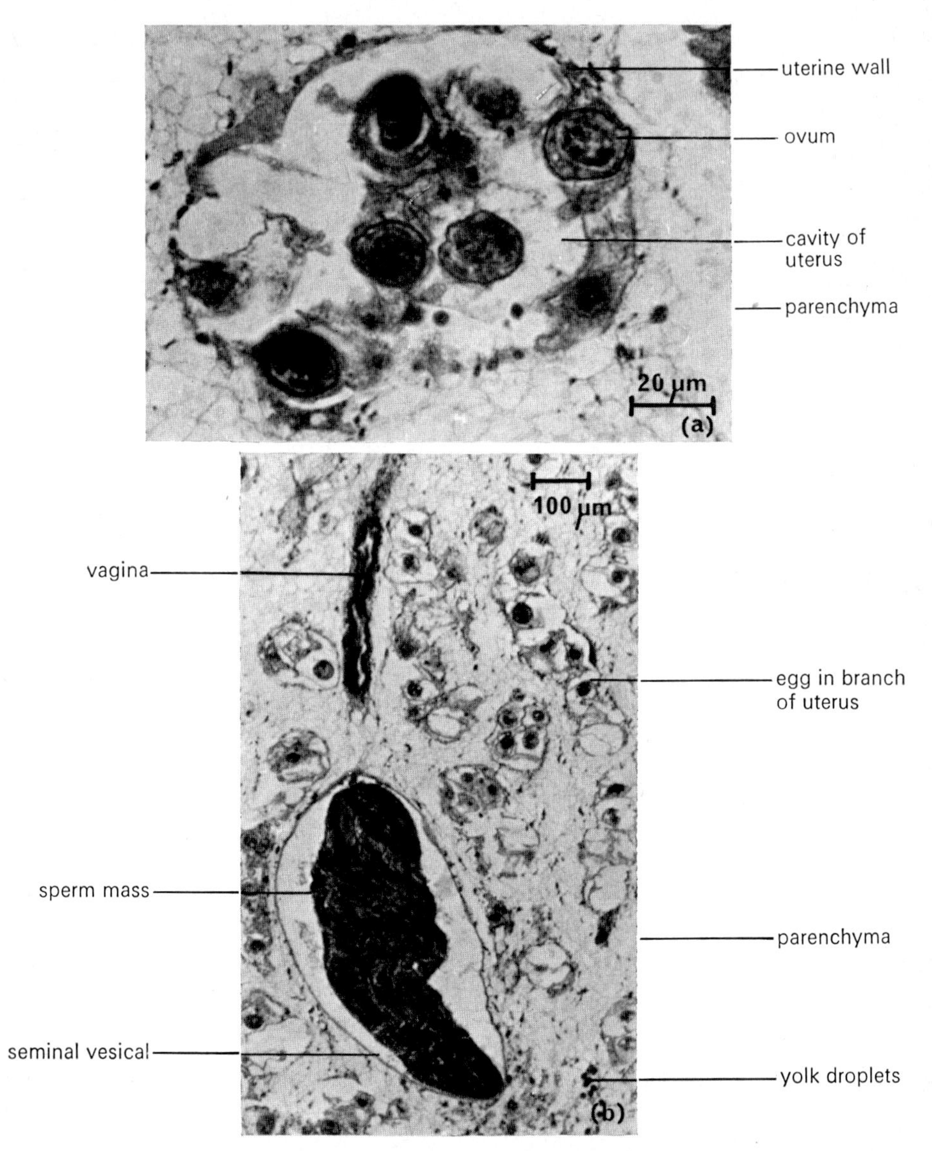

Fig. 1.39 (a) Detail of developing ova in a branch of a uterus of *M. expansa*; (b) section through the vagina, seminal vesicle and parts of the uterus of one proglottid of *M. expansa*.

Chapter 2

Phylum Acanthocephala

Acanthocephalans form a small group of parasitic worms showing similarities to both platyhelminths and nematodes. They all require two hosts: the larvae living mainly in arthropods and the adults parasitizing vertebrates. They are bilaterally symmetrical pseudocoelomates, and possess a characteristic proboscis covered with rows of curved hooks or spines with which they attach to the host tissues. The body consists of a small anterior presoma bearing the proboscis, a short neck sometimes thickened anteriorly to form a neck-bulb, and a trunk which may be covered with spines or divided into superficial annuli. Like the cestodes, the acanthocephalans do not possess a gut but absorb all nutrient material they require through the body wall. There are no excretory organs except in one order (Archiacanthocephala) in which modified paired protonephridia are associated with the reproductive system. Since they live in partially or wholly anaerobic conditions, acanthocephalans rely on anaerobic metabolism, and no respiratory structures are present. There are insufficient differences between acanthocephalans to split them into classes.

Example: *Pomphorhynchus* (*laevis*)

P. laevis is a commonly occurring acanthocephalan parasitizing *Gammarus pulex* and a variety of freshwater fish. The adult has a short trunk divided superficially into annuli, an elongate neck bearing a neck-bulb and a conical proboscis covered with rooted hooks (Fig. 2.1).

The body wall consists of a mucopolysaccharide epicuticle overlying a thick syncytial hypodermis, and lined with circular and longitudinal muscle layers. The hypodermis contains a closed network of lacunar vessels in which absorbed food materials are thought to circulate (Fig. 2.2). In the neck region the radial layer containing these lacunae is reduced in thickness and circular muscle is absent (Fig. 2.3). The longitudinal muscle fibres shown are typical of the rather unusual muscles that occur in this phylum, which possess peripheral fibrils and a cytoplasmic core bearing the nucleus. Just above the cuticularized partition which separates the neck and trunk, two long bulges of hypodermis (lemnisci) project from the base of the neck into the pseudocoelom (Fig. 2.4). Their function is not really known but they are presumed to act as reservoirs for lacunar fluid and to assist proboscis movements.

The proboscis, shown in transverse section in Fig. 2.5(a), is radially symmetrical and is also covered by an epicuticle overlying a thin hypodermis. It is armed with cuticular hooks which have their roots in the hypodermis [Fig. 2.5(b)] and which are arranged in alternating radial rows (quincunxial arrangement). The proboscis can be withdrawn by large retractor muscles into the proboscis receptacle (Fig. 2.6). This is a cup-shaped structure opening from the base of the proboscis and is composed of a double layer of muscles.

Nervous system

The nervous system is much reduced, consisting only of a cerebral ganglion and its associated nerves. There is no sign of a nerve net. The cerebral ganglion is located on the inside ventral wall of the proboscis receptacle (see Fig. 2.6). An enlarged picture of this ganglion shows a few large peripheral neurones surrounding a web of axons, the neuropile (Fig. 2.7). Two main nerve trunks, sheathed in muscle cells (retinaculae,) run from the ganglion through the receptacle wall to the body wall where they lie in parallel with the longitudinal muscle layer. In the male paired genital ganglia are found at the base of the penis.

As in other parasites, there are very few sense organs in the Acanthocephala, and these are mostly simple tactile receptors on the proboscis.

Reproduction

The sexes are separate in the Acanthocephala, and, as is often the case in parasites, the female is larger than the male.

In the male (Fig. 2.8) two testes are suspended one behind the other in a nucleated strand of connective tissue (ligament strand) which is believed to represent the endoderm and which stretches from the base of the receptacle to the hind end of the body. Testes and cement glands empty their products into a common sperm duct. These structures are all enclosed in a thin-walled ligament sac, a hollow tube of connective tissue peculiar to the Acanthocephala (see also Fig. 2.1). The ends of the sperm ducts and cement glands are enclosed by a muscular genital sheath continuous with the ligament sac (Fig. 2.9). The two sperm ducts unite to form a common sperm duct which discharges the thread-like sperm into a penis. The penis projects into a cup-shaped bursa that can be everted through the gonopore during copulation. Once the penis is inserted into the female, the bursa becomes sealed round the female gonopore by cement gland secretions, which remain as a sealing plug after copulation.

In the female (Fig. 2.10) paired ovaries, also derived from the ligament strand, break up during development into ovarian balls (Fig. 2.11). Each ball consists of a central multi-nucleate syncytium from which the surrounding ovogonia are formed and move outwards (Fig. 2.12). These ovarian balls lie within the ligament sac, or freely in the pseudocoel if (as in the case of *P. laevis*) the ligament sac ruptures. Internal fertilization

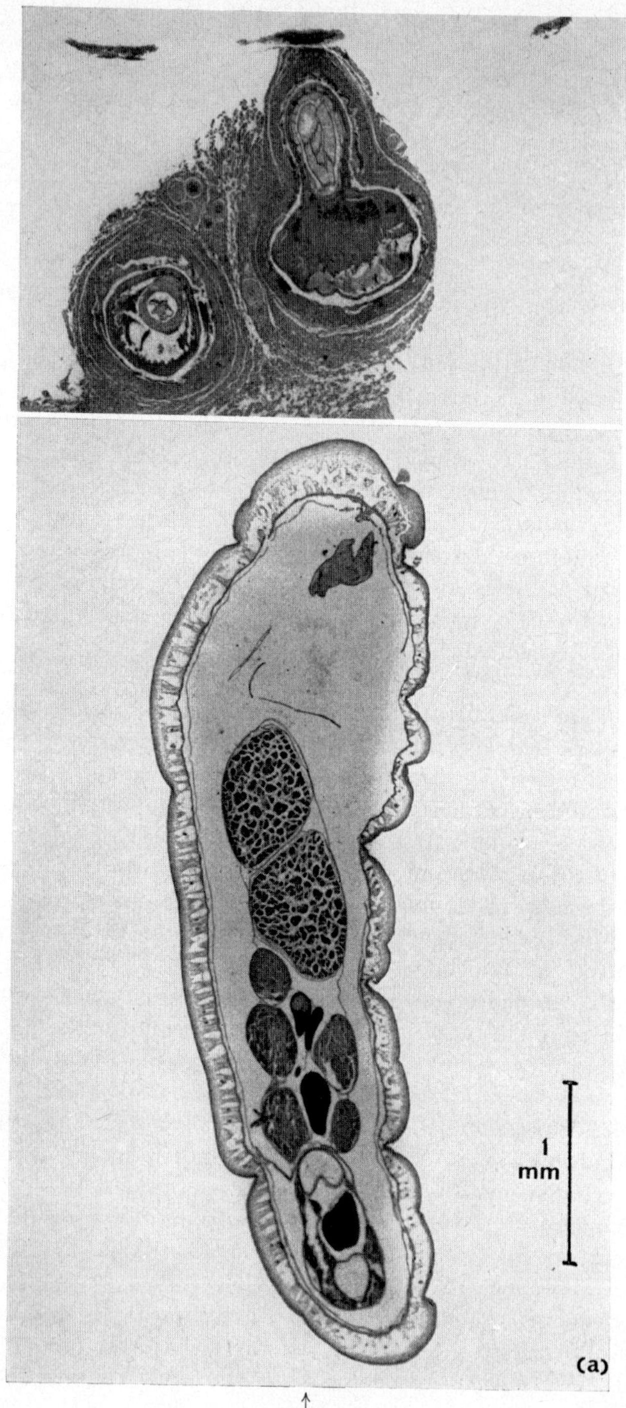

Fig. 2.1

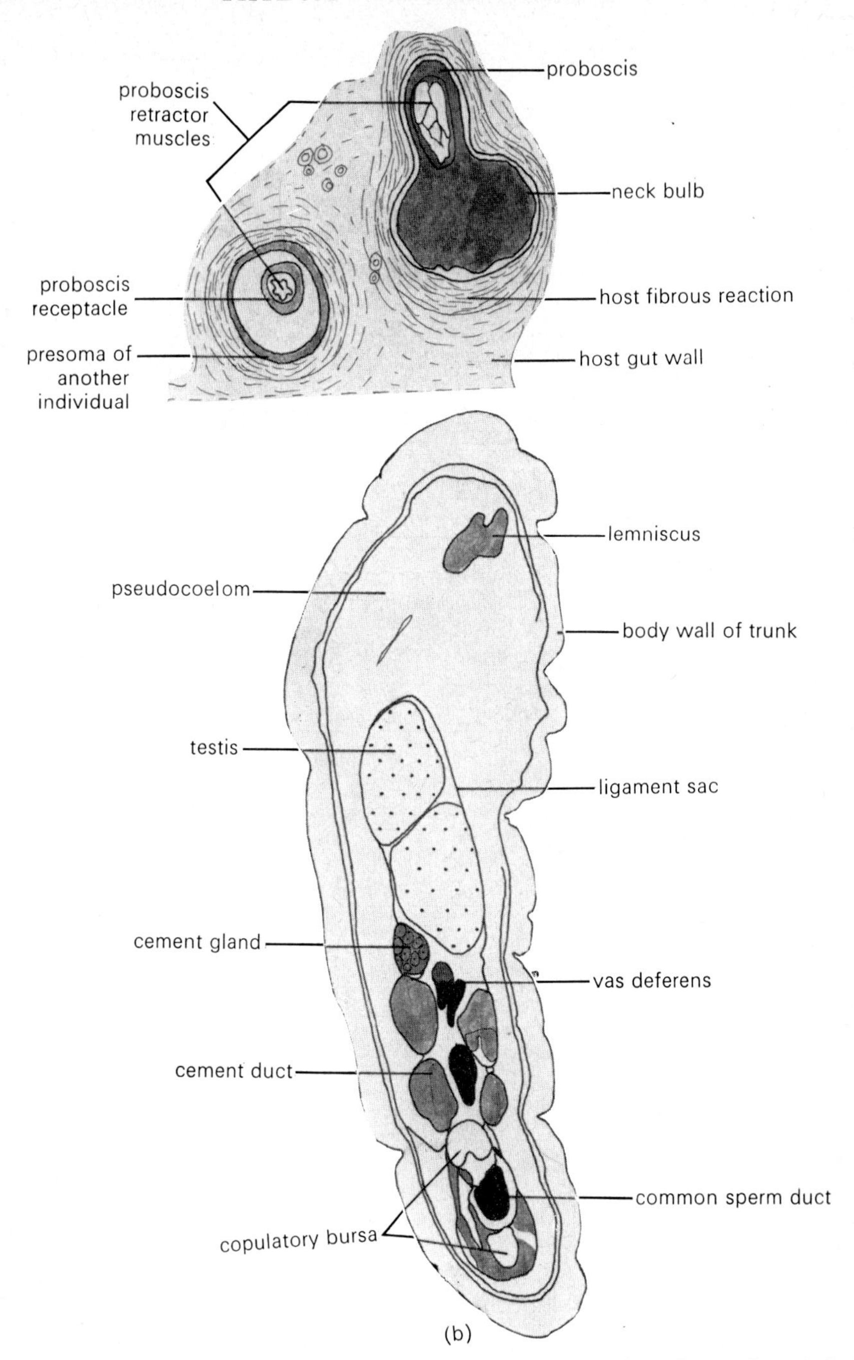

FIG. 2.1 (a) Composite oblique LS and (b) explanatory drawing of a male acanthocephalan *Pomphorhynchus laevis*, showing the relationship of the trunk to the proboscis and neck-bulb which are embedded in the gut wall of a fish host. The proboscis of a second individual is also seen in the picture.

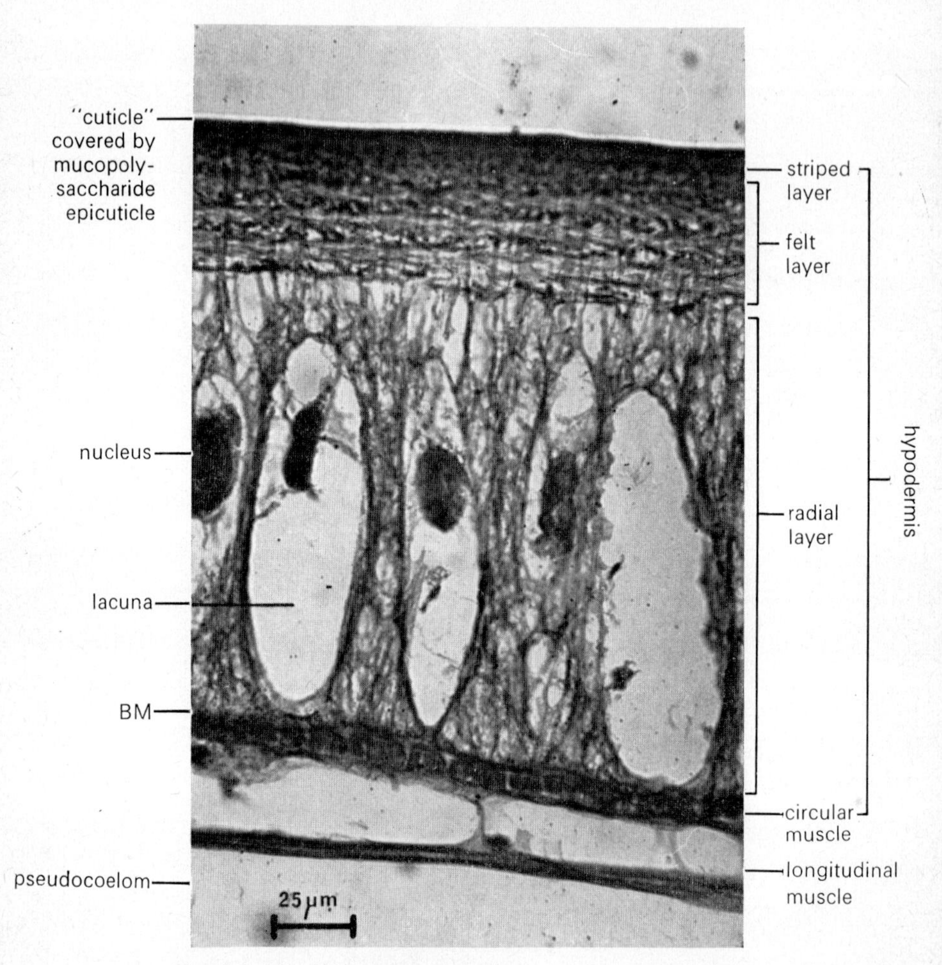

FIG. 2.2. LS of the body wall of *P. laevis* in the trunk region.

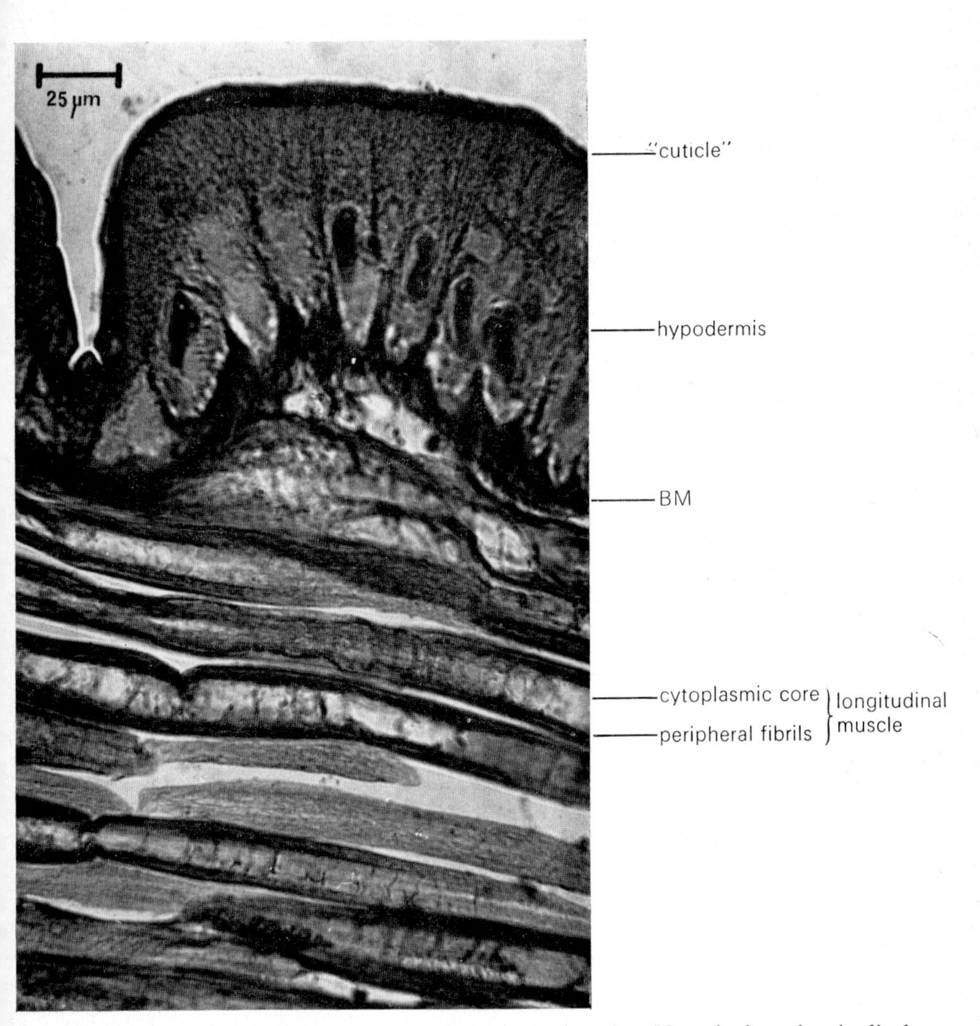

FIG. 2.3. LS of the body wall of *P. laevis* in the neck region. Note the large longitudinal muscle fibres with their peripheral fibrils and central cores of cytoplasm.

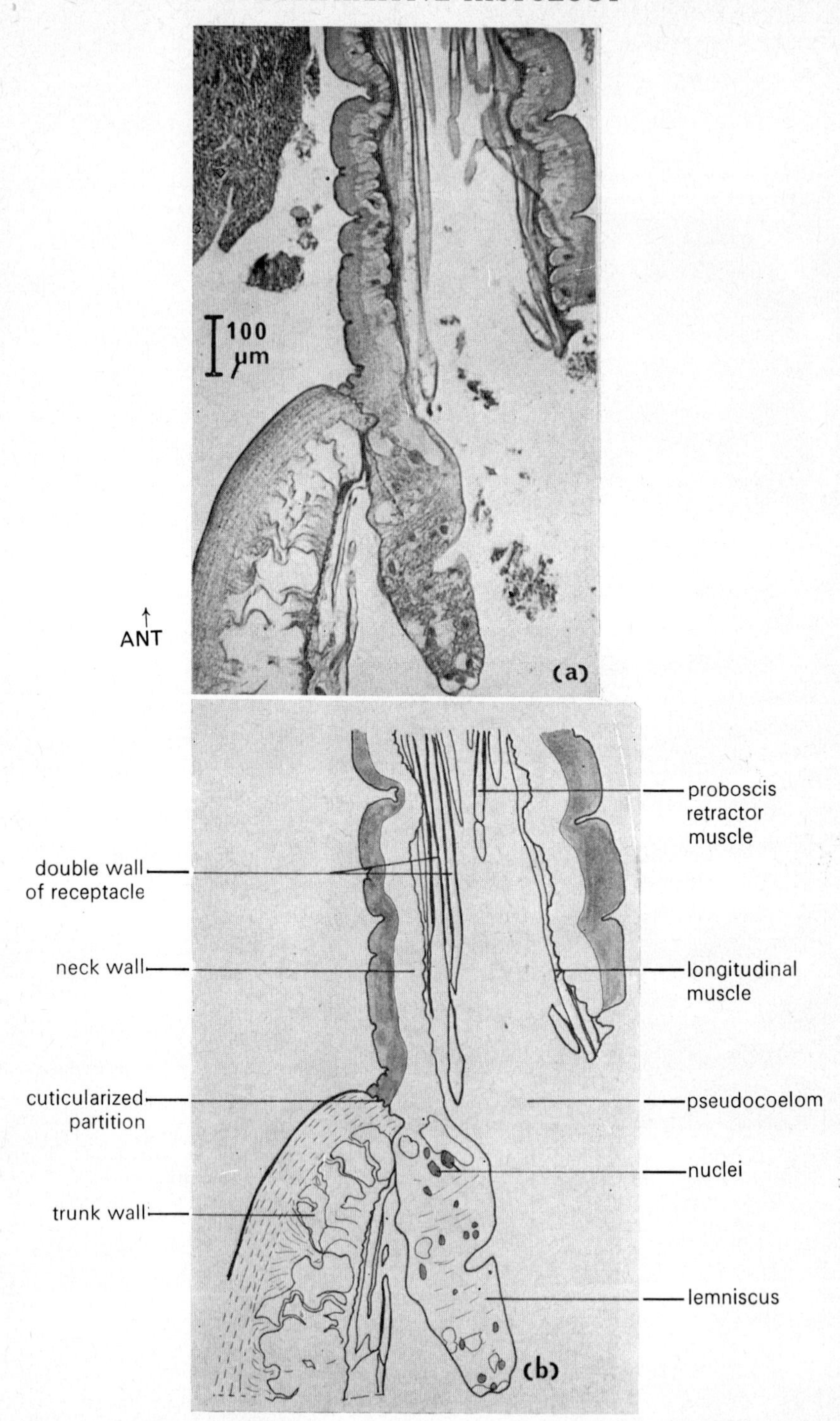

FIG. 2.4 (a) LS and (b) explanatory drawing of the neck and anterior trunk region of *P. leavis* showing a lemniscus.

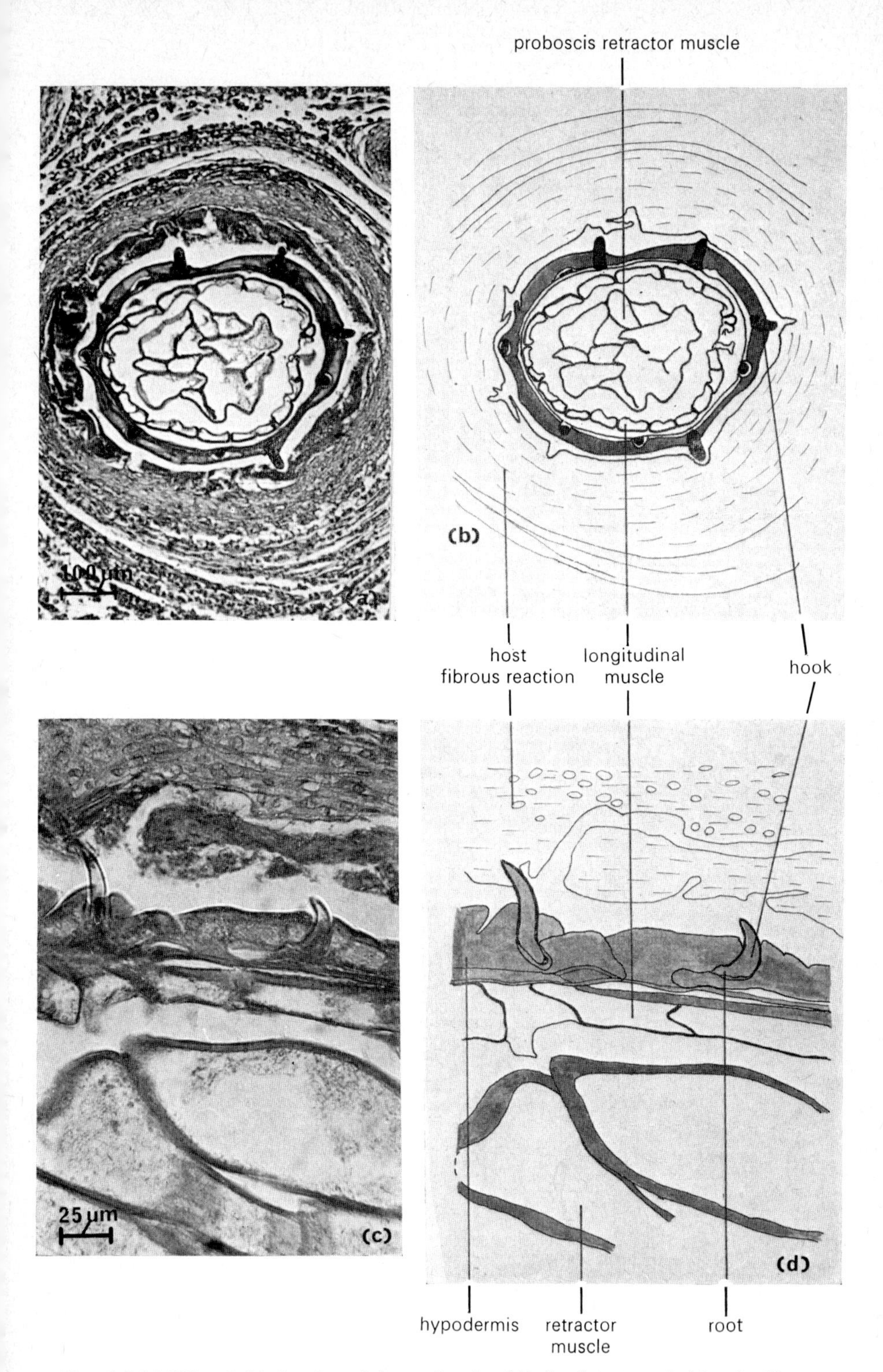

FIG. 2.5 (a) TS and (b) drawing of the proboscis of *P. laevis* surrounded by the fibrous reaction in the host gut wall; (c) detail and (d) drawing of the proboscis wall of *P. laevis*.

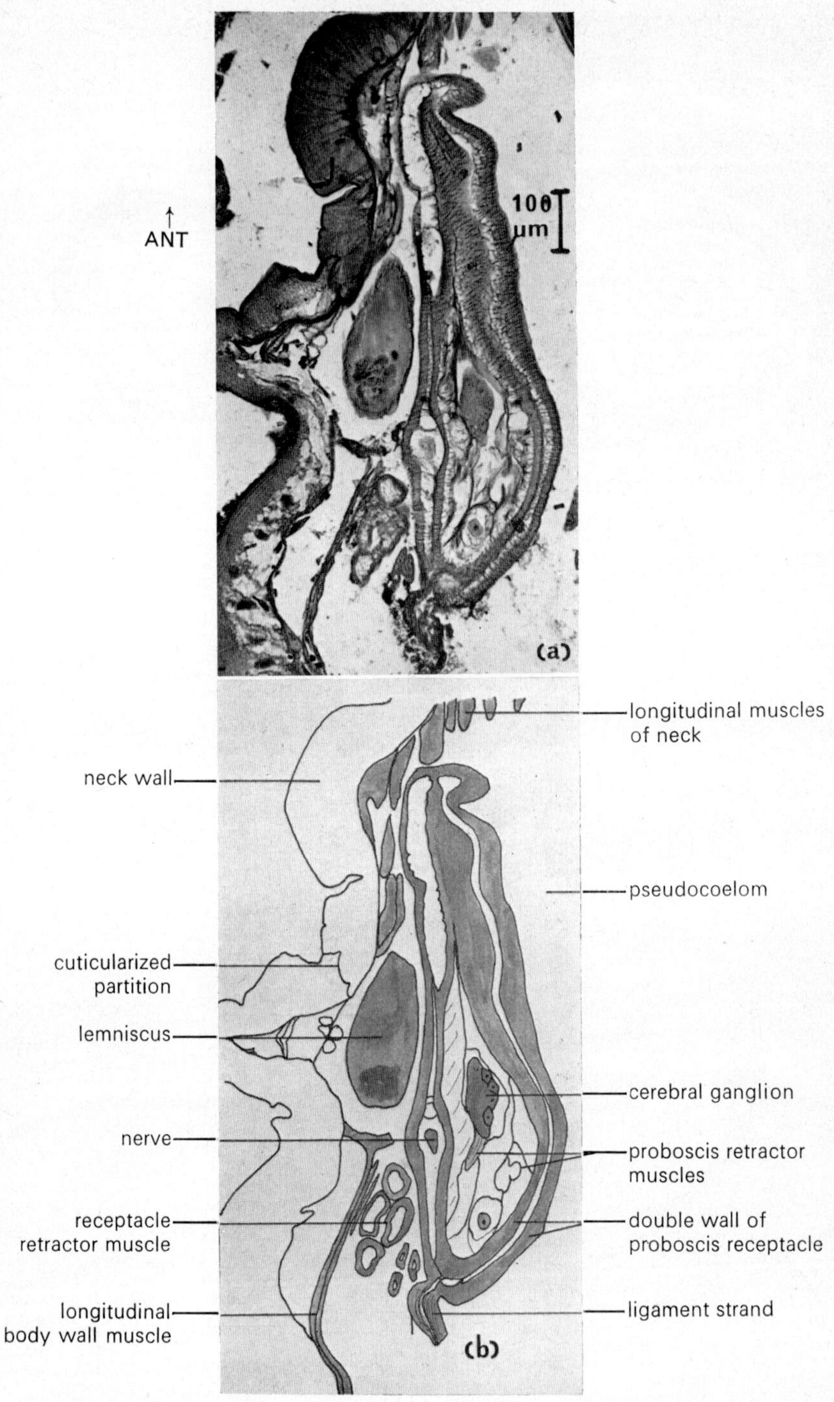

FIG. 2.6. (a) LS and (b) drawing of the proboscis receptacle of *P. laevis*. Note the position of the cerebral ganglion inside the receptacle.

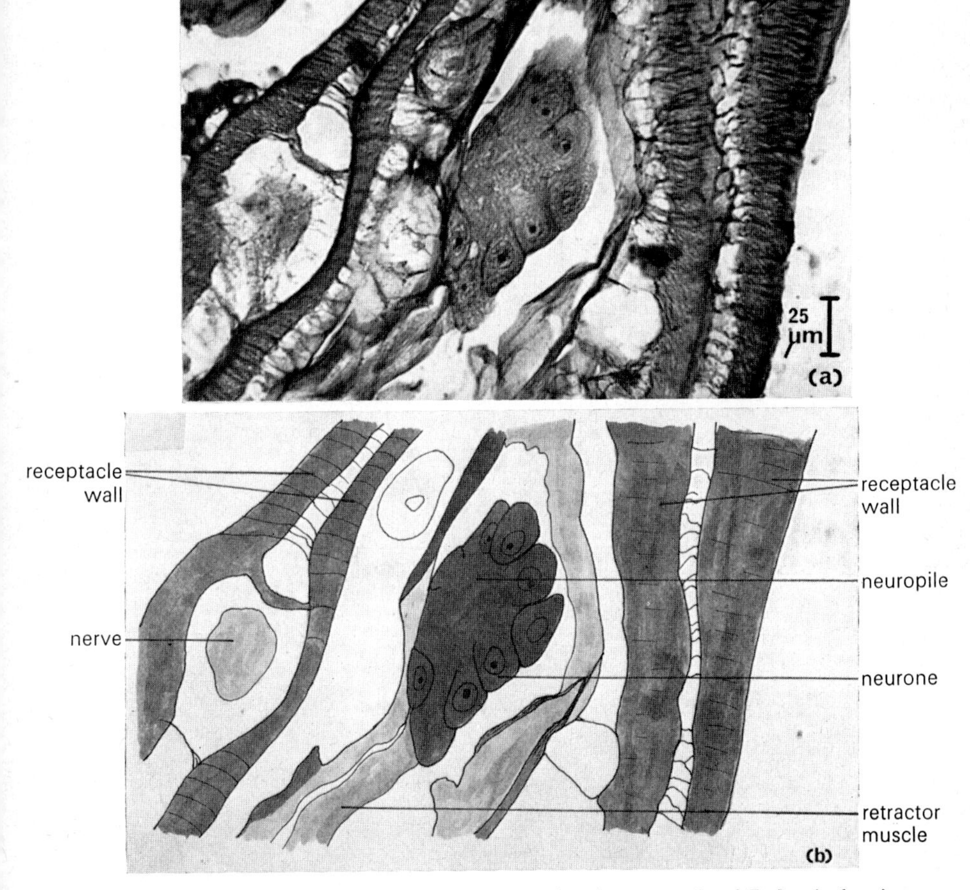

FIG. 2.7. (a) LS and (b) drawing of part of the proboscis receptacle of *P. laevis* showing the structure and position of the cerebral ganglion.

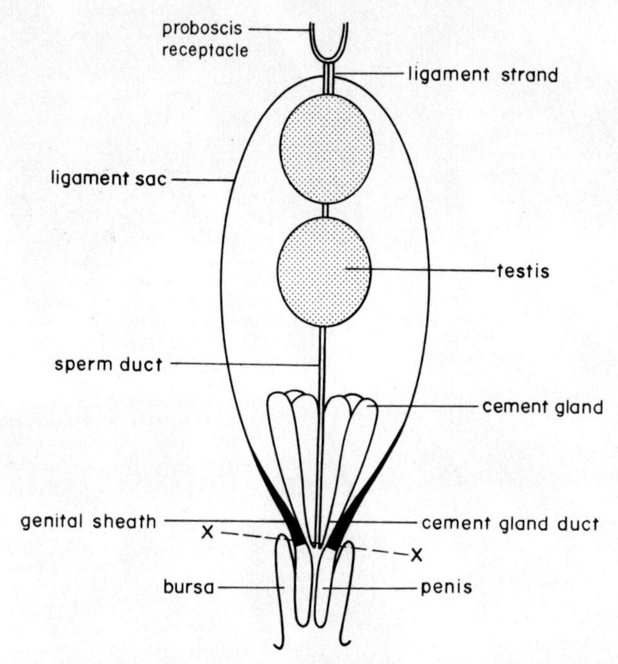

FIG. 2.8. Plan of the male reproductive system of *P. laevis* (X-X is the section shown in Fig. 2.9).

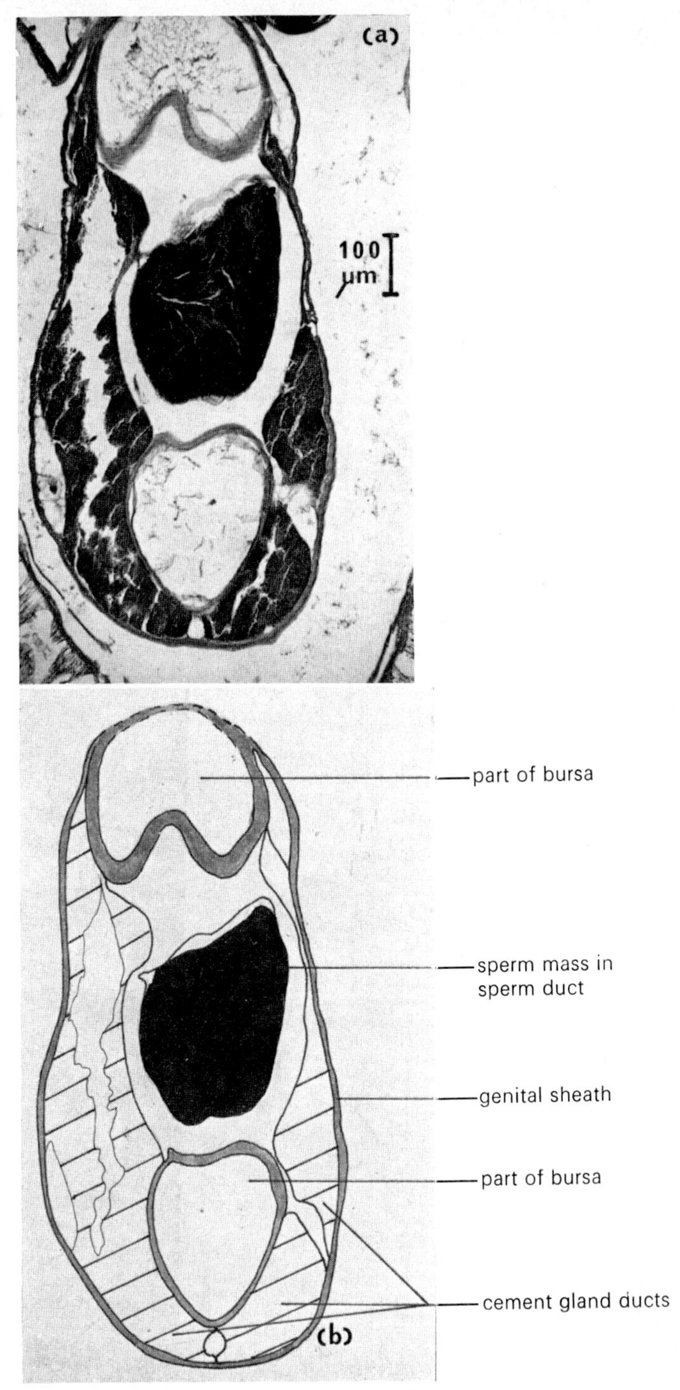

FIG. 2.9. (a) Oblique section and (b) explanatory drawing of the posterior end of the male reproductive system of *P. laevis* (see X-X in Fig. 2.8).

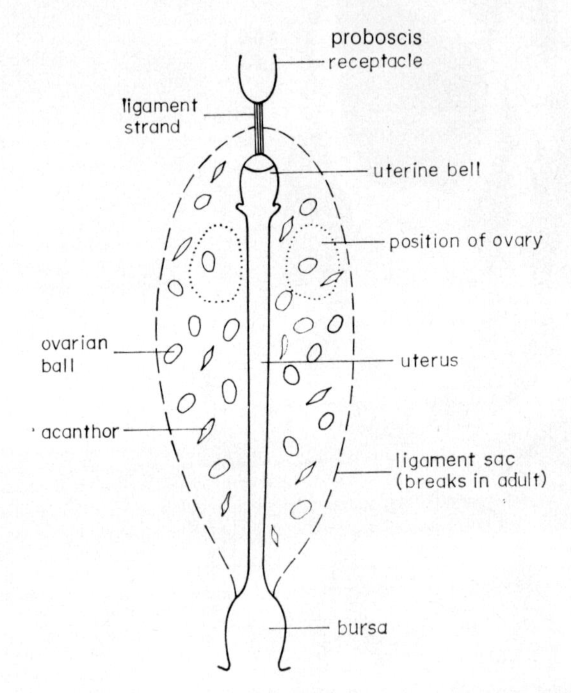

FIG. 2.10. Plan of the female reproductive system of *P. laevis*.

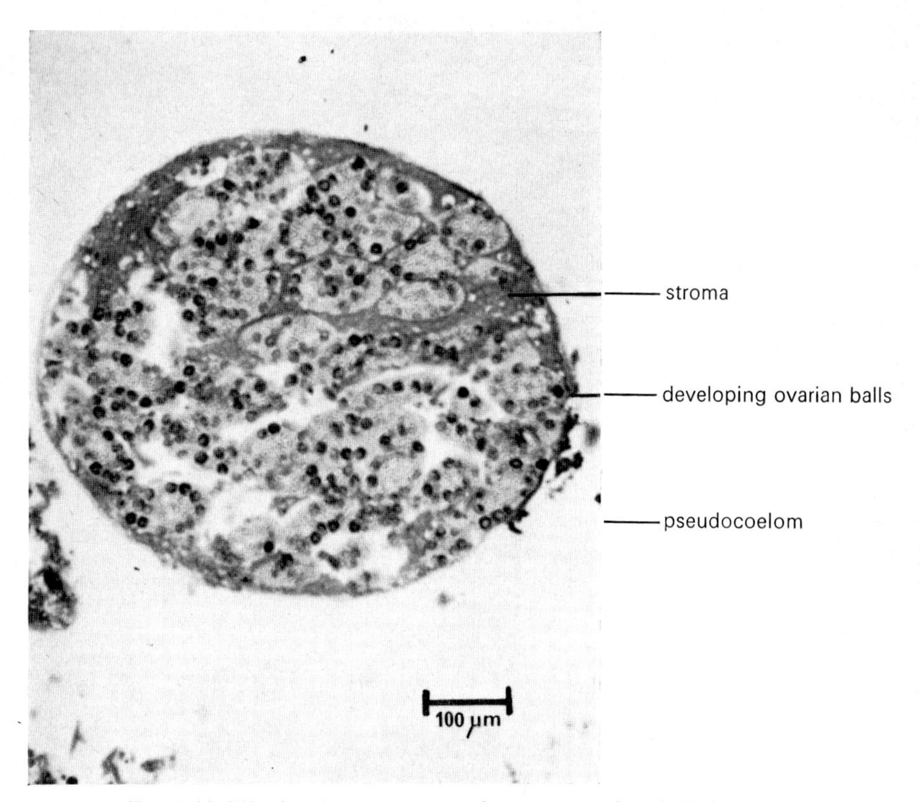

FIG. 2.11. TS of an immature ovary from a young female *P. laevis*.

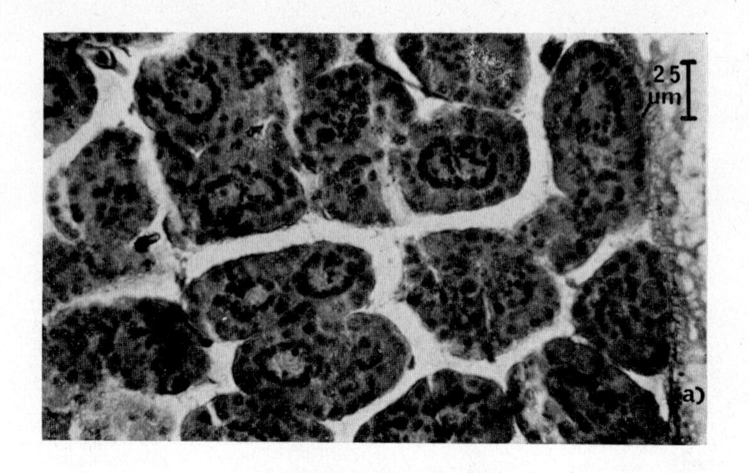

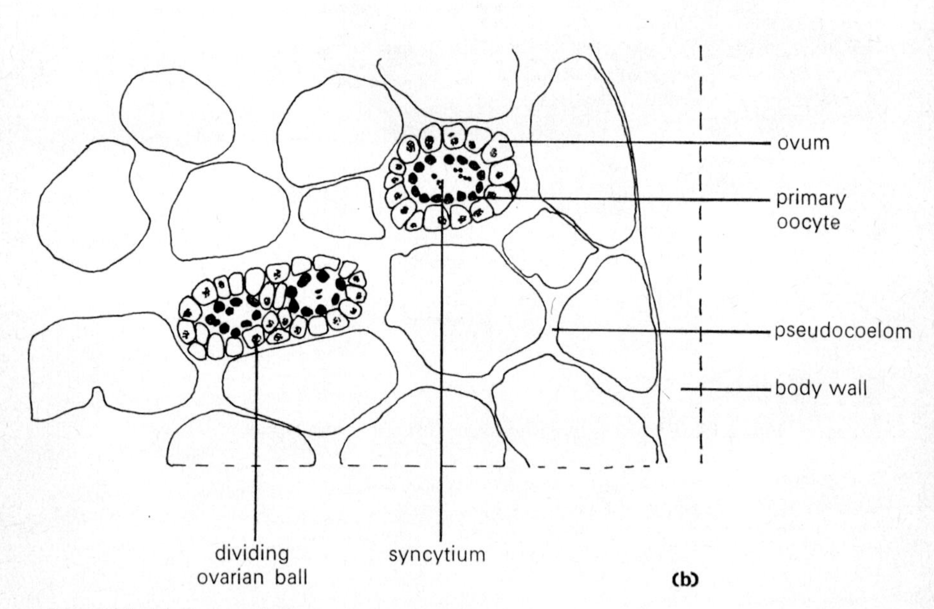

FIG. 2.12. (a) Ovarian balls in the pseudocoelom of an adult female *P. laevis*; and (b) an explanatory drawing.

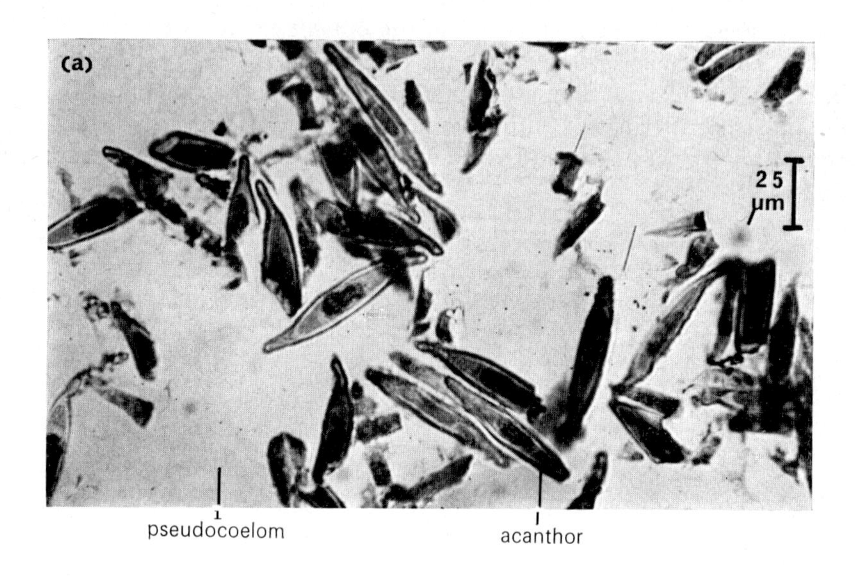

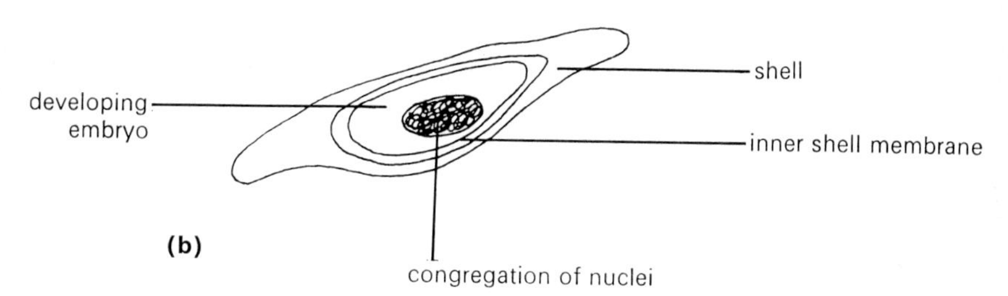

FIG. 2.13. (a) Acanthors in the pseudocoelom of an adult female *P. laevis*; (b) diagram of an individual acanthor.

takes place and sperm cluster round the ovarian balls. As the fertilized embryo develops in the pseudocoel it becomes enclosed in a hard shell. This forms the boat-shaped acanthor (Fig. 2.13) which is the resistant resting stage and which is passed to the outside in the vertebrate host's faeces. A muscular sorting organ, the uterine bell, is suspended from the ligament strand and allows acanthors to pass from the pseudocoel into the muscular uterus and through the thin-walled vagina to the outside. Immature eggs are rejected by the uterine bell and returned to the body cavity.

Chapter 3

Phylum Nematoda

The nematodes or thread-worms form a large group of pseudocoelomates. They are ubiquitous, having colonised almost all aquatic and terrestrial habitats. Many are free-living and very small, while some of the more important and better known members are large and parasitic. However, regardless of mode of life, their structure is remarkably uniform.

Nematodes are bilaterally symmetrical, unsegmented worms with long spindle-shaped bodies that are not subdivided into clear regions. There is often a marked hexamerous symmetry of structures round the mouth (e.g. lips, papillae). Many nematode features are thought to have arisen as adaptations to the changes in body length and high internal pressures created by their hydrostatic skeleton acting against the body muscles. Thus the body is covered by a resilient cuticle; the thin-walled collapsible gut is sealed off at both ends by stout muscles and can be dilated by dilator muscles; and the delicate excretory tubules are protected by being embedded in lateral cords of hypodermal tissue. The pseudocoelom often has fenestrated membranes stretched across it, and is filled with hypotonic fluid which keeps it distended. There are no circulatory or respiratory systems, all oxygen being obtained across the gut or body wall. The sexes are separate and the gonads are tubular and continuous with their ducts. The single male gonad opens with the gut into a cloaca, while the paired female gonads open directly (via their ducts) to the outside. Eggs of all species hatch to larvae which resemble the adult.

The phylum Nematoda is sometimes divided into a small group of Aphasmids which do not possess phasmids (caudal sense organs) and a larger group (Phasmids) which do. Since, however, all nematodes have a similar internal structure, I shall only describe one representative of the phylum.

Example: *Ascaris* (*suum*)

A. suum is a large nematode parasitic in pigs. Ascaroids are characterized by having three large lips (one dorsal and two ventro-lateral) round the mouth. They possess phasmids which are fairly complex. Figure 3.1 shows a transverse section through the mid-region of an adult female *A. suum*, and demonstrates the general arrangement of internal structures.

The body surface is covered by an elastic proteinaceous multilayered cuticle overlying a thin hypodermis containing deposits of fat and glycogen and traversed by fine fibres (Fig. 3.2). The hypodermis bulges into the

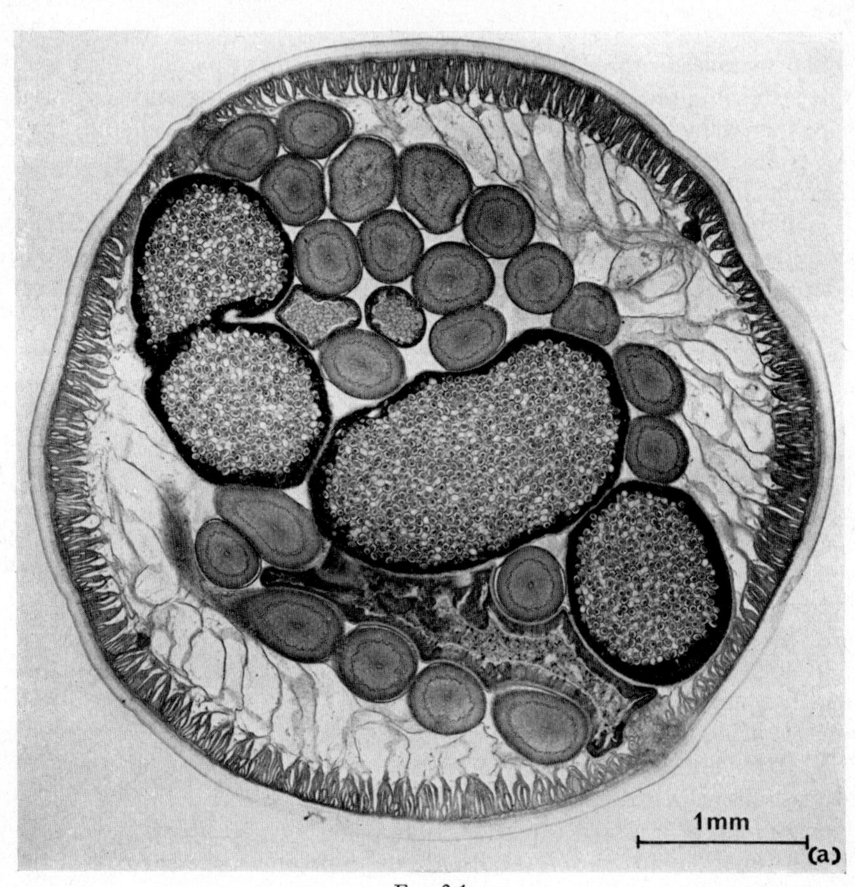

Fig. 3.1

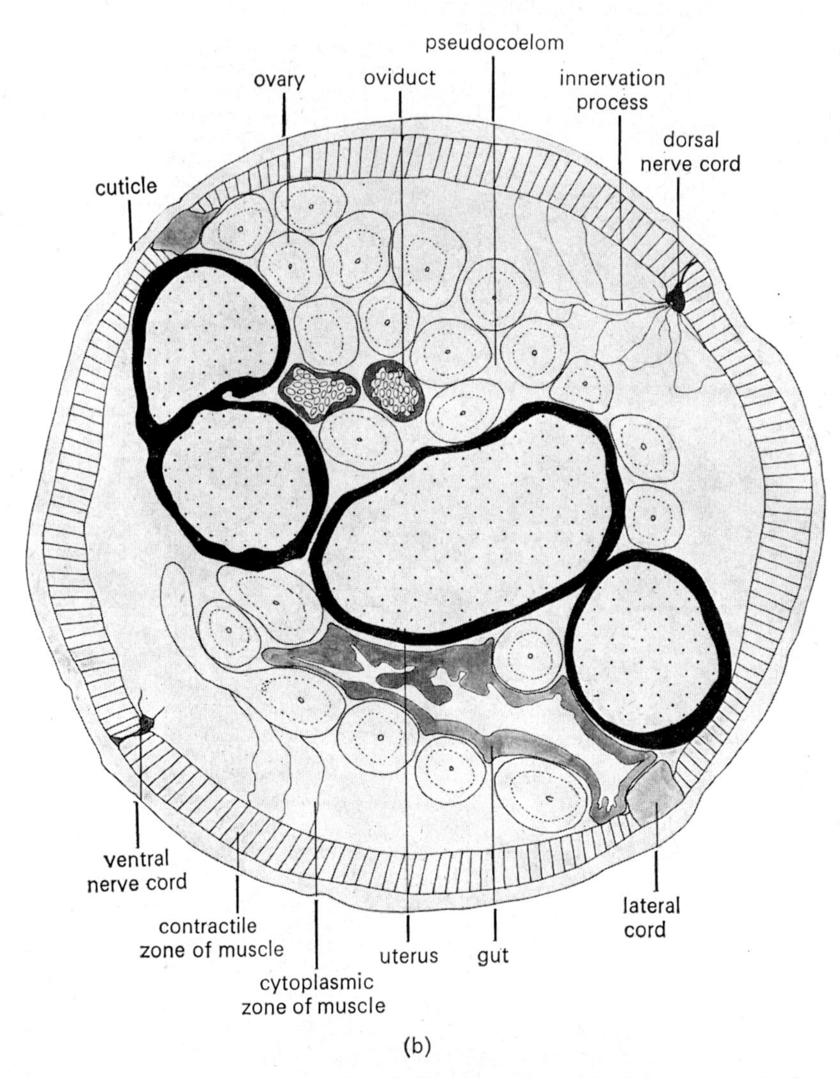

FIG. 3.1 (a) TS and (b) drawing of the middle region of an adult female nematode *Ascaris suum*.

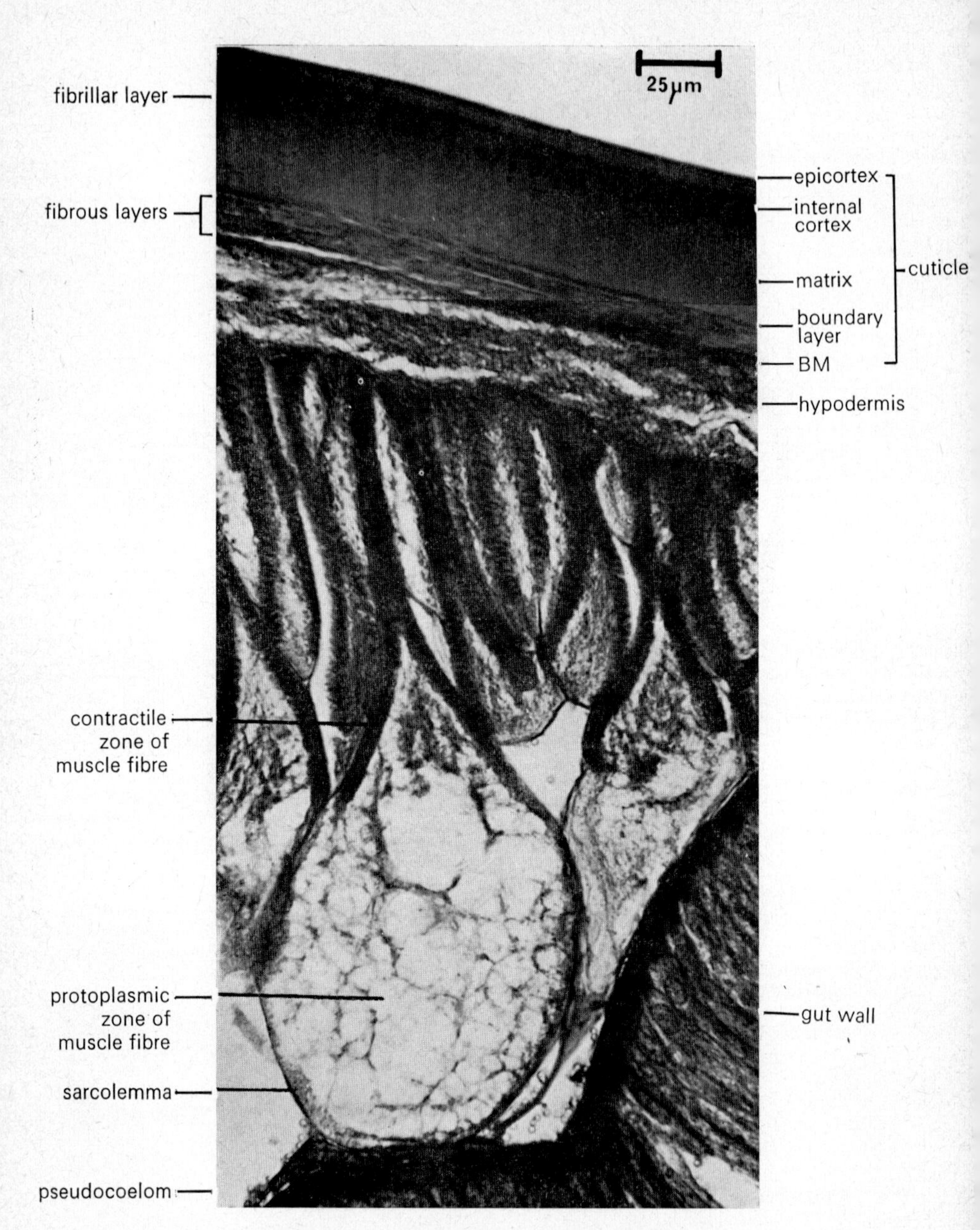

FIG. 3.2. TS through the body wall of *A. suum*. Note the longitudinal muscle fibres with their restricted fibrillar zones and bulging protoplasmic zones.

body cavity to form thickened dorsal, ventral and lateral longitudinal cords (see Fig. 3.1), which divide the body musculature up into four blocks. The cords contain longitudinal nerves and excretory canals (see also Fig. 3.8). Attached to the hypodermis is a single layer of longitudinal muscle fibres of a type peculiar to nematodes. One end of each fibre, adjacent to the hypodermis, consists of longitudinal bands of contractile filaments while the other forms a nucleated sarcoplasmic bulge into the pseudocoel. This end contains storage depots of fat and glycogen, and sends out processes to join the nervous system (Fig. 3.3). There are no circular muscles, movement generally being effected by alternating contractions of dorsal and ventral longitudinal muscles.

Nervous system

The nervous system is simple, consisting of a circumpharyngeal ring of fibres and associated paired lateral ganglia, a single ventral ganglion and other smaller ganglia connected to the ring. Six nerves supply the head end, while towards the posterior end run a small non-ganglionated dorsal nerve cord, a large ganglionated ventral nerve cord which joins an anal ganglion at the hind end of the worm, and two lateral ganglionated nerves which run in the lateral cords (see Fig. 3.8). The longitudinal nerve fibres are joined at intervals by ventro-lateral commissures, and connect with processes from the protoplasmic portions of muscle cells (Fig. 3.3).

Sense organs

Photoreceptors are not common although they do occur in some aquatic species. The chief sense organs in parasitic nematodes are papillae, sensory bristles, phasmids and amphids. Papillae, probably tactile receptors, occur in large numbers round the mouth and a few round the cloaca in males. In *Ascaris* these are small evaginations of the cuticle provided with nerves. Phasmids are tiny pits in the cuticle just behind the anus containing clusters of nerve cells and receiving openings from unicellular glands. They are probably chemoreceptive and are better developed in parasitic than in free-living species. The latter, however, are better equipped with amphids which are similar in structure but which occur on each side of the head. *Ascaris* has two amphids which are simple nerve fibres embedded in supporting cells.

Digestion

Ascaris lives in the intestine of its host and is therefore bathed in semi-digested food. It does, however, possess its own digestive enzymes. The gut is a straight tube with its extreme anterior and posterior ends lined with cuticle. The terminal mouth is surrounded by six lips in free-living nematodes, but in *Ascaris* these have become reduced to three. The mouth

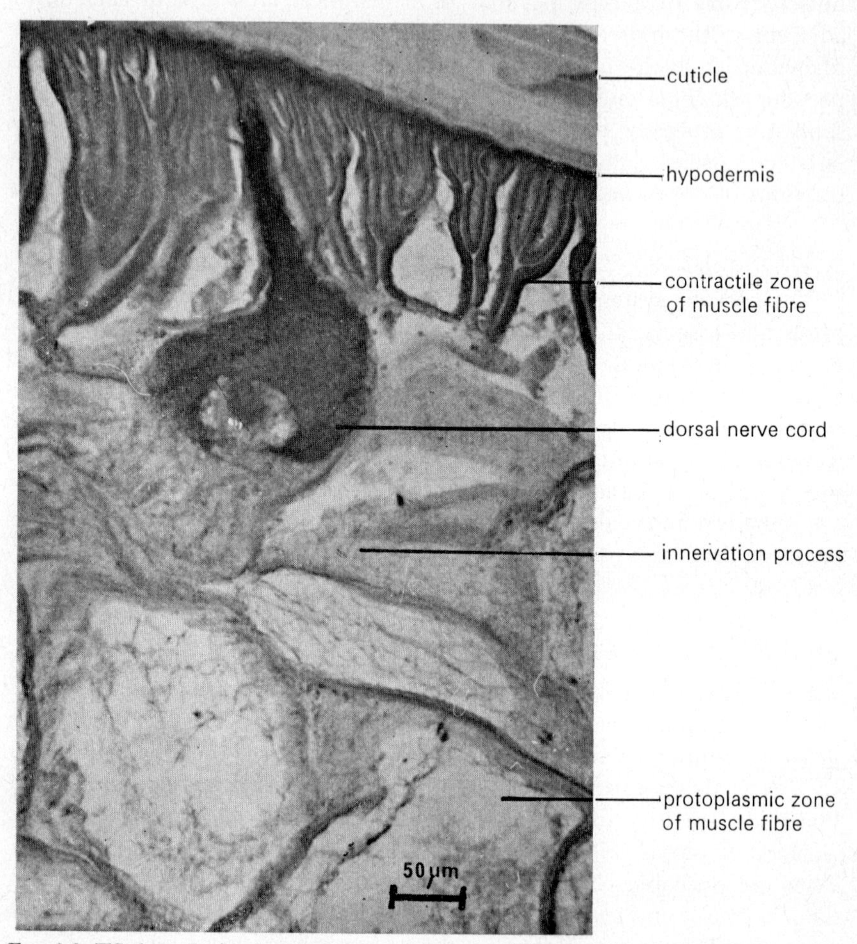

FIG. 3.3. TS through the dorsal nerve cord and muscle innervation processes of *A. suum*.

leads by a short thin-walled buccal region to a muscular pharynx which acts as a valve and powerful suction pump. It is cylindrical with a triradiate lumen, and has a thick wall composed of radial muscle fibres and long pharyngeal glands, which open by ducts into the pharynx and buccal cavity [Fig. 3.4(a, b)]. Food is pumped by the pharynx into the endodermal mid-gut. This is lined by a single layer of columnar epithelial cells which have a well-defined inner border of microvilli, previously known as the bacillary layer (Fig. 3.5). The cells are both secretory and absorptive in function. There is no gut musculature, food being moved along the intestine by body movements. The mid-gut is separated by a muscular intestino-rectal valve from a short, cuticle-lined hind-gut (Fig. 3.6) which is provided with large unicellular rectal glands. The hind-gut opens to the outside in females by the slit-like anus, and into the cloaca in males.

Excretion

There are no nephridia in *Ascaris*, but there is a well-developed excretory system consisting of two longitudinal tubules running in the lateral cords (Figs 3.7 and 3.8). These canals are connected anteriorly by a network of cross-ducts from which arises a short canal to the excretory pore. There are also two small anterior canals, so the excretory system is H-shaped. The duct cavities are intracellular as the whole excretory system is derived from only three cells. Water is taken up by the tubules from the pseudocoel and actively pumped out of the body.

Reproduction

Nematodes are generally dioecious, the male normally being smaller than the female and equipped with accessory copulatory structures.

In the male *Ascaris* there is a single, very long, convoluted testis (Fig. 3.9). Spherical, non-flagellate sperm are produced at the blind end (germinal zone) of the testis [Fig. 3.10(a)] and gradually pass down through a growth zone [Fig. 3.10(b)] into a vas deferens [Fig. 3.11(a)]. The end of the vas deferens is expanded to form a large seminal vesicle lined with columnar epithelium bearing characteristic long villar processes [Fig. 3.11(b)]. Finally an ejaculatory duct with puckered muscular walls joins the rectum in a common cloaca. Evaginations of the cloacal walls form paired pouches containing copulatory spicules which are protruded by muscular action during copulation. The male also bears numerous accessory copulatory papillae and a pair of phasmids near the cloacal opening.

The female is oviparous, the egg being covered with a resistant coat from several different sources. The paired ovaries are long coiled tubes packed tightly with maturing oogonia (Fig. 3.12). The ovary also possesses a germinal zone and a growth zone in which oogonia are attached to a central strand of cytoplasm (rachis) until they mature (Fig. 3.13). They pass down a short oviduct into the paired uteri where sperm are stored

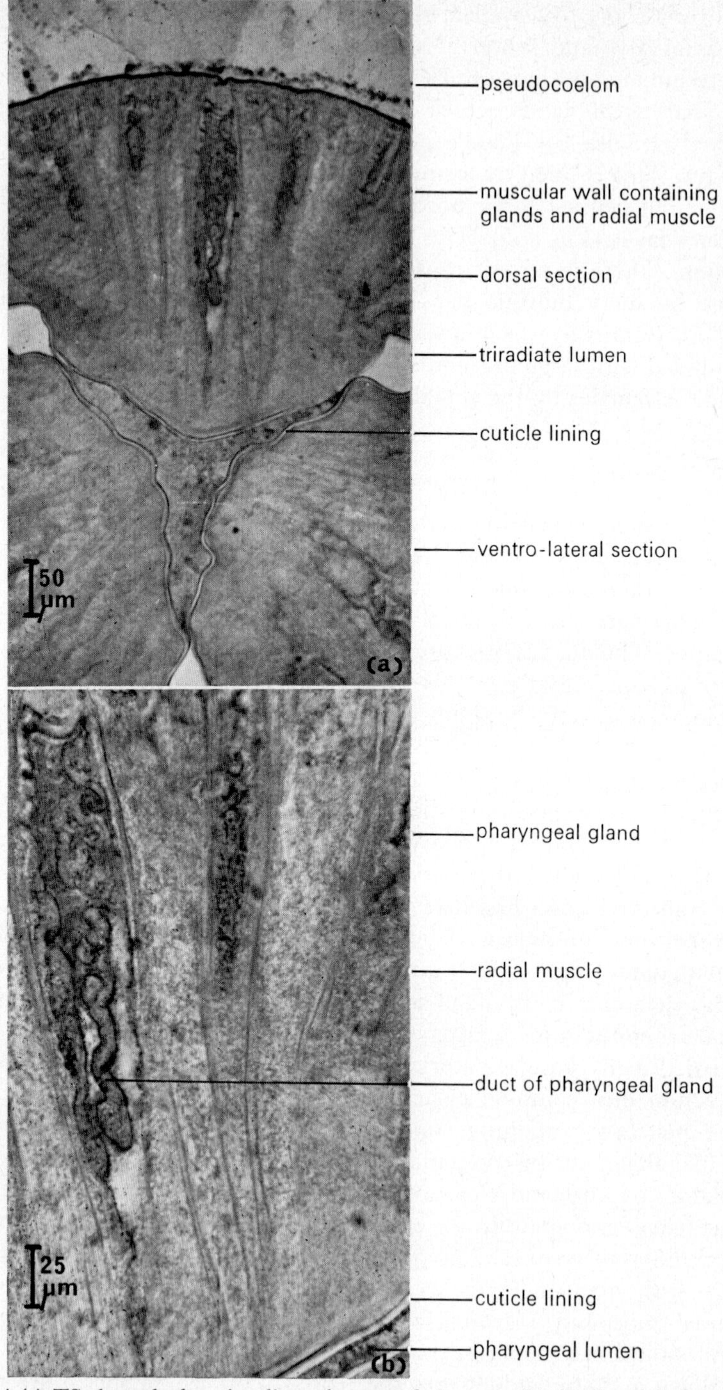

FIG. 3.4 (a) TS through the tri-radiate pharynx of *A. suum*; (b) details of the structure of the pharynx wall shown in (a).

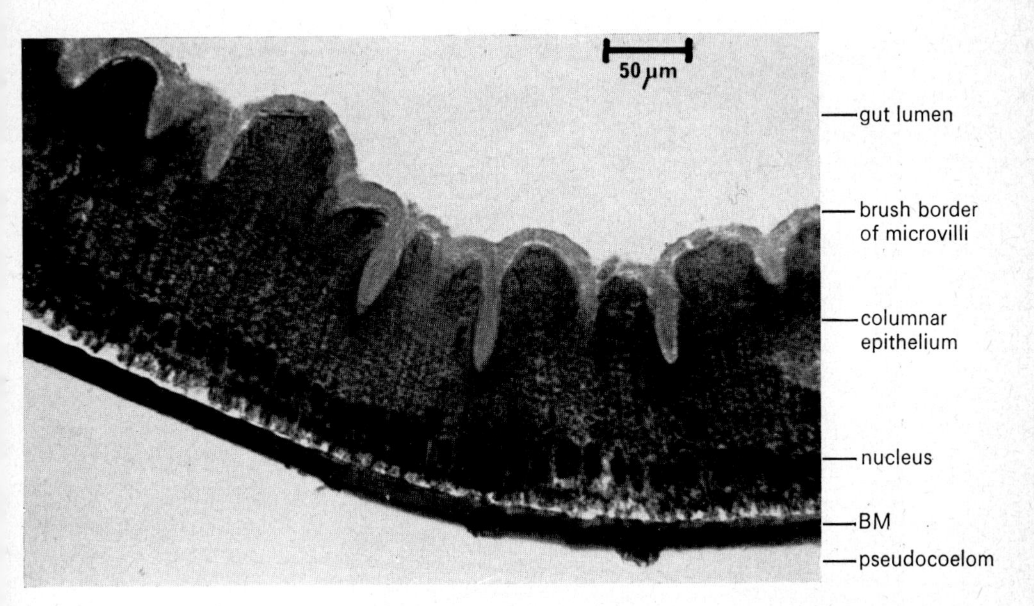

FIG. 3.5. TS though the intestine wall of *A. suum*. Note the lack of gut musculature.

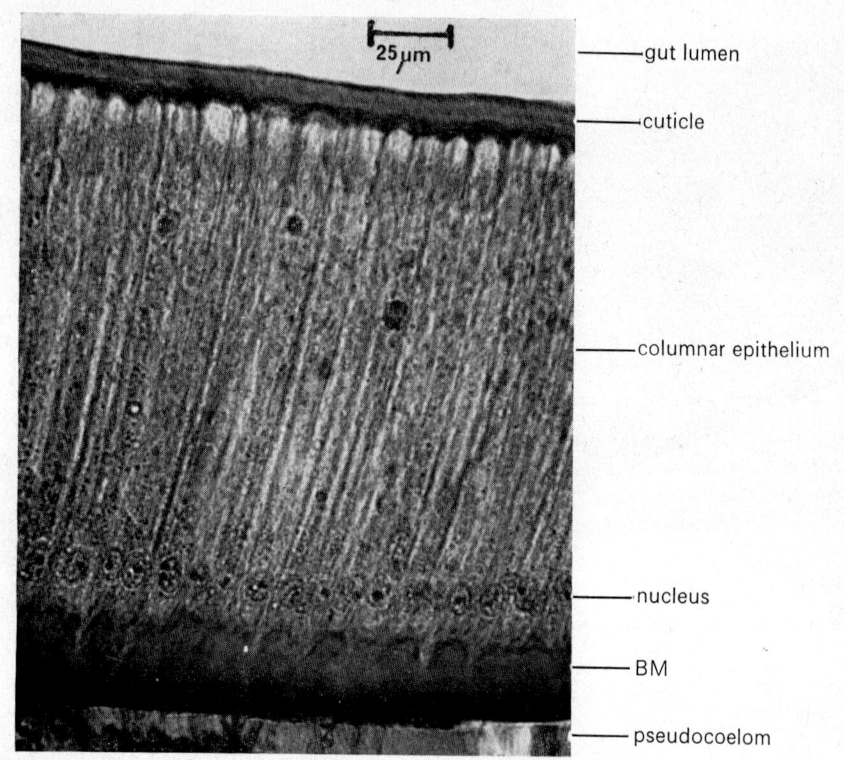

FIG. 3.6. TS through the rectum wall of *A. suum*.

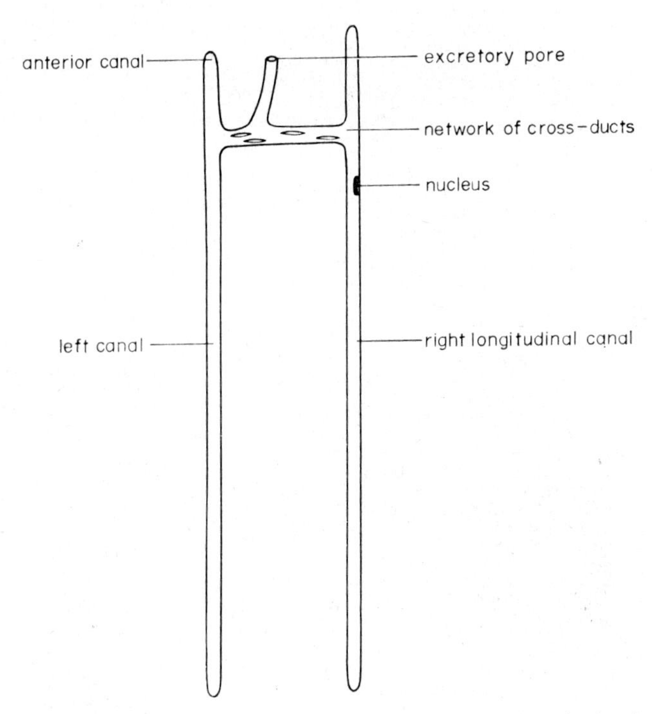

FIG. 3.7. Plan of the excretory system of *Ascaris*. Note the asymmetry in the size and shape of the longitudinal canals.

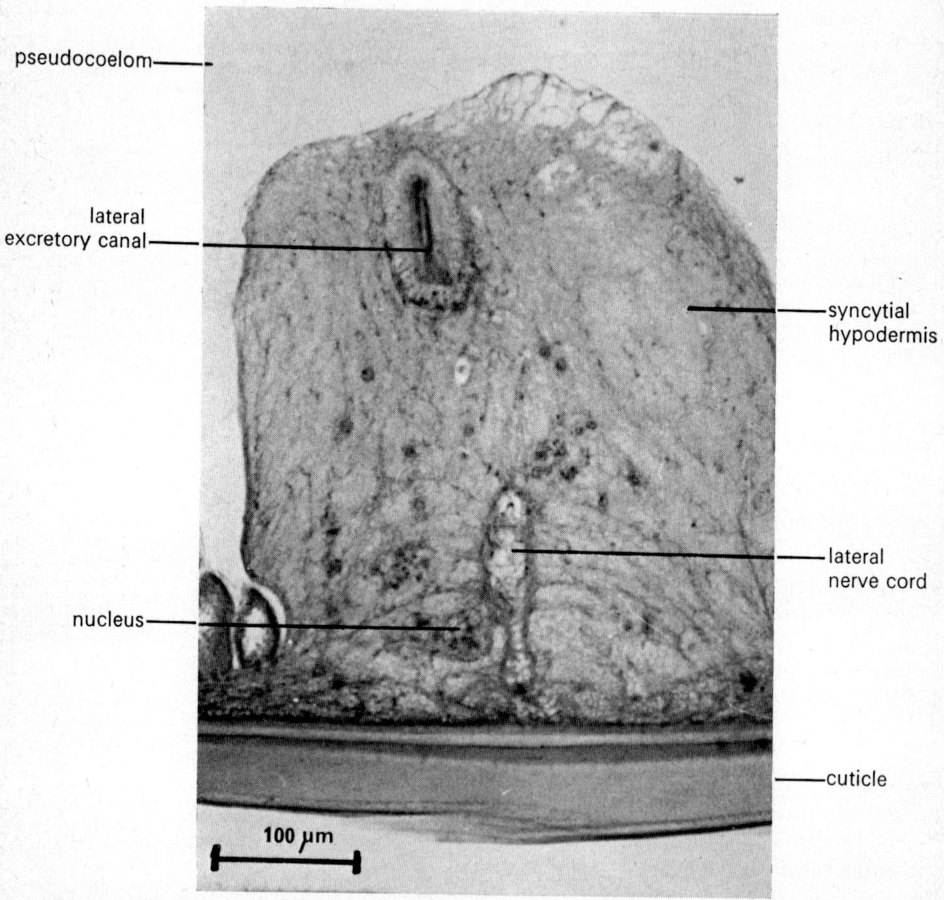

FIG. 3.8. TS through a lateral cord of *A. suum*, showing the positions of the nerve cord and excretory canal.

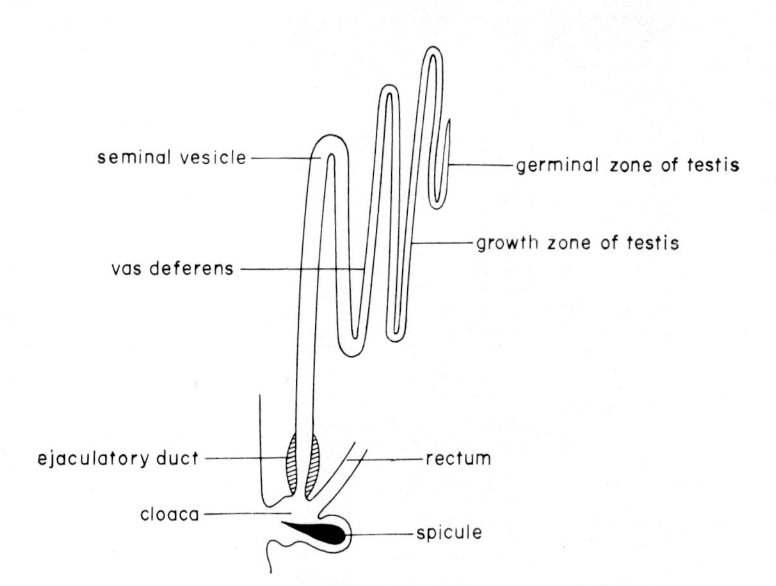

FIG. 3.9. Plan of the male reproductive system of *Ascaris*.

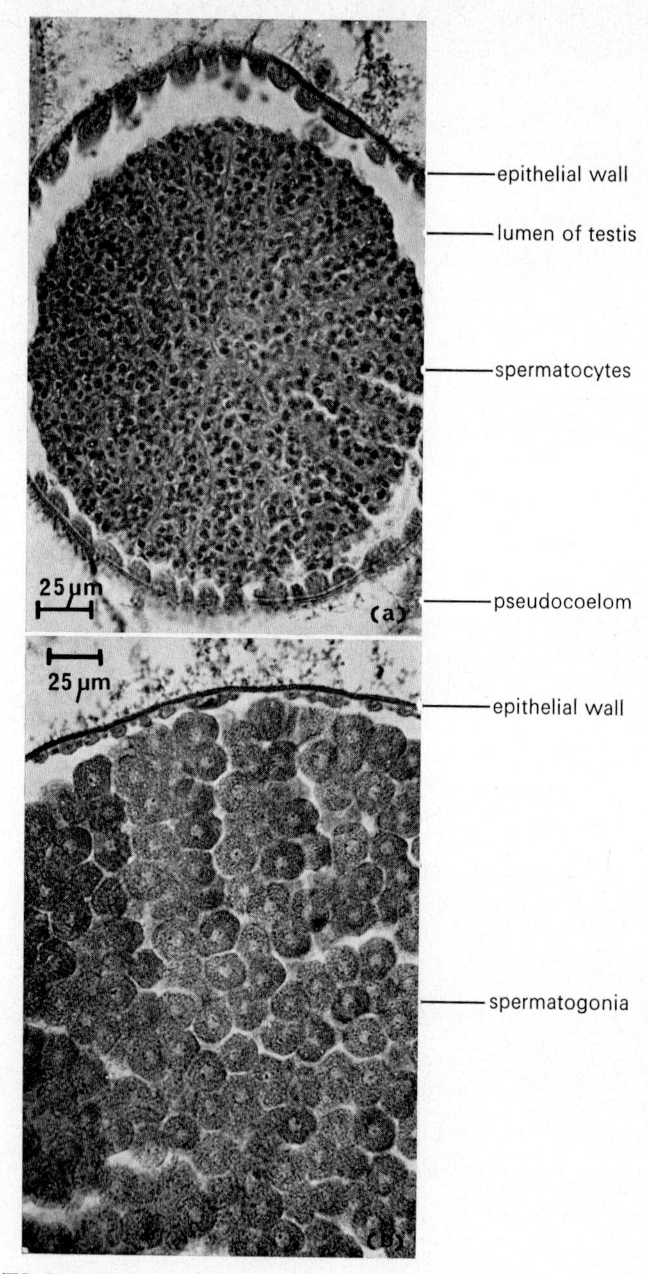

FIG. 3.10. TS through the testis of an adult male *A. suum* in (a) the germinal zone; and (b) the growth zone.

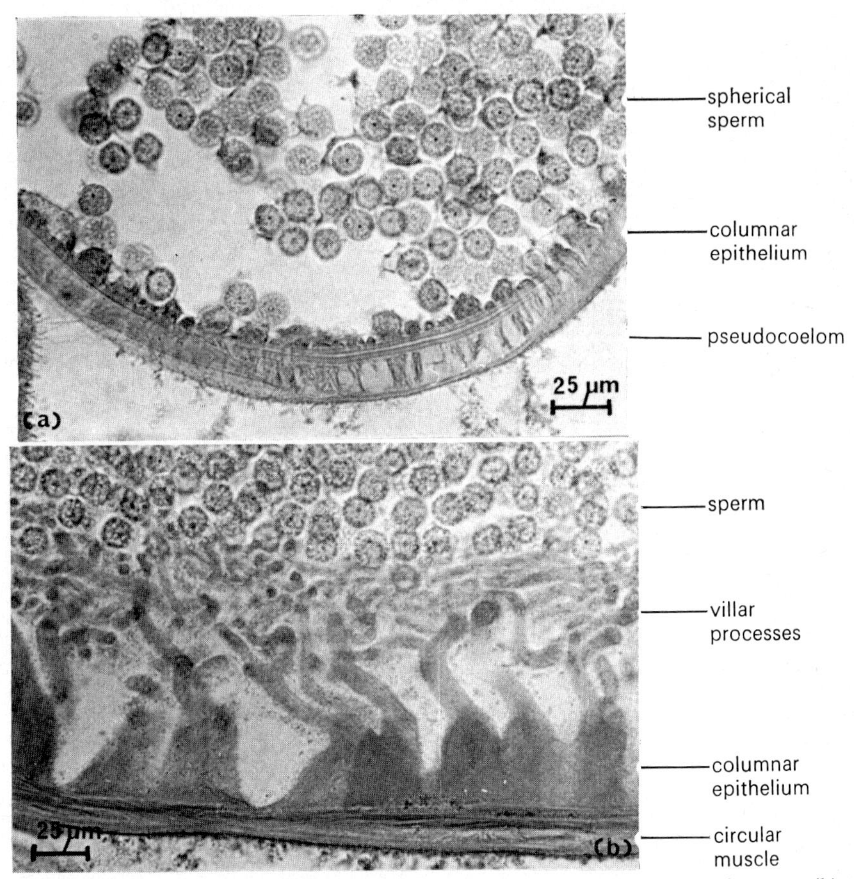

FIG. 3.11 (a) Part of the vas deferens of a male *A. suum*. Note the spherical sperm; (b) section through the wall of the seminal vesicle of a male *A. suum*. Note the large epithelial cells with their long villar processes.

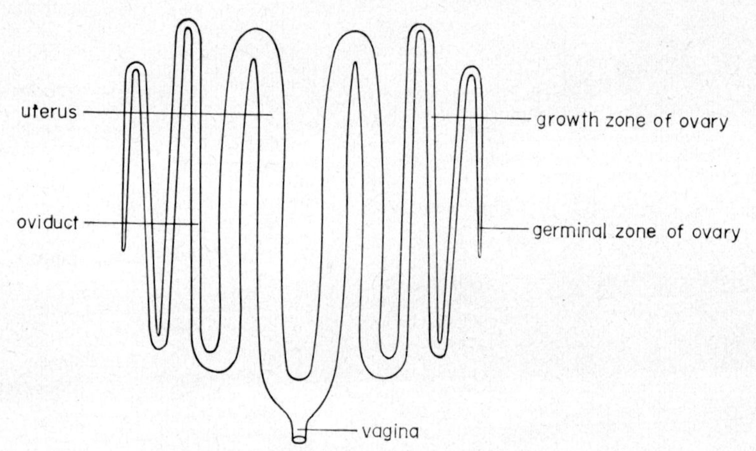

FIG. 3.12. Plan of the female reproductive system of *Ascaris*.

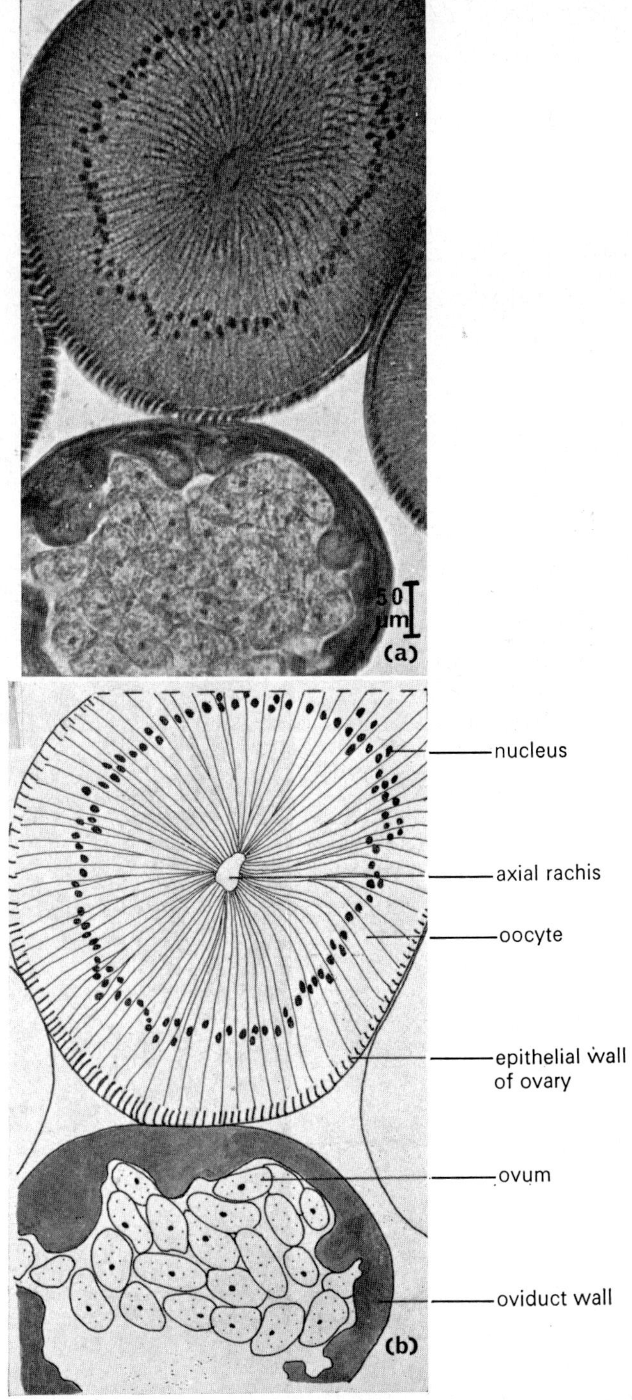

FIG. 3.13 (a) TS and (b) drawing of the ovary (growth zone) and oviduct of a female *A. suum*.

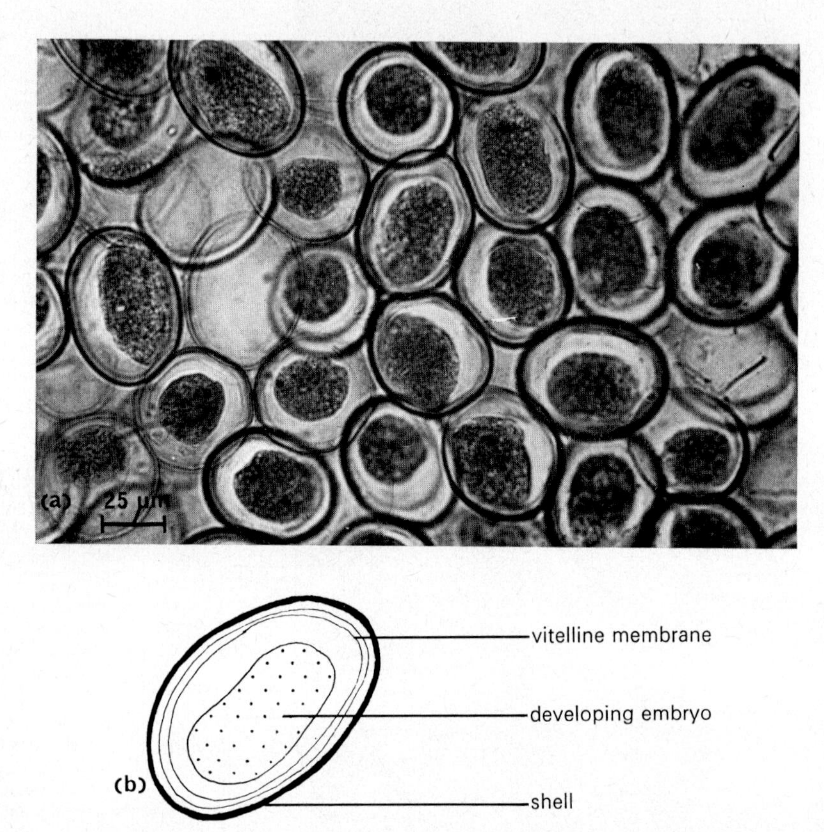

FIG. 3.14 (a) Mass of fertilized eggs in the uterus of a female *A. suum*; (b) diagram of a single fertilized egg.

temporarily and where fertilization occurs (see Fig. 3.1). The fertilized eggs remain in the uteri until their protective shell has been added by the ovum itself and by the uterine wall (Fig. 3.14). When the eggs are ripe they are shed to the outside via a cuticle-lined vagina. The gonopore is controlled by radial dilator muscles.

Chapter 4

Phylum Annelida

Annelids are segmented worms with a perivisceral coelom and a single pre-oral segment (prostomium). The body wall is composed of an external layer of circular muscle and an internal layer of longitudinal muscle. It is protected by a thin, non-chitinous cuticle. The nervous system consists of paired cerebral ganglia connected by commissures to a pair of ventral nerve cords containing giant nerve fibres. These nerve cords are expanded in each segment to form ganglia which give rise to segmental nerves. Neuro-secretory cells are common in the nervous system, and are probably the source of many hormones. The beginnings of a circulatory system is found. Nephridia and coelomoducts are both present.

Class Oligochaeta

Oligochaetes live in terrestrial and freshwater habitats. They show marked segmentation and the large coelom is normally divided into compartments by intersegmental septae. It is this arrangement, together with the body wall muscles, that gives the animal a most efficient hydrostatic skeleton. There is generally a distinct prostomium, but there are no appendages or limbs and only a few chaetae. All oligochaetes are hermaphrodite and possess only a few pairs of gonads; the ovaries always being posterior to the testes. Special genital ducts open into the coelom by funnels. In the mature adult there are present spermathecae (for storage of sperm from another individual) and a clitellum.

Example: *Lumbricus* (*terrestris*)

The earthworm is a terrestrial member of the class. It lives in soil through which it burrows, largely by ingesting it. *Lumbricus* has a wide-spread distribution and is nocturnal in its habits. The segmental arrangement of the anterior end is seen clearly in Fig. 4.1. Figure 4.2. is a TS of an earthworm, and shows the arrangement of the body wall musculature and the positions of the chaetae.

The body wall consists of a thin cuticle covering the unicellular epidermis. This is well supplied with mucous cells and overlies a thinnish layer of circular muscle and a very thick layer of longitudinal muscle. The circular muscle is reduced in thickness at the junction between two segments, but the longitudinal layer is not (Fig. 4.3). The normal muscle arrangement is lost in the region of segments 32–37 in the mature adult. In this area (clitellum) the epidermis is very swollen with unicellular glands (Fig. 4.4). They produce mucus to aid copulation, albumin to nourish the eggs and also secrete a cocoon round the developing eggs.

Earthworms have four pairs of short chaetae in each segment which assist locomotion. Each chaeta is secreted by sunken epidermal cells in a chaetal sac and is moved by protractor and retractor muscles (Fig. 4.5).

Nervous system

The basic plan of annelid nervous systems is shown in Fig. 4.6. Paired cerebral ganglia lying dorsally to the gut at the anterior end of the body connect by large commissures with ventral sub-oesophageal ganglia and a double ventral nerve cord bearing segmental ganglia. The ganglia show the typically invertebrate arrangement of a central mass of axons surrounded by the nerve cell bodies (Fig. 4.7). A cross-section of the ventral nerve cord (Fig. 4.8) shows three dorsal giant nerve fibres which run the length of the animal and which mediate fast escape reactions. The very large median giant fibre is linked to the two smaller lateral fibres by transverse connections. Giant fibres receive inputs from several giant neurones in each segment, and a few of these can be distinguished on the ventral side of the segmental ganglion. The two small ventral giant fibres are not apparent.

Earthworms possess only a few sense organs of a simple type.

Circulation

Although respiratory exchange occurs across the body wall, there is a simple open circulatory system to help transport oxygen round the body. Blood, containing a respiratory pigment (haemoglobin), is pumped anteriorly by a contractile dorsal blood-vessel assisted by five pulsatile vessels (pseudohearts) which surround the oesophagus (see Fig. 4.1). It passes into a large ventral vessel in which it flows posteriorly and is distributed to the tissues. The position of the dorsal and ventral vessels can be seen in Fig. 4.2.

Digestion

The earthworm is a non-specific omnivorous feeder, extracting nutritional requirements from vegetation and from soil which passes through the gut during burrowing. The gut is an open-ended tube showing some specialization of structure in different regions (Fig. 4.9). A muscular pharynx leads into a thin-walled oesophagus with its associated pouches and calciferous glands (Fig. 4.10). The calciferous glands secrete calcium carbonate into the lumen of the gut, and are believed to help maintain acid-base balance and control water and electrolyte balance in the worm. The thin-walled crop is used for filtration and storage, while the muscular

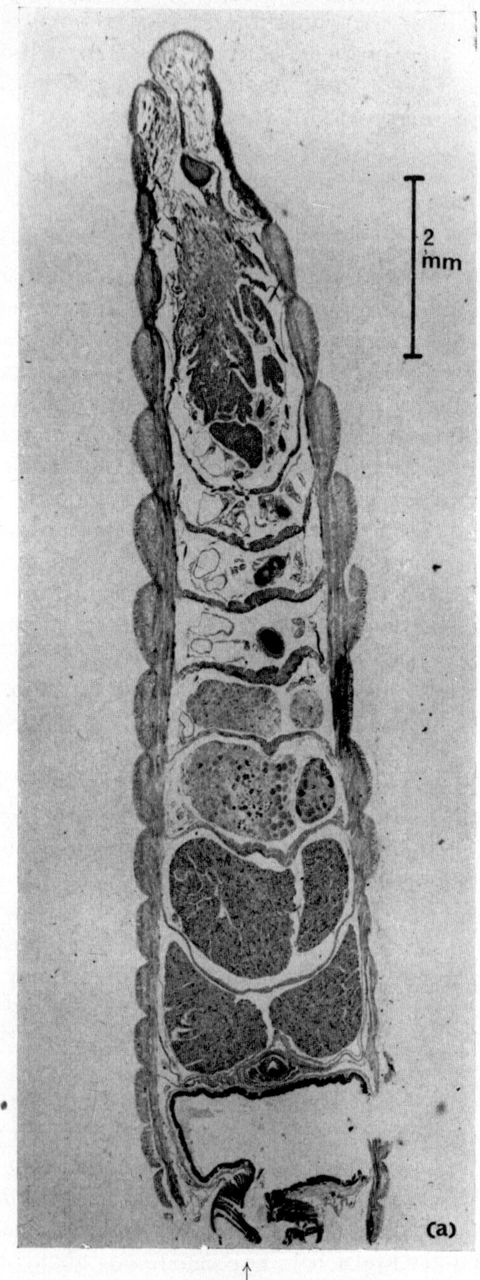

ANT

Fig. 4.1

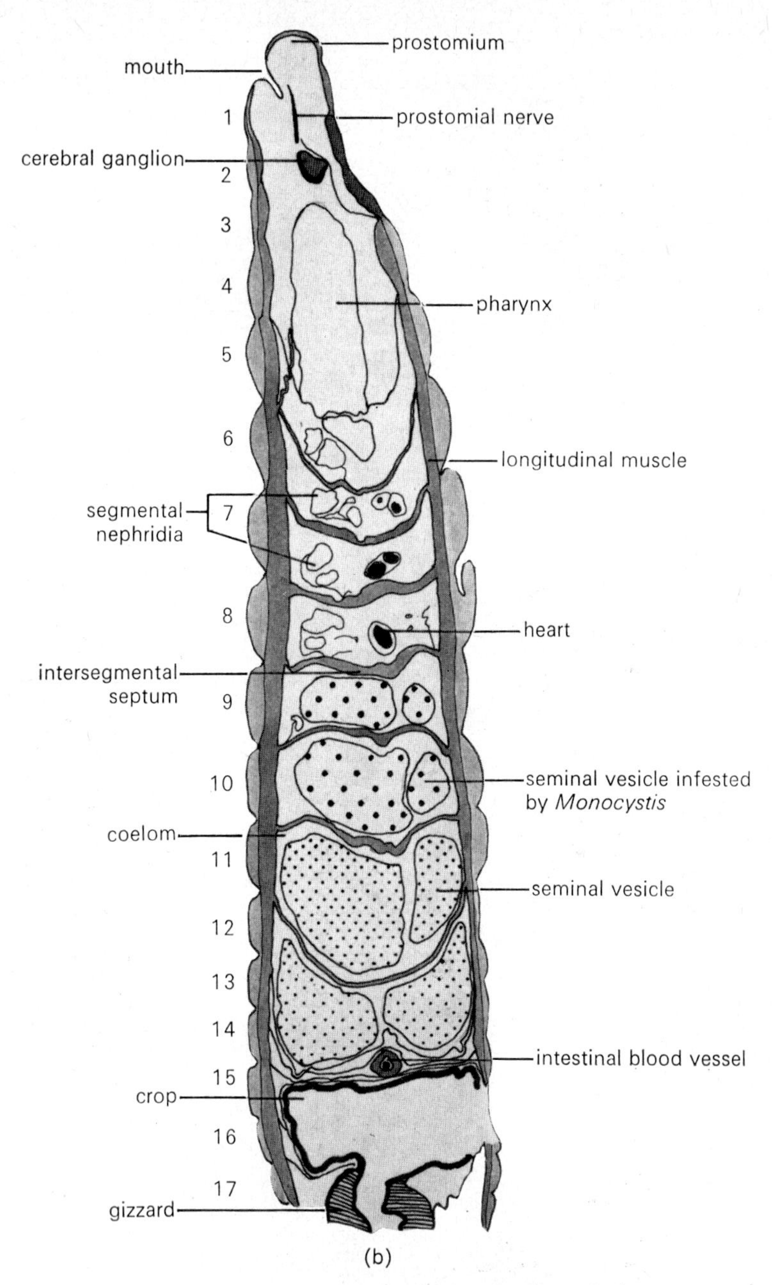

FIG. 4.1 (a) VLS and (b) explanatory drawing of the anterior end of a young **earthworm** (*Lumbricus terrestris*). The segments are numbered in Arabic numerals.

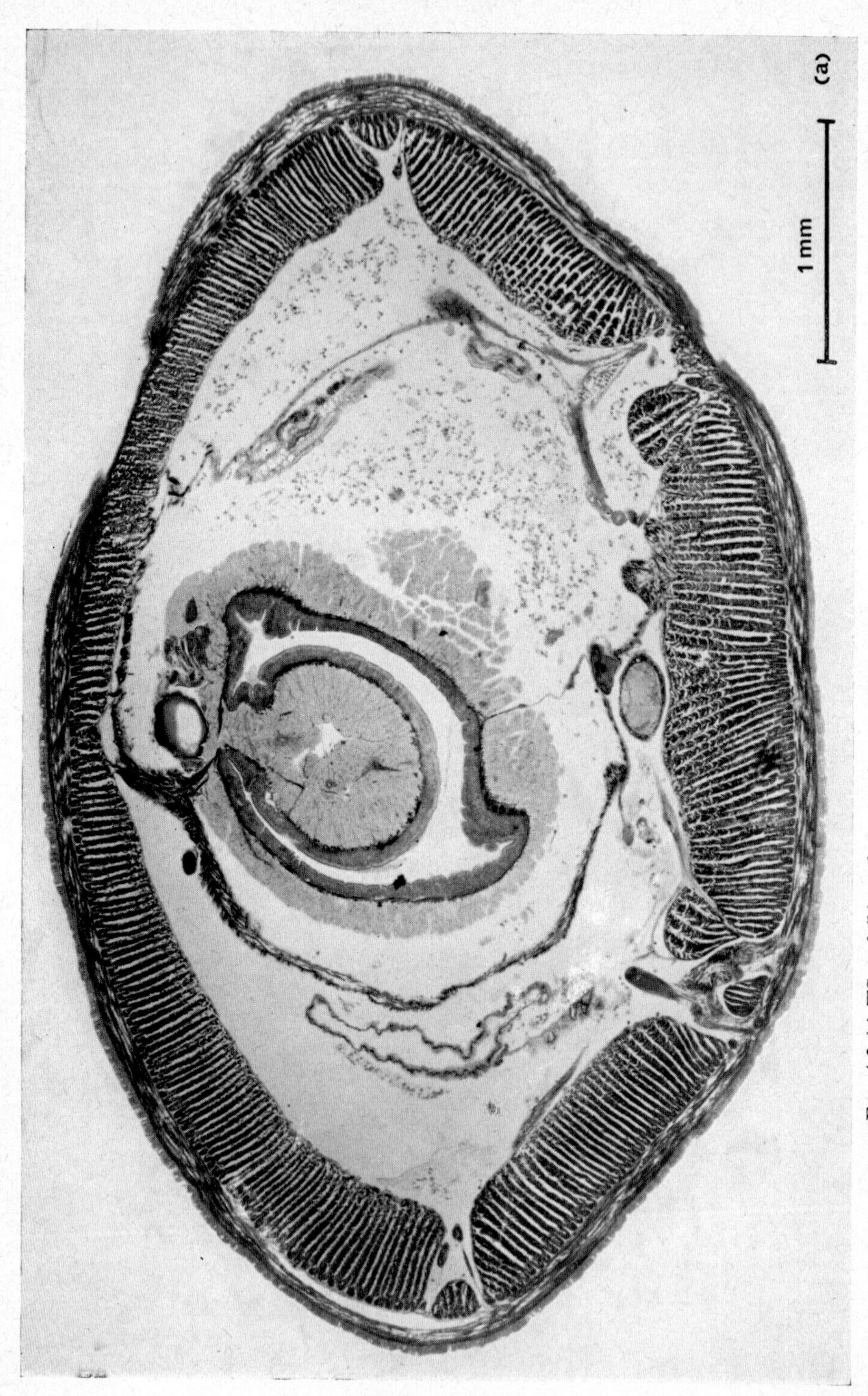

FIG. 4.2 (a) TS and (b) explanatory drawing of *L. terrestris* in the intestinal region.

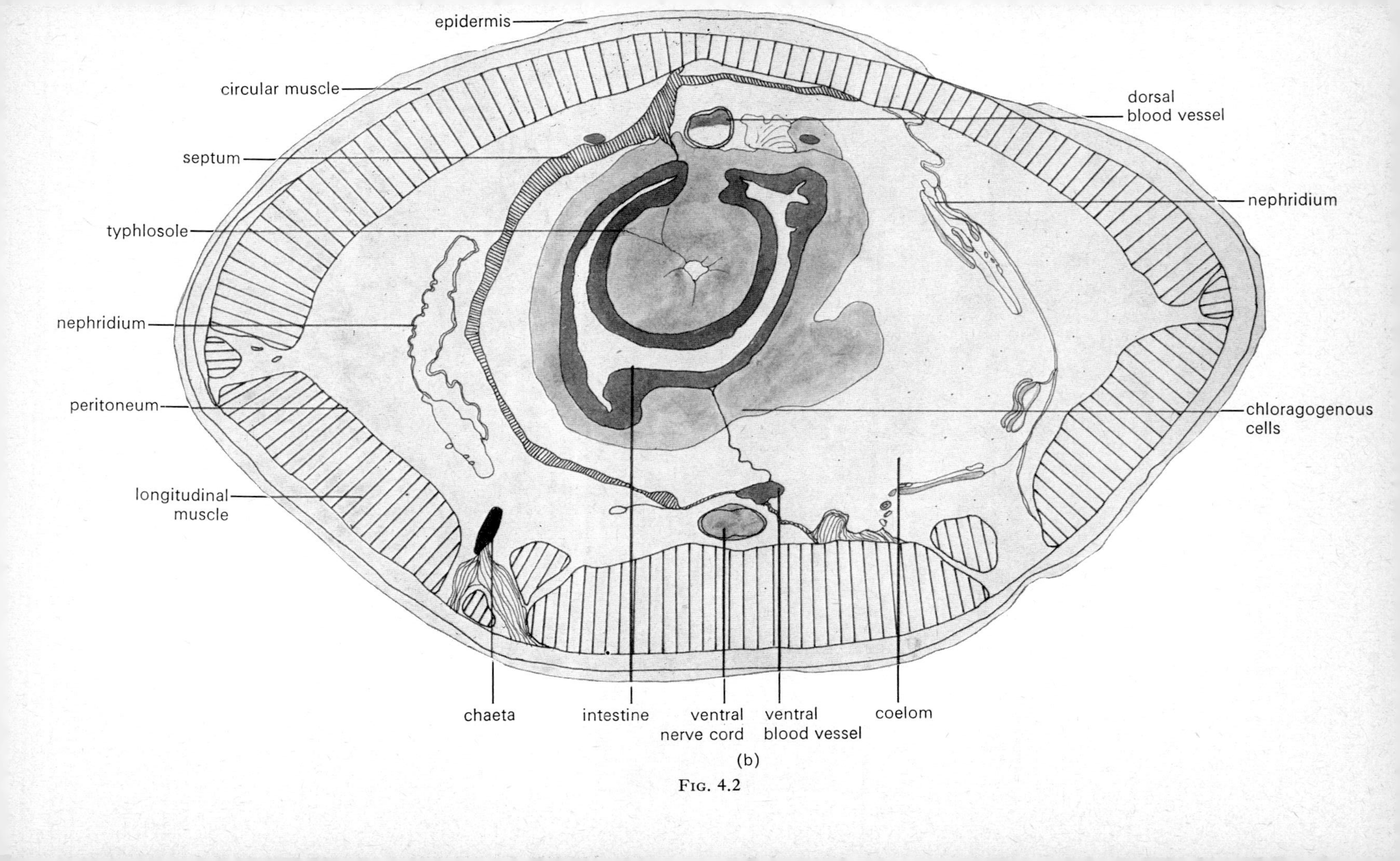
epidermis
circular muscle
dorsal blood vessel
septum
nephridium
typhlosole
nephridium
peritoneum
chloragogenous cells
longitudinal muscle
chaeta
intestine
ventral nerve cord
ventral blood vessel
coelom

(b)

FIG. 4.2

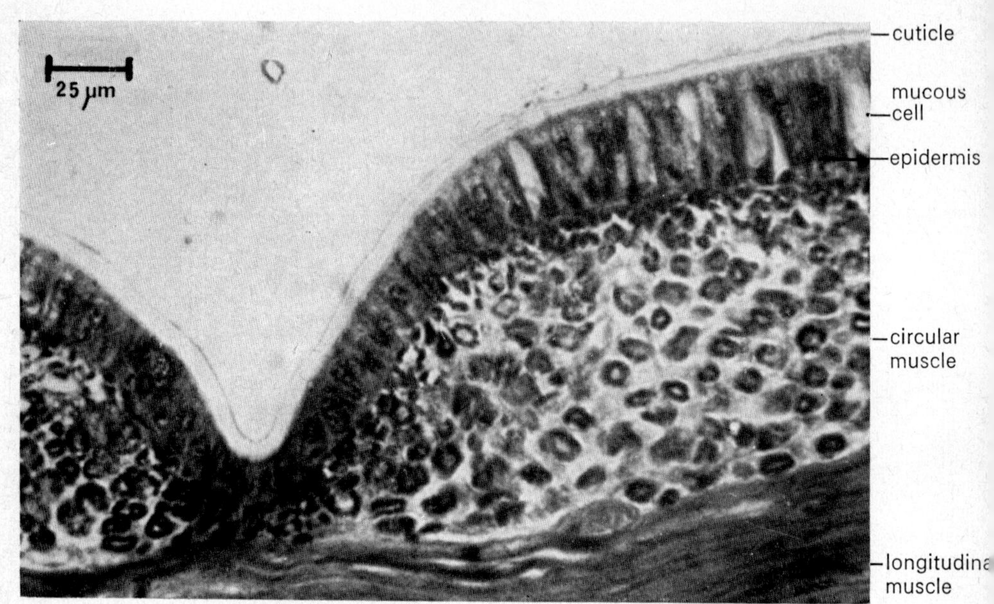

FIG. 4.3. LS through part of the body wall of *L. terrestris*.

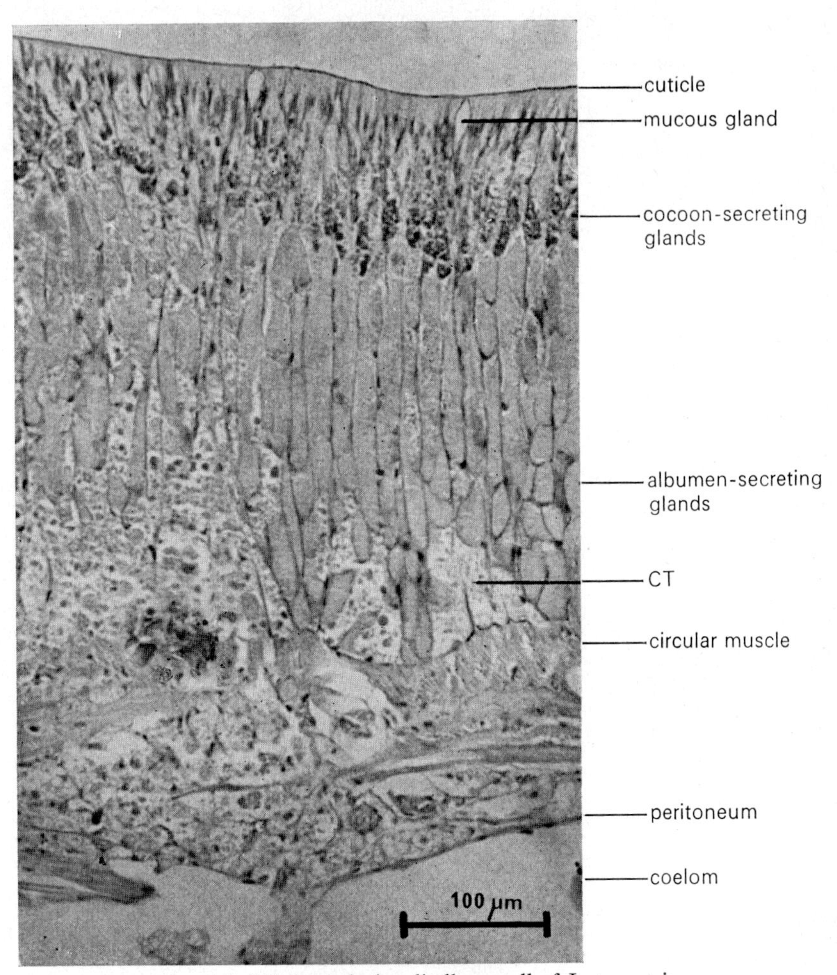

FIG. 4.4. LS through the clitellum wall of *L. terrestris*.

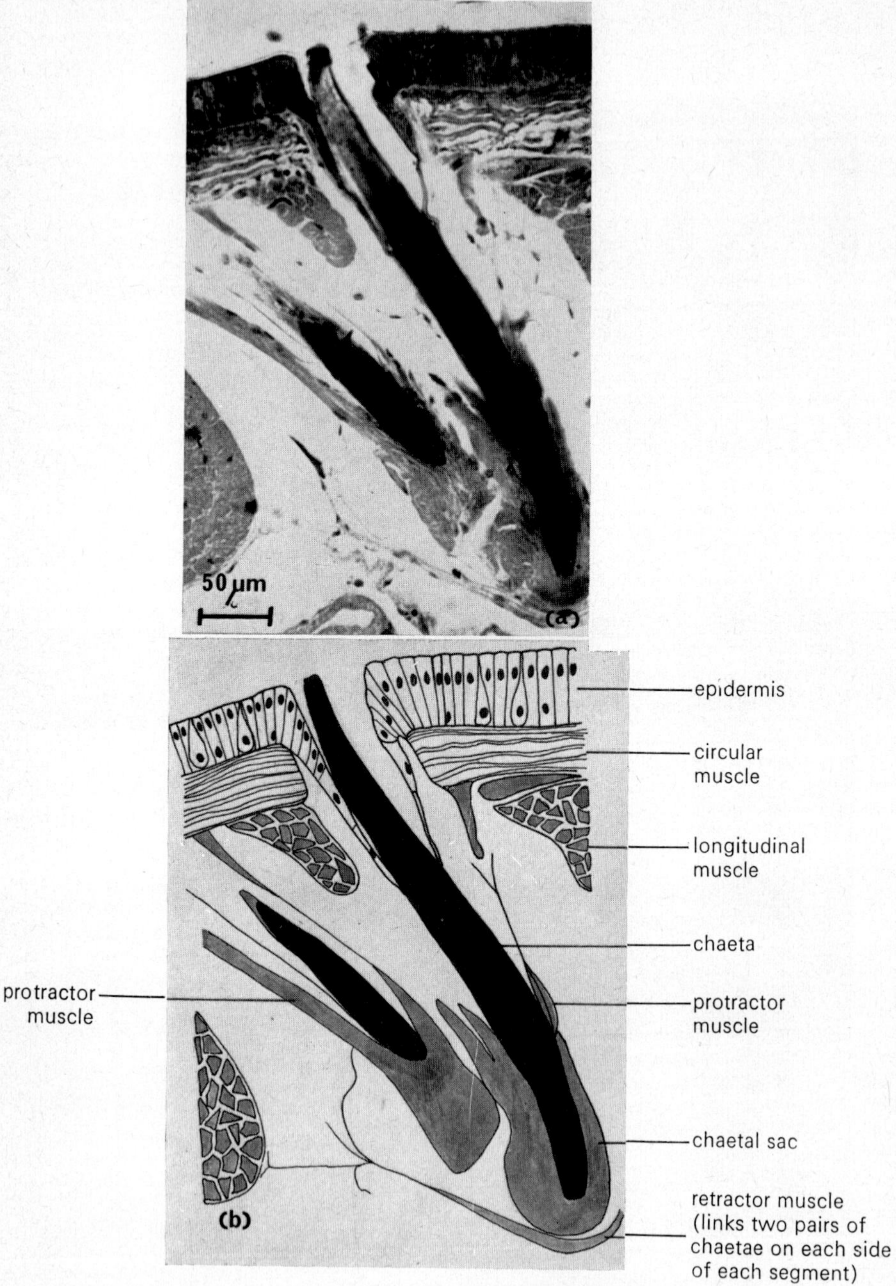

Fig. 4.5 (a) TS and (b) drawing of part of the body wall of *L. terrestris*, showing two chaetae and their associated muscles.

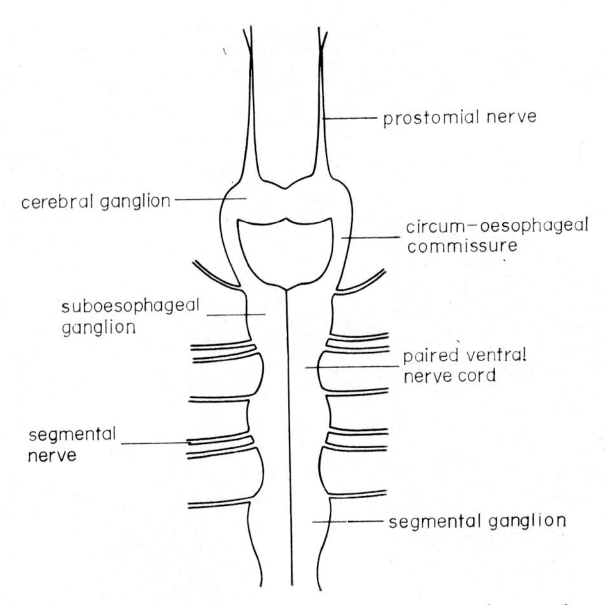

FIG. 4.6. Plan of the anterior end of the nervous system of an earthworm.

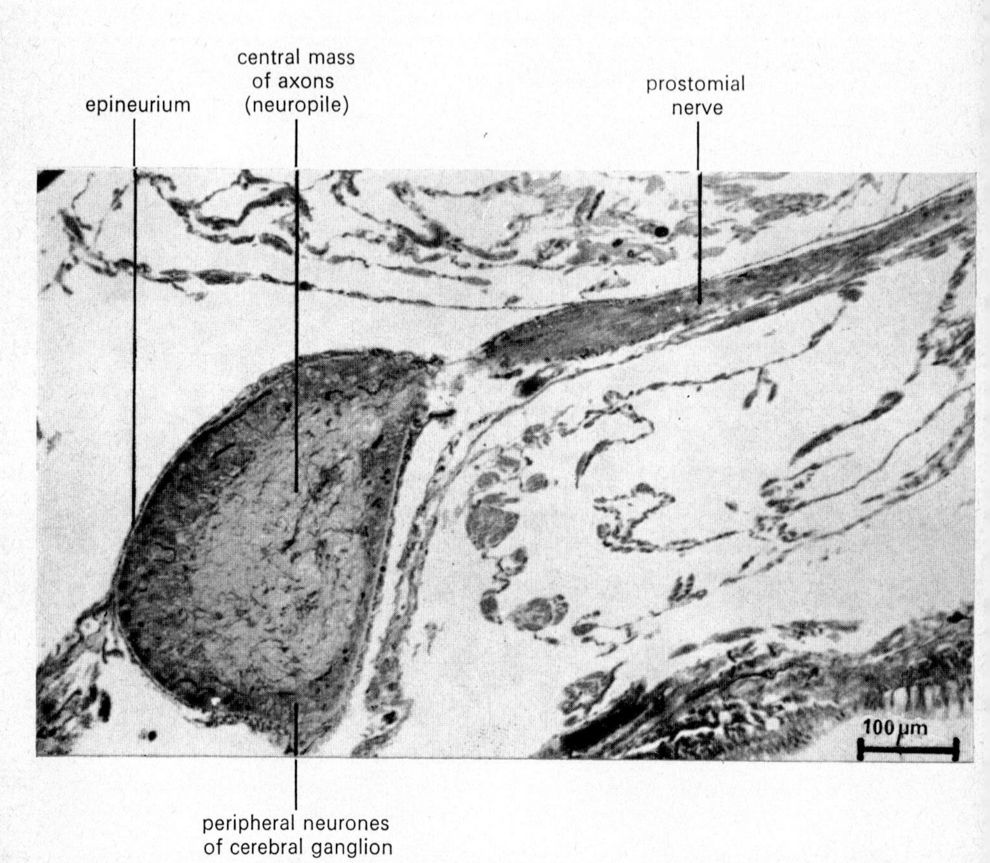

FIG. 4.7. LS through one cerebral ganglion and prostomial nerve of *L. terrestris*.

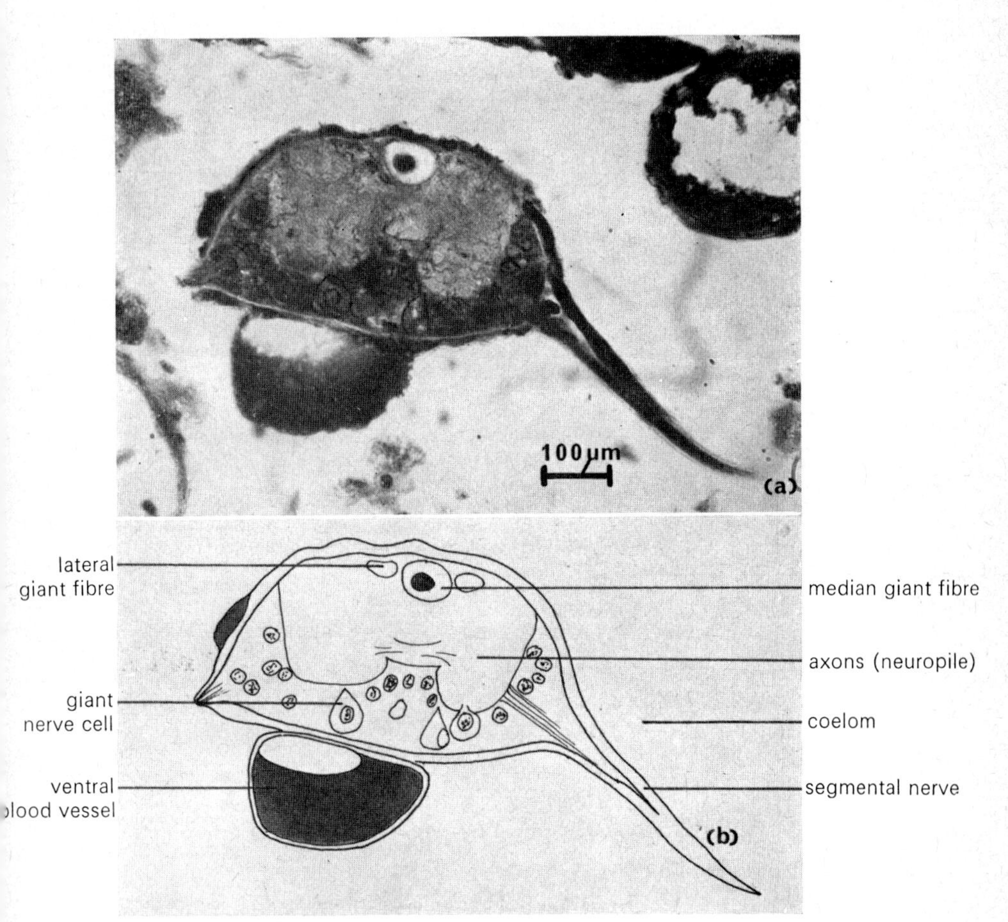

FIG. 4.8 (a) TS and (b) drawing of the ventral nerve cord of *L. terrestris* in the region of a segmental ganglion. Note the large median giant nerve fibre.

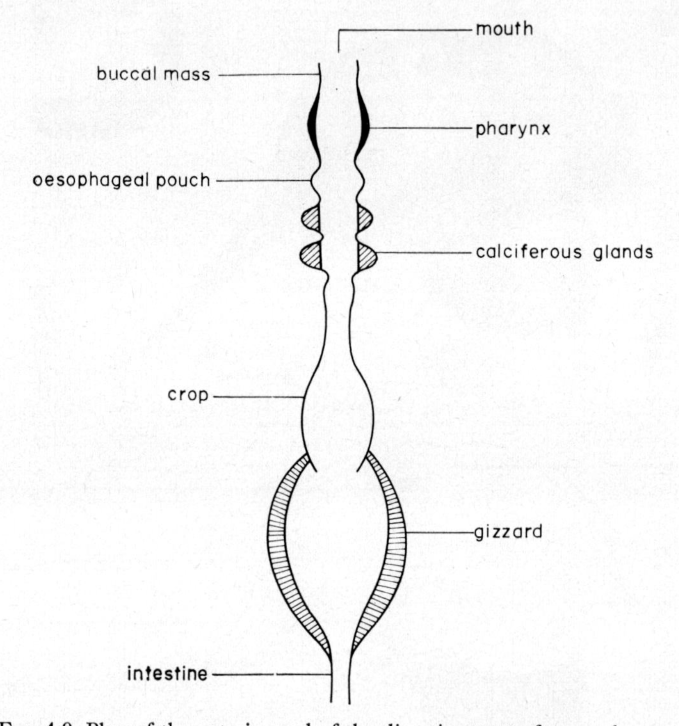

FIG. 4.9. Plan of the anterior end of the digestive tract of an earthworm.

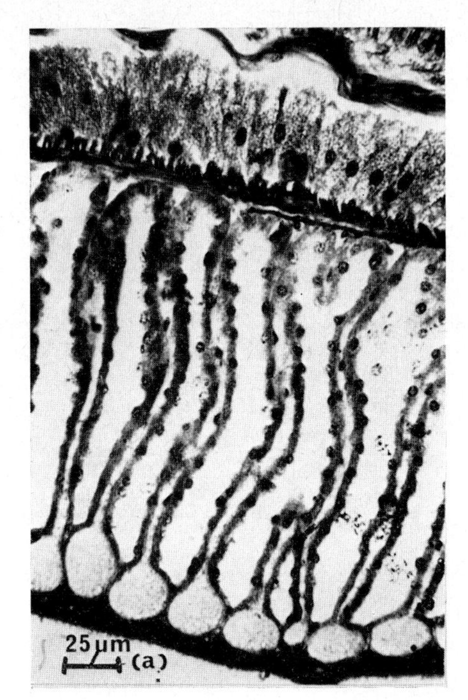

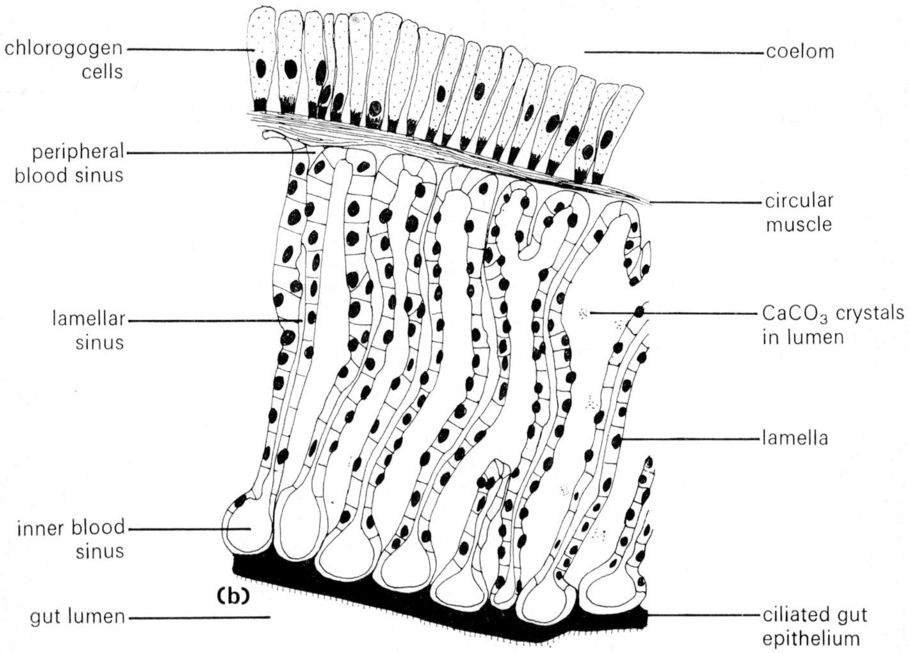

FIG. 4.10 (a) TS and (b) drawing of calciferous glands in the oesophagus of *L. terrestris.*

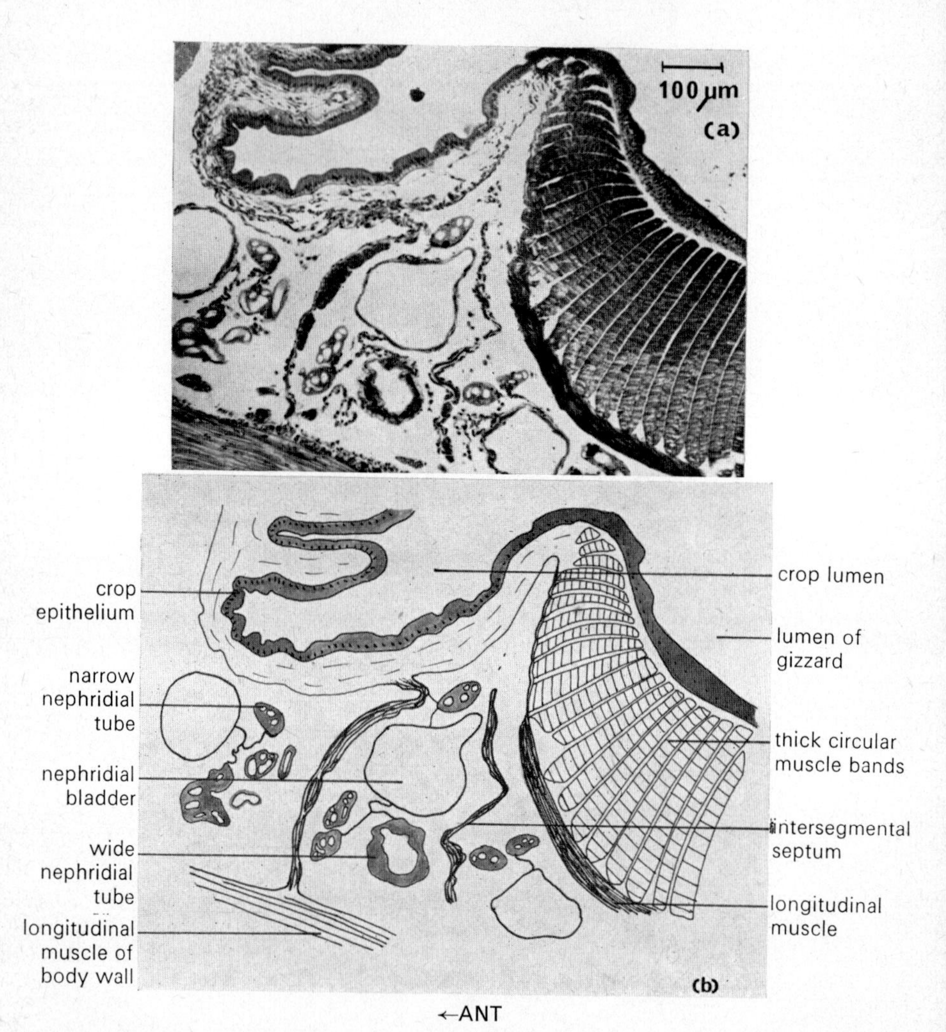

FIG. 4.11 (a) LS and (b) drawing of part of *L. terrestris* showing the crop and gizzard, and parts of the nephridial excretory system.

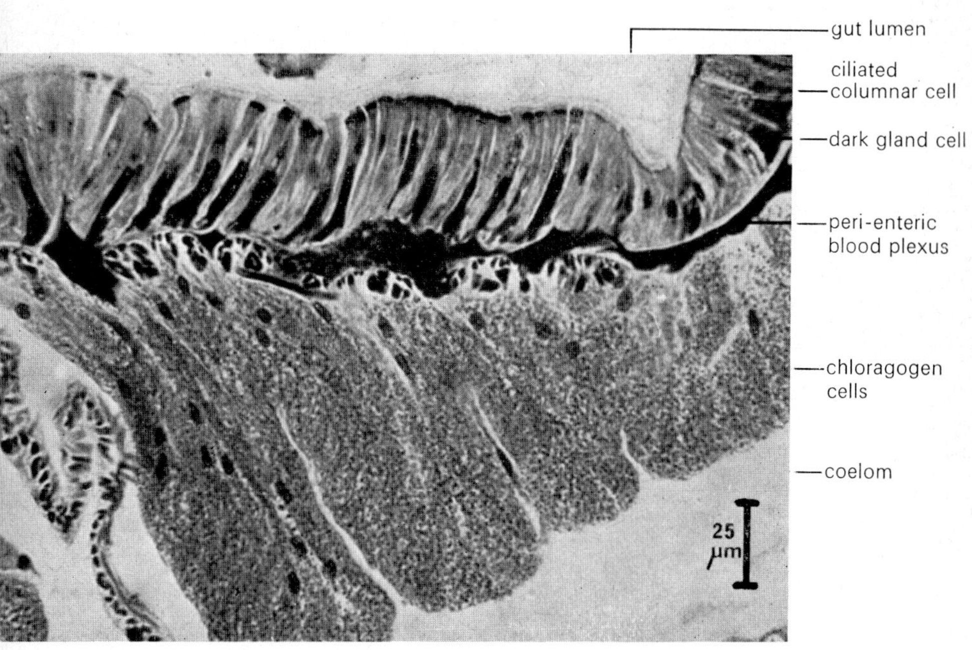

FIG. 4.12. TS through the intestine wall of *L. terrestris*, showing the digestive epithelium and chloragogen tissue.

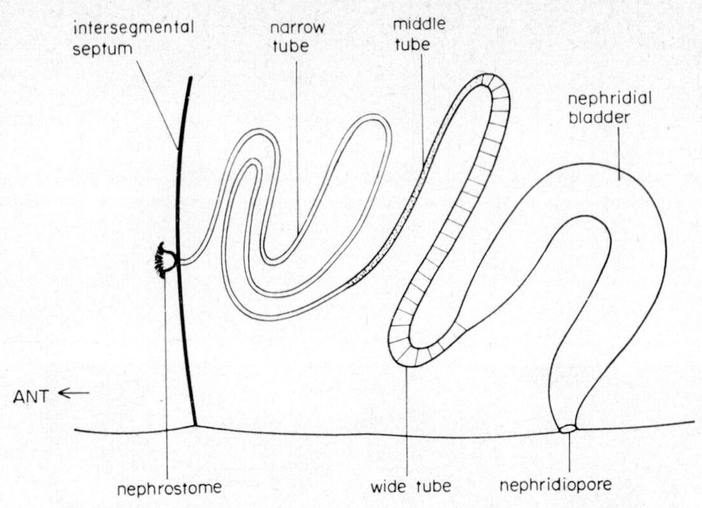

FIG. 4.13. Diagram showing the arrangement of an oligochaete metanephridium.

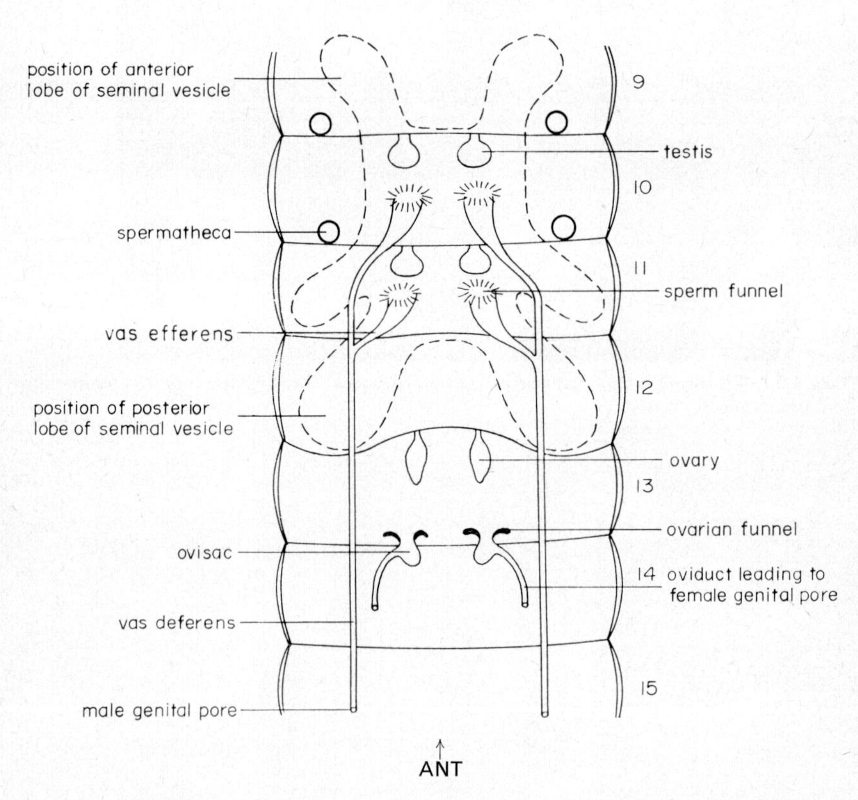

FIG. 4.14. Plan of the hermaphrodite reproductive system of an earthworm.

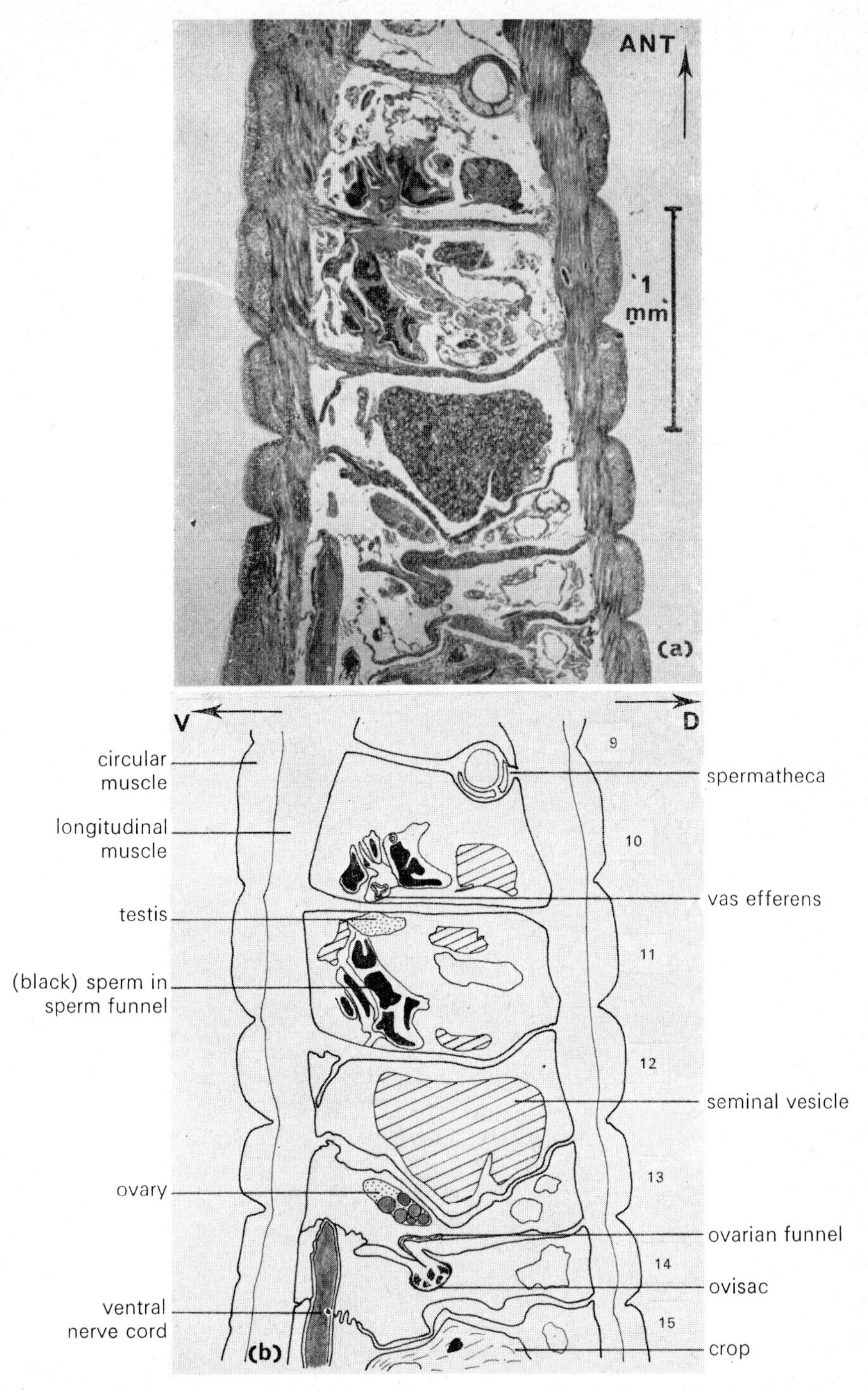

FIG. 4.15 (a) Oblique LS and (b) drawing of segments 9–15 of *L. terrestris*, showing many of the reproductive organs.

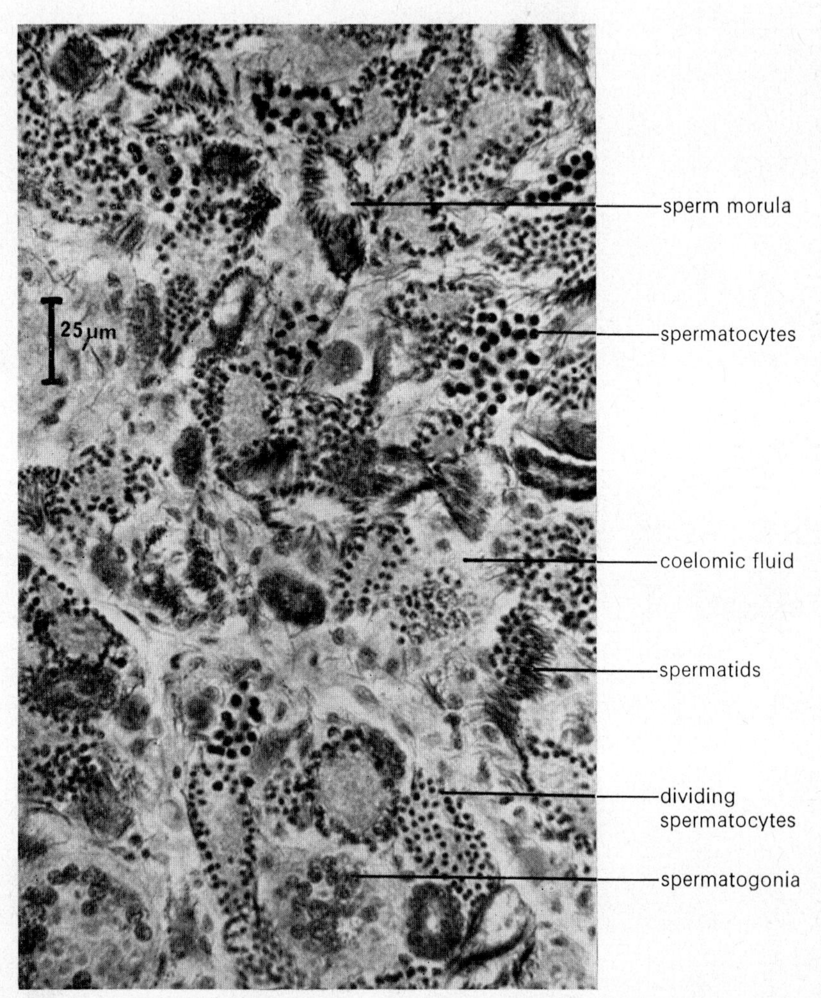

FIG. 4.16. Masses of developing sperm in the seminal vesicles of *L. terrestris.*

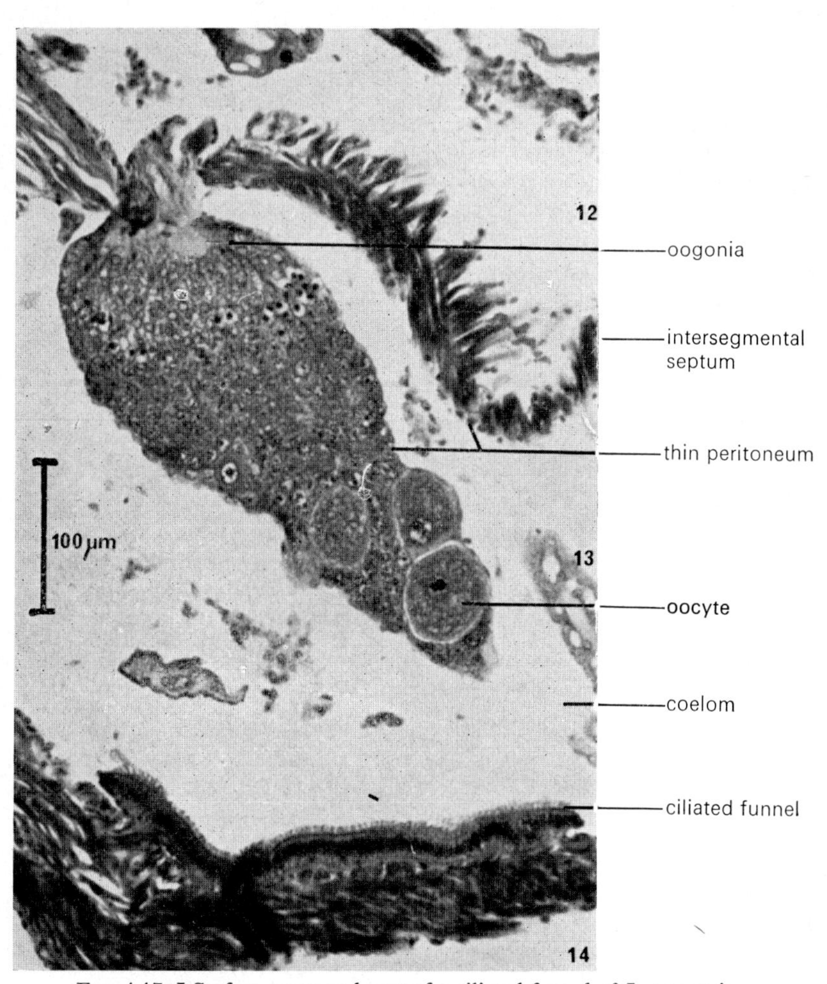

FIG. 4.17. LS of an ovary and part of a ciliated funnel of *L. terrestris*.

gizzard is used for trituration (Fig. 4.11). The intestine is the main site of digestion and absorption, and has an intucked wall (typhlosole) to increase surface area (see Fig. 4.2). It is lined by simple columnar epithelium (Fig. 4.12). The voluminous chloragogen cells surrounding the gut (see also Fig. 4.2 and 4.10) have a liver-like function involving storage of fat and glycogen, and detoxification and excretion of nitrogenous materials.

Excretion

The chief excretory organs in earthworms are metanephridia. Paired nephridia occur in most segments of the body, each consisting of a ciliated funnel (nephrostome) in one segment with a long coiled tube leading back into the adjacent posterior segment. This tube begins as a narrow ciliated duct widening out into a middle and wide tube and finally into a non-ciliated storage bladder which leads directly to a nephridiopore on the ventro-lateral surface (Fig. 4.13). Parts of several nephridia can be seen in Figs 4.1, 4.2 and 4.11.

Reproduction

Earthworms are hermaphrodite, each possessing two pairs of testes and one pair of ovaries. A plan of the reproductive structures is shown in Fig. 4.14, while many of them can be seen in Fig. 4.15.

The testes release spermatogonia into enclosed portions of the coelom (seminal vesicles) where they develop freely in the coelomic fluid (Fig. 4.16). Mature sperm travel from the seminal vesicles through paired coelomoducts (ciliated sperm funnels), vasa efferentia and vasa deferentia to the male genital pore on segment 15. From here they are passed to the spermathecae (between segments 9 and 10) of another individual where they are stored.

The small pear-shaped ovaries (Fig. 4.17) release oocytes into the coelom. They are captured by ciliated ovarian funnels and transferred into the ovisac or receptaculum ovarum, a coelomic pouch in which they develop (see Fig. 4.15). Mature eggs are extruded through oviducts to the female genital pore on segment 14, and laid in a cocoon secreted by the clitellum (see Fig. 4.4). Sperm are discharged into the cocoon as it passes the spermathecal openings. Fertilization is therefore external, and live young emerge from the cocoons.

Class Polychaeta

Polychaete worms live in marine or estuarine habitats. They show great diversity of form, many being modified for burrowing or dwelling in tubes. In many respects polychaetes resemble oligochaetes: i.e. they show obvious segmentation; the large perivisceral coelom is usually divided up by inter-segmental septae; the body wall layers, circulatory system and nervous system are similar; and segmental nephridia are the main excretory organs. However, polychaetes are generally characterized by the possession of a distinct head bearing appendages, although this may be reduced or modified in burrowing or sedentary forms. There is no clitellum present in the mature adult. Polychaetes also bear lateral projections from the body wall (parapodia) which are provided with blood-vessels and chaetae, and which act as limbs and gills. True gills are also found in some species.

Example: *Nereis* (*virens*)

This nereid lives in estuarine conditions and is one of the least specialized of the polychaetes. It can swim rapidly, and crawls and burrows in the sand using its parapodia and body movements. Nereids are errant (free-living) polychaetes and do swim about but they normally live in mucous tubes through which they circulate the water by undulations of the body. *N.virens* has a well-developed head with a prostomium bearing antennae, eyes and palps, and a peristomium formed by the fusion of two segments (Fig. 4.18). Although the coelom is divided up by septae, these are perforated so that the coelomic cavity extends throughout the body. Figure 4.19 is a transverse section through a nereid showing the arrangement of longitudinal muscle blocks and the position of the laterally-compressed segmental parapodia. Each parapodium consists of two fleshy lobes (notopodium and neuropodium) supported by chitinous acicula (Fig. 4.20). The parapodia are moved by muscles attached to the body wall, and possess tactile cirri on the dorsal and ventral surfaces. The parapodia move in metachronal rhythm; each one describing an elipse as it moves backwards on the power stroke and then forwards in a raised position on the return.

Nervous system

The nervous system of nereids is very similar to that of oligochaetes (see Fig. 4.6) with the two ganglionated ventral nerve cords held together by connective tissue. Polychaetes are well-known for their large giant nerve fibres (Fig. 4.21). Those of nereids show an arrangement somewhat similar to that of earthworms (see Fig. 4.8), but some polychaetes have only one very large fibre.

Sense organs

Polychaetes have a more plentiful supply of sense organs than oligochaetes, associated with their more active mode of life. Many of these receptors are simple nerve endings, or ciliated sensory pits on the head region (nuchal organs). Light receptors range from eyespots to highly-developed focusing structures in some pelagic species.

Nereis possesses two pairs of eyes on the prostomium (see Fig. 4.18.) They are of the direct retinal cup type with rod-like receptors directed

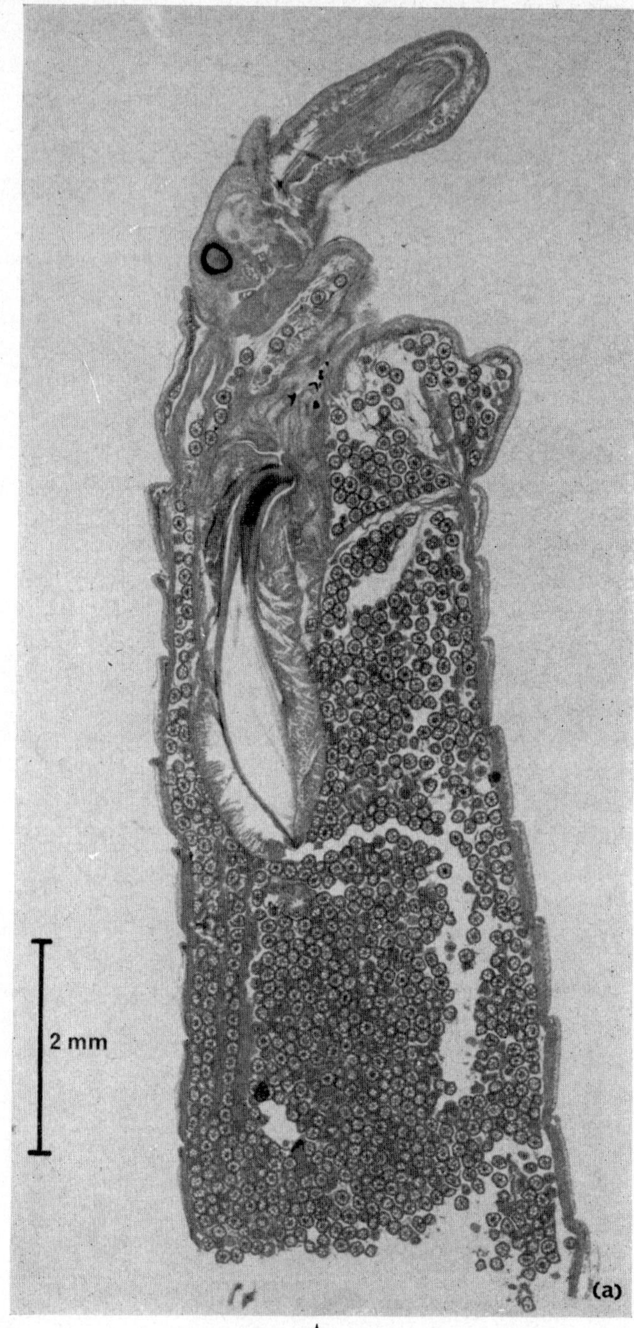

Fig. 4.18

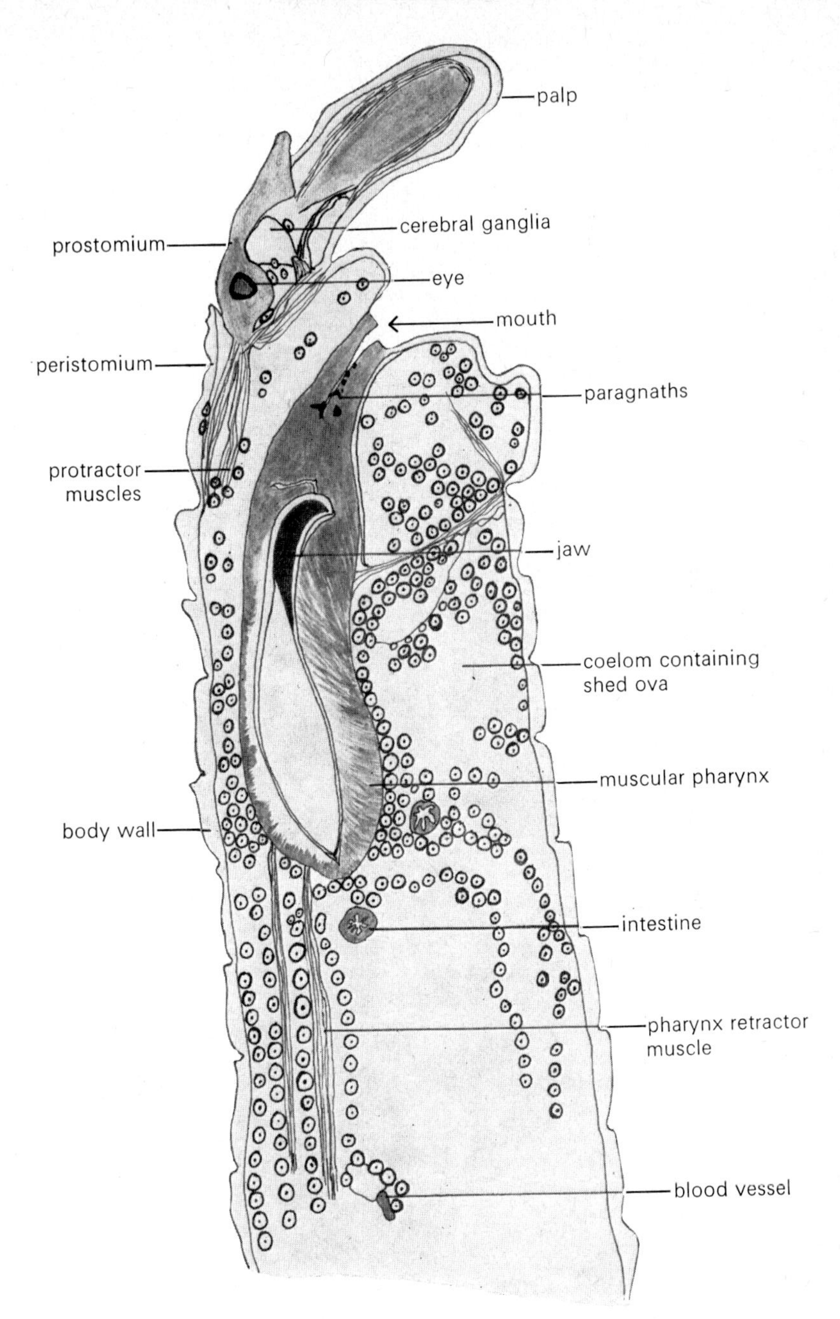

FIG. 4.18 (a) VLS and (b) explanatory drawing of the anterior end of an adult female nereid (*Nereis virens*).

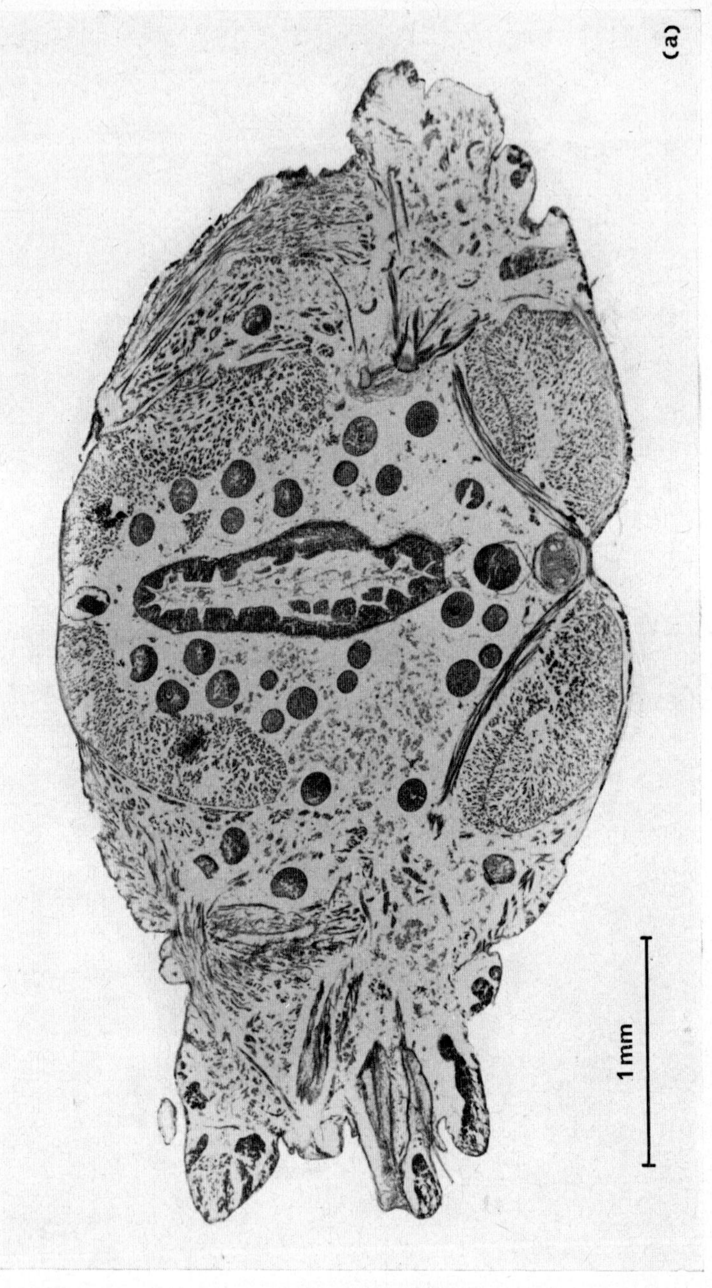

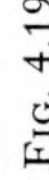

FIG. 4.19

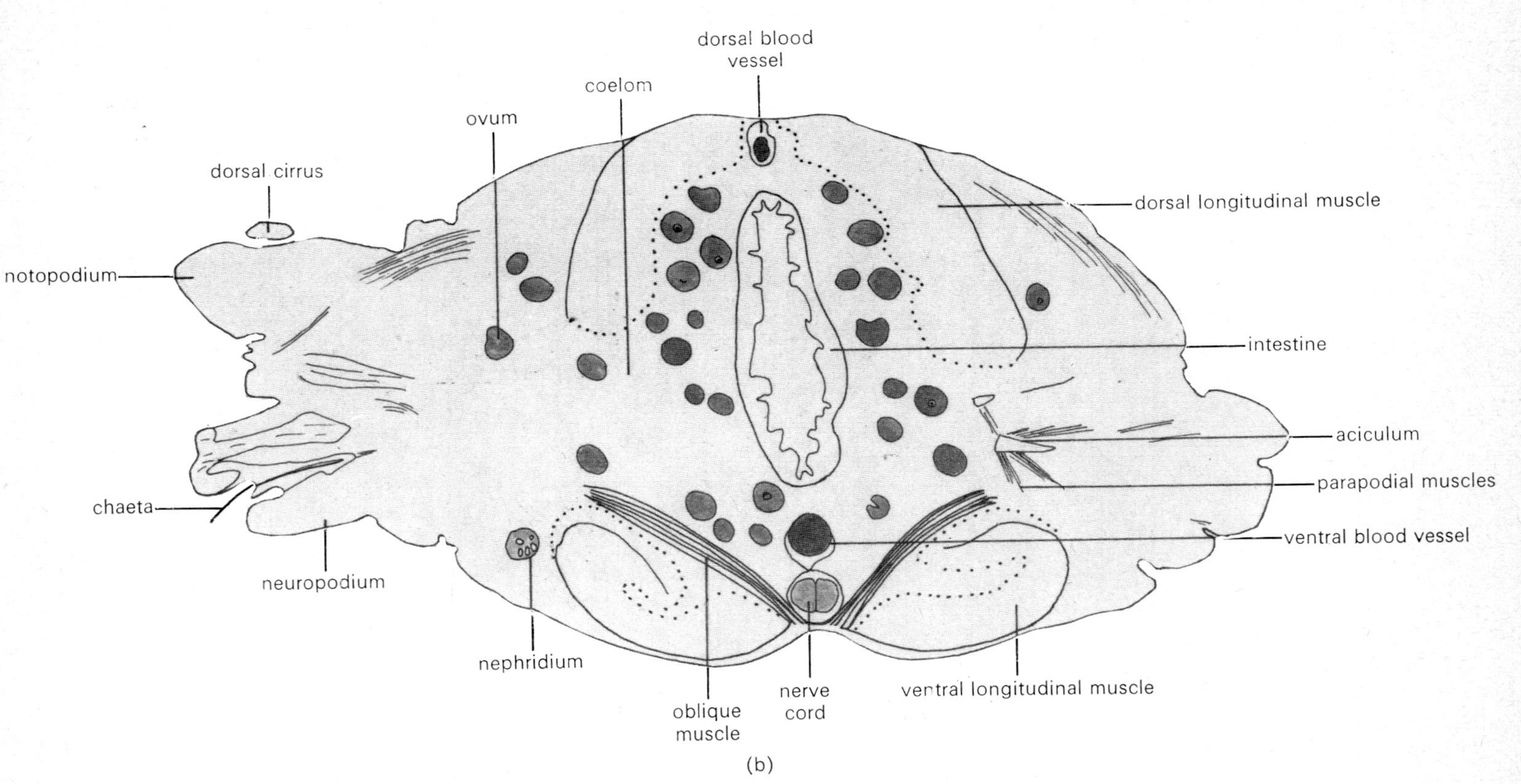

FIG. 4.19 (a) TS and (b) explanatory drawing of an adult female *N. virens* in the mid-body region.

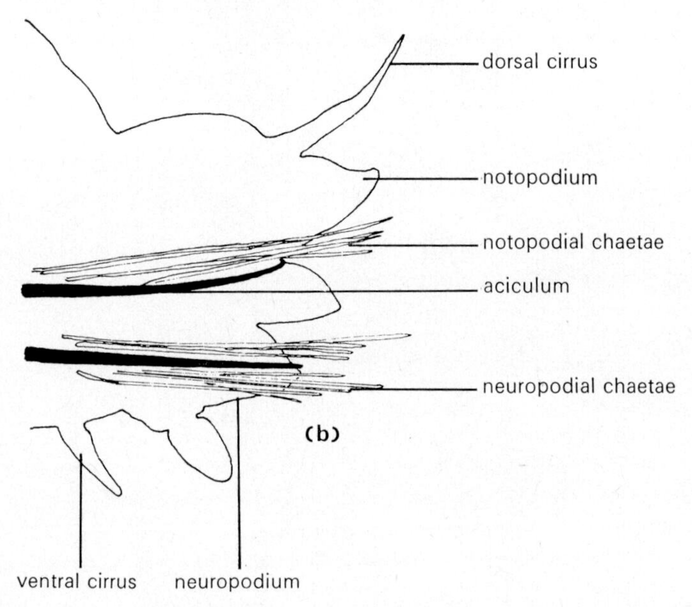

FIG. 4.20 (a) Whole mount of a nereid parapodium; and (b) explanatory drawing. Note the chaetae, acicula and sensory cirri.

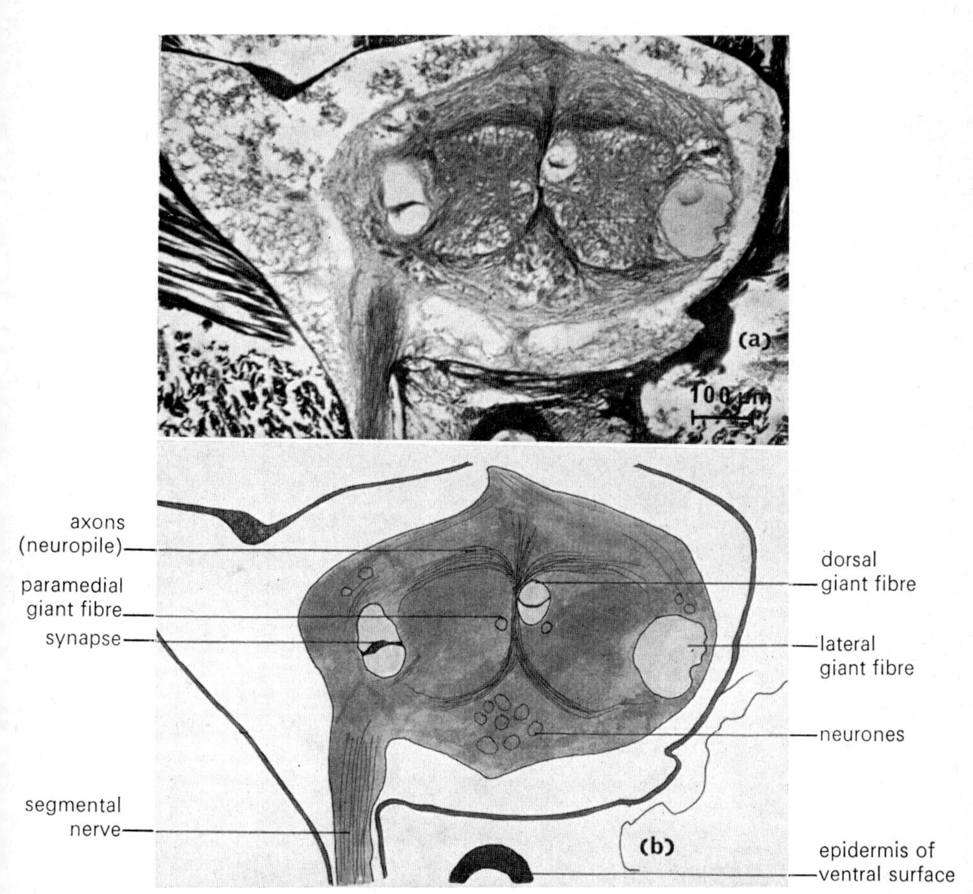

FIG. 4.21 (a) TS and (b) drawing of a segmental ganglion from the ventral nerve cord of *N. virens*. Note the very large lateral giant nerve fibres (one of them showing a synaptic junction between two sequential fibres).

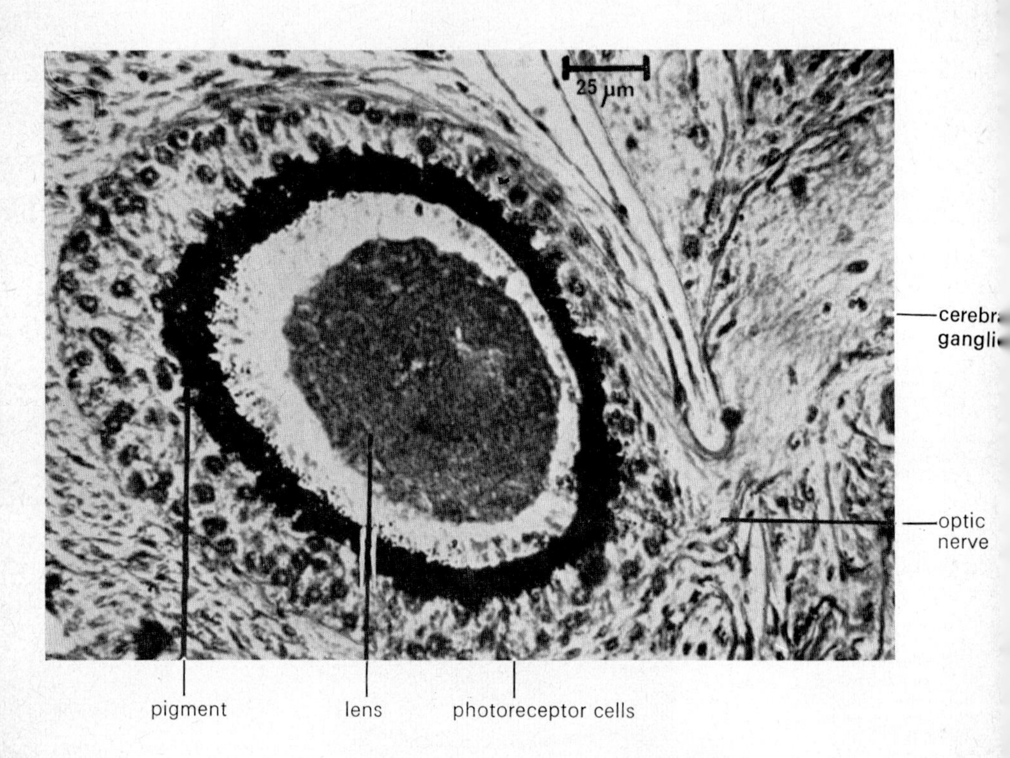

FIG. 4.22. Section through the eye of *N. virens*.

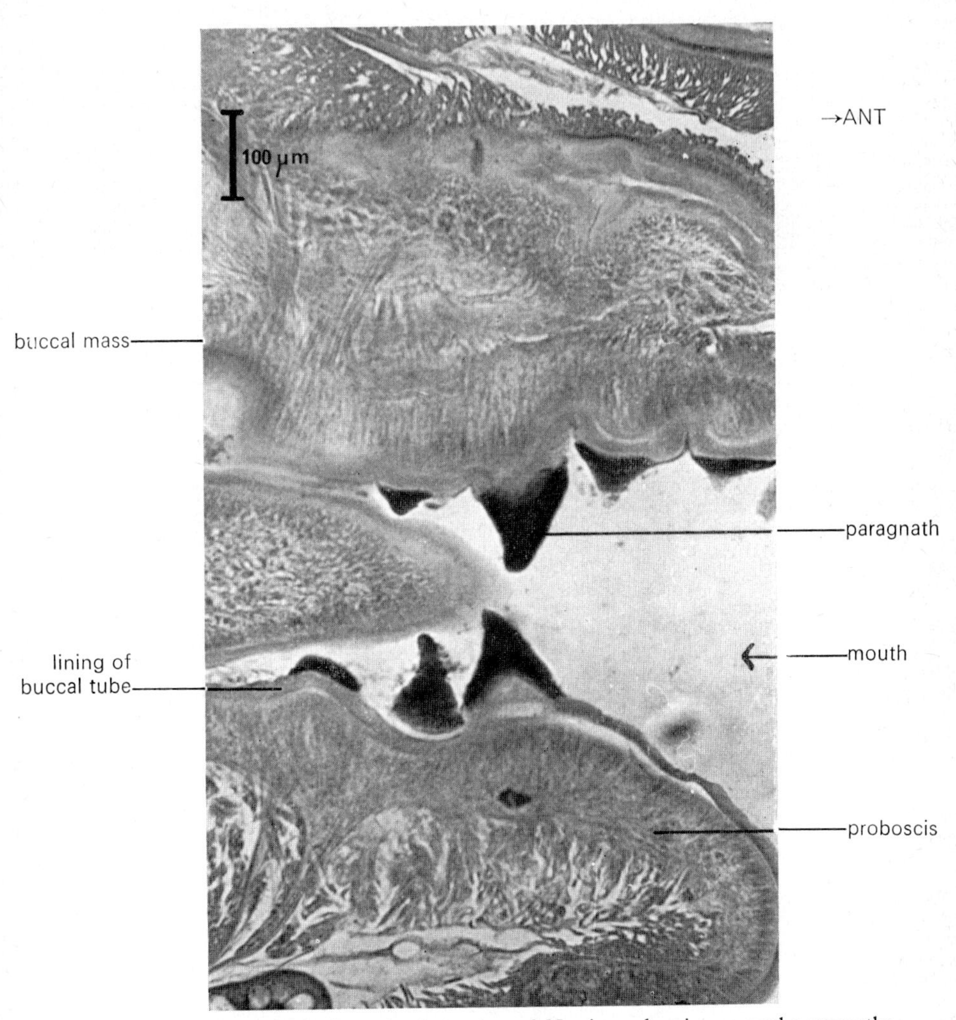

FIG. 4.23. VLS through the buccal tube region of *N. virens* showing several paragnatha. The proboscis is in the resting (inverted) position.

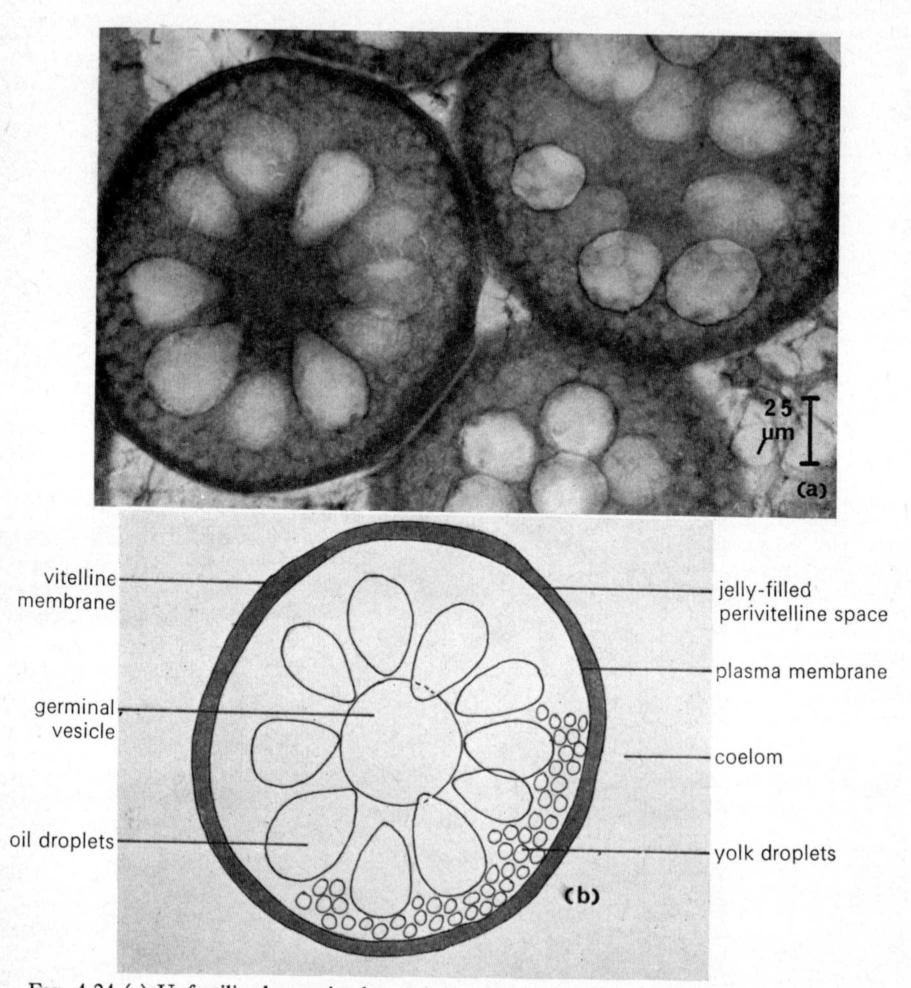

FIG. 4.24 (a) Unfertilized eggs in the coelom of *N. virens*; (b) drawing of one egg seen in (a).

towards the lumen of the eye and connected to the brain by short optic nerve fibres (Fig. 4.22). The eye is lined with supporting cells and pigment, is filled with a refractive lens, and is covered externally by a cuticular "cornea". Such eyes probably function only in light detection.

Digestion

N. virens is carnivorous, eating small invertebrates and algae. A muscular pharynx bearing chitinized jaws (see Fig. 4.18) is linked to the mouth by a short buccal tube lined with chitinous teeth or paragnaths (Fig. 4.23), This whole structure (proboscis) can be partially or wholly everted when feeding, with the aid of longitudinal retractor and protractor muscles. The rest of the gut is similar to that of oligochaetes (see Fig. 4.9), except that there is no crop or gizzard.

Excretion

As in earthworms, the main excretory organs are metanephridia and chloragogen tissue. Being an estuarine animal, *N. virens* exhibits great tolerance to salinity changes. This is probably achieved by active salt uptake at low external salinities combined with an ability of the tissues to function at a wide range of osmotic concentrations.

Reproduction

Some polychaetes reproduce asexually, but in most species sexual reproduction occurs, and the sexes are separate. In nereids the simple gonads arise by proliferation of the coelomic epithelium and extend throughout the body. Developing gametes are shed into the body cavity where they mature in the coelomic fluid (see Figs 4.18 and 4.19). Ripe gametes are released through the nephridioducts. The eggs (Fig. 4.24) are provided with a plentiful supply of yolk and oil droplets (for buoyancy), and are protected by a layer of jelly. Fertilization is external, and the eggs develop into free-swimming trochophore larvae.

At breeding time many nereids exhibit a sexually-active (epitokous) form in which whole or part of the worm differs morphologically from the normal (atokous) individual. Epitokous forms (heteronereids) are generally adapted for a more active mode of life than normal. They have bigger eyes, and larger fan-like parapodia to allow for an increased respiratory rate and faster swimming movements.

Class Hirudinea

Most of the Hirudinea (leeches) live in freshwater, but there are also marine and terrestrial species. Many of them are parasitic and show adaptations to this mode of life. The body is dorso-ventrally flattened, and, unlike other annelids, has a fixed number of segments (33) which are divided up externally into rings or annuli. The original segmentation is thus obscured, and septae are lost. The prostomium is reduced, while several segments at the anterior and posterior ends of the body are modified to form ventrally placed suckers. The body surface is otherwise smooth with no appendages, parapodia or chaetae. The coelom is reduced by ingrowth of mesenchyme, and forms several longitudinal sinuses with transverse connections containing blood. The foregut is specialized consisting of a proboscis (acanthobdellids and rhyncobdellids) or a muscular sucking pharynx (pharyngobdellids and gnathobdellids) well supplied with salivary glands. The gnathobdellids also possess jaws and secrete anticoagulants from their salivary glands. All leeches are hermaphrodite, possessing several pairs of testes and a single pair of ovaries continuous with their ducts. There are single male and female genital pores, and a clitellum is developed at maturity.

Example: *Erpobdella* (*testacea*)

E. testacea is a small leech living in shallow ponds. It belongs to the Pharyngobdellidae, and therefore has a muscular pharynx but no jaws.

The coelomic cavity is packed with mesenchyme traversed by longitudinal and radial muscles with a few large dorso-ventral muscle fibres (Fig. 4.25). All that remains of the coelom in this species is the single ventral sinus and two lateral sinuses with their transverse (latero-lateral) connections; the dorsal sinus having disappeared. The ventral blood sinus encloses the double nerve cord. In the posterior region of the body (Fig. 4.26) the longitudinal muscle is divided up into blocks and is confined to the outer margin of the animal; the rest of the body being occupied by the gut and gonads.

The body wall consists of a thin cuticle and single layer of epidermis overlying a thin dermis which contains sunken epidermal mucous glands (Fig. 4.27). The typical leech muscle structure is shown in this photograph, with a central core of sarcoplasm containing the nucleus and surrounded by contractile fibres (see also Fig. 4.29).

Nervous system

The nervous system of leeches is more specialized than in other annelids. The six most anterior segmental ganglia are fused to form the cerebral and sub-pharyngeal ganglia. While in other annelids the cerebral ganglia consist of a single pair of prostomial ganglia, in leeches they are composed of a pair of prostomial ganglia and one pair of segmental ganglia which have migrated dorsally to the pharynx. The seven most posterior ganglia are also fused in the posterior sucker region. Each segmental ganglion consists of six fibrous capsules enclosing the nerve cell bodies (two antero-dorsal, two postero-dorsal and two ventral) (Fig. 4.28). There are two giant Retzius cells in each segmental ganglion, which are thought to innervate epidermal mucous cells.

Leeches have many simple sense organs, and eyes consisting of pigment cups surrounding clusters of photoreceptive cells.

Digestion

E. testacea is carnivorous and swallows small invertebrates whole. The anterior third of the gut is a muscular pharynx well supplied with unicellular salivary glands (Fig. 4.29). The pharynx is triangular in cross-section (see Fig. 4.25), while in the jawed leeches there are three jaws arranged in a triangle (one dorsal and two ventro-lateral). The salivary glands secrete enzymes or, in some blood-sucking species, an anticoagulant (hirudin).

The pharynx leads, via a short oesophagus, into the crop or stomach, which forms a storage organ and often possesses diverticula or caecae. The stomach is usually separated from the intestine by a pyloric sphincter. The intestine is lined by a simple epithelium and is the site of digestion and absorption (Fig. 4.30). There is a noticeable lack of muscle in the gut wall; food being moved through the gut by pumping action of the pharynx and by contractions of the body muscles. In many species the internal surface of the intestine is increased by a spiral folding of the epithelium, and sometimes caecae are present here also. The intestine passes by a short rectum to the dorsal anus, which lies just in front of the posterior sucker and is remarkable for its plentiful supply of mucous cells (Fig. 4.31).

Excretion

The excretory organs are metanephridia, and a special tissue (botryoidal tissue) which is probably analogous to the chloragogen cells of oligochaetes.

The nephridia of leeches differ from those of other annelids in that the nephrostome is usually separated from the nephridial canal. The ciliated funnel of each nephridium projects into a coelomic blood sinus, and opens into a capsule which forms a phagocytic organ and acts as a site for the manufacture of coelomic corpuscles. In erpobdellid leeches there are two funnels and capsules to each nephridium [Fig. 4.32(a)]. The nephridial canal ends blindly near this ciliated organ: excretory material enters it by filtration through the wall. The canal leads eventually into a bladder or vesicle which opens to the outside by a nephridiopore [Fig. 4.32(b)].

Botryoidal tissue consists of swollen pigment-containing cells surrounding capillary channels which arise from the blood sinuses. Bits of such tissue are seen in Figs 4.26, 4.27, 4.32 and 4.33. It is thought to assist in excretion.

Reproduction

The cavities of the gonads are of coelomic origin and therefore form true gonocoels.

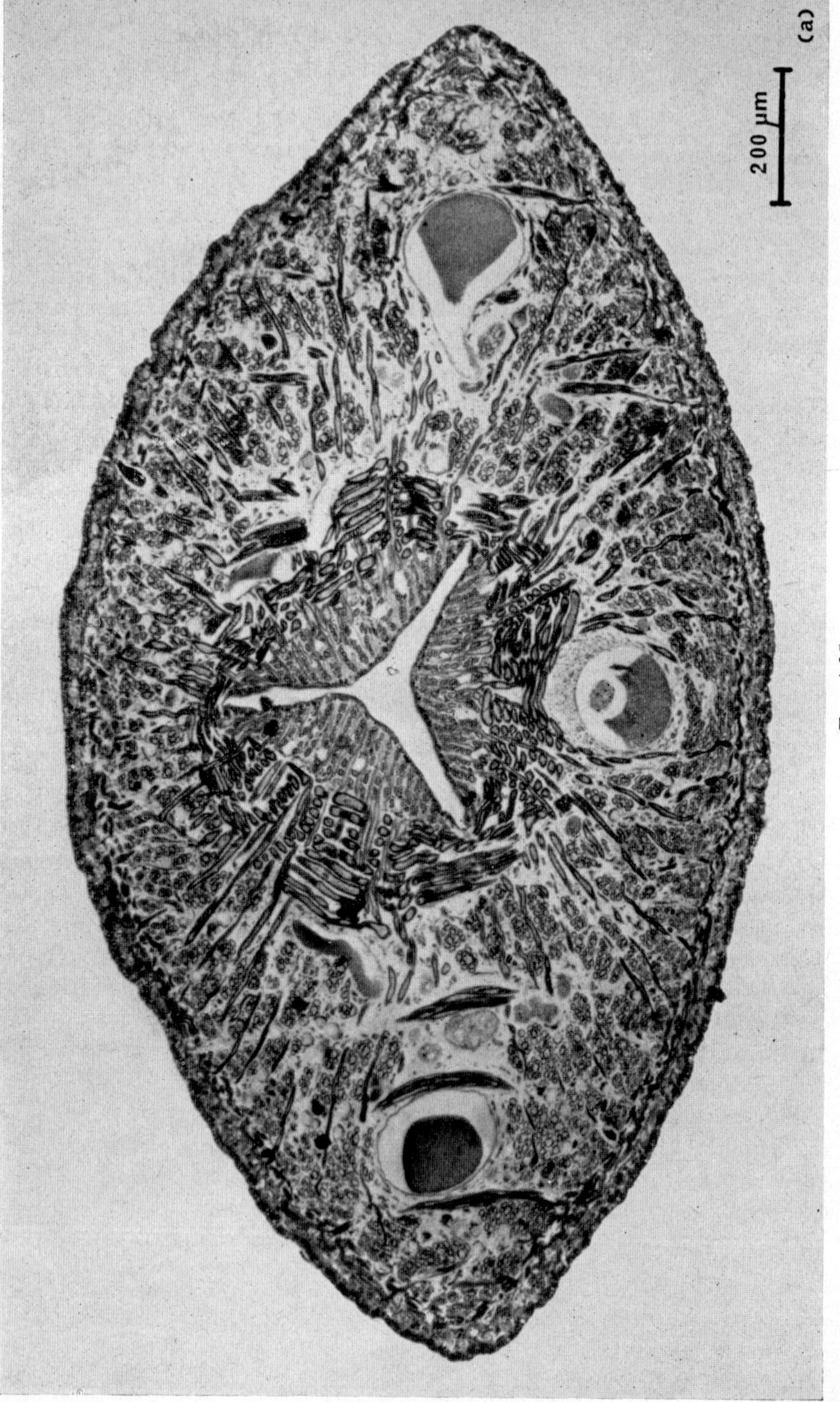

FIG. 4.25

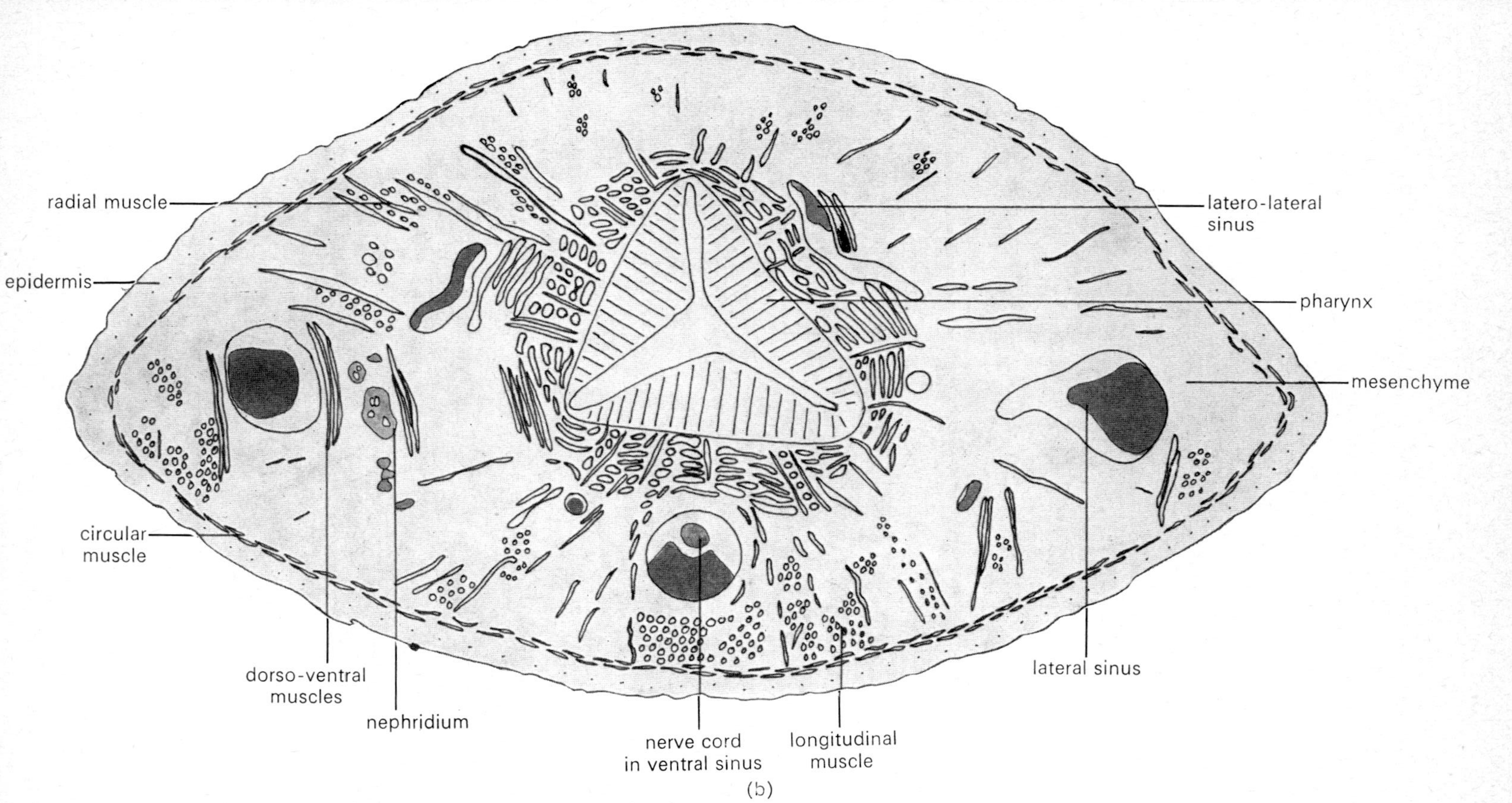

Fig. 4.25 (a) TS and (b) explanatory drawing of a leech (*Erpobdella testacea*) in the pharyngeal region.

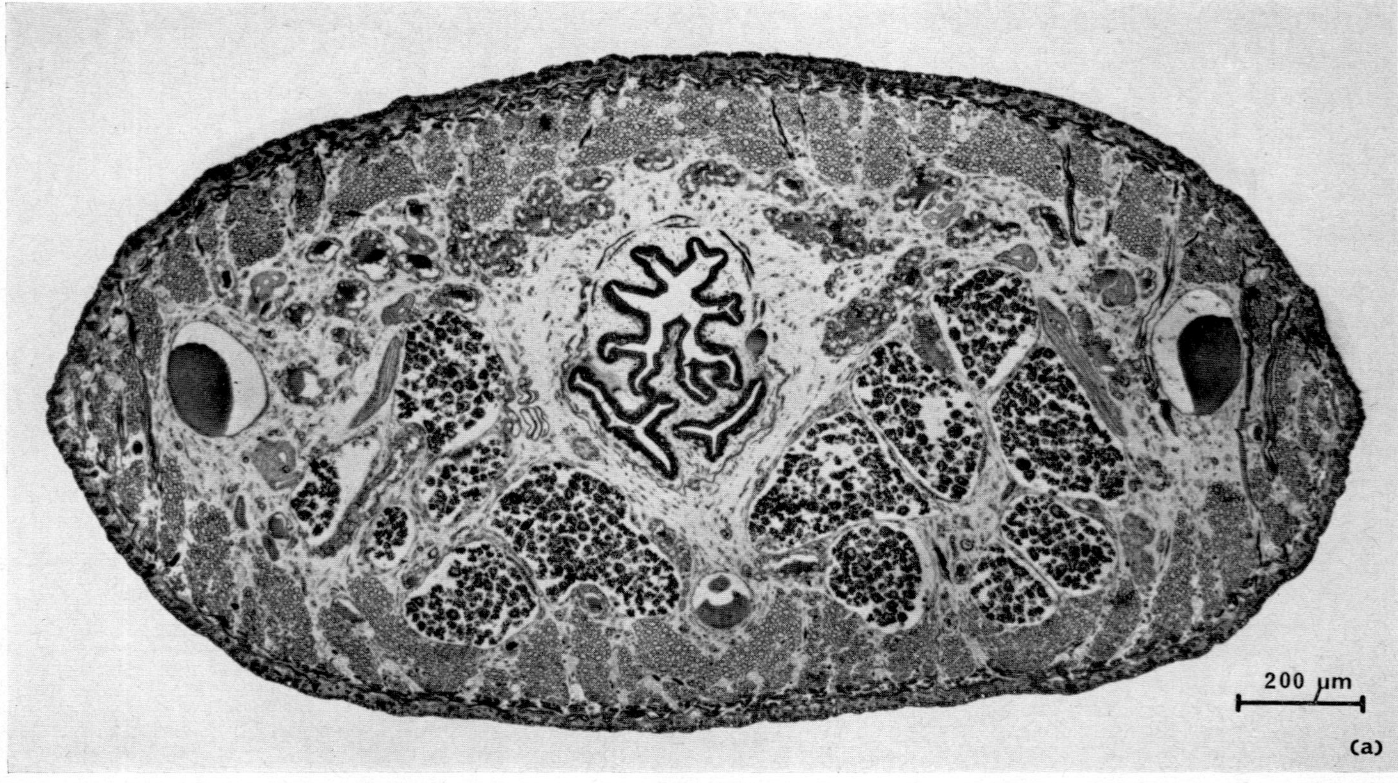

Fig. 4.26

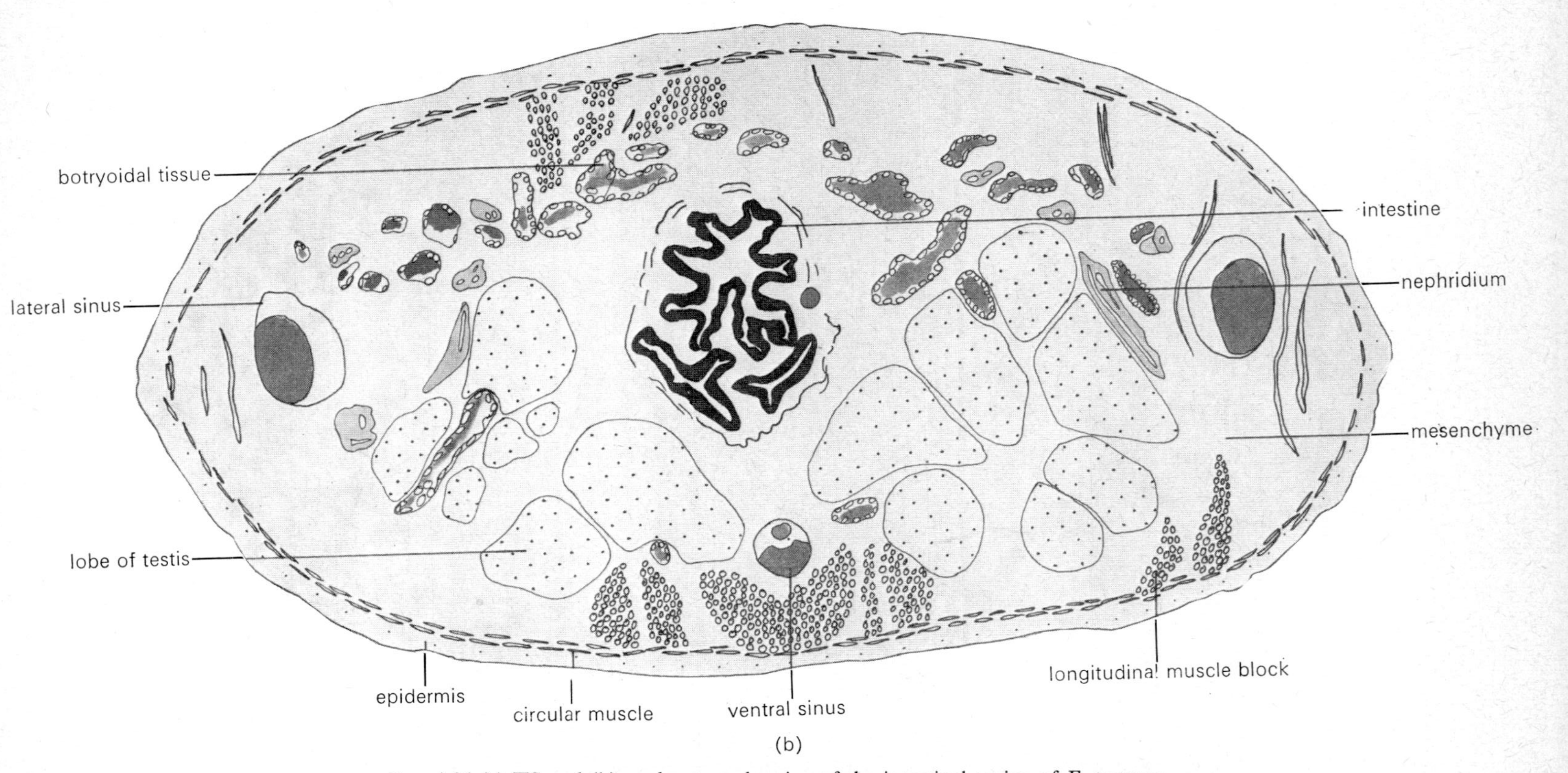

FIG. 4.26 (a) TS and (b) explanatory drawing of the intestinal region of *E. testacea*.

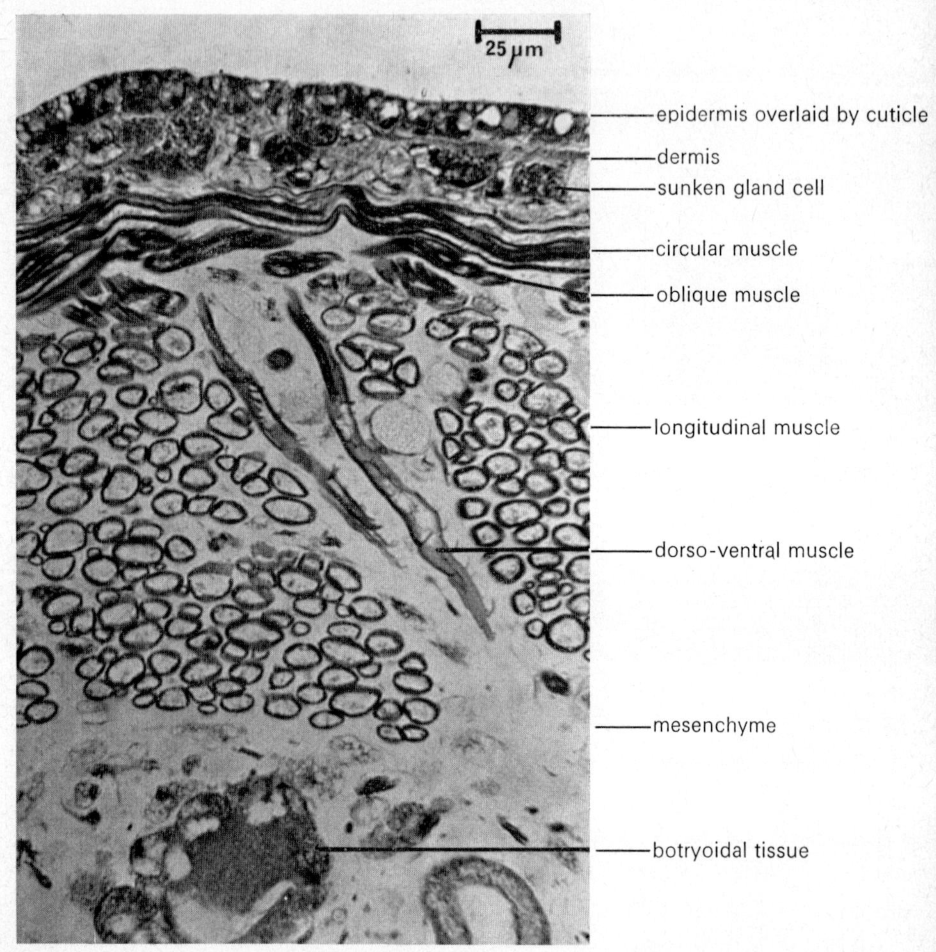

FIG. 4.27. TS through the body wall of *E. testacea*. Note the muscle fibres with their peripheral fibrils and central cytoplasmic core.

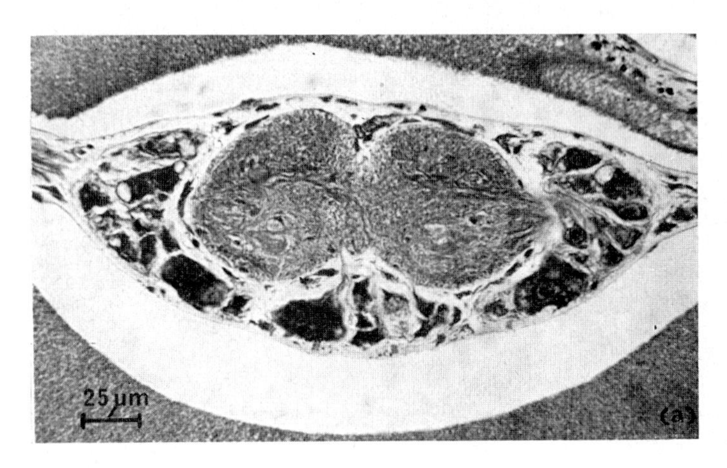

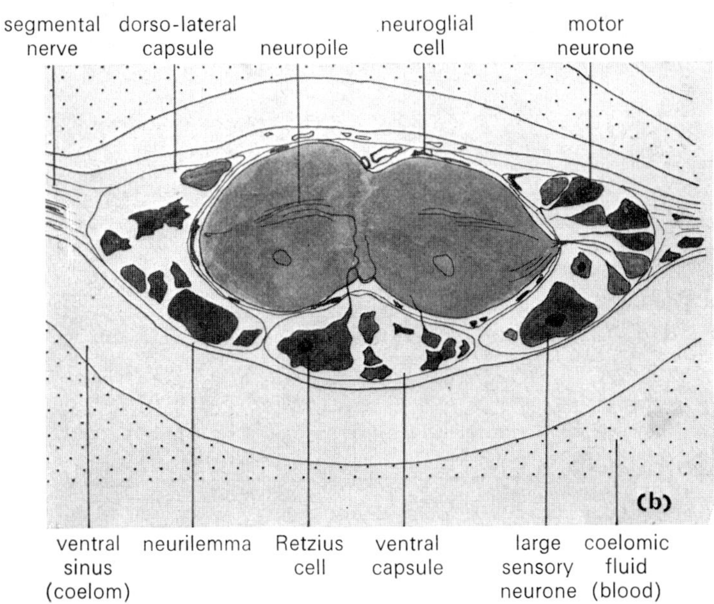

FIG. 4.28 (a) TS and (b) drawing of a segmental ganglion in the ventral nerve cord of *E. testacea*. Note that the nerve cord lies inside the ventral coelomic (blood) sinus.

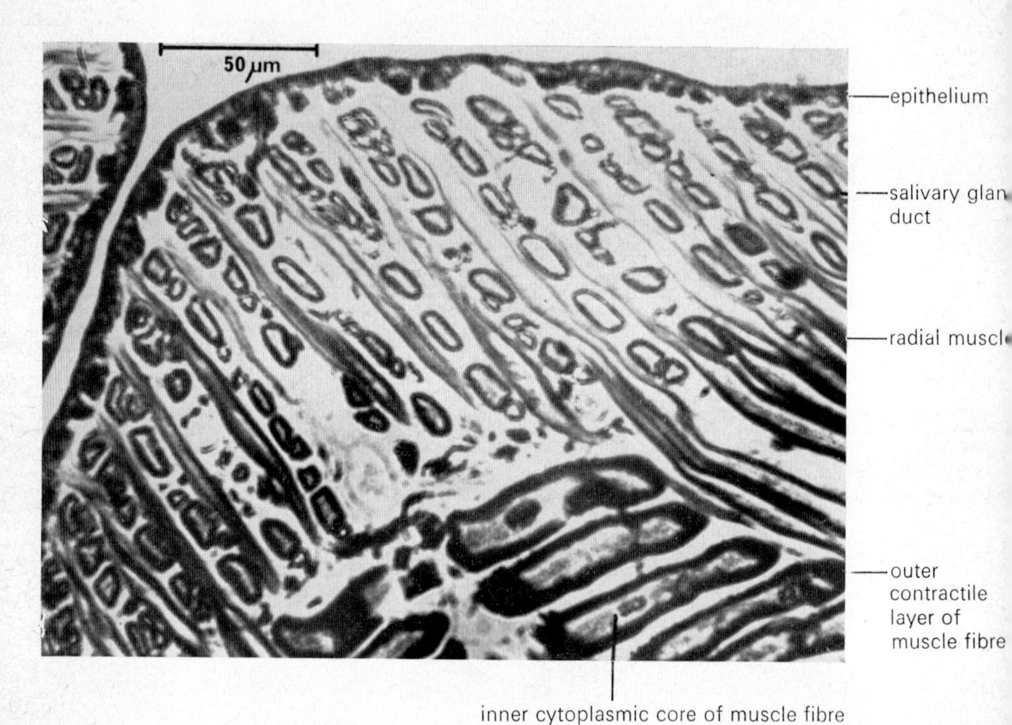

Fig. 4.29. TS through the pharynx wall of *E. testacea*.

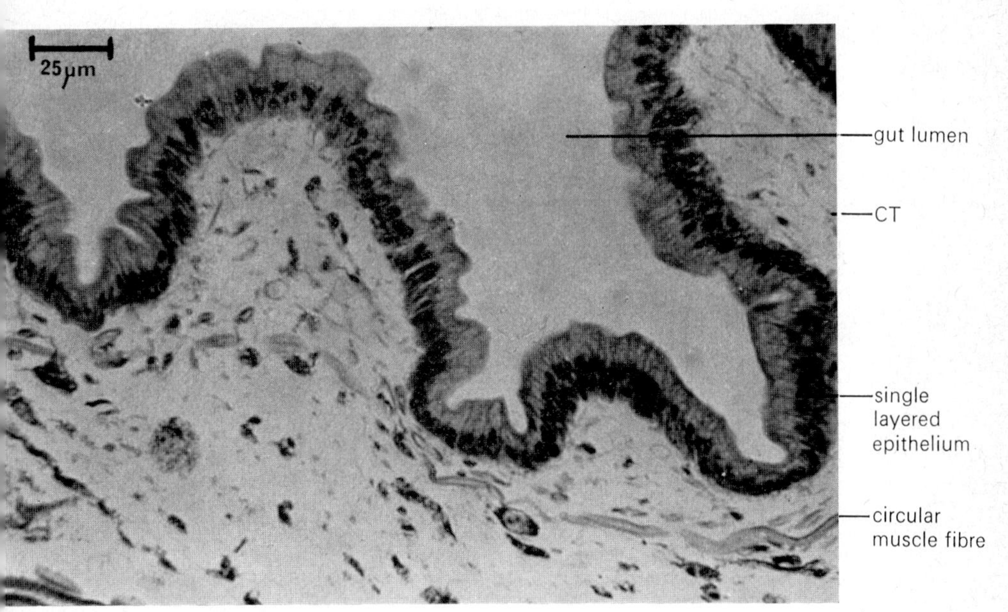

FIG. 4.30. TS through the intestinal wall of *E. testacea*. Note the paucity of gut musculature.

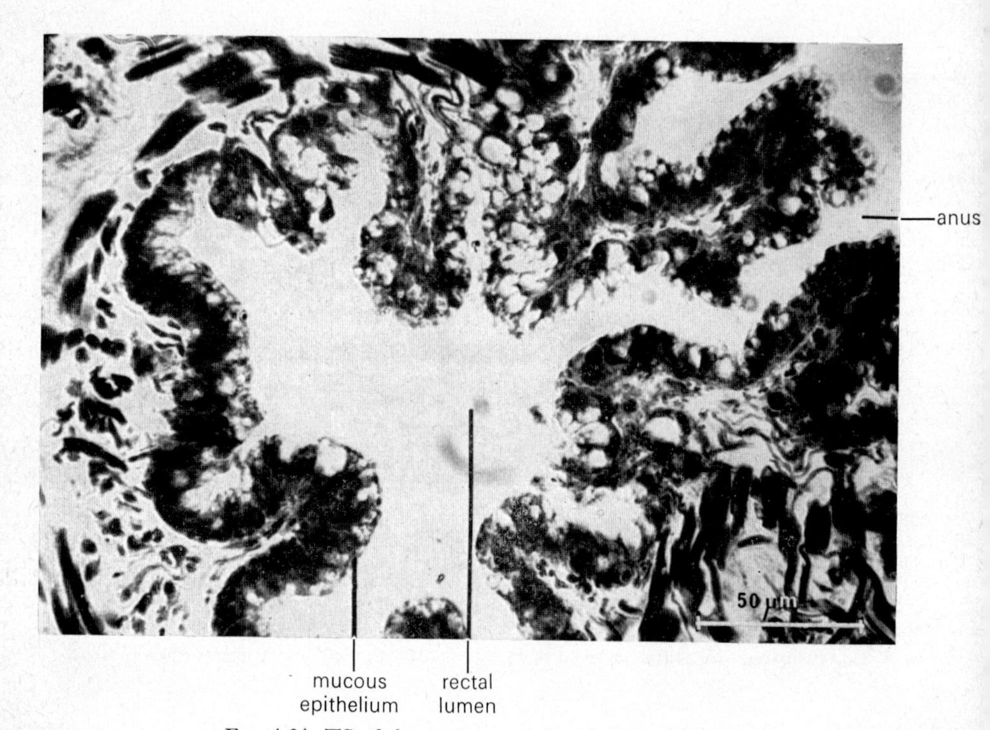

FIG. 4.31. TS of the rectum and dorsal anus of *E. testacea*.

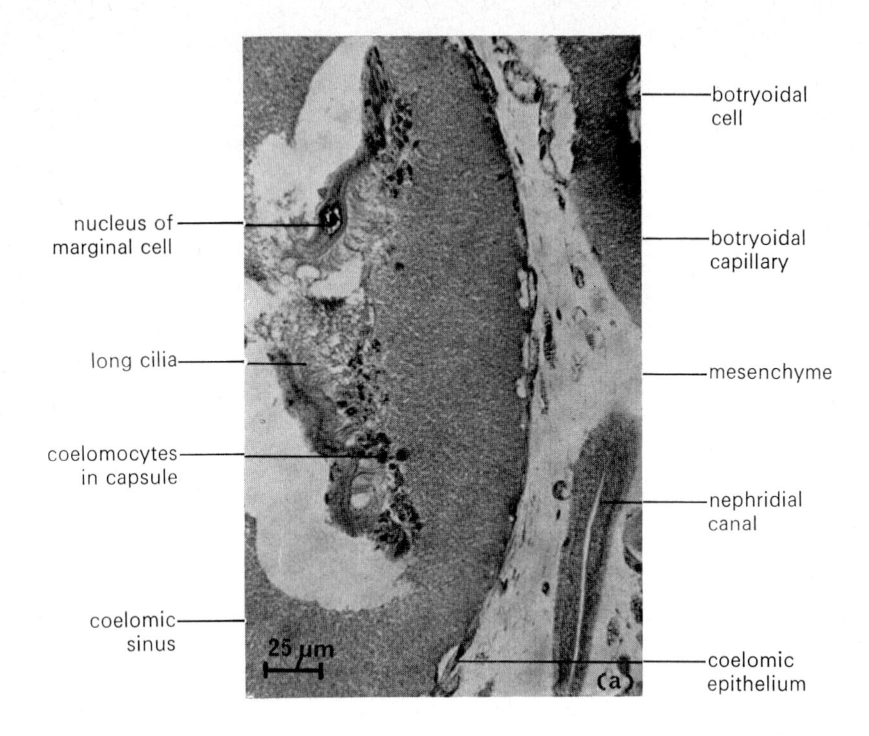

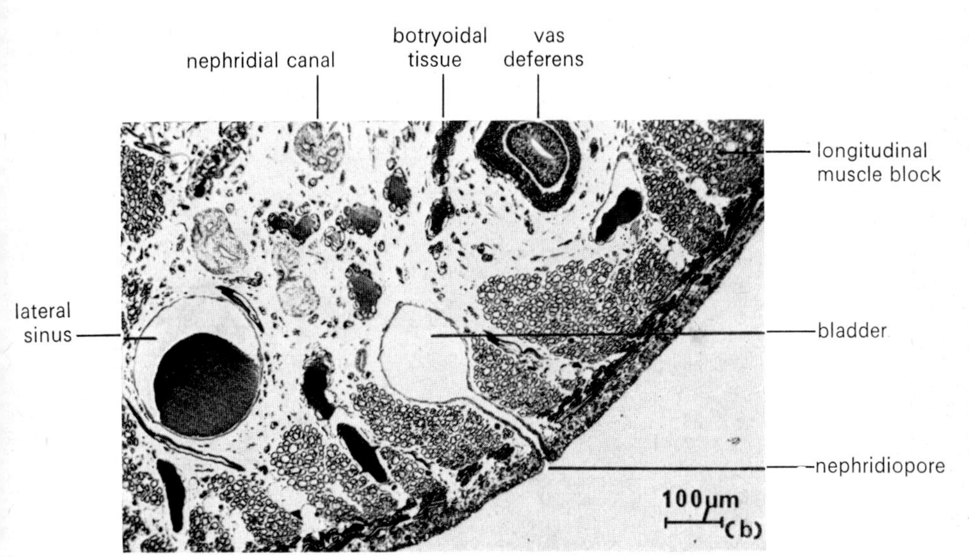

FIG. 4.32 (a) Section through part of the ciliated organ of *E. testacea*. Note the long cilia of the two nephridial funnels; (b) TS through part of *E. testacea* showing portions of the nephridial excretory system.

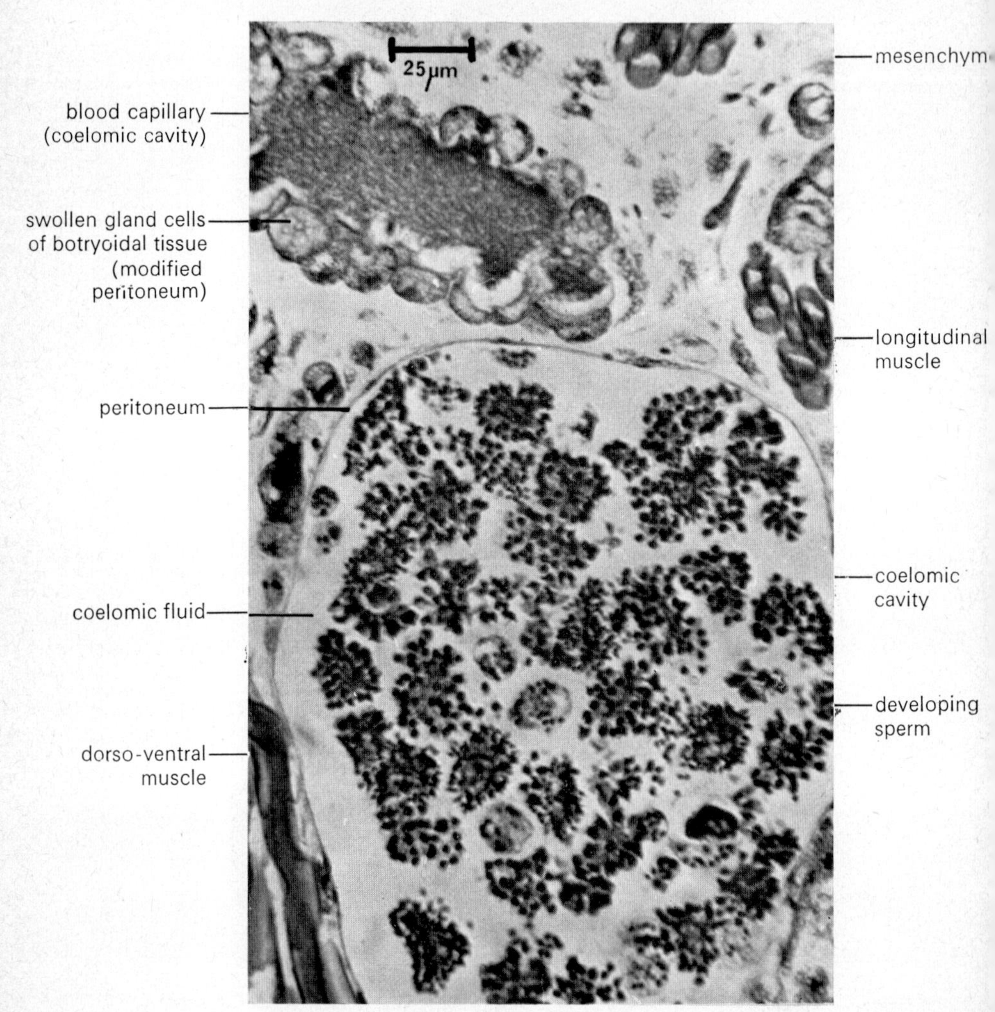

FIG. 4.33. Testis and botryoidal tissue of *E. testacea.*

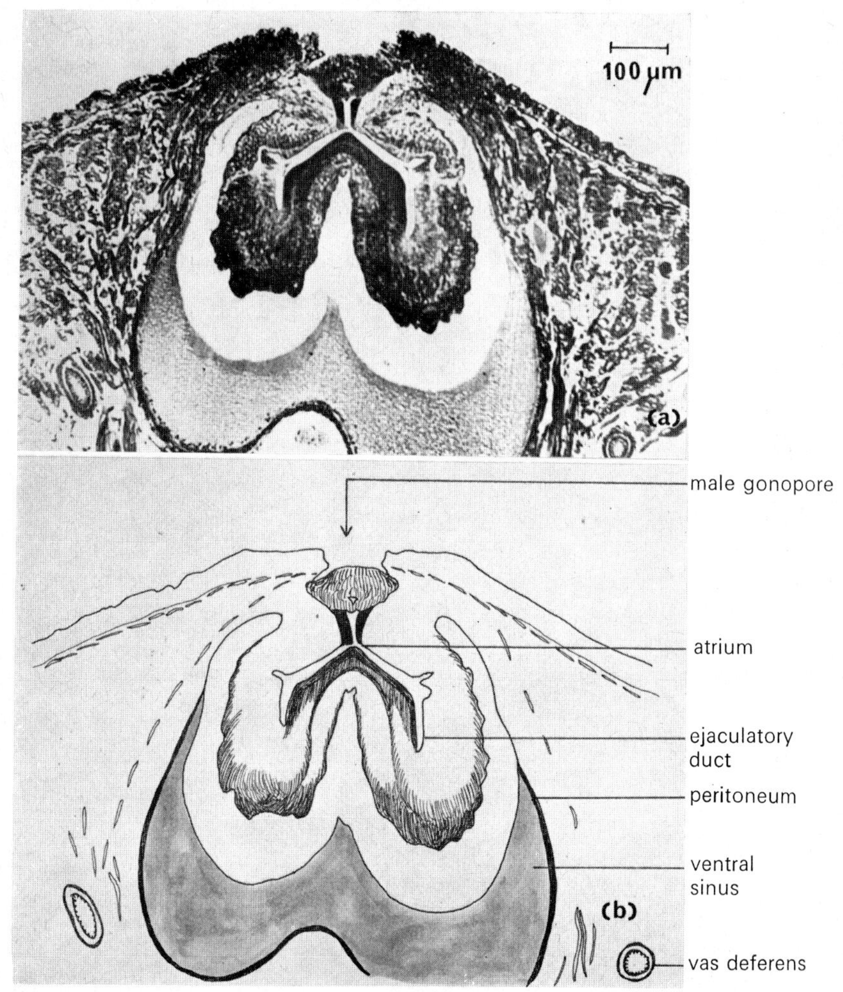

FIG. 4.34. (a) Section and (b) drawing of the genital atrium of *E. testacea*. The male gonopore opens in segment 10.

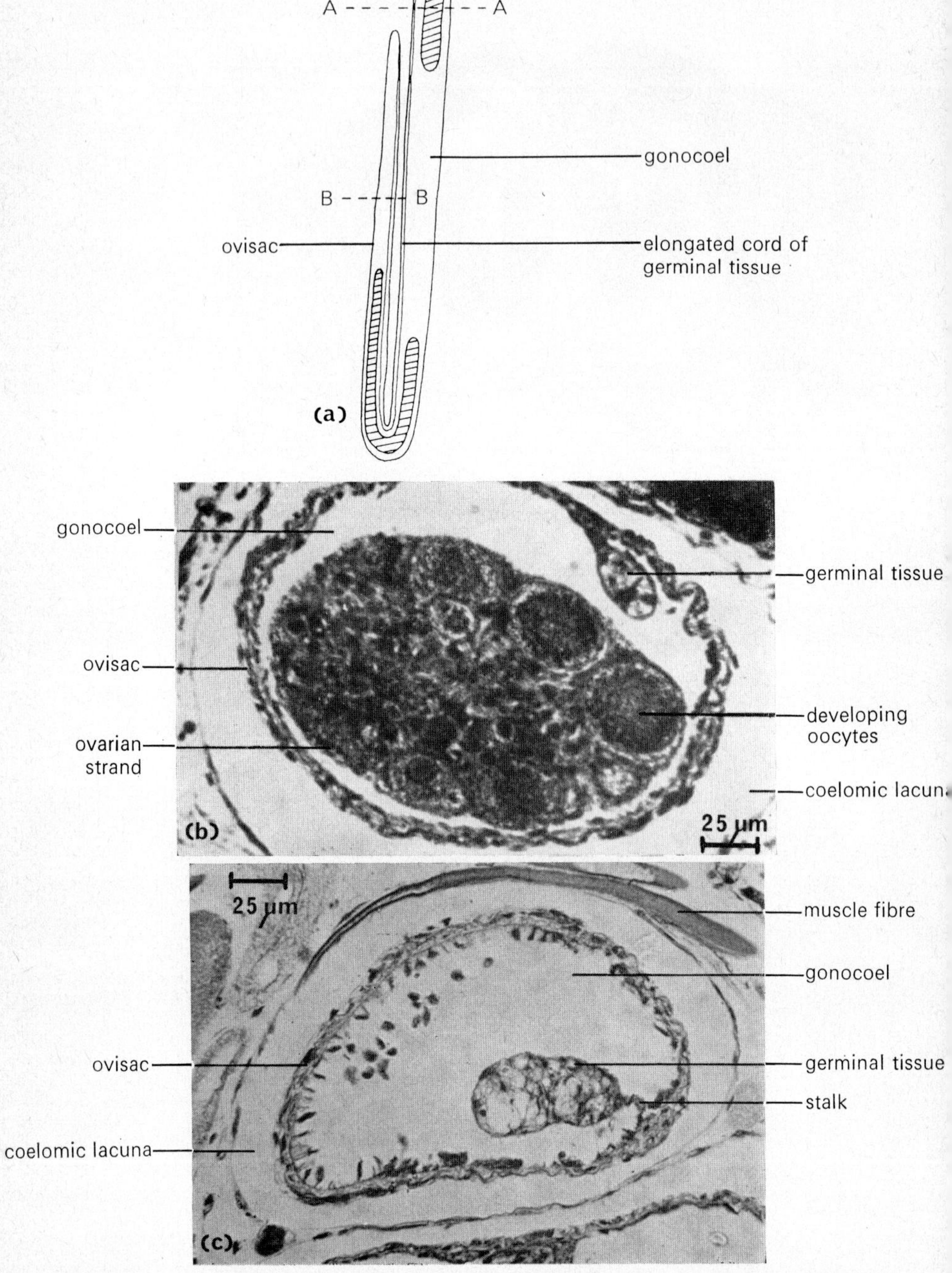

FIG. 4.35 (a) Plan of the female reproductive system of *E. testacea*; (b) TS of the ovisac and ovarian strand (A-A in Fig. 4.35a); (c) TS of the ovisac (B-B in Fig. 4.35a).

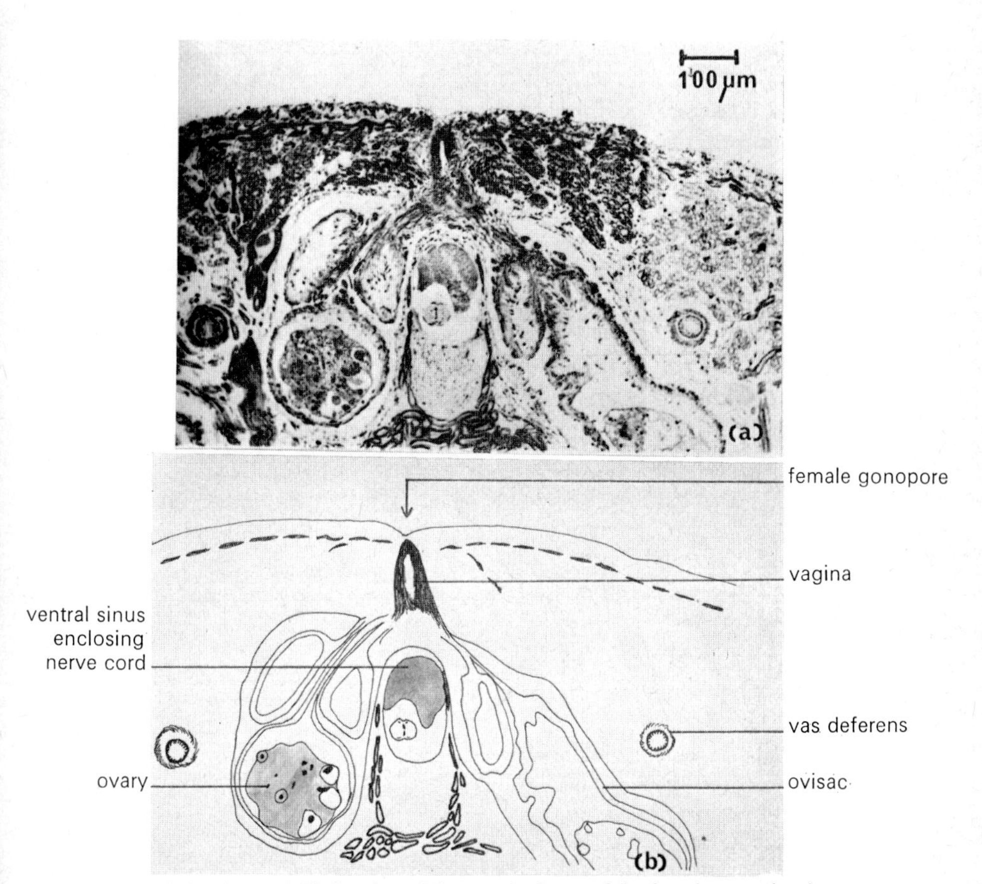

FIG. 4.36 (a) Section and (b) drawing of the terminal part of the female reproductive system of *E. testacea*. The female gonopore opens in segment 11.

The male reproductive organs consist of 4–10 pairs of branched testes which lie behind the female organs. Note the similarity of the testes (Fig. 4.33) to the seminal vesicles of earthworms (Fig. 4.16); both of them being coelomic cavities. Spermatogenesis occurs in the coelomic fluid in the testis lumen, and sperm travel thence by short vasa efferentia to the long vasa deferentia [seen in Fig. 4.32(b)]. These become coiled and enlarged at their ends to form seminal vesicles. From here thick-walled ejaculatory ducts run to a single median atrium which opens to the exterior through a male gonopore on segment 10 (Fig. 4.34). The genital atrium in most leeches is a muscular chamber. Gland cells in the atrial wall secrete the spermatophore walls that surround sperm masses pumped out by the ejaculatory ducts. In the gnathobdellids the muscular atrium is eversible and forms a penis.

The female organs comprise a single pair of ovaries, each composed of a mass or a number of masses of germinal tissue forming elongated cords with club-shaped endings lying freely in the coelomic sac (ovisac). Each ovisac is elongated and often folded forward on itself, as in *E. testacea* (Fig. 4.35). Short oviducts run from the anterior ends of each ovisac, pass beneath the ventral sinus, and join to form a muscular vagina which opens to the surface through the female gonopore in segment 11 (Fig. 4.36).

Chapter 5

Phylum Onychophora

The Onychophora is a small phylum, once classified with the Arthropoda but more recently separated from them. Onychophorans are terrestrial animals with a discontinuous distribution in temperate and tropical regions. They illustrate characters of both the Annelida and the Arthropoda, thus supporting the idea that annelids and arthropods have advanced along the same evolutionary line.

The onychophoran head consists of only three segments and bears simple appendages (one pair each of pre-antennae, jaws and oral papillae) and a pair of eyes at the base of the pre-antennae. The remaining body segments are alike and have an annulated appearance similar to that of leeches. Each segment carries a pair of non-jointed legs, but these limbs, unlike those of annelids, have their own extrinsic muscles and are moved ventrally, thus lifting the body off the ground during locomotion. Other annelidan features include a soft body wall covered on the outside by a thin flexible, non-jointed cuticle; a fairly primitive nervous system; simple sense organs; and a complete set of segmental coelomoducts.

On the other hand, onychophorans show several obviously arthropodan features which have arisen as adaptations to a terrestrial mode of life. The respiratory system is a tracheal one similar to that of insects. The vascular and digestive systems have affinities with those of arthropods. The coelom is reduced to a renocoel and gonocoel, and is replaced as the main body cavity by a perivisceral haemocoel. The sexes are always separate, and each individual generally possesses a single pair of gonads and paired ducts (formed from modified coelomoducts) which fuse terminally before reaching the ventral gonopores.

Example: *Peripatus* (*capensis*)

P. capensis is a native of South Africa, and is found in damp secluded places (e.g. under stones and logs). Its organization is shown in Fig. 5.1 which is a longitudinal section of the anterior end, and Fig. 5.2, a transverse section in the middle region of the body.

There is no exoskeleton such as is found in arthropods, and so the body wall, like that of annelids, is composed of complete layers of circular, diagonal and longitudinal muscles covered by an epidermis and thin cuticle [Fig. 5.3(a)]. The cuticle, however, is similar in composition to the arthropod cuticle and is shed periodically. The body surface is thrown

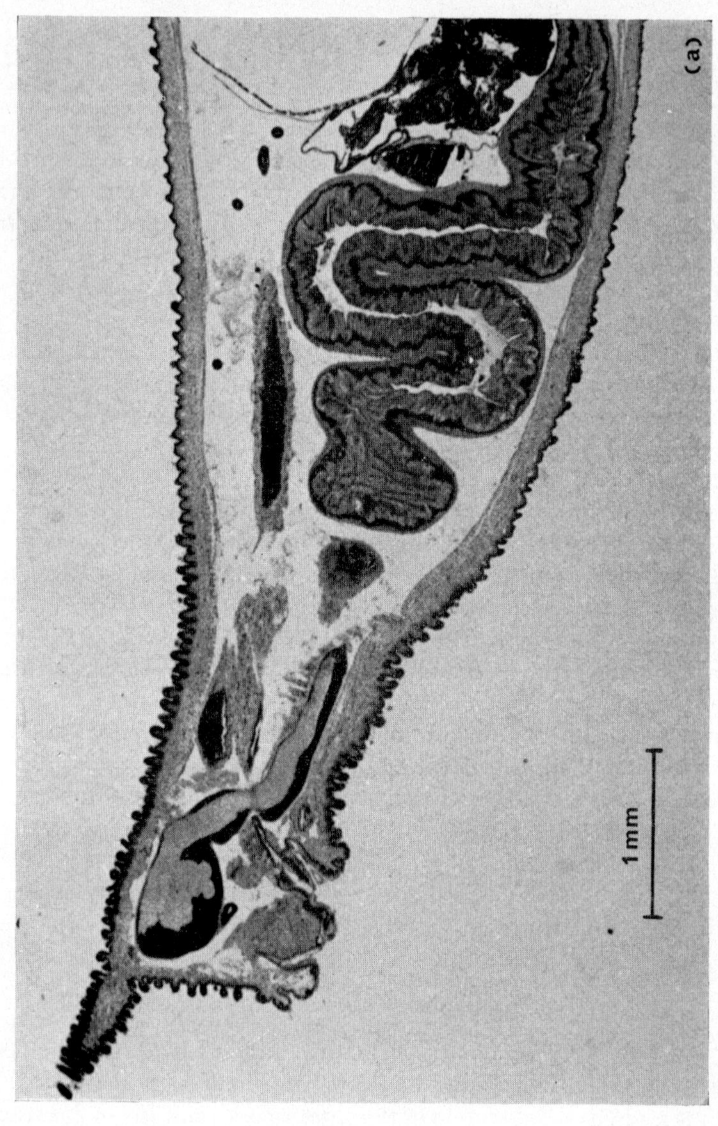

FIG. 5.1

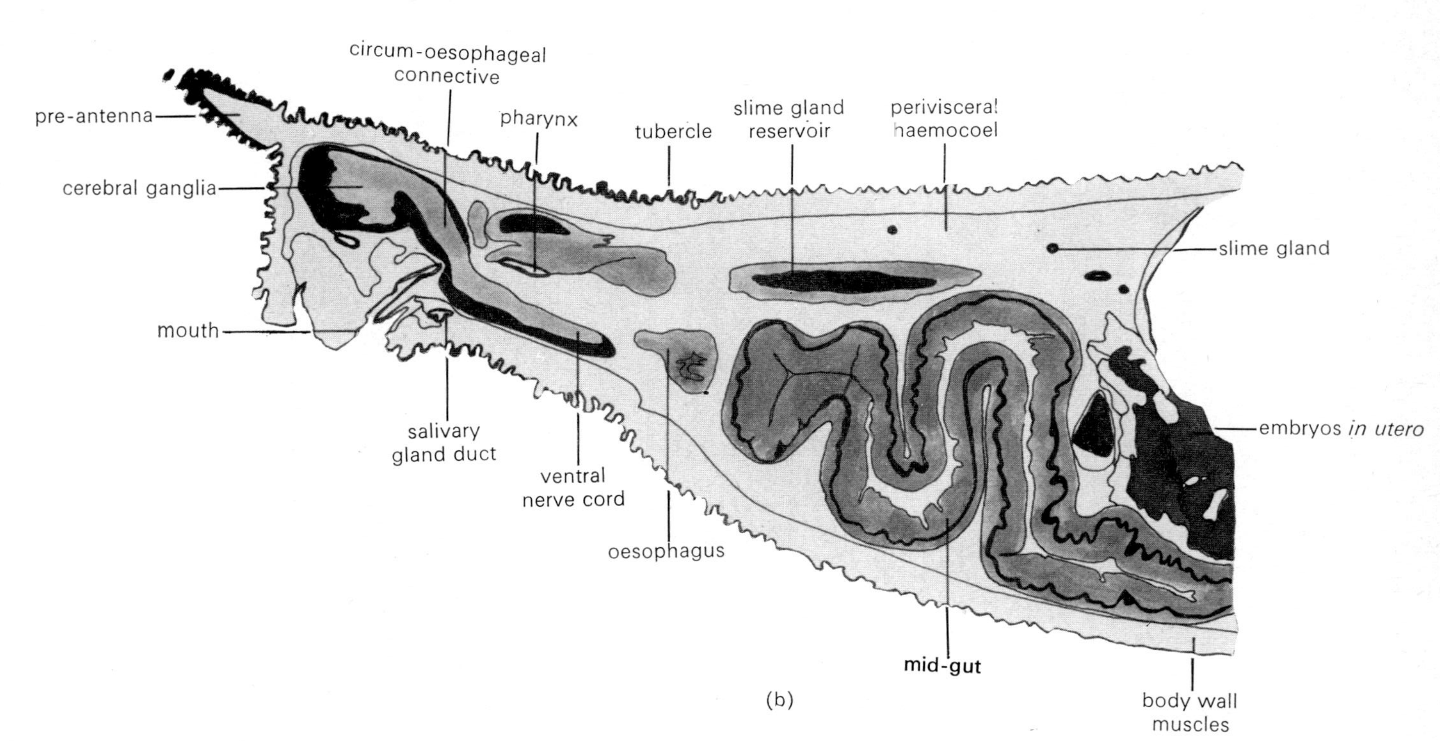

FIG. 5.1 (a) VLS and (b) explanatory drawing of the anterior end of an adult female onychophoran (*Peripatus capensis*).

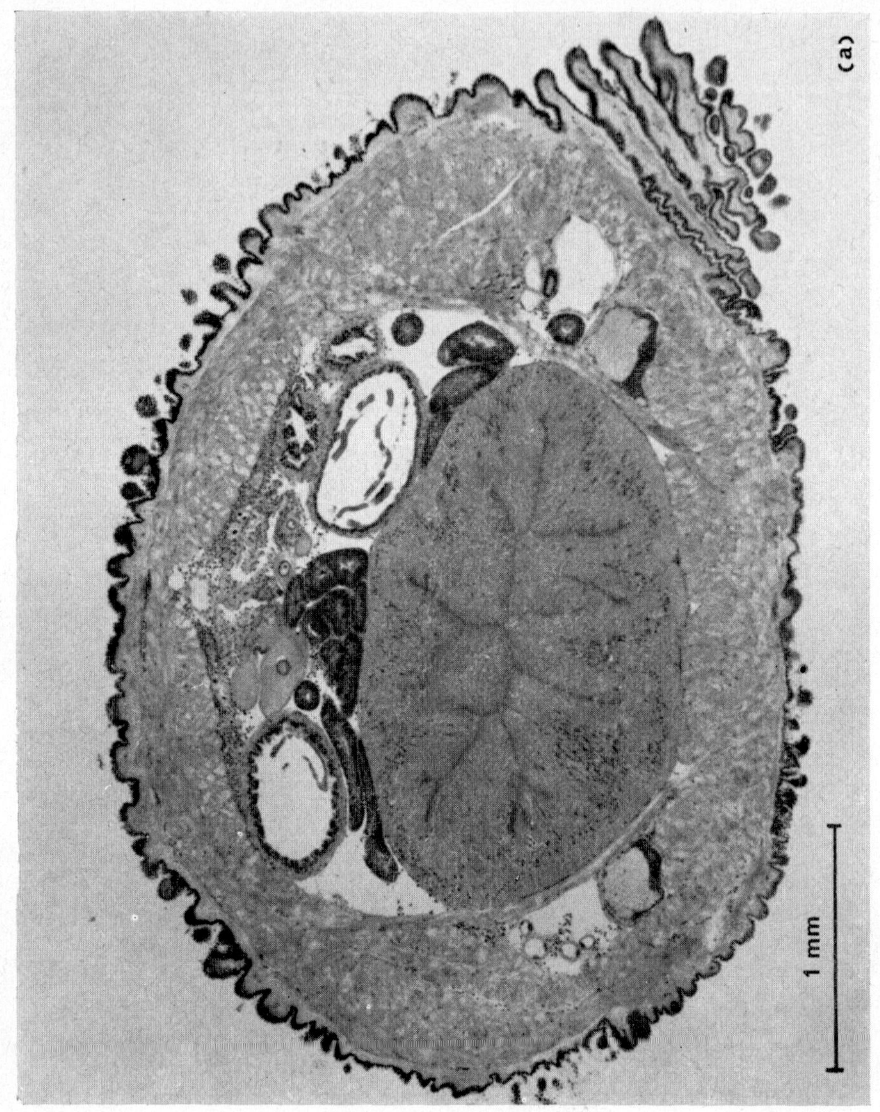

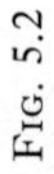
Fig. 5.2

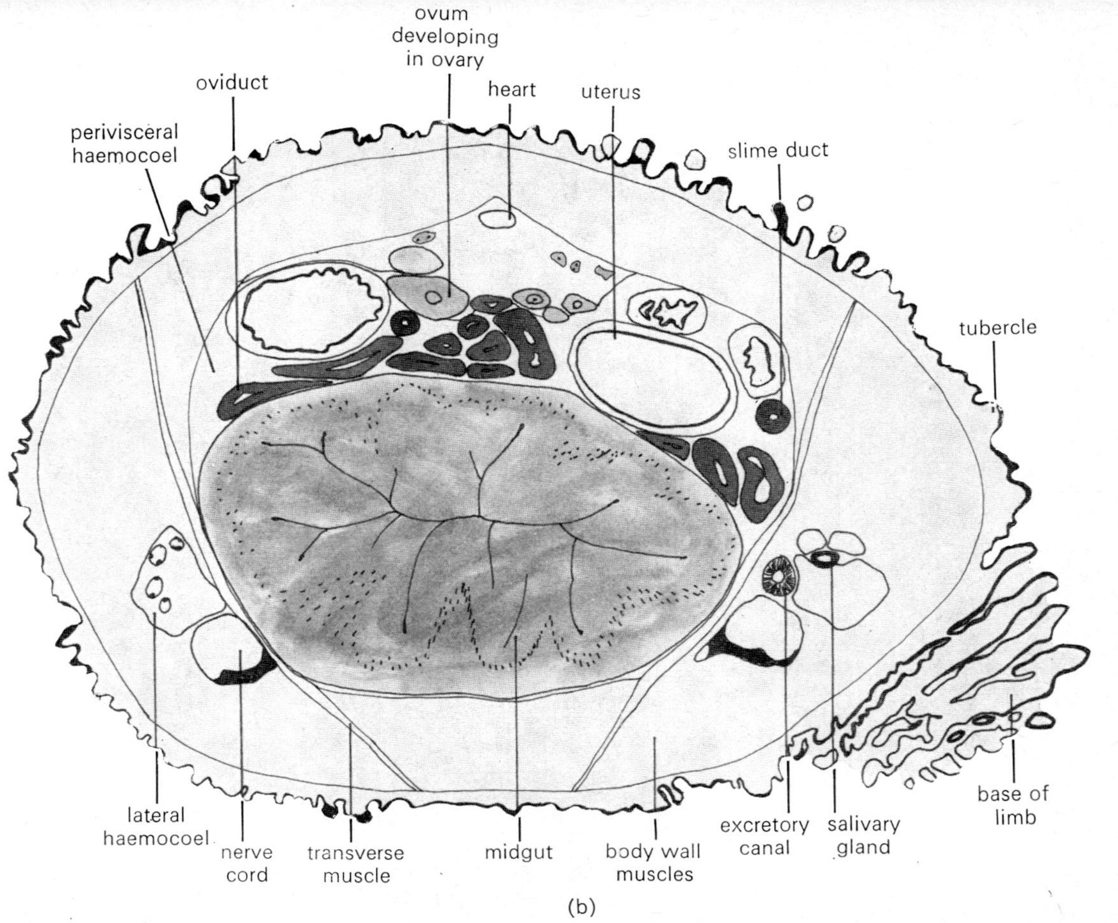

FIG. 5.2 (a) TS and (b) explanatory drawing of a female *P. capensis* in the intestinal region.

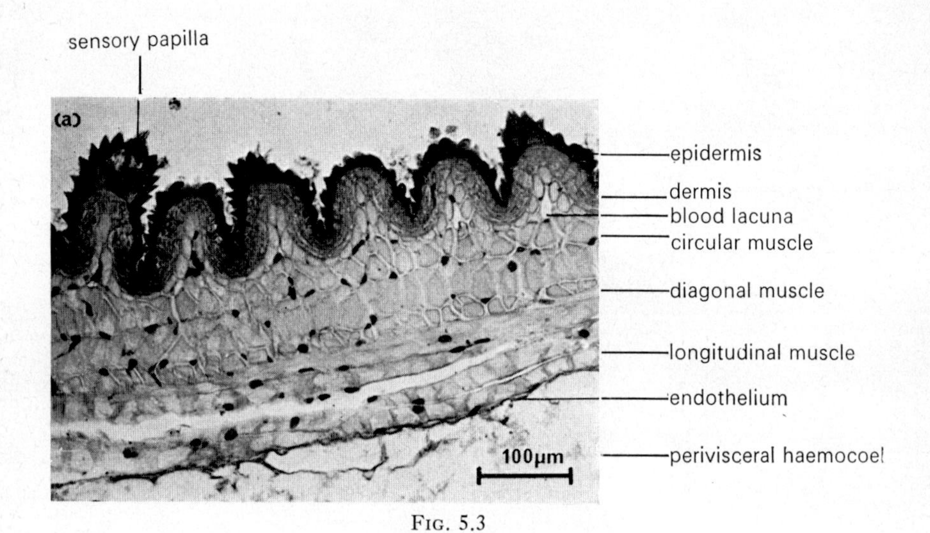
sensory papilla
(a)
epidermis
dermis
blood lacuna
circular muscle
diagonal muscle
longitudinal muscle
endothelium
perivisceral haemocoel
100μm

FIG. 5.3

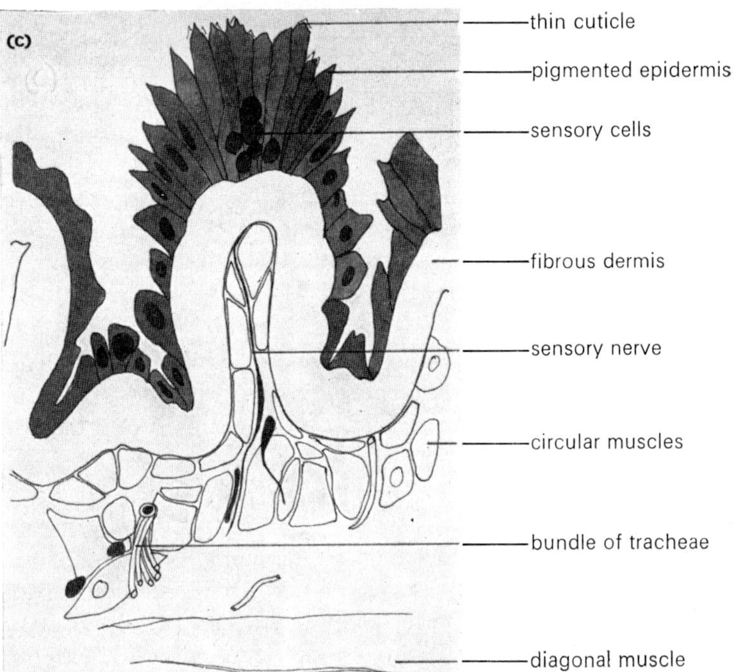

FIG. 5.3 (a) LS through the body wall of *P. capensis* showing a sensory papilla; (b) detail of a sensory papilla of *P. capensis*. Note the fine bundle of tracheae (which open into a pit at the base of the papilla); (c) explanatory drawing of (b).

into papillae and tubercles covered in scales, and with a sensory bristle at the end [Fig. 5.3(b, c)]. The conical limbs end in chitinous claws and are well supplied with sensory structures (Fig. 5.4).

Nervous system

The nervous system is relatively simple. A pair of supraoesophageal ganglia supply nerves to the eyes and head appendages, and are joined by stout connectives (see Fig. 5.1) to a pair of widely separated ventral nerve cords (see Figs 5.2 and 5.15). These are of uniform structure along their whole length, with ventro-lateral bands of neurones, but they are thickened in each segment to form the so-called segmental ganglia (Fig. 5.5).

Sense organs

Peripatus, living as it does in damp situations, is well-provided with hygroreceptors in the skin, and there are a number of other simple sense organs including chemosensory and tactile receptors (Fig. 5.6, see also Figs 5.3, 5.4 and 5.8). The eyes resemble those of annelids in being pigment lined cups (ocelli) each with a transparent chitinous lens (Fig. 5.7). They are thought to be capable of forming images although *Peripatus* is a nocturnal animal and tends to avoid light.

Respiration

The respiratory system comprises a series of tubules (tracheae), bundles of which open into randomly arranged pits on the body surface. They are not easy to see in light microscope sections, but a bundle of fine tracheae is visible in Fig. 5.3(b, c). Unlike the insects, there is no spiracular closing mechanism and so the onychophoran respiratory system is a source of water loss, another feature which confines them to moist habitats.

Circulation

The vascular system is an open haemocoel with a thin-walled dorsal tubular heart running nearly the entire length of the body, with openings (ostia) into it in most segments. The heart is enclosed in a membranous pericardium. Its position can be seen in Fig. 5.2.

Digestion

Peripatus is omnivorous, feeding on decaying vegetation and small animals, some of which may be rather hard. The endodermal mid-gut region, like that of insects, is protected by a cuticular peritrophic membrane which is shed at intervals. The fore-gut (stomodaeum) and hind-gut (proctodaeum) are of ectodermal origin and are therefore lined with cuticle.

The buccal cavity encloses a pair of cuticular mandibles (Fig. 5.8)

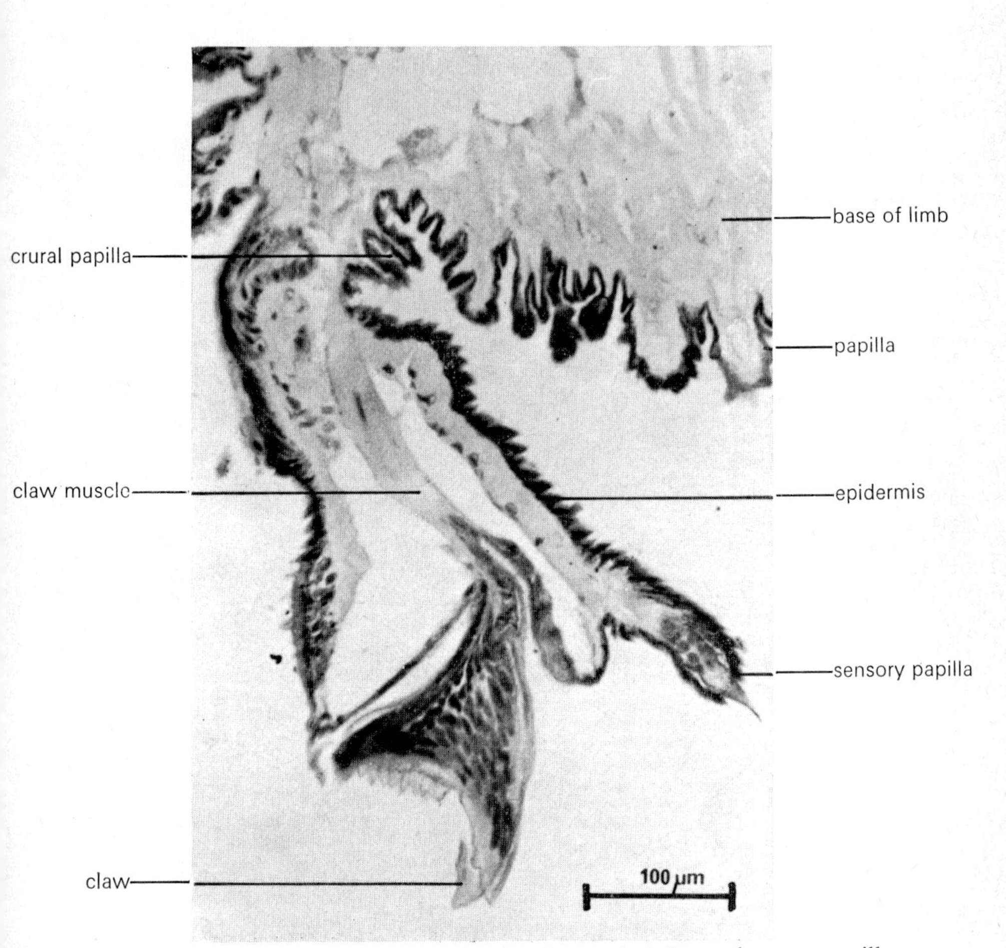

FIG. 5.4. LS of a limb of *P. capensis* showing the terminal claw and sensory papilla.

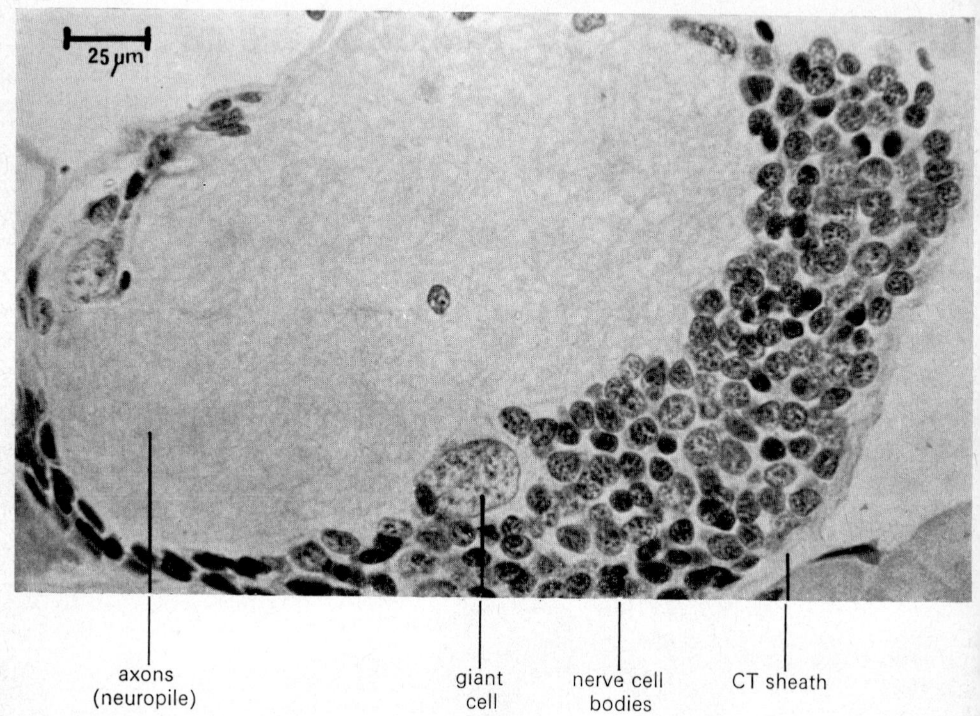

FIG. 5.5. TS through a segmental ganglion of the ventral nerve cord of *P. capensis*.

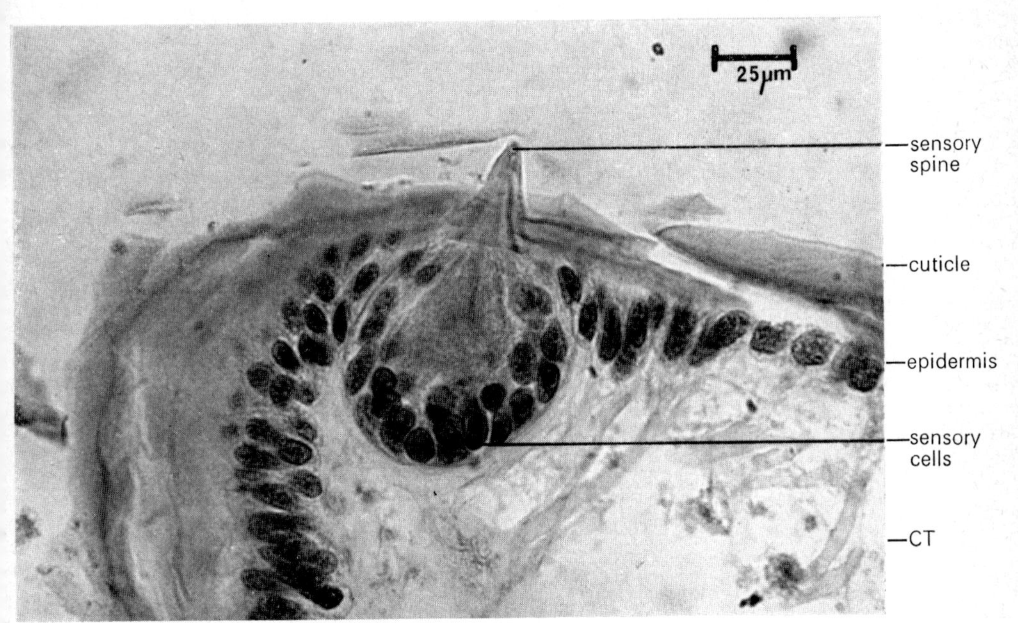

FIG. 5.6. Sensory bourgeon on the inner lip of *P. capensis*.

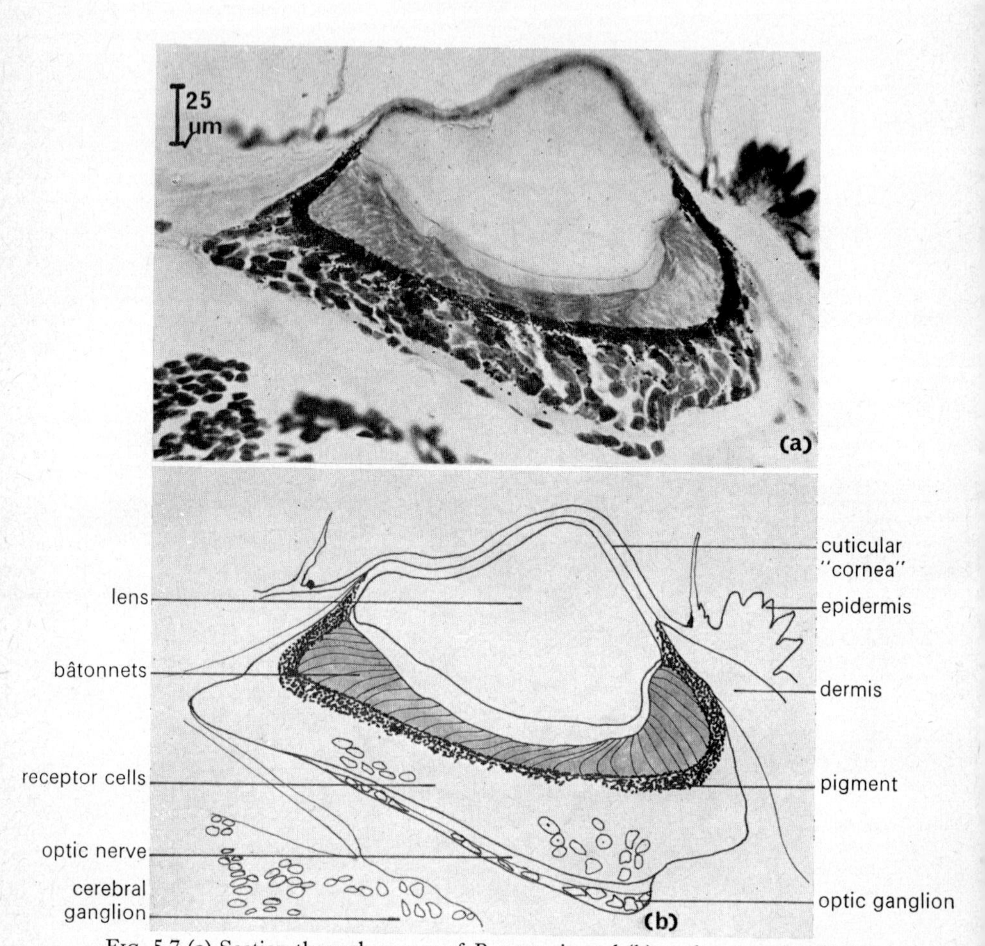

FIG. 5.7 (a) Section through an eye of *P. capensis*; and (b) explanatory drawing.

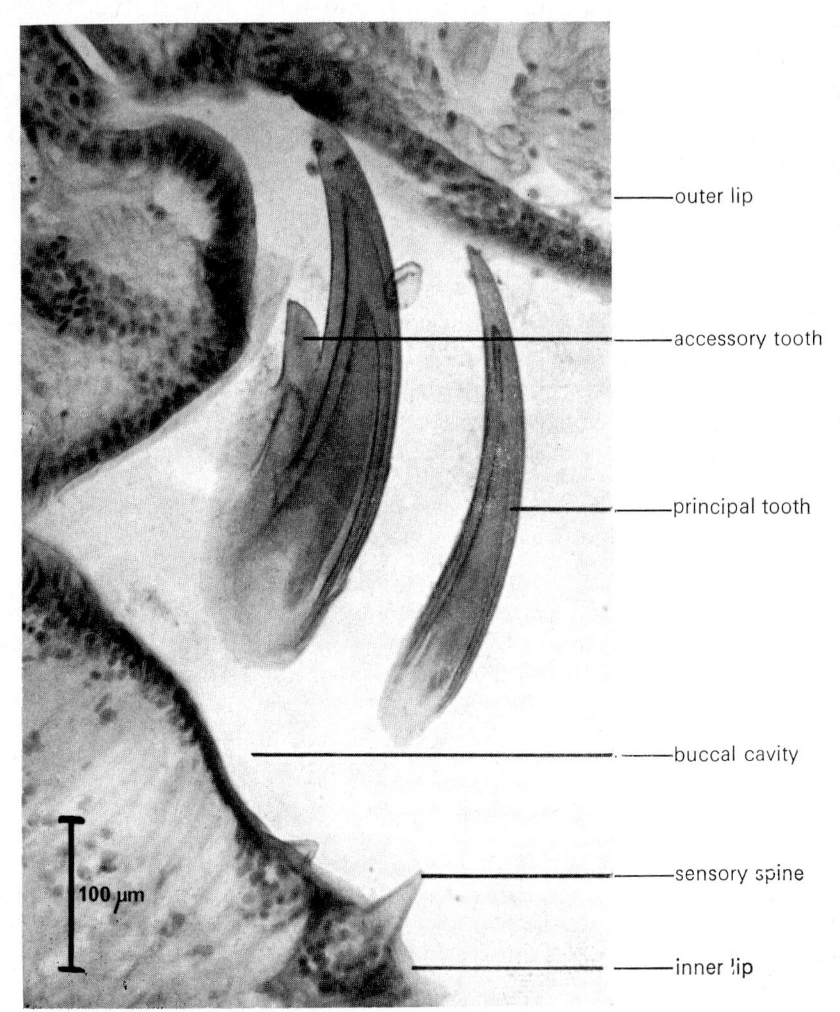

Fig. 5.8. LS through the buccal region of *P. capensis*, showing the teeth and two sensory bourgeons on the inner lip.

formed from modified limbs. Compare the teeth in this figure with the claws on the end of the limb in Fig. 5.4. Food is passed from the buccal cavity via a muscular suctorial pharynx [Fig. 5.9(a)] to the oesophagus, which has a very folded wall and is used for storage of food [Fig. 5.9(b)]. As in arthropods, the principal site of digestion and absorption is the mid-gut which is lined with a tall columnar epithelium that takes on a completely different appearance at various stages of digestion (Fig. 5.10). All the cells are apparently at the same stage at the same time. Uric acid is secreted by the mid-gut cells into the lumen as a means of excretion of nitrogenous waste, and is removed with the peritrophic membrane at regular intervals.

There are no digestive glands associated with the mid-gut. Paired tubular salivary glands, formed from the modified first pair of coelomoducts, are located in the lateral haemocoels (see Fig. 5.2) and may assist with preliminary extracellular digestion. They discharge into the buccal cavity. Opening at the tip of the oral papillae are paired adhesive (slime) glands, each with a branched secretory portion (Fig. 5.11) and a reservoir (see Fig. 5.1). The slime is probably, however, used for trapping enemies rather than food.

Excretion

The paired excretory organs (coxal glands), although segmental, are formed from coelomoducts rather than from nephridia, and bear strong resemblance to certain arthropod excretory organs (particularly the coxal glands of some arachnids). Each coxal gland (Fig. 5.12) consists of a coelomic end-sac which filters off materials from the tissues, a ciliated duct continuous with an excretory canal which leads to a storage vesicle or bladder, and a terminal duct opening at the base of a limb.

Reproduction

The sexes are separate in *Peripatus*, and, although there is no obvious morphological differences between them, the females are generally larger than the males. Both sexes possess saccular crural glands which open via crural papillae near the bases of the limbs (see Fig. 5.4) and which are thought to have a sexual function.

In the male a single pair of testes, situated close to the dorsal body wall, release thread-like sperm into long coiled vasa deferentia where they are compacted into spermatophores (Fig. 5.13). The spermatophores leave the body through a muscular ejaculatory duct (Fig. 5.14) and are deposited on the ventral surface of the female, from where they are said to make their way through the body wall and haemocoel to the dorsal ovary.

The female tract of *P. capensis* consists of a single ovary formed by fusion of a pair, two long coiled oviducts (see Fig. 5.2) parts of which are used as seminal receptacles, and two extensive thin-walled uteri which fuse

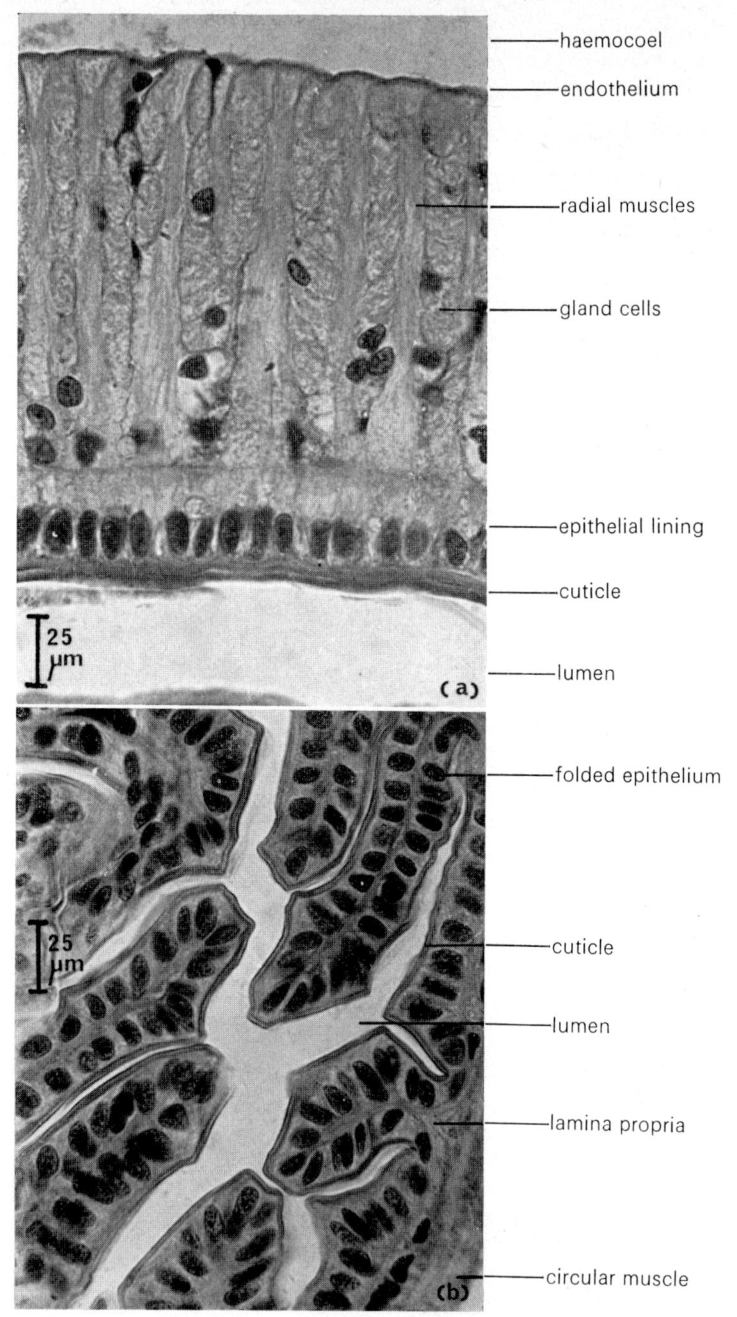

FIG. 5.9 (a) LS through the pharynx wall of *P. capensis*; (b) TS through part of the oesophagus of *P. capensis*. Note the folded wall and cuticular lining.

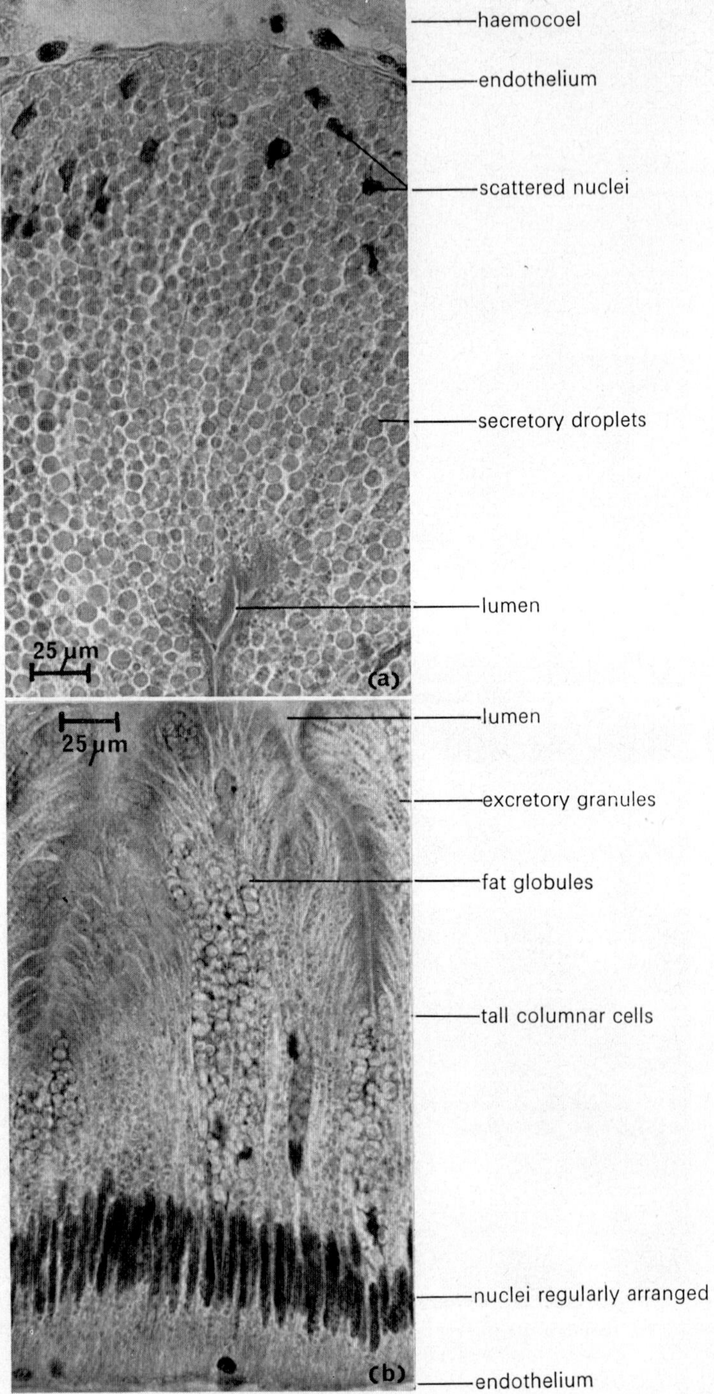

FIG. 5.10. TS through the mid-gut wall of *P. capensis* showing the epithelium in (a) a secretory phase; and (b) an absorptive phase. The two sections were taken from different individuals.

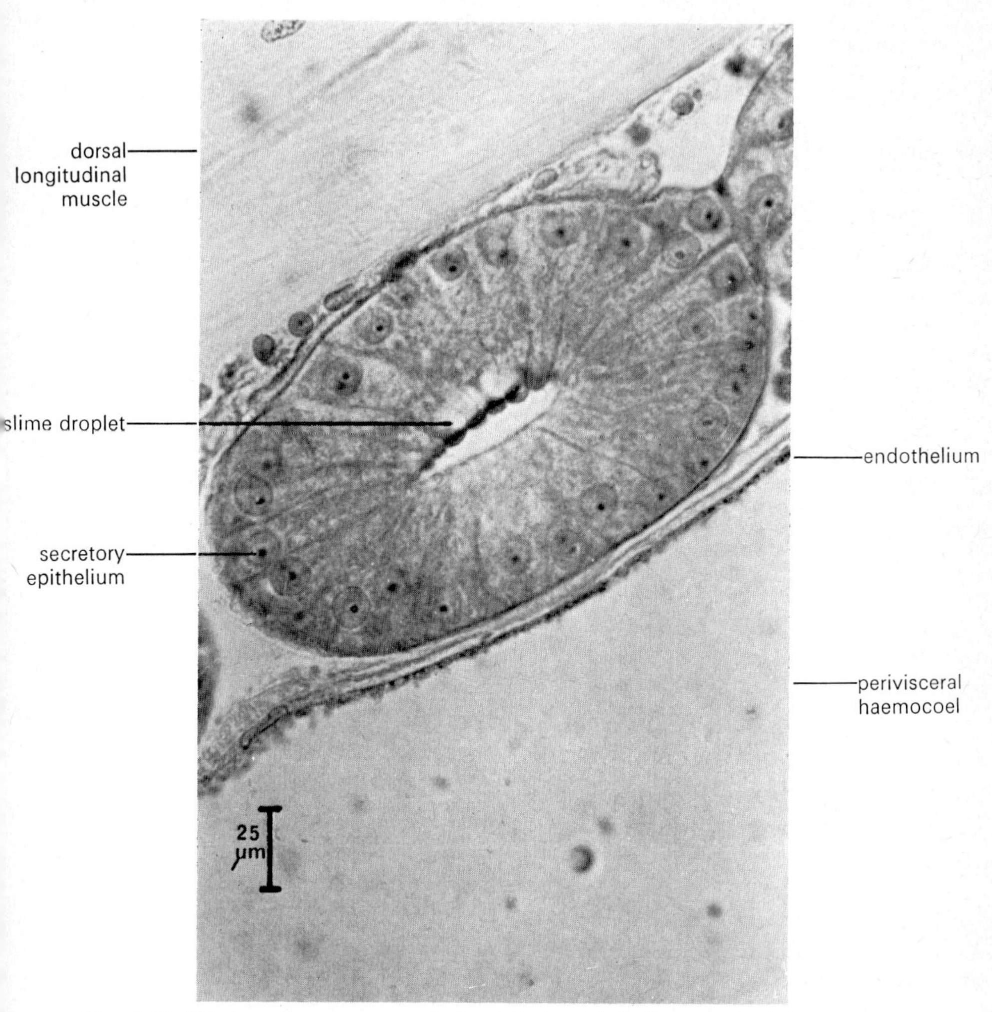

FIG. 5.11. TS through a secretory tubule of the branched slime-gland of *P. capensis*.

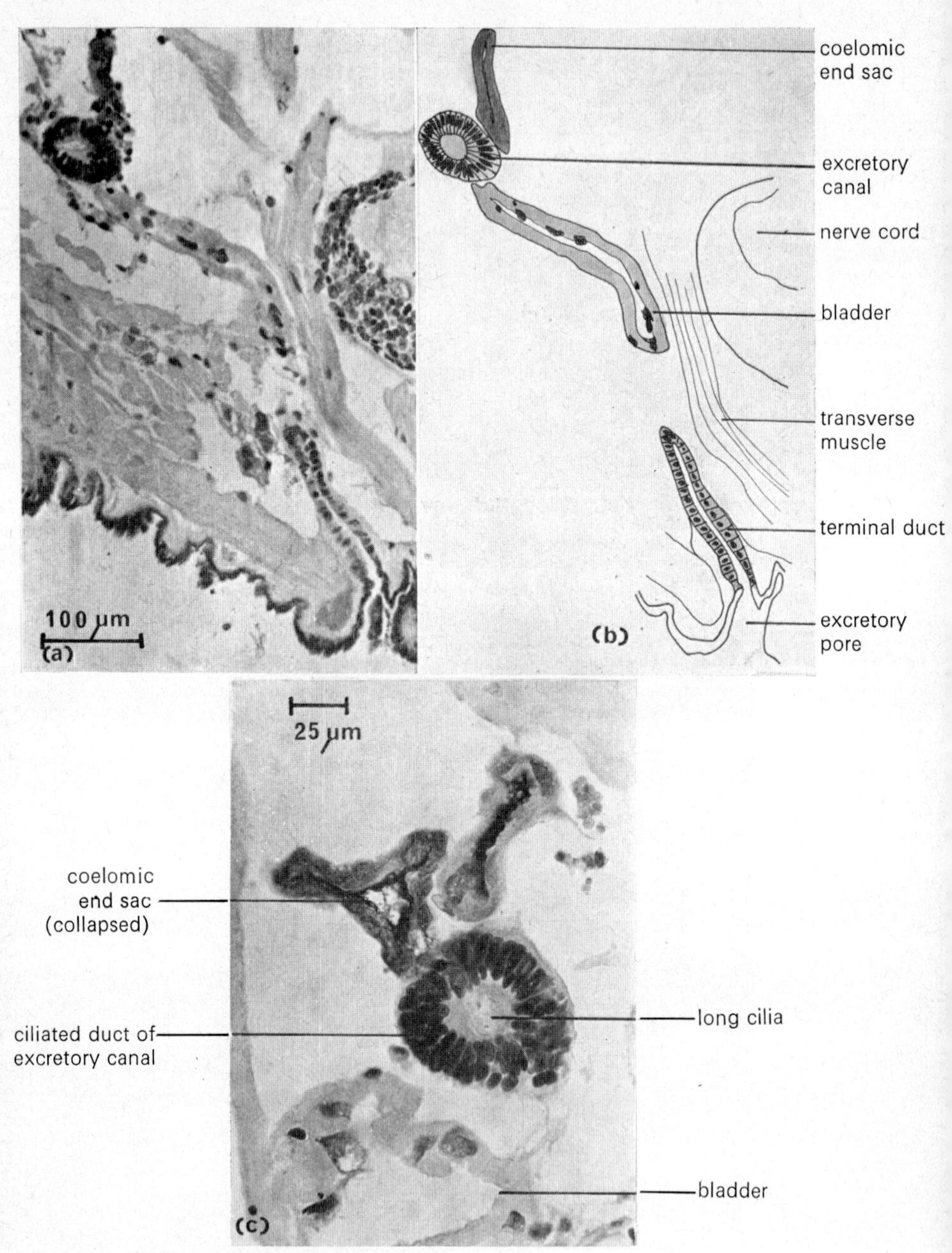

FIG. 5.12 (a) Part of a TS of the ventral region of *P. capensis*; and (b) explanatory drawing, showing a coxal gland opening at the base of a limb; (c) details of part of the coxal gland.

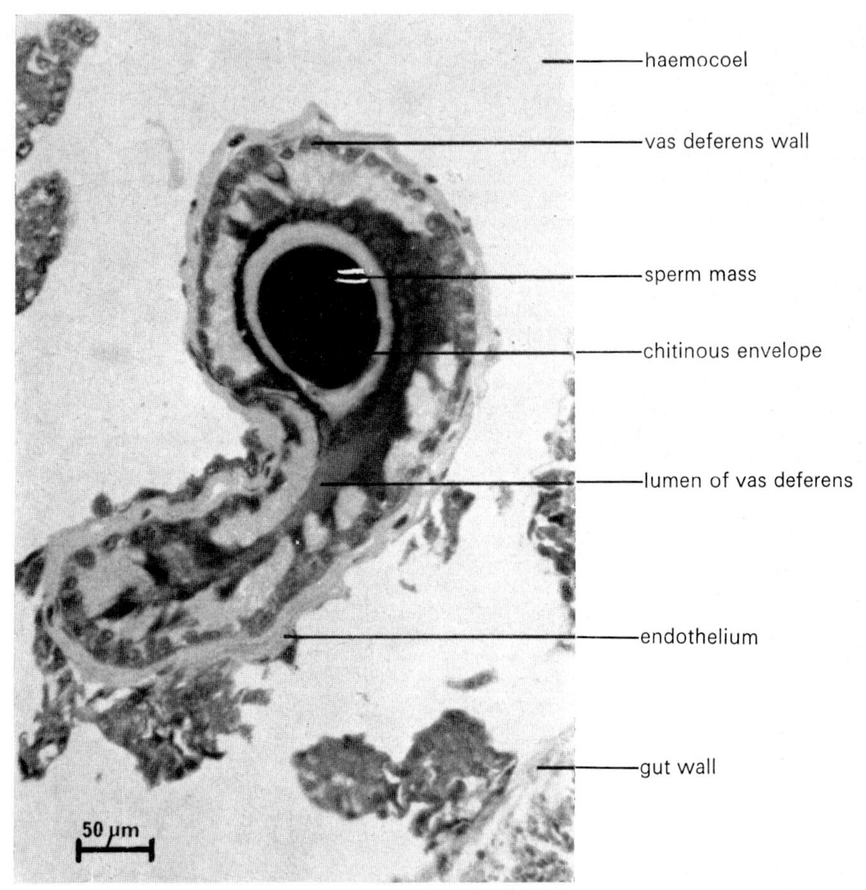

FIG. 5.13. Section through a spermatophore in the vas deferens of a male *P. capensis*.

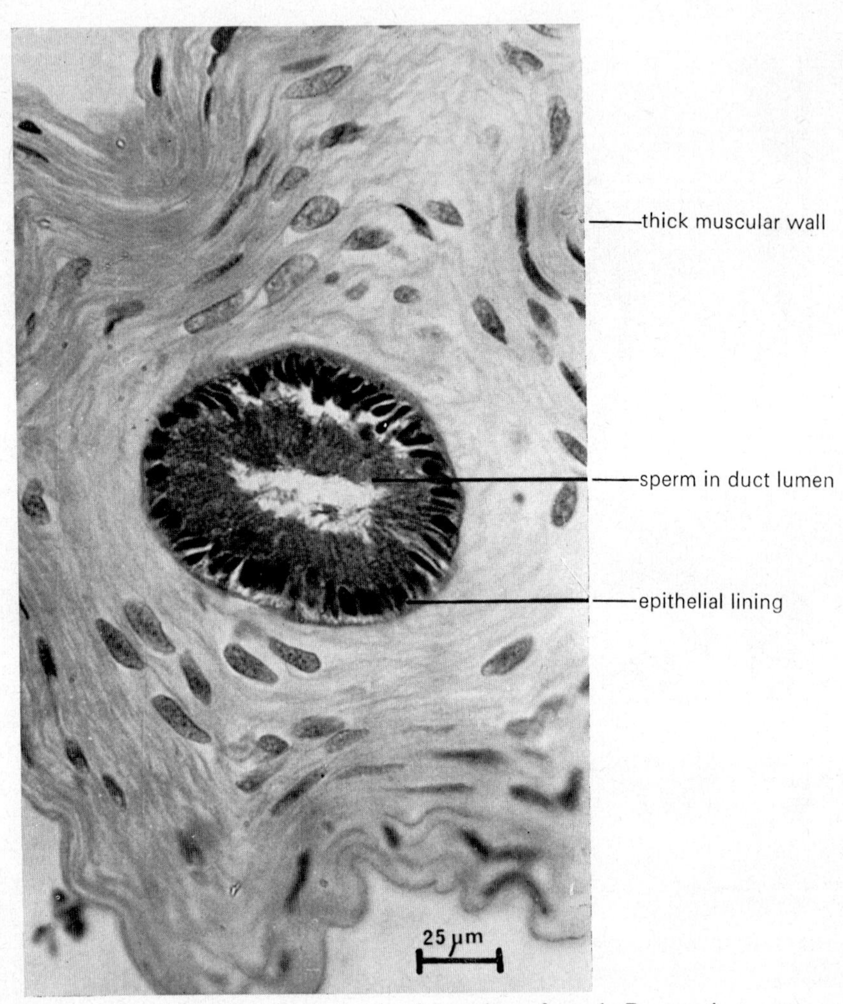

FIG. 5.14. TS through the ejaculatory duct of a male *P. capensis*.

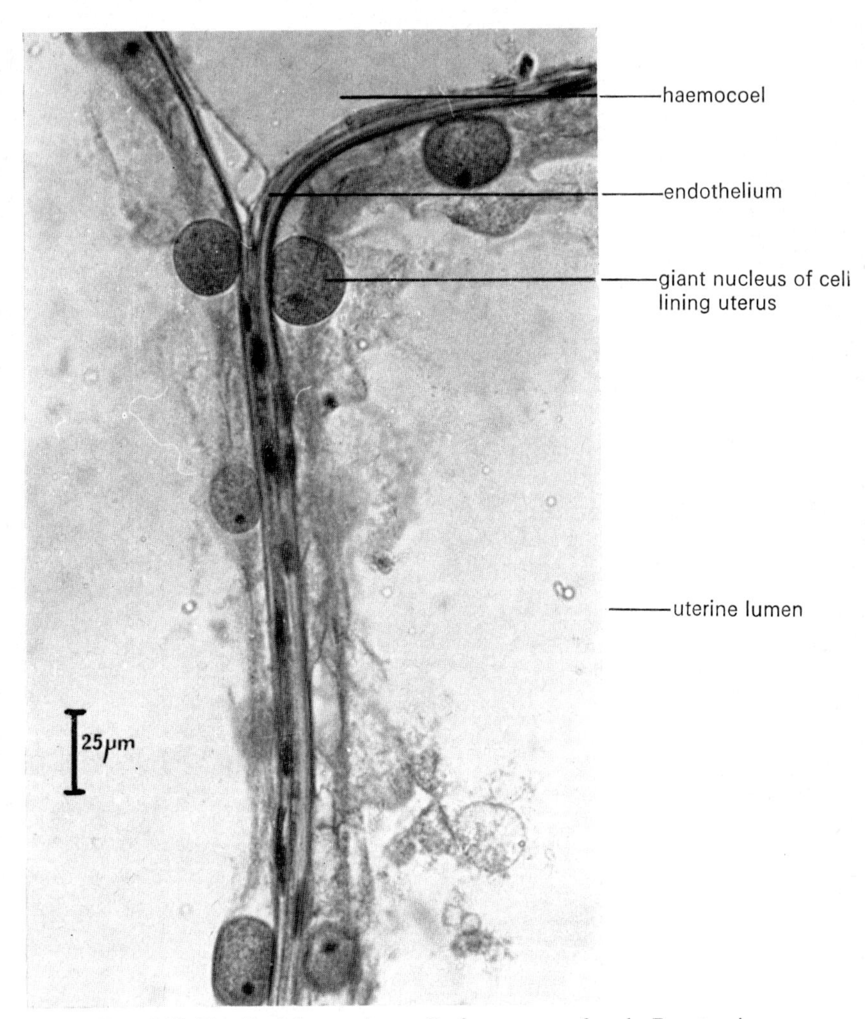

FIG. 5.15. Detail of the uterine wall of a pregnant female *P. capensis*.

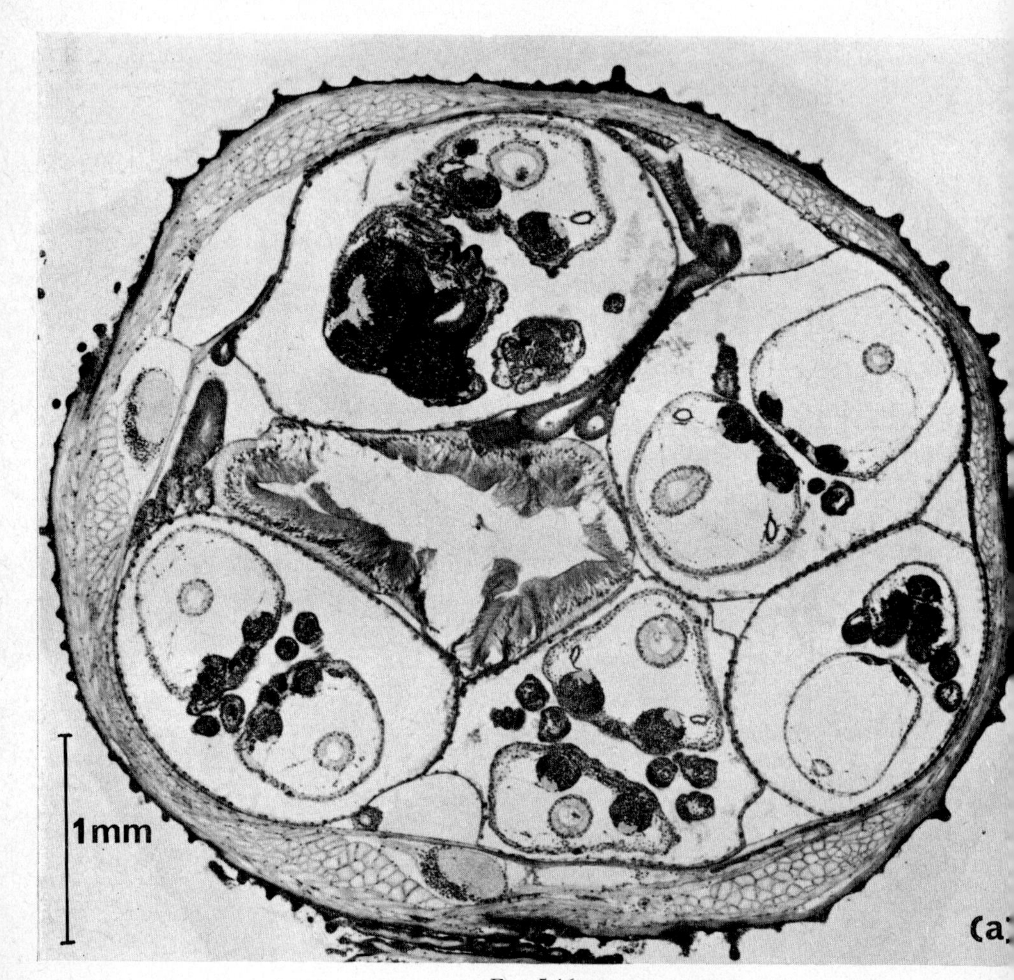

Fig. 5.16

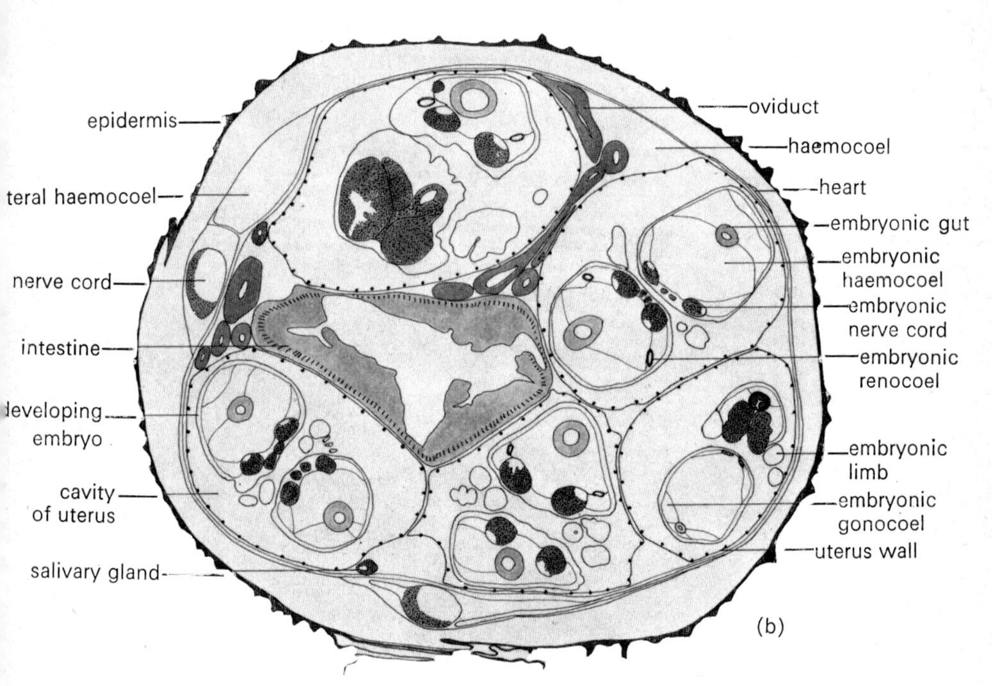

FIG. 5.16 (a) TS and (b) explanatory drawing of a pregnant female *P. capensis*, showing many embryos developing in the distended uterus.

just before the genital opening. Fertilization is internal. Many onychophorans are oviparous, but *P. capensis* is viviparous, and the young develop for long periods in the uterus which becomes very distended and occupies most of the body cavity in a gravid female (Figs 5.15 and 5.16).

Chapter 6

Phylum Arthropoda

Arthropods constitute the largest invertebrate phylum. Members of this phylum have colonized a great variety of habitats, some of them even thriving in extreme conditions such as are found in deserts, and some taking to the air.

Arthropods are bilaterally symmetrical animals which have certain features in common with annelids and onychophorans. They are basically metamerically segmented, but this metamerism has secondarily disappeared in some forms. They possess paired appendages on some or all of the segments (somites): certain of the anterior ones have been modified to form sensory structures and feeding organs. The nervous system is based on the same plan as that of annelids although arthropods show an increasing degree of cephalization correlated with the development of anterior sense organs. The "brain" is divisible into three regions (proto-, deutero- and tritocerebrum). Nerve cell bodies, however, occur only in ganglia and not scattered along the connectives as in onychophorans.

On the other hand, arthropods differ from annelids in several major respects. Perhaps the most important difference is that the hydrostatic skeleton has been replaced by a hard cuticular exoskeleton. The cuticle is divided into hard plates with thin, flexible joints between them so that free movement is possible. The exoskeleton is shed periodically (at ecdysis) so that growth of the soft tissues can occur. A new cuticle, formed beneath the previous one, then hardens by a process known as sclerotization. The cuticle covers the entire body surface including any intucked ectodermal surfaces: thus the fore- and hind-guts are lined with cuticle as well as the majority of the respiratory system and parts of the reproductive tract. The evolution of a hard exoskeleton has been accompanied by a variety of other modifications. The continuous muscle layers characteristic of the annelid body are replaced by a system of separate cross-striated muscles attached to the inner surface of the exoskeleton. The coelom is greatly reduced, being retained only as a gonocoel, and in a few members as a renocoel; and there is concomitant reduction of the coelomoducts. The coelom is replaced as the main body cavity by an enlarged haemocoel, formed by breakdown of blood vessels. The circulation is therefore open and the dorsal vessel has evolved as a muscular heart pumping blood through arteries into a series of open sinuses among the tissues. Due to the impermeability of the cuticle, respiration does not take place across the general body surface in most species, and special respiratory structures have evolved. There are no chaetae or external cilia, locomotion being undertaken by limbs adapted for that purpose.

Arthropods are classified into two main groups, primarily on their limb function. The largest group, the Mandibulata, which includes insects, crustaceans and myriapods, have their first (and second if present) pairs of post-embryonic limbs modified as sensory antennae, their third pair as jaws bearing mandibles and their fourth (and sometimes fifth and sixth) pairs as additional jaws (maxillae). Mandibulates are often characterized by the possession of compound eyes.

The second group, the Chelicerata, including the scorpions and spiders, has a completely different arrangement whereby there are no antennae and any of the first six pairs of post-embryonic limbs can show a variety of functions from acquisition of food to the transfer of sperm to another individual. The first pair are usually chelate (chelicerae) and used as feeding structures, the second pair (pedipalps) are mostly palp-like and tactile but they may also be chelate or leg-like, and the third-sixth pairs generally serve as legs. In addition to this the Chelicerata do not generally have true compound eyes.

Class Crustacea

The Crustacea is a large class, most of whose members are aquatic although there are a few terrestrial forms. The first six segments fuse to form a head: however, the most anterior one only appears briefly during development. The rest of the body is normally divided into thorax and abdomen, though this arrangement is not always found in some of the more simple members. The thorax is usually covered dorsally and laterally by a posterior fold from the head exoskeleton which fuses with skeletal plates behind it to form a carapace. The head bears two pairs of antennae, one pair of mandibles and two pairs of maxillae. The appendages of the thorax and abdomen are typically biramous and show adaptive radiation for a variety of functions (e.g. walking, swimming, respiration, feeding, reproduction).

Example: *Carcinus* (*maenas*)

Carcinus maenas is the common shore crab. It belongs to the subclass Malacostraca, which comprises the most highly evolved crustaceans, showing the greatest degree of tagmosis (see p. 5). The head and thorax are fused and the body is dorso-ventrally flattened; features which can be seen in Fig. 6.1 which shows an HLS of a young crab. Also visible are the compound eyes on stalks. There is a large carapace covering the thorax, which is composed of eight segments all bearing appendages: three pairs of maxillipeds and five pairs of legs (order Decapoda), the most anterior of which are modified as claws (chelae). The abdomen is short and tucked in underneath the thorax. It is composed of six segments all carrying "swimming" appendages (pleopods) while the terminal segment bears the tail-fan (telson).

The exoskeleton of crabs is normally calcified. It is formed of a lipoprotein epicuticle overlying a chitinous endocuticle where calcification occurs (Fig. 6.2). This layer lies on an uncalcified chitinous membrane which adjoins the epidermis. The epidermis possesses chromatophores, and is the site of secretion of the new cuticle during the moulting cycle. Beneath the epidermis is a thin dermis containing more chromatophores and secretory tegmental glands which open by pores onto the surface of the body. The powerful cross-striated muscles are inserted onto the basement membrane below the epidermis.

Nervous system

The basic arrangement of the arthropod nervous system is the same as that of annelids (see Fig. 4.6), but it is modified in crabs due to fusion of the ganglia (Fig. 6.3). The dorsal "brain" (Fig. 6.4) is composed of three pairs of pre-oral (proto-, deutero- and tritocerebral) ganglia. The ventral nerve cord and segmental (thoracic and abdominal) ganglia are fused together with the suboesophageal ganglia to produce a single ventral ganglionic mass. Giant nerve fibres, arising in the brain, are present in Crustacea, but their greatest development occurs in the lobsters and shrimps where they innervate the abdominal flexor muscles and mediate fast escape reactions. Neurosecretory cells are common in Crustacea, and produce hormones which are released from neurohaemal (storage-and-release) organs directly into the blood.

Sense organs

The advent of an exoskeleton is accompanied by a reduction in the number of sensory nerve endings on the body surface, and the more specialized development of internal sense organs. The arthropods show a major advance in the development of the compound eye, a device for increasing visual discrimination.

Simple ocelli occur throughout the Crustacea, but in crabs there are paired lateral compound eyes on mobile stalks (see Fig. 6.1). Each eye is composed of a large number of visual units (ommatidia), each screened from adjacent ones by pigment cells. Thus each ommatidium can operate as an isolated unit, and stimulation of many such units will lead to a mosaic pattern of vision. Each ommatidium (Fig. 6.5.) consists of two focusing lenses (corneal lens and crystalline cone) and a group of receptor cells (retinular cells) connected by optic nerve fibres to the optic ganglia, which are extensions of the brain located in the eyestalks (see Fig. 6.1). The outer corneal lens is formed of thickened cuticle secreted by the underlying layer of epidermal corneagenous cells, while the crystalline cone [Fig. 6.6(a)] is secreted by the vitrellar cells. In the crab there are seven retinular cells each of which secretes a fibrillar rhabdomere containing visual pigment along its inner edge: these seven rhabdomeres fuse to form a central

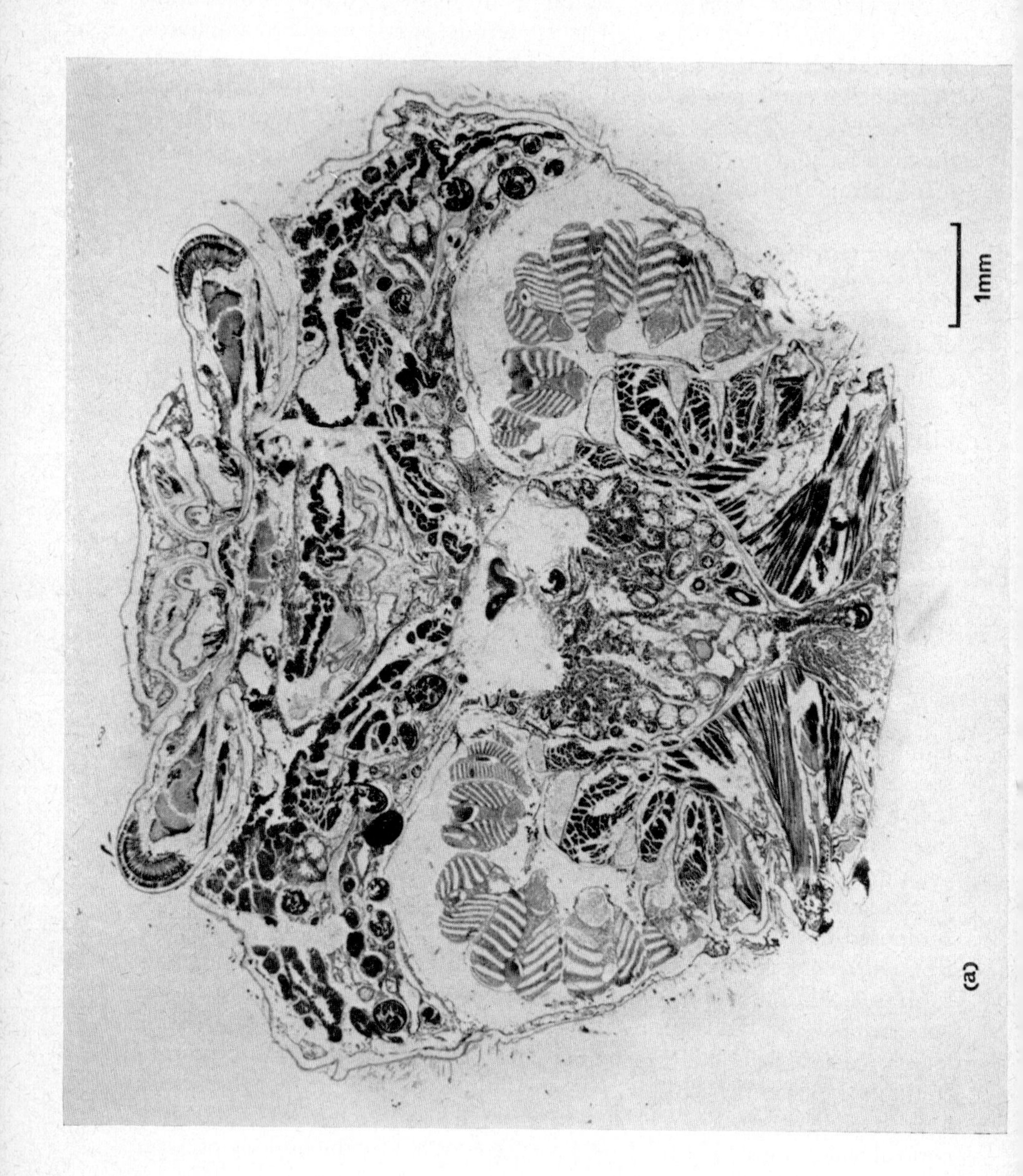
1mm
(a)

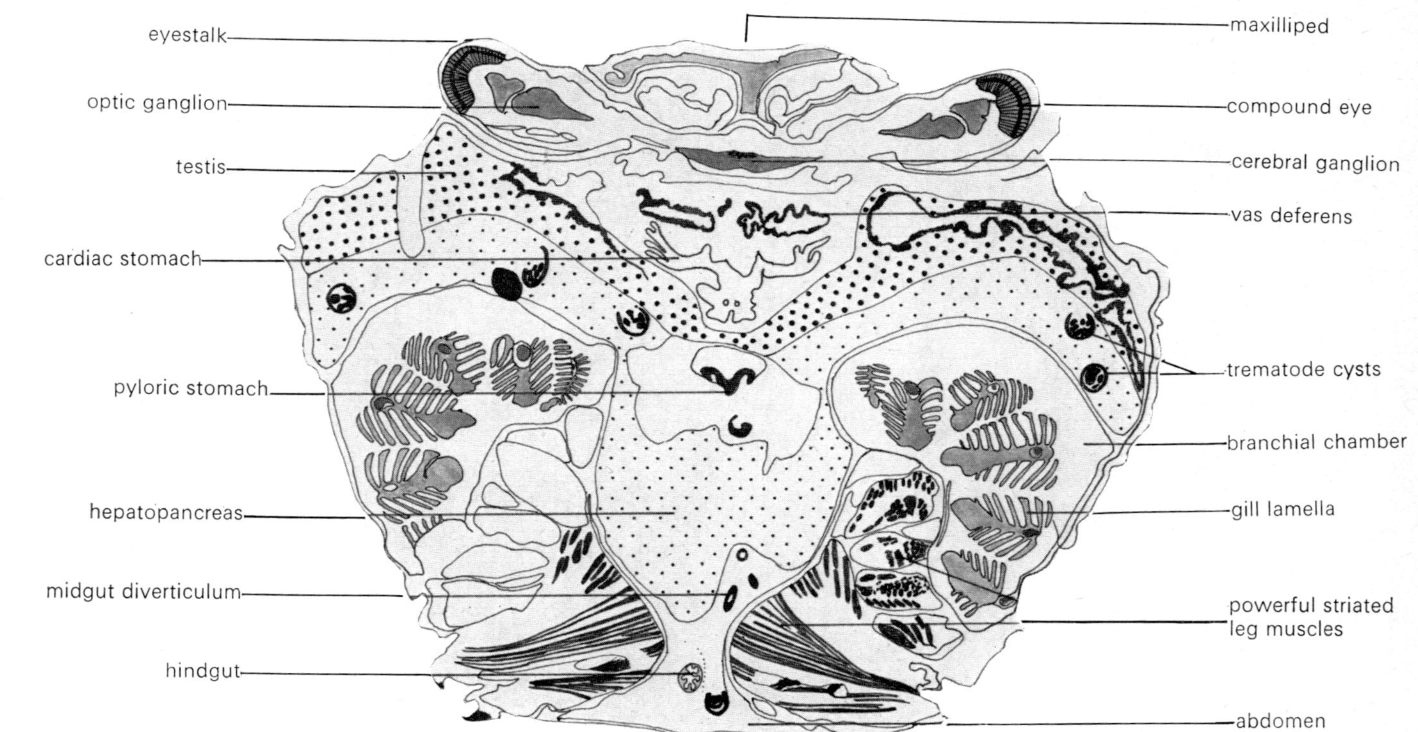

FIG. 6.1 (a) HLS and (b) explanatory drawing of a young crab (*Carcinus maenas*) in the thorax region. This crab is lightly infected with encysted larvae of a trematode.

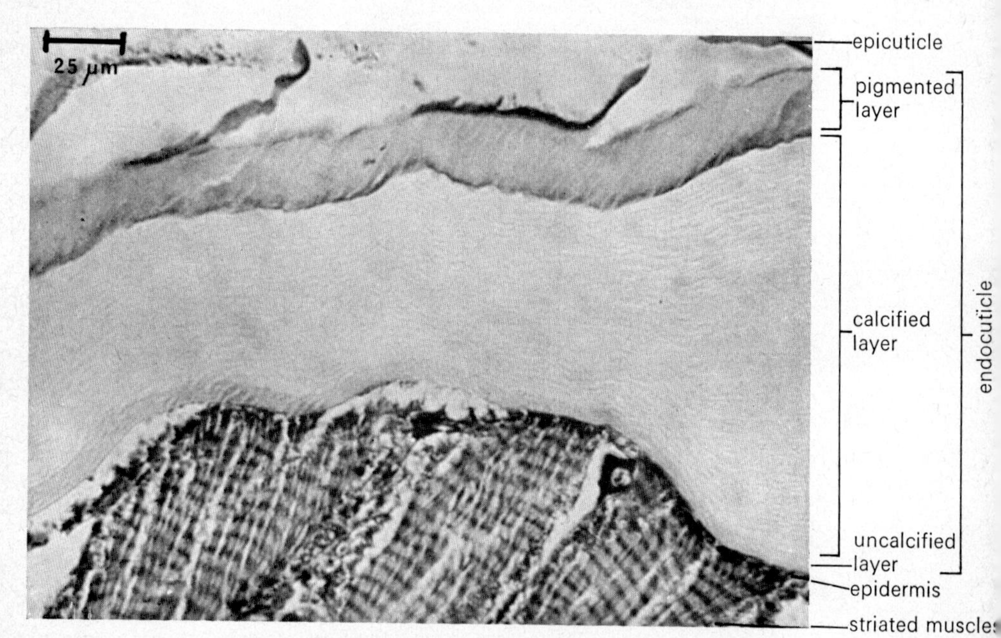

Fig. 6.2. Section through the body wall of *C. maenas*.

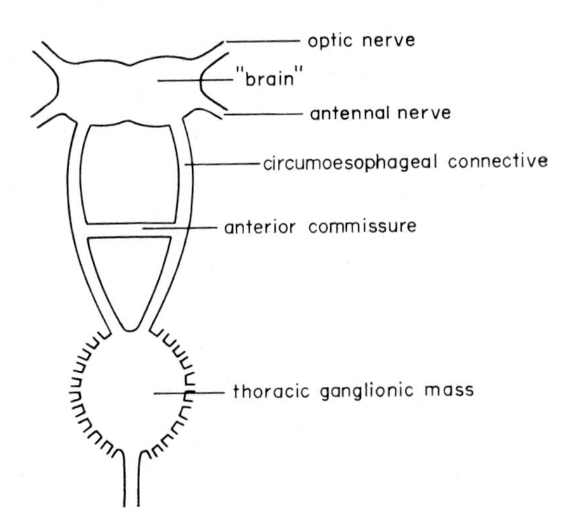

FIG. 6.3. Diagram showing the arrangement of the cerebral ganglia ("brain") and thoracic ganglionic mass of the crab.

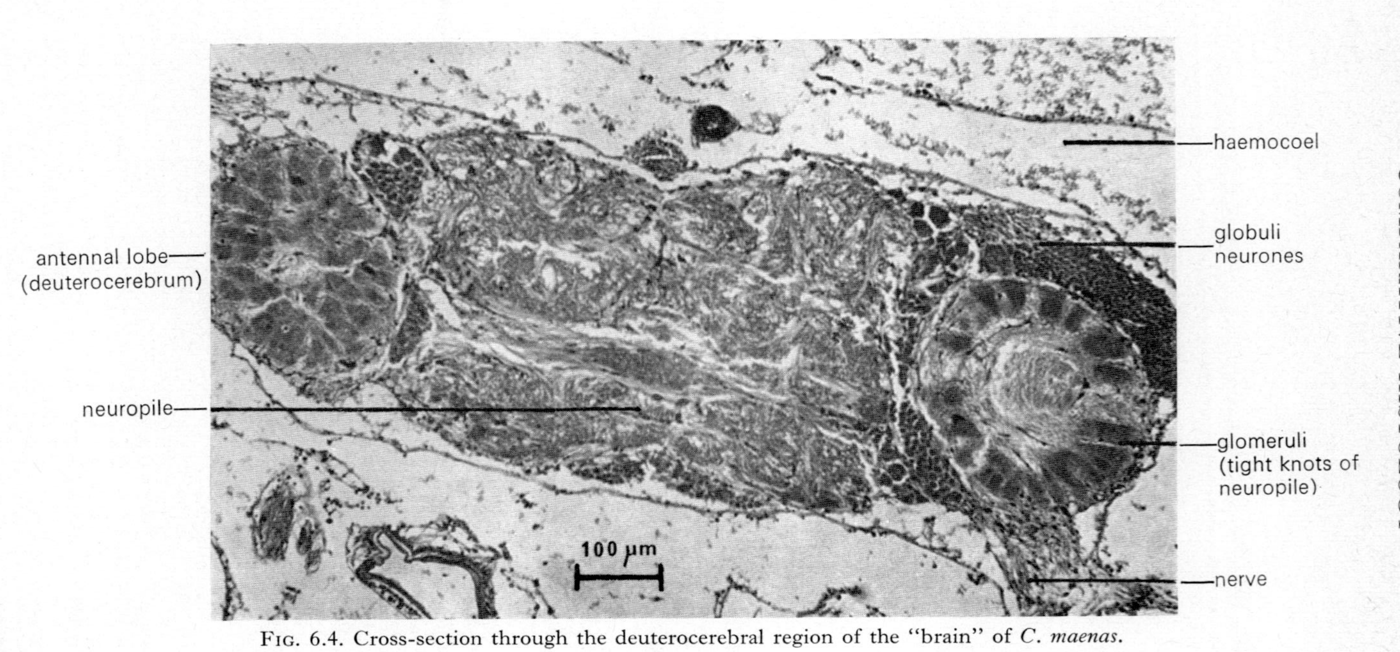

FIG. 6.4. Cross-section through the deuterocerebral region of the "brain" of *C. maenas*.

rhabdom [Fig. 6.6(b)]. The pigment round the ommatidia can be moved distally or centrally depending on whether the eye is dark or light adapted. In the crab, which is diurnal, the arrangement is such that light from a point source is confined to a single ommatidium by movement of the pigment screen distally to surround the visual cells (apposition eye). Compare Fig. 6.5 with Fig. 6.26, which shows an eye from the nocturnal cockroach.

Crabs possess many proprioreceptors including muscle stretch receptors and myochordotonal organs at the joints of the appendages, and a variety of sensory hairs on the body surface (Fig. 6.7). Statocysts occur at the base of the 2nd antennae. These are gravity or vibration receptors consisting of a chamber enclosing sensory hair cells which are deformed by movements of a heavy statolith.

Respiration

The special respiratory organs in crabs are finely divided gills, formed as outgrowths from the base of the thoracic limbs. They are situated inside a branchial chamber created by a downgrowth of the carapace at each side of the body (see Fig. 6.1). Water is circulated through this branchial chamber by paddle-shaped projections (scaphognathites) from the 2nd maxillae. In *Carcinus* there are nine gills on each side. They have a phyllobranchiate (lamellar) structure, are well vascularized and supplied with afferent and efferent blood vessels (Fig. 6.8). Each lamella consists of a thin epidermal covering surrounding a blood-filled cavity (lamellar sinus) which connects the afferent and efferent vessels in the central axis of the gill with an outer lamellar sinus running round the outside edge of the lamella. Scattered cells with brown contents occur in the branchial septum and are thought to be excretory in function.

Circulation

Blood, carrying a blue respiratory pigment (haemocyanin) drains from spaces among the tissues into the afferent branchial sinuses of the gills. It is oxygenated in the lamellar sinuses and passes through efferent branchial veins back to the pericardium, which in arthropods is part of the haemocoel and not coelomic. Inside the pericardium lies the contractile heart, a sac-like structure with thin walls composed of cross-striated muscle fibres. Blood flows from the pericardium into the heart, via four dorsal and two lateral valved ostia, whence it is pumped through five anterior and two posterior arteries to the tissues. Heartbeat is controlled by neurosecretions released from the pericardial organs positioned over the ostia, and by cardiac ganglia.

Digestion

Crustaceans have a wide range of diets. Most small species are filter feeders, but crabs generally are carnivorous scavengers. The mouth-parts

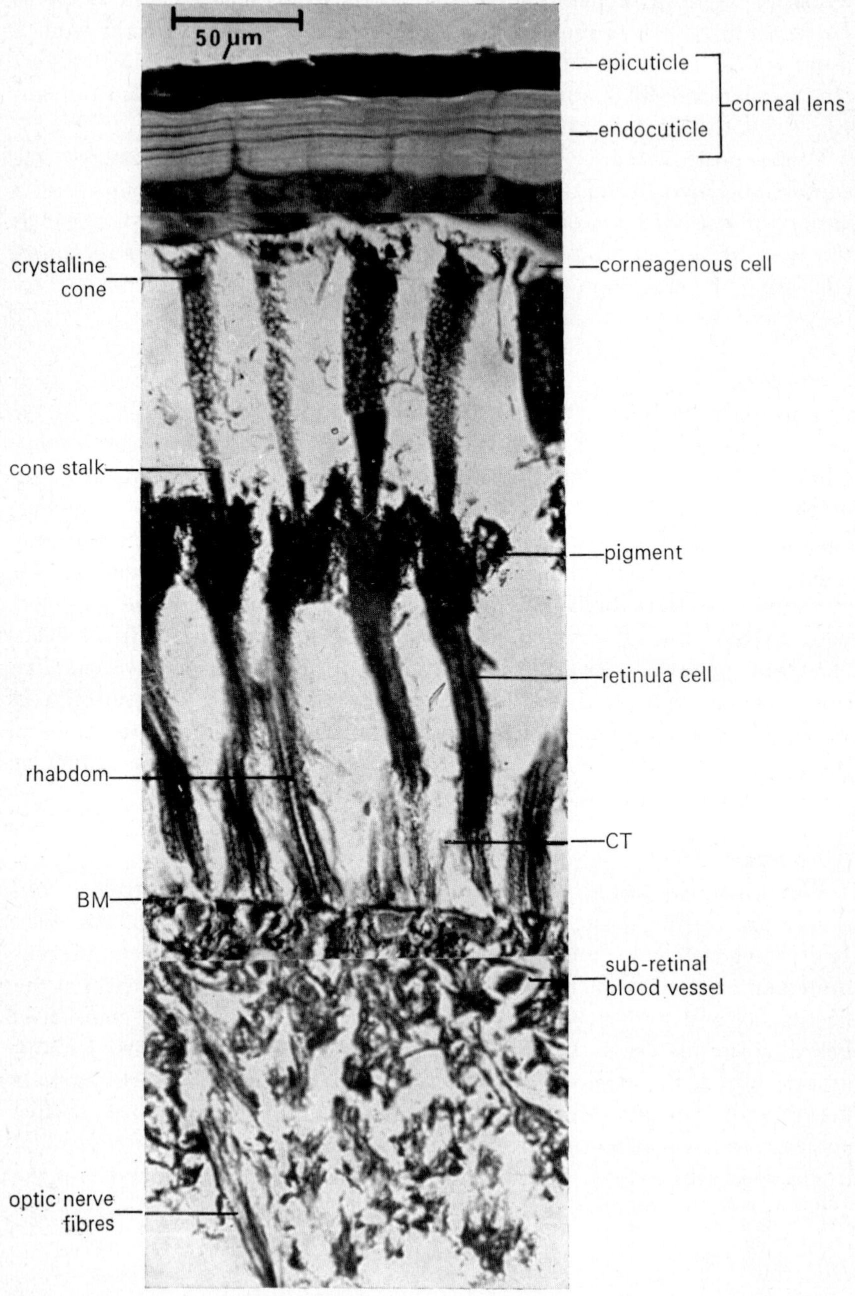

FIG. 6.5. Section through the apposition eye of *Carcinus maenas*. Note that all the pigment is clustered around the retinula cells.

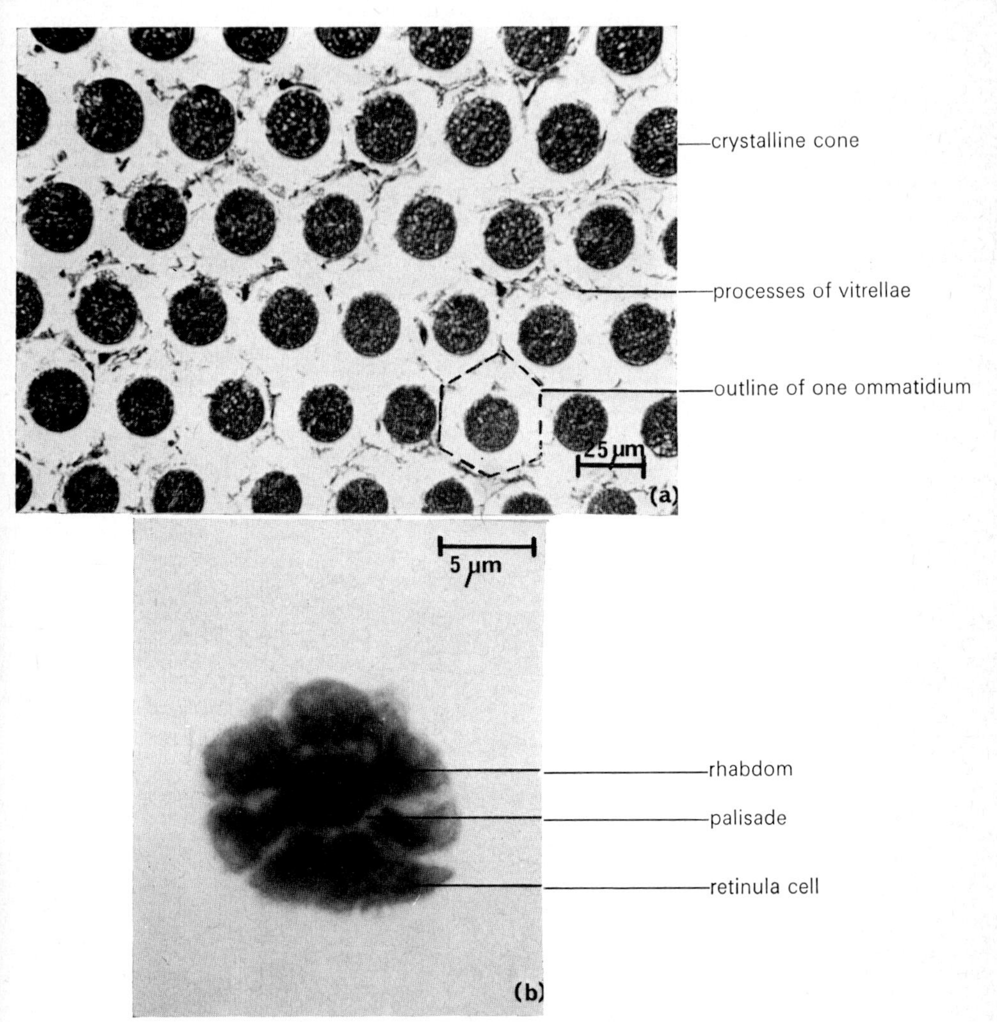

FIG. 6.6 (a) TS of the crab eye at the level of the crystalline cones, showing the regular arrangement of the ommatidia; (b) TS of the photoreceptive region of one ommatidium from the eye of *C. maenas*. Note the central rhabdom and seven retinula cells.

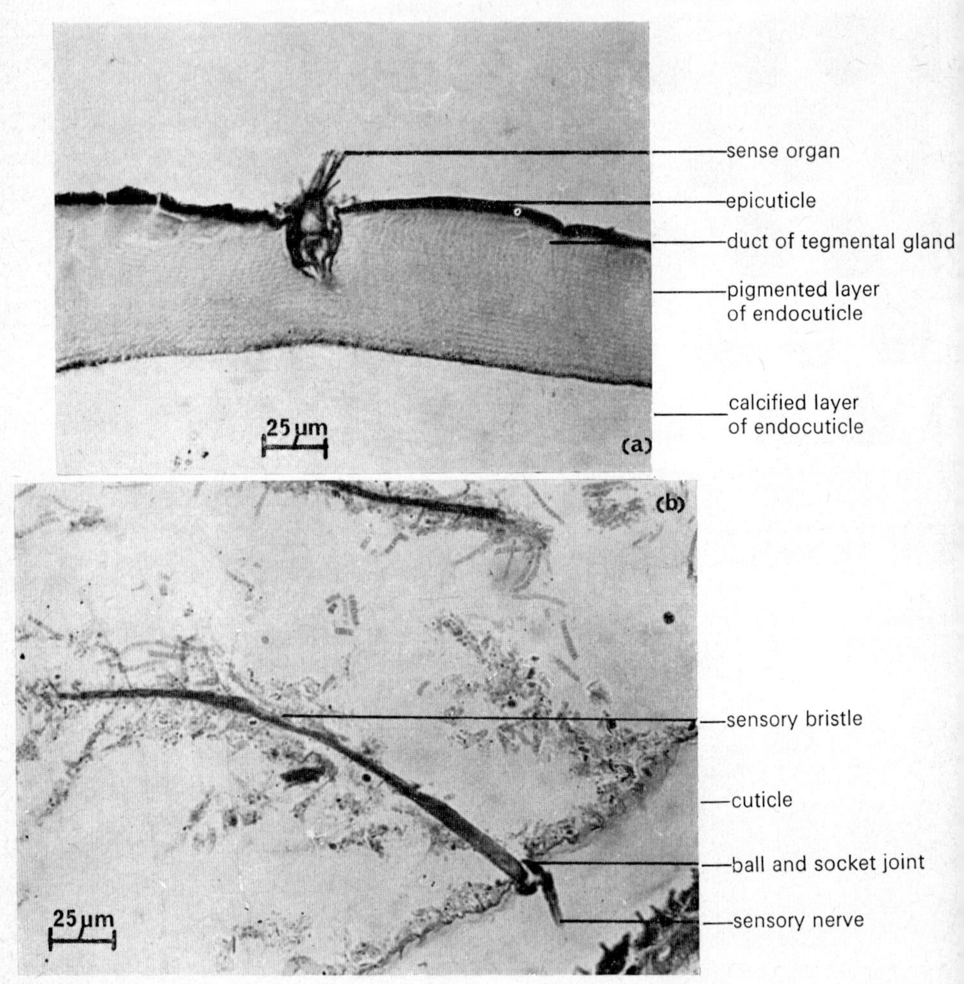

FIG. 6.7 (a) Cuticular sense organ; and (b) sensory bristle from the head region of *C. maenas*.

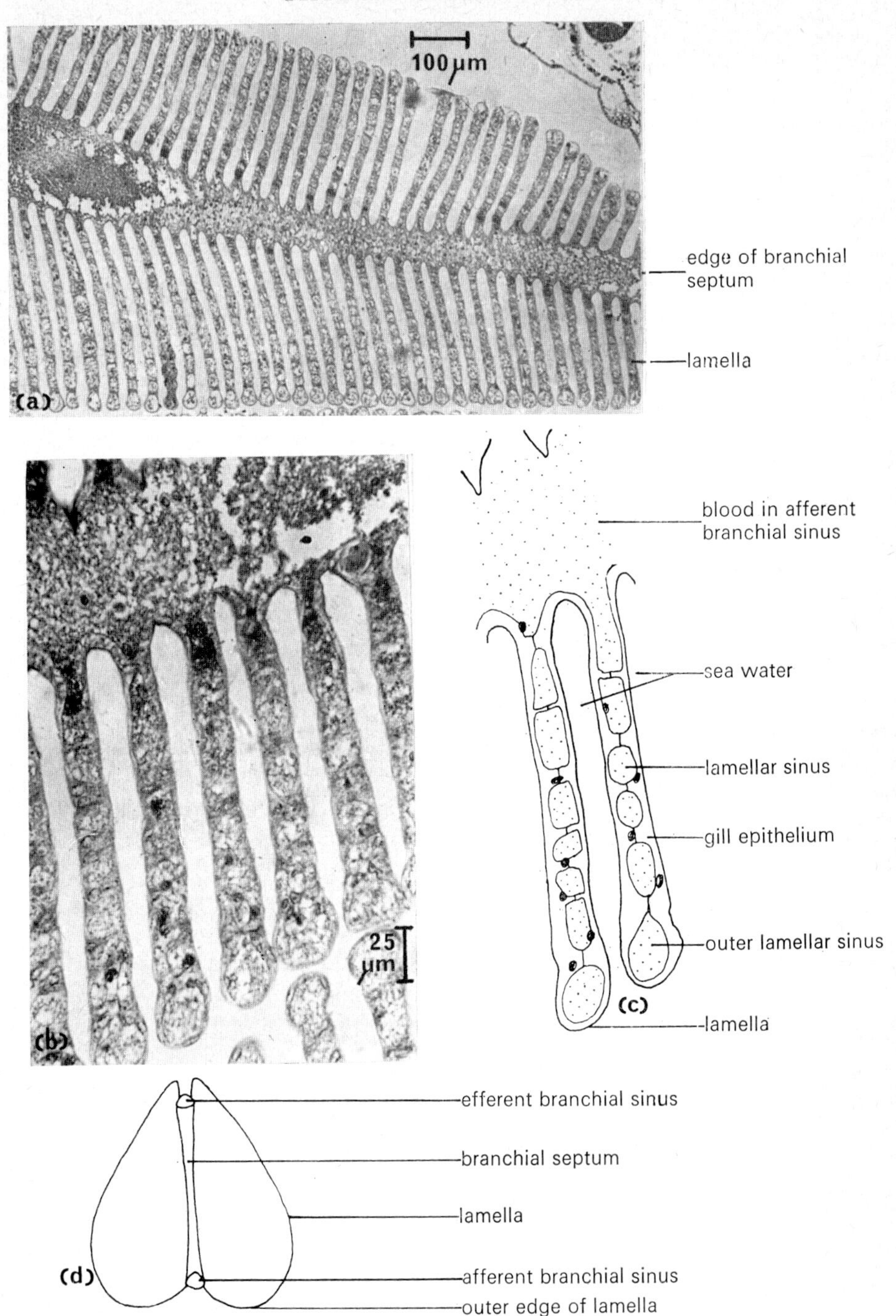

FIG. 6.8 (a) LS of a lamellar gill from *C. maenas*; (b) detail of the gill lamellae; (c) drawing of two gill lamellae shown in (b); (d) diagram of a crab gill in transverse section.

are well adapted for this purpose. Round the mouth the upper lips (labrum), developed from the body wall, and the lower lips (labium), formed from the fore-gut, assist food manipulation. The gut (Fig. 6.9) is also adapted for such a diet: all parts are provided with powerful striated muscles which move food onwards by their peristaltic action.

Food passes from the mouth through a cuticle-lined oesophagus possessing salivary glands (Fig. 6.10) to the stomach complex (see Fig. 6.1). The cardiac stomach (proventriculus) is a large sac-like chamber with muscular walls and a folded lining (Fig. 6.11). Its chief function is storage of food while some preliminary digestion by salivary amylases occurs, but the lining is calcified in certain regions to form ossicles which act as a gastric mill for trituration. This function is further assisted by chitinous teeth close to the junction with the pyloric stomach. The two parts of the stomach are separated by a muscular valve. The pyloric stomach (Fig. 6.12) has complex folded walls strengthened by plates and chitinized hairs (setae) which form a filtering apparatus. This filter, combined with a complicated valve arrangement, allows only fluid containing small particles through channels into the ducts of the digestive gland. Solid matter is passed to the mid-gut and compacted into faeces.

The digestive gland (hepatopancreas) consists of a mass of fine, blunt-ended tubules lined with absorptive cells, cells secreting enzymes and cells which store fat, glycogen and calcium (Fig. 6.13). This is the main site of extracellular digestion and absorption in crabs. The tubules join in paired hepatopancreatic ducts which open at the point where the pyloric stomach meets the mid-gut, and which are guarded by a fine filter of hairs.

Further digestion and absorption occurs in the mid-gut. This is a straight tube lacking a cuticular lining and bearing a pair of dorsal caecae at its anterior end, which serve to increase surface area. Both mid-gut and caecae are lined with a single layer of absorptive epithelium, that of the caecae often giving the appearance of being ciliated (Fig. 6.14). This is probably an artifact as it is generally reported that the arthropod gut is devoid of cilia. The hind-gut possesses a single coiled caecum of similar structure to the mid-gut caecae, and has a cuticular lining (Fig. 6.15).

Excretion

The main excretory organs in crabs are a pair of modified glandular coelomoducts in the antennary segment, immediately behind the eye sockets. These antennal or green glands (Fig. 6.16) consist of a terminal end-sac of coelomic origin, an excretory tubule and a storage bladder which opens to the exterior. Ultrafiltration of dissolved substances (including ammonia) in the blood of the haemocoel occurs across the wall of the end-sac. Since *Carcinus* occurs in littoral and estuarine habitats, it has the ability to regulate the salt and water content of the body over a wide range of environmental salinities. This is achieved partly by absorption

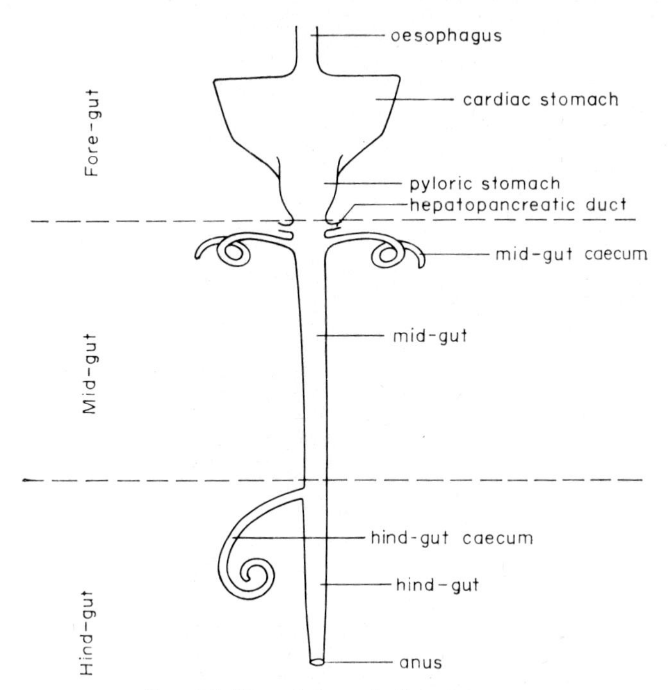

FIG. 6.9. Plan of the crab digestive tract.

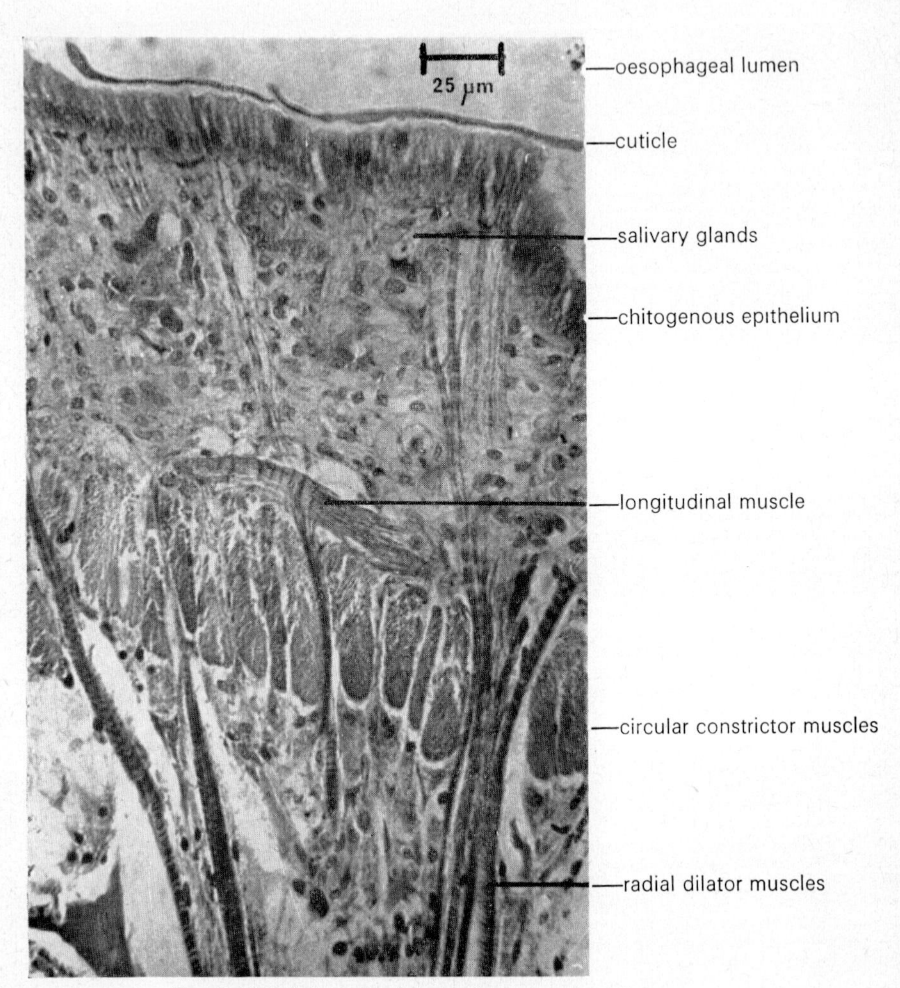

FIG. 6.10. LS through the oesophageal wall of *C. maenas*.

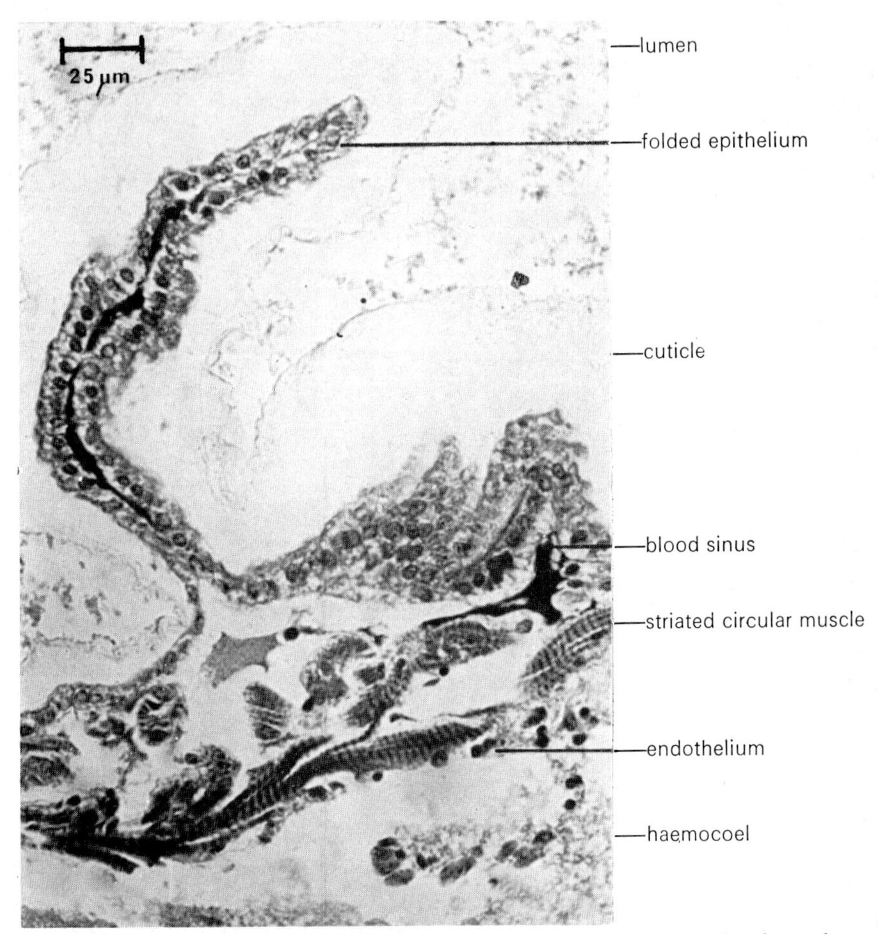

FIG. 6.11. TS of the cardiac stomach wall of *C. maenas*. Note the circular striated muscles.

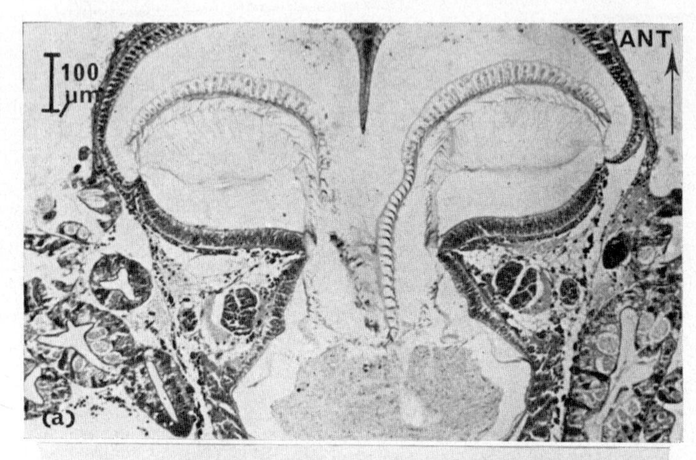
100 µm
ANT
(a)

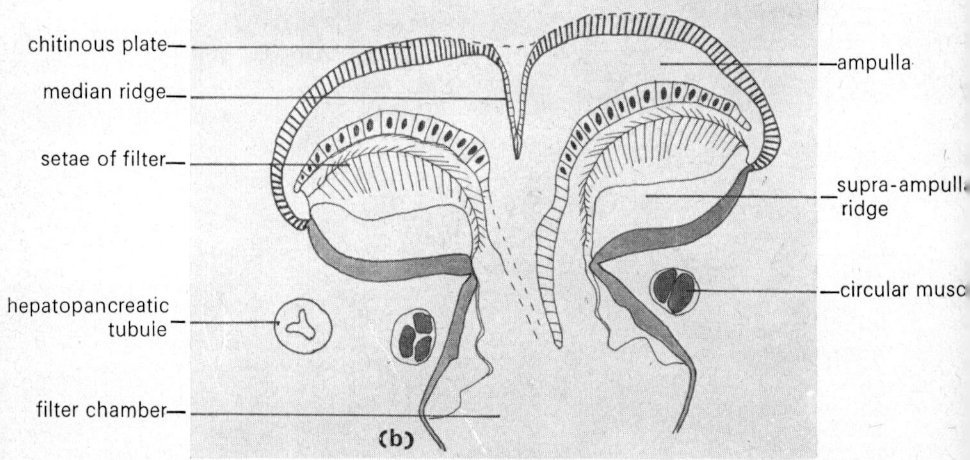
chitinous plate
median ridge
setae of filter
hepatopancreatic tubule
filter chamber
ampulla
supra-ampull
ridge
circular musc
(b)

FIG. 6.12

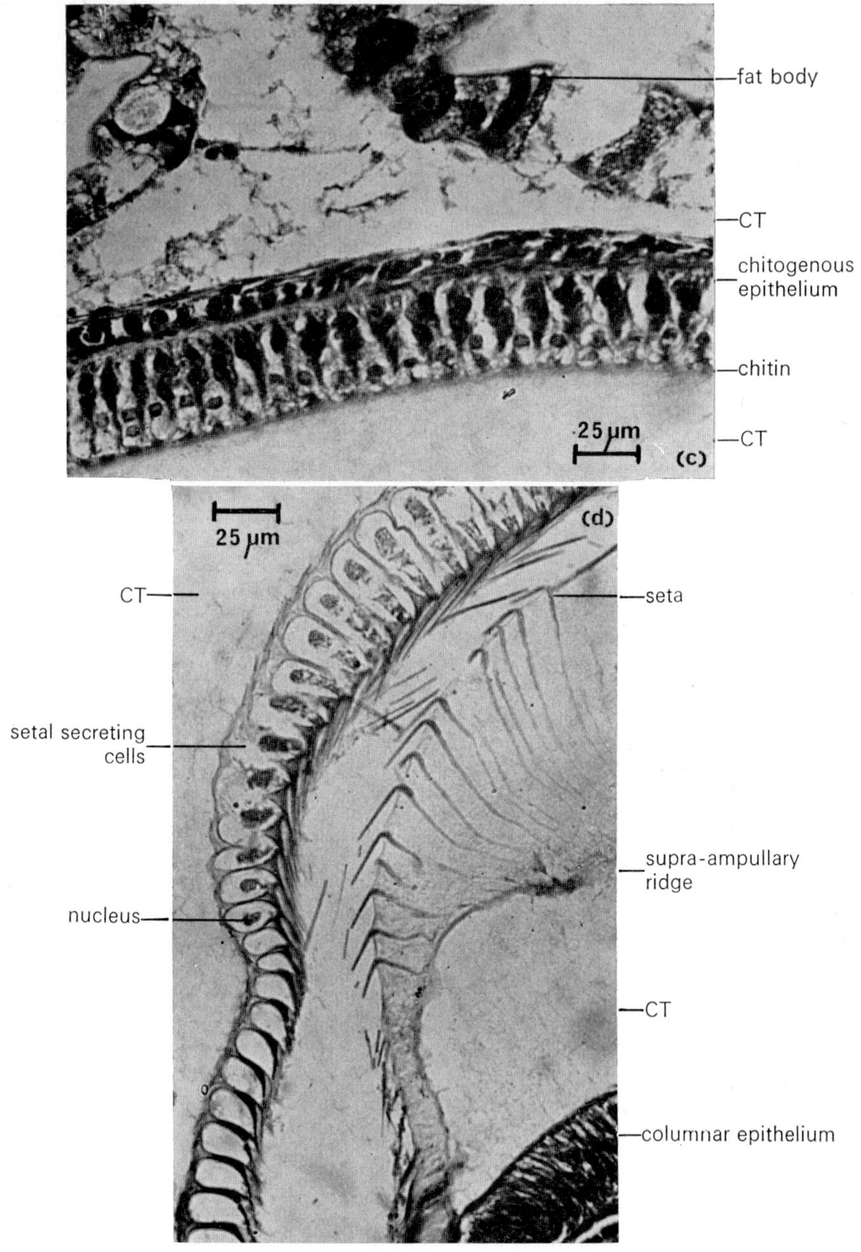

FIG. 6.12 (a) LS and (b) drawing of the pyloric stomach of *C. maenas*, showing the filtering apparatus; (c) detail of the chitogenous plate; and (d) detail of the setae of the filtering apparatus in the crab pyloric stomach.

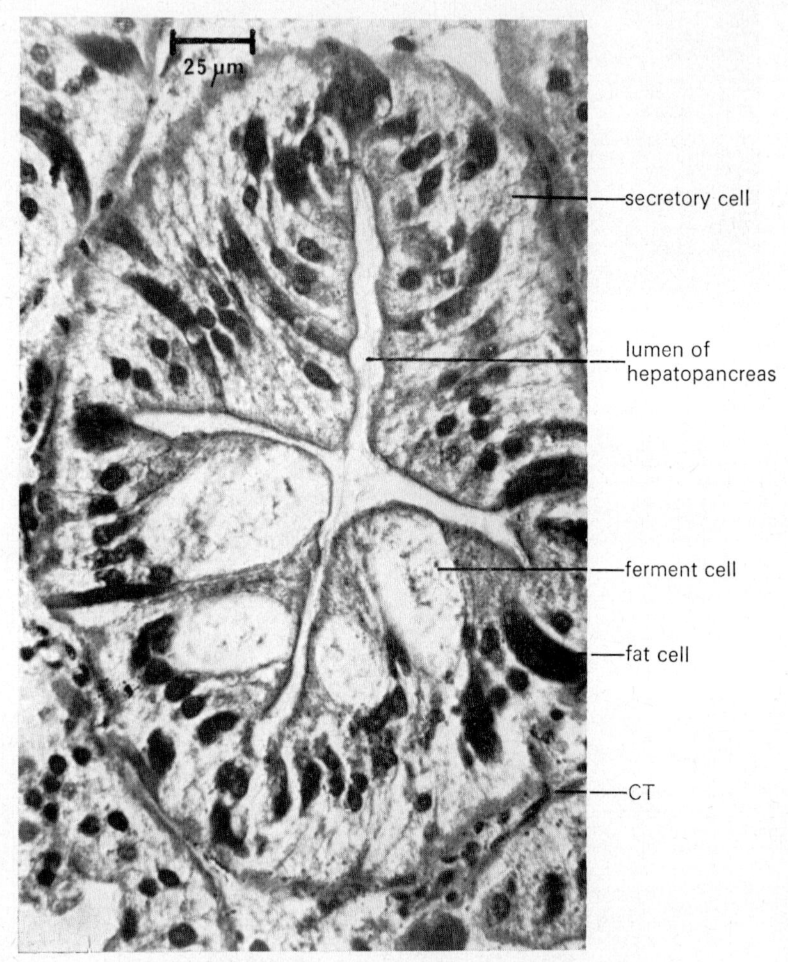

FIG. 6.13. Section through an hepatopancreatic tubule of *C. maenas*.

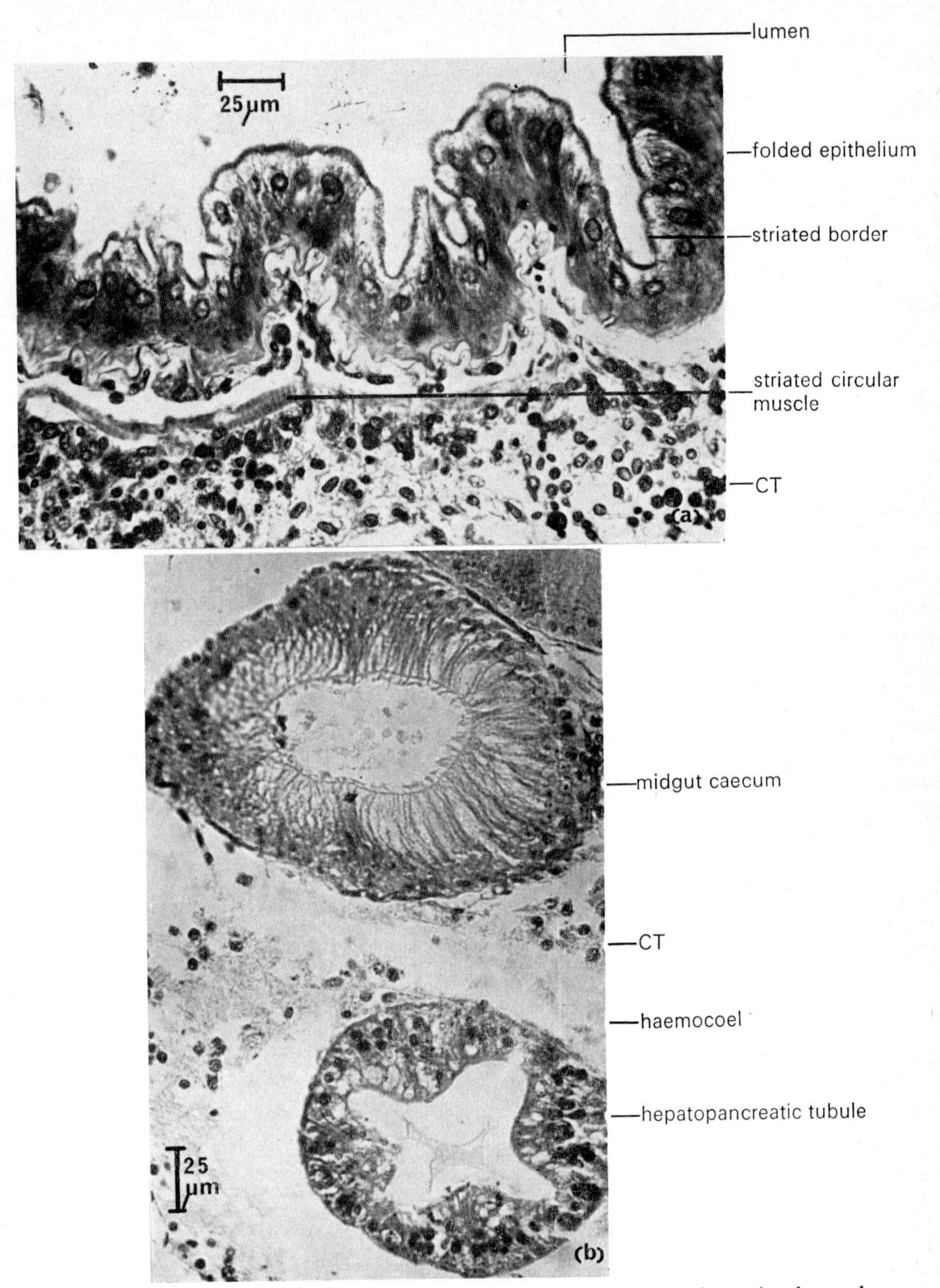

FIG. 6.14 (a) TS of the mid-gut wall of *C. maenas*. Note the circular striated muscle; (b) TS through a mid-gut caecum of *C. maenas*.

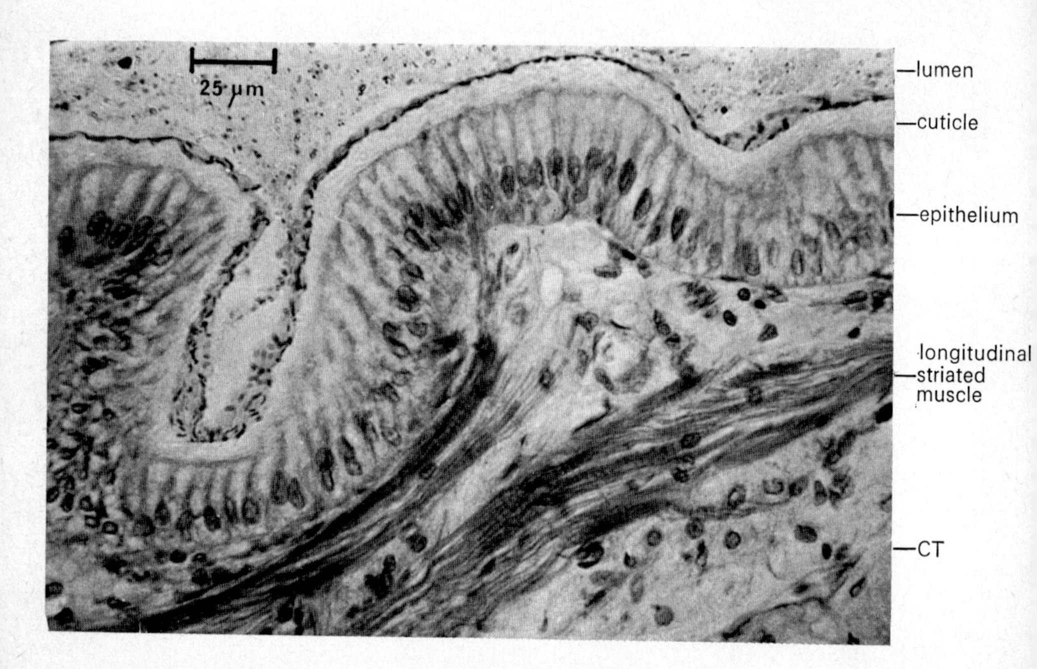

Fig. 6.15. Section of the hind-gut wall of *C. maenas*. Note the powerful striated muscles.

of salts across the gill surface, and partly by reabsorption of salts and water across the wall of the excretory tubule. The anterior portion of this tubule is expanded to form a labyrinth, the walls of which are extremely folded and glandular. The bladder is a thin-walled sac with lateral extensions which lie among the viscera.

Reproduction

The sexes are separate in crabs and can be distinguished by the shape of the abdomen. Elongated gonads on each side of the body join across the mid-line and extend laterally into the carapace fold. They are hollow organs continuous with ducts which lead directly to the outside.

The testes [Fig. 6.17(a)] are compact structures composed of tightly packed follicles lined with germinal epithelium. Sperm develop in the lumina of these follicles; all the cells in one follicle being at the same stage of development at the same time. Ripe sperm are passed into the coiled vasa deferentia which have glandular walls [Fig. 6.17(b)] specialized for the formation of spermatophores. The terminal parts form muscular ejaculatory ducts leading to paired tubular penes which fit into grooves on the first pair of pleopods. Males possess certain appendages modified for clasping the females.

In the female there is a pair of small ovaries which become multilobulate in mature adults and packed with large numbers of enormous ova filled with yolk particles (Fig. 6.18). Eggs are released into paired oviducts, the ends of which are modified as glandular seminal receptacles (spermathecae). Two vaginas open on the ventral surface of the sixth thoracic segment. Fertilization is internal and development is by pelagic larvae.

Class Insecta

Insects form the largest class of animals and are found in most terrestrial habitats, with relatively few being secondarily aquatic. They are characterized by the possession of a hard exoskeleton, and a body showing complete tagmosis with a head of four segments (six in the embryo), a thorax of three segments, and an abdomen of nine to eleven segments. The head bears a pair of antennae, a pair of mandibles and one pair of maxillae: the labium is formed by fusion of a second pair of maxillae, while the labrum is part of the head capsule. There is normally a pair of compound eyes, and often ocelli are present. The thorax possesses three pairs of legs and, usually, two pairs of wings, although the latter may be primitively or secondarily reduced or absent in some groups. The abdomen carries no appendages except for the sensory cerci on the terminal segment. Respiration is typically by a tracheal system. With a few exceptions, insects are dioecious and the gonoducts open at the posterior end of the abdomen.

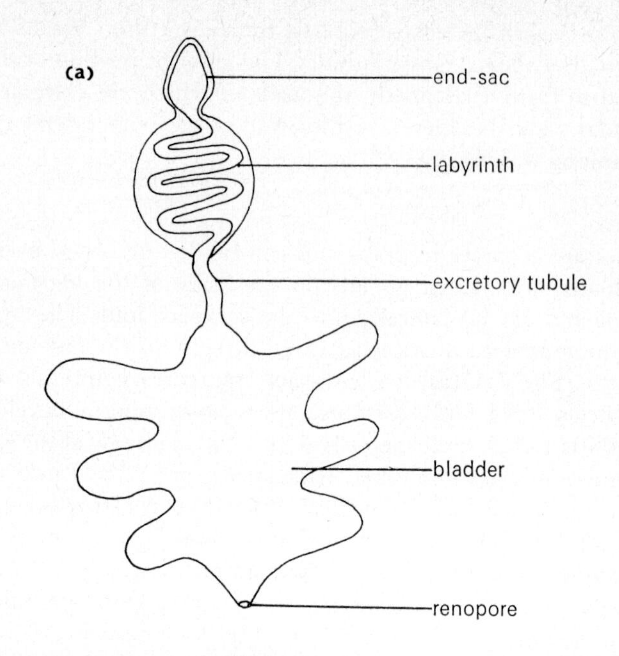
(a)
end-sac
labyrinth
excretory tubule
bladder
renopore

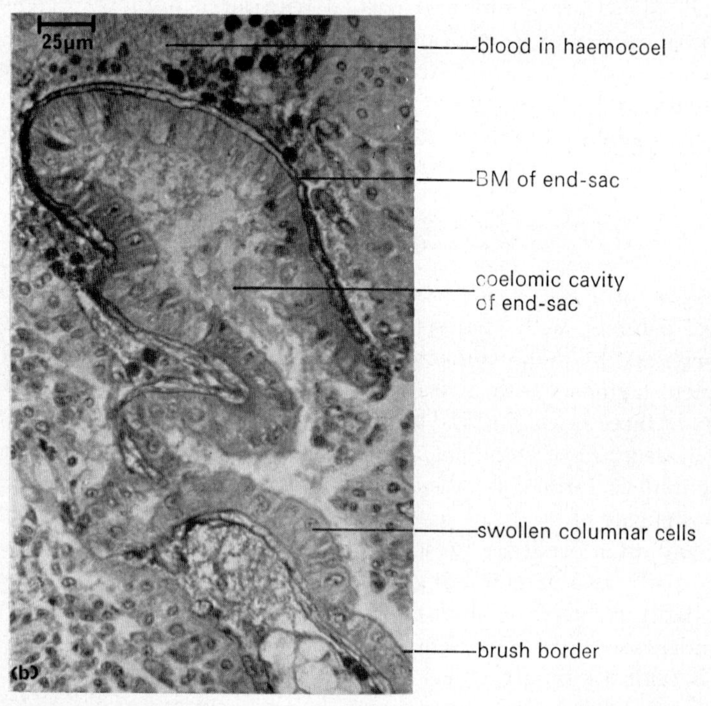
25μm
blood in haemocoel
BM of end-sac
coelomic cavity
of end-sac
swollen columnar cells
brush border
(b)

FIG. 6.16

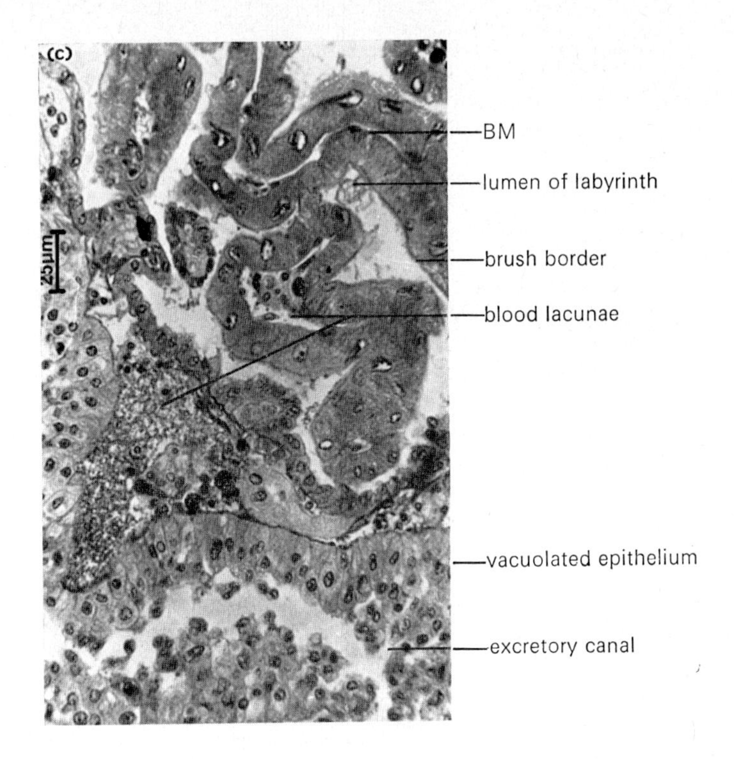

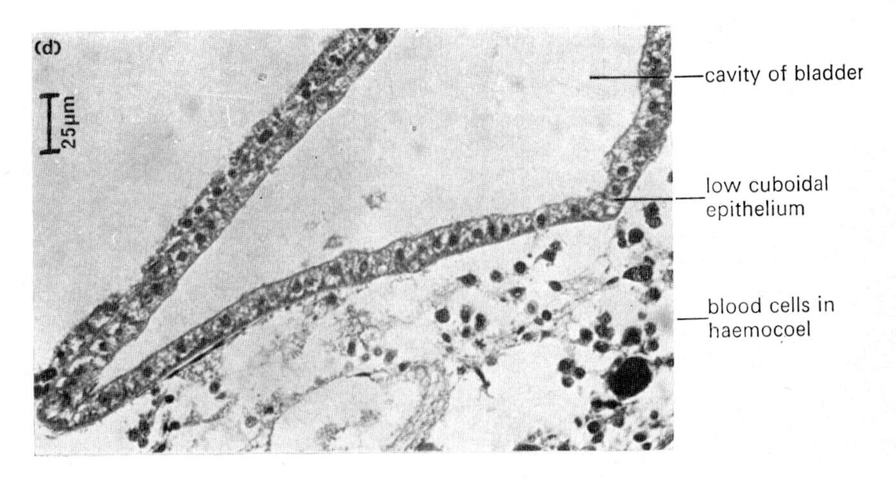

FIG. 6.16 (a) Plan of a crab antennal (green) gland. Details of (b) the end-sac and part of the labyrinth; (c) the labyrinth and excretory canal; and (d) the storage bladder of the antennal gland of *C. maenas*.

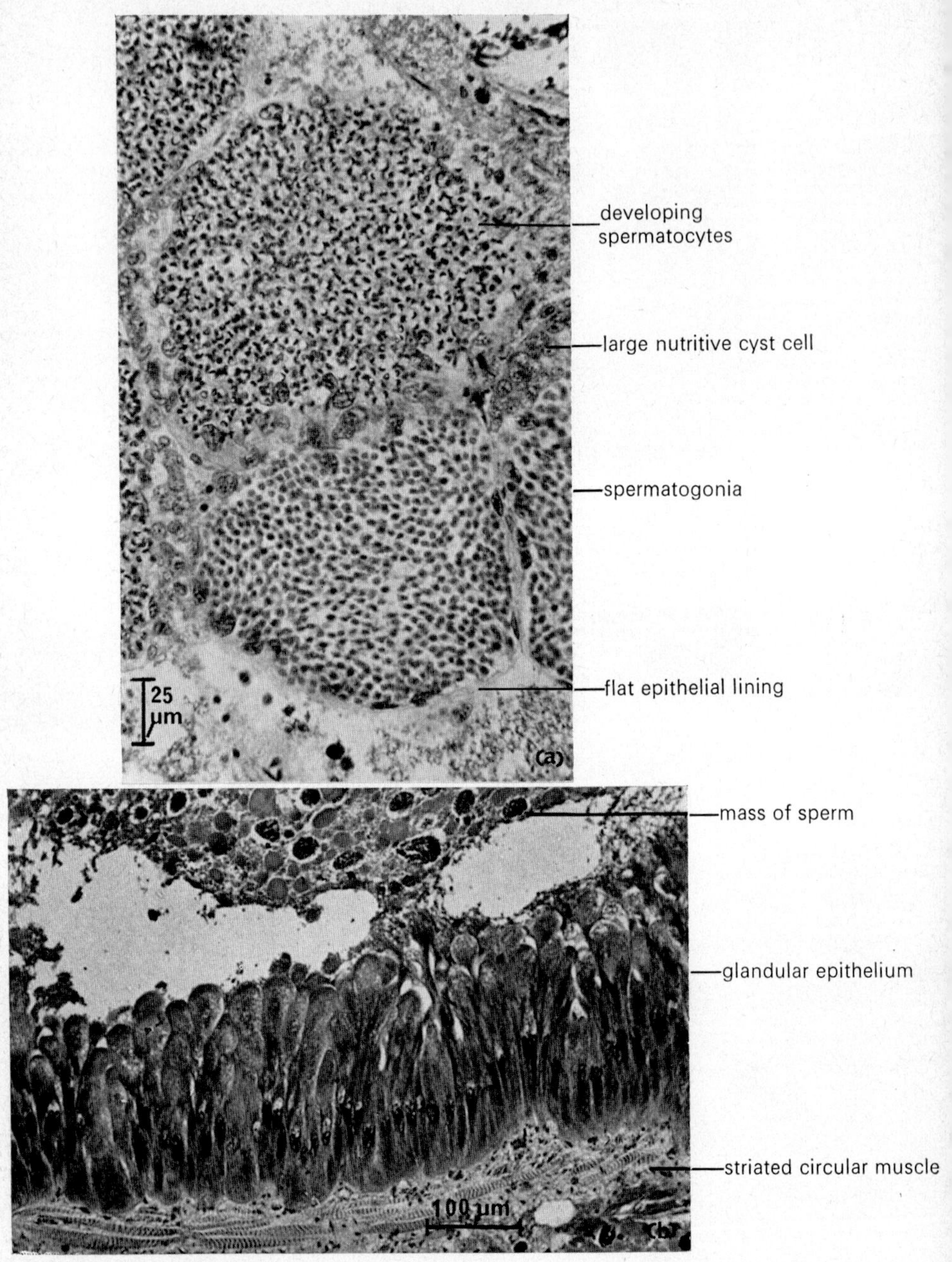

FIG. 6.17 (a) Section through the testis of a male *C. maenas*, showing two follicles. Note that all the gametes in each follicle are at the same stage of development; (b) TS through the vas deferens wall of an adult male *C. maenas*.

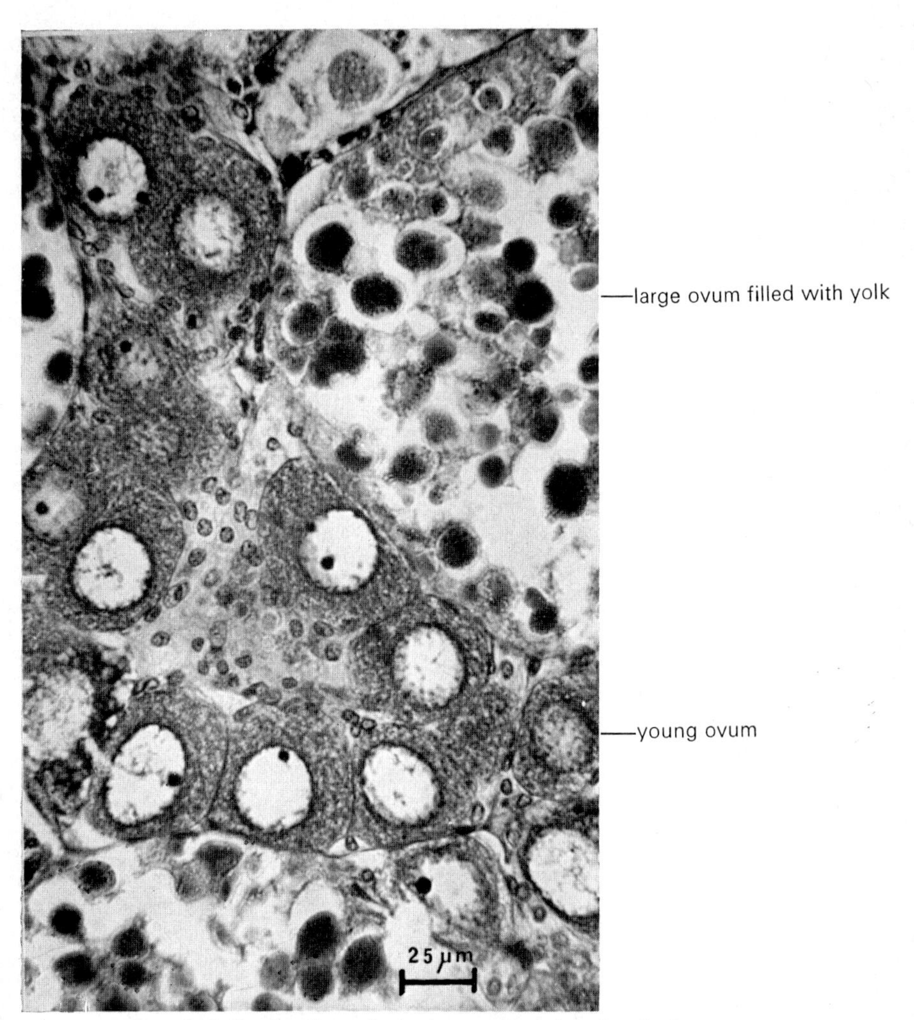

FIG. 6.18. Section through the ovary of an adult female *C. maenas*.

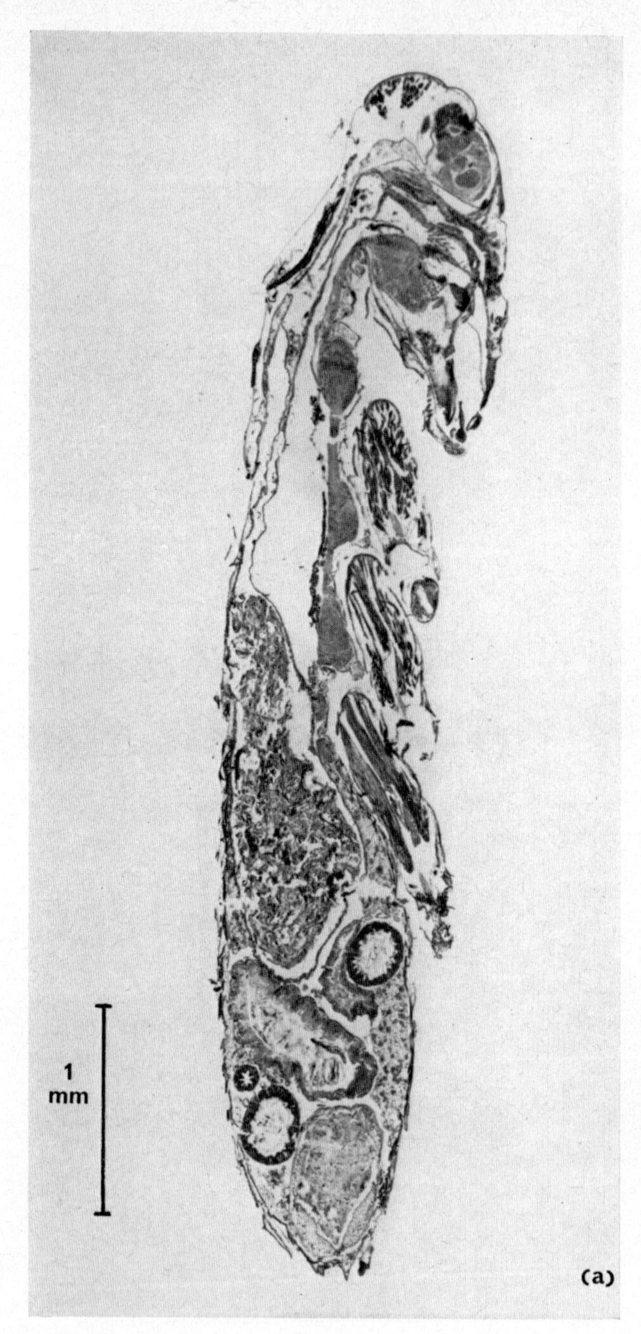

Fig. 6.19

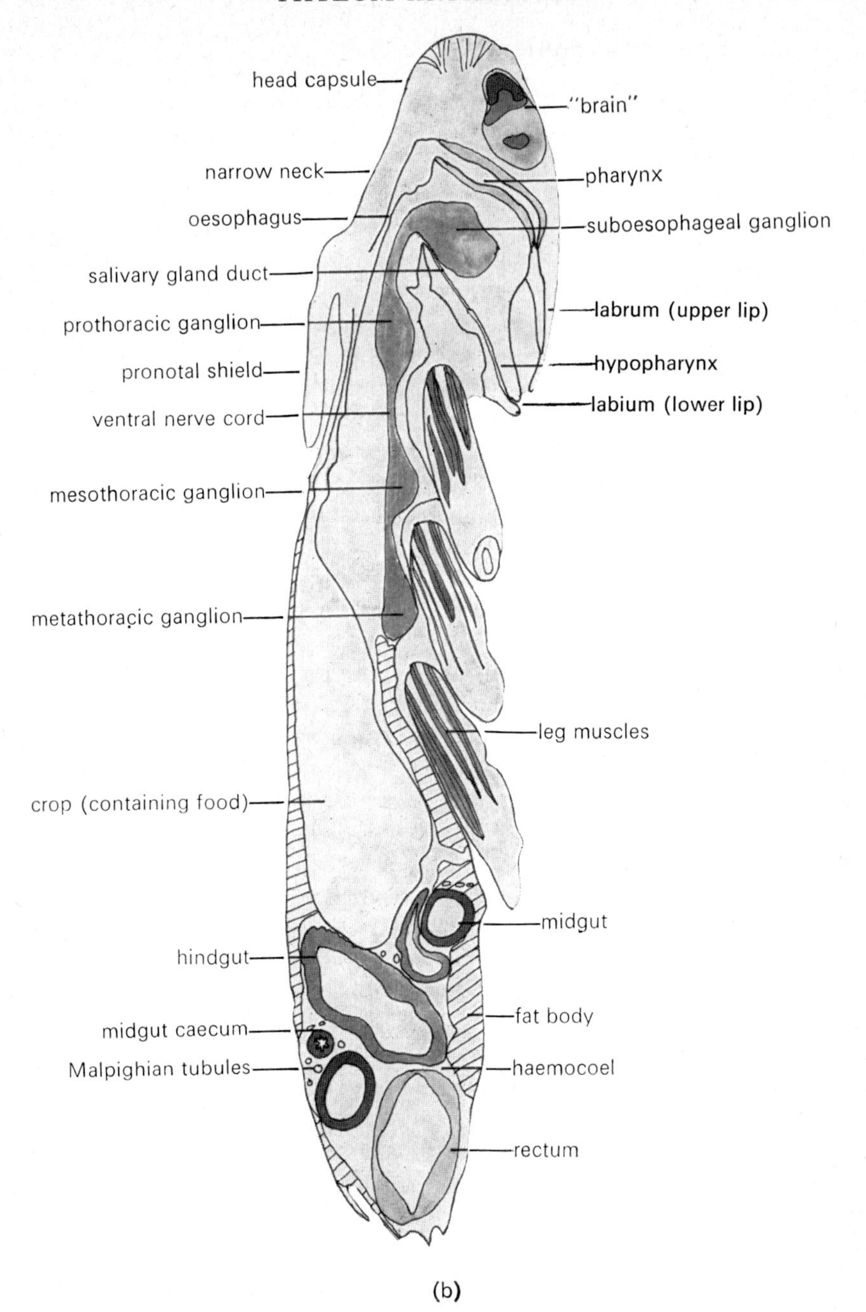

FIG. 6.19 (a) VLS and (b) explanatory drawing of a young cockroach (*Periplaneta americana*).

Example : *Periplaneta* (*americana*)

The American cockroach is one of the most widely distributed of the cockroaches and often occurs in close association with man. The body is dorso-ventrally flattened and consists of head, thorax and an abdomen of ten segments with a reduced eleventh segment bearing multi-articulate sensory cerci. The wings are fully developed in the adult and folded when at rest. There are strong running legs and powerful biting mouthparts. Figure 6.19 shows a VLS of a young cockroach, before it has acquired wings. Note the narrow neck joining the head and thorax: this means that the head is very mobile but that the gut must of necessity be very constricted at this point. Because of movement in this region the "brain" is supported on a chitinized platform (tentorium).

As in the Crustacea, the body wall of insects is covered by an epidermis secreting a cuticle which, however, is thin and non-calcified (Fig. 6.20). The cuticular plates (sclerites) contain chitin and are hardened, while the joints between sclerites remain flexible. Powerful striated skeletal muscles insert onto the basement membrane at the base of the epidermis.

Nervous system

The nervous system of cockroaches follows the basic arthropodan plan: the arrangement of the anterior end being illustrated in Fig. 6.21. The cerebral ganglia or "brain", formed by fusion of the three anterior head ganglia (proto-, deutero- and trito-cerebrum), is linked by commissures to the suboesophageal ganglion, created by fusion of the last three head ganglia (Fig. 6.22). The paired ventral nerve cords lie close to one another, and the paired segmental ganglia are fused in the mid-line, thus giving the appearance of single ganglia (Fig. 6.23). Compare this with the situation in *Peripatus* (see Figs 5.2 and 5.15). There are three large (pro-, meso- and meta-) thoracic ganglia and six abdominal ganglia, the last one of which is composed of ganglia from several terminal segments. Giant nerve fibres occur in the ventral nerve cord (Fig. 6.23), but they are not as common in insects as they are in crustaceans. The "brain" and suboesophageal ganglion are, however, a major source of neurosecretions, which play a role as hormones (Fig. 6.24).

Closely associated with the brain are the corpora cardiaca and corpora allata (see Fig. 6.21). These are paired bodies, which in many species are fused and which release neurosecretions produced by the nervous system. The corpora cardiaca are elongated structures closely applied to the walls of the aorta. They are neurohaemal (storage-and-release) organs consisting of axons, swollen nerve endings, glial cells and blood sinuses [Fig. 6.25(a)]. The corpora allata [Fig. 6.25(b)] are small, round, paired bodies connected by nerves to the corpora cardiaca. They possibly also release neurosecretions, but in addition to this they are stimulated by brain

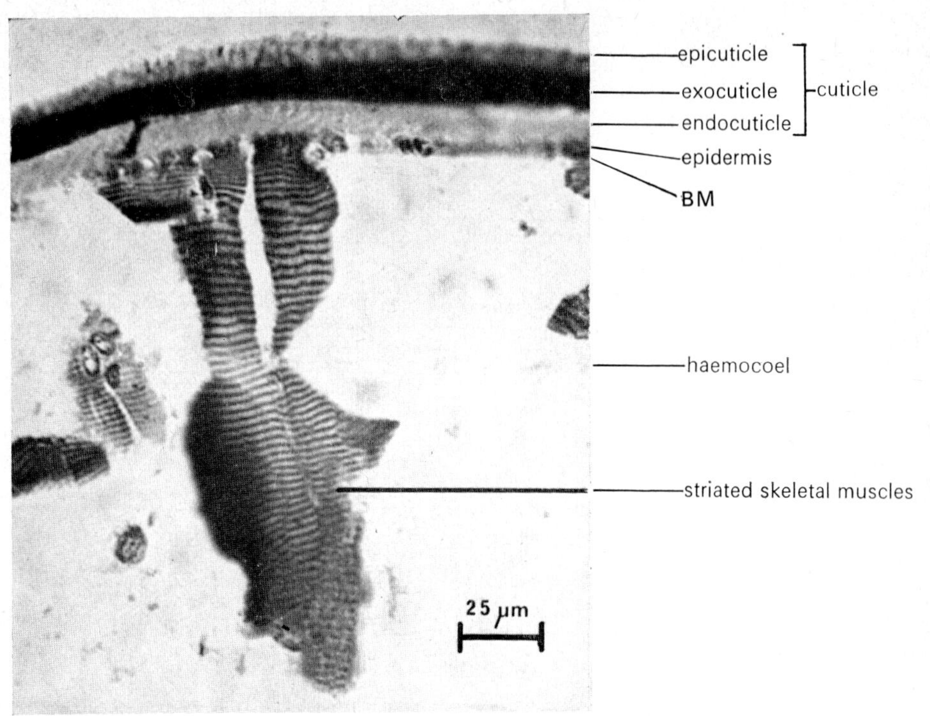

FIG. 6.20. Section through the body wall of *P. americana*, showing the powerful striated skeletal muscles.

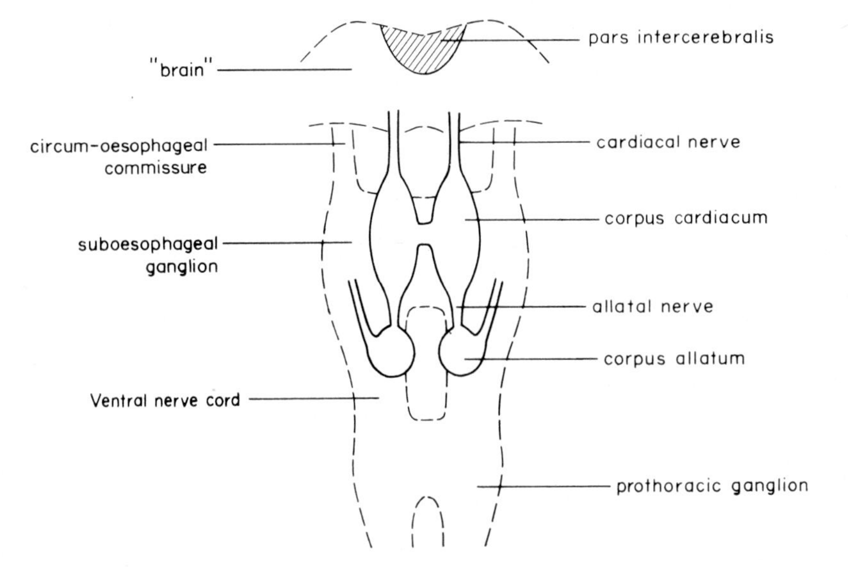

FIG. 6.21. Diagram to show the arrangement of the anterior nervous system of a cockroach, and its relation to the corpora cardiaca and corpora allata (viewed from the dorsal side).

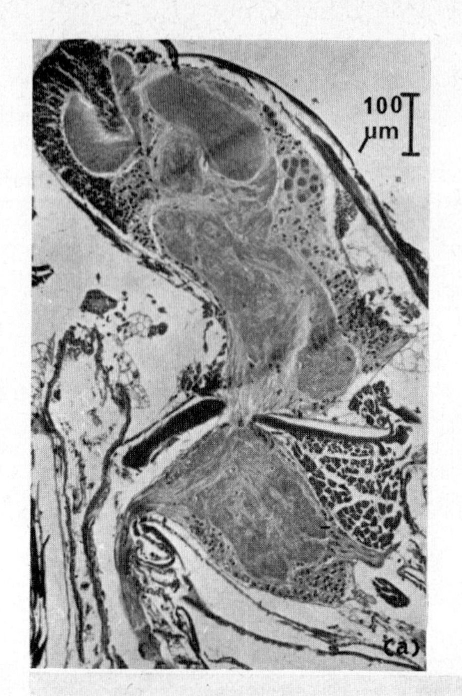

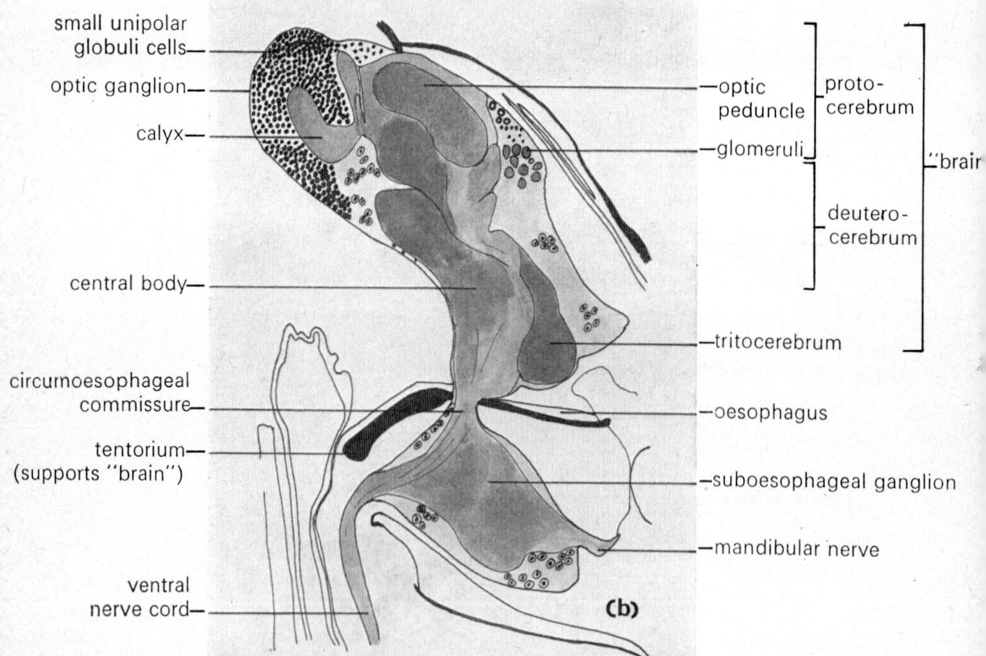

FIG. 6.22 (a) LS and (b) drawing of the anterior end of the nervous system of *P. americana*.

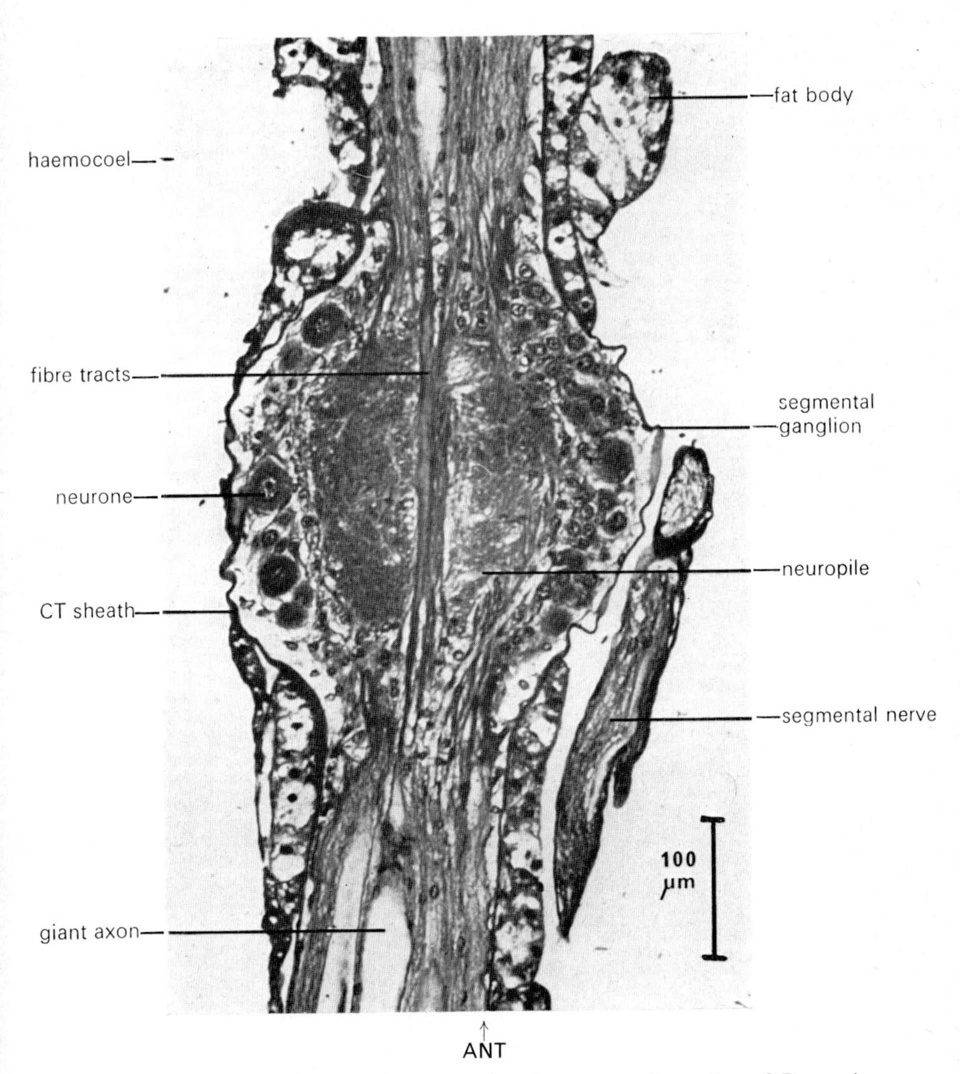

FIG. 6.23. LS through the ventral nerve cord and a segmental ganglion of *P. americana*.

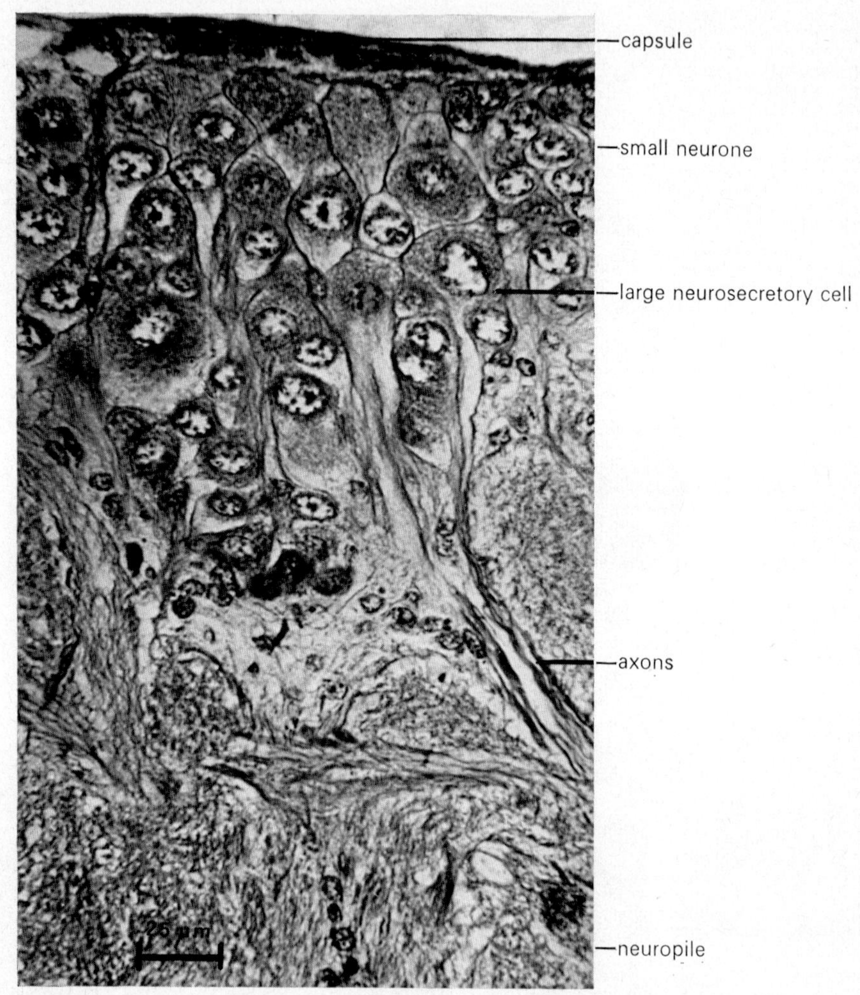

FIG. 6.24. Detail of neurosecretory cells in the pars intercerebralis of the brain of *P. americana*.

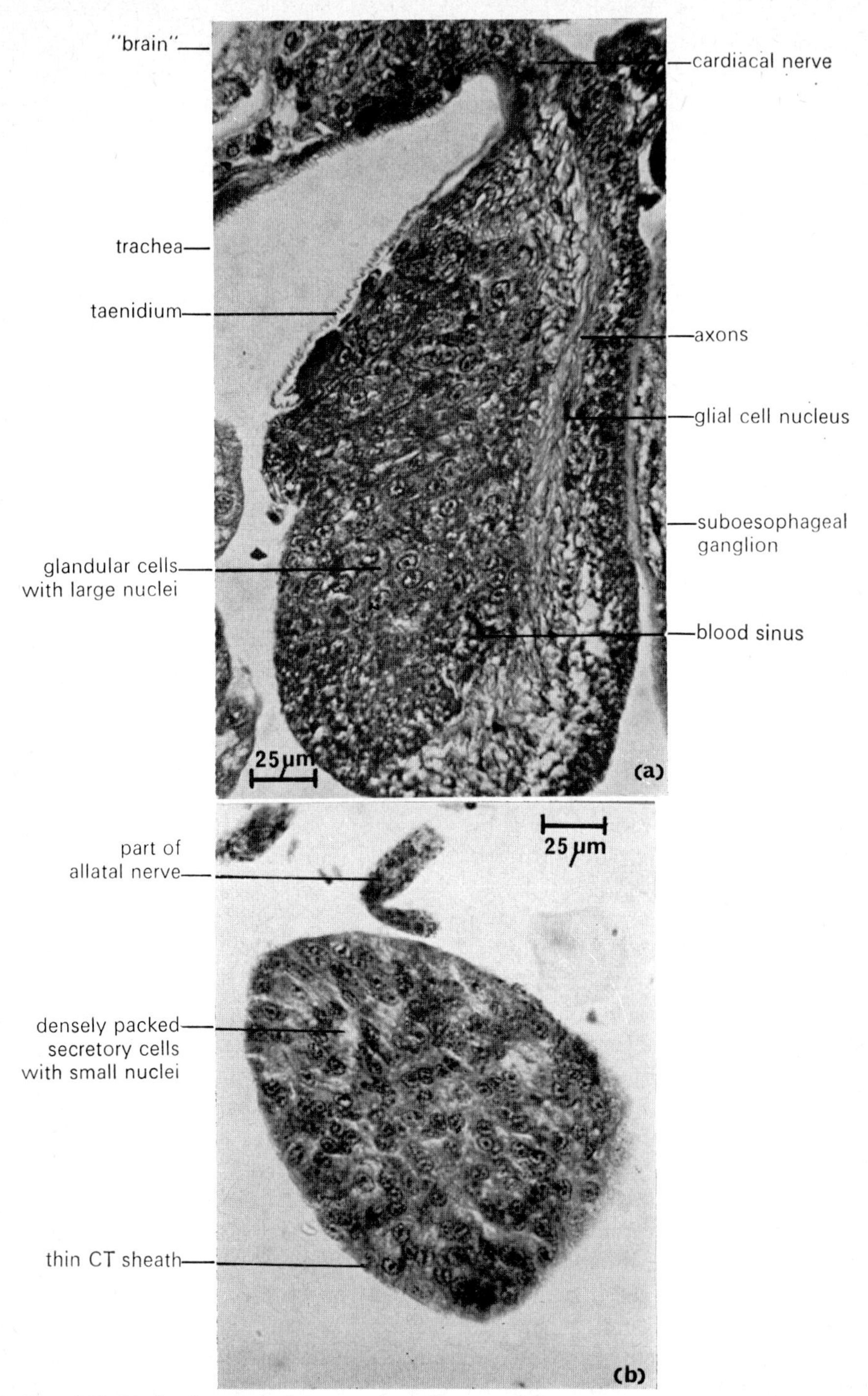

FIG. 6.25 (a) Section through a corpus cardiacum of *P. americana* showing its position close to the cerebral and suboesophageal ganglia; (b) section through a corpus allatum of *P. americana*.

hormones and act as endocrine organs in their own right. They are composed of densely packed glandular cells with large nuclei. Both glands are involved in the hormonal control of metamorphosis as well as other functions. The moulting hormone (ecdysone) is produced by a purely endocrine organ, the prothoracic gland, which is controlled by neurosecretions from the brain that are released by the corpora cardiaca.

Sense organs

Cockroaches are fast moving, terrestrial insects and have well-developed sense organs. There are large numbers of small, single-celled sense organs (sensilla) which take the form of tactile hairs and spines, chemoreceptors round the mouth, campaniform sensilla for detecting strains in the cuticle and chordotonal organs attached to the inside of the cuticle which register body movements. Multicellular sense organs include tactile antennae, anal cerci for appreciating air movements, tympanic organs for hearing, and eyes.

The eyes are of two types: the simple light-sensitive ocelli, which in the cockroach are reduced to unpigmented areas of head cuticle (fenestrae); and the compound eyes which are specialized for movement perception. As in the Crustacea, the compound eyes of insects are made up of a number of ommatidia, each comprising a biconvex corneal lens formed of transparent cuticle, a crystalline cone and seven retinular cells surrounding a rhabdom (Fig. 6.26). The cockroach is nocturnal and possesses a superposition eye where light from a point source falls on more than one ommatidium at once. This is achieved by movement of pigment to the outer regions of the eye around the cone. Compare this with the crab where the pigment is normally moved towards the inner regions of the eye around the receptors (see Fig. 6.5).

Respiration

The respiratory system of cockroaches consists of a ramifying network of tubular invaginations of the ectoderm (tracheae) lying in the haemocoel. The tracheae send branches to all the tissues, where they end intracellularly. They connect with the outside by protective closing mechanisms (spiracles) of which there are ten pairs in the cockroach (2 thoracic and 8 abdominal). Figure 6.27 is a surface view of an insect spiracle and shows the filter of fringe processes typical of terrestrial insects. A muscular closing mechanism, controlled by carbon dioxide concentration and humidity, prevents excess water loss through the respiratory system. The large longitudinal tracheal trunks (see Figs 6.25 and 6.43d) and their smaller cross-connections are lined with cuticle which is folded and thickened to produce strengthening rings and/or spirals (taenidia). The trachea branch repeatedly into tracheoles which have thin, permeable walls and are often fluid-filled. Oxygen reaches the tissues by diffusion, assisted by ventilatory

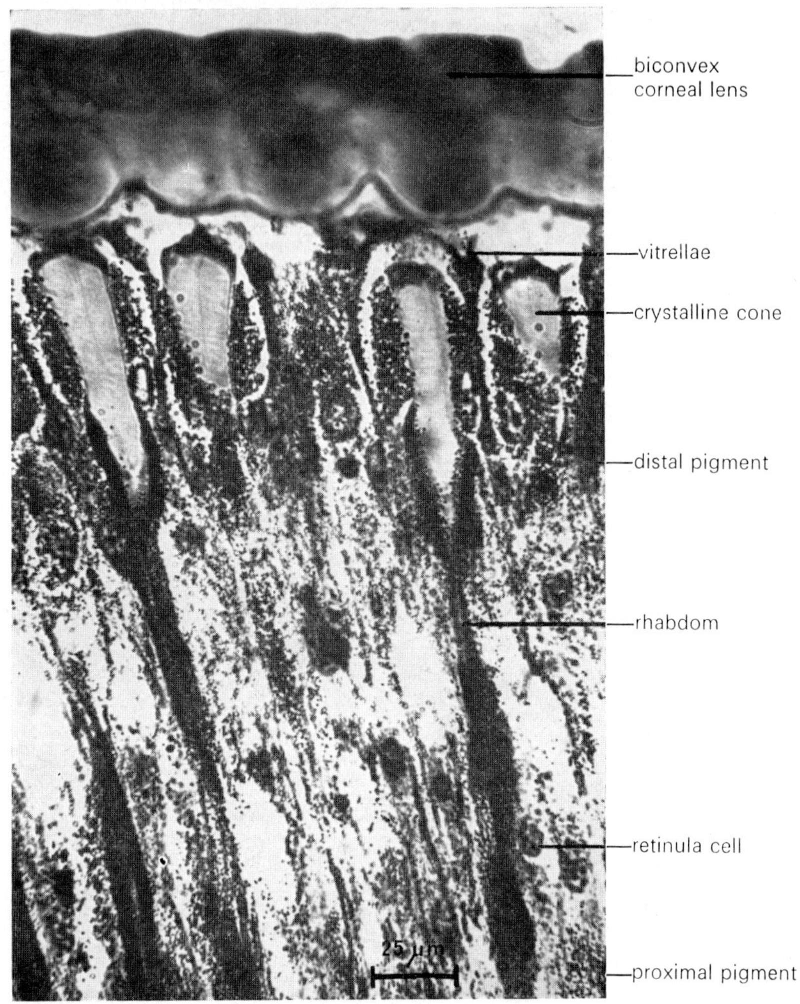

Fig. 6.26. Section through the superposition eye of *P. americana*. Note that the pigment is distributed around the cones as well as around the lower portions of the photoreceptive cells.

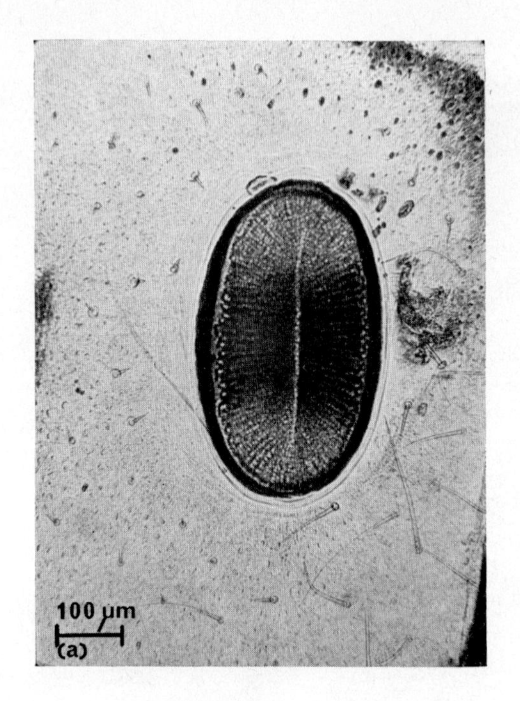

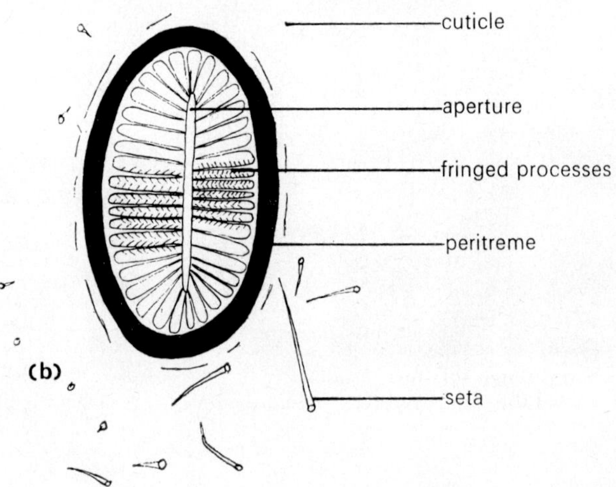

FIG. 6.27 (a) Surface view and (b) drawing of a lepidopteran insect spiracle.

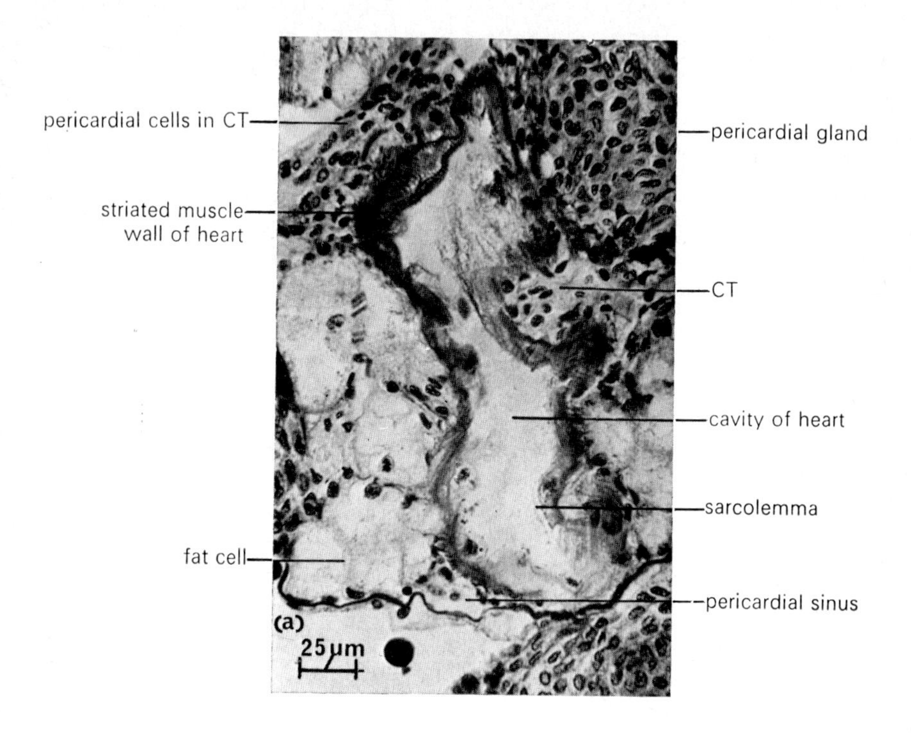

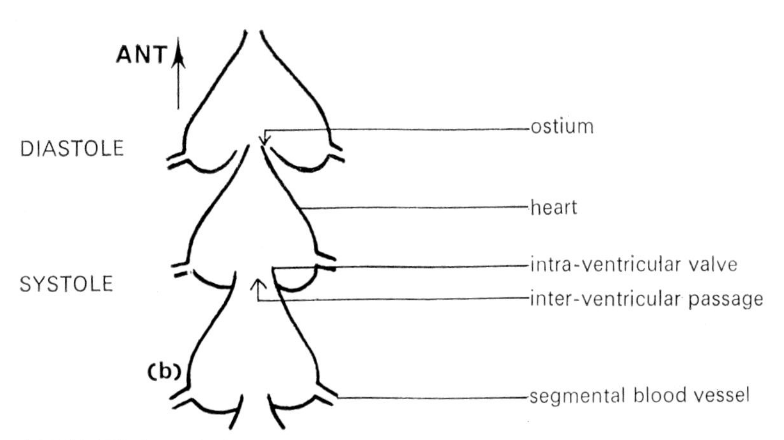

FIG. 6.28 (a) TS through the heart of *P. americana.* Note the pericardial glands and pericardial cells lying close to the heart; (b) diagram of part of the abdominal region of the cockroach heart.

body movements. In some insects the tracheae are widened to form distensible air-sacs which lack taenidia. Insect spiracles are usually protected against wetting, either by hydrofuge hairs or by a lipid secretion produced by peristigmatic glands round the spiracle.

Circulation

The blood of insects acts as a transport medium for nutrients and hormones but is not responsible for carriage of oxygen, which is taken straight to the cells by tracheoles. Respiratory pigments are not normally present. The circulatory system is open and the haemocoel is divided into dorsal (pericardial), visceral and ventral (perineural) sinuses. In the cockroach there is a dorsal tubular heart which is composed of a single layer of striated muscle fibres, and which is controlled by ganglia and secretions from pericardial cells [Fig. 6.28(a)]. The heart lies in the thoracic and abdominal segments, is closed posteriorly, and merges anteriorly with the aorta which later divides first into two cephalic arteries and then into a number of smaller vessels. The heart is constricted between each segment and blood enters through a pair of ostia at each constriction [Fig. 6.28(b)]. These twelve pairs of ostia are guarded by valves which not only prevent escape of blood from the ostia but which also ensure its forward movement inside the heart by acting as "intraventricular" valves as waves of contraction pass forwards along the heart. Six pairs (two thoracic and four abdominal) of segmental vessels leave the heart. Accessory contractile vesicles at the bases of the wings and antennae aspirate blood through these structures which would tend to miss the normal body circulation.

Digestion

Cockroaches are omnivorous and their mouth-parts are adapted for dealing with a wide range of foodstuffs. As in crustaceans the gut can be divided into ectodermally-derived fore- and hind-guts which are lined with cuticle that is shed at each moult, and a non-chitinized endodermal mid-gut. Figure 6.29 is a plan of the cockroach gut and defines these regions.

Opening at the base of the hypopharynx (see Fig. 6.19) is a duct from the paired salivary glands. These glands consist of a network of fine tubules and bunches of acini comprising two types of cell: granular zymogen cells, which appear to have different densities depending on their phase of secretory activity, and ductule-containing cells which pass the secretions into the salivary gland ducts (Fig. 6.30). Each gland is associated with a large reservoir for the storage of saliva. The saliva contains enzymes responsible for the initial stages of digestion.

The muscular pharynx gives way to a narrow oesophagus passing through a small hole in the tentorium (see Fig. 6.22). The crop is a large saccular chamber with thin, folded, cuticle-lined walls to allow for

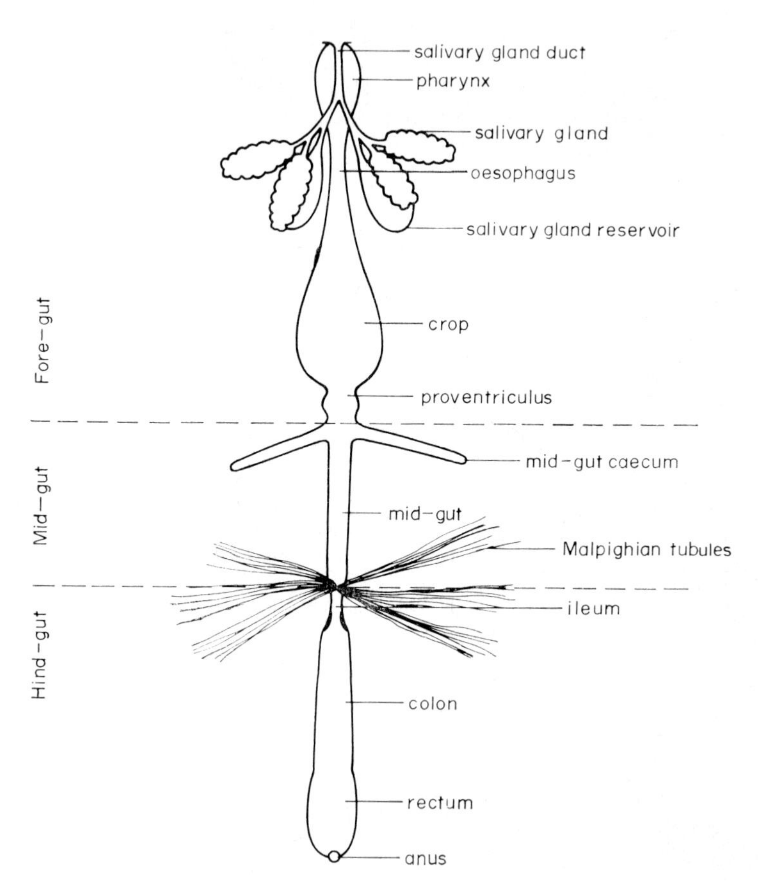

FIG. 6.29. Plan of the digestive tract of a cockroach.

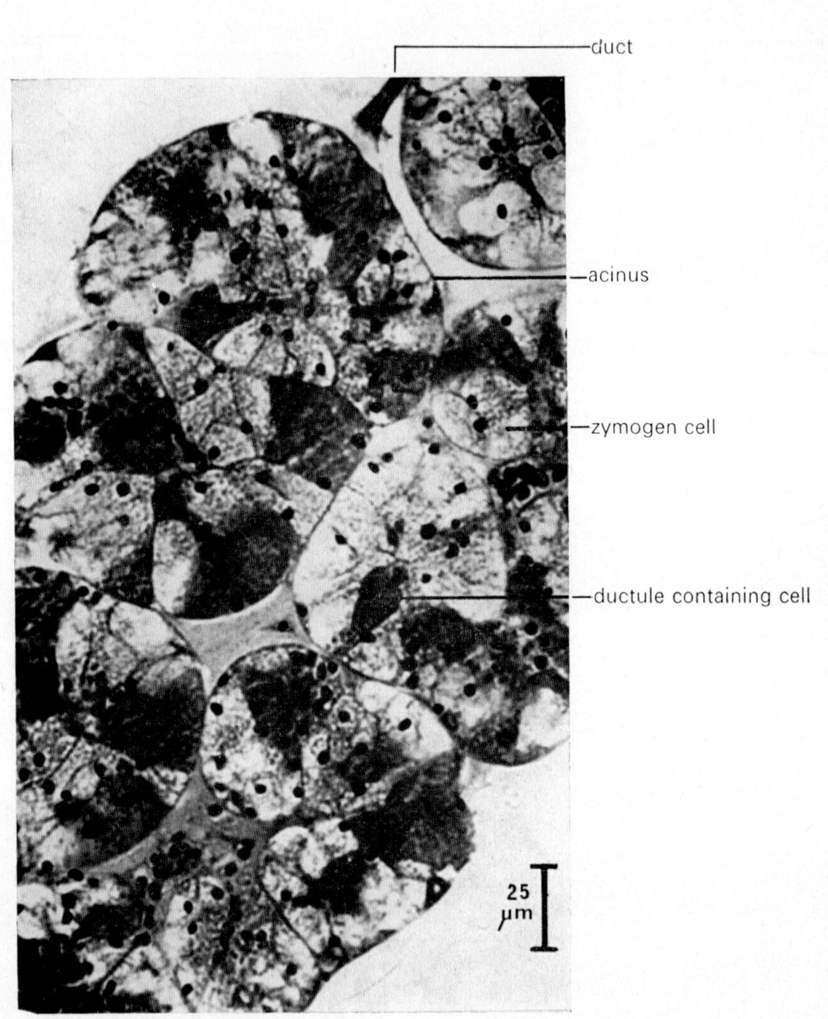

FIG. 6.30. Section through the salivary gland of *P. americana*.

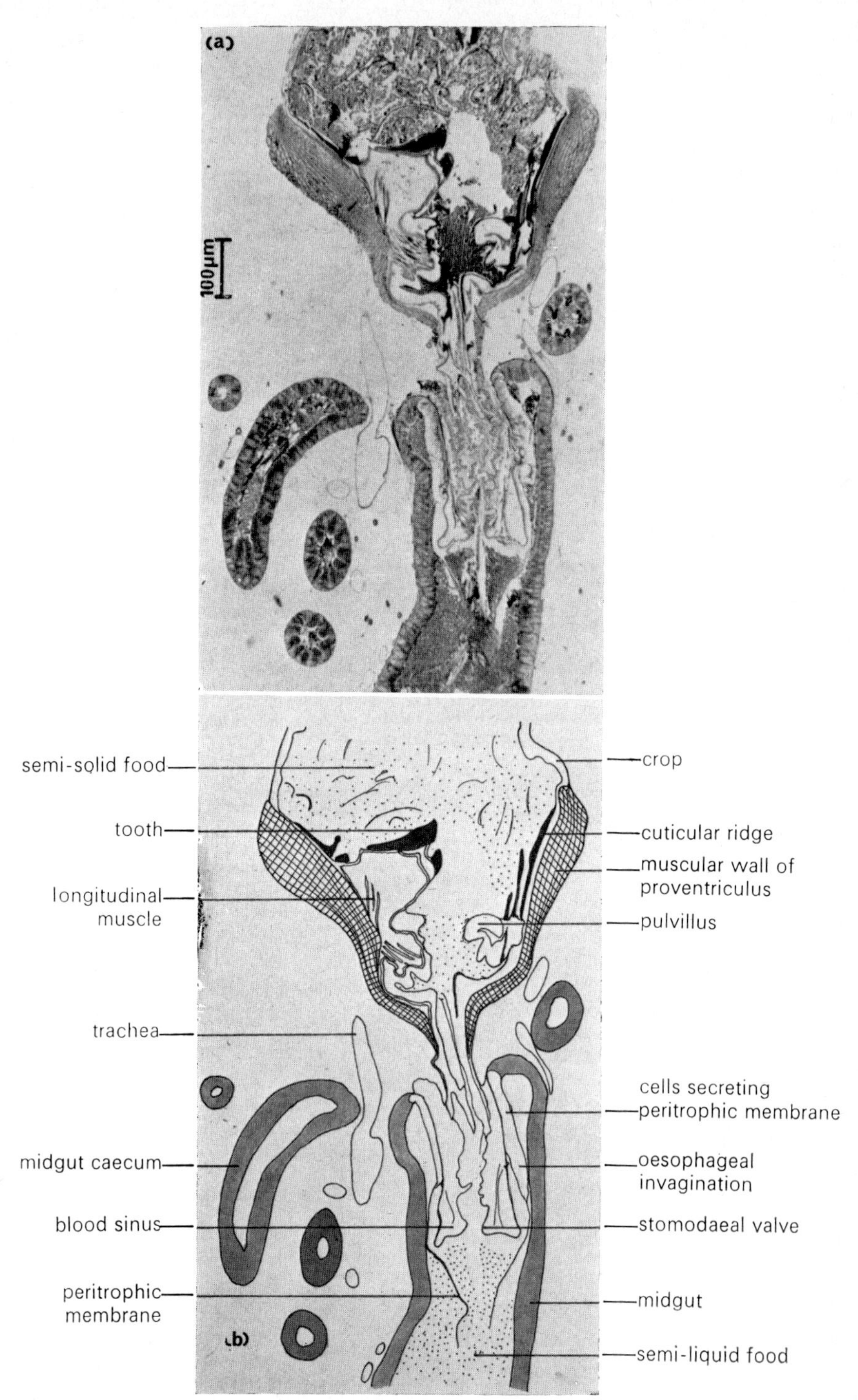

FIG. 6.31 (a) LS and (b) drawing of part of the digestive tract of *P. americana* showing the crop, proventriculus, midgut and midgut caecae.

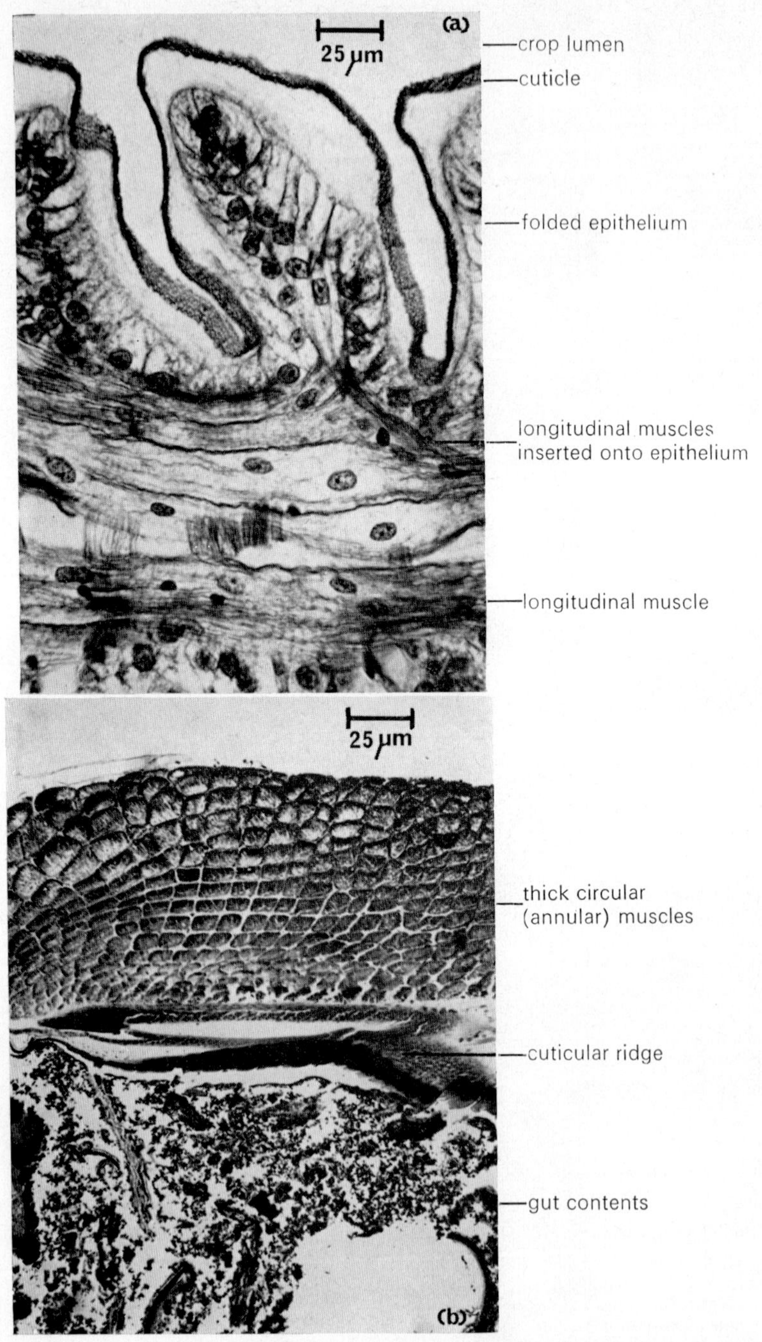

FIG. 6.32 (a) Detail of the crop wall of *P. americana*; (b) detail of the proventricular wall of *P. americana*.

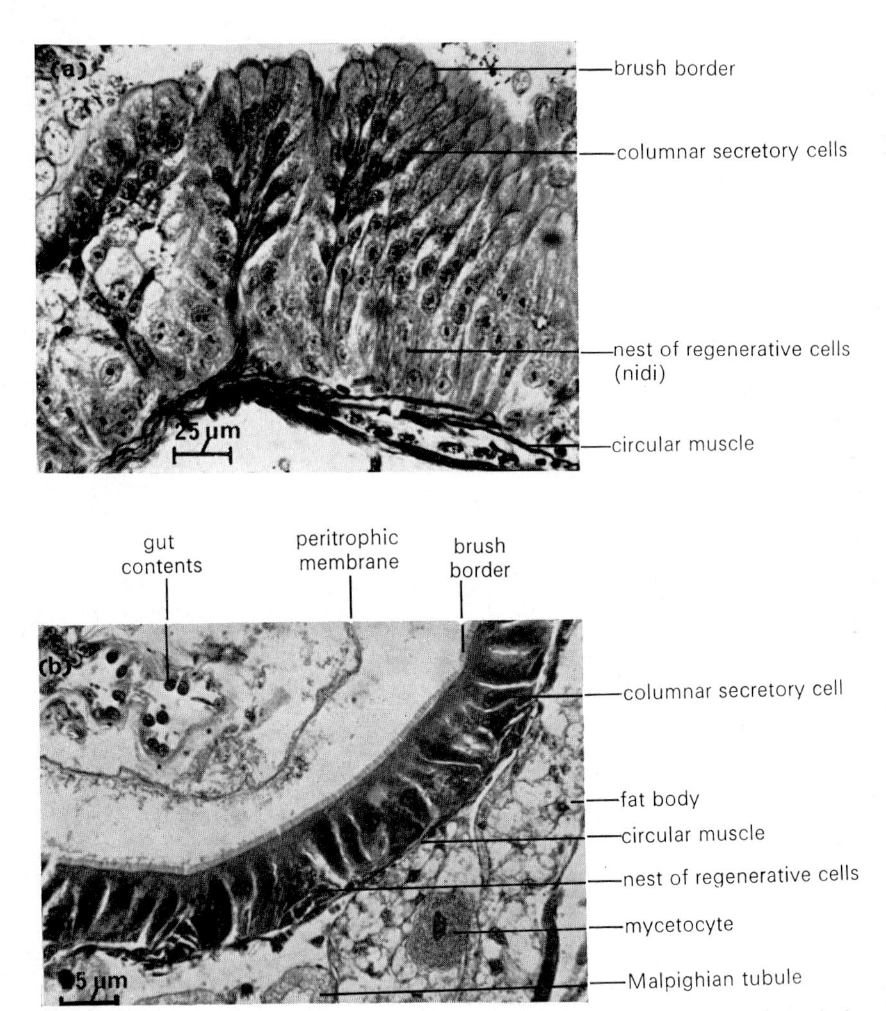

FIG. 6.33 (a) TS through the wall of a midgut caecum of *P. americana*; (b) TS through the midgut wall of *P. americana* showing the brush border and peritrophic membrane.

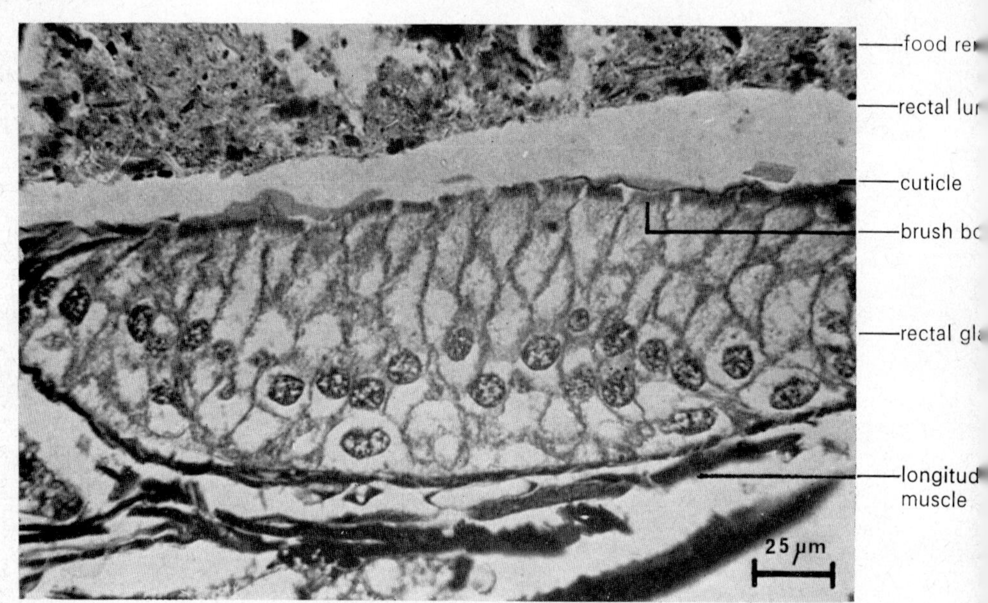

FIG. 6.34. LS of the rectal glands lining the rectum of *P. americana*.

stretching [Figs 6.31 and 6.32(a)]. Food is held in the crop while a certain amount of extracellular digestion occurs. Lipid can be absorbed directly across the crop walls. Food passes from the crop to the proventriculus (gizzard), where trituration takes place. This region [Figs 6.31 and 6.32(b)] is armed with ridges and teeth, and has a thick wall of annular muscles and a filter system for allowing small particles to pass on to the mid-gut for digestion. The sclerotized portion of the proventriculus (armarium) has six cuticular folds each armed with a tooth for crushing food and a pulvillus bearing hairs which forms a fine filter. The posterior part acts as a soft valve to allow liquids and small particles down a channel into the mid-gut, which is the chief site of digestion. There is no digestive gland in insects.

The mid-gut, ventriculus or stomach, is a tubular structure bearing eight caecae at its anterior end (Fig. 6.31). The structure of this region in the cockroach is characteristic of that in many of the more primitive insects. The columnar epithelial lining, which secretes digestive enzymes, is thrown up into evaginations with nests (nidi) of regenerative cells at their bases [Fig. 6.33(a)]. The degenerating epithelial cells are gradually replaced by the regenerative cells. The mid-gut epithelium [Fig. 6.33(b)] is protected against hard food by a peritrophic membrane which is a transparent layer of cuticle-like material formed continuously by a group of epithelial cells at the anterior end of the mid-gut (see Fig. 6.31). It is permeable to enzymes and digested food, and is shed at regular intervals. Bacteria are often found in the mid-gut which aid cellulose digestion. The muscle layers of the mid-gut (inner circular and outer longitudinal) are reversed with respect to those of the fore-gut, and are very thin.

The hind-gut is lined with cuticle-covered epithelium and possesses muscular walls for expelling the dry faeces. This is the region where water is absorbed, a function that is chiefly carried out by the rectal glands in the terminal portion. The rectal glands (Fig. 6.34) are six longitudinal thickenings of the epithelial lining of the rectum which project into the lumen and which function in reabsorption of water, salts and fats from the faeces.

Excretion

At the junction between the mid-gut and hind-gut are located large numbers of fine tubules of ectodermal origin (see Fig. 6.29). These are the Malpighian tubules which are the chief excretory organs in insects. In the cockroach there are six groups, each containing 15–20, of very long blind-ended tubules lying freely in the haemocoel. They are lined by cuboidal epithelial cells with prominent nuclei and a brush border (Fig. 6.35). Malpighian tubules work on a secretory principle rather than by ultrafiltration. They undergo peristalsis and actively absorb water, salts, amino acids and nitrogenous products from the blood. Much of the water

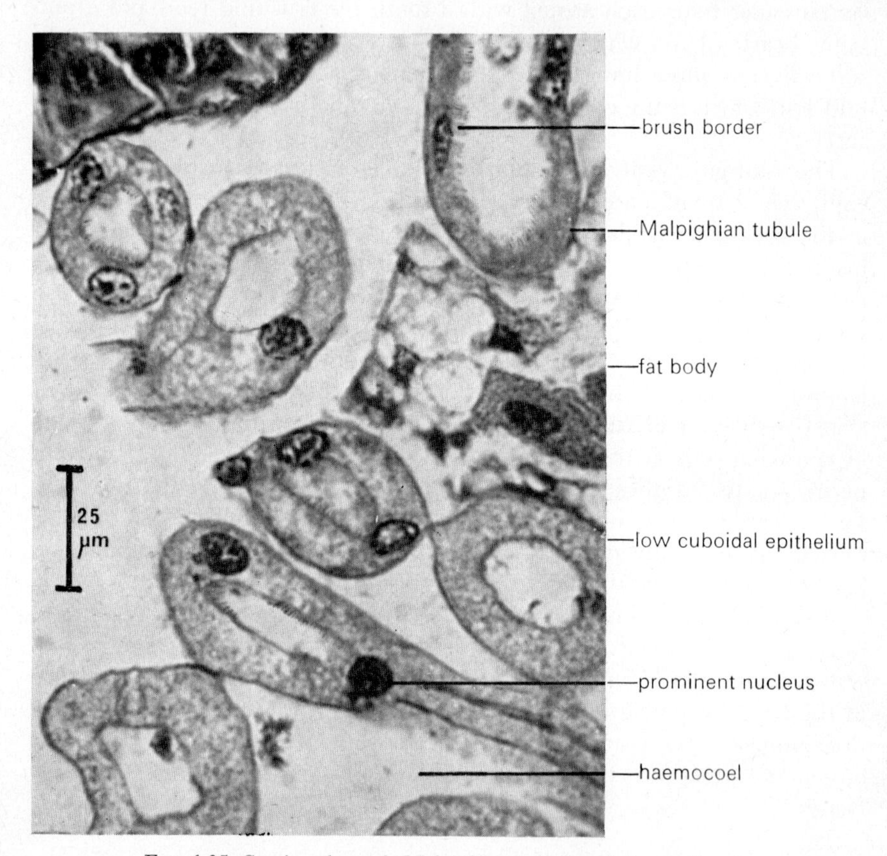

FIG. 6.35. Section through Malpighian tubules of *P. americana*.

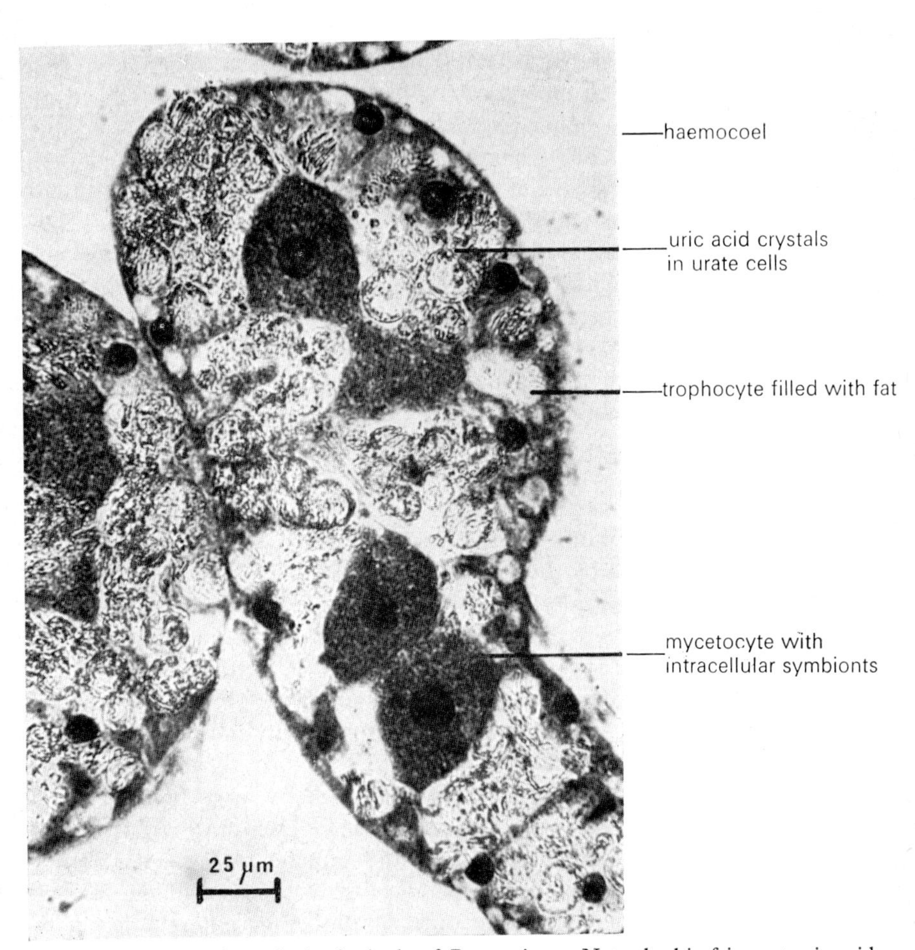

FIG. 6.36. Section through the fat body of *P. americana*. Note the birefringent uric acid crystals and the dark myetocytes filled with intracellular symbionts.

and some dissolved substances can be reabsorbed at the proximal ends of the tubules, while yet more reabsorption occurs across the rectal walls. In some species living in dry environments, the Malpighian tubules can be seen to lie close to the rectum and it is suggested that a counter-current concentrating system is set up. In addition to their excretory function, the Malpighian tubules may well play a part in nutrition as they often contain large amounts of vitamins.

The role of the cockroach Malpighian tubules in nitrogen excretion is probably secondary to their osmoregulatory function. The other structure which may play a major part in excretion is the fat body associated with the gut. The fat body has a number of functions associated with metabolism, and is composed of irregular lobes of vacuolated cells (trophocytes) which are packed with fat, protein and glycogen, and are therefore chiefly concerned with food storage (Fig. 6.36). Scattered among the trophocytes are the yellow urate cells, specialized for the formation and storage of insoluble urates and uric acid. The cockroach fat body is characterized by the occurrence of dark cells (mycetocytes) filled with symbiotic rod-shaped bacteroids. The symbionts are thought to be involved in sulphur incorporation and to metabolize uric acid formed in the urate cells.

Reproduction

The sexes are separate in the majority of insects, and the reproductive tracts are characterized by the possession of one or more accessory glands.

A plan of the male reproductive system of the cockroach is shown in Fig. 6.37(a). The paired lateral testes, which are embedded in fat body, are composed of masses of long seminiferous tubules or follicles (Fig. 6.38). There is progressive gamete development along each follicle from an area containing germ cells, through zones of growth, division and reduction to the zone of transformation where spermatozoa are formed. From the spermatocyte stage the germ cells are nourished by large testicular cyst cells. Ripe sperm are released into paired vasa deferentia which unite and lead into the ejaculatory duct and to the outside by a penis. Large, lobulated accessory ("mushroom") glands (Fig. 6.39) and a single, ventral, conglobate gland (Fig. 6.40) contribute a number of different materials involved in formation of the spermatophores. Seminal vesicles are present as small sacs on the ventral surface of the ejaculatory duct.

In the female [Fig. 6.37(b)] paired ovaries occur, each composed of eight panoistic ovarioles (chains of developing ova with no nutritive cells) united posteriorly into a calyx [Fig. 6.41(a)]. This is a primitive condition: many other insects show the more advanced meroistic condition (with nutritive cells). Each ovariole is divided into an anterior germarium containing oogonia, and a posterior vitellarium containing developing oocytes and mature ova. The oocytes [Fig. 6.41(b)] are surrounded by follicular

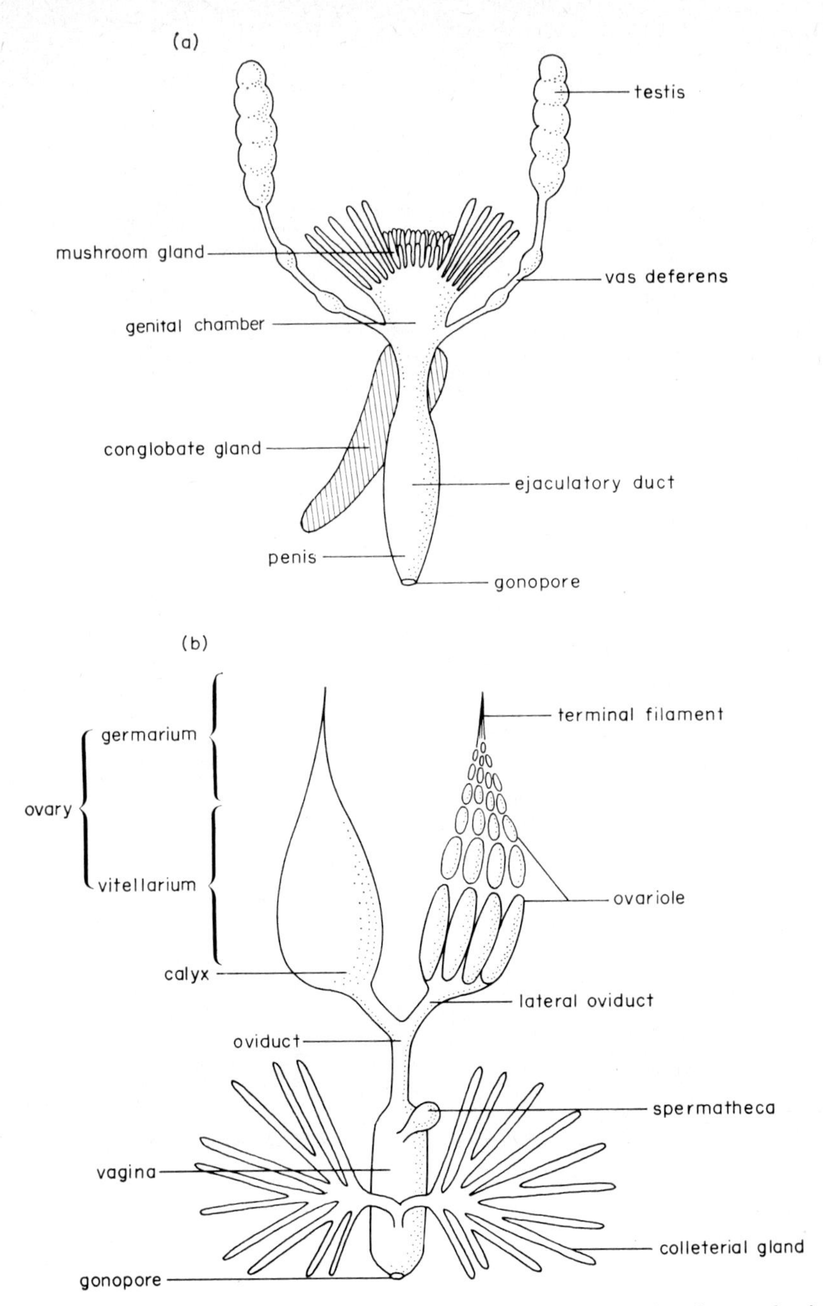

FIG. 6.37 (a) Plan of the male reproductive organs; and (b) plan of the female reproductive organs of the cockroach.

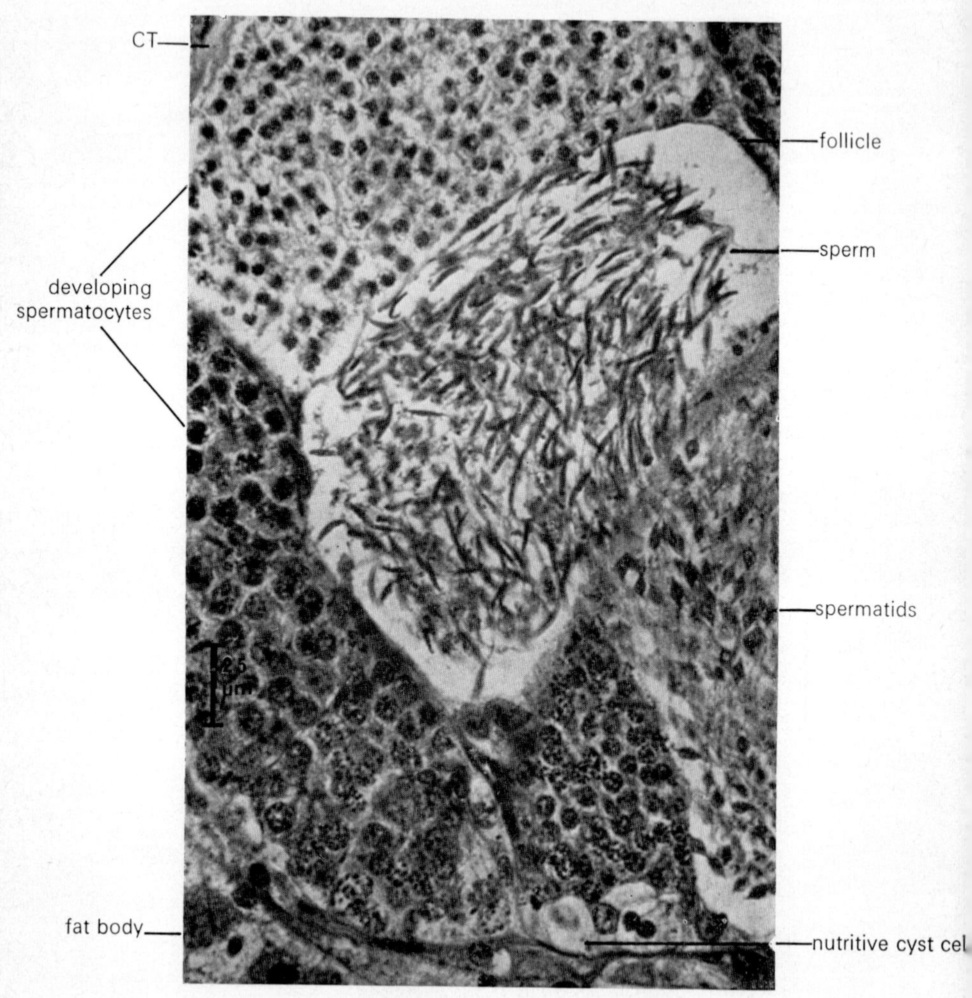

FIG. 6.38. Section through the testis of an adult male *P. americana*, showing different stages of spermatogenesis.

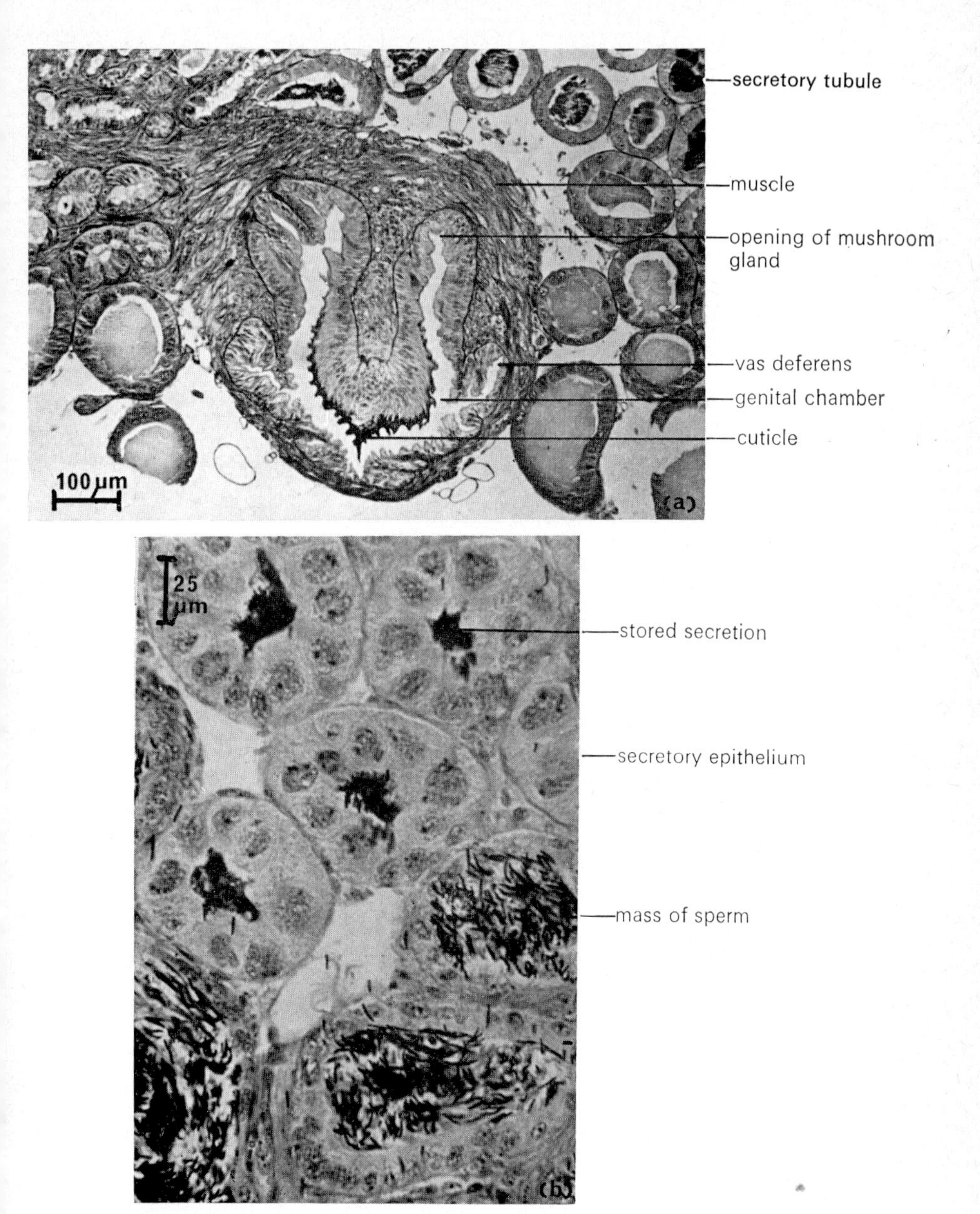

FIG. 6.39(*a*) Section through the base of the accessory ("mushroom") gland of a male *P. americana*, showing the cuticle-lined genital chamber and several different types of tubules; (*b*) details of two types of tubule, containing stored secretions and sperm.

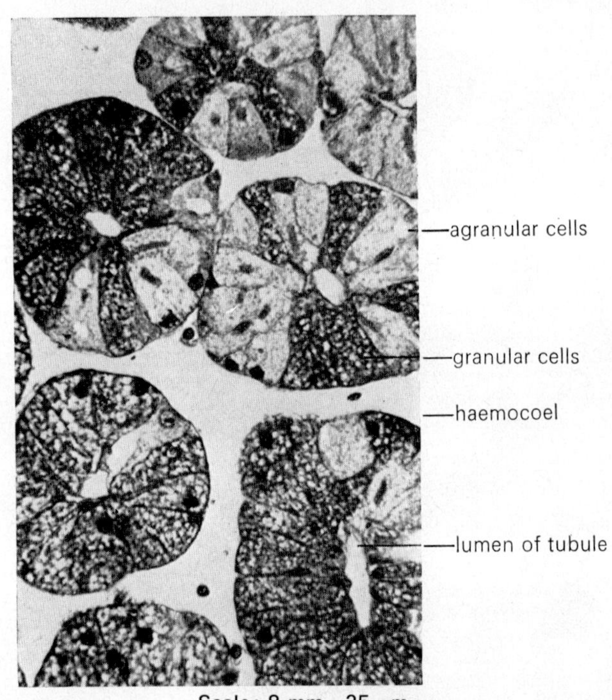

FIG. 6.40. TS through secretory tubules of the conglobate gland of a male *P. americana*.

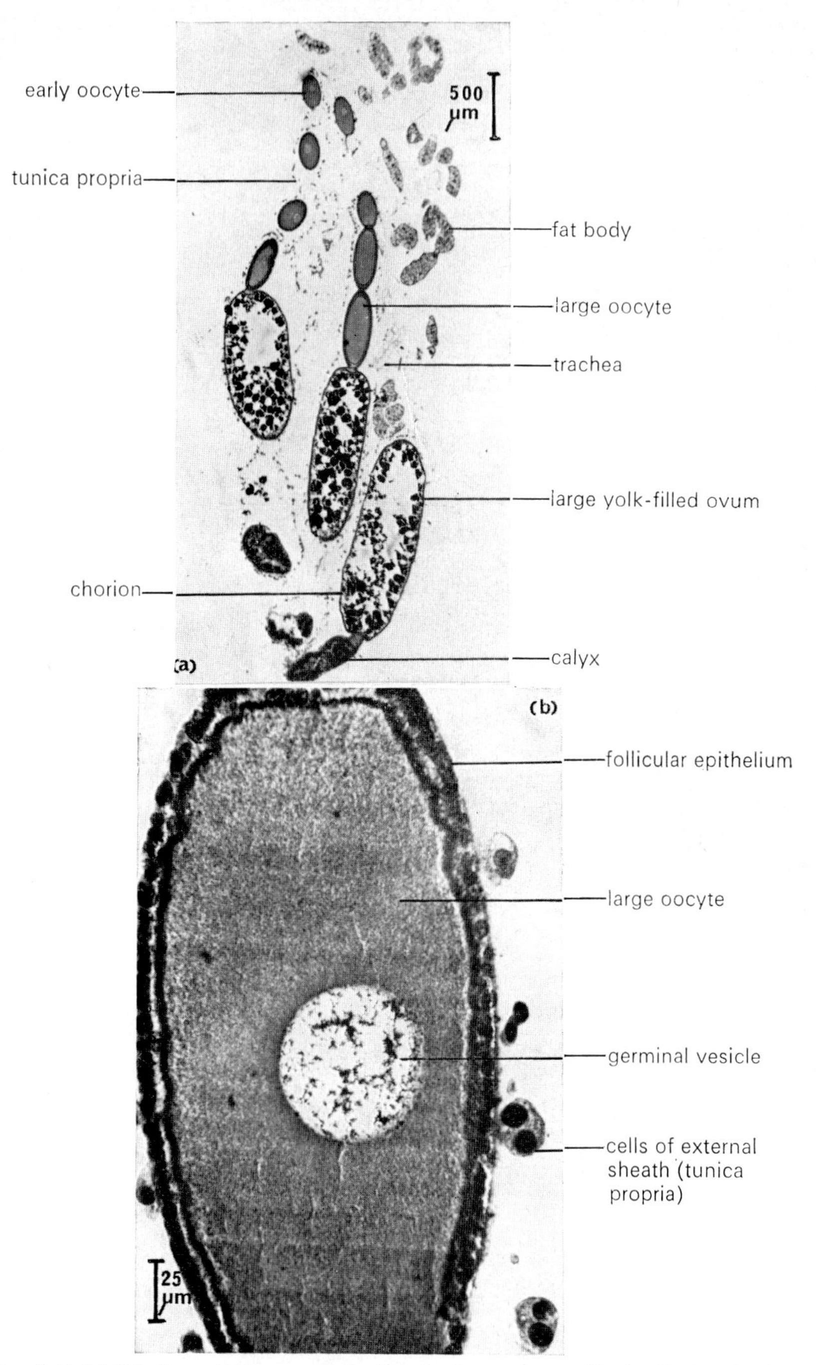

Fig. 6.41 (a) LS through the ovary of a female *P. americana*; (b) detail of a developing panoistic ovum of *P. americana*.

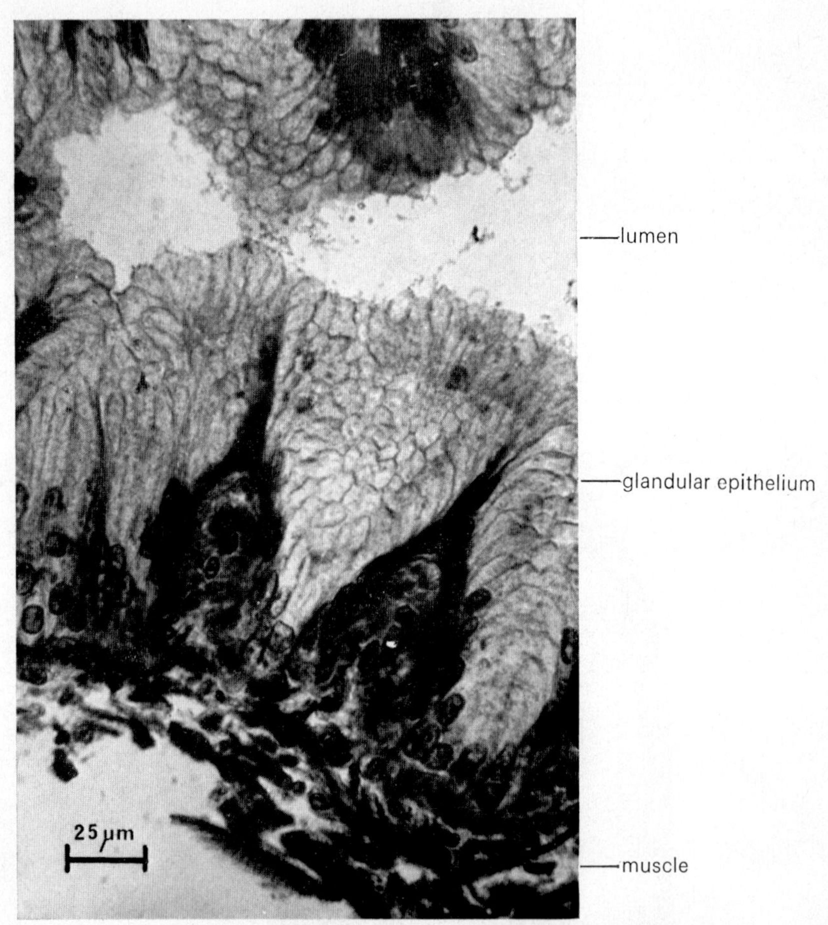

Fig. 6.42. Section through the oviduct wall of a female *P. americana*.

epithelium which secretes materials for yolk formation as well as the egg shell (chorion). The mature ova are large, yolk-filled cells surrounded by an egg-shell which has a small canal (micropyle) through which fertilization can be effected. Ova are shed into paired lateral oviducts (Fig. 6.42) which join in a common oviduct leading into a vagina which opens on the ventral side of the eighth abdominal segment. A small cuticle-lined spermatheca opens by a short neck into the dorsal wall of the vagina and stores sperm received from the male. Branched accessory or colleterial glands (Fig. 6.43), concerned with the formation of a hard egg-case (ootheca), also open into the vagina. The two sides of the gland secrete different components (carbohydrates, enzymes and other proteins), which, when mixed, form a tanned protein of which the ootheca is composed.

Fertilization is internal, with the male introducing spermatophores into the vagina, close to the spermathecal duct. Eggs are fertilized as they pass the spermatheca on their way to the vagina, and are then covered by the ootheca which is closed up before they are laid. Development of the cockroach is holometabolous (i.e. via a number of nymphal instars) while in other insects it may be hemimetabolous (i.e. via larval and quiescent pupal stages).

Class Arachnida

Arachnids are generally terrestrial arthropods, although some (e.g. *Limulus*) are secondarily aquatic. They are characterized by the division of a basically segmented body into an anterior prosoma (cephalothorax) consisting of six segments, and a posterior opisthosoma comprising up to thirteen segments and a telson. The prosoma bears six pairs of appendages; the first (preoral) pair of which are chelicerae, the second sensory or prehensile pedipalps and the remaining four pairs are walking legs. The opisthosoma has few appendages and possesses intucked respiratory structures in the form of lung-books and/or tracheae. The exoskeleton is completely chitinized. The sexes are separate in arachnids, and the genital orifices occur on the opisthosoma some distance from the terminal anus.

Example: *Epeira* (*diadematus*)

Epeira belongs to the largest order of arachnids, the Araneida (spiders). It is a commonly occurring garden spider, notable for its elaborate webs spun from silk secreted by opisthosomatic silk glands. Figure 6.44 is of a VLS of *Epeira* and shows that the "head" is indicated by a transverse groove and that the prosoma is joined to the opisthosoma by a narrow pedicel. The opisthosoma is soft, patterned and slightly pointed. It is foreshortened, does not carry a telson and shows no sign of segmentation. Small chitinous spinning organs (spinnerets), probably representing modified appendages, occur on the opisthosoma just below the anus.

Spiders have a flexible waterproof cuticle covering the body surface

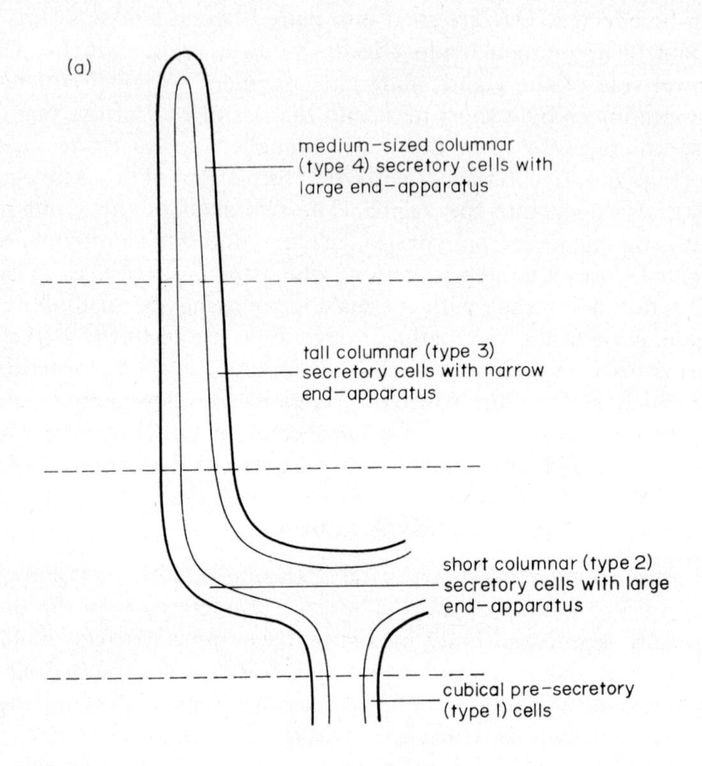
(a)
medium-sized columnar (type 4) secretory cells with large end-apparatus
tall columnar (type 3) secretory cells with narrow end-apparatus
short columnar (type 2) secretory cells with large end-apparatus
cubical pre-secretory (type 1) cells

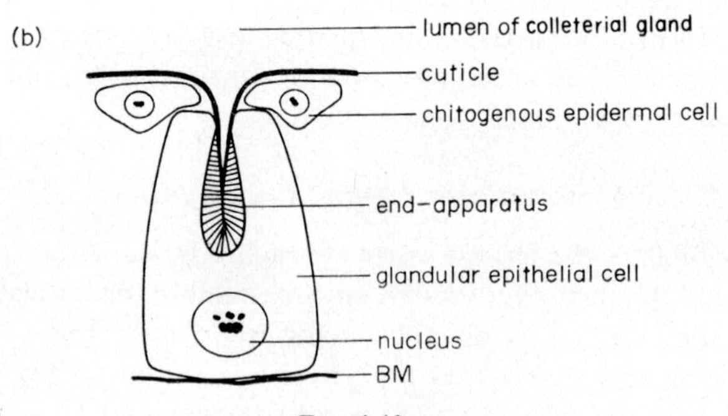
(b)
lumen of colleterial gland
cuticle
chitogenous epidermal cell
end-apparatus
glandular epithelial cell
nucleus
BM

FIG. 6.43

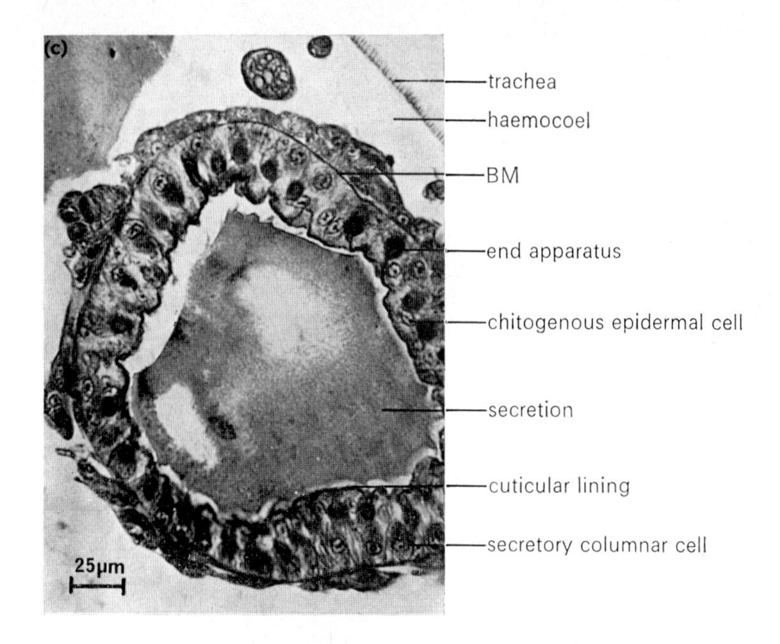

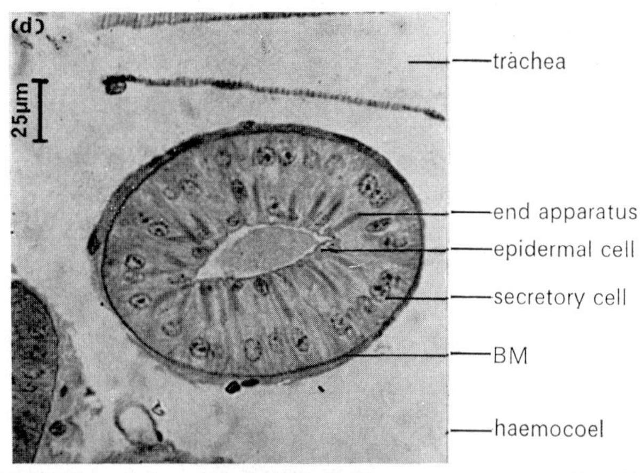

FIG. 6.43 (a) Plan of the arrangement of cell types in the left colleterial gland of a female cockroach. The different types of cell secrete the various components involved in spermatophore formation. The smaller right gland is composed of only two types of cell; (b) diagram of a single generalized secretory cell from the cockroach colleterial gland; (c) TS through the proximal end of a secretory tubule from the colleterial gland of *P. americana*. Note the wide tubule, low columnar (type 2) secretory cells and the wide dark end-apparatus; (d) TS through the middle region of a colleterial gland secretory tubule from *P. americana*. Note the small tubule, the tall (type 3) secretory cells and narrow end-apparatus.

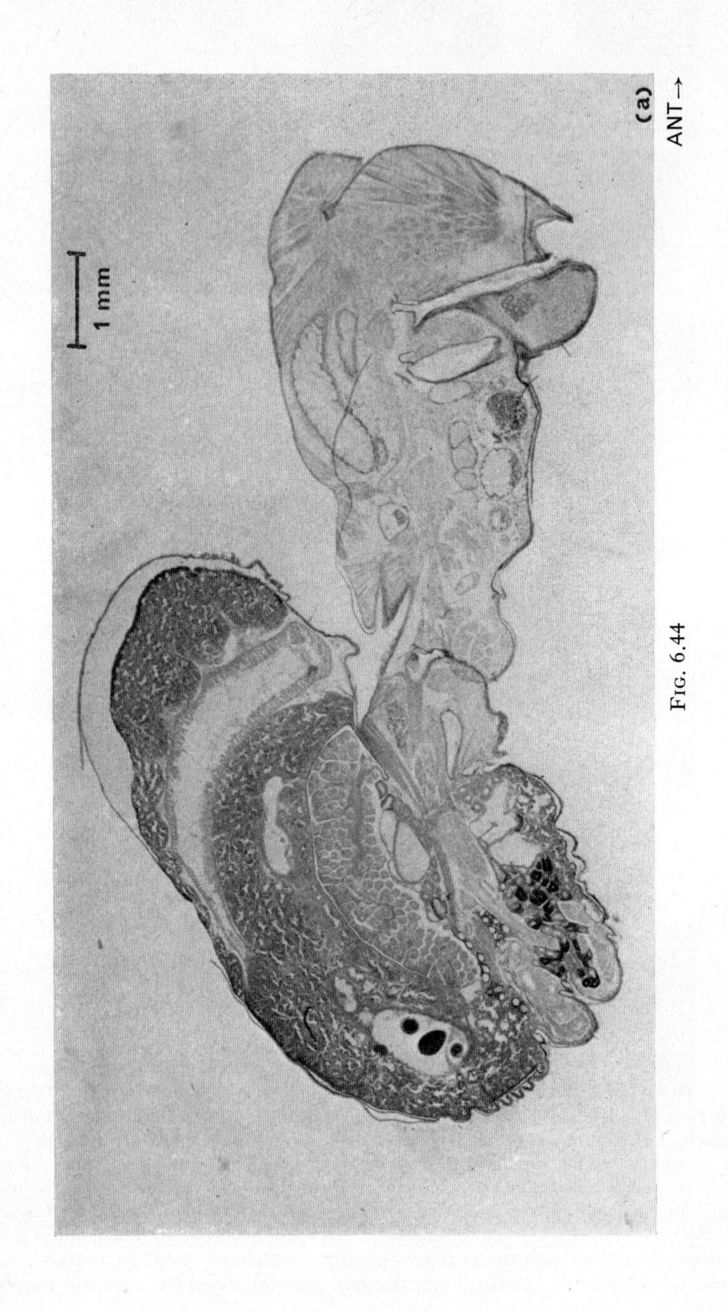

FIG. 6.44

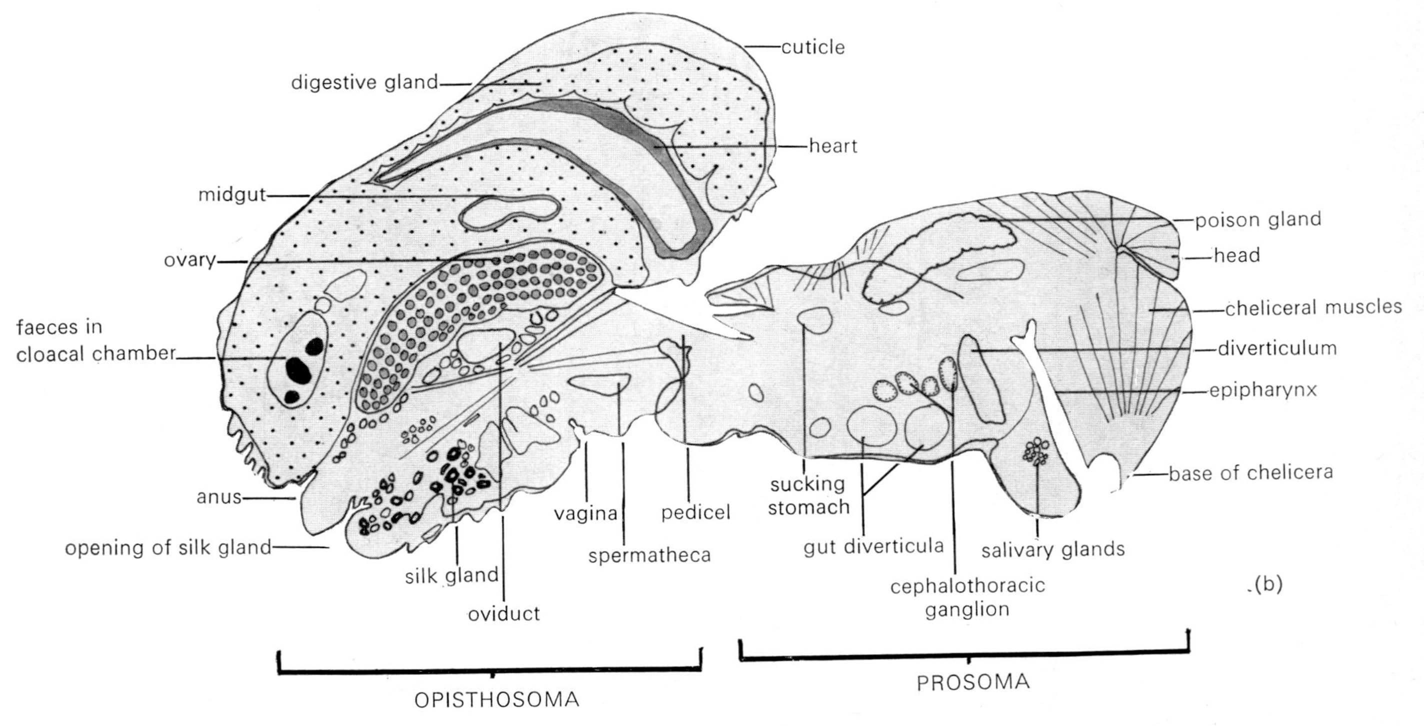

FIG. 6.44 (a) VLS and (b) explanatory drawing of an adult female spider (*Epeira diadematus*).

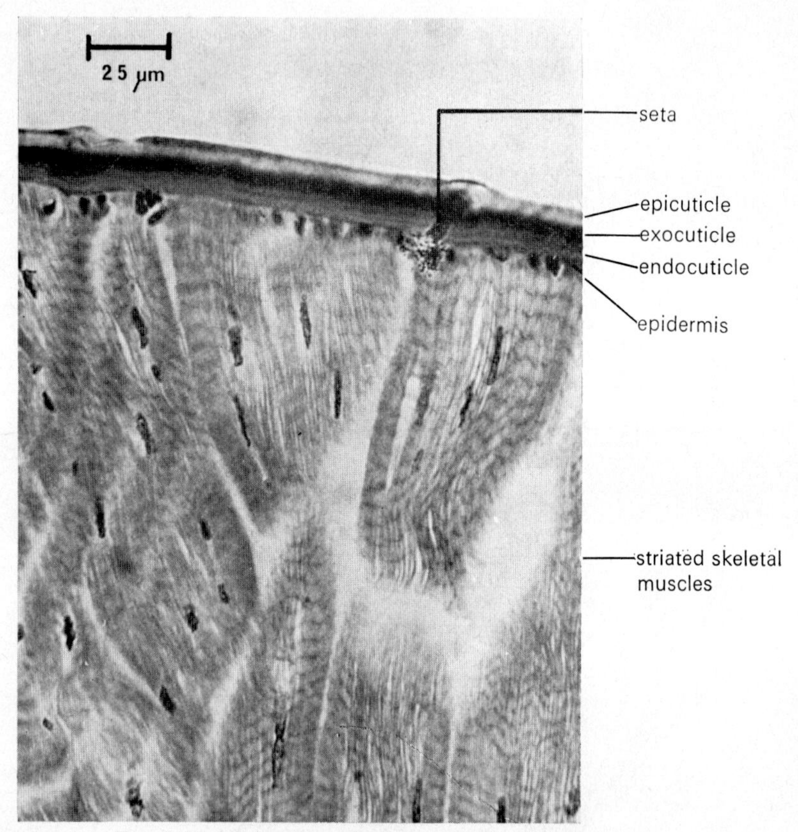

FIG. 6.45. Section through the body wall of *E. diadematus*.

and serving as a support for the extensive body muscles (Fig. 6.45). It is thinner than that of other arthropods, but the epicuticle contains a layer of wax; a feature which reduces the risk of desiccation.

Nervous system

The basic plan of arthropod nervous systems is modified in spiders. In *Epeira* the cerebral ganglia are linked by short connectives to a large cephalothoracic ganglion, consisting of suboesophageal ganglia fused with thoracic and abdominal ganglia which have moved forward during development (Fig. 6.46). Although there are separate ganglia associated with the digestive system, there is no ventral nerve cord or associated ganglia in the opisthosoma. The cerebral ganglia or "brain" lacks a deuterocerebrum, a feature probably associated with the absence of antennae.

Sense organs

Spiders are active predators and as such are well supplied with sense organs. These include tactile setae and spines all over the body surface, acoustic hairs (trichobothria) on the legs and pedipalps, chemoreceptive tarsal organs on the legs, slit-like lyriform organs which are widely distributed, and eyes. The eyes are atypical among arthropod eyes in being simple ocelli, not compound eyes. They are generally only used for light appreciation, although the eyes of some hunting spiders are capable of image formation. *Epeira* has simple indirect-type eyes with a cuticular cornea and lens, and photoreceptor cells backed by a pigmented membrane (Fig. 6.47).

Respiration

Respiratory organs in arachnids include "gill books", but the more advanced members have characteristic lung-books and tracheae. The tracheae appear to have evolved from lung-books, both being ectodermal invaginations. *E. diadematus* possesses posterior tracheae and a pair of lung-books on the anterior ventral side of the opisthosoma. The lung-books consist of sunken pockets containing stacks of parallel leaflets (lamellae) formed from the pocket wall [Fig. 6.48(a and b)]. The inside of the pocket and edges of the lamellae are cuticularized [Fig. 6.48(c)] while the thin, vascularized lamellae [Fig. 6.48(d)] are held apart by cuticular spines, thus allowing air to circulate freely amongst them. Air is drawn into the cavity and through the leaflets by the action of a muscle attached to the dorsal side of the cavity. The tracheae found in spiders are similar to those of insects, with spiracles which can be opened and closed.

Circulation

A powerful pump is required to circulate blood through the numerous narrow vascular channels in the lamellae of the lung-books. The dorsal

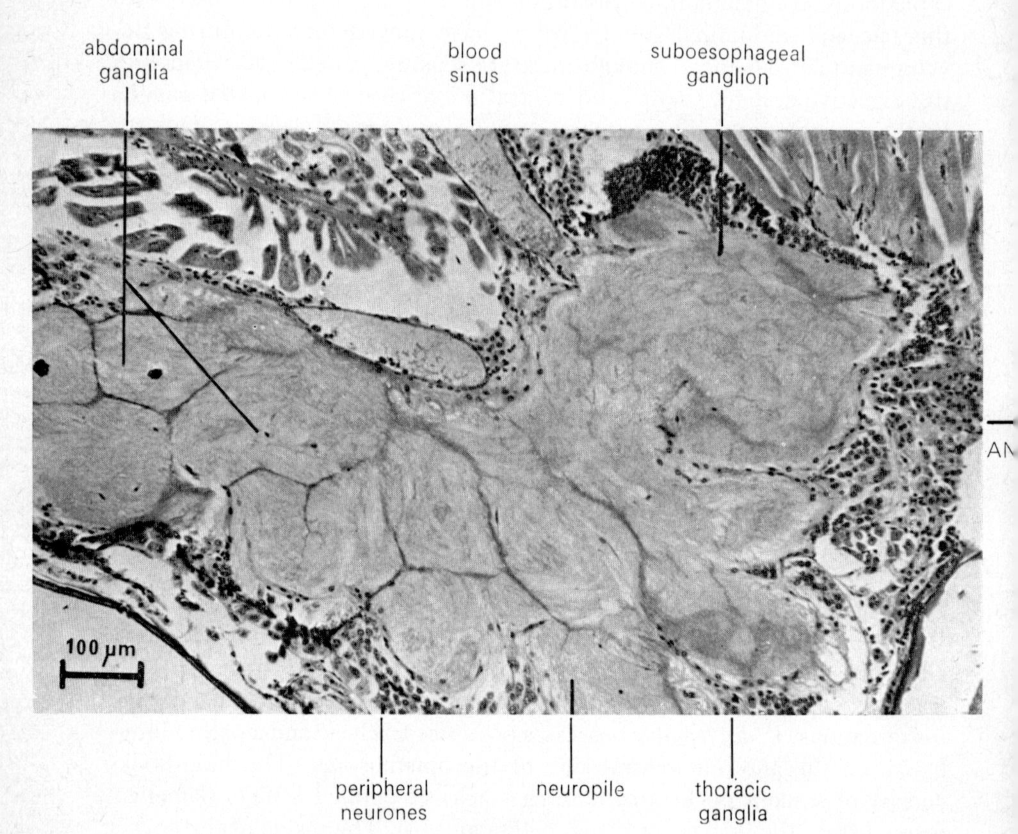

FIG. 6.46. VLS through the cephalothoracic ganglion of *E. diadematus*.

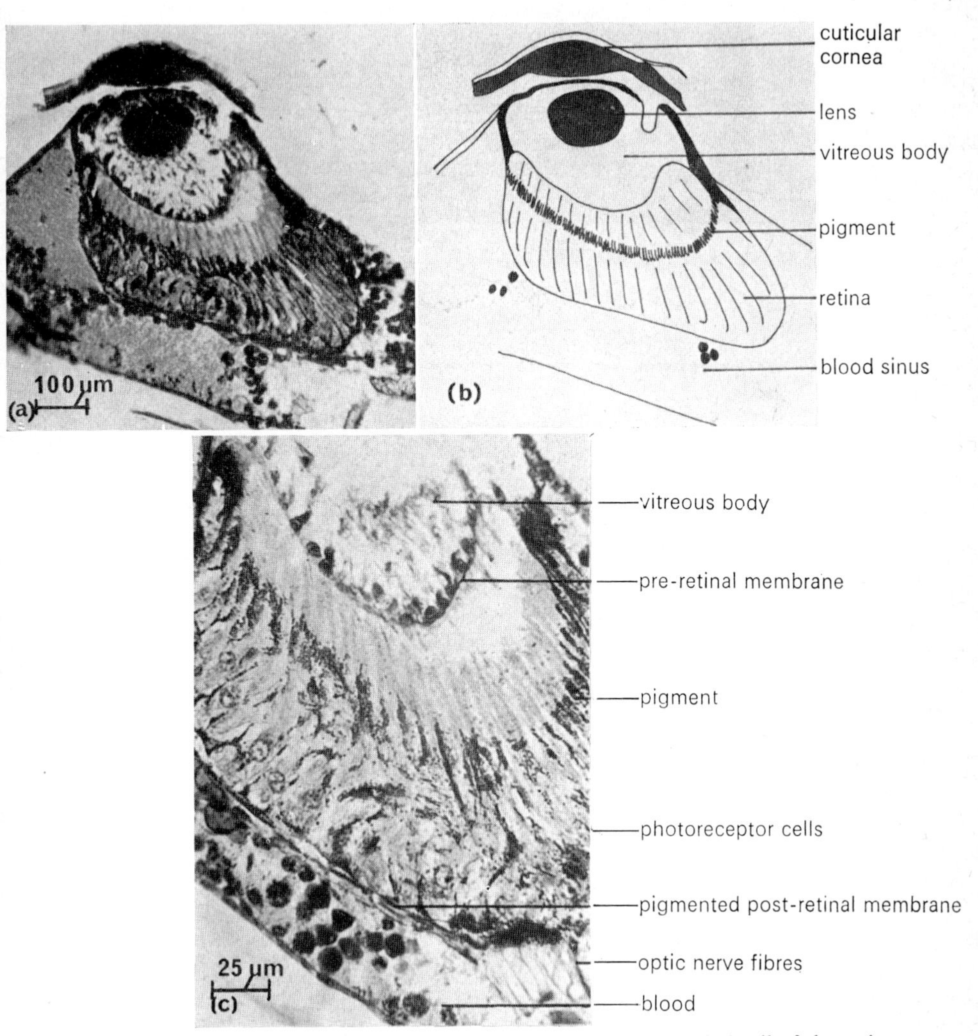

FIG. 6.47 (a) Section and (b) drawing of the eye of *E. diadematus*; (c) detail of the retina from the spider eye.

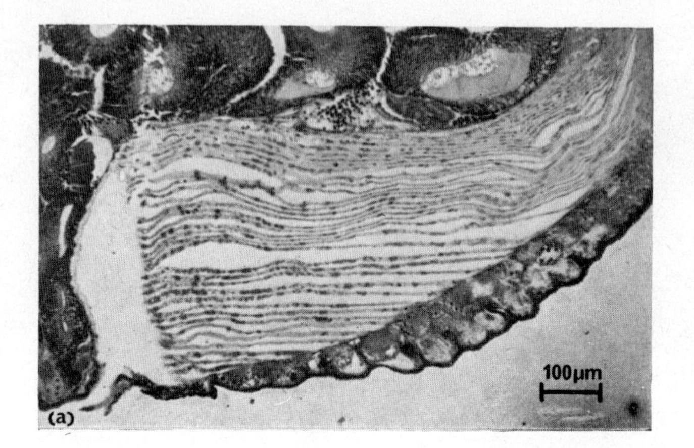
100µm
(a)
→ANT

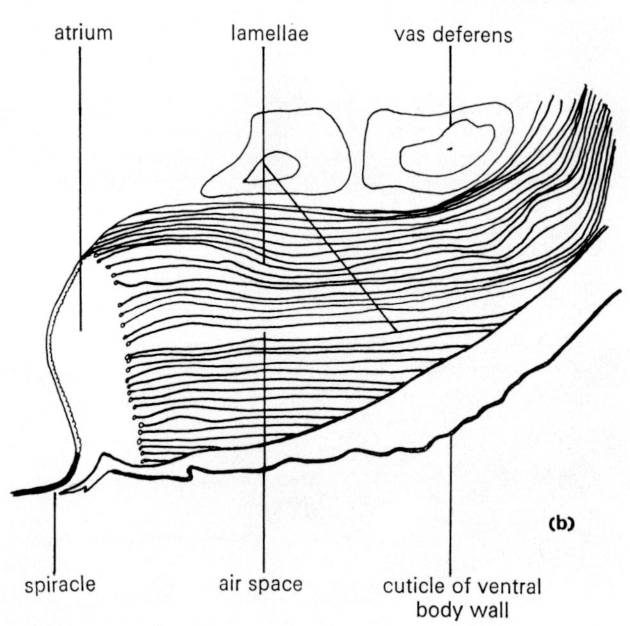
atrium
lamellae
vas deferens
spiracle
air space
cuticle of ventral body wall
(b)

Fig. 6.48

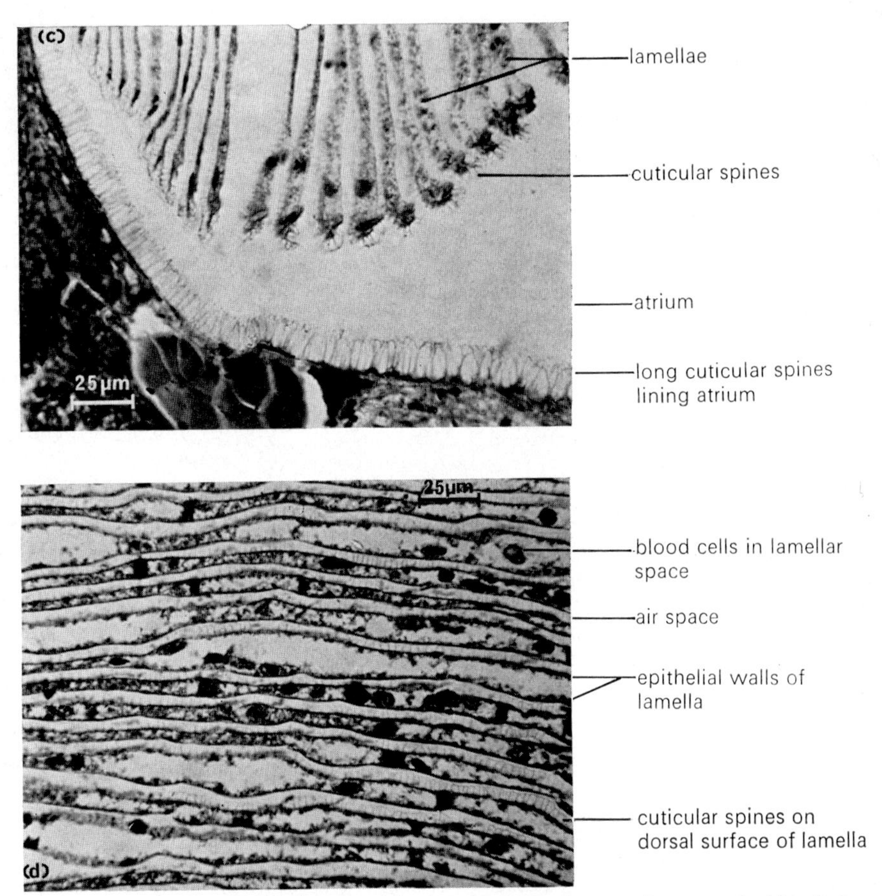

FIG. 6.48 (a) VLS and (b) drawing of a lungbook on the anterior ventral side of the opisthosoma of *E. diadematus*; (c) details of the cuticular spines lining the atrium and lamellae of the spider lungbook; (d) details of the lungbook lamellae of *E. diadematus*.

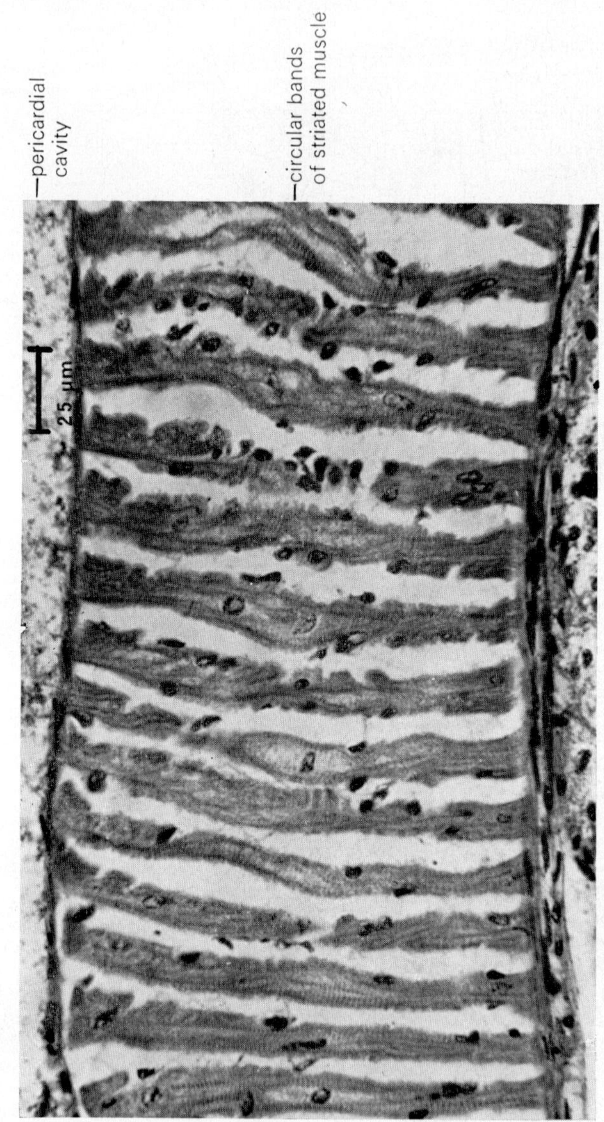

FIG. 6.49. VLS through the heart of *E. diadematus*. This section is cut towards the side of the heart where the thick bands of circular muscle are continuous.

tubular heart of spiders is, therefore, large and provided with strong circular muscle bands (Fig. 6.49). A pair of abdominal arteries leave the heart in each segment, while a large anterior aorta supplies the cephalothorax and a small posterior aorta runs to the hind end of the opisthosoma. Blood from the tissues collects in a large ventral sinus, flows into the pericardium and thence into the heart via ostia.

Digestion

Spiders are generally carnivorous and much of the digestion process occurs externally. The chelicerae bear ducts from poison glands, consist of only two segments, and are for piercing rather than biting. Prey is captured in the silken web and immobilized by injection of poison from the poison gland [Fig. 6.50(a)]. It is manipulated by the limbs and the upper (rostrum) and lower (labium) lips, and torn apart by the chelicerae. Extracellular enzymes are poured out onto the food from the salivary glands in the epipharynx [Fig. 6.50(b)]. Spiders are thus really liquid feeders as they suck up the liquified food, and filter off large particles by a sieve of stiff hairs on the mouth-parts and in the hypopharynx [Fig. 6.50(c)]. Partially digested food is sucked through the expansible pharynx and narrow oesophagus into the "sucking stomach" which is actually a modified end of the oesophagus and therefore also chitin-lined as well as provided with powerful muscles. Digestion occurs in the mid-gut, the surface area of which is increased by numerous finely branching diverticula forming a digestive gland that occupies most of the opisthosoma (see Fig. 6.44). The digestive gland [Fig. 6.50(d)] is lined with secretory and absorptive epithelium, and acts as a digestive and a storage organ, allowing spiders to exist for months or years without food. The hind-gut is similar to that of other arthropods except that it possesses an evagination forming a cloacal chamber in which faeces can accumulate (see Fig. 6.44). The anus is closed by a sphincter muscle.

Excretion

The excretory organs present in spiders are Malpighian tubules and coxal glands. The Malpighian tubules are similar to those of insects (see Fig. 6.35) except that they are of mesodermal origin and open into the posterior end of the mid-gut. They ramify among the organs of the opisthosoma absorbing waste material from the haemocoel which they excrete as insoluble guanine, a compound related to uric acid. The coxal glands are derived from coelomoducts and are very like those of onychophorans (see Fig. 5.12). In *Epeira* there is only one pair of simple coxal glands opening at the bases of the first pair of legs. Each one possesses a coelomic end-sac and straight tubule but has no storage bladder. Other coxal glands are modified as silk glands (Fig. 6.51), which secrete proteinaceous silk. This is stored in the lumen of the glands as a viscous

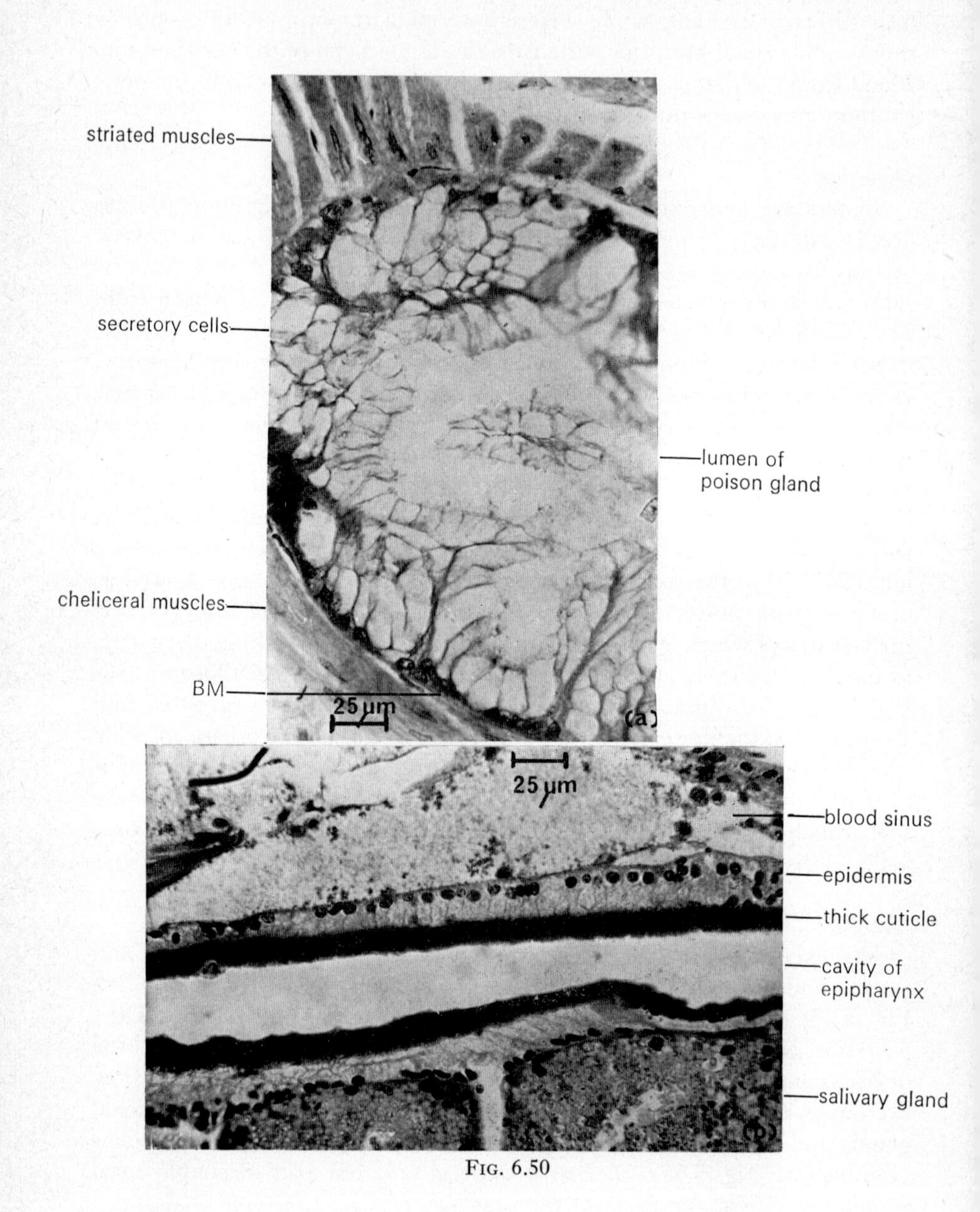
striated muscles
secretory cells
lumen of poison gland
cheliceral muscles
BM
25 µm
(a)
25 µm
blood sinus
epidermis
thick cuticle
cavity of epipharynx
salivary gland
(b)

Fig. 6.50

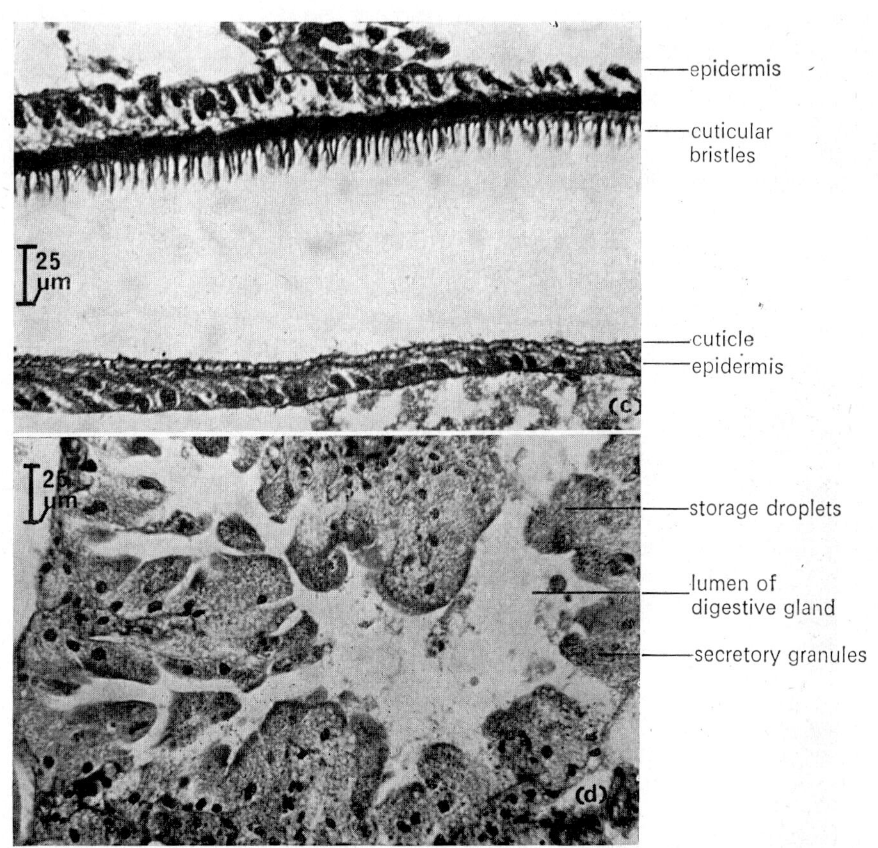

FIG. 6.50 (a) Section through the poison gland in the cephalothorax of *E. diadematus*; (b) LS through the epipharynx and salivary glands of *E. diadematus*; (c) filter of bristles on the side of the mouthparts of *E. diadematus*; (d) section of a digestive gland tubule of *E. diadematus.*

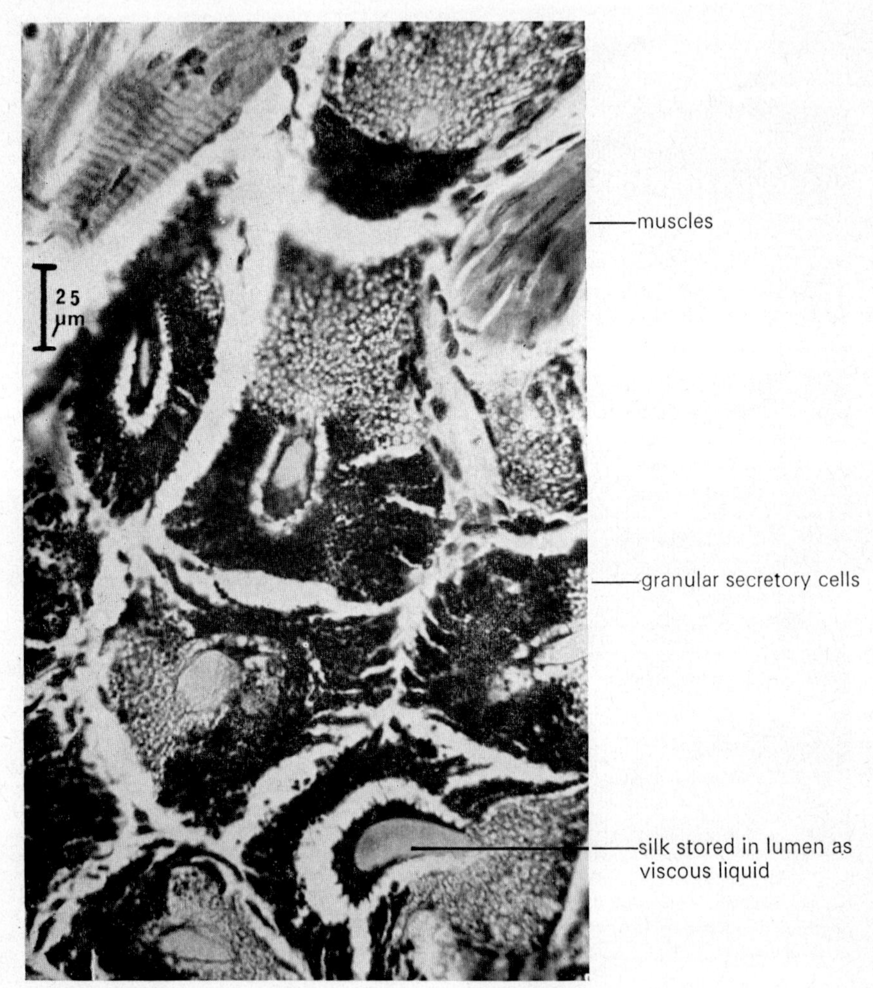

Fig. 6.51. Section through the tubular silk glands of *E. diadematus*.

liquid, and is spun out by the spinnerets, whereupon it is converted by this extension into a strong elastic thread.

Reproduction

Spiders are dioecious, and the pedipalps of the male are modified as intromittent organs. The arrangement of gonads is very variable in arachnids, but generally in spiders there is a single pair of gonads. In the male *Epeira* tubular testes lead by paired vasa deferentia (see Fig. 6.48) to a single orifice on the ventral side of the 8th body segment. In the female two sac-like ovaries packed with eggs (see Fig. 6.44) lead by paired oviducts to a single large vagina. Grooves guide the male intromittent organ into the vagina, and sperm are stored in paired spermathecae which open off the vagina. Fertilization is internal, and eggs are laid which develop directly into minute spiders.

Chapter 7

Phylum Mollusca

The Mollusca is the second largest invertebrate phylum and its members have become adapted to a wide range of environments. Molluscs are unsegmented coelomates with, usually, a well-developed head, a ventral muscular foot and a dorsal visceral hump covered by a soft skin (mantle). This mantle often secretes a partly calcareous shell from its margin, and is extended away from the visceral hump to enclose a mantle cavity into which the gut and kidneys discharge their waste. All that remains of the coelom are the interconnecting pericardial, renal and gonadal cavities. There is an open blood system with a heart, arterial system and venous system which normally expands into a haemocoel. Many molluscs use the large volume of blood in the haemocoel as a hydrostatic skeleton. The nervous system consists of a ganglionic ring round the oesophagus with separate pedal ganglia and visceral loops. Sense organs are plentiful on the head and in the mantle cavity. The gut is generally provided with a crystalline style and ciliary sorting mechanisms or a muscular buccal mass and chitin-covered tongue (radula). There is always a stomach and a digestive gland. Molluscs are classified on body shape.

Class Gastropoda

The Gastropoda is the largest molluscan class. Its members show considerable variation, and many of their characteristics are associated with the fact that they all show torsion at some stage of their development. Initially the larva is bilaterally symmetrical with an anterior visceral hump and a mantle cavity that opens posteriorly and ventrally. Torsion is the 180° anticlockwise rotation of the viscera and mantle cavity in relation to the head and foot, so that all the structures behind the "neck" are reversed. This is due to contraction of asymmetrical retractor muscles from the shell to the head/foot, and is quite independent of any spiralling of the shell and visceral hump. The mantle cavity is thus twisted to face the water currents and assumes respiratory and sensory functions. Detorsion may occur secondarily (in some Opisthobranchs). There is also unequal growth of the visceral mass into a right-handed spiral which is usually covered by a single shell, although the shell may be secondarily absent. In aquatic gastropods the anterior mantle cavity typically has paired gills although bilateral symmetry of other mantle organs is often lost due to torsion. In terrestrial species (pulmonates) the mantle is converted into a vascularized air-breathing "lung" and the gills are lost. The head of gastropods is well-developed and bears tentacles, eyes and a radula for feeding. There is a large flattened ventral foot for attachment to the substratum and for locomotion.

Example: *Helix* (*aspersa*)

H. aspersa is the common garden snail which belongs to the subclass Pulmonata, or terrestrial gastropods possessing lungs. A young specimen is shown in VLS in Fig. 7.1. The snail shows torsion and also possesses a spiral shell which affords it considerable protection on land. In the snail photographed, the shell was removed before the animal was sectioned. The snail can withdraw into the shell using the specially developed columellar (pharynx retractor) muscle, and can close the shell opening with an epiphragm of mucus hardened with calcium which is secreted by the mantle edge. Thus the shell can be effectively sealed off during periods of drought or cold. Locomotion is achieved by waves of contraction in the longitudinal muscles of the foot. The foot is lubricated by mucus produced by the pedal mucous gland which lies dorsally to it, and opens by a duct just below the mouth (Fig. 7.2.)

The head and foot of the snail are covered by an elastic skin composed of epidermis and muscle fibres, while the visceral hump is protected permanently by the shell. The mantle which lines the shell forms the roof of the mantle cavity. It is a thin, moist epidermal sheet with numerous "pulmonary" blood-vessels on its inner surface [Fig. 7.3(a)] and acts as the respiratory surface. Its anterior rim is thickened to form a glandular collar [Fig. 7.3(b)] which is involved in shell secretion.

Nervous system

The nervous system of snails develops asymmetrically as a result of torsion, but in adults it shows a secondary bilateral symmetry. Primitively in the gastropod class the nerve ganglia are all separate, but in snails the nervous system comprises paired cerebral ganglia situated dorsally to the oesophagus and joined by large circumoesophageal connectives to a sub-oesophageal ganglionic mass formed by the fusion of paired pleural, parietal and pedal ganglia and a single visceral ganglion [Fig. 7.4(a)]. The whole collection of ganglia is often called the "brain". Some of the structures mentioned can be seen in Fig. 7.4(b) which shows an oblique section through the snail "brain". Nerves run from the suboesophageal mass to the viscera, mantle and muscles of the body and foot. The ganglia are covered by a thick connective tissue sheath often containing calcium deposits. Neurones are situated peripherally around a central neuropile of axons, and very large nerve cells are common (Fig. 7.5).

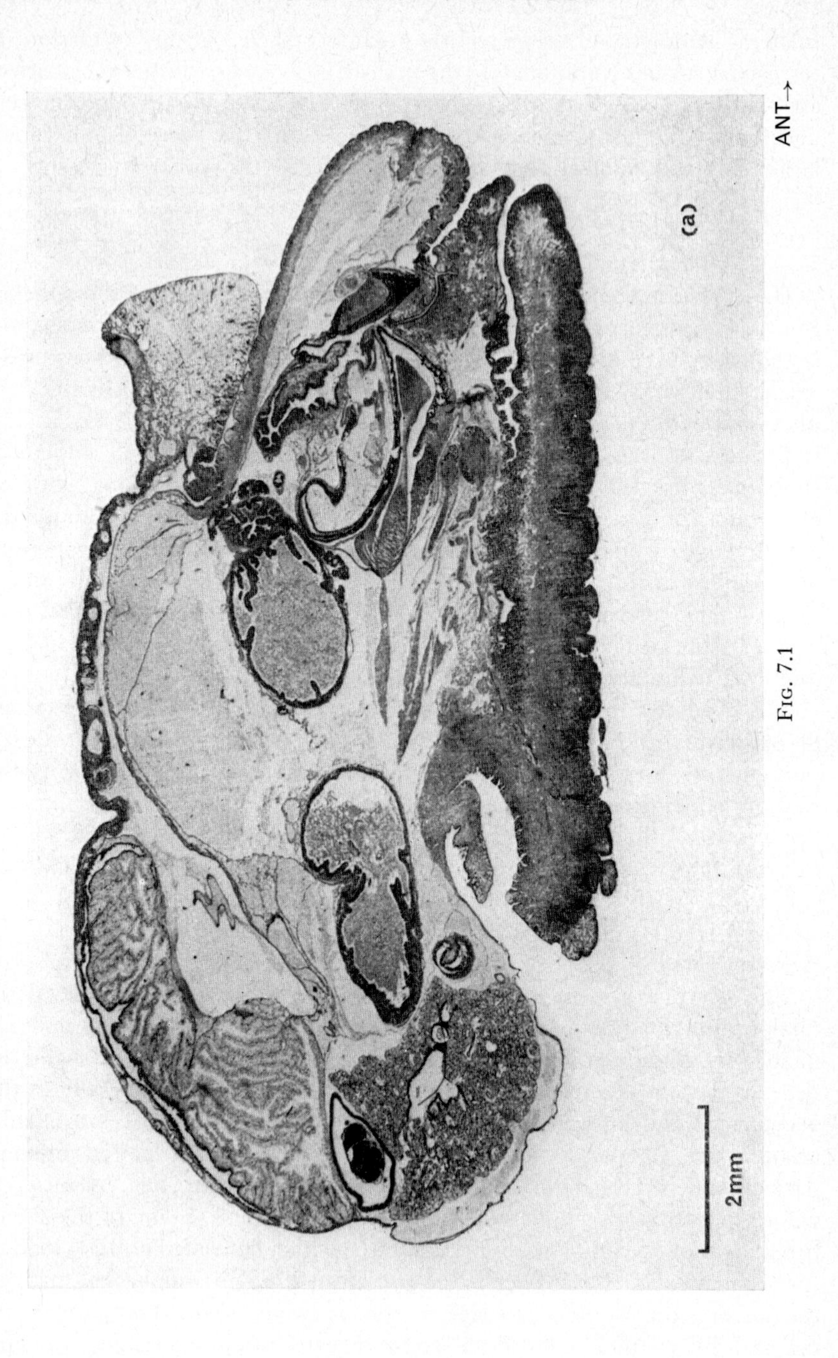

FIG. 7.1

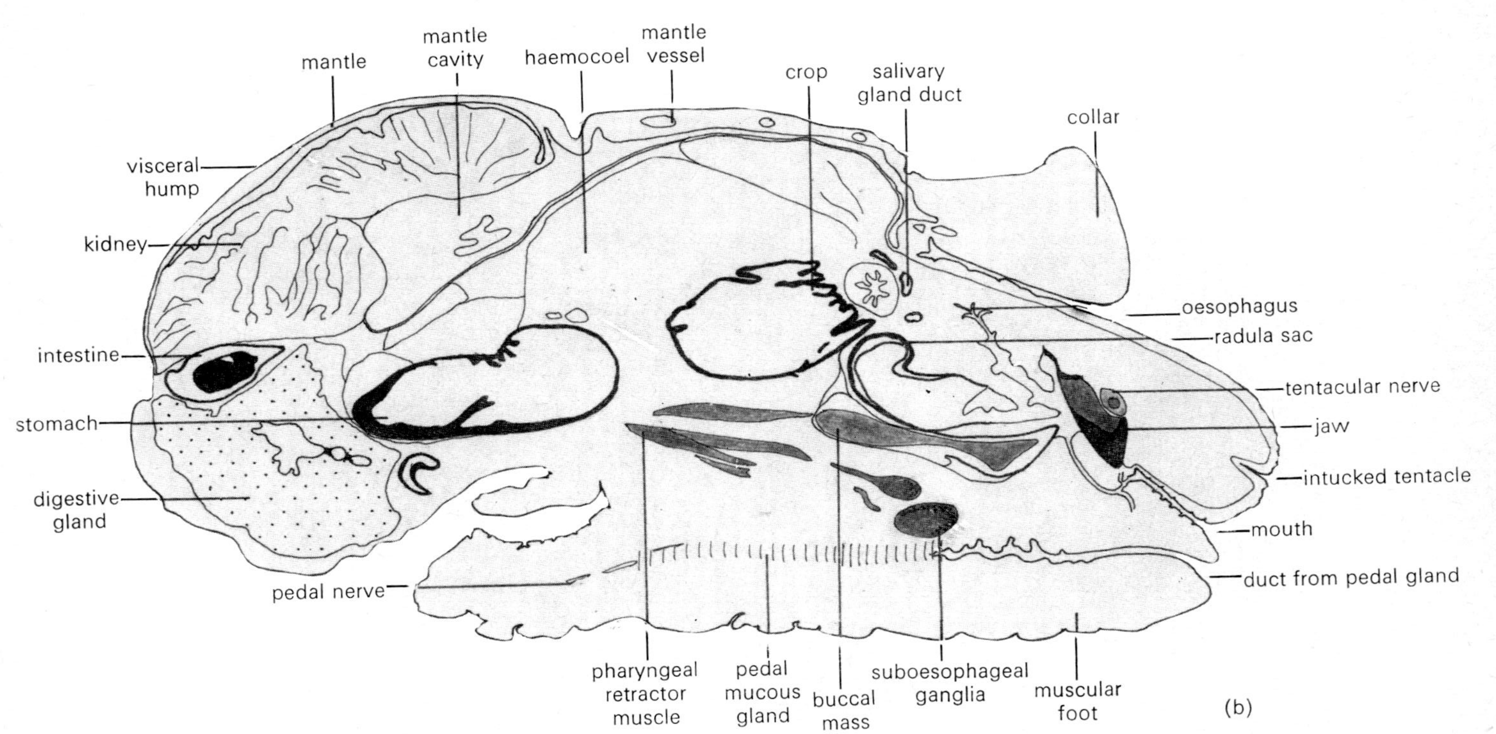

FIG. 7.1 (a) VLS and (b) explanatory drawing of a young garden snail (*Helix aspersa*). Note that the shell has been removed.

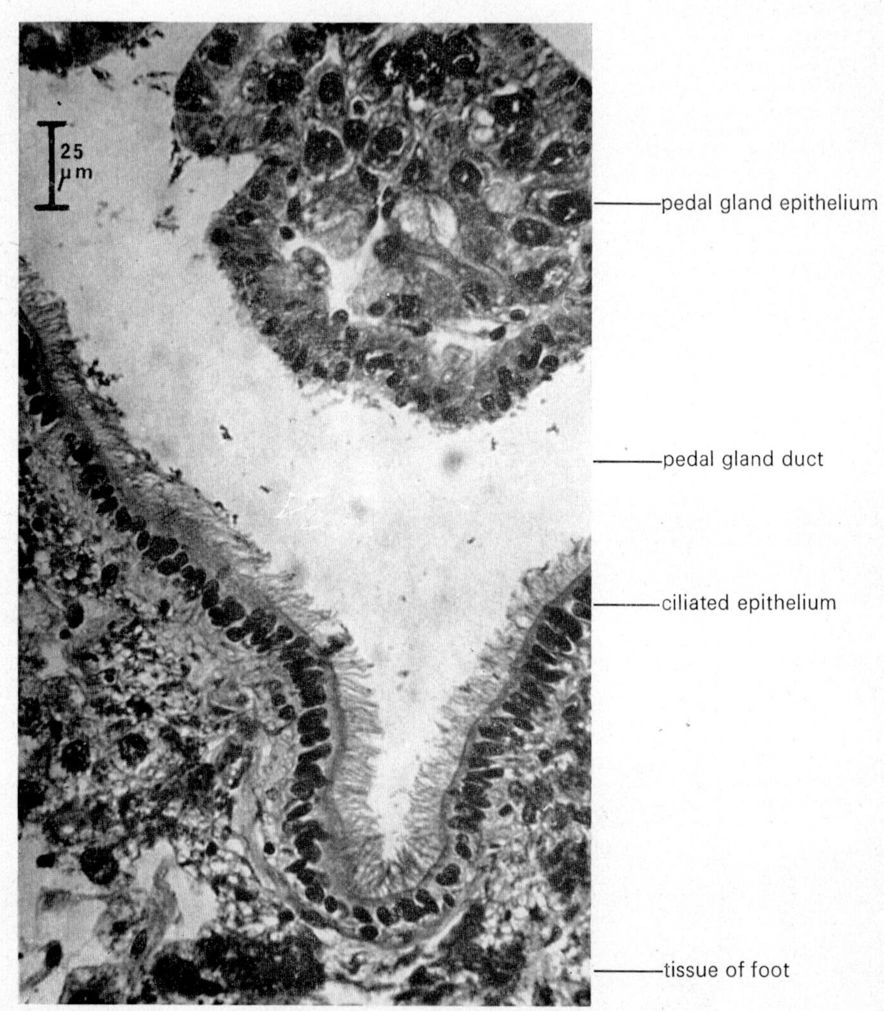

FIG. 7.2. Section through part of the foot of *H. aspersa*, showing the pedal mucous gland and ciliated duct.

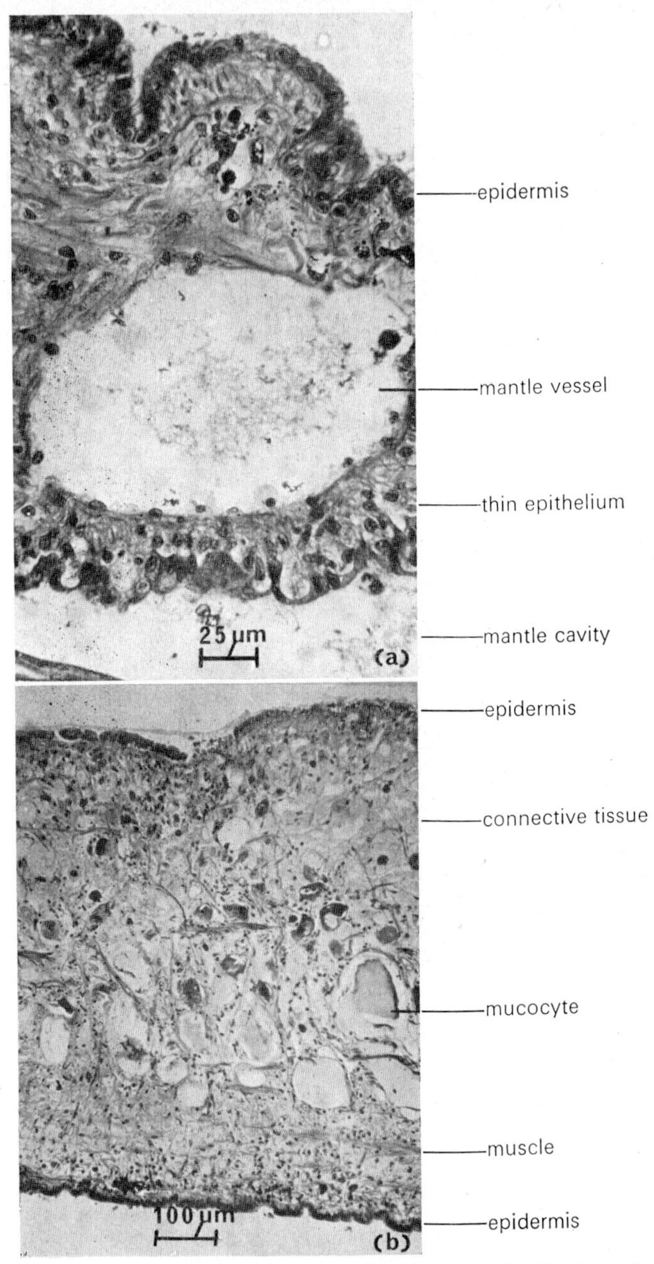

FIG. 7.3 (a) LS through the mantle of *H. aspersa* showing a mantle blood vessel; (b) section through the collar of *H. aspersa*.

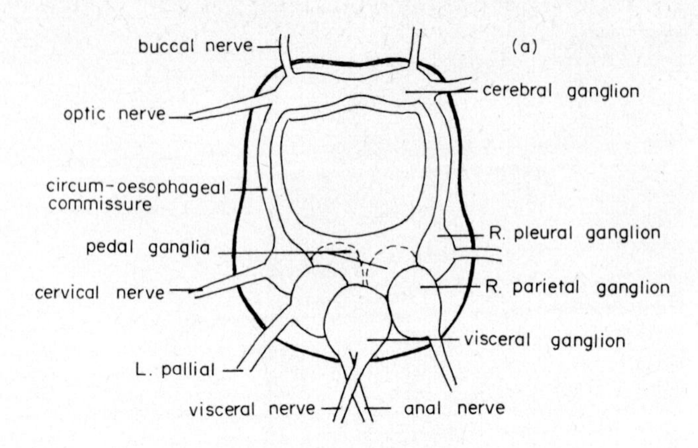

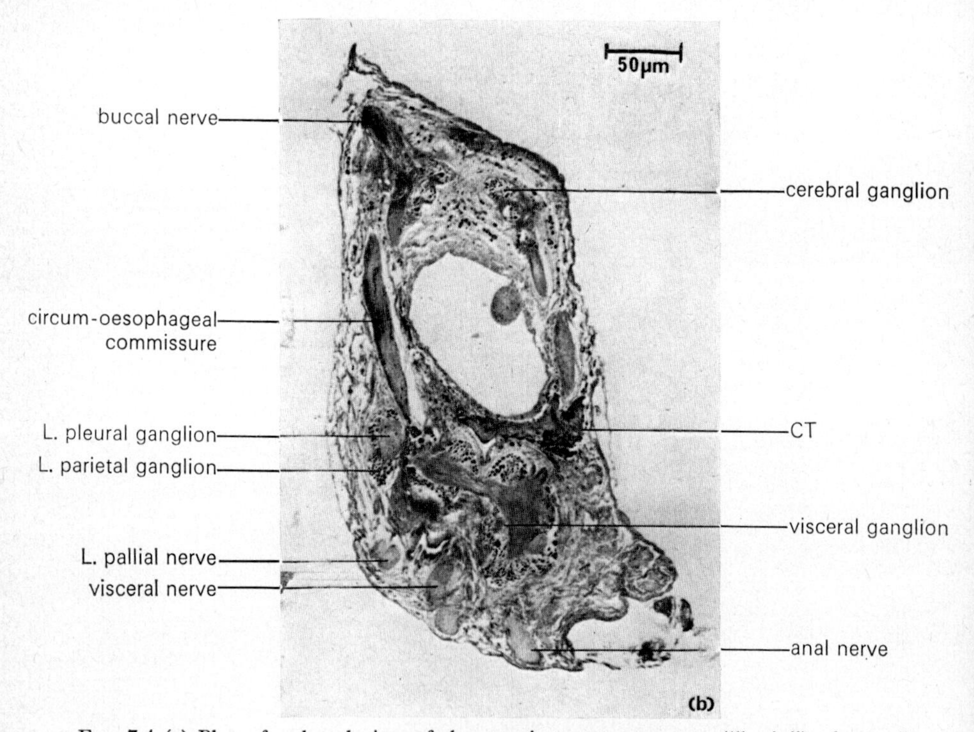

FIG. 7.4 (a) Plan of a dorsal view of the anterior nervous system ("brain") of a snail; (b) oblique HLS through the "brain" of *H. aspersa*. The right pleural and parietal ganglia have not been cut in this section, and the pedal ganglia lie ventrally to the plane of section.

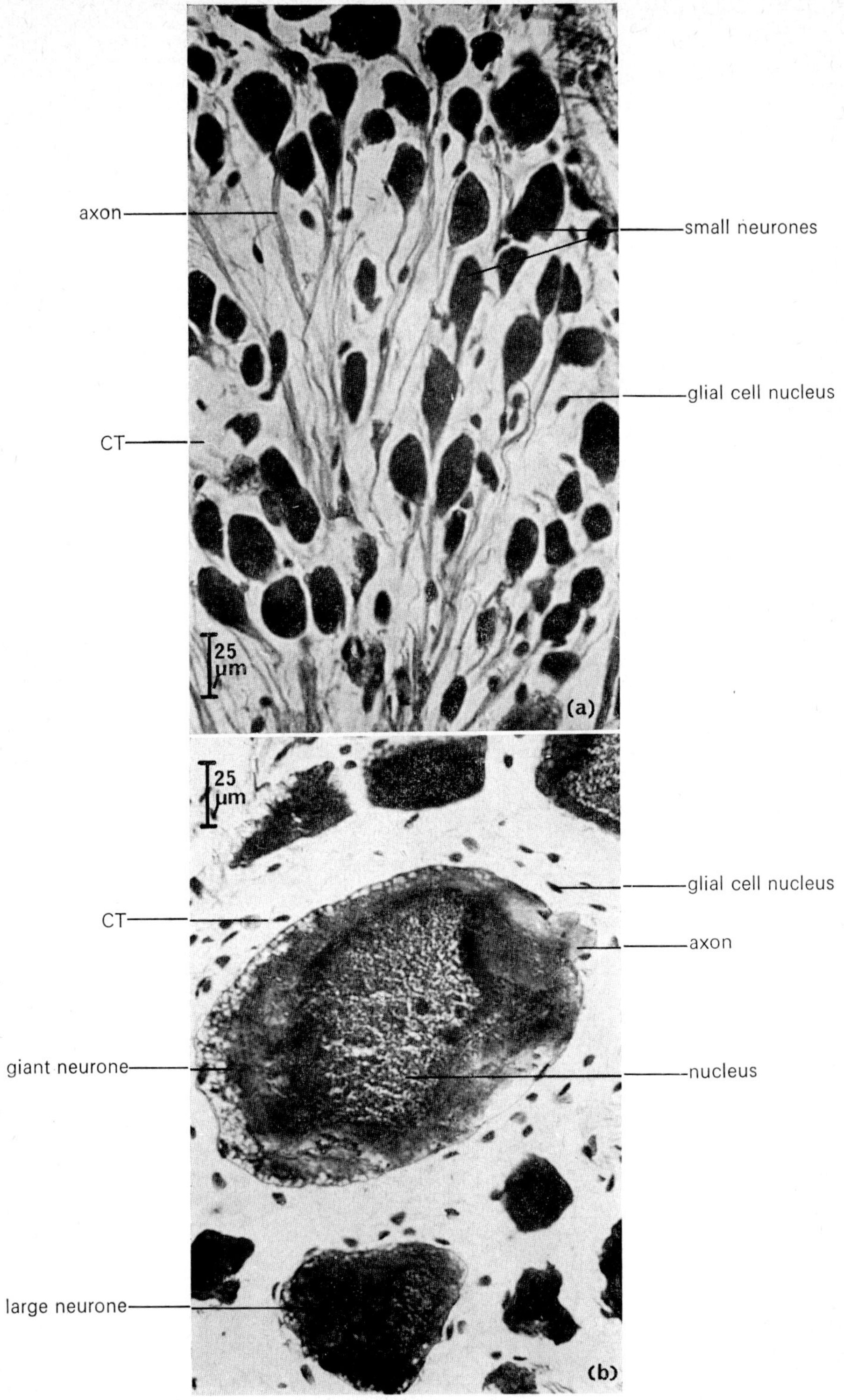

FIG. 7.5 (a) Small neurones from the R. parietal ganglion of the "brain" of *H. aspersa*; (b) giant neurones from the R. parietal ganglion of *H. aspersa*.

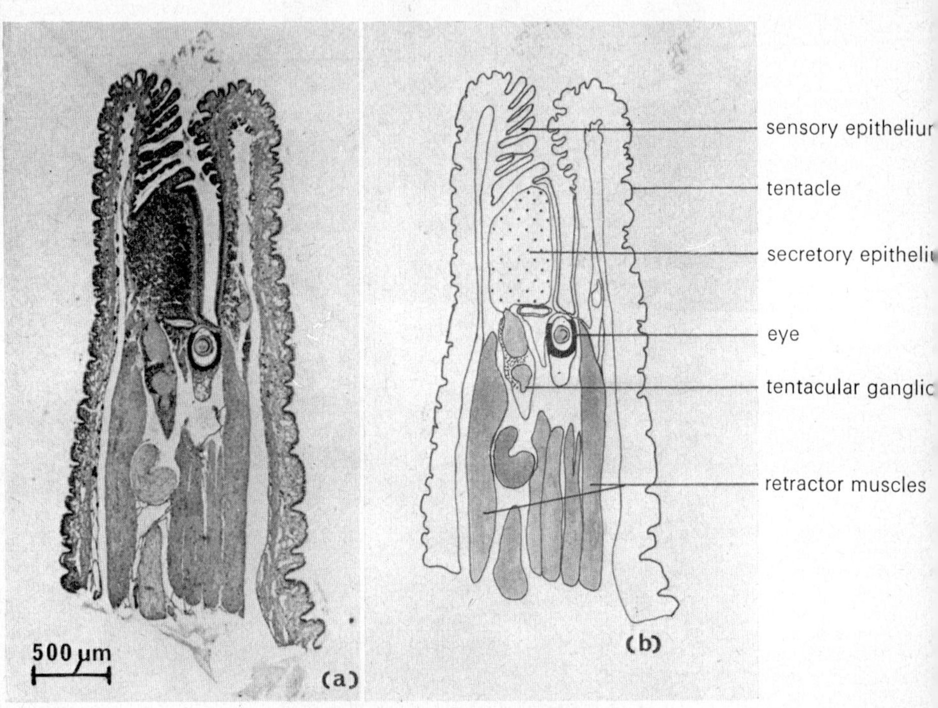

FIG. 7.6 (a) LS and (b) drawing of a partially retracted eyestalk of a snail (*H. pomatia*).

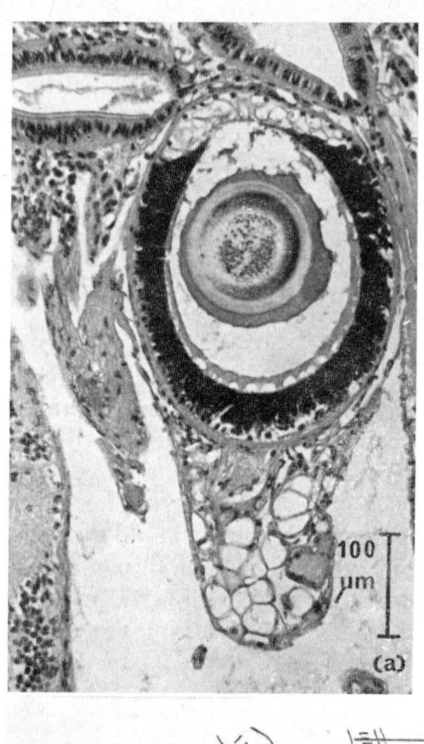

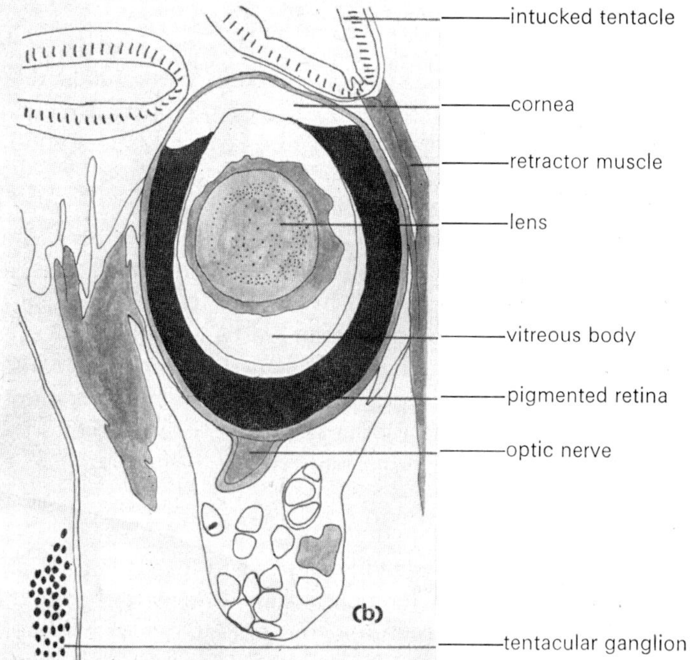

FIG. 7.7 (a) Detail of the snail eye shown in Fig. 7.6, and (b) explanatory drawing.

Sense organs

Snails are well supplied with sense organs. The smaller pair of tentacles on the head are equipped with tactile and chemoreceptor cells. *Helix* belongs to the order Stylommatophora which possess eyes at the tip of the larger pair of retractile tentacles. The eyes are fairly complex in structure with a focusing lens to concentrate the light, but probably they only function in light detection (Fig. 7.6). They are eyes of the direct type with the photoreceptor cells facing the source of light (Fig. 7.7). Some pelagic carnivorous species have much more complex eyes probably capable of image formatio ı and accommodation.

Respiration

The respiratory surface in snails is the vascular lung formed by the mantle [see Fig. 7.3(a)]. The mantle edge is joined onto the foot, thus forming an enclosed pulmonary sac with only a small opening, the pneumostome. The floor of this cavity is muscular and acts like a diaphragm. During inspiration muscular contractions flatten the floor of the mantle cavity and increase its volume so that air is drawn in through the pneumostome. The pneumostome is then closed by a small valve, and the muscles of the mantle cavity floor relax, thereby compressing the air within the pulmonary sac and facilitating diffusion of oxygen into the pulmonary vessels.

Circulation

The circulatory system of snails is partially open. Blood that is oxygenated in the pulmonary vessels lining the mantle cavity is returned to the heart, which consists of a single auricle and ventricle (Fig. 7.8). The heart is composed of a network of branching muscle fibres covered by an epithelial epicardium. The muscle fibres are myogenic (i.e. show inherent rhythmicity) and are innervated by ordinary nerve fibres and neurosecretory axons. The thin-walled auricle is separated from the ventricle by a pair of valves. Blood is pumped by the ventricle into an aorta which later splits into anterior and posterior aortae supplying the head and visceral hump respectively. These vessels divide many times into smaller arterioles which empty blood into the haemocoel. This is a spongy network of sinuses and lacunae surrounding the body tissues, from which blood eventually passes into small veins, and returns to the heart.

Digestion

Helix aspersa eats green vegetable matter and its gut is typical of herbivorous molluscs (Fig. 7.9). The mouth leads directly to the buccal mass, a strong muscular organ surrounding the buccal cavity and possessing a ventral diverticulum (radula sac) in which the radula is formed. The

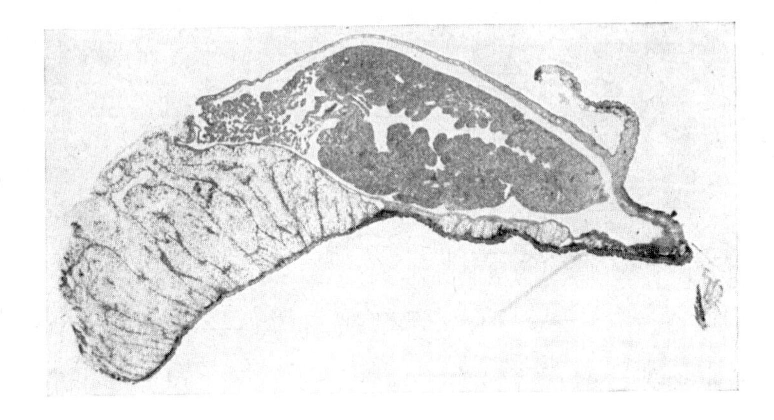

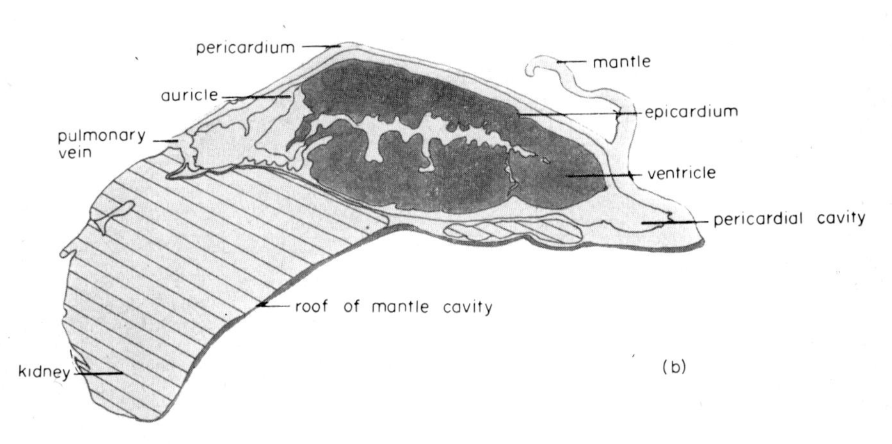

FIG. 7.8 (a) VLS and (b) drawing of the dorsal part of the visceral hump of *H. aspersa*, showing the heart and its relationship to the kidney.

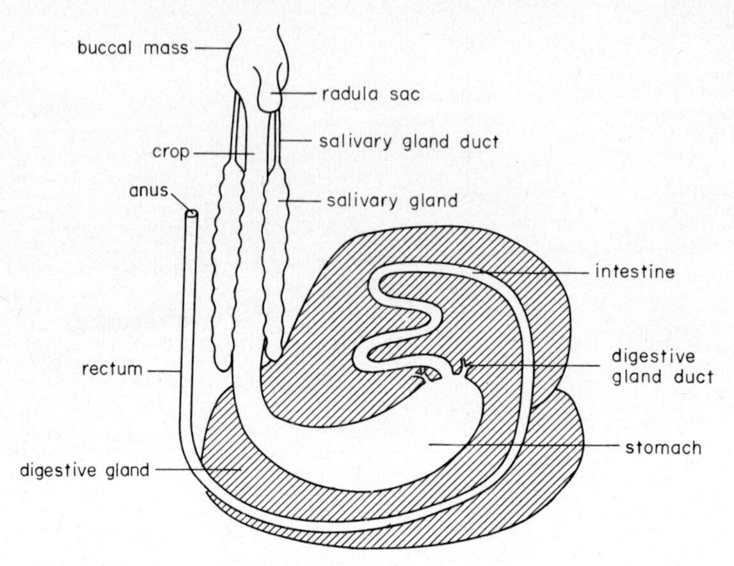

FIG. 7.9. Plan of the digestive tract of the snail, from the ventral side.

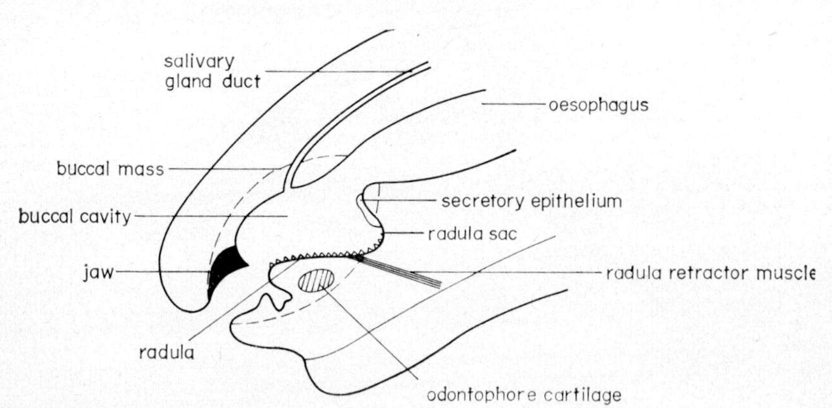

FIG. 7.10. Diagram of a VLS through the snail head, showing the structures associated with feeding.

radula is a chitinous membrane bearing rows of tiny file-like teeth, and is spread over the surface of a tongue-like structure (odontophore) which is supported by cartilage (Fig. 7.10). The odontophore moves forwards, and then backwards into the mouth taking strips of food with it. The rasping action of the radula against the chitinized jaw removes small bits of food.

Salivary glands (Fig. 7.11) secreting mucus and amylases lie attached to the outside of the crop, a thin-walled dilation of the oesophagus used for storage and for the first stages of digestion (Fig. 7.12). The crop and stomach are filled with a brown fluid secreted by the digestive gland. It contains cellulases of bacterial origin which break down the cell walls of plant food and release the contents which are then taken up into the digestive gland for intracellular digestion. The progressive development of a macrophagus habit and extracellular digestion have led to a simplification of the snail stomach as compared with more primitive gastropods. The stomach is lined by a simple epithelium (Fig. 7.13) which is not clearly differentiated from the intestinal epithelium. As food passes down the intestine, mucus is added and moisture extracted so that fairly dry faeces are discharged from the anus as a faecal string.

Very small food particles are moved by cilia into the fine tubules of the digestive gland. The ends of the tubules are expanded into alveoli lined with secretory (ferment) cells producing enzymes, lime-containing cells storing reserves of calcium carbonate, and phagocytic (digestive) cells responsible for the final intracellular stages of digestion (Fig. 7.14).

Excretion

Terrestrial snails excrete mainly uric acid. The excretory organ is a single kidney which represents a part of the coelom. It lies close to the pericardium (see Fig. 7.8) and is connected to it by a very small renopericardial canal. The internal walls of the kidney are thrown into complex folds covered by vacuolated secretory cells (nephrocytes) containing uric acid crystals (Fig. 7.15). The kidney connects with the outside by a thin-walled ureter which runs along the edge of the mantle cavity parallel to the rectum and opens close to the anus just outside the pneumostome. Only small amounts of urine are produced.

Reproduction

All pulmonates are hermaphrodite and there are a number of elaborate mechanisms to ensure cross-fertilization between two individuals. The genital system of *Helix* is shown in Fig. 7.16. The gametes develop consecutively in a common gland, the ovotestis, seen in Fig. 7.17 to consist of oocytes intermingled with developing sperm. The sperm mature first, early in the summer, and are passed via a thin-walled coiled hermaphrodite

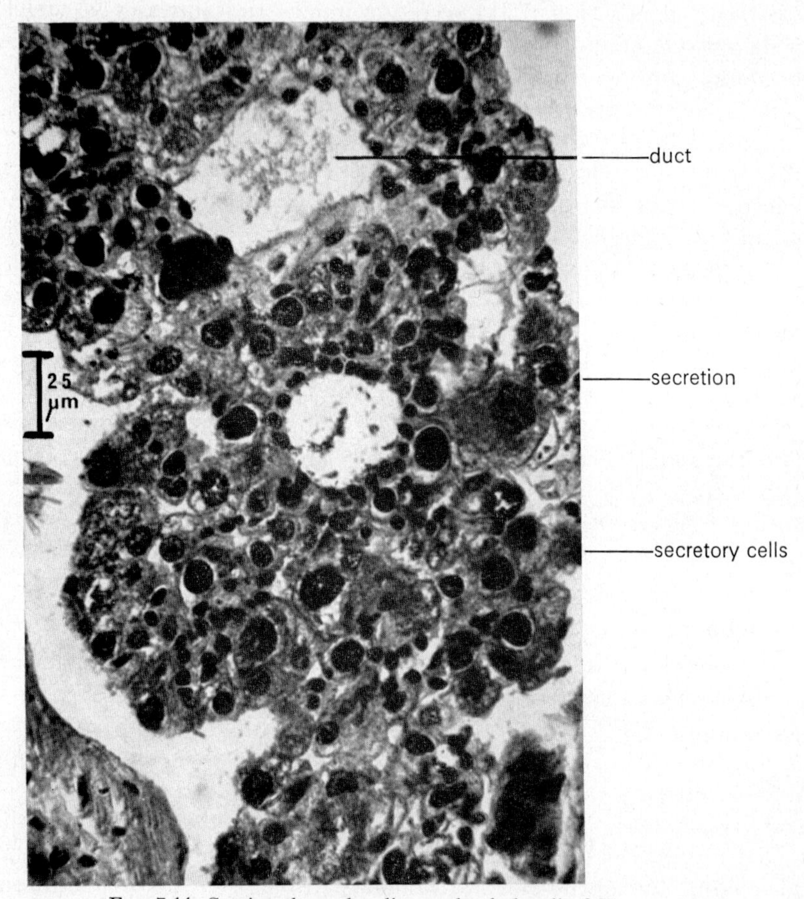

FIG. 7.11. Section through salivary gland alveoli of *H. aspersa*.

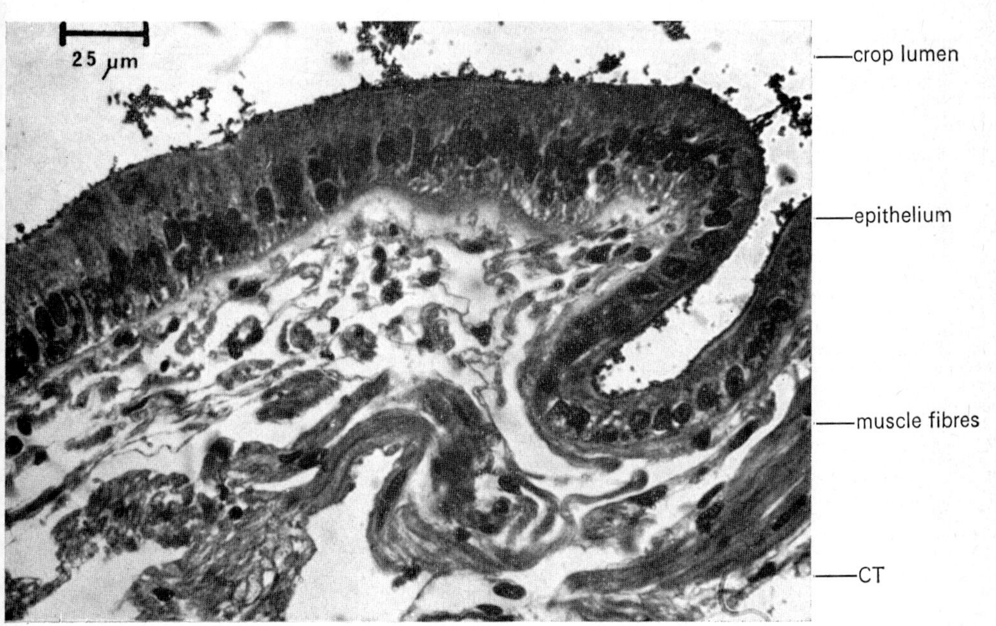

FIG. 7.12. Section through the crop wall of *H. aspersa*.

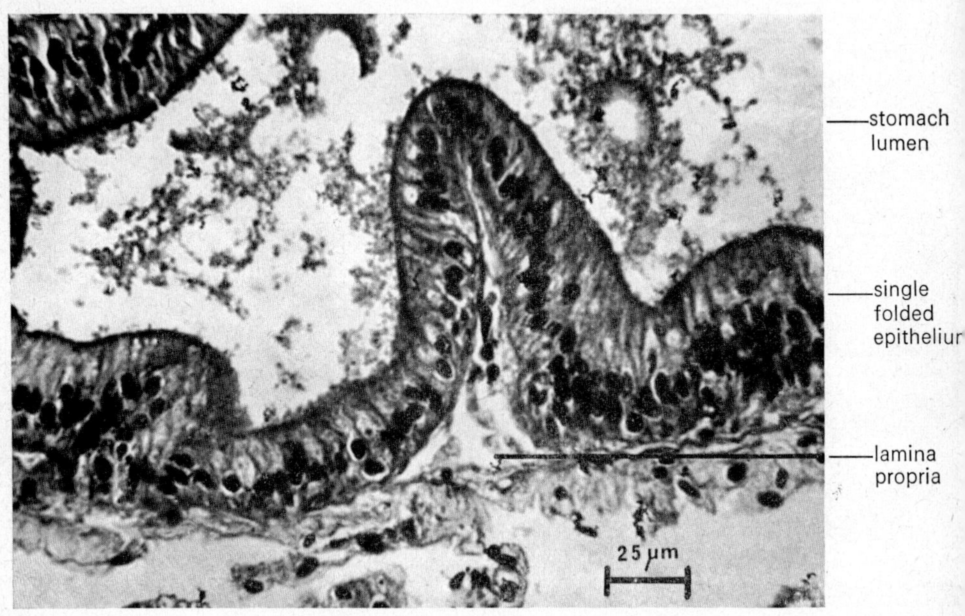

FIG. 7.13. TS through the stomach wall of *H. aspersa*.

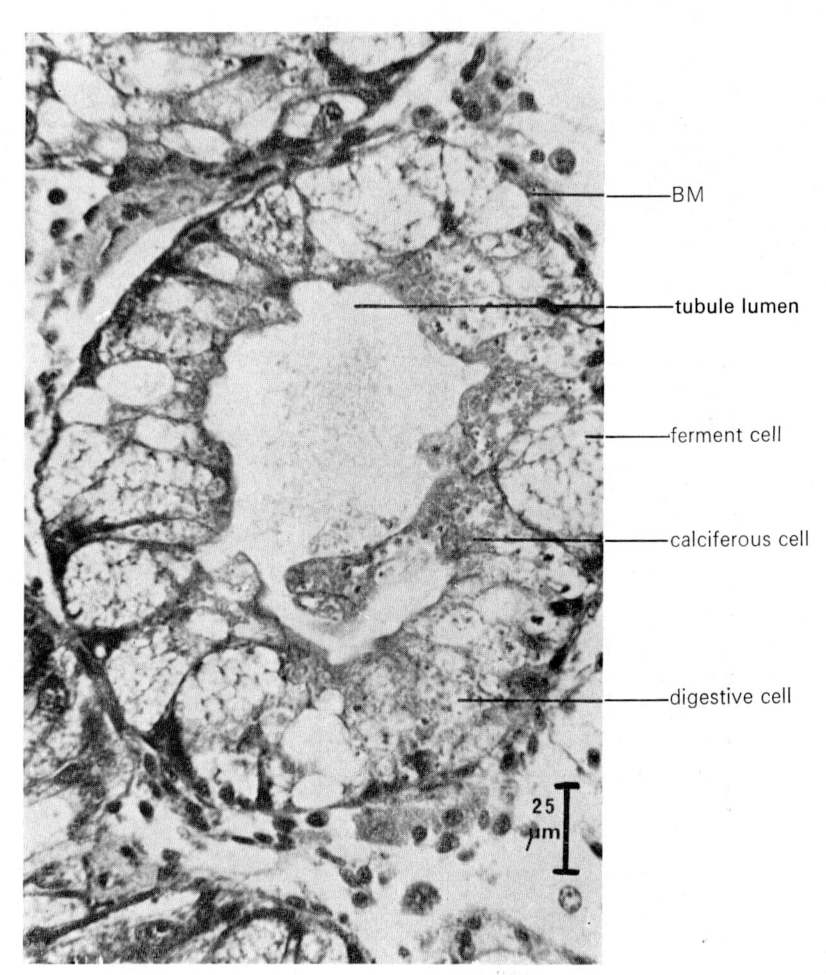

FIG. 7.14. Section through a digestive gland tubule of *H. aspersa*.

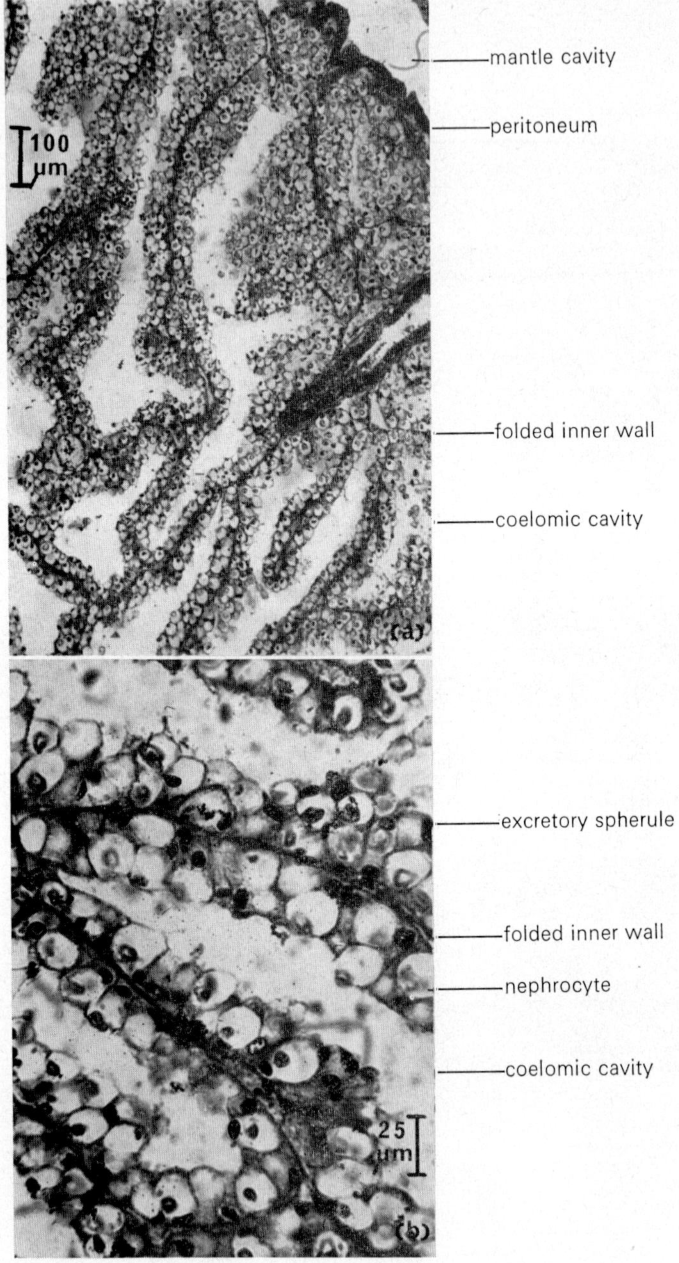

FIG. 7.15 (a) Section through the kidney of *H. aspersa*; (b) detail of the folded internal walls of the snail kidney.

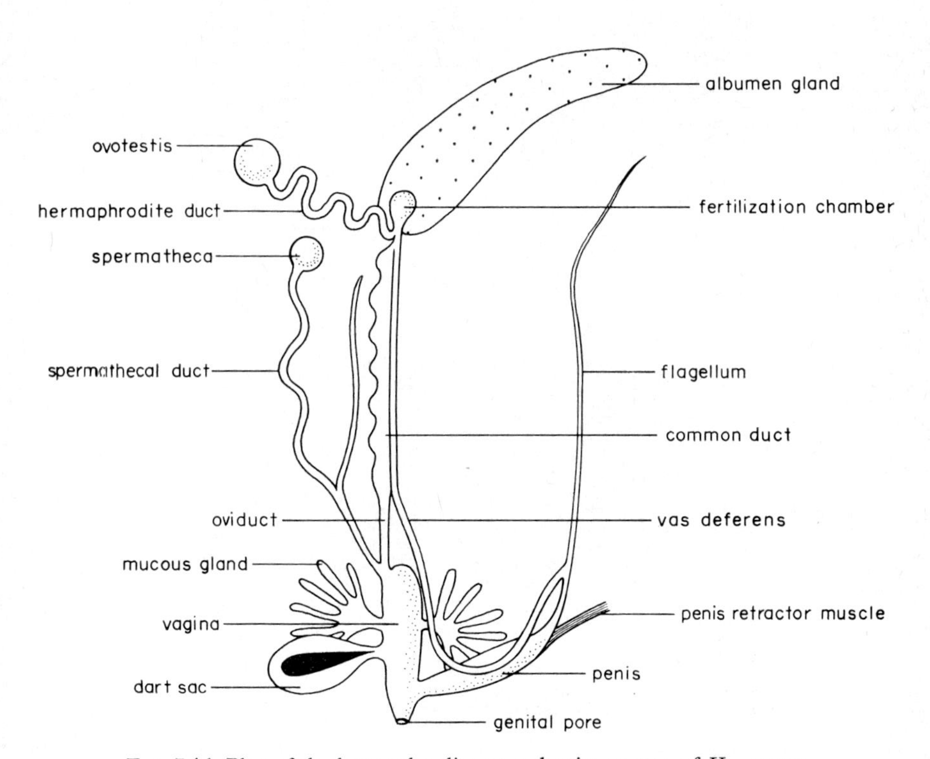

FIG. 7.16. Plan of the hermaphrodite reproductive system of *H. aspersa*.

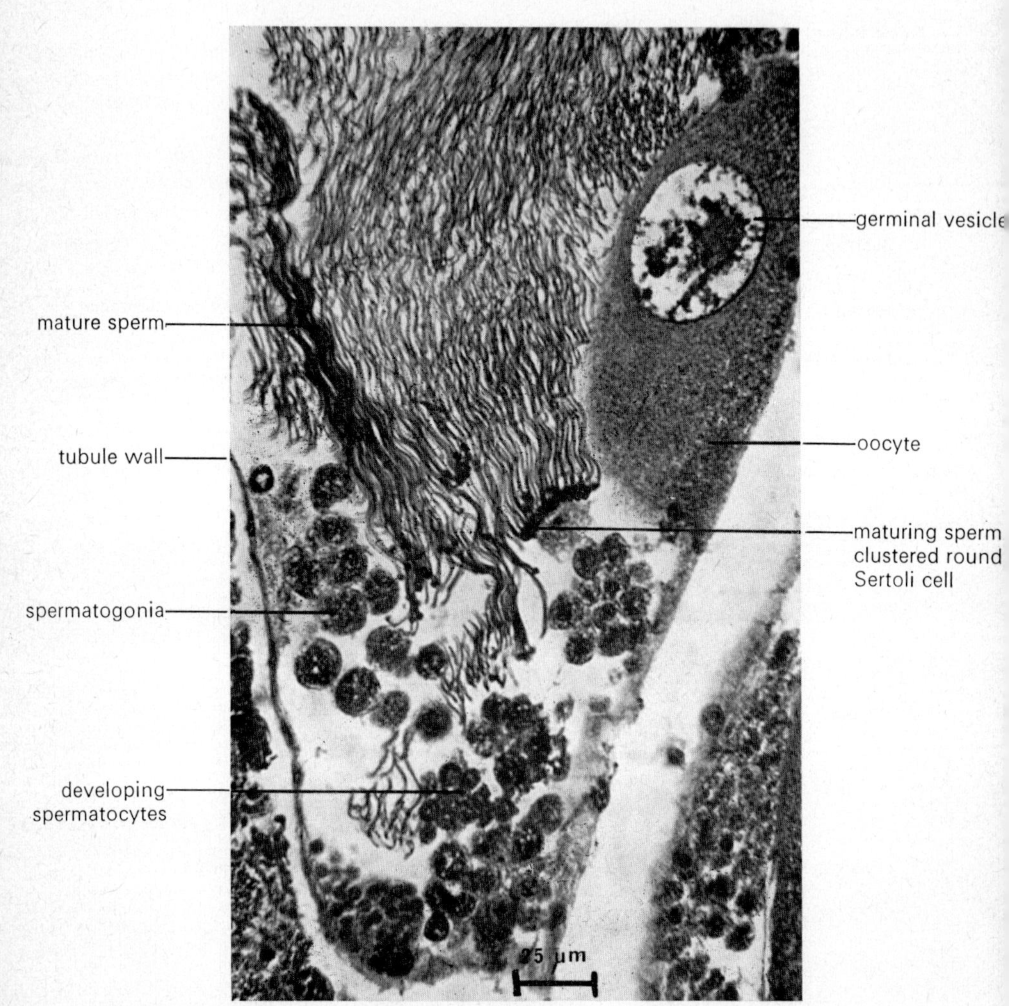

Fig. 7.17. Section through the ovotestis of *H. aspersa*, showing mature sperm and a young oocyte (tissue fixed in April).

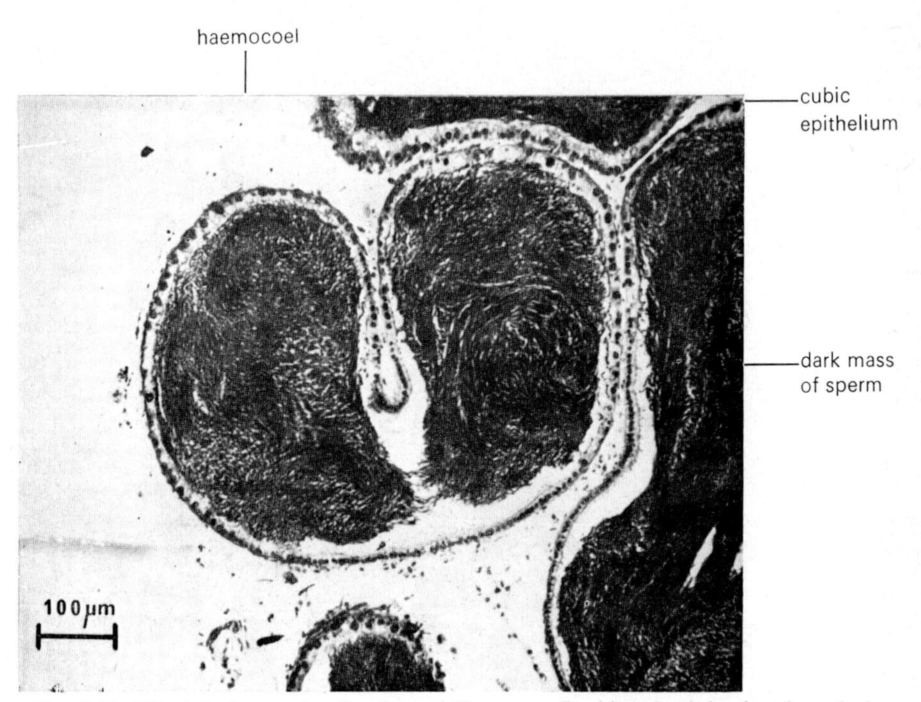

FIG. 7.18. TS of the hermaphrodite duct of *H. aspersa*. In this animal the duct is packed with a mass of ripe sperm

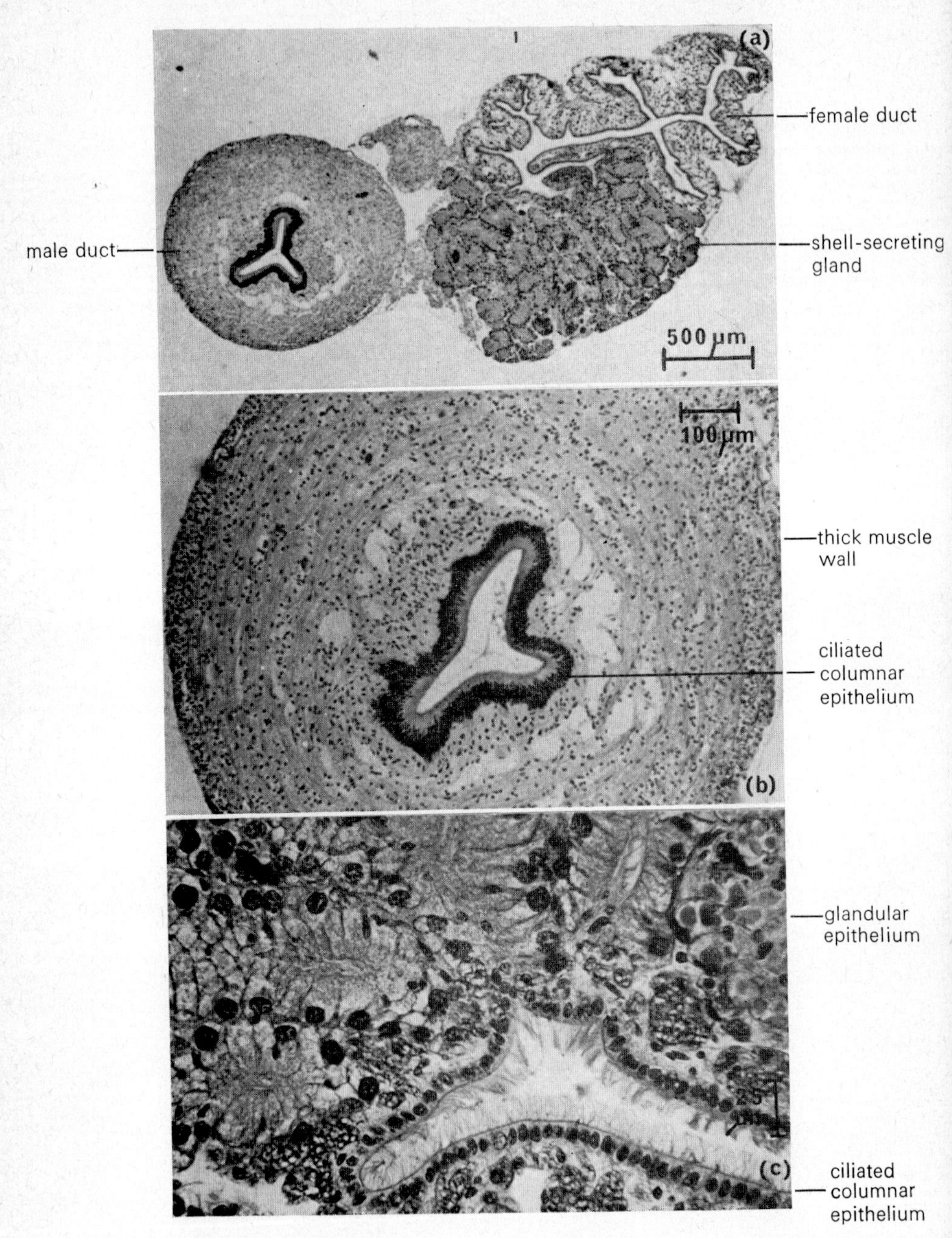

FIG. 7.19 (a) TS through the common duct of *H. aspersa*; (b) detail of the male half of the common duct of *H. aspersa*; (c) detail of the female portion of the common duct of *H. aspersa*.

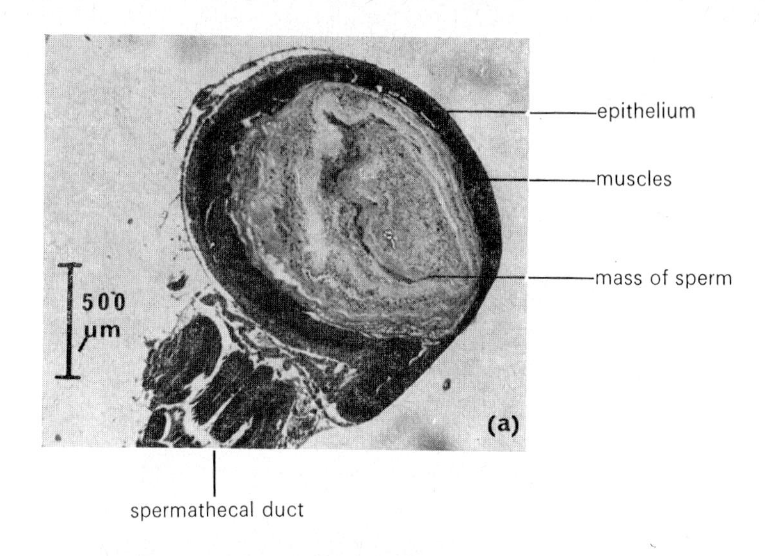

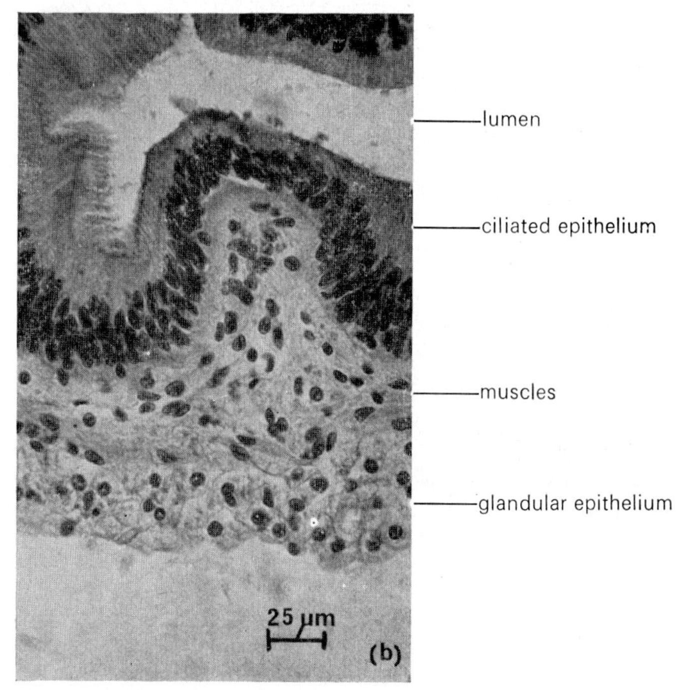

FIG. 7.20 (a) Section of a spermatheca of *H. aspersa*, packed with sperm from a second individual; (b) section through the wall of the spermathecal duct of *H. aspersa*.

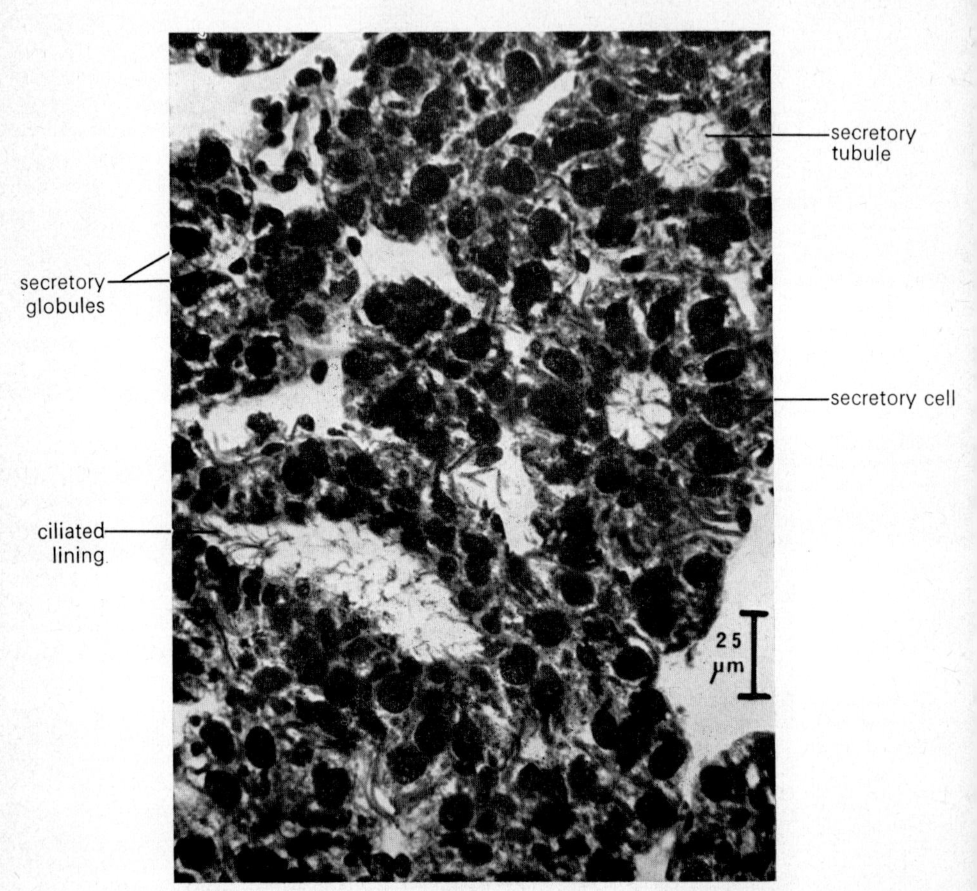

Fig. 7.21. Section through the albumen gland of *H. aspersa*.

duct (Fig. 7.18) to the male half of the common duct [Fig. 7.19(a, b)], and thence via the vas deferens to a muscular penis. During their passage the sperm are compacted into spermatophores and covered by a chitinous envelope which is secreted by the flagellum. At copulation the penis is everted into the vagina of another individual and spermatophores are discharged. They pass from the vagina through a ciliated spermathecal duct into the spermathecae [Fig. 7.20(a, b)] where their chitinous coat is dissolved and sperm are then stored.

Ova, produced later in the summer, pass from the ovotestis through the hermaphrodite duct to a fertilization chamber in the albumen gland. Here they are fertilized by sperm from another individual which are passed up from the spermathecae when required. The fertilized eggs are coated with albumen secreted by the albumen gland (Fig. 7.21) and passed down to the oviduct and vagina via the female half of the common duct [Fig. 7.19(a, c)], the walls of which secrete a thin calcareous shell round them. Ova and mucus, produced by branched mucous glands, are passed to the outside through the common genital pore, move down towards the ground along the genital groove and are laid in covered damp situations. They hatch directly into small snails, although some aquatic species develop into pelagic larvae.

The vagina also receives the opening of a muscular dart sac. This secretes a ridged calcareous "love dart" which is fired by strong muscular contractions of the dart sac at the onset of mating, causing (in the second snail) marked tissue damage which is believed to result in sexual stimulation.

Class Lamellibranchiata (Bivalvia)

Lamellibranchs are mainly marine with a few freshwater and estuarine representatives. They are mostly sedentary and often show adaptations for burrowing or boring. The body is bilaterally symmetrical and laterally compressed, and is enveloped by extensive mantle lobes and a bivalved shell which is hinged dorsally and can be closed ventrally by large adductor muscles. These muscles contain a special contractile protein (paramyosin) which enables them to remain in a contracted state for long periods of time. The head is rudimentary and is withdrawn permanently into the mantle cavity. It bears lateral palps formed from drawn-out lips, but there are no tentacles or eyes. Lamellibranchs live enclosed within their shell valves, and must either grope for food particles or filter-feed; both buccal mass and radula have been lost. The foot is typically large, wedge-shaped and laterally compressed. It is used for locomotion and burrowing and is operated by a haemoskeleton and foot musculature which allows alternate lengthening and shortening. There is a very large mantle cavity and the two gills (ctenidia) are often enlarged to form a sieve for food collection in addition to their respiratory function.

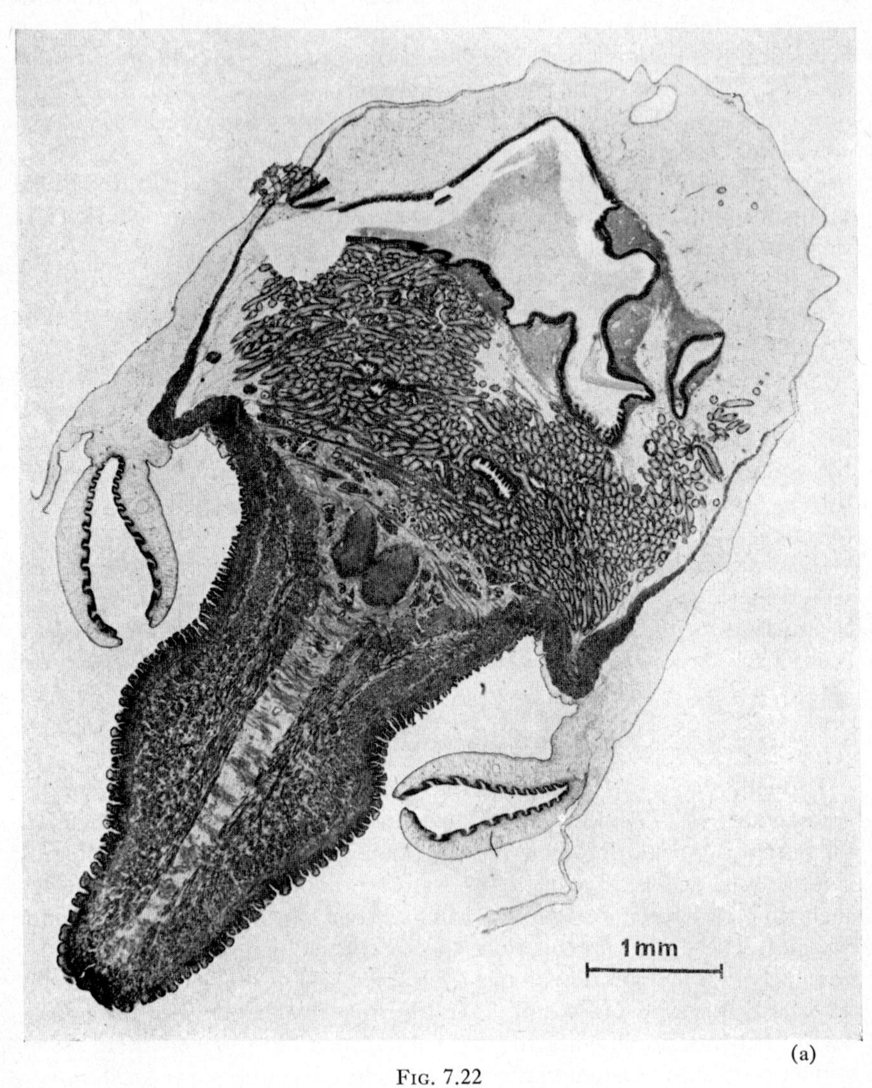

(a)

Fig. 7.22

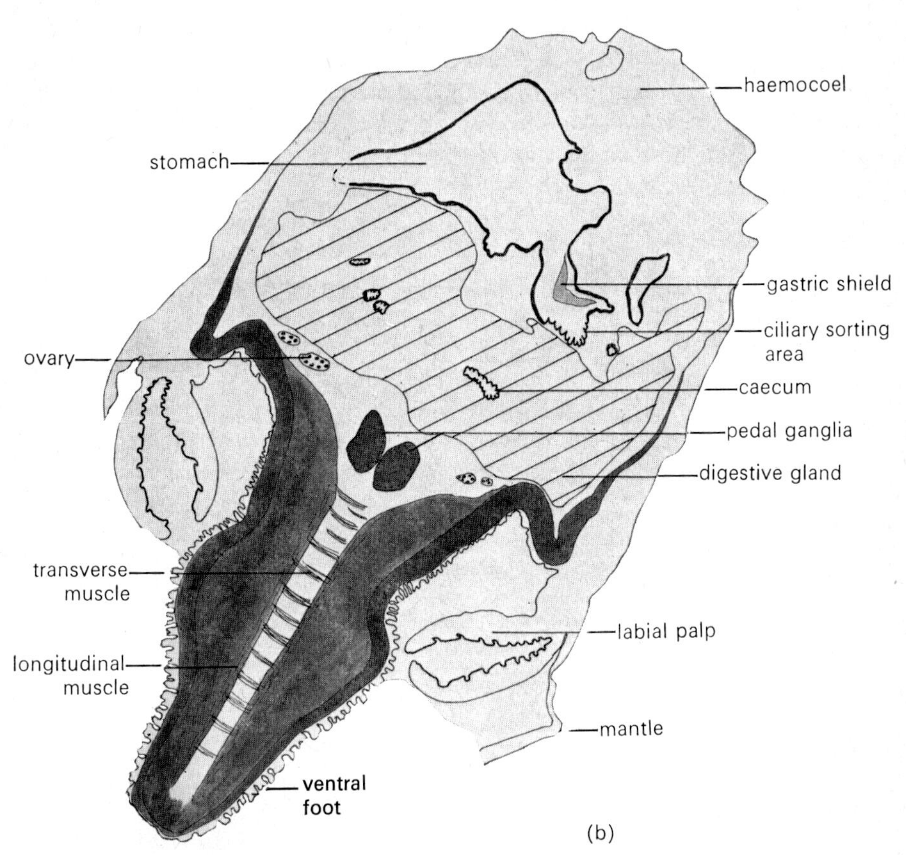

FIG. 7.22 (a) TS and (b) explanatory drawing of the anterior region of a young lamellibranch (*Anodonta cygnea*).

Example: *Anodonta* (*cygnea*)

Anodonta cygnea (the freshwater clam or swan mussel) occurs widely in this country in sluggish freshwater ponds and streams where it lies buried in the mud on the bottom. It moves slowly by muscular action of the foot combined with fluctuations in blood pressure within the foot vessels. The foot musculature can be seen clearly in Fig. 7.22 which shows a section of the anterior region of a young *A. cygnea*.

The large bivalved shell is secreted by the margin of the mantle. This is a thin epithelial membrane thickened along its free edge to form a series of glandular folds, and modified along its posterior border to form two siphons, with the mantle fused between them. The dorsal (exhalent) siphon has smooth walls while the ventral (inhalent) siphon is incompletely closed and has a sensory papillated edge. Even when the shell is open, the muscles along the mantle border allow its edges to be pressed tightly together so that water only moves through the siphons.

Nervous system

The nervous system of bivalves is relatively simple (Fig. 7.23). It shows bilateral symmetry, with three pairs of ganglia (cerebral, visceral and pedal) separated from each other by large connectives containing scattered nerve cells (see also Fig. 7.22). The visceral ganglia are almost fused.

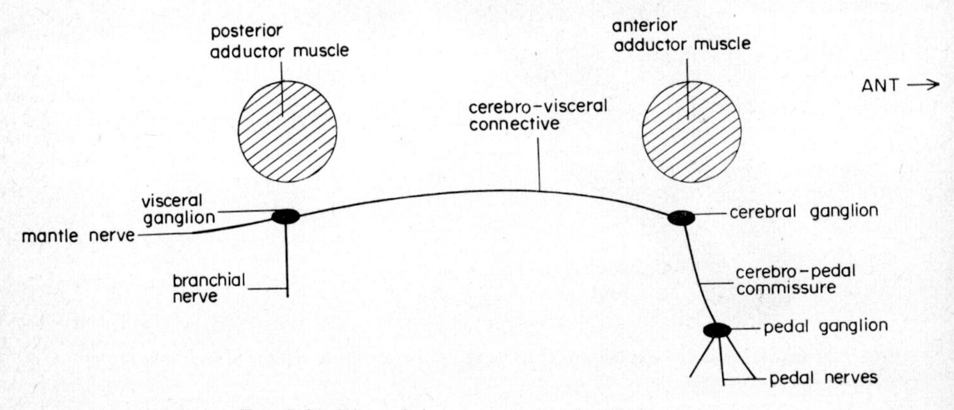

Fig. 7.23. Plan of the nervous system of *A. cygnea*.

Sense organs

Bivalves possess simple receptors (chemoreceptors, tactile receptors and photosensory cells) which are sometimes aggregated to form sense organs. They are usually situated at the mantle margins (e.g. the eyes in *Pecten*) especially around the inhalent siphon. An osphradium, which is a patch of sensory epithelium, is located beneath the posterior adductor muscle in the exhalent chamber. It was thought to be chemosensitive but would be unable to test incoming water in that position, so it is now thought to be primarily a tactile receptor for appreciation of sediment passing through or into the mantle cavities, although it may also play a part in regulating water flow.

Respiration

The gills of lamellibranchs are composed of very elongated filaments which are folded over to form a double lamellar structure on each side of the body (Fig. 7.24). Thus the mantle cavity is divided by the gills into a small dorsal exhalent chamber and a larger ventral inhalent chamber. The gill lamellae are formed by apposition of numerous filaments. The two limbs of each filament are joined by vascular junctions, while the folded edge of each filament is indented so that a marginal food groove has developed along each lamella and is used for transporting food particles towards the mouth. Lamellibranch gill structure falls into two main groups (filibranch and eulamellibranch); other gill types being modifications of these two patterns. *Anodonta* belongs to the Eulamellibranchiata which possess the most specialized gills. In this order adjacent gill filaments are connected by vascular and skeletal interfilamentary junctions so that the lamellae consist of fairly rigid sheets of tissue joined by a large number of interlamellar junctions (Fig. 7.25, see also Fig. 7.34). In the filibranchs, of which *Mytilus* is the best known example, adjacent gill filaments remain discrete, bridged only by lateral cilia. There are also fewer interlamellar junctions so the whole gill is a much more fragile, flexible structure (Fig. 7.26).

Each gill filament possesses several tracts of cilia, each of which performs a different function. The lateral cilia are responsible for drawing the main respiratory current into the mantle cavity. Water is drawn in from the inhalent chamber, between the gill filaments into the interlamellar spaces and out into the exhalent chamber whence it is expelled through the exhalent siphon. Water currents are directed in opposition to the blood flow in the gill vessels to allow maximum oxygen extraction from the water. The frontal gill cilia convey food particles; the short frontals beating continuously and transporting fine particles dorsally to tracts at the base of the lamella, while the longer frontals beat in the opposite direction and move slightly bigger pieces ventrally towards the marginal groove and thence towards the mouth. The latero-frontals are peculiar to bivalve gills and deflect particles onto the face of the filaments. Any large particles are completely rejected and fall out into the water.

Circulation

The circulatory system follows much the same plan as in gastropods. The heart consists of a pair of lateral auricles and a single ventricle which

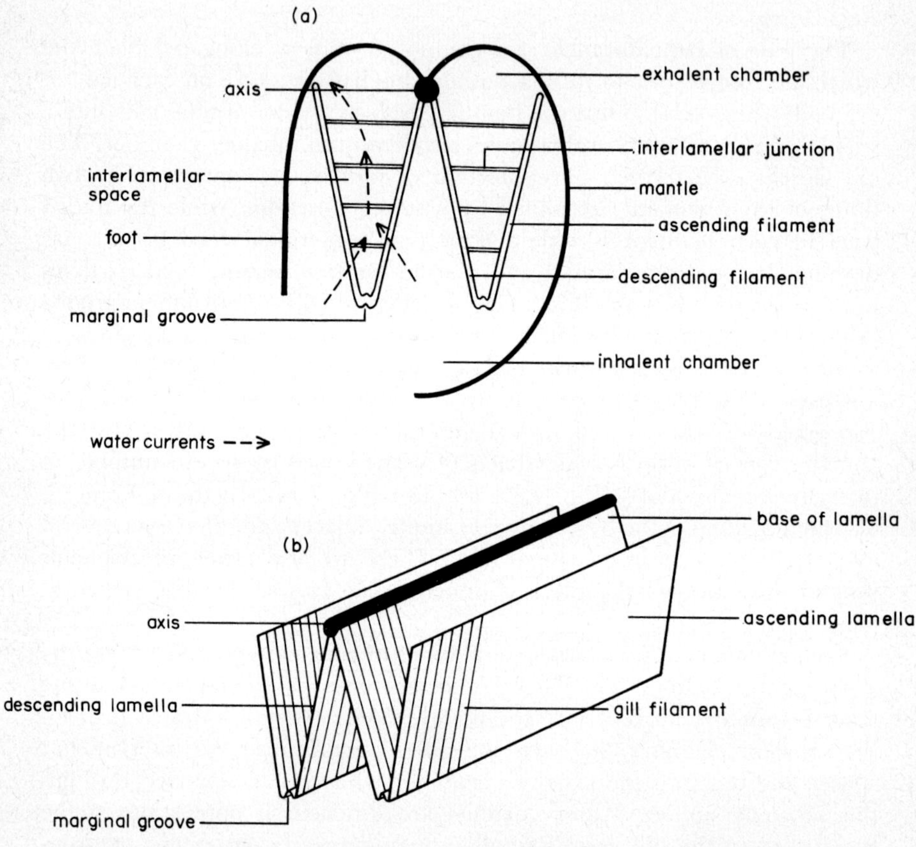

FIG. 7.24 (*a*) Basic plan of a transverse section through one side of the mantle cavity of a filibranch or eulamellibranch mussel, showing the position of the gills, inhalent and exhalent siphons, and the direction of water flow; (*b*) diagram of the basic gill structure in a filibranch or eulamellibranch mussel.

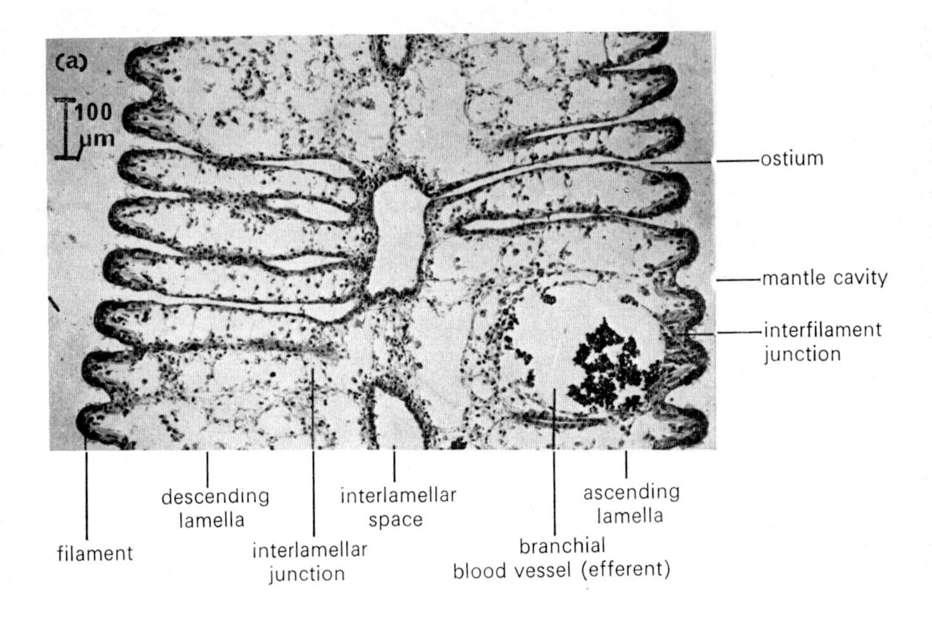

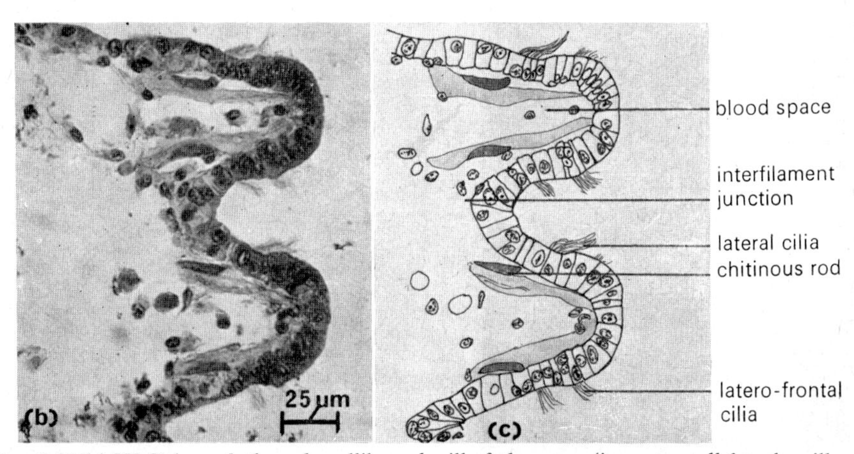

FIG. 7.25 (a) HLS through the eulamellibranch gill of *A. cygnea* (i.e. cut parallel to the gill axis). Note the solid inter-filament junctions and the numerous thick interlamellar junctions; (b) detail and (c) drawing of gill filament structure of *A. cygnea.*

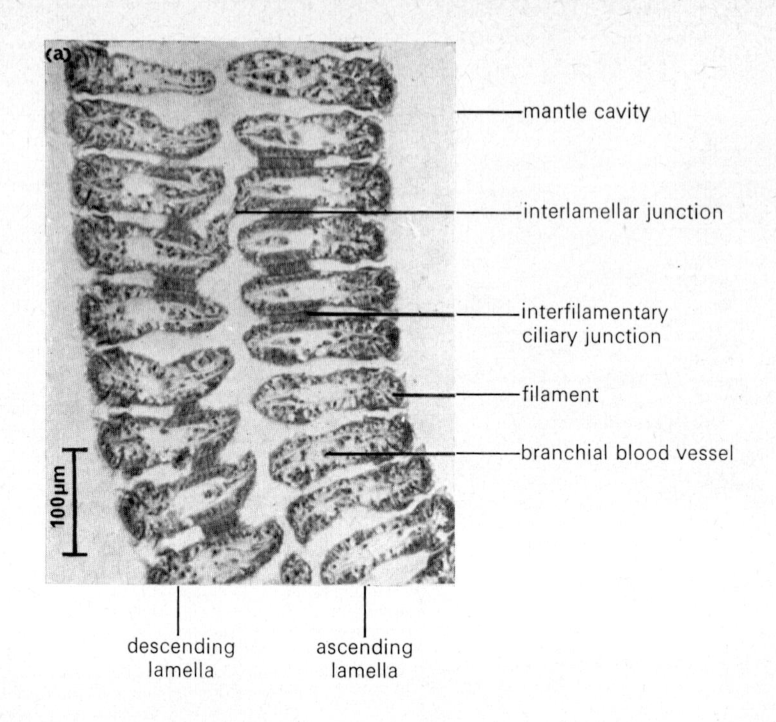

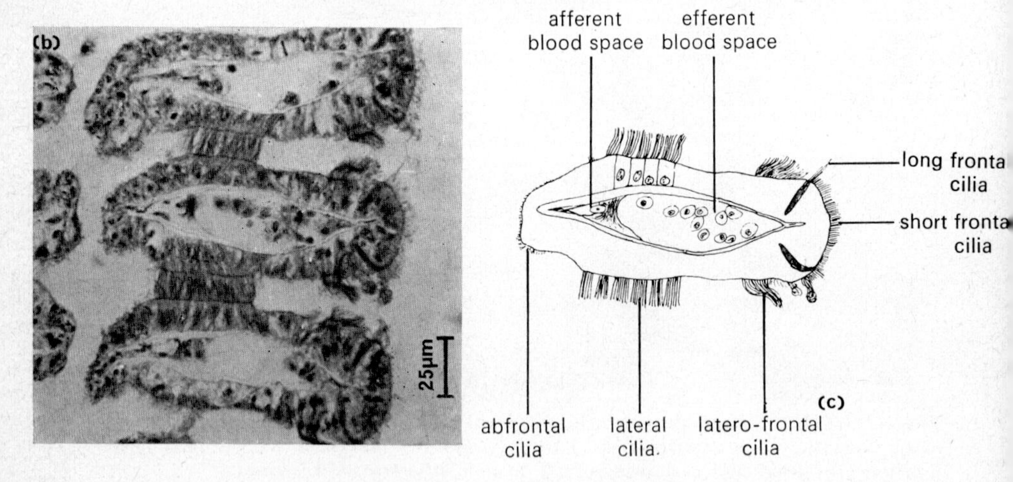

FIG. 7.26 (a) HLS through the filibranch gill of a marine mussel (*Mytilus edulis*), cut in the same plane as that of *A. cygnea* shown in Fig. 7.25. Note the fragile ciliary inter-filament junctions, and the paucity of interlamellar junctions; (b) detail and (c) drawing of gill filament structure of *M. edulis*.

pumps blood into an anterior aorta supplying the viscera and foot, and paired posterior aortae which run to the mantle and hind end of the body. The heart ventricle in *Anodonta* is characteristically folded round the rectum (see Fig. 7.31). Blood from the body collects in irregular lacunae which form the haemocoel and is passed to the kidneys before being returned to the heart.

Digestion

Anodonta, like the majority of lamellibranchs, feeds by sieving microscopic organisms and bits of organic matter from the water. Food collection is performed by the gills and labial palps. Ciliary tracts on the gills sort food particles and pass those of a suitable size up the marginal food grooves to an oral groove and thence to the mouth. Further sorting of particles occurs on the palps which are ridged and ciliated on their inner surfaces (see Fig. 7.22).

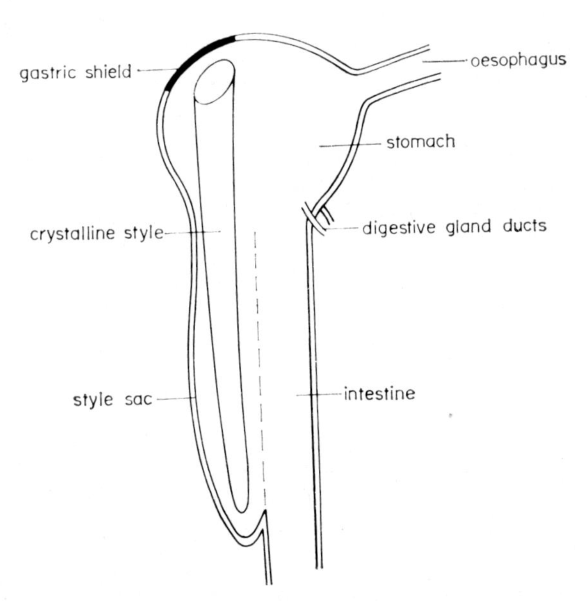

FIG. 7.27. Plan of the stomach and style sac of *A. cygnea*

The mouth leads by a short tubular oesophagus into a complex saccular stomach (Fig. 7.27). In the stomach food is mixed with digestive enzymes, small particles are channelled into the ducts of the digestive gland for further digestion, and unwanted particles are directed into the intestine from where they are passed to the anus. The lamellibranch stomach therefore possess a complex pattern of ridged ciliary sorting areas [Fig. 7.28(a)] as well as a chitinous gastric shield [Fig. 7.28(b)]. A diverticulum of the anterior intestine (style sac) lined with secretory cells and cilia secretes the characteristic crystalline style of bivalves. This is a mucopolysaccharide rod with amylase adsorbed onto it. Beating of the cilia lining the style sac rotates the style against the gastric shield. This action liberates the enzymes which are then mixed with the food. Styles are permanent in many bivalves (e.g. *Mya*), but in *Anodonta* they are only produced in response to the presence of food. Figure 7.29 shows the developing style secreted by glandular cells at the margins of the typhlosoles.

Partially digested food is passed by ciliated ducts into the digestive gland where it is digested intracellularly. Figure 7.30 shows small spheres beginning to break off the tips of a few absorptive cells in the digestive gland. These contain fragments of waste material which will be returned to the stomach.

Particles of food rejected by the digestive gland are wound up in a mucus string and passed by cilia down the intestine where they are compacted into faeces. A ridge of tall epithelial cells extends the length of the intestine forming a "typhlosole" or caecum which is probably involved in faeces formation. A short ciliated rectum passes through the ventricle of the heart (Fig. 7.31) to an anus which empties into the exhalent chamber. Closely applied to the outside of the intestine wall is some granular chloragogen tissue packed with glycogen, phospholipids and pigment, which appears to function as an organ of storage and excretion.

Excretion

Excretory functions have been proposed for the chloragogen tissue attached to the outside of the intestine, and osmotic and ionic regulation may occur across the gill surface. However, the chief excretory organs in *Anodonta* are a pair of pericardial (Keber's) organs and the kidneys. The pericardial organs are formed from thickened pericardial epithelium and are composed of a system of tubes surrounded by blood lacunae. Urine, containing dissolved ammonia, is filtered off from the blood into the anterior end of the pericardium. It moves thence to the kidneys via a reno-pericardial opening (nephrostome) [Fig. 7.32(a)]. The kidneys are paired, each one being U-shaped with a lower glandular secretory limb [Fig. 7.32(b)], and an upper storage end or bladder which opens by a wide duct into the exhalent chamber.

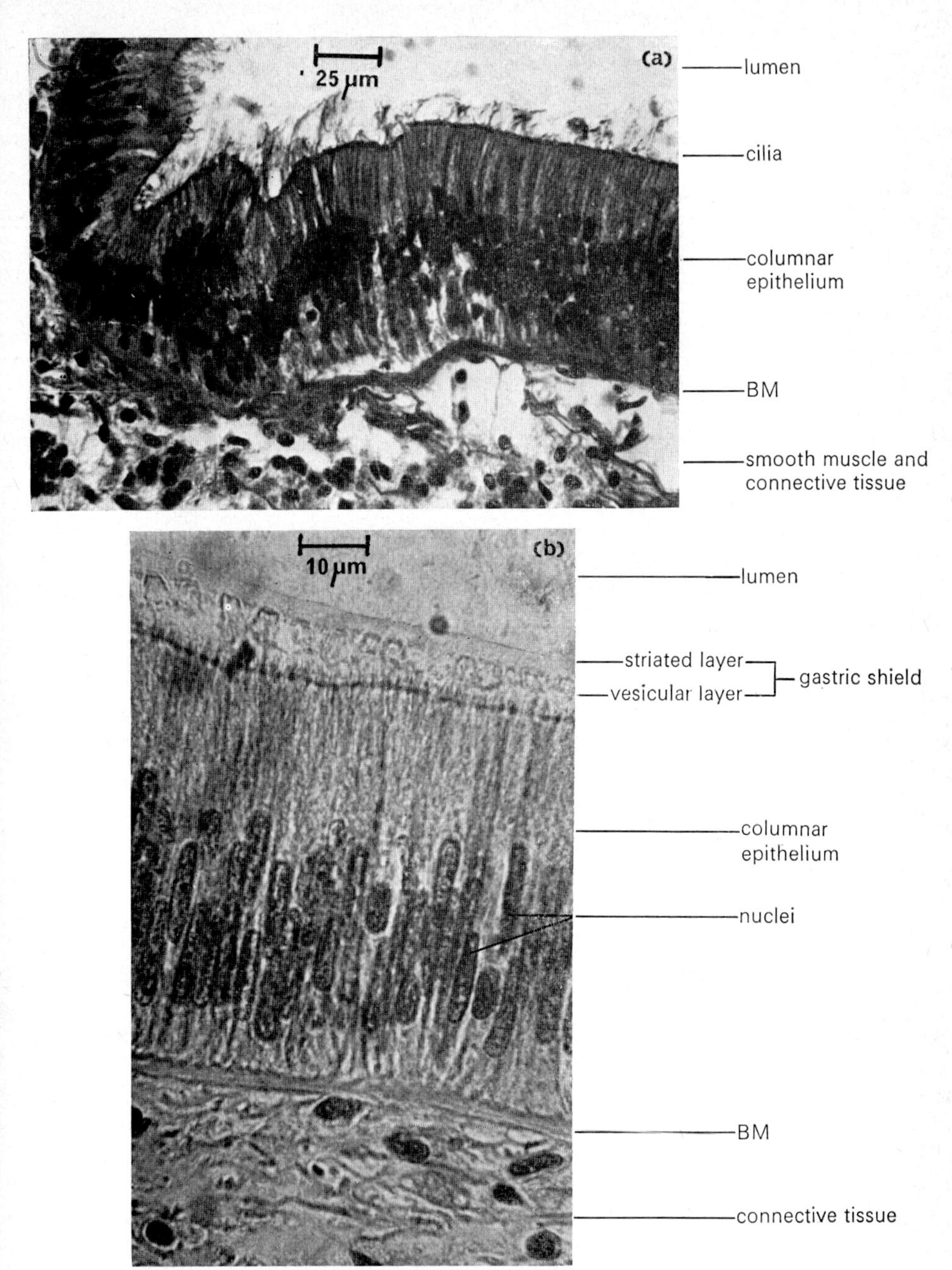

FIG. 7.28 (a) Section through the stomach wall of *A. cygnea*, showing part of a ciliary sorting area; (b) section through the stomach wall and gastric shield of the lamellibranch *Dreissena polymorpha*, a species closely related to *A. cygnea*.

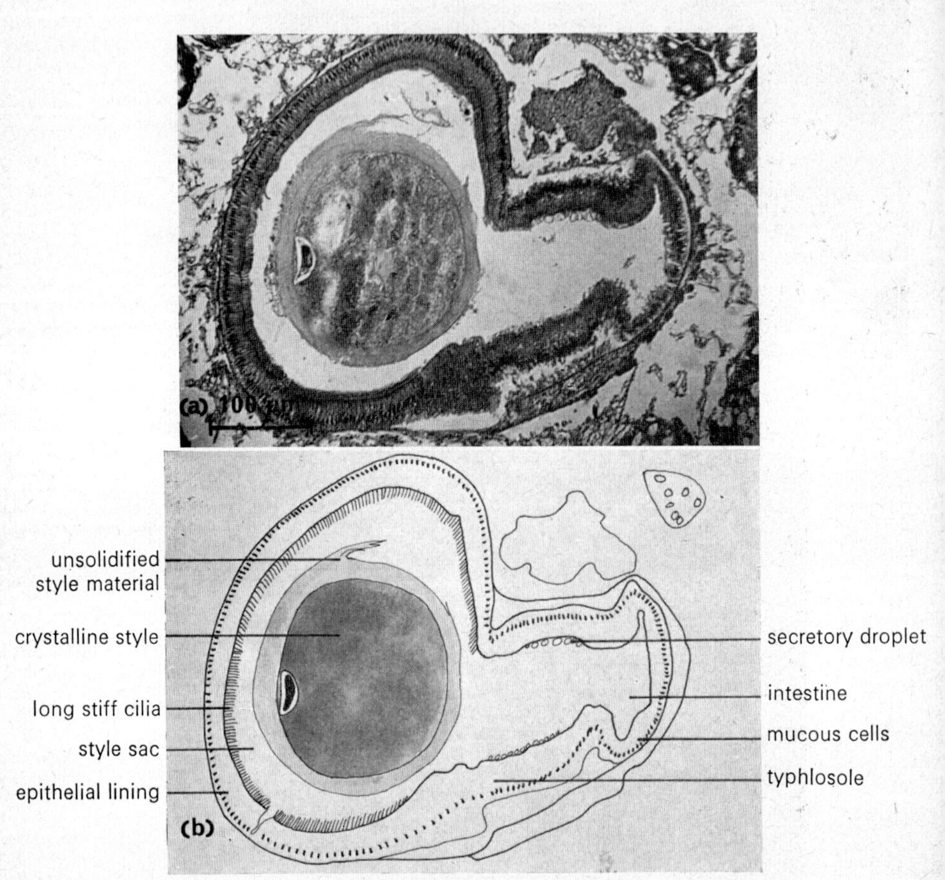

FIG. 7.29 (a) TS and (b) drawing of the intestine of *A. cygnea*, showing a crystalline style developing in the intestinal diverticulum (style sac).

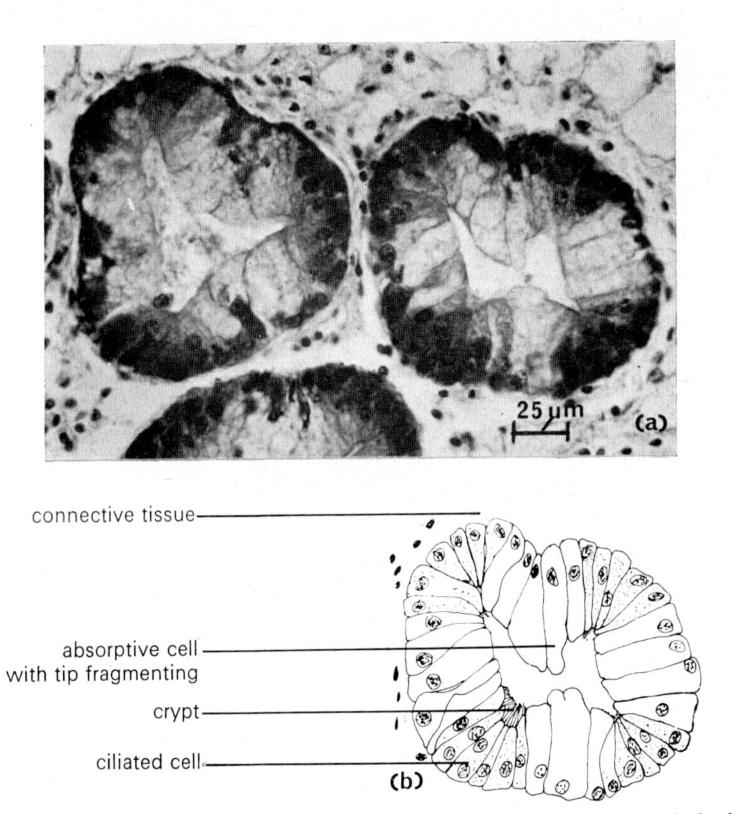

FIG. 7.30 (a) TS of digestive gland tubules of *A. cygnea*; (b) drawing of one tubule shown in (a).

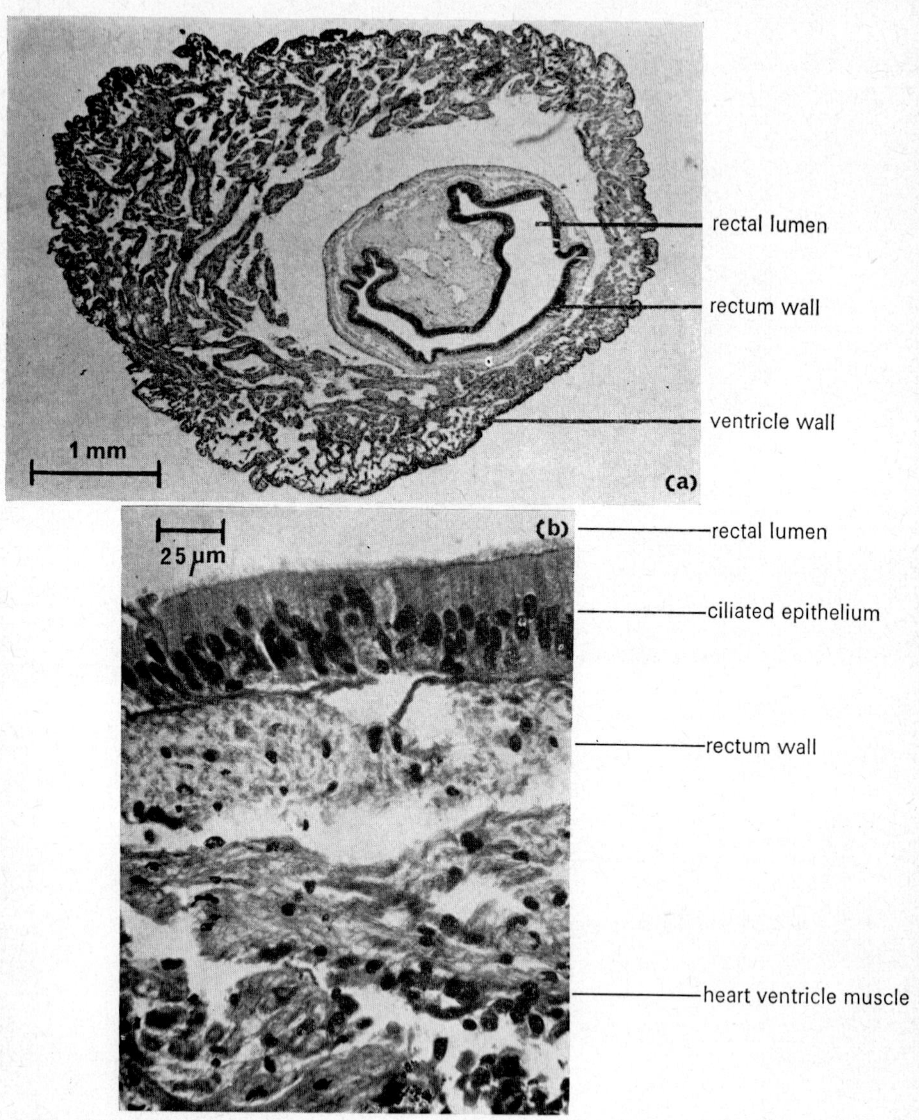

FIG. 7.31 (a) TS of the heart ventricle and rectum of *A. cygnea*; (b) detail of the rectum wall and ventricular muscle.

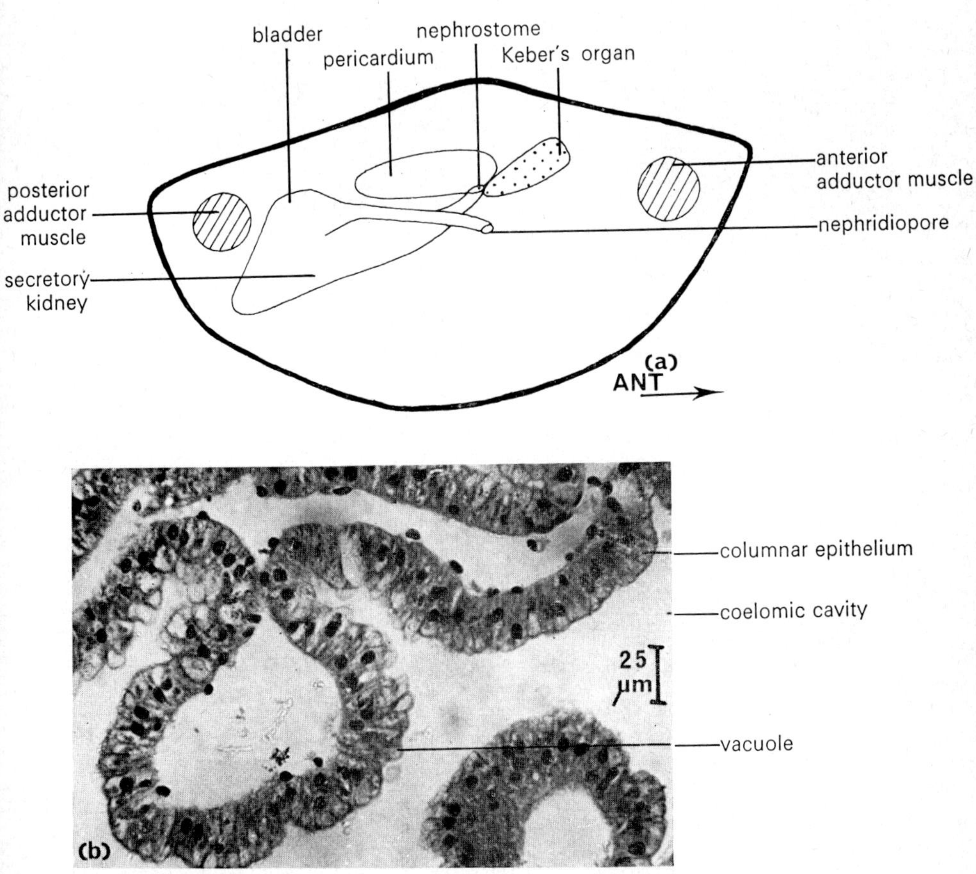

FIG. 7.32 (a) Plan of the excretory system of *A. cygnea*; (b) detail of the folded wall of the glandular secretory limb of the kidney of *A. cygnea*.

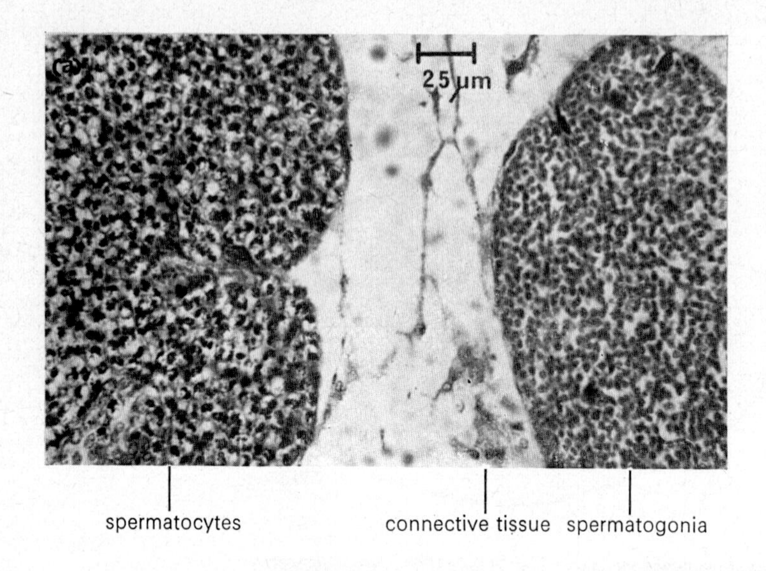

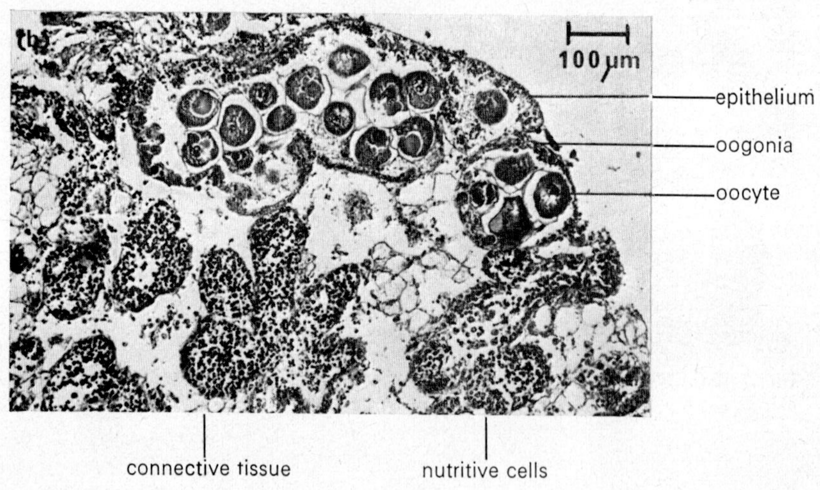

FIG. 7.33 (a) Section through the testis of a male *A. cygnea*, showing two tubules containing sperm at different stages of development; (b) section through the ovary of a female *A. cygnea*.

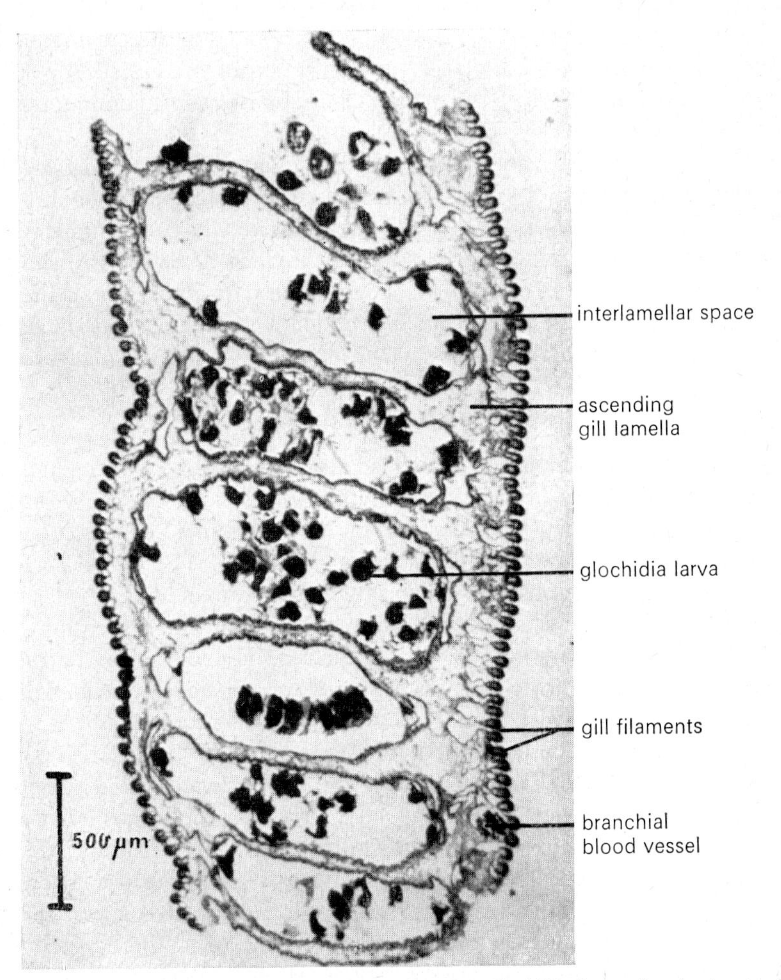

Fig. 7.34. LS of the gill of a female *A. cygnea*, showing glochidia larvae incubating in the interlamellar spaces.

Reproduction

Bivalves can be either hermaphrodite or dioecious: in *Anodonta* the sexes are generally separate although some hermaphrodites have been reported. There are simple paired gonads which are fused in the mid-line to form apparently single organs subdivided into lobes. The gonads develop as evaginations of the coelom, and are lined with germinal epithelium. Maturing gametes lie freely in the lumen of the gonad (Fig. 7.33). They are shed directly into the mantle cavity by two short gonoducts which open close to the kidney ducts.

In the female, ova are fertilized in the mantle cavity. They are then incubated in between the gill lamellae, in the interlamellar spaces (Fig. 7.34). The small glochidia larvae which develop have triangular shells bearing spines with which they attach to the fins of freshwater fish, when they are released from the parent. They live a parasitic mode of life until they are large enough to lead an independent existence. This is atypical of lamellibranchs which, in general, produce free-swimming larvae.

Class Cephalopoda

Cephalopods are the most specialized and highly evolved of the molluscs. They are often very large and are active, predacious marine carnivores which are particularly well-adapted for swimming. They are bilaterally symmetrical about an elongated dorso-ventral axis. The dorsal surface is drawn out into a blunt cone which is generally directed forwards during locomotion and therefore becomes the functional anterior end. A well-developed head occurs on the ventral surface, surrounded by a crown of prehensile tentacles which may be formed during development from the basic molluscan foot, although there still seems to be some doubt as to their origin. Just behind the head lies a muscular spout or funnel, also possibly formed from the foot, which controls outflow of water from the mantle cavity and can produce directional jets of water to be used in fast swimming. The mantle cavity (in the primitively posterior position) thus lies at the lower side of the body, in an effectively ventral location. It contains paired, non-ciliated gills; respiratory currents are directed across them by mantle muscles. Primitively in the class there was a chambered shell which was probably a buoyancy mechanism. This still exists in *Nautilus*, but in other present-day species the shell is normally reduced and internal or secondarily absent. Cephalopods are further characterized by a large complicated brain and well-developed sense organs, especially the eyes. They nearly all discharge black ink from an ink-sac to produce a decoy.

Example: *Sepia* (*officinalis*)

Cuttlefish are abundant in shallow coastal waters. They normally live on the sea bottom but feed at night near the surface. *Sepia* belongs to the order Dibranchiata which possess a single pair of gills, and it is a decapod, *i.e.* the head bears a ring of ten arms or tentacles. These are provided with suckers, although in the male one of the arms is modified as a copulatory organ, and on this the suckers are reduced. The body is cylindrical with a long pointed visceral hump bearing two lateral fins which are muscular extensions of the mantle and are used for slow swimming. These features can be seen in Fig. 7.35 which shows a dorso-ventral section of a young cuttlefish shortly after hatching. The mantle has lost its thin, vascular nature and has developed thick muscles (Fig. 7.36). It now forms the body wall and is used in respiration and locomotion. The shell is reduced in size, and is modified for use as a buoyancy device although it still serves as a supporting structure. Its internal position can be seen in both photographs, although the shells have been dissolved away. The body wall is covered by a single-layered epidermis overlying a dermis which contains large chromatophores operated by radial muscle fibres that are used in fast colour changes. Beneath the dermis lie sheets of longitudinal, circular and radial muscle fibres (Fig. 7.37).

Nervous system

The nervous system of dibranchiate cephalopods (cuttlefish, squid and octopods) is very highly developed with the typical molluscan ganglia fused to form a distinct brain. The brain is almost completely enclosed in a capsule of a hard packing tissue, "cartilage" (Fig. 7.38) and is buffered by gelatinous tissue. Figure 7.39 shows a plan of the cuttlefish brain. The pleural, brachial, pedal and visceral ganglia are fused round the oesophagus while the cerebral ganglia remain practically distinct although they connect with the other ganglia. There is a high degree of differentiation of function within the ganglia; the cerebral ganglia being divided into many lobes and acting as higher controlling centres. The brain has a plentiful blood supply and follows the normal invertebrate pattern of peripheral neurones surrounding the neuropile: giant neurones are also common (Fig. 7.40). The neurones of the dorsal vertical lobe of the brain are numerous and small. These are characteristics of neurones from integrative brain centres (see also Figs. 13.5, 15.7 and 17.6).

Since the eyes are exceptionally well-developed there are two very large optic ganglia with many optic nerve fibres. The ganglia have a highly organized structure with an outer cortex and an inner medulla containing large multipolar cells (Fig. 7.41). The pedal ganglia send nerves to the funnel and tentacles, while the visceral ganglia supply the visceral hump and mantle. The mantle nerves contain giant nerve fibres (reaching diameters of 1 mm in the squid) which mediate fast escape reactions. They have their cell bodies in the stellate ganglia, which supply the new mantle muscles.

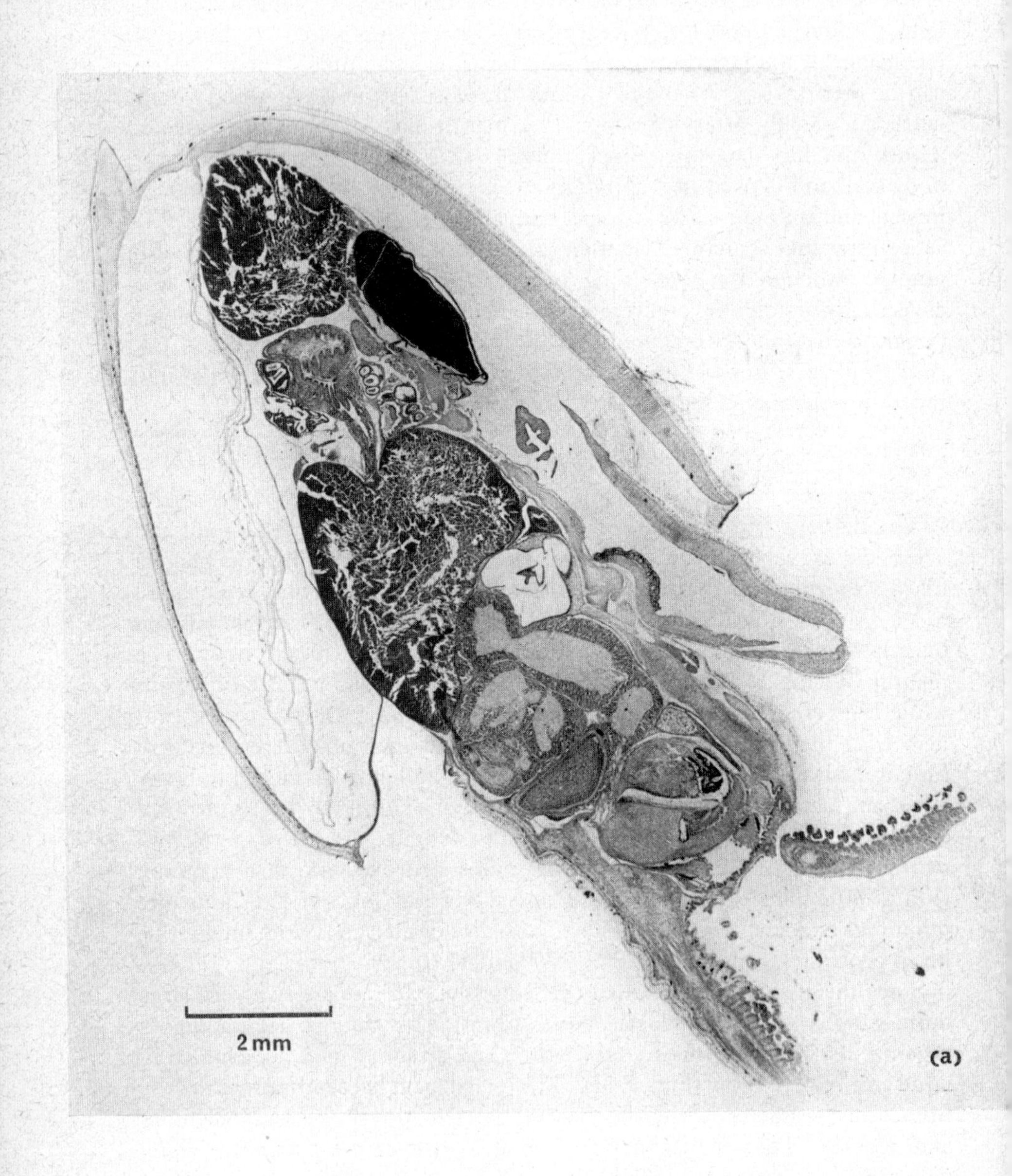
2 mm
(a)

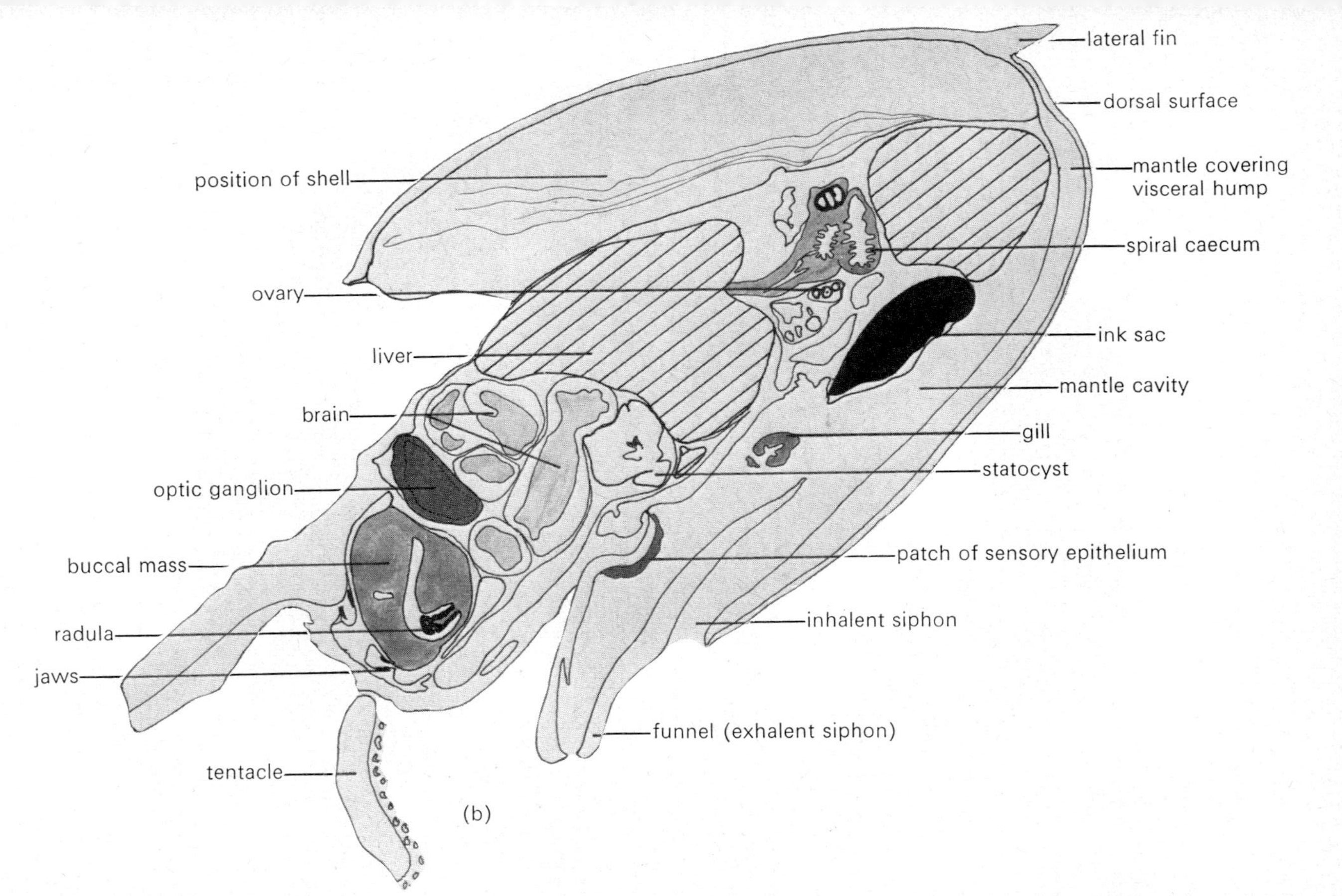

FIG. 7.35 (a) Dorso-ventral LS; and (b) explanatory drawing of a young cuttlefish (*Sepia officinalis*). Note that the shell has been dissolved away.

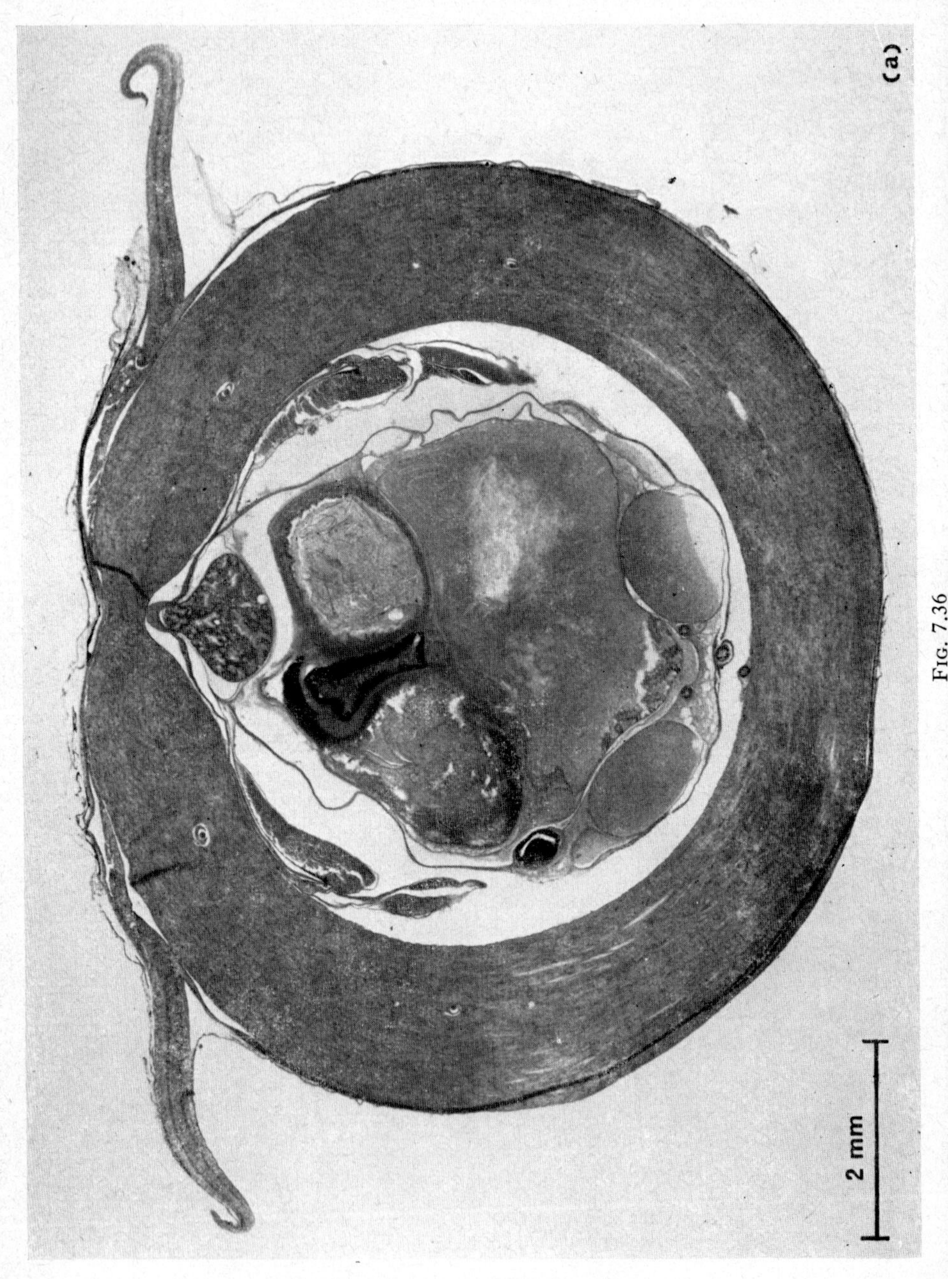

Fig. 7.36

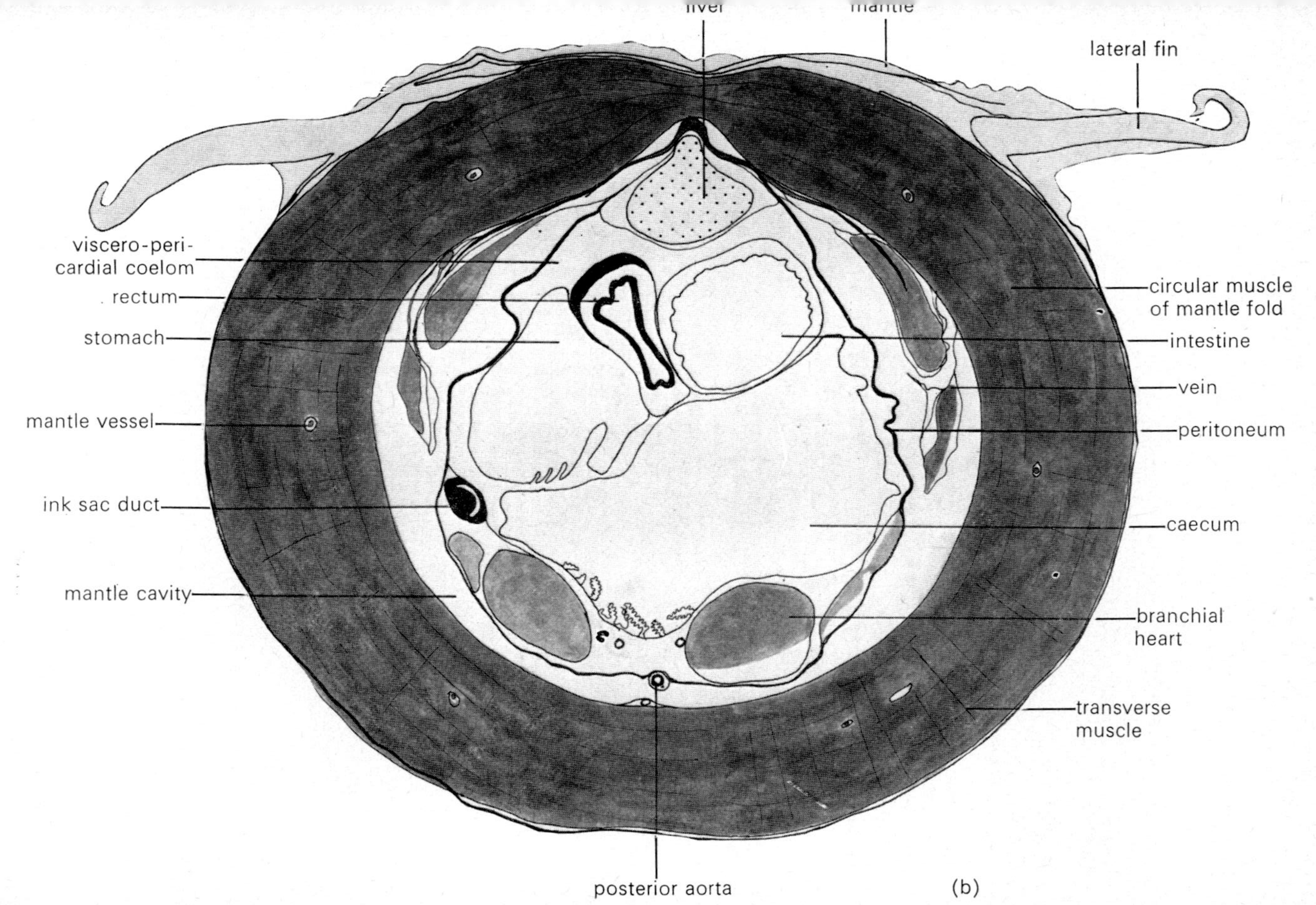

FIG. 7.36 (a) TS and (b) explanatory drawing of a young *S. officinalis* in the visceral hump region. Note that the shell has been dissolved away.

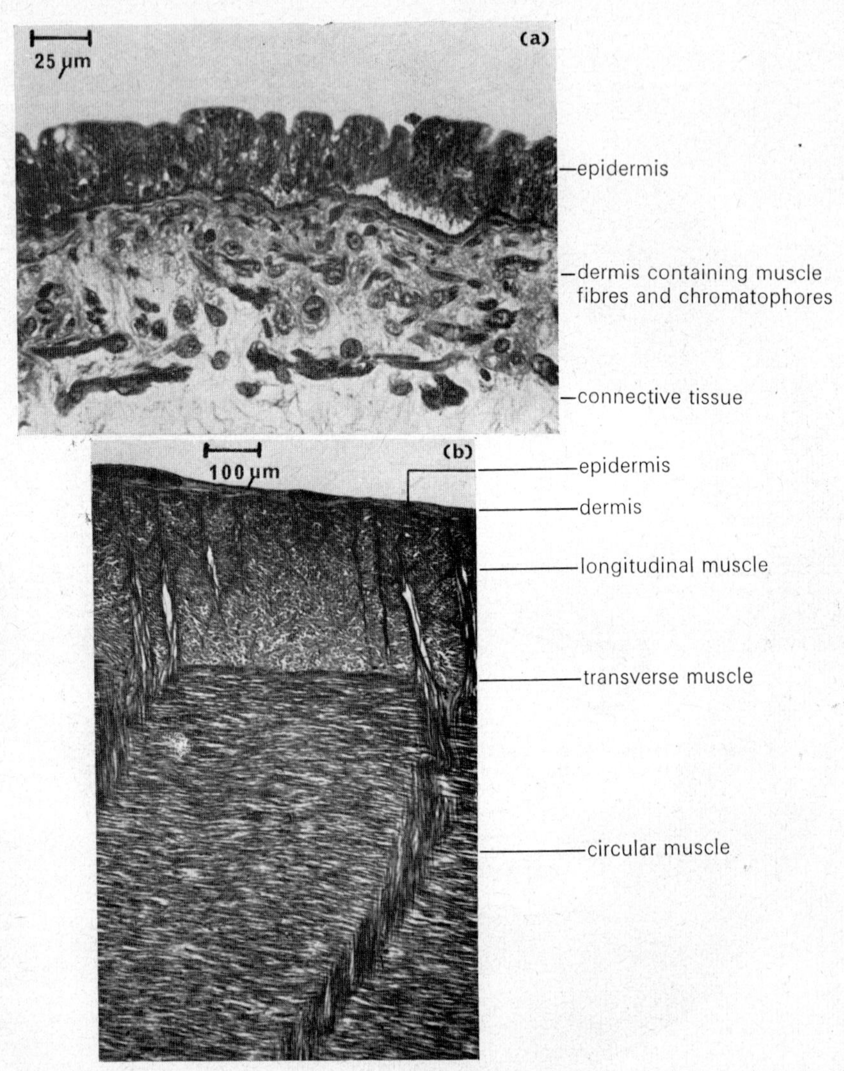

FIG. 7.37 (a) Section through the skin of *S. officinalis*; (b) TS through the mantle of *S. officinalis*. Very thick muscle layers convert the mantle into a tough muscular body wall.

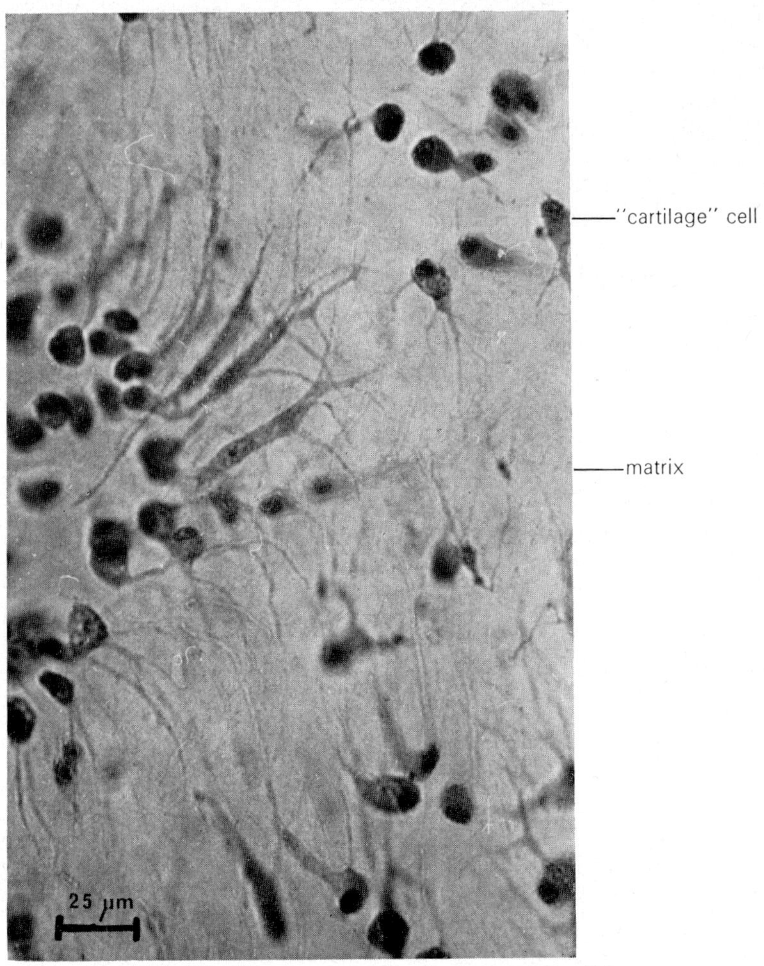

FIG. 7.38. "Cartilage" surrounding the brain of *S. officina*

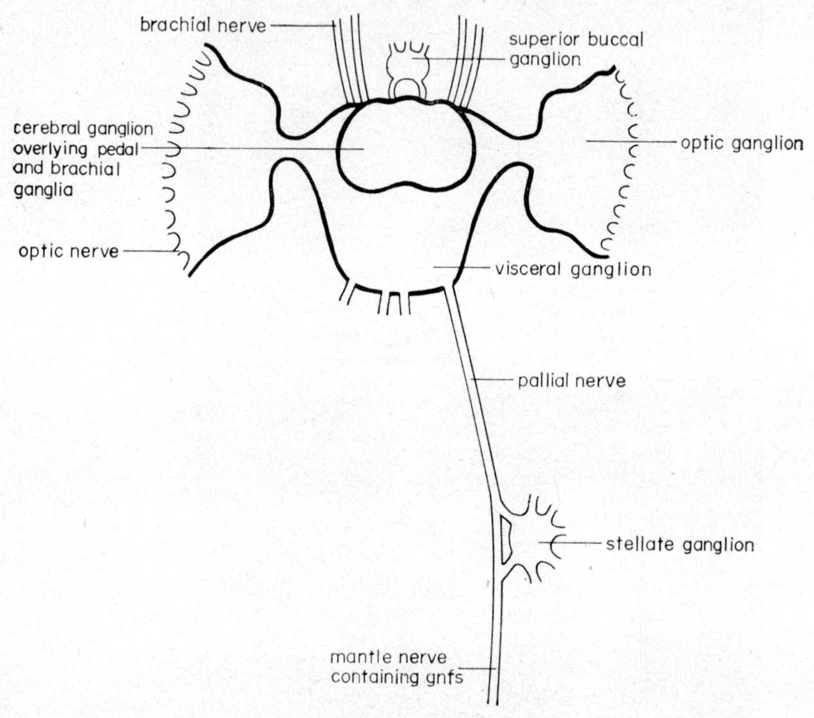

FIG. 7.39. Plan of the anterior part of the cuttlefish nervous system.

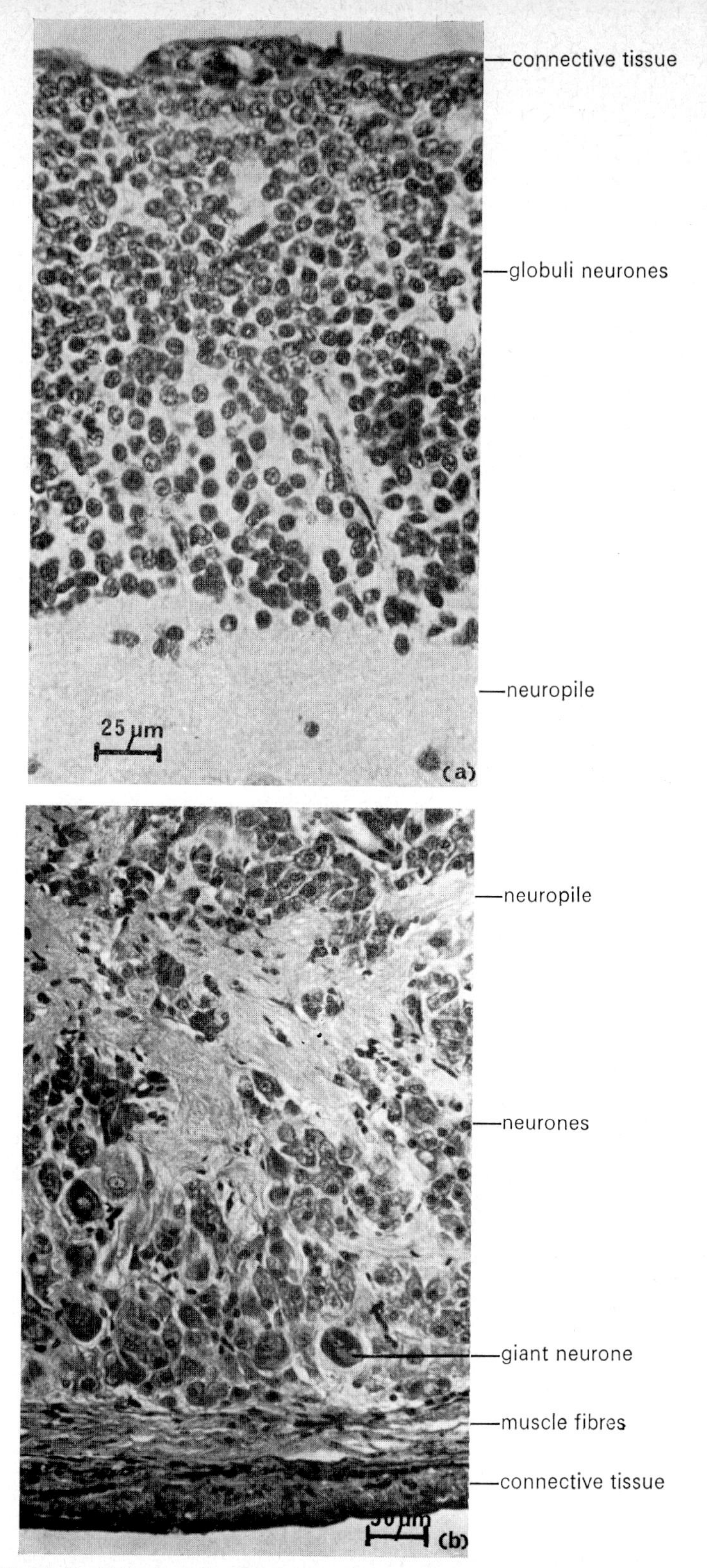

FIG. 7.40 (a) Section through the dorsal surface of the vertical lobe of the brain of *S. officinalis*; (b) section through the ventral surface of the basal lobe of the cuttlefish brain.

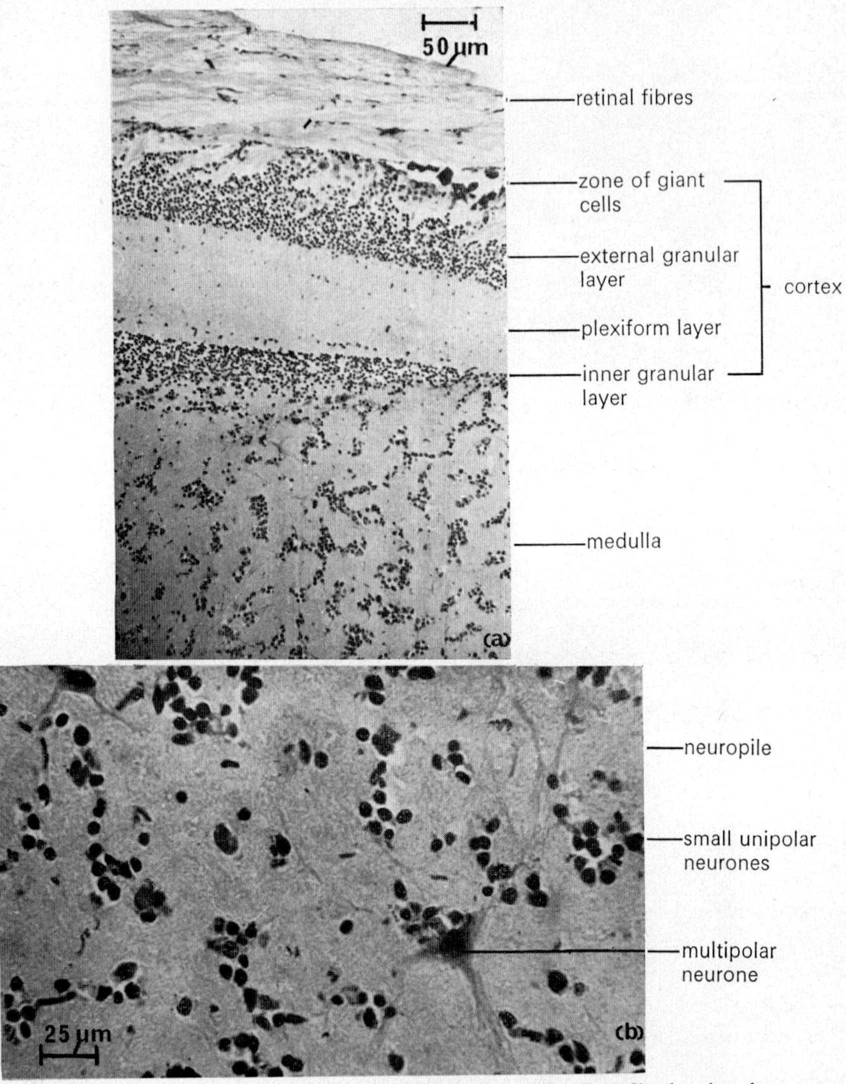

FIG. 7.41 (a) Section through the optic lobe of the brain of *S. officinalis*, showing the cortex and medulla; (b) detailed structure of the medulla of the optic lobe.

Sense organs

Cephalopods require advanced sense organs for their mode of life. The sense of smell is poorly developed but the eyes are large and complex. They are similar in structure to the vertebrate eye and are capable of forming images (Fig. 7.42). There are two refractive surfaces, the cornea [Fig. 7.43(a)] and lens. The lens is secreted and supported by a ciliary body [Fig. 7.43(b)]. The eye itself can be moved by extrinsic eye muscles, while the lens can also be shifted backwards and forwards by ciliary muscles to focus an image on the retina. The extensive and heavily pigmented retina is of the direct type, i.e. with the photoreceptors directed towards the source of light and with underlying nerves (Fig. 7.44). The photoreceptors are connected to retinal cells which send back fibres to the optic ganglion. There is evidence that most cephalopods have colour vision, but it has not yet been established if there are separate photoreceptors responsible for colour vision as in vertebrates. The amount of light entering the eye is controlled by a muscular iris diaphragm, while the amount of light reaching the visual cells can be further regulated by movement of pigment round the receptors. The shape of the eye is maintained by a scleroid coat reinforced by cartilage.

Other sense organs include small olfactory pits behind the eyes and a pair of large statocysts or auditory organs (see Fig. 7.35). Each statocyst is a fluid-filled vesicle bearing a patch of sensory epithelium. A large calcareous otolith suspended in the fluid is moved over the sensory hairs by sound waves. It thus works on a principle similar to that of the vertebrate ear (see p. 455).

Respiration

The increase in size and activity of the cephalopods has led to a raised metabolic rate which has necessitated a modification in the structure of the gills to provide more oxygen for the tissues. The two gills, suspended one on either side of the visceral mass, are composed of fleshy, non-ciliated filaments with an extensive capillary circulation. The gills are bipectinate (with two rows of filaments) and their surface area has been greatly increased by secondary and tertiary folding (Fig. 7.45).

Water currents, created by the action of mantle muscles, flow from the afferent to the efferent side of the gills. Relaxation of the circular mantle muscles causes an enlargement of the mantle cavity and movement of the mantle edge away from the head; water is thus drawn in round the base of the funnel (see Fig. 7.35). Contraction of the mantle muscles decreases the size of the mantle cavity and forces the mantle against the head, thereby closing the inhalent siphon and leading to the expulsion of water out through the funnel, which thus acts as the exhalent siphon. The funnel is mobile and can produce directional jets of water for rapid forward or backward movements.

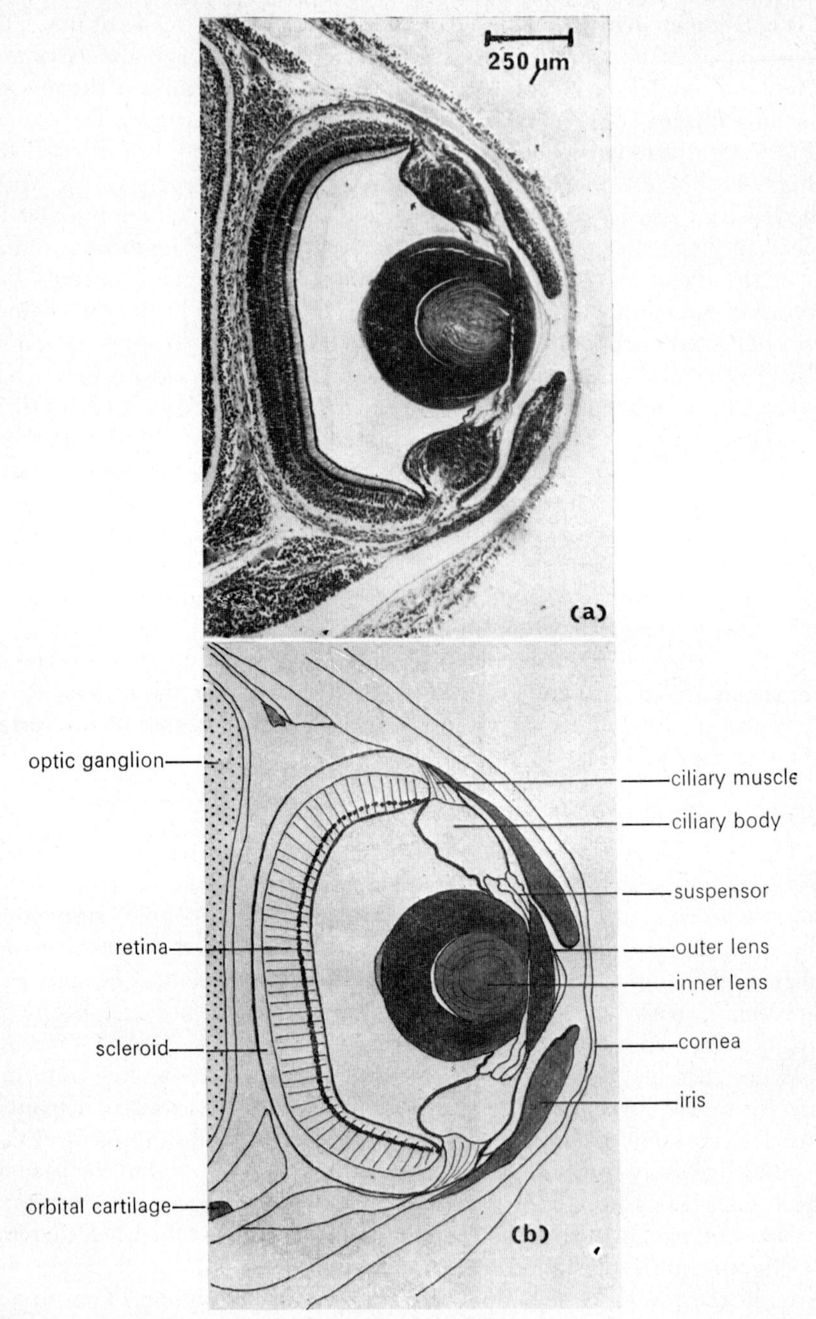

FIG. 7.42 (a) Vertical section and (b) drawing of the eye of an embryo *S. officinalis* shortly before hatching.

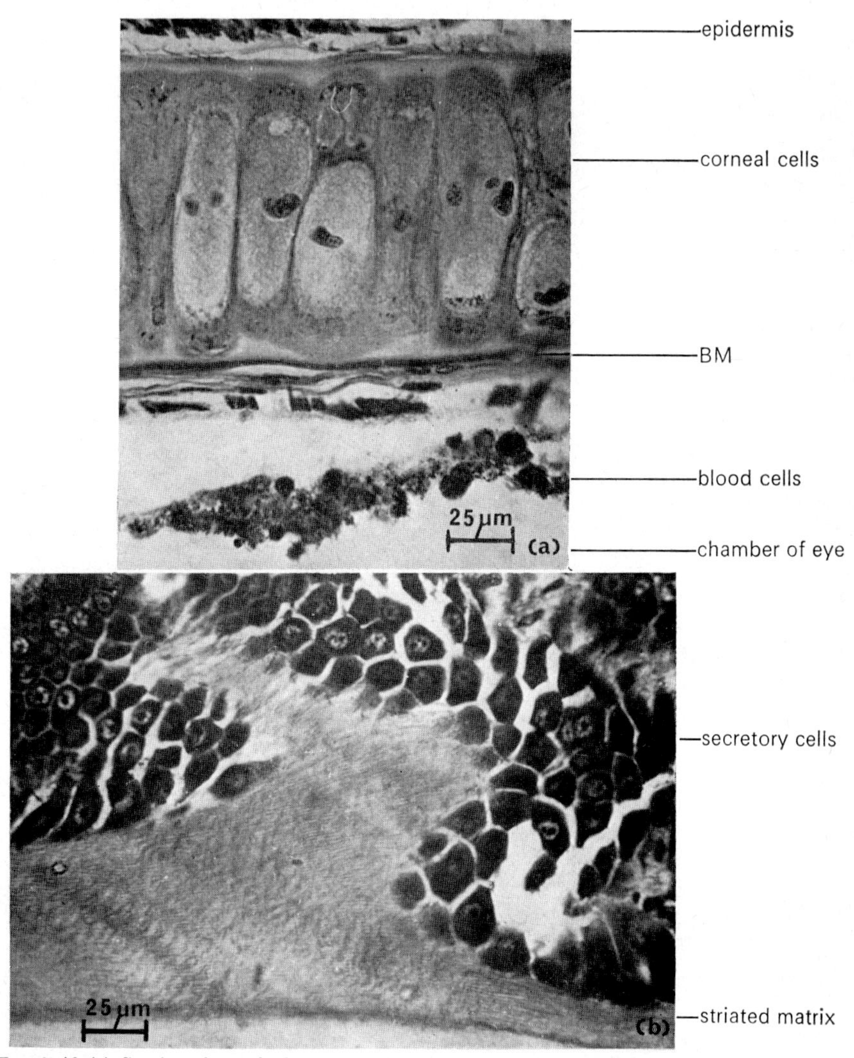

FIG. 7.43 (a) Section through the cornea from the eye of an adult *S. officinalis*; (b) ciliary body from the eye of an adult cuttlefish.

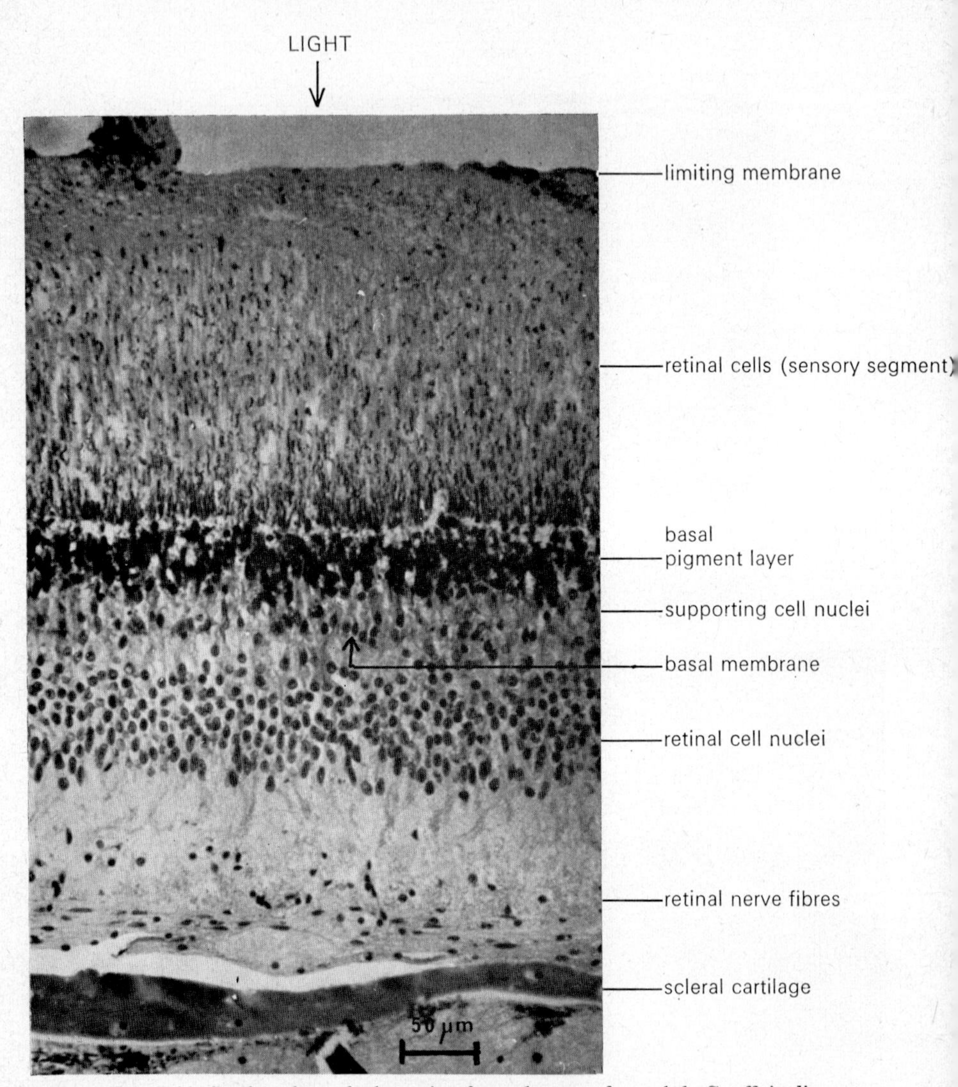

FIG. 7.44. Section through the retina from the eye of an adult *S. officinalis*.

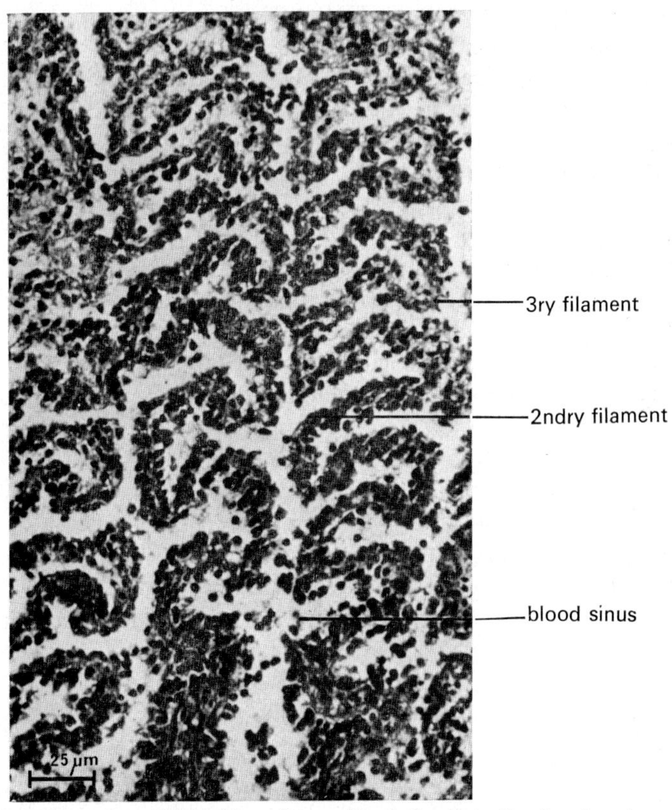

FIG. 7.45. Section through the bipectinate gill of *S. officinalis*, showing the secondary folding of the gill surface. This tissue shows signs of necrosis.

Circulation

Because of the nature of the gills and the high metabolic rate, cephalopods have lost the haemocoel of lower molluscs, and their circulatory system is almost completely closed with a high-pressure supply to the gills. A large heart, consisting of paired lateral auricles and a single ventricle, pumps oxygenated blood through anterior and posterior aortae to minute capillaries in the tissues. Deoxygenated blood is returned from the tissues via the kidneys to contractile branchial hearts at the base of the gills which force blood through the gills and back to the auricles (Fig. 7.46). Circulation of blood through this relatively high resistance circuit is assisted by elastic walls of the aortae and vessels of the liver, mantle, gills and arms.

Digestion

Cuttlefish prey on small fish, prawns and other crustaceans. This is a highly digestible protein-rich diet, and is rapidly assimilated. The digestive tract is modified accordingly, food being passed through the gut by powerful muscles rather than only by cilia.

The prehensile arms round the mouth are equipped with suckers for catching moving food. These suckers are muscular cup-shaped organs with special cells round the rim which secrete mucus for adhesion, and horny material to assist gripping and even rasping (Fig. 7.47). The buccal mass, enclosing a pair of strong jaws embedded in muscle, lies within a blood sinus which thus allows it to move when the animal is biting. Muscles surrounding the sinus allow extrusion of the buccal mass. The buccal cavity is filled by an odontophore bearing a radula (see Fig. 7.35), while opening into it are two pairs of salivary glands. The anterior pair within the buccal mass secrete mucus, while the second pair, lying in front of the digestive gland, secrete mucus, proteolytic enzymes and poison (Fig. 7.48).

The arrangement of the digestive tract is shown in Fig. 7.49. The oesophagus is lined with chitin for protection against hard food [Fig. 7.50(a)]. It leads into a sac-like stomach ("gizzard") with thick muscular walls and a ciliated epithelial lining which also secretes a chitinous coat [Fig. 7.50(b and c)]. Food passes from the stomach to a large spirally coiled caecum (Fig. 7.51) possessing ciliary sorting areas which channel fine particles into a groove and up to the digestive gland. This groove is separated by a columellar ridge (=typhlosole), from a second mucus-collecting groove, which collects the string of waste material from the caecum. Much extracellular digestion occurs in the stomach and caecum, only very small particles reaching the digestive gland for intracellular digestion. Food rejected by the caecum is passed out of the body by a short, cuticle-lined intestine and rectum to the anus which is situated just behind the funnel. The ink-sac, which is a rectal diverticulum, also ejects ink through the anus.

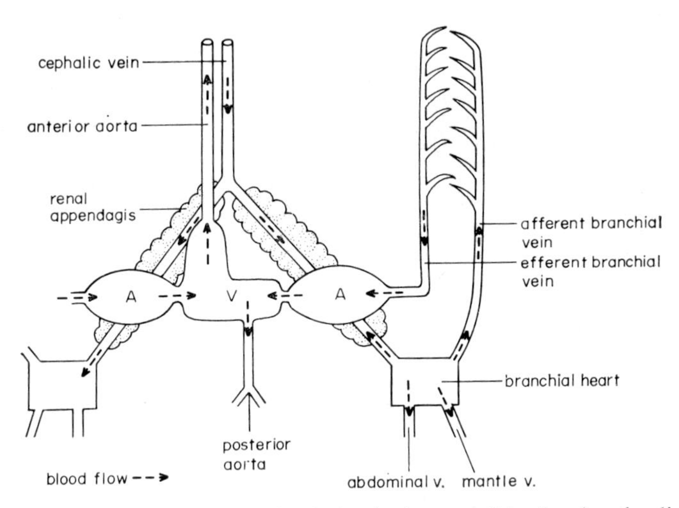

FIG. 7.46. Plan of the heart and gill circulation in the cuttlefish, showing the direction of blood flow. (A=auricle; V=ventricle.)

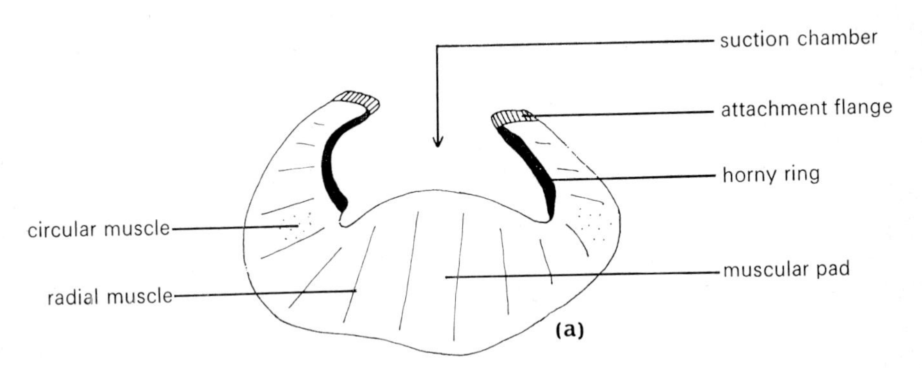

FIG. 7.47 (a). See p. 306 for legend.

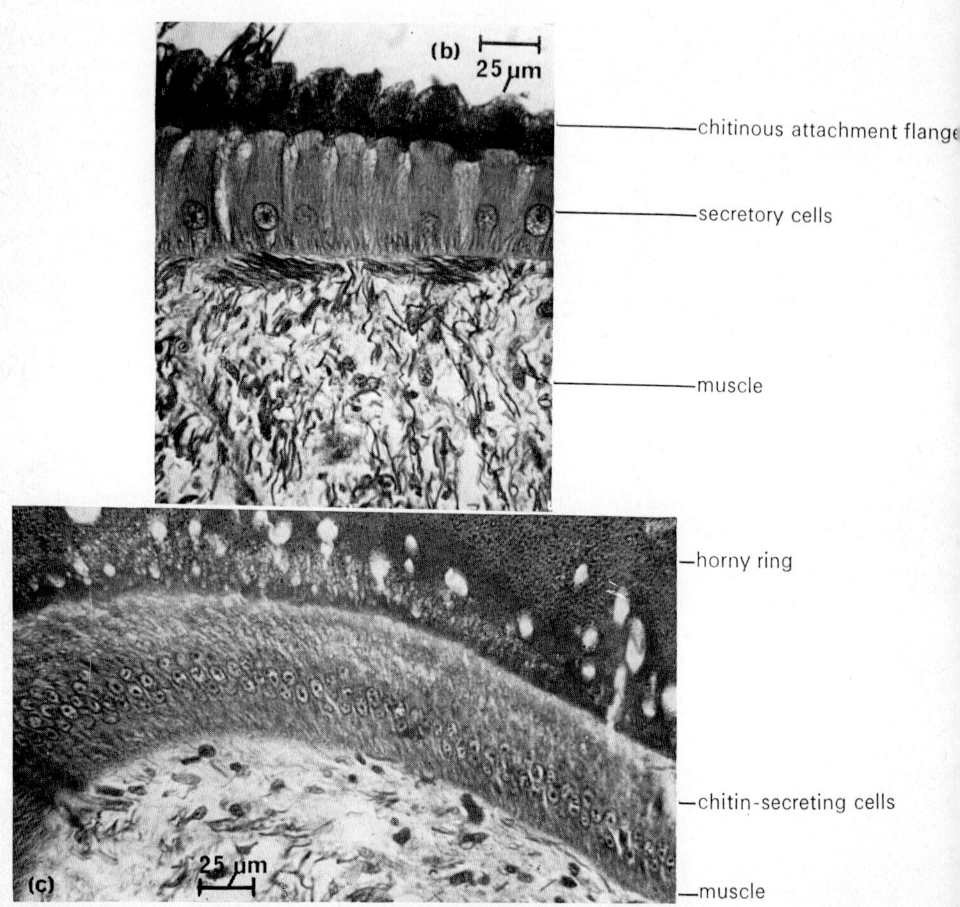

FIG. 7.47 (a) Diagram of a vertical section through a sucker of *S. officinalis*; (b) detail of the cells round the rim of the sucker, and the attachment flange; (c) detail of the horny ring and chitogenous cells lining the sucker.

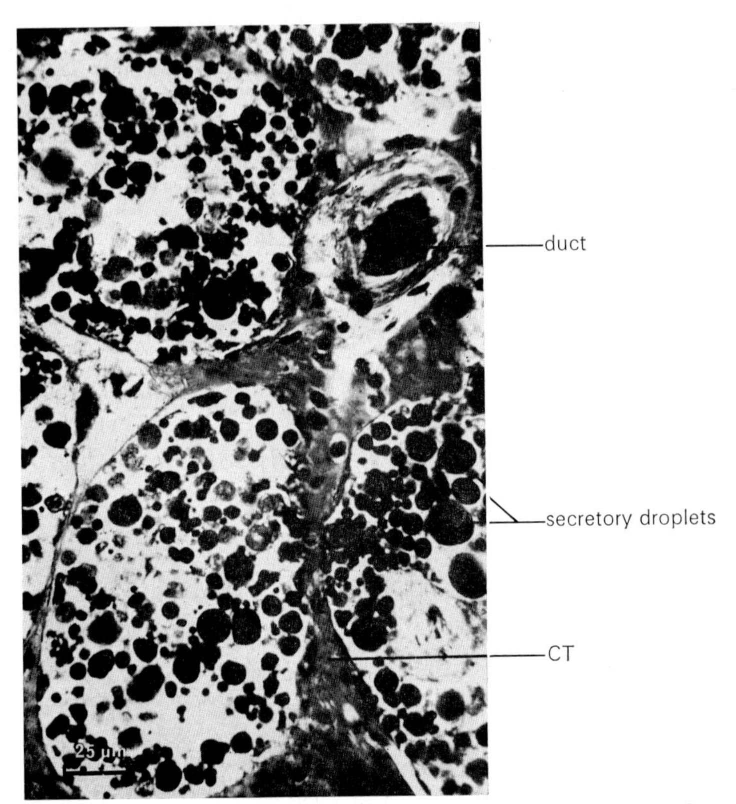

FIG. 7.48. Section through the posterior salivary gland of *S. officinalis*.

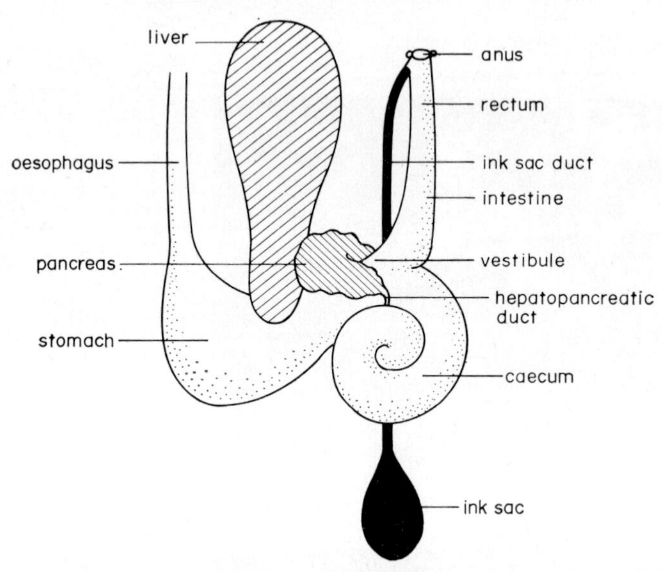

FIG. 7.49. Plan of the digestive tract of the cuttlefish.

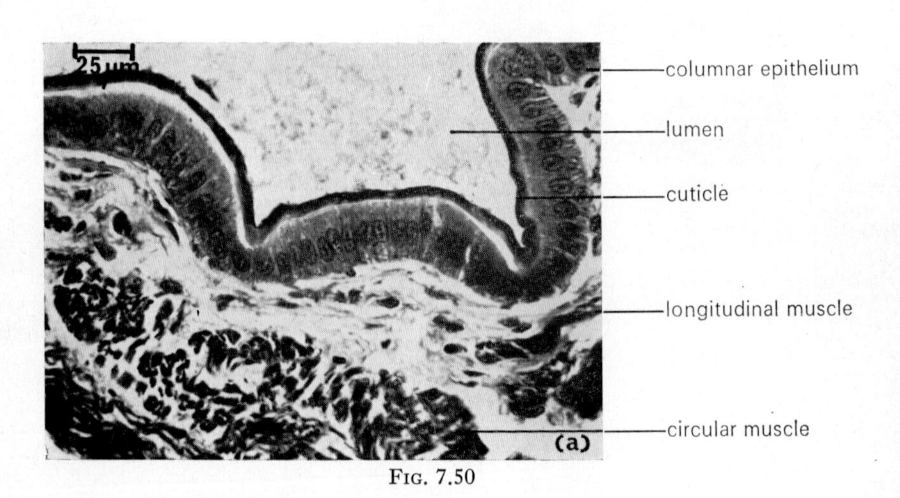

FIG. 7.50

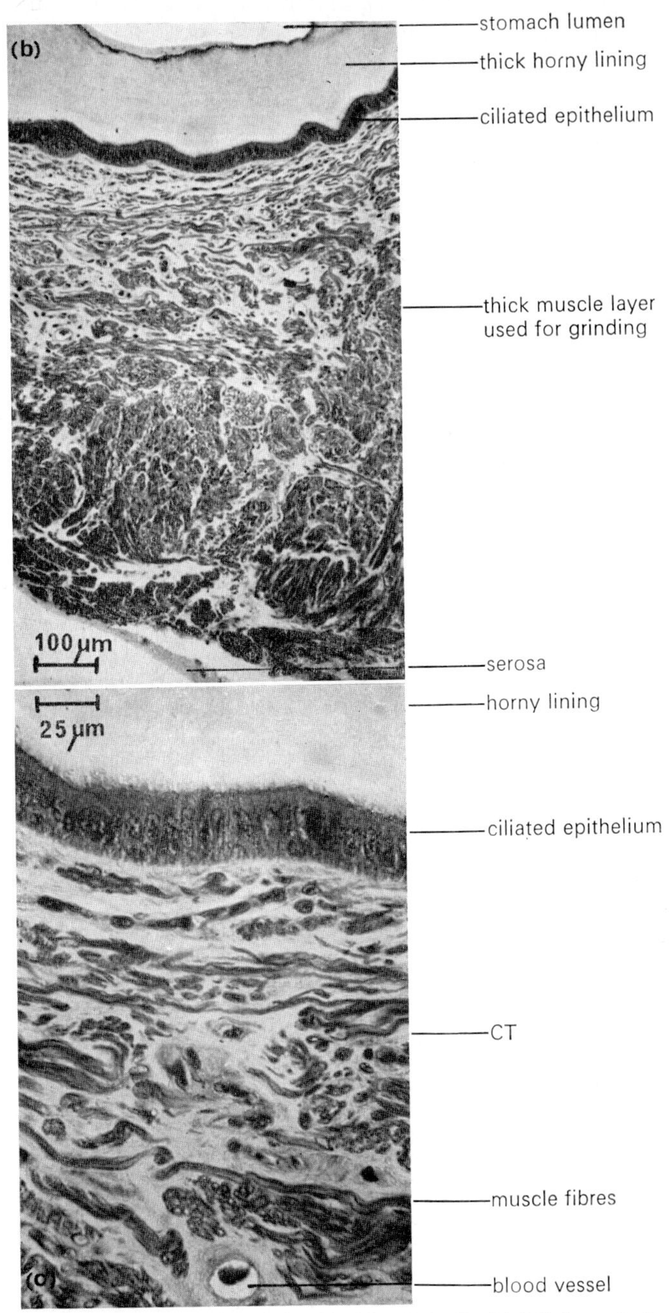

FIG. 7.50 (a) LS through the oesophagus wall of *S. officinalis*; (b) TS through the stomach wall of *S. officinalis*. Note the thick horny lining and powerful grinding muscles; (c) detail of the stomach wall shown in (b). Note the horny lining and the ciliated epithelium.

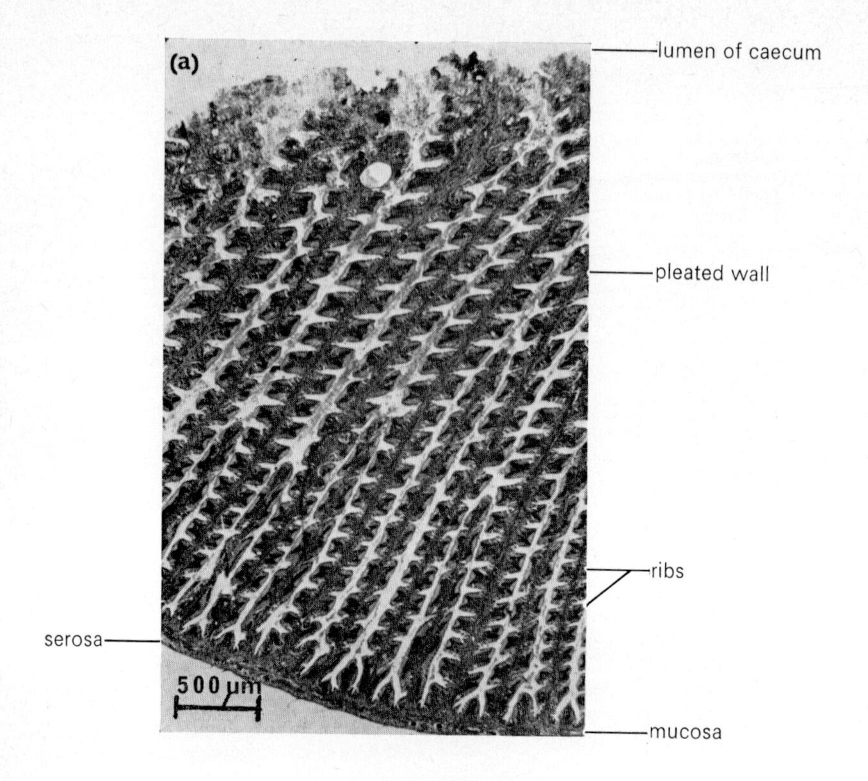

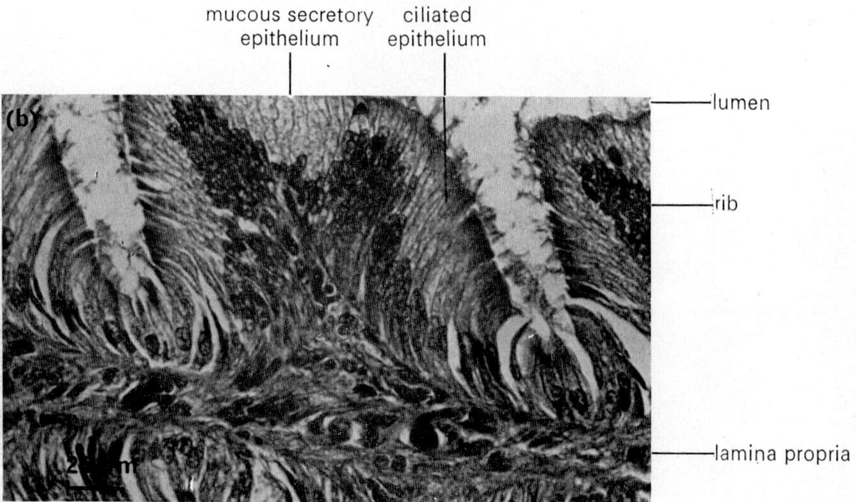

FIG. 7.51 (a) TS through the caecum of *S. officinalis*; (b) detail of the pleated wall of the cuttlefish caecum.

The digestive gland (hepatopancreas) can be divided into two morphologically and functionally distinct parts served by the same ducts (see Fig. 7.49).

The anterior part (liver) is a compact organ consisting of densely packed tubules opening into wide tubules continuous with the hepatopancreatic duct (Fig. 7.52). The tubules are lined with columnar cells which show phases of secretion, absorption and excretion (Fig. 7.53). Calcium-secreting cells often occur between the bases of the other cells. The liver is chiefly a storage organ but it also secretes enzymes involved in the final stages of intracellular digestion.

The smaller posterior portion is the pancreas. This consists of a loose mass of follicles opening by narrow ducts into the hepatopancreatic ducts. The epithelial cells are lower (Fig. 7.54) and secrete enzymes which are released into the stomach and caecum and are involved in the preliminary stages of digestion. The pancreas is also thought to have an excretory function, the tubules having a similar structure to those of the kidney [Fig. 7.55(b)].

Excretion

The chief excretory organ in cephalopods is a kidney or renal organ [Fig. 7.55(a)]. This possesses a large dorsal sac communicating with paired ventral sacs, and is connected with the mantle cavity by two renal papillae and with the pericardium by reno-pericardial canals. The thin-walled ventral chambers have afferent branchial veins attached to the inner dorsal surface. These veins are surrounded by voluminous glandular appendages composed of loosely packed tubules lined with tall columnar epithelium [Fig. 7.55(b)]. The cavities of the appendages are continuous with those of the veins so that blood is pumped into them and waste products are filtered off into the renal sacs and passed to the mantle cavity.

The pericardial organs, which are appendages of the branchial hearts, are accessory excretory organs. They filter off material from the body cavity into the pericardial cavity, whence it passes with the blood filtrate into the renal sacs and out through the renal papillae [see Fig. 7.55(a)]. Ammonia is the main excretory product in cephalopods.

Reproduction

Cephalopods are dioecious: in each individual there is generally a single median gonad representing part of the coelom, and paired genital ducts leading to gonopores situated in the mantle cavity. In *Sepia* only one (left) duct and pore are retained.

The testis is a solid mass of seminiferous tubules [Fig. 7.56(a)] showing continuous development of motile sperm. The sperm are released into the coelom through a longitudinal slit on the ventral surface of the testis

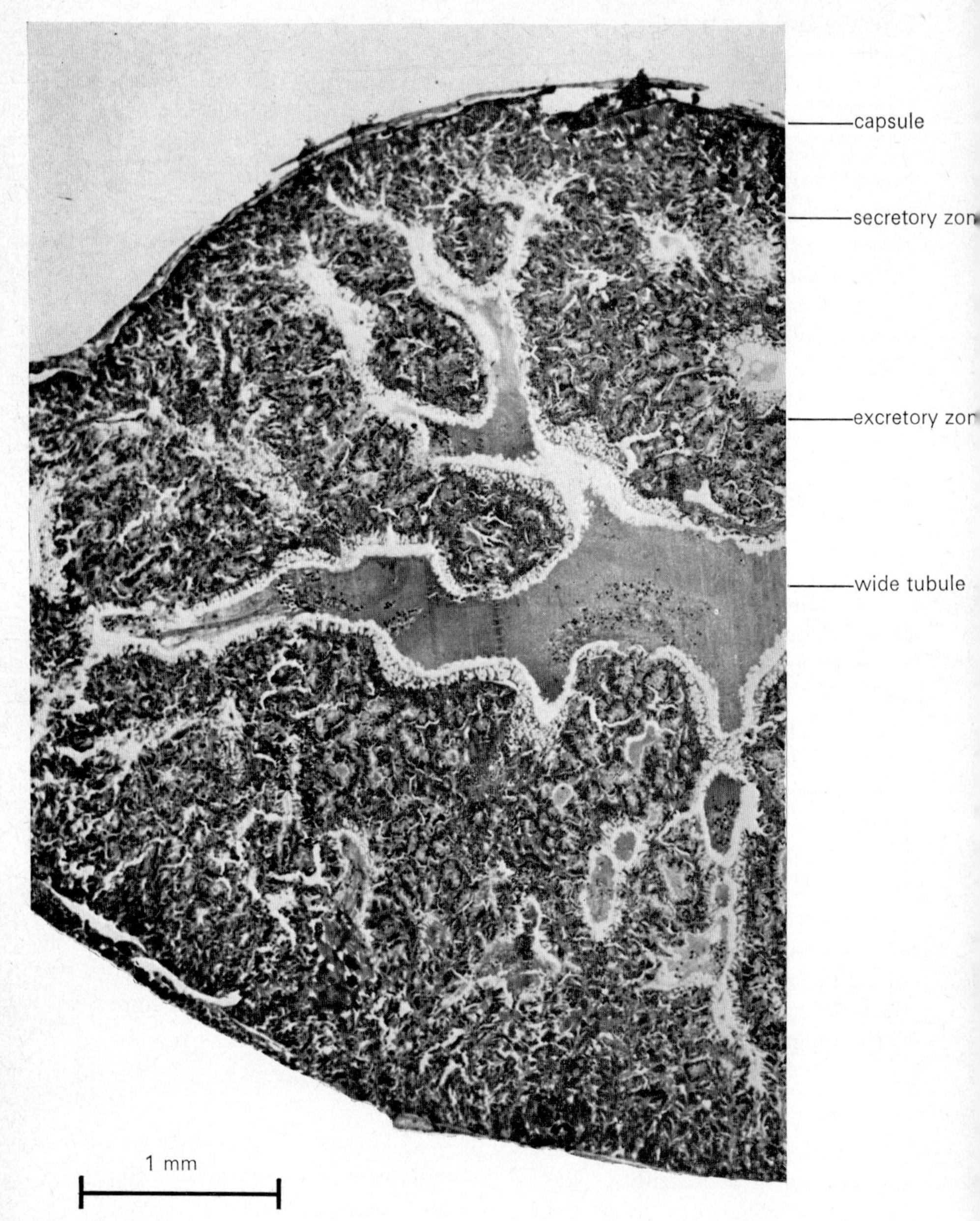

FIG. 7.52. Section of the liver of *S. officinalis*, showing the wide central tubule which joins the hepatopancreatic duct.

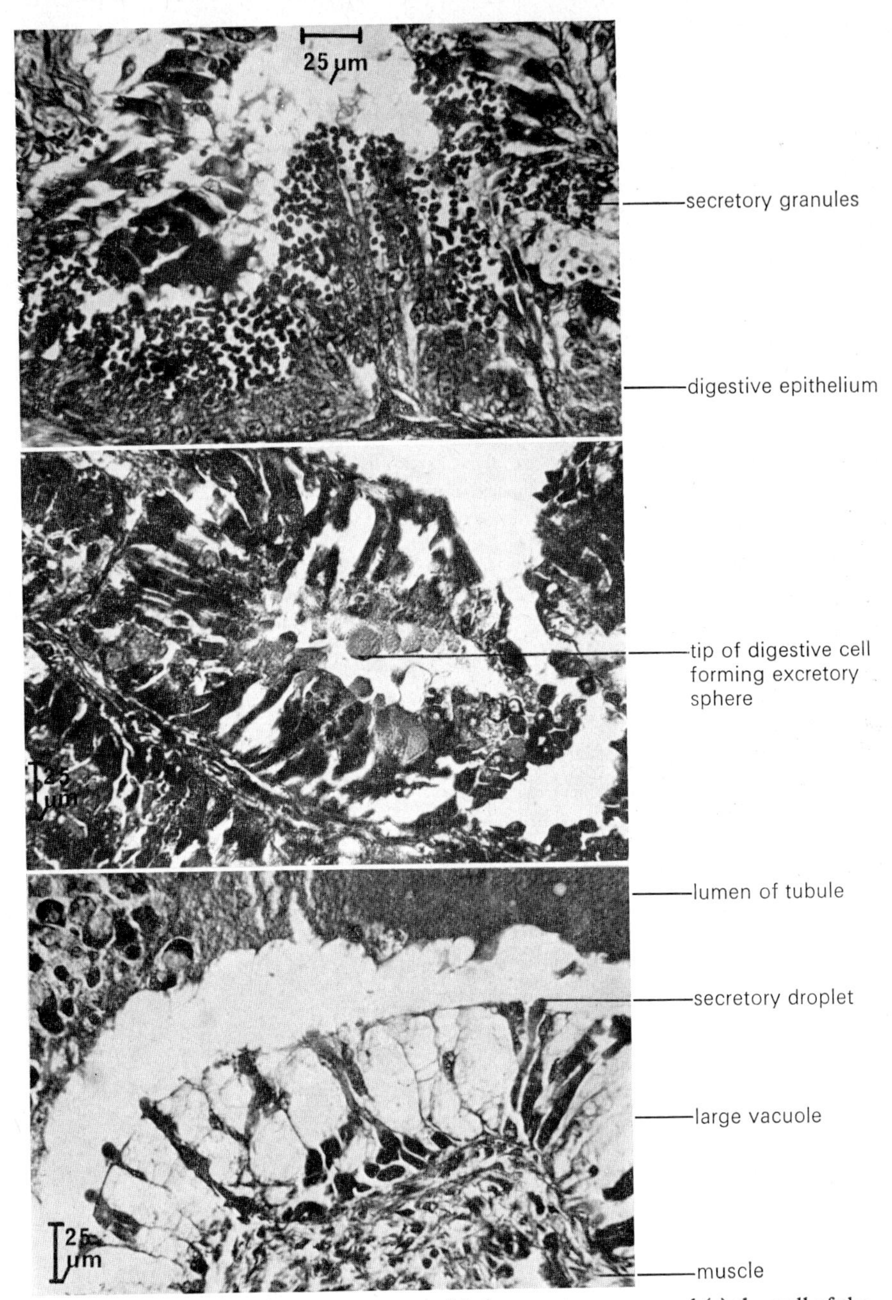

Fig. 7.53. Details of (a) the secretory zone; (b) the excretory zone; and (c) the wall of the wide tubule of the liver of *S. officinalis*.

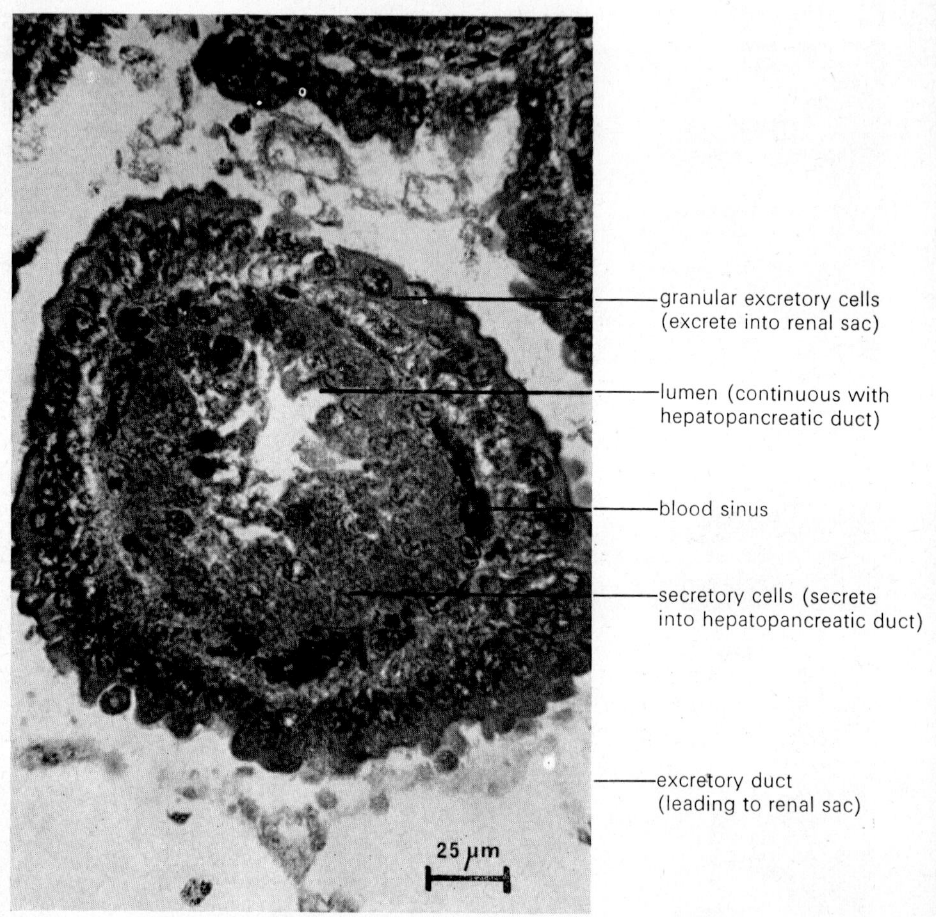

FIG. 7.54. Section through the pancreas of *S. officinalis* showing one follicle.

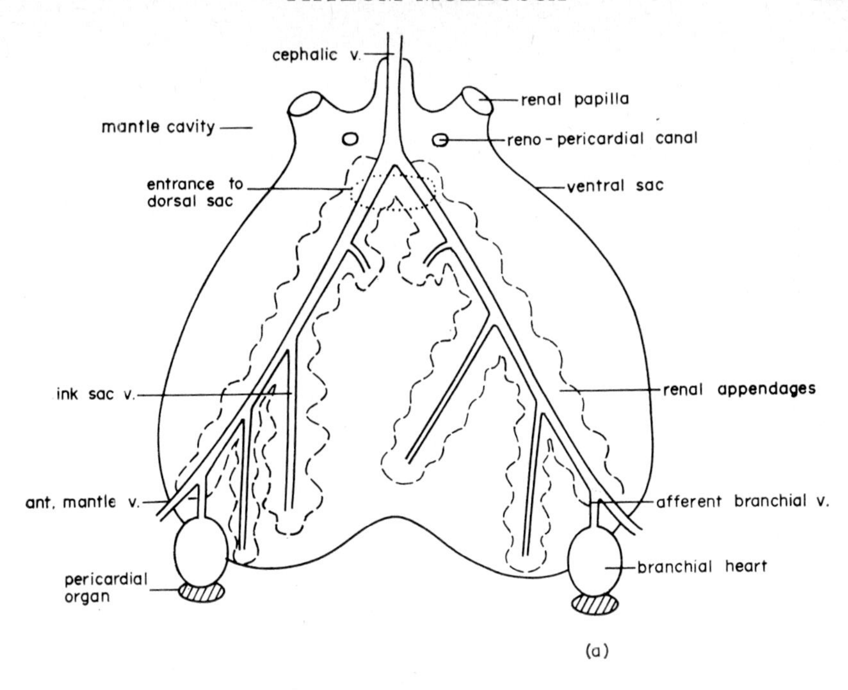

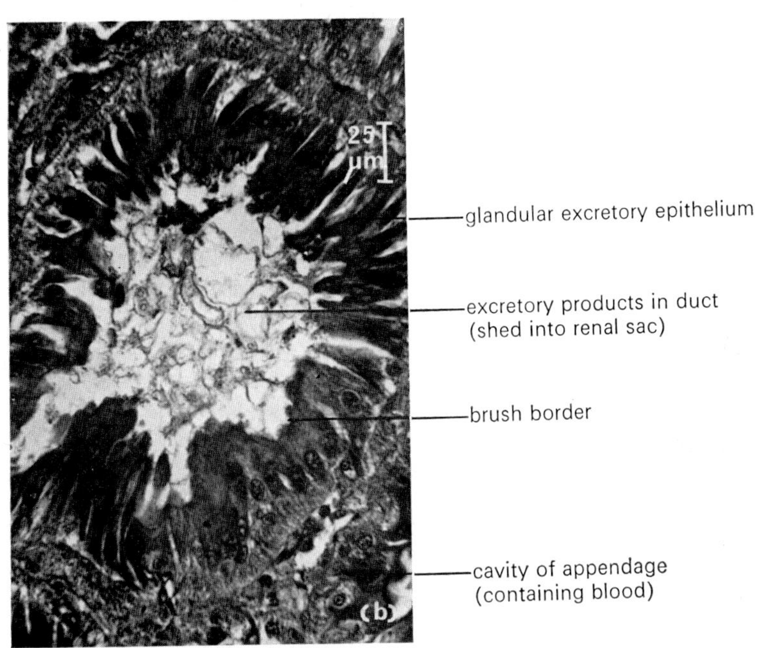

FIG. 7.55 (a) Plan of the cuttlefish renal organ (ventral view). The dorsal sac is therefore obscured by the overlying ventral sacs; (b) section through the renal appendages in the renal organ of *S. officinalis*.

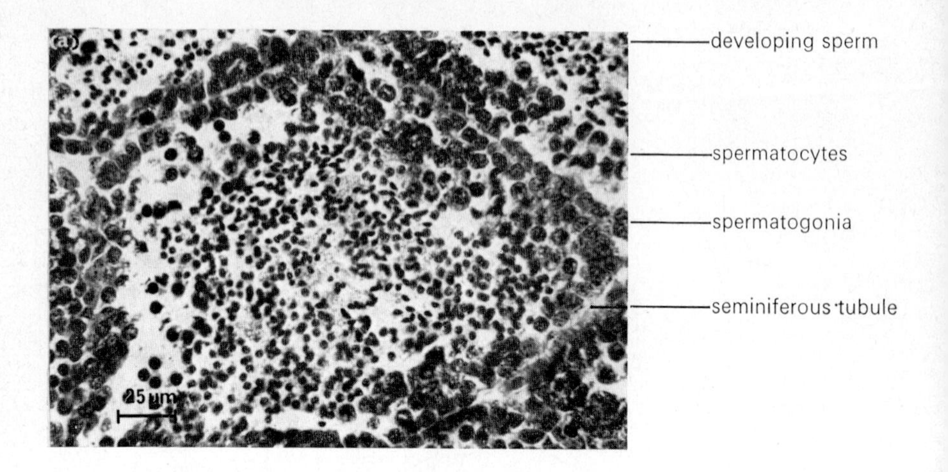

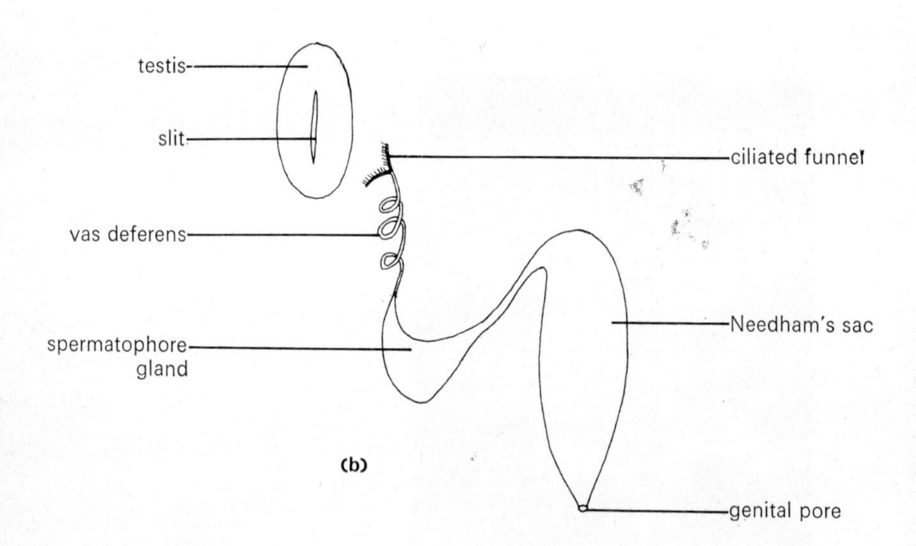

FIG. 7.56 (a) TS through the testis of a male *S. officinalis*; (b) plan of the male reproductive system of the cuttlefish.

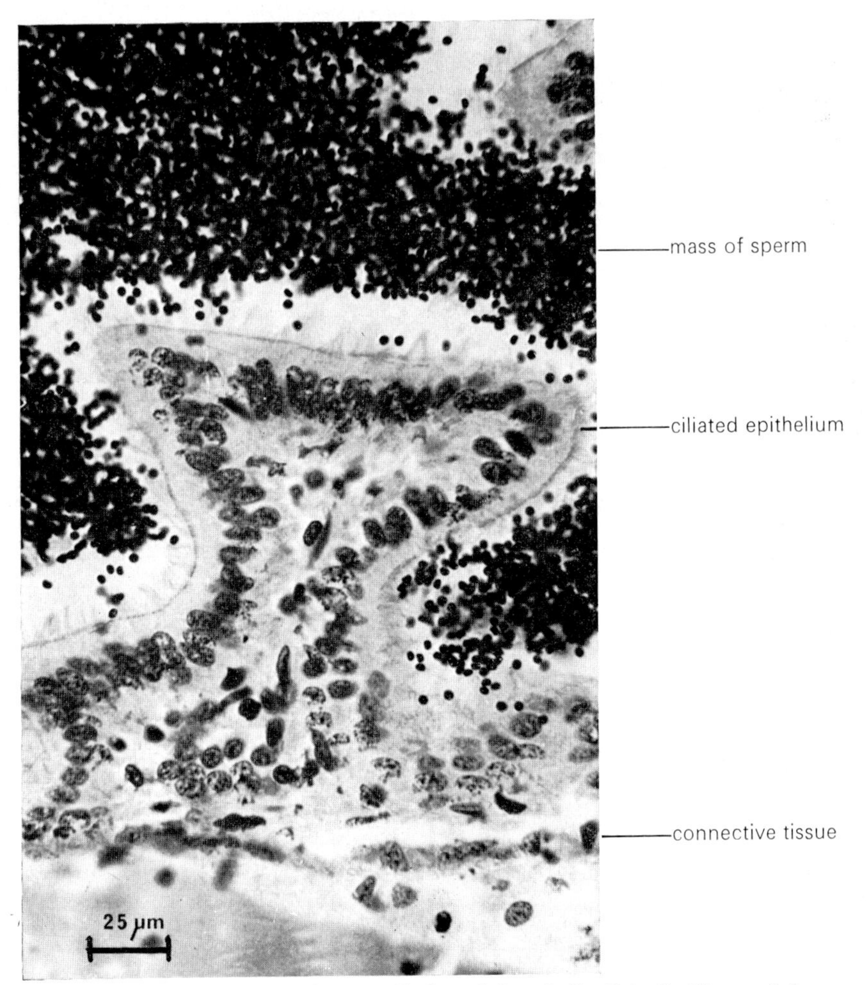

FIG. 7.57. TS through the vas deferens wall of an adult male *S. officinalis*. The vas deferens is packed with sperm.

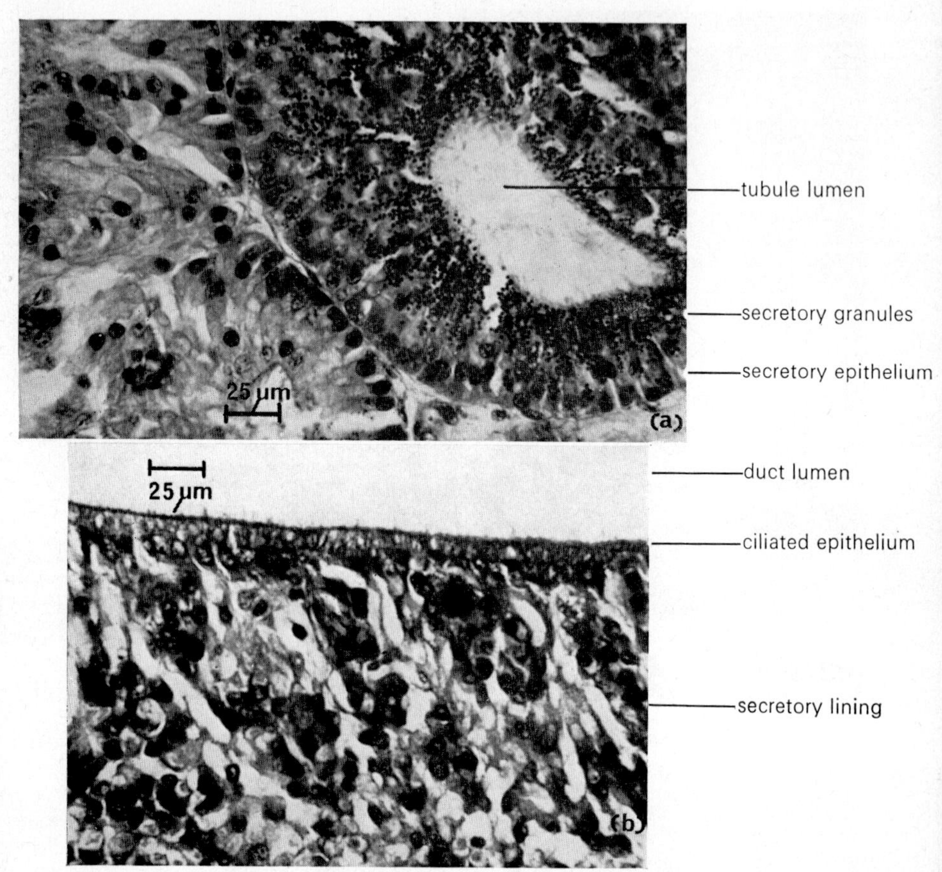

FIG. 7.58. Section through (a) the secretory tubules; and (b) the large ciliated duct of the spermatophore gland from an adult male *S. officinalis*.

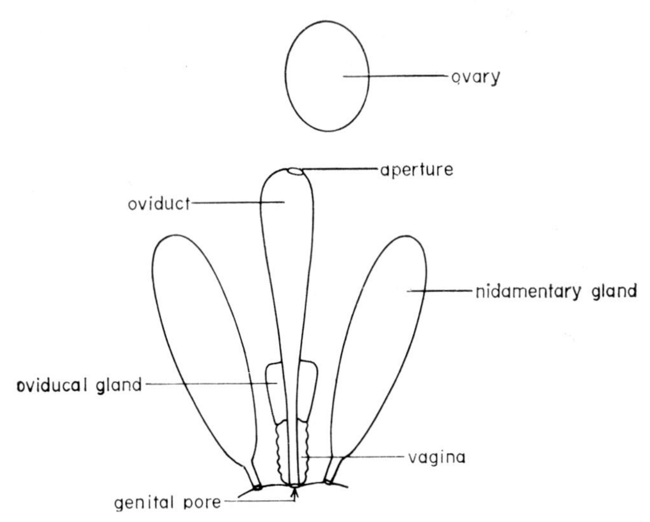

FIG. 7.59. Plan of the female reproductive system of the cuttlefish.

[Fig. 7.56(b)]. They are then captured by a ciliated funnel and passed down a coiled ciliated vas deferens (Fig. 7.57). Part of the genital tubule is modified as a glandular region (spermatophore gland) for the production of complex spermatophores which surround the sperm (Fig. 7.58). These spermatophores are stored in the dilated end of the gonoduct (spermatophore sac or Needham's sac) and at copulation they are removed by one of the arms (hectocotylus) which has been specially modified for this purpose. They are deposited in the mantle cavity of a female, near the genital opening.

In the female, eggs develop surrounded by nutritive follicles in the large saccular ovary. They too are released into the coelom and pass through a small aperture into an oviduct (Fig. 7.59). The terminal part of the oviduct has thick glandular walls which form an oviducal gland and secrete layers of albumen and yolk round the ova. As the ova are passed out into the mantle cavity they are covered with a layer of ink from the ink sac, and an elastic protein secretion from the nidamentary glands. This forms a tough protective coat on exposure to seawater. The large yolky eggs are fertilized in the mantle cavity, and then laid attached in a mass to the substrate. Development is direct in all cephalopods.

Chapter 8

Phylum Echinodermata

Echinoderms form a large group of marine coelomates, and are the only major invertebrate phylum to show a pattern of development similar to that of the chordates (see Introduction, p. 12). They are usually bottom dwellers or burrowers. The free-swimming pelagic larvae are bilaterally symmetrical, but the adults become secondarily radially symmetrical and more often than not exhibit pentamerous symmetry. This changes the positions of all the organs of the body; and the left side of the larva becomes the oral surface of the adult, bearing the mouth. Five radii carrying organ systems diverge from the mouth to the aboral surface, which thus corresponds to the right side of the larva. The coelom in adults is divided into a large perivisceral cavity and a series of tubular coelomic systems: a hydrocoel (water vascular system); a haemal system of lacunae containing blood and coelomocytes; and a perihaemal system which surrounds the haemal system. The water vascular system contains water which generally connects with the external medium by a perforated plate (madreporite), a device for equalizing the pressure within the water vascular system with that of the seawater. Along each radius of the body (ambulacrum) the water vascular system gives rise to rows of soft hollow appendages, the podia or tube-feet. They are used as a hydrostatic skeleton in locomotion, and also function in respiratory exchange and nutrition. Echinoderms possess a calcareous endoskeleton taking the form of plates or ossicles of magnesium calcite, often with projecting spines and tubercles. The nervous system remains elementary and in contact with the ectoderm from which it originates. There is no excretory system and generally no respiratory structures. Echinoderms are usually dioecious, possessing simple reproductive tracts which discharge gametes directly to the exterior by special ducts. No sexual dimorphism is apparent.

The echinoderms are divided into two sub-phyla. The Pelmatozoa are sessile and stalked at some stage of their life although they may be secondarily free as adults. Many of them are known only as fossils. A small primitive class, the Crinoidea (sea lilies and feather stars), are alive today represented by *Antedon* and a few other species. Most of the present day echinoderms fall into the sub-phylum Eleutherozoa. These are free-living echinoderms which have lost stalks: they have jaws or oral appendages for feeding and a means of locomotion for moving after their food. Three classes have been chosen for a detailed examination on the basis that they represent the three main body types (star-shaped, cushion-shaped and cylindrical) found among eleutherozoans. In fact, internal structures are remarkably uniform throughout the echinoderms.

Class Asteroidea

The asteroids (sea stars), which are the most primitive of the free-living echinoderms, are star-shaped with five or ten arms containing caecae of the alimentary canal. These arms are not constricted off from the central disc of the body. Asteroids live on the sea bottom with their oral surface downwards and have an aboral madreporite. There is generally a well-developed system of tube-feet with suckers for locomotion and feeding. On the oral surface of each arm is a groove (ambulacral groove) extending from the mouth to the tip of the arm. This is termed "open" in asteroids as all the radial components of the nervous and tubular coelomic systems lie in the groove but outside the skeleton and are covered only by epithelium. The groove is flanked by tube-feet.

Example: *Asterias* (*rubens*)

Asterias is a common starfish in Britain and lives near the low tide mark. It possesses five long flexible arms and is covered by a rough orange integument bearing spines, tubercles, small finger-like gills (papulae) and pincer-like pedicellariae for protection against settling larvae and detritus. Figure 8.1 is a section through one arm of *A. rubens* and shows the ambulacral groove containing two double rows of tube-feet and guarded by moveable spines.

The body wall of *Asterias* is covered by tall columnar flagellated epidermal cells surmounted by a thin cuticle (Fig. 8.2). Lying scattered among the epithelial cells are a large number of mucous and muriform (mulberry) cells as well as pigment and neurosecretory cells. Beneath the epidermis lies the main nerve plexus and, under this, the dermis with its reticulated skeleton of calcareous ossicles embedded in a connective tissue framework (Fig. 8.3). These ossicles sometimes project above the surface as spines. Finally, bordering the coelom, are layers of circular and longitudinal muscles and a coelomic epithelium (peritoneum) of flagellated cuboidal cells. The muscle sheets are thickened in some places to form discrete muscles.

Between the ossicles (Fig. 8.3), conical evaginations of the body wall (papulae) have a very thin dermal layer and enclose a fragment of coelom. They act as gills in respiratory exchange and play a part in excretion. Phagocytic coelomocytes are circulated in the coelom by the flagellated peritoneum. They pick up waste material and make their way, together with other soluble waste products, to the tips of the papulae which constrict off and thus release waste to the outside. Some coelomocytes may pass directly through the external epithelium (e.g. in slime on the tube-feet) while others may travel to the gut cavity.

Nervous system

The nervous system of echinoderms follows the elementary pattern of two nerve nets thickened in certain regions to form tracts. There is no concentration of ganglia such as is found in most other invertebrates. Nevertheless echinoderms are capable of performing very complicated movements involving a high degree of co-ordination of different parts of the nervous system.

In asteroids the nervous system is divided into superficial and deep parts, although the two make contact at various points. The superficial (ectoneural) nerve net consists of a diffuse plexus of interconnecting axons lying at the base of the epidermis (see Fig. 8.2). The net is thickened into a circumoral nerve ring round the mouth and five large radial nerve cords running along the bases of the ambulacral grooves (Fig. 8.4, see also Fig. 8.8). The plexus is also thickened along the edges of the ambulacral grooves into paired marginal nerve cords which run the length of the arms. The superficial nerve net is predominantly sensory and is particularly well developed underneath specialized areas of the body such as ciliary tracts or collections of sensory receptors.

The second deep nerve net, which is closely associated with the peritoneum, is primarily motor and is thickened in the vicinity of special muscles. It consists chiefly of Lange's nerves which are aggregations of motoneurones in the oral walls of the circumoral and radial perihaemal canals, separated from the ectoneural nerve cords by a layer of connective tissue (Fig. 8.4). They innervate the ampullae and tube-feet, the ambulacral muscles and the body wall. The body wall and ampullar musculature is also supplied by the lateral centres, which too are part of the deep nervous system.

Sense organs

There are no specialized sense organs in asteroids except for the eye-spots. These are composed of numerous ocelli, each consisting of a cup of epidermal cells containing red pigment filled with elongated receptor cells. These receptors connect with each radial nerve cord in a small "cushion-shaped" thickening at the base of the sensory terminal tentacle on the oral side of each arm. Light is focused on the receptors by a thickened cuticle which acts as a lens.

There are numerous sensory cells in the epidermis, particularly on the suckers of the podia at the bases of the spines and pedicellariae, along the margins of the ambulacral grooves and on the terminal tentacles. These receptors are connected to the superficial nerve plexus.

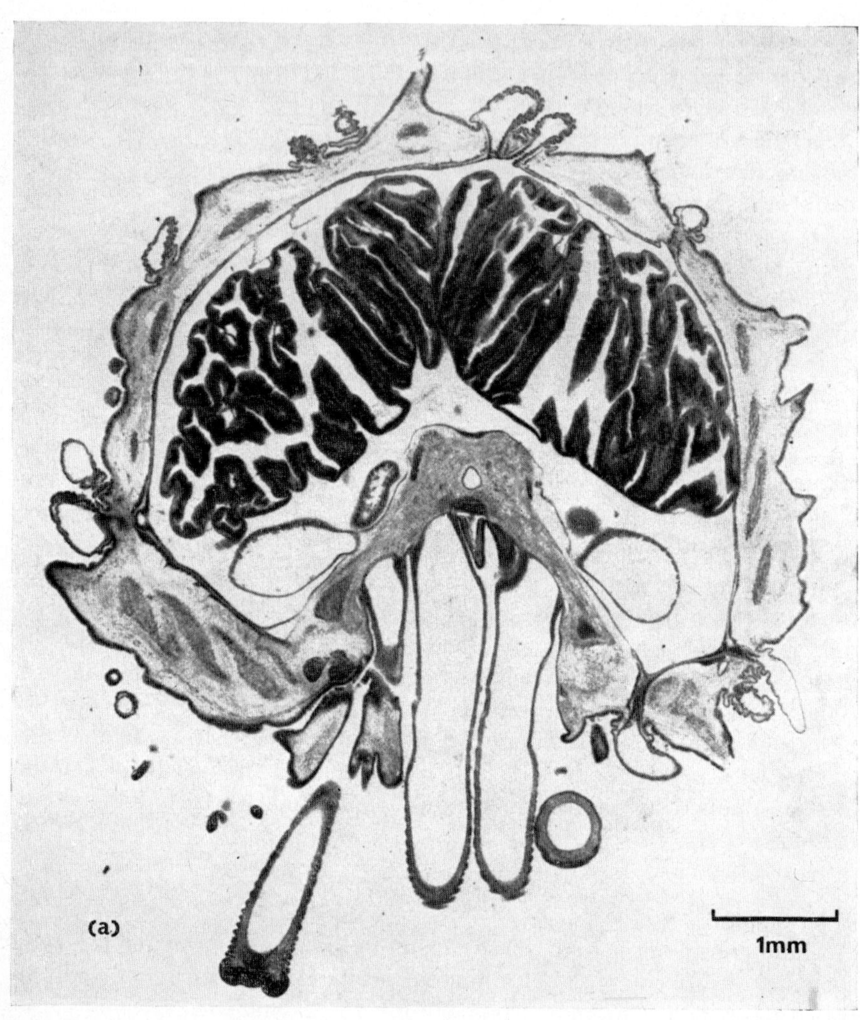

FIG. 8.1

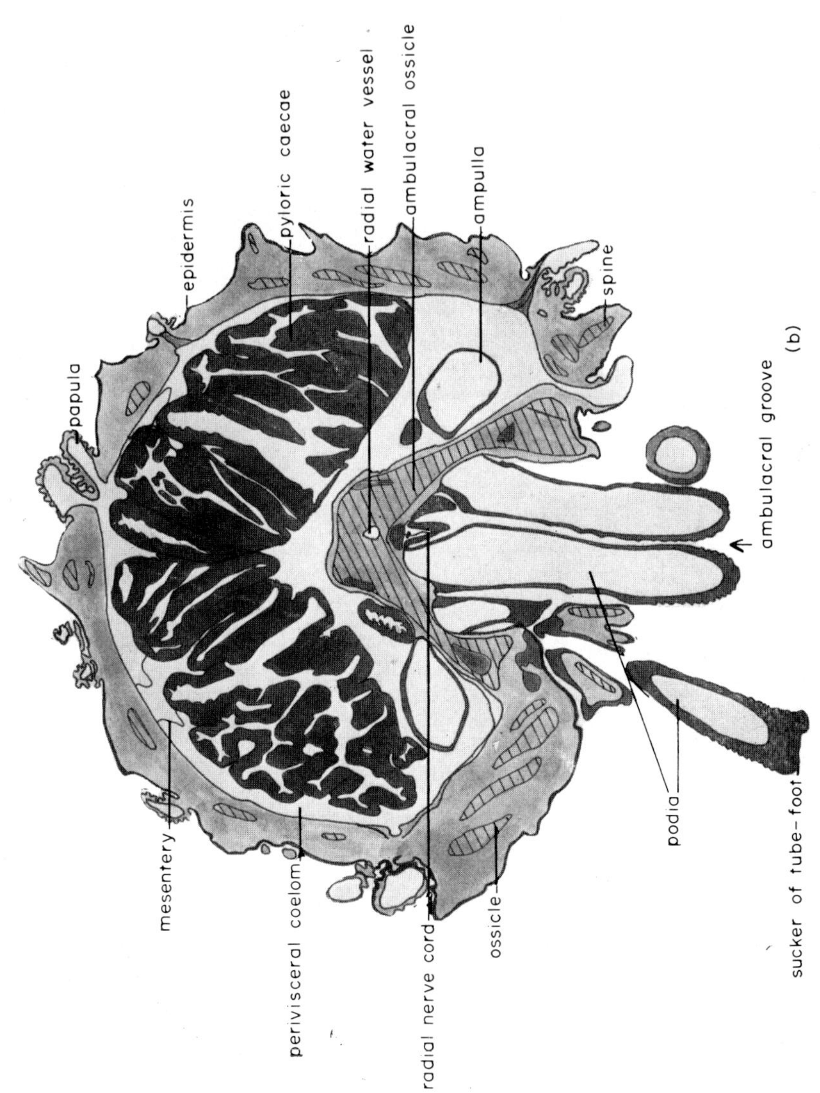

FIG. 8.1 (a) TS and (b) explanatory drawing of an arm of the starfish *Asterias rubens*.

FIG. 8.2. Section through the integument of *A. rubens*.

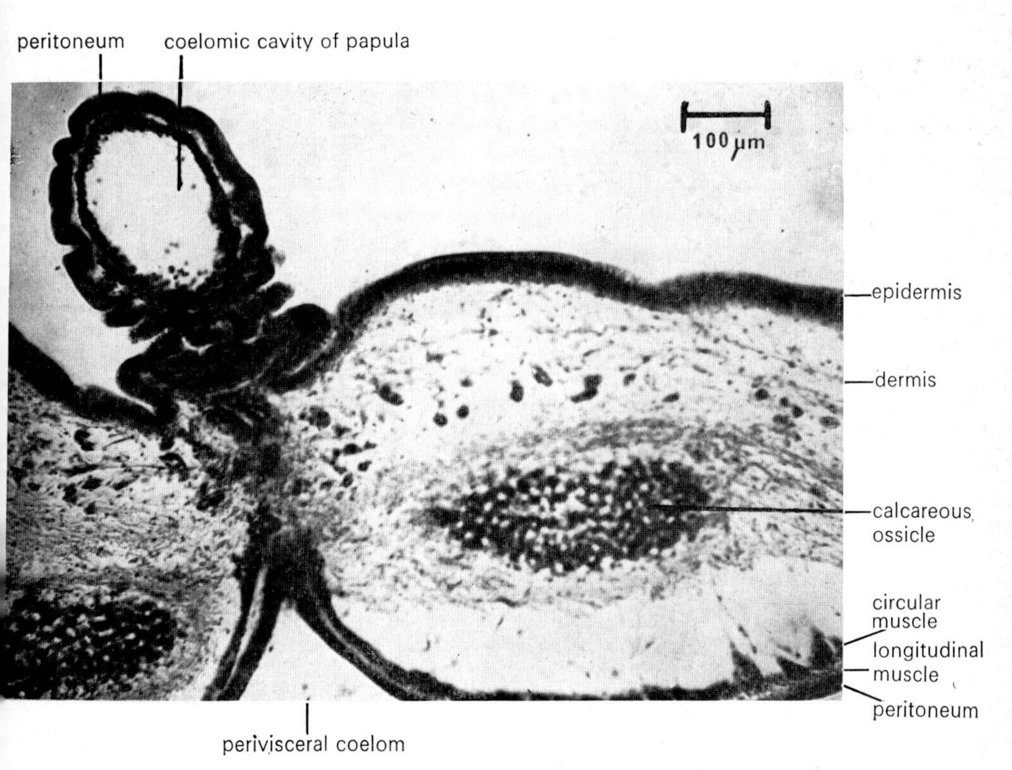

FIG. 8.3. Section through the body wall of *A. rubens* showing a papula and calcareous ossicles.

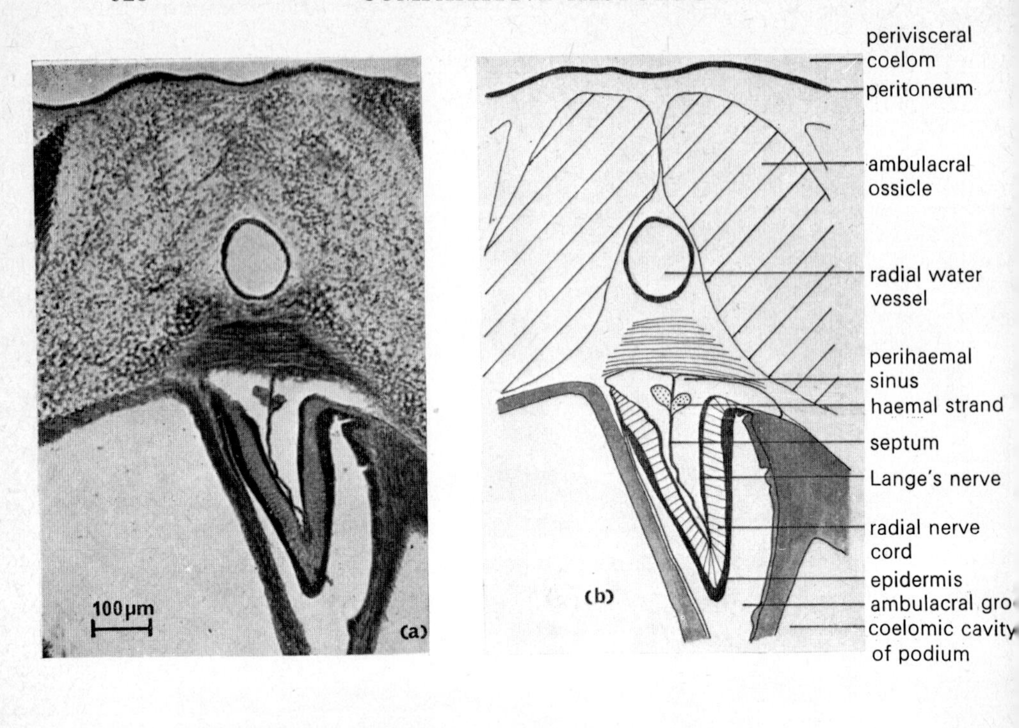

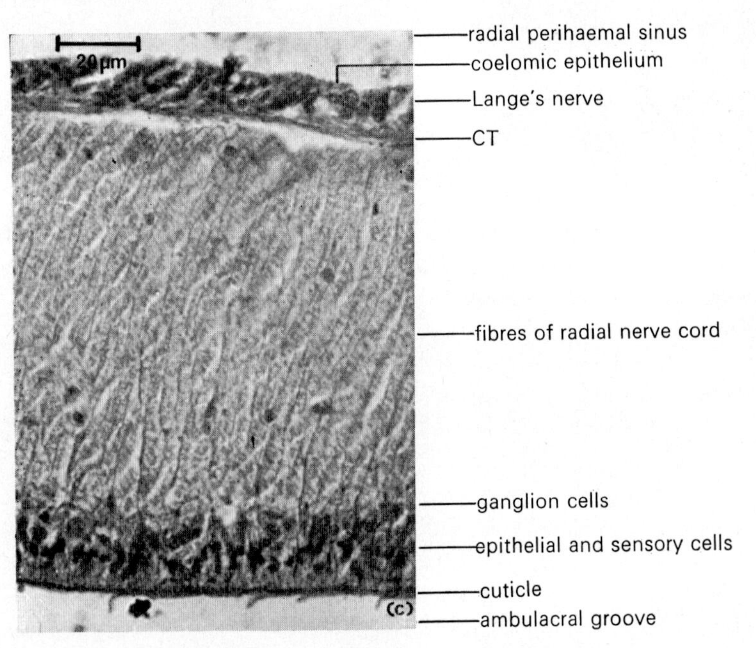

FIG. 8.4 (a) TS and (b) drawing of part of the arm of *A. rubens*, showing the position of the radial nerve cord; (c) detail of the radial nerve cord and Lange's nerve.

Water vascular system

Locomotion and feeding are brought about chiefly by one of the tubular coelomic systems, the water vascular system. In asteroids this consists of five radial canals bearing tube-feet and a circumoral ring vessel, the cavity of which communicates with the seawater by way of a short stone canal and porous madreporite (Fig. 8.5).

The stone canal is a ciliated tube supported by calcareous rings, and, in asteroids, possesses a scroll-shaped projection from one side to increase surface area (Fig. 8.6). This partition has ciliated walls, the cilia being more abundant on the concave surface and being responsible for circulation of fluid within the water vascular system. The stone canal is accompanied along its length by part of the haemal system (the axial organ) and surrounded by the axial sinus with which it communicates aborally.

The madreporite is a calcareous sieve-like structure opening onto the aboral surface. It allows contact between fluid within the water vascular system and the outside medium. The fine pore canals are also lined with long cilia or flagella (Fig. 8.7). The cavity of the madreporite bears a small muscular pocket, the madreporic ampulla, which is thought to pulsate and thus assist circulation of fluid within the stone canal.

The circumoral ring vessel gives rise to a radial vessel running along each arm, with small lateral branches which pass through the skeletal plates (ambulacral ossicles) bordering the ambulacral groove and supply the globular ampullae and tube-feet (Fig. 8.8). There are valves at the junctions of the radial and lateral vessels which can close and allow pressure to build up within the tube-feet. The tube-feet are the locomotory appendages and work on a hydraulic principle.

All parts of the water vascular system have muscular walls but those of the ampullae and tube-feet are particularly well developed. The ampulla wall [Fig. 8.9(a)] consists of layers of circular and longitudinal muscle covered on both sides by coelomic epithelium. The longitudinal muscle fibres are rather peculiar in that they possess an outer layer of contractile filaments and an inner layer of cytoplasm rich in glycogen. Once the valves in the lateral branch vessels have closed, the muscular walls of the ampullae contract against the incompressible coelomic fluid within them and force the tube-feet to extend. The walls of the podia are similar to those of the ampullae except that there is no circular muscle [Fig. 8.9(b)]. There are, however, thick longitudinal retractor muscles which can contract, shortening the podia and forcing fluid back into the ampullae. In this way the tube-feet can be alternately extended and withdrawn as well as moved from side to side. Large numbers of podia operating in this way provide a remarkably efficient locomotory system. The tube-feet end in suckers [Fig. 8.9 (c, d)] composed of radial and levator muscles and well supplied with nerves, sensory receptors and mucous glands. These strong suction pads assist locomotion and are also used in feeding.

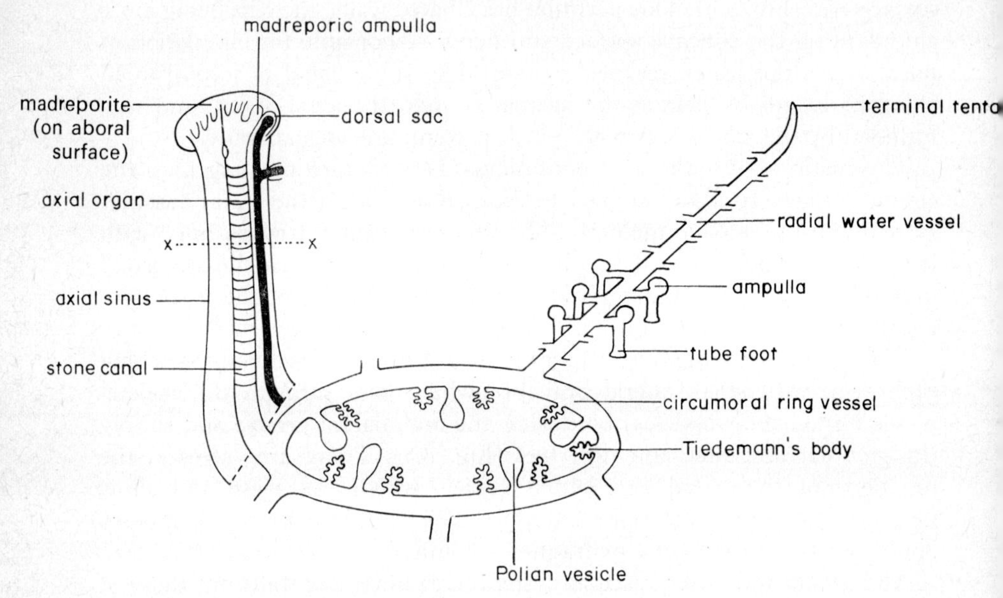

FIG. 8.5. Plan of the water vascular system of *A. rubens* (X-X is the section shown in Fig. 8.6).

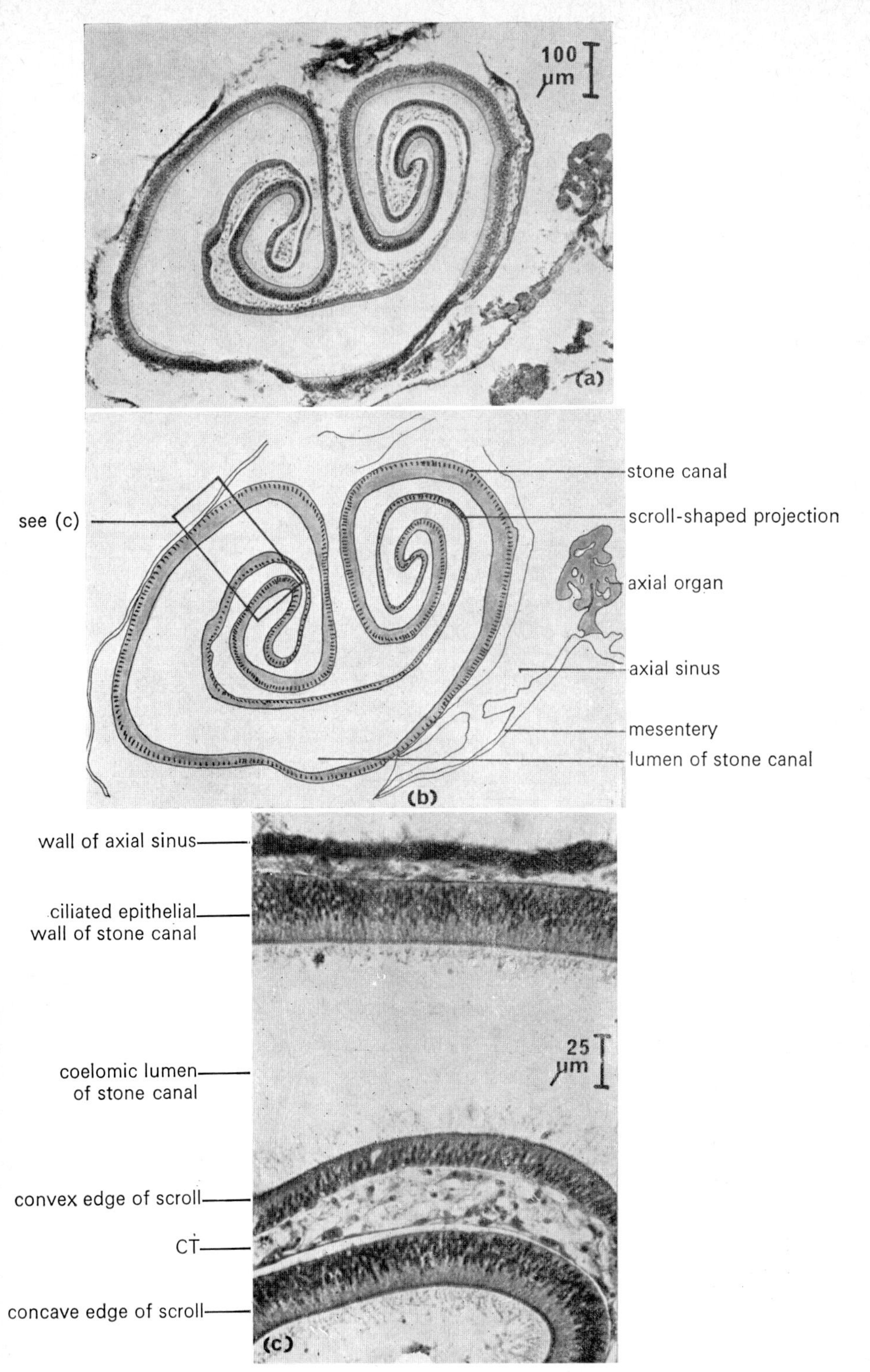

Fig. 8.6 (a) TS and (b) drawing of the stone canal of *A. rubens* (X-X in Fig. 8.5); (c) detail of the stone canal wall [of the area outlined in (b)].

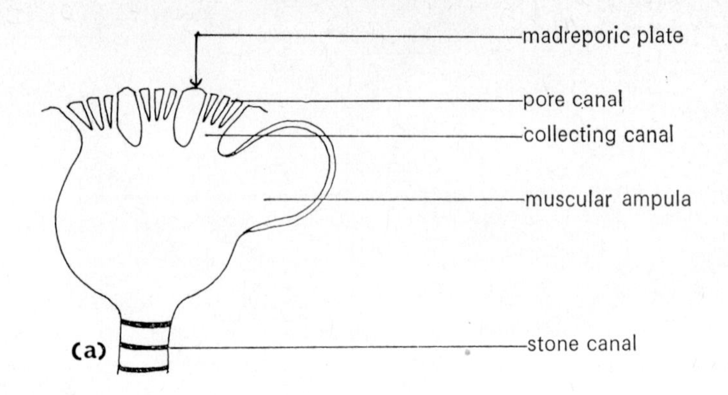

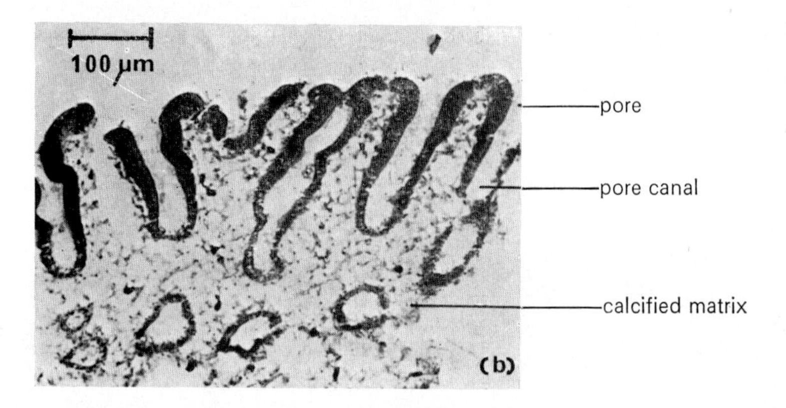

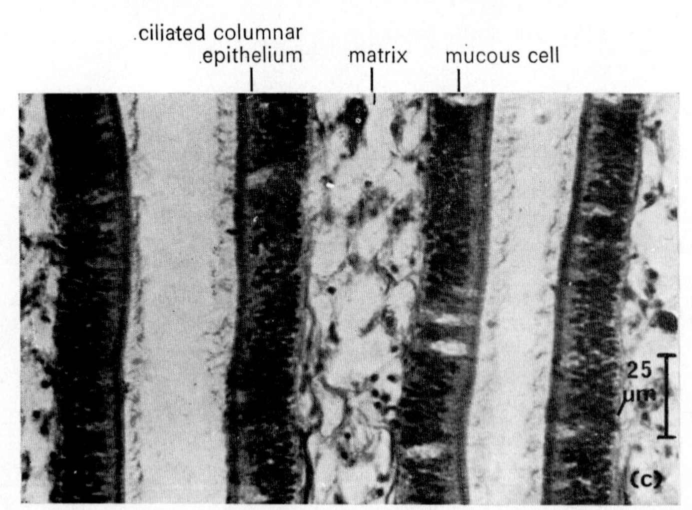

FIG. 8.7 (a) Diagram of the madreporite structure of asteroids; (b) vertical section through the madreporite of *A. rubens*; (c) detail of pore canals in the madreporite plate.

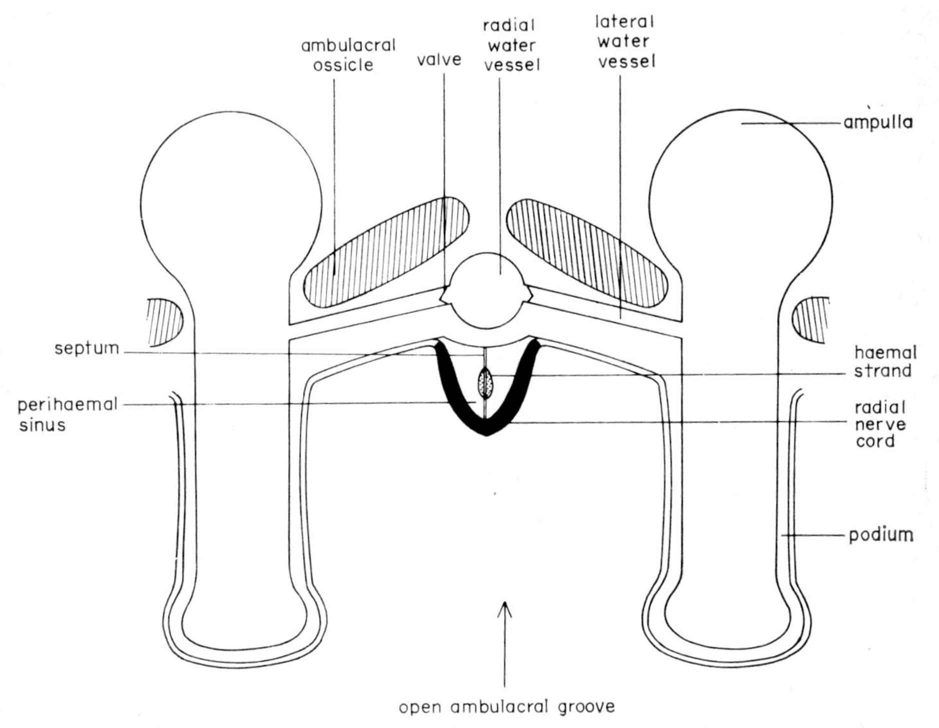

FIG. 8.8. Diagram of the radial tubular coelomic systems as seen in a TS of a starfish arm.

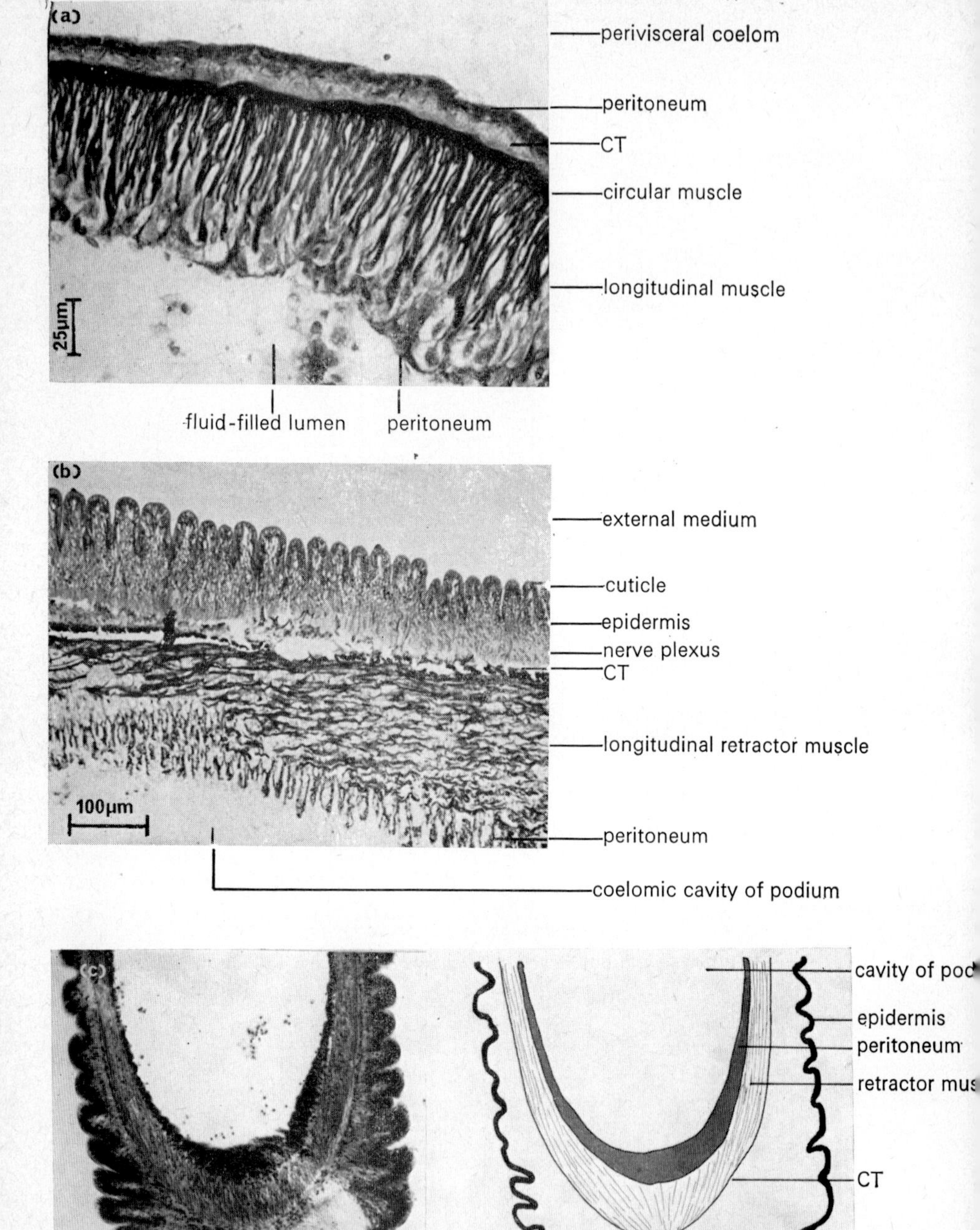

FIG. 8.9 (a) Section through the ampulla wall of the water vascular system of *A. rubens*; (b) LS through the podium wall of *A. rubens*; (c) LS; and (d) drawing of the end of a tube-foot of *A. rubens* showing the sucker musculature.

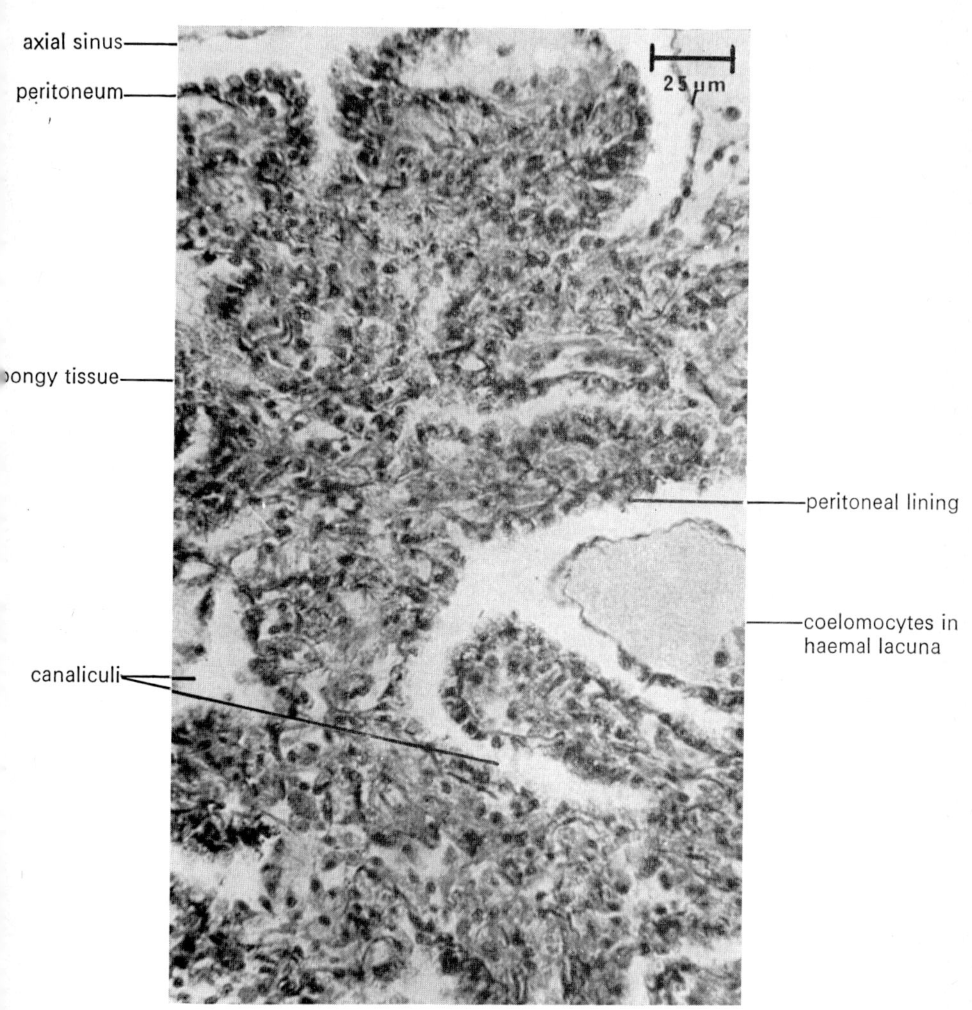

FIG. 8.10. Section through the axial organ of *A. rubens*.

Each radial water vessel ends in a sensory terminal tentacle which represents a tube-foot without its ampulla. Paired Tiedemann's bodies, having the form of folded pouches lined with columnar epithelium, project from the inner side of the circumoral ring vessel and are thought to manufacture coelomocytes. Interspersed between these bodies are pear-shaped muscular sacs, the Polian vesicles. These have the same basic structure as the tube-feet although they possess a very thick connective tissue layer, and are thought probably to act as low-pressure reservoirs for fluid within the water vascular system (see Fig. 8.5).

Haemal and perihaemal systems

The water vascular system is accompanied along part of its length by two more tubular coelomic systems (the haemal and perihaemal systems) which are also in contact with the perivisceral coelom.

The haemal system consists of branching glandular tissue and muscle fibres. A diffuse axial organ (Fig. 8.10) is thought to be involved in the production of coelomocytes and the defence mechanism of the body against invading organisms and injury. It connects orally with a circumoral haemal ring which gives rise to radial haemal strands in each arm. Aborally, the axial organ sends a small projection into the dorsal sac, a coelomic remnant lying under the madreporite (see Fig. 8.5). It connects with another (aboral) ring of lacunar tissue which sends branches to the gonads and hepatic caecae, and a pyloric haemal ring round the pyloric stomach with haemal tufts supplying the cardiac stomach.

All the elements of the haemal system, except some of the aboral branches, are suspended in a vertical septum of the perihaemal system (see Figs. 8.4 and 8.8). A wide axial sinus surrounds the axial organ and stone canal and connects aborally with the stone canal in the madreporite. Orally it joins a perihaemal ring which sends a radial sinus into each arm. There is also an aboral perihaemal canal surrounding the aboral haemal ring and sending branches to the gonads, but this forms a separate cavity from the perihaemal system, having lost its connection with the axial sinus. The perihaemal system probably has a vascular function.

Digestion

Starfish are normally carnivorous, prising open bivalve molluscs and sucking out their contents, but they will also scavenge on dead animal matter. The mouth, positioned in the middle of the peristomial membrane, is separated by a muscular sphincter from a short oesophagus and a large complex stomach. The stomach can be everted through the mouth by the action of body wall muscles and coelomic fluid pressure, and is used as a suction pump. It is divided horizontally into a large pouched cardiac stomach and a smaller flattened, star-shaped pyloric stomach (Fig. 8.11).

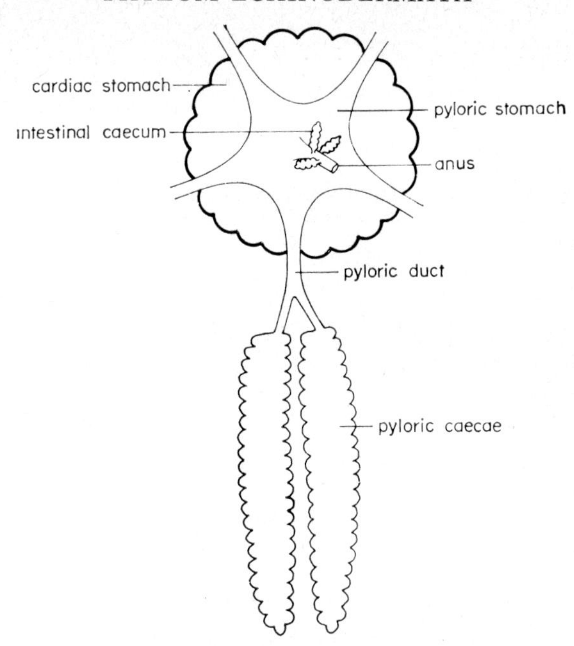

FIG. 8.11. Plan of the digestive tract of *A. rubens*, as seen from the aboral side.

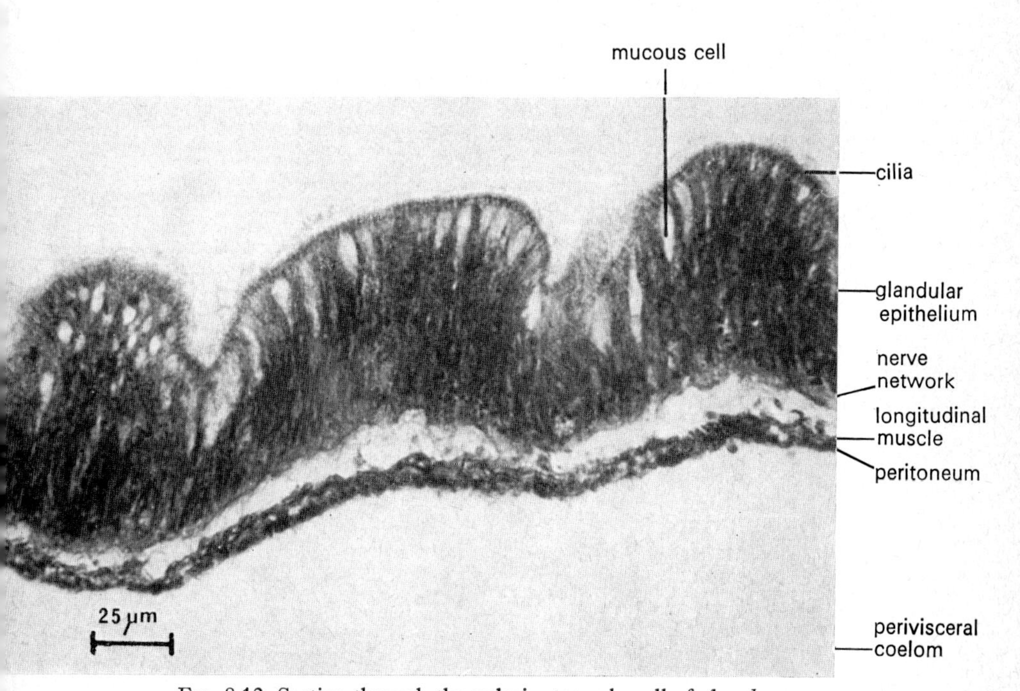

FIG. 8.12. Section through the pyloric stomach wall of *A. rubens*.

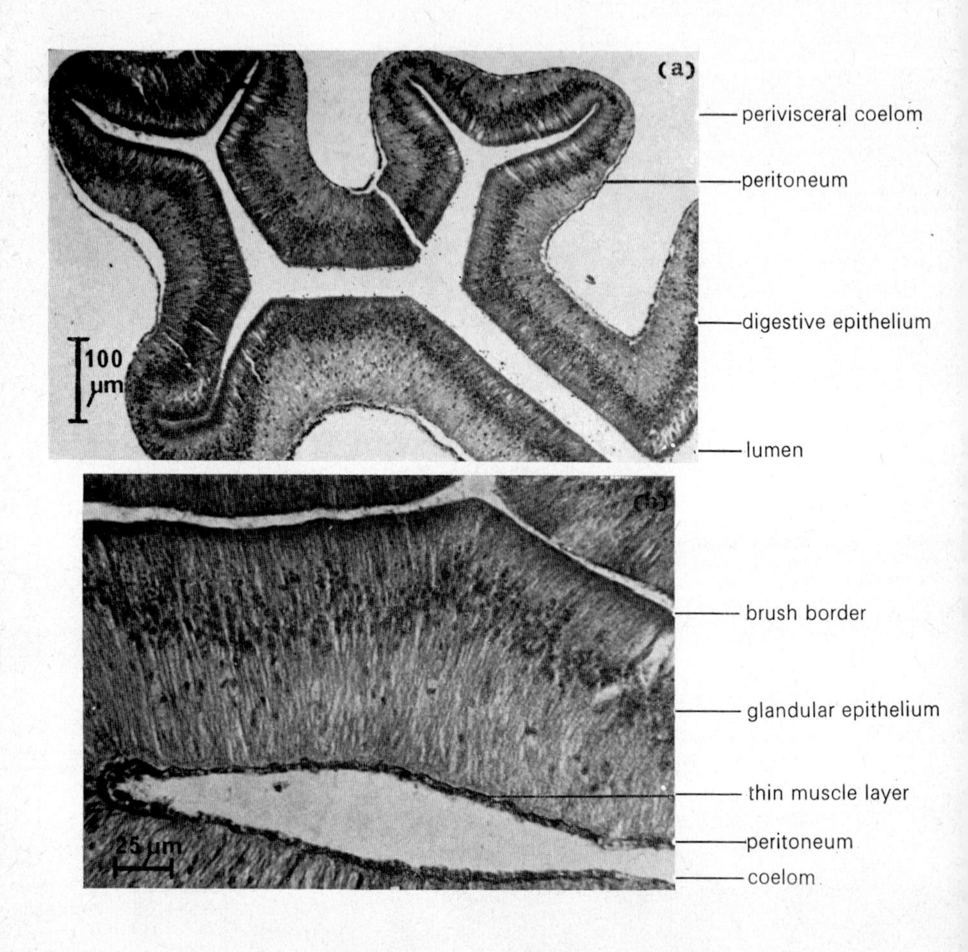

FIG. 8.13 (a) Section through the pyloric caecae of *A. rubens*; (b) detail of the mucosal epithelium of a pyloric caecum.

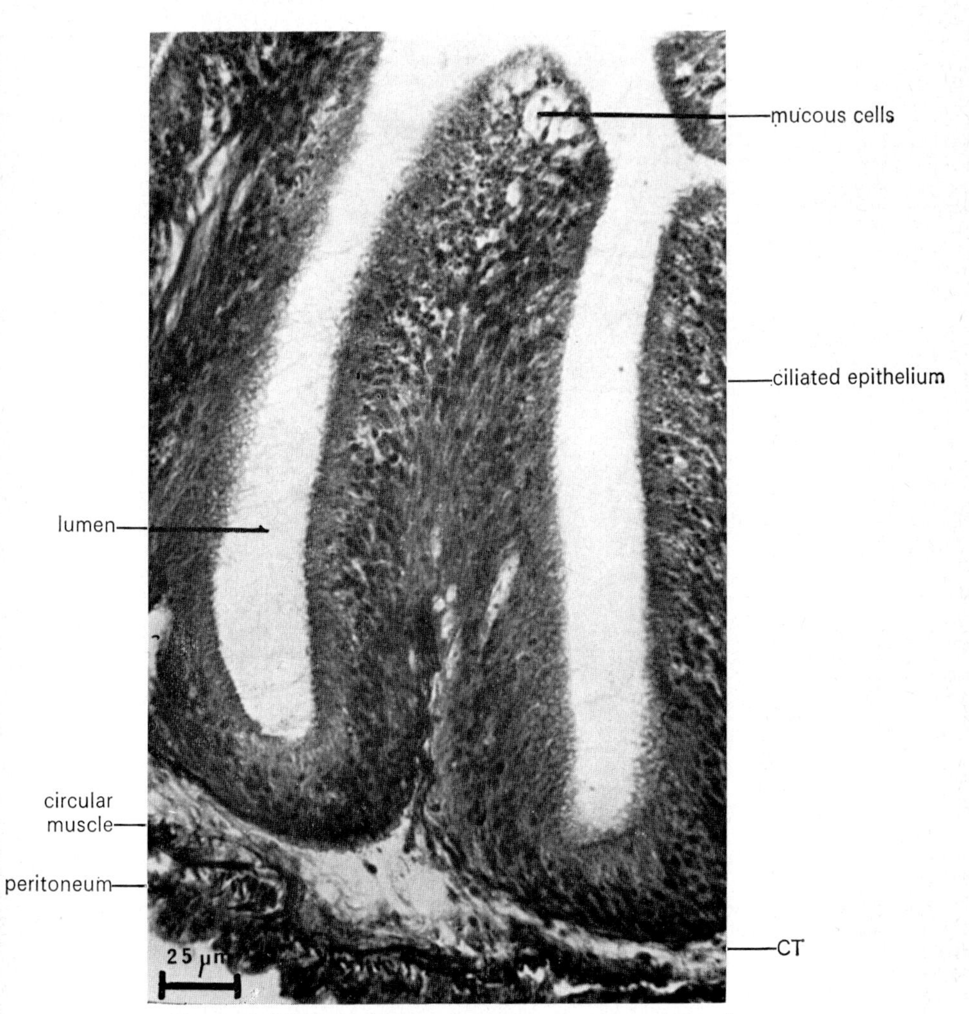

Fig. 8.14. TS through the wall of a rectal caecum of *A. rubens*.

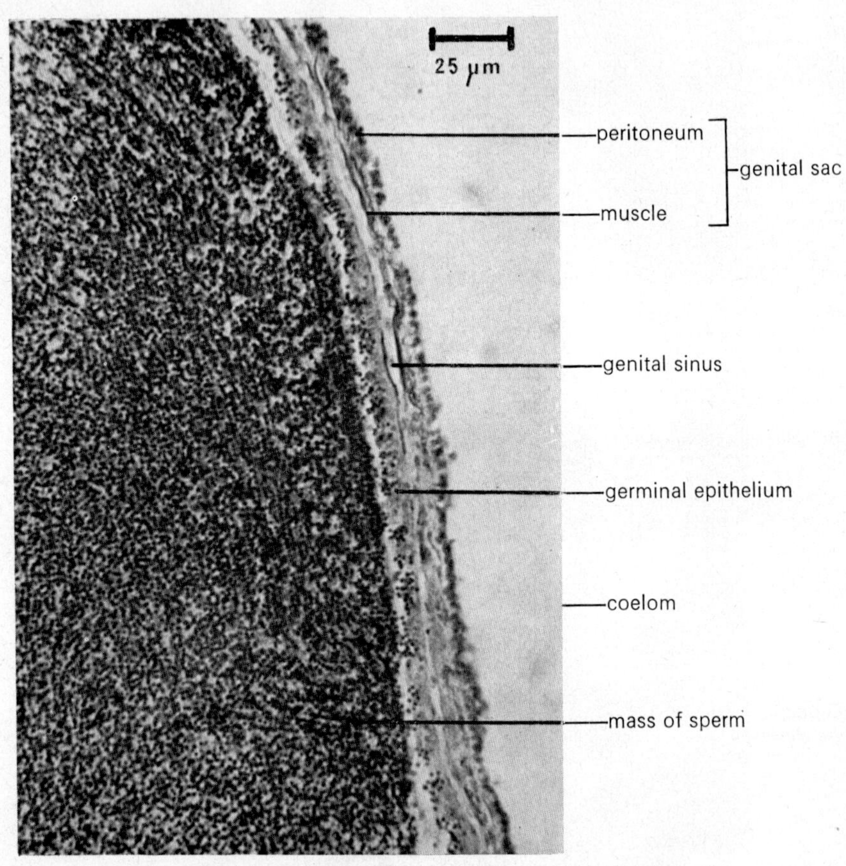

Fig. 8.15. Section through part of the testis of *A. rubens*.

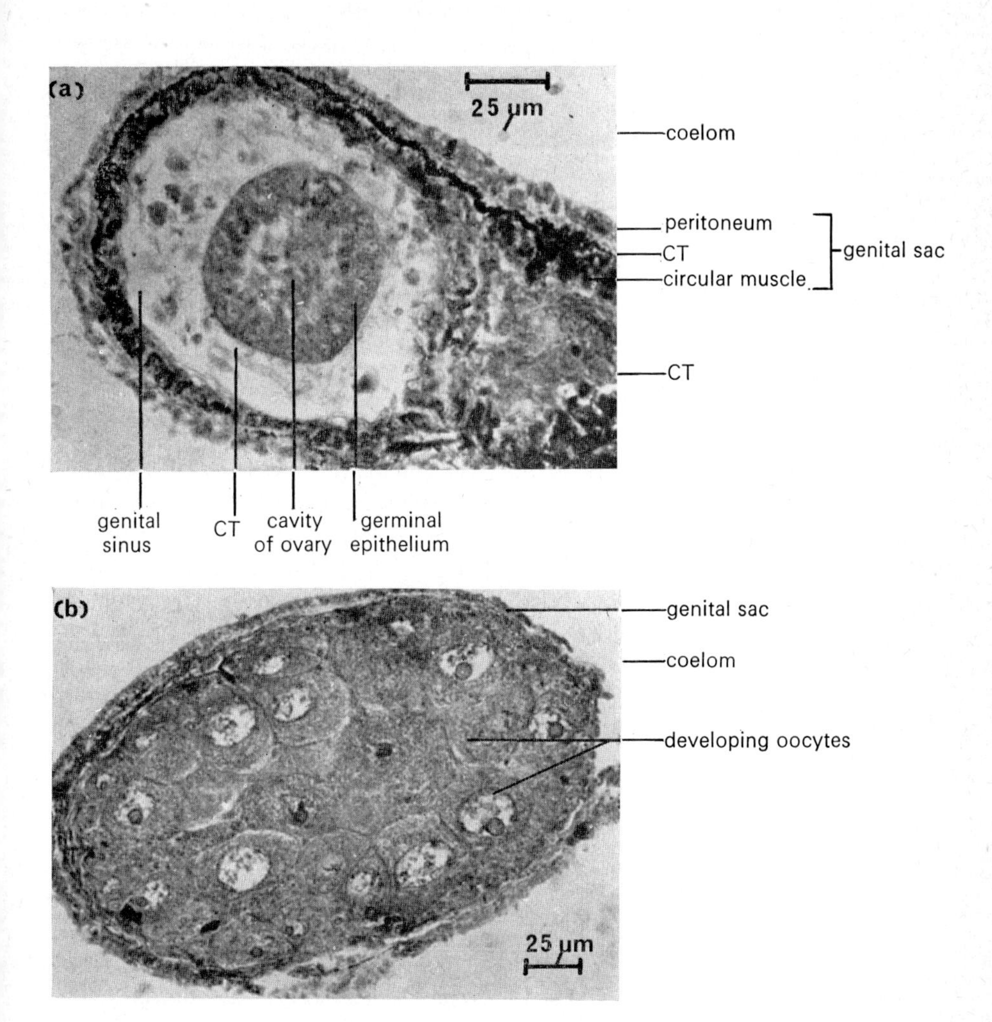

FIG. 8.16 (a) Section through an immature ovary of *A. rubens*; (b) section through a developing ovary of *A. rubens*. Note that the genital sinus and cavity of the ovary have been obscured by the large oocytes as they develop.

The stomach has a dense nerve network and a lining of ciliated epithelium and mucous cells (Fig. 8.12). This lining is thick in the cardiac stomach and thinner in the pyloric stomach. No enzymes are produced in this region, but extracellular digestion occurs using enzymes produced in the digestive gland.

The digestive gland consists of ten pyloric caecae; two suspended in each arm by a double dorsal mesentery, and opening by a duct into the pyloric stomach (see Fig. 8.11). The caecae are composed of masses of tubules lined with tall glandular columnar cells which have a secretory and digestive function (Fig. 8.13). A short tubular intestine, lined with an epithelium similar to that of the stomach, leads to the aboral anus and gives rise, on one side, to a few small branched rectal or intestinal caecae. These caecae have a very folded epithelium containing mucous and gland cells (Fig. 8.14).

Reproduction

In asteroids there is a pair of tuft-like gonads in each arm. Each gonad is attached to the body wall by a mesentery near the junction between arm and central disc; and lies freely in the coelom covered externally by peritoneum. Gametes are discharged from each gonad into a short gonoduct and to the outside by a gonopore situated between the bases of the arms. Fertilization is external.

The gonads lie each within a genital sac of peritoneum and connective tissue, and are lined with connective tissue and germinal epithelium (Figs. 8.15 and 8.16). Germ cells in the male give rise to spherical sperm. In the female the germ cells enlarge enormously as they mature, so that the ovary of an asteroid immediately before spawning is vast. Development is through pelagic larvae.

Class Echinoidea

Echinoids are globose or cushion-shaped echinoderms which live with their flattened oral surface downwards on the sea floor. They do not have arms, and the whole body is covered by a solid test, formed of fused skeletal ossicles. This test is perforated along the radii by five double rows of tube-feet, and is normally covered by numerous long moveable spines and pedicellariae. Echinoids have closed ambulacral grooves, with all the radial tubular coelomic systems and nerve cords internal to the skeleton. There is a vast fluid-filled coelom divided by septae into a large perivisceral coelom, a small oral peripharyngeal cavity and several minor aboral cavities (peri-anal, peri-proctal and genital sinuses). The madreporite is always aboral. Echinoids can be either pentamerous (regular) or secondarily bilateral (irregular).

Example: *Echinus* (*esculentus*)

E. esculentus is a common sea urchin usually occurring on the sea bottom in deep water although it is sometimes found in the littoral zone. It is a regular or radially pentamerous echinoid, nearly spherical in shape and covered by strong spines. *Echinus* is omnivorous and is equipped with powerful jaws arranged in a complex pattern in the extrusible Aristotle's lantern. The structure and arrangement of the body systems is very similar to that of asteroids and therefore will not be dealt with in great detail. Figure 8.17 is an oral/aboral section through a small echinoid (*Psammechinus miliaris*) which is very common round the coasts of Britain and which resembles *E. esculentus* but is more convenient to section. The enormous perivesceral coelom and the musculature associated with the large Aristotle's lantern are both clearly visible.

The body wall is built on the same plan as that of asteroids except that the calcareous ossicles are much larger and fused into a continuous case (test). Long cylindrical spines, which can be moved by dermal muscles, are used for protection and to assist in locomotion. Numerous pedicellariae of four different types are present, some of which are shown in Fig. 8.18. There are no papulae, so most of the respiratory exchange and excretion occurs across the walls of the tube-feet and their ampullae and across the peristomial gills. These are five pairs of thin-walled evaginations from the peristomial membrane, and are lined inside and out with ciliated epithelium. Their internal cavities are filled with fluid and connect with the peripharyngeal coelom (see Fig. 8.17). Certain muscles of the lantern act as a pump for squeezing fluid in and out of the gill cavities.

Echinoid tube-feet are often very powerful and show great adaptive radiation in accordance with habitat and mode of life. Instead of radial muscle fibres, they possess a complex calcite skeleton which is in two parts; a solid plate-like "rosette" round the base of the sucker, and a series of smaller ossicles forming a frame [Fig. 8.19(a)]. Five pairs of tube-feet round the mouth (buccal podia) lack this skeleton, are foreshortened and are chemosensory in function for finding food [Fig. 8.19(b)].

The nervous and sensory systems are similar to those of asteroids, with pedicellariae, spines and tube-feet as the chief sensory structures. Small spherical modified spines (sphaeridia) occur in tiny pits in the ambulacral areas of the test round the mouth and are thought to be organs of balance. A few species possess eyespots, but not *Echinus*.

Tubular coelomic systems

The water vascular and haemal systems of echinoids follow the same plan as in asteroids, except that the stone canal is non-calcified (Fig. 8.20). The external surface of the madreporite is covered by epidermis, like the rest of the test, although the pores are lined with ciliated columnar cells

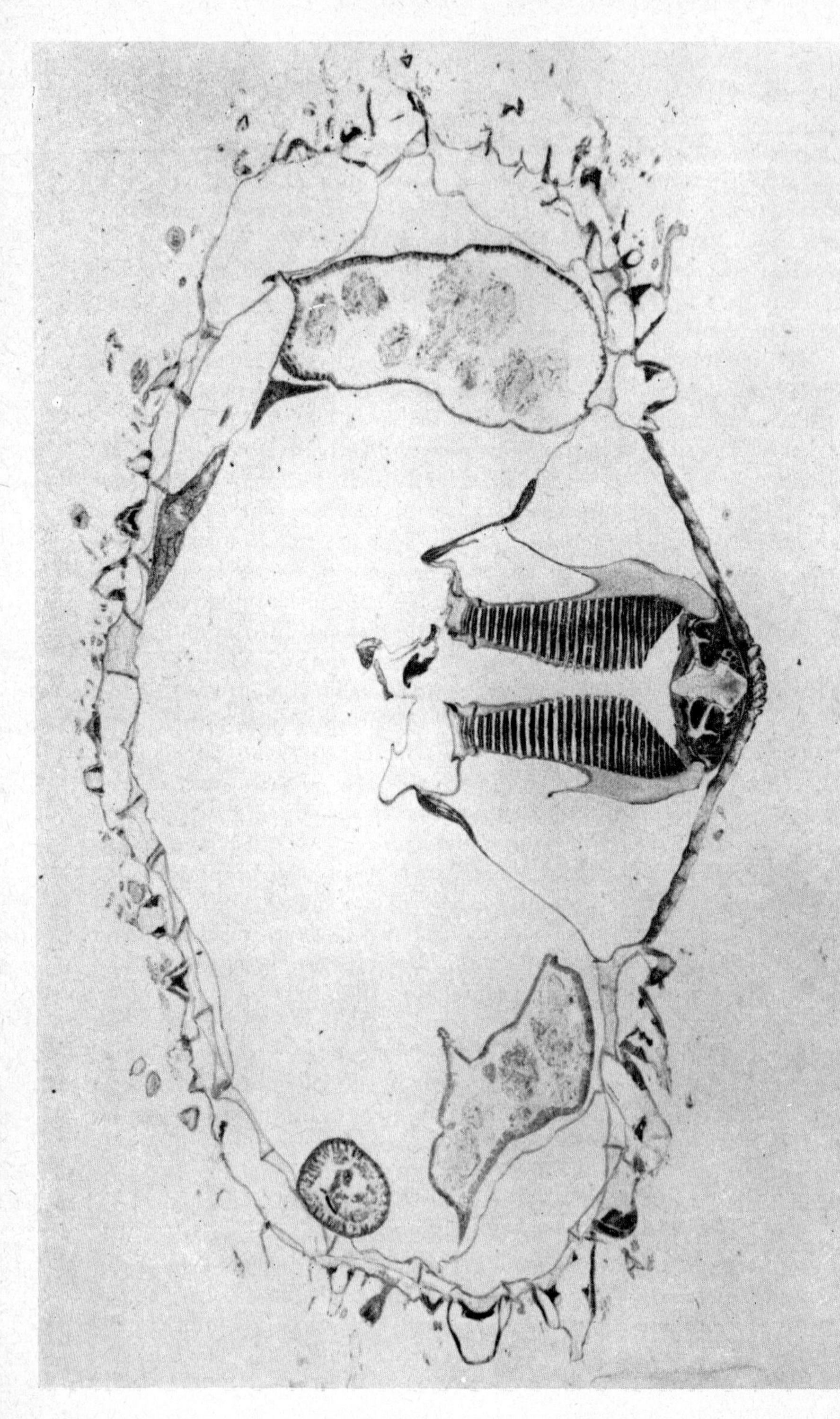
2 mm
(a)

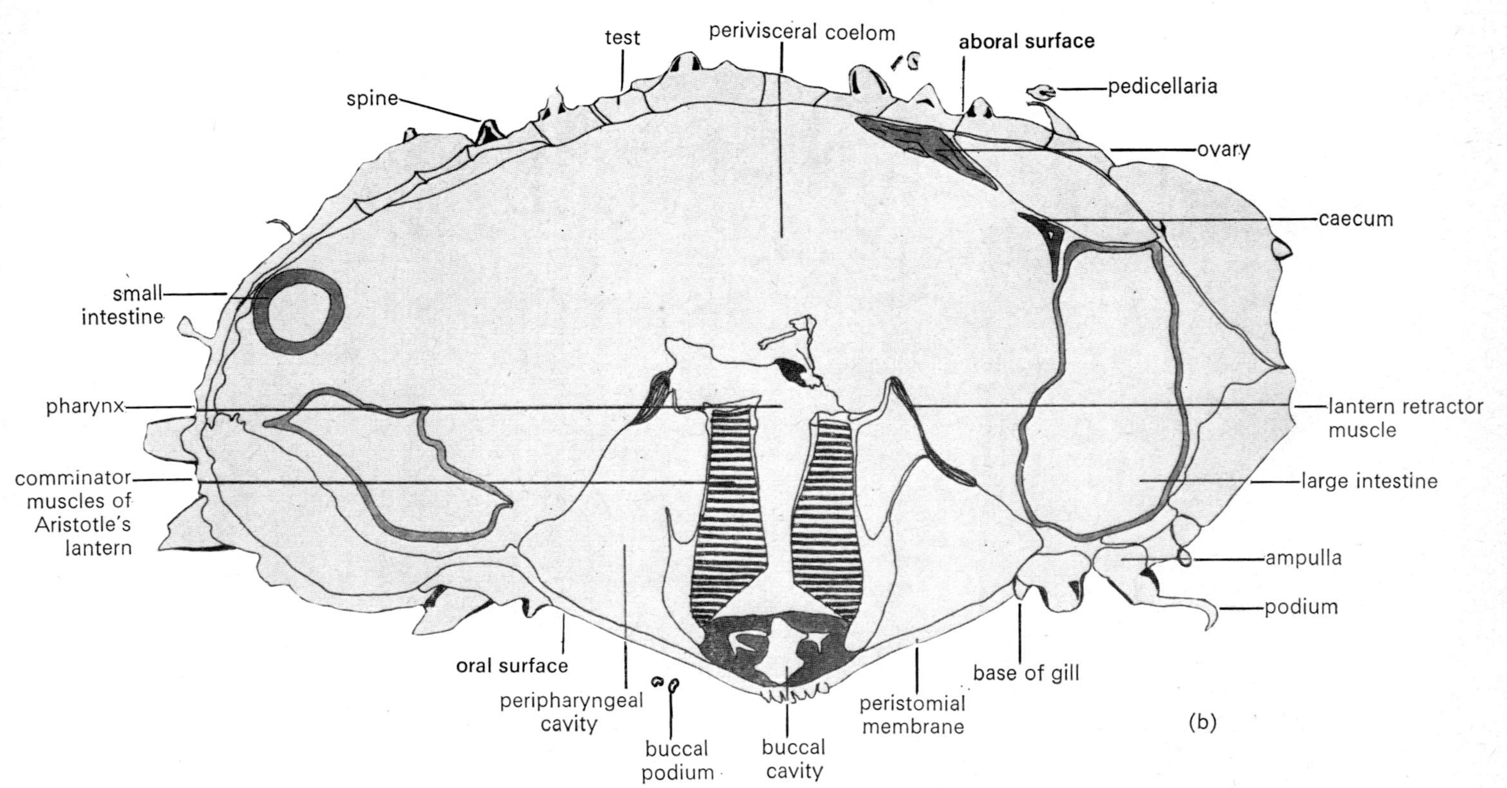

FIG. 8.17 (a) Oral/aboral section; and (b) explanatory drawing of a small regular sea-urchin (*Psammechinus miliaris*).

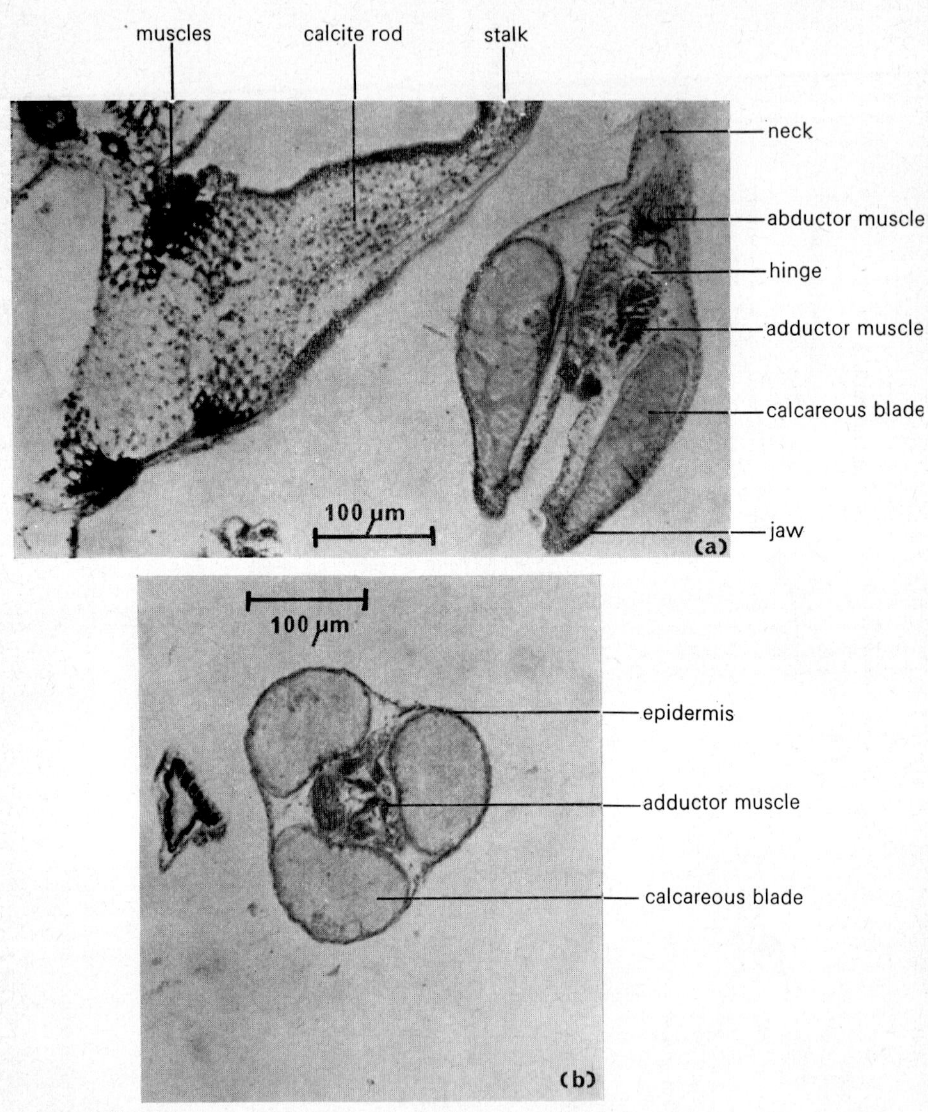

FIG. 8.18 (a) Part of a tridentate pedicellaria of an echinoid, showing the positions of the jaws, hinge, adductor and abductor muscles; (b) TS through the jaw region of a tridentate pedicellaria.

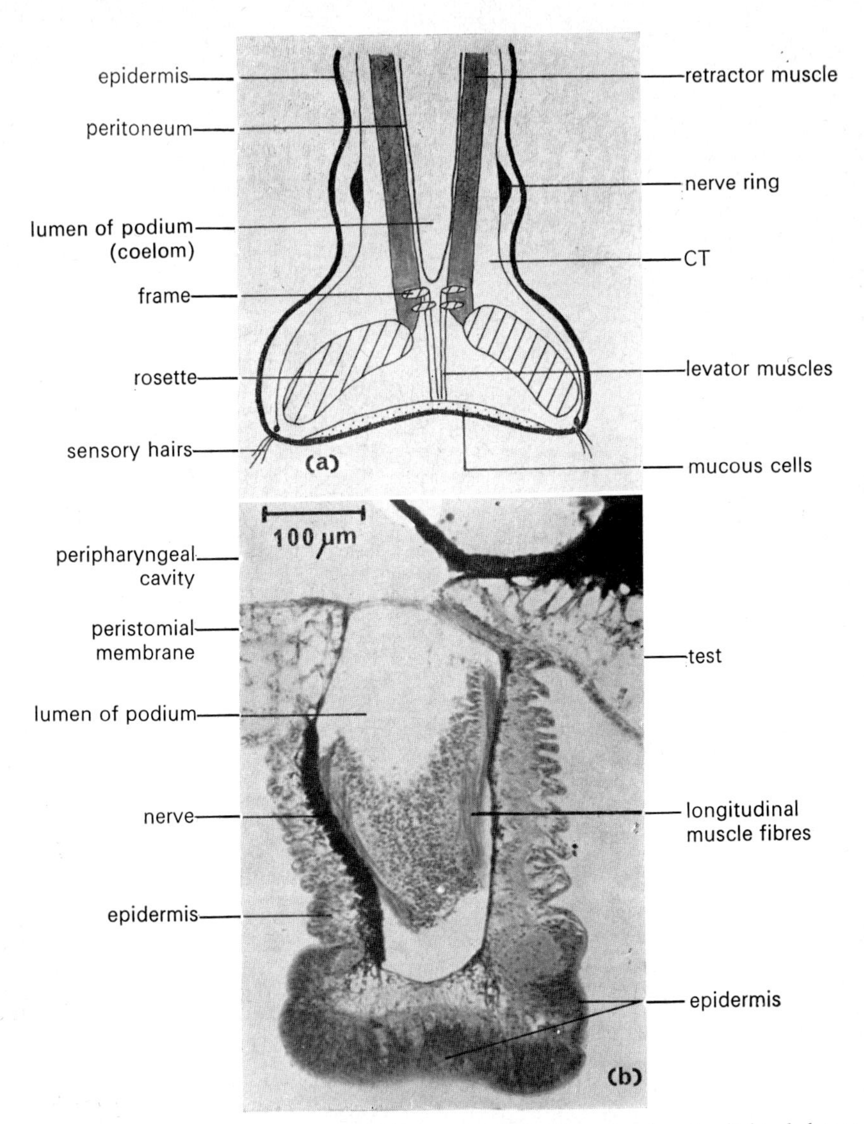

FIG. 8.19 (a) Diagram of an echinoid locomotory tube-foot showing the light skeleton; (b) LS through a buccal tube-foot of an echinoid. Note the absence of skeleton.

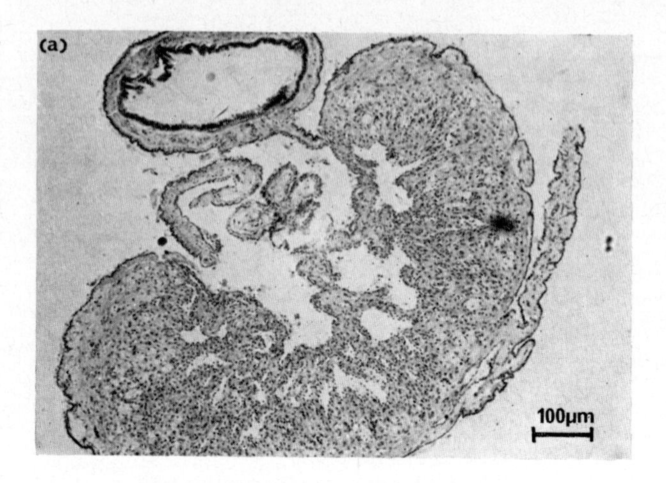
(a)
100µm

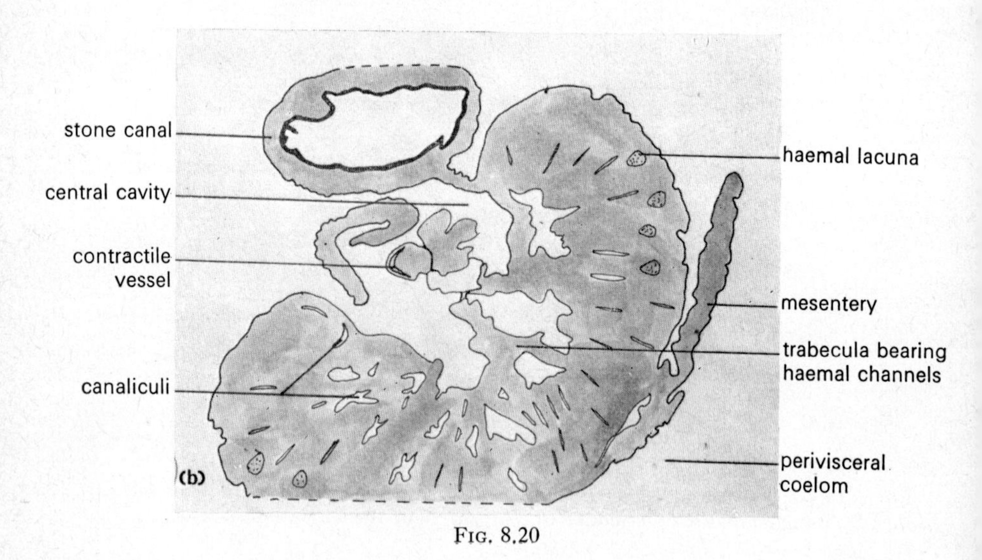
stone canal
central cavity
contractile vessel
canaliculi
(b)
haemal lacuna
mesentery
trabecula bearing haemal channels
perivisceral coelom

FIG. 8.20

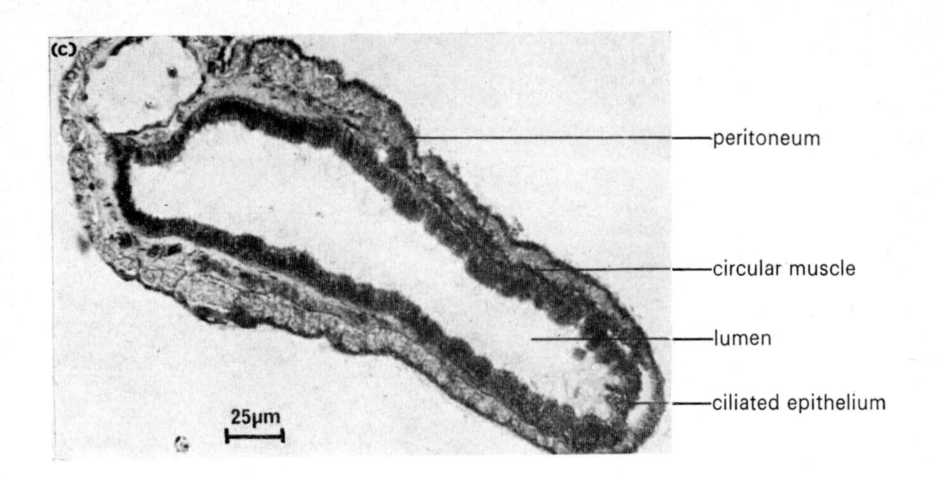

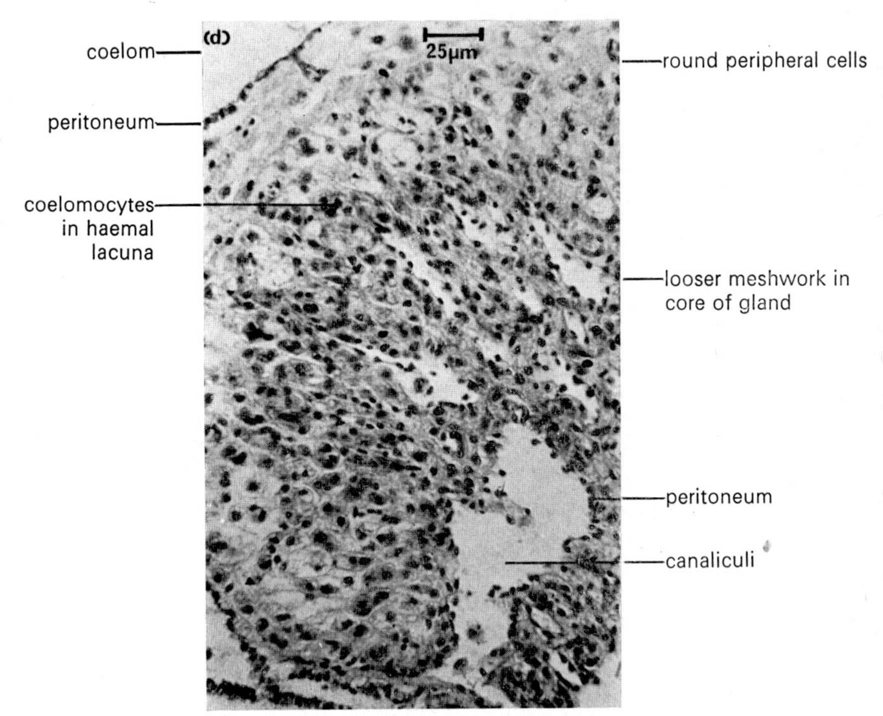

FIG. 8.20 (a) TS and (b) drawing of the stone canal and axial organ of *Echinus esculentus*; (c) detail of the stone canal of *E. esculentus*; (d) detail of the axial organ. Note the coelomocytes in the interstices of the tissue.

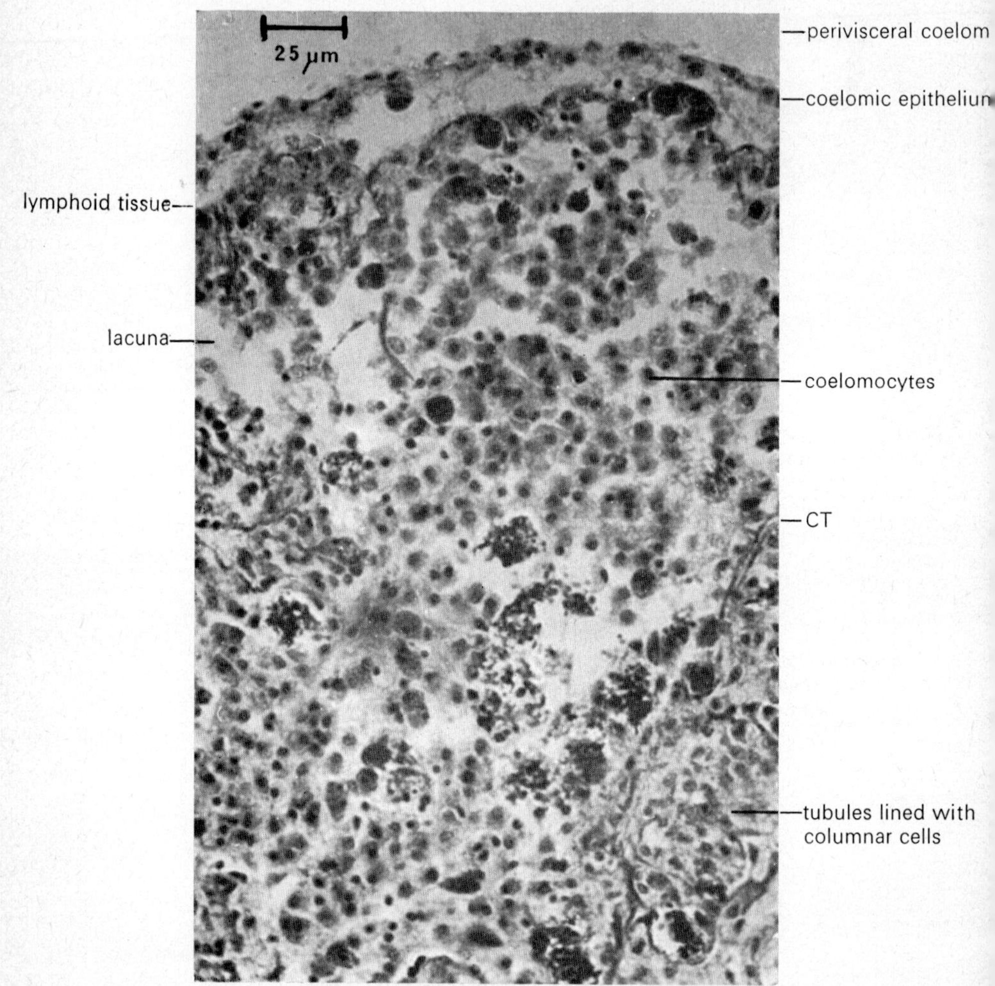

FIG. 8.21. Section through the spongy body of *E. esculentus*. Note the numerous coelomocytes.

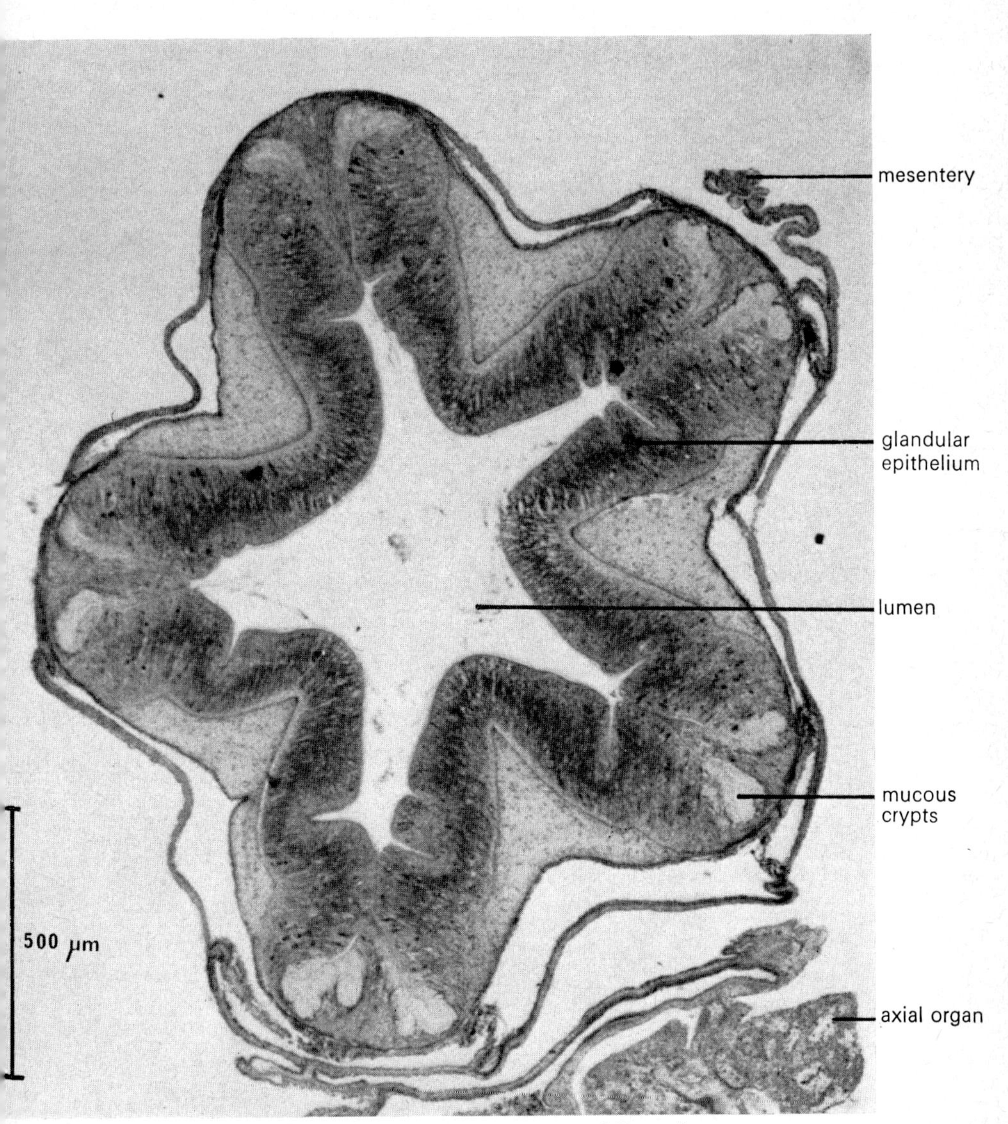

FIG. 8.22. TS through the pharynx of *E. esculentus*.

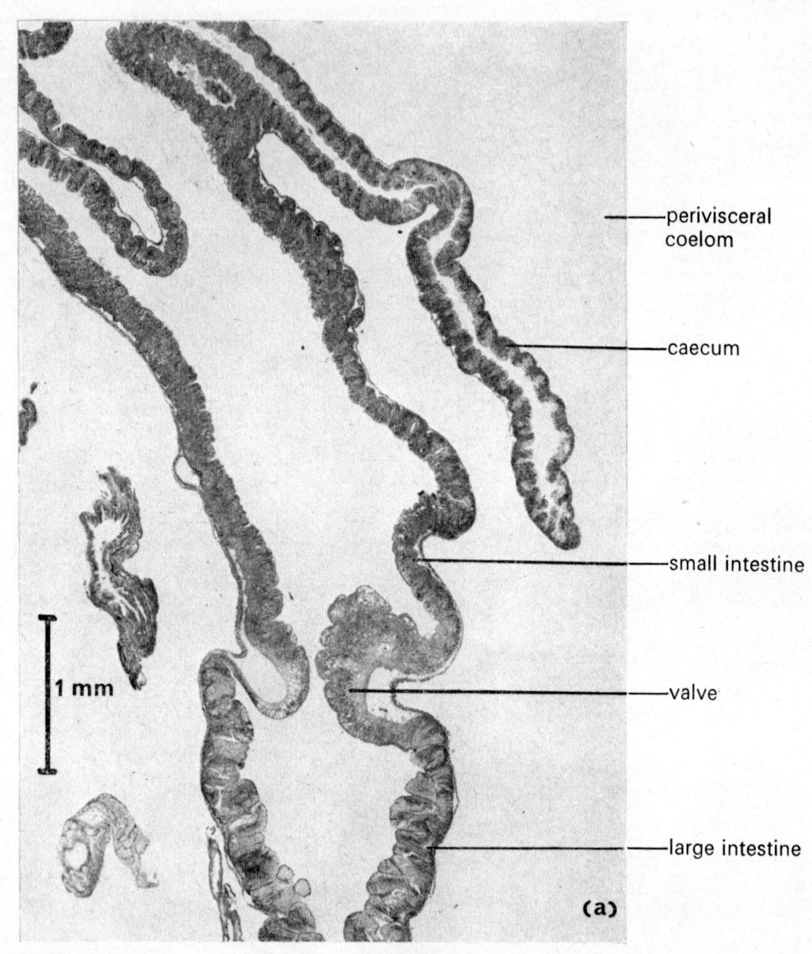
perivisceral coelom
caecum
small intestine
1 mm
valve
large intestine
(a)

FIG. 8.23

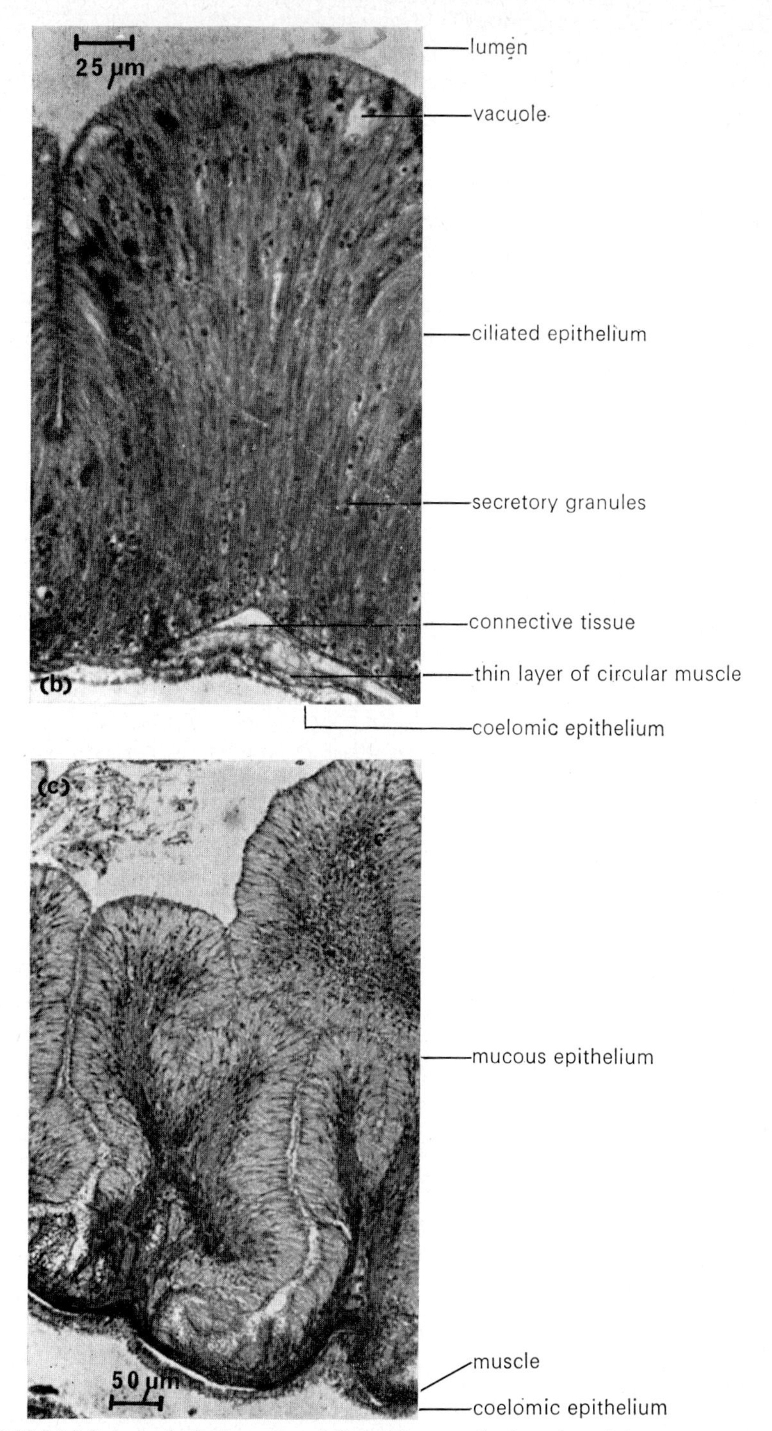

FIG. 8.23 (a) LS through the intestine of *E. esculentus* at the junction of the small and large intestines; (b) detail of the small intestine wall; (c) detail of the large intestine wall.

as in asteroids. Five spongy lymphoid glands, which have often been called Polian vesicles but are now thought to be Tiedemann's bodies, sprout from the ring vessel in the interradii. They consist of numerous branched tubules lined with columnar cells and bound together with connective tissue (Fig. 8.21). They are part of the haemal system and appear to produce amoeboid coelomocytes which are stored in lacunae. The stone canal is accompanied by an extensive axial organ although there is no axial sinus. The axial gland has been implicated in excretion on the grounds that its internal cavity represents part of the coelom and its walls contain coelomocytes which almost certainly play some part in excretion (see Fig. 8.20). The perihaemal system is somewhat reduced as there is no axial sinus, and the circumoral ring does not apparently connect with the radial canals.

Digestion

Echinus is omnivorous, feeding mainly on algae and small bits of animal material. Food particles are rasped off by the jaws of the Aristotle's lantern and passed back into a short grooved pharynx where they are mixed with mucus. Figure 8.22 is a TS of the pharynx and shows clearly the zones responsible for mucus production as well as the marked pentamerous symmetry of *Echinus*. The pharynx leads via a short oesophagus into a small intestine which lies coiled round the inside of the test, attached to it by peritoneal mesenteries. It has a characteristic lining of tall columnar cells bearing cilia and a brush border and containing rows of secretory granules and clear vacuoles [Fig. 8.23(a, b)]. A narrow non-ciliated siphon, of similar structure, leaves the gut at the junction between oesophagus and small intestine and joins it again at the point where the small and large intestines meet. At this point too there are many caecae, mostly from the small intestine. The siphon is probably used as a fluid by-pass to help concentration of food, while the caecae provide a greater surface area for digestion. On either side of the small intestine lie large blood sinuses linked by a dense haemal network. The large intestine also lies coiled round the inside of the test, but in the opposite direction. It secretes mucus [Fig. 8.23(c)] but not digestive enzymes, and functions only in absorption and the storage of food. A short narrow rectum leads to an aboral anus.

Reproduction

Echinoids are dioecious. In regular echinoids there are five gonads: a single sacculated gonad is suspended in each interradius by mesenteries, and leads by a short gonoduct to a gonopore opening on one of five genital plates on the aboral surface. The testes are similar to those of asteroids (see Fig. 8.15). The ovaries are divided up into follicles, each covered by a thin genital sac and lined with germinal epithelium. The developing eggs

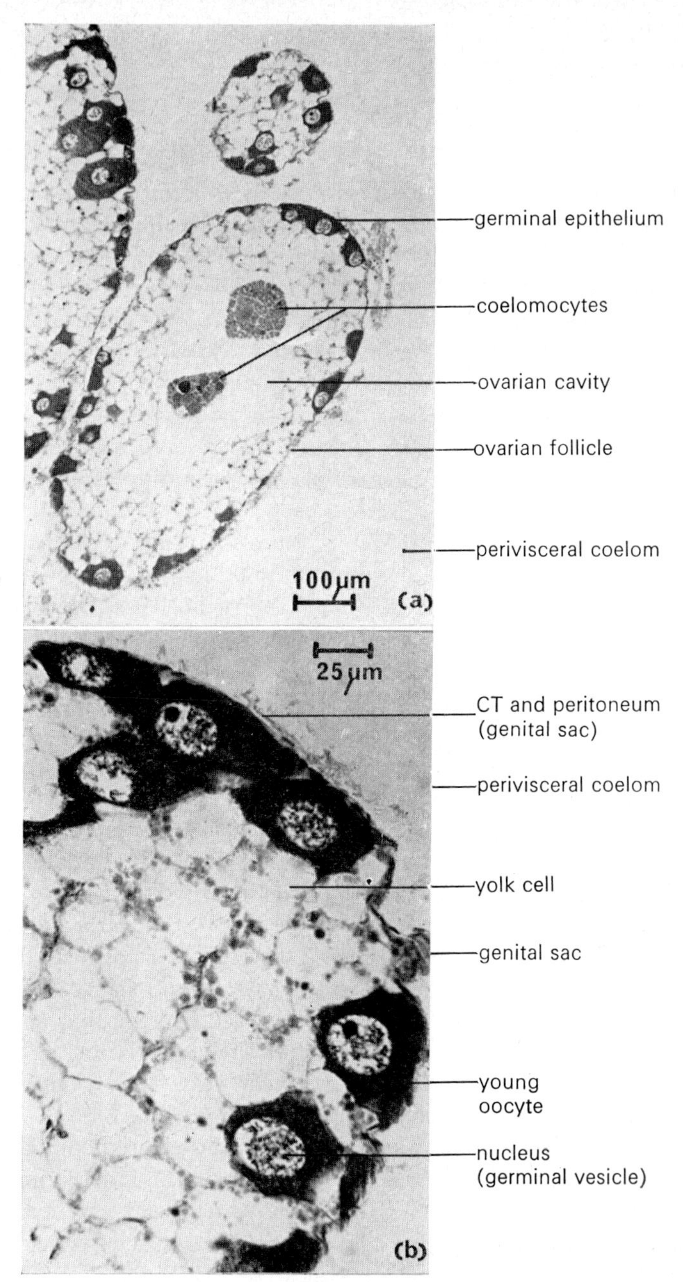

FIG. 8.24 (a) TS through the ovary of a female *E. esculentus*; (b) detail of part of one ovarian follicle.

are provided with nutrients by yolk cells (Fig. 8.24). Fertilization is external. In some species the fertilized eggs are held close to the test by spines, and develop in this protected position.

Class Holothuroidea

The holothuroids, or sea cucumbers, are free-living cylindrical echinoderms, elongated in an oral-aboral direction and lacking arms. They show pentamerous symmetry but with a superimposed bilateral symmetry, i.e. they move oral end first and show some dorso-ventral differentiation in that they generally move on the ventral surface. The aboral anus is directed backwards. There are no recognizable ambulacral and adambulacral areas and the ambulacral grooves are closed. The tube-feet in contact with the substrate are normal but many of the others are reduced and some round the mouth have become modified to form branched or pinnate buccal tentacles. They generally occur in multiples of five and are used in food capture. The skeleton is reduced to minute calcareous ossicles embedded in the dermis, but there is a ring of well-developed ossicles round the mouth and oesophagus to which muscles of the buccal podia and body wall are attached. Spines and pedicellariae are absent, and the madreporite is internal.

Example: *Holothuria* (*forskäli*)

H. forskäli is a sluggish, bottom-living sea cucumber which burrows into cracks and crevices. Its ventral surface is flattened, and covered by tube-feet provided with suckers and bearing a light skeleton in place of levator muscles. They assist locomotion which is effected by undulations of the thick body wall muscles. The dorsal surface is arched and covered with a tough, slimy, papillated skin with the podia reduced to warts and tubercles scattered randomly over the surface. Fig. 8.25 is a vertical transverse section of a small holothuroid (*Cucumaria elongata*), since even young *Holothuria* were found to be too large to section. In this species the pentamerous arrangement is retained and the dorsal tube feet are not greatly reduced. Holothuroids possess a very large body cavity which is sub-divided by mesenteries into four compartments (Fig. 8.26).

The body wall has a thin cuticle covering a non-ciliated epidermis well-supplied with pigment, mucous and sense cells and overlying a sub-epidermal nerve net [Fig. 8.27(a)]. The thick leathery dermis, also containing a nerve plexus, covers a layer of circular muscles which connects with five bands of longitudinal muscle at the radii (see Fig. 8.25). The superficial dermal layers contain small scattered ossicles which represent the remains of an endoskeleton. The body wall is bordered on its inner edge by a thick layer of coelomic epithelium [Fig. 8.27(b)]. The body wall is so thick that very little exchange of oxygen or metabolites can occur across it.

The nervous system follows the general echinoderm plan. The body surface is sensitive to light, implying the occurrence of scattered simple photoreceptors. The main sensory receptors are tactile and chemosensory cells present in the epithelium [see Fig. 8.27(a)], especially round the mouth and anus.

Tubular coelomic systems

The tubular coelomic systems are similar to those of asteroids except that the perihaemal system is barely represented.

The madreporite is generally attached to a very short non-calcified stone canal and lies freely in the coelom as the madreporite body, with no connections to the body surface. Thus the water vascular system communicates with coelomic fluid in the body cavity rather than with seawater. There are no Tiedemann's bodies attached to the circumoral water ring, and in *Holothuria* there is only one long Polian vesicle to act as a fluid reservoir. The buccal tube-feet or tentacles can be shut off from the rest of the water vascular system by muscular valves, thus preventing backflow of fluid. The podia on the dorsal side are reduced and have very small ampullae, while the ventral ones are strengthened by calcareous spicules and are well provided with mucous glands.

The haemal system of holothuroids is well developed although the axial organ appears to be reduced or absent. There is a circumoral haemal ring, five radial haemal strands accompanying the water vascular canals and two large vessels or lacunae running one dorsal and one ventral alongside the gut. These lacunae extend masses of small rete (haemal tufts) over the surface of the intestine wall. They are lined with connective tissue and a thick coat of coelomic epithelial cells stacked with chains of granules and vacuoles (Fig. 8.28). The cells are thought to secrete digestive enzymes and fatty acids, which are then picked up by wandering amoebocytes and transferred to the gut.

Digestion

Holothuroids are plankton or detritus feeders. They trap food in the mucus produced by the ten buccal tentacles, and then use the tentacles to stuff it into the mouth. The mouth lies in the middle of a buccal membrane guarded by a sphincter muscle, and leads to a short pharynx which is generally enclosed in a ring of calcareous ossicles possibly analogous to the echinoid lantern. The stomach is not well defined in *H. forskäli* although in most species it is a muscular organ with the usual muscle layers reversed so that there is a longitudinal inner layer and a circular outer layer. The pharynx and stomach are lined with cuticle and often have longitudinally folded walls, while the tall columnar epithelium is well supplied with

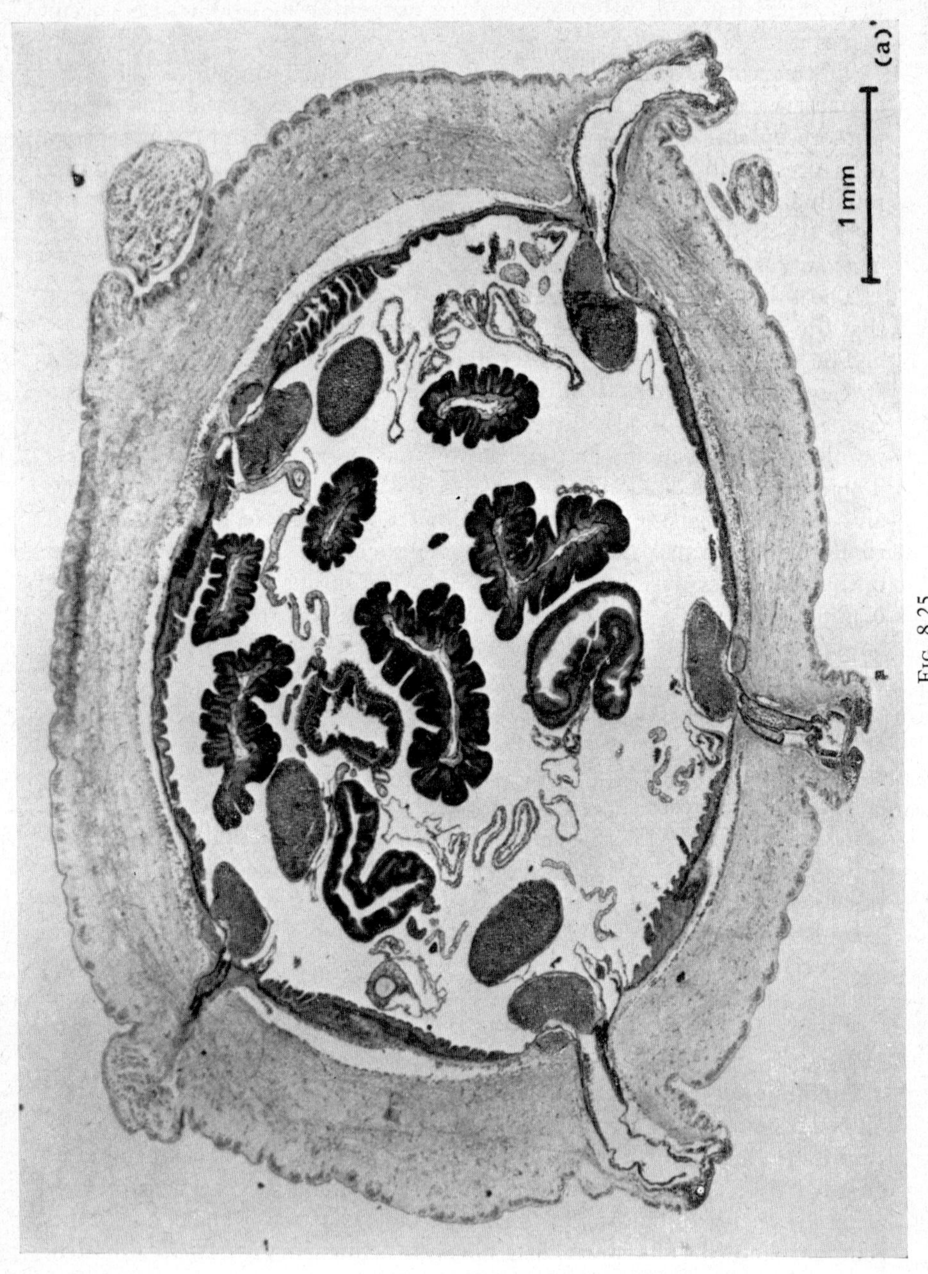

FIG. 8.25

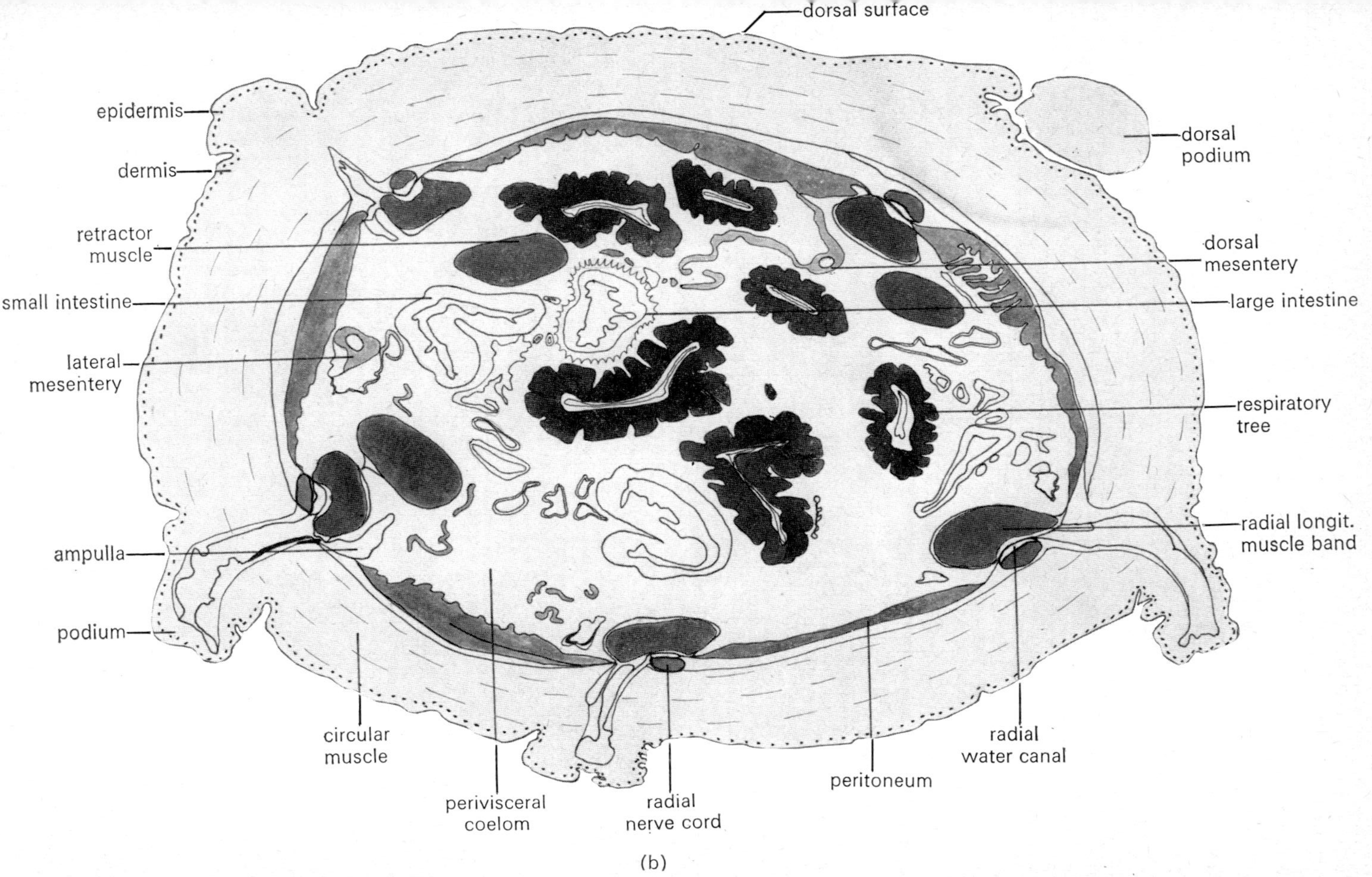

FIG. 8.25 (a) TS and (b) explanatory drawing of a small holothuroid (*Cucumaria elongata*).

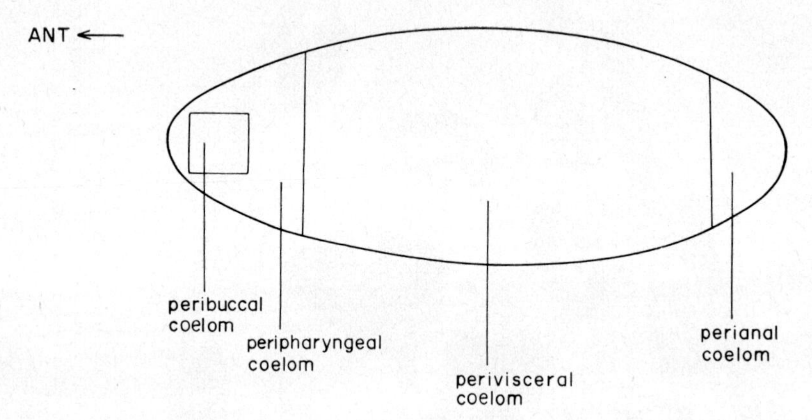

FIG. 8.26. Diagram to show the divisions of the body cavity in holothuroids.

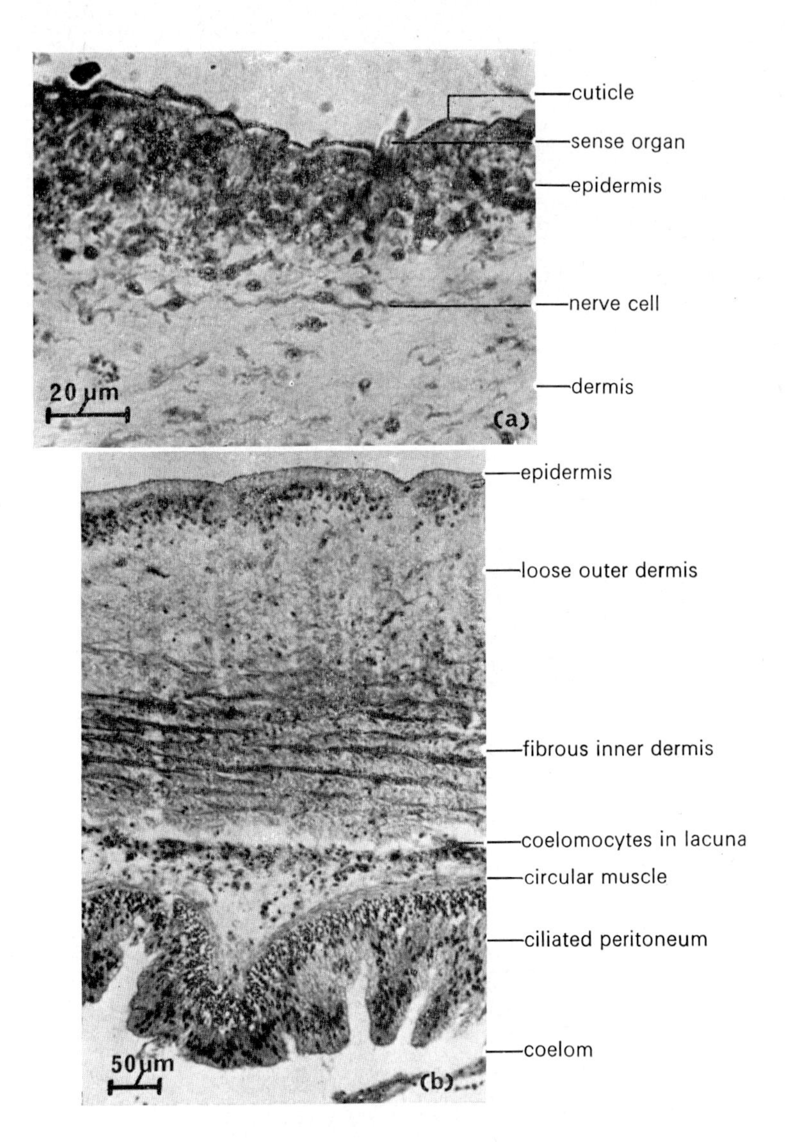

FIG. 8.27 (a) Section through the integument of *Holothuria forskäli*, showing the position of the subepidermal nerve net; (b) section through the holothuroid body wall.

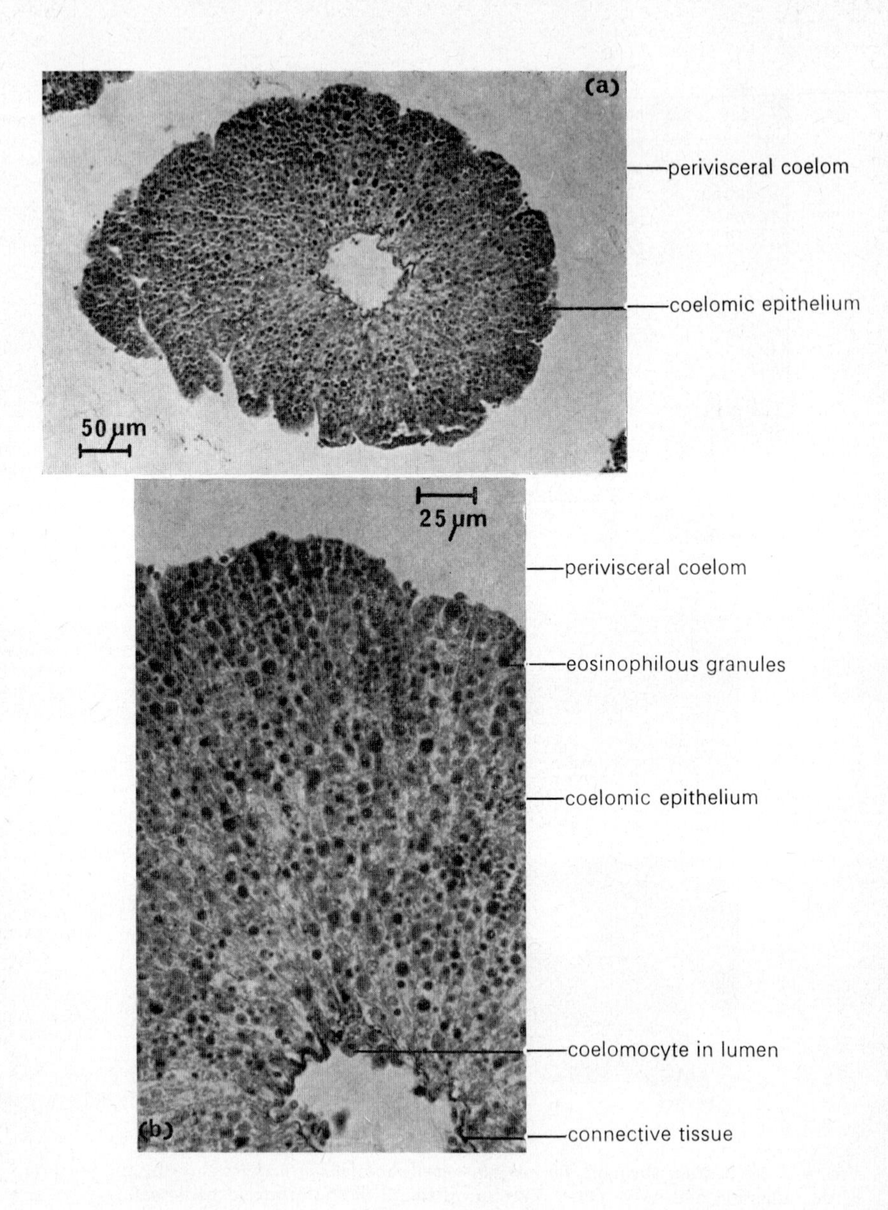

FIG. 8.28 (a) TS through the haemal rete of *H. forskäli*; (b) detail of the epithelial wall of the haemal rete.

gland and mucous cells. Most digestion occurs in the long intestine which is looped round the inside of the body wall in mesenteries. The first part, the small intestine, is well vascularized and lined with a tall ciliated epithelium containing gland cells overlying thin layers of muscle (Fig. 8.29). Wandering amoebocytes are often found within the intestine as they have the ability to cross the gut wall in either direction, carrying enzymes into the gut and waste products out into the body cavity. The large intestine is distended and lined with a progressively thinner epithelium well supplied with mucous cells (Fig. 8.30). The gut musculature in this region shows the more usual arrangement of a circular inner layer and an outer longitudinal layer. It is suggested that the cross-over point provides a weak place where the gut can split and allow either the hind end of the gut or the Cuverian organs (see next section) to be extruded from the body to confuse or entrap an aggressor. The large intestine widens at its distal end to form a cloaca (Fig. 8.31) which plays a part in this extrusion and in respiration. The anus is guarded by a sphincter muscle.

Respiration

Opening into the cloaca are ducts from the chief respiratory structures, the respiratory trees [Fig. 8.32(a)]. These are long, much-branched tubules ending in small thin-walled vesicles, into which water is forced by contractions of the cloacal wall muscles and anal sphincter. Water is later expelled by contractions of the muscular walls of the trees themselves. The tubule walls are similar in structure to the gut wall from which they are derived [Fig. 8.32(b)]. They are closely intermingled with the intestinal haemal tufts, thereby allowing exchange of respiratory gases between the tubules, the coelom and the vascular system. Excretory material is often carried by wandering coelomocytes into the cavities of the respiratory trees (see Fig. 8.32) as well as into the gut and gonadal tubules, whence it is removed to the outside.

The lower branches of the respiratory trees of *H. forskäli* are modified to form Cuverian organs [Fig. 8.32(a)]. These are long tubules with their walls composed of a thick layer of muscle and collagen fibres covered by muscles, and an external glandular layer producing an adhesive secretion [Fig. 8.32(c)]. The Cuverian organs form a very effective defence mechanism against predators. When danger threatens, water is forced into the tubules by the respiratory trees, and the proximal ends of the tubules are occluded by muscles. Powerful contractions of the body wall muscles cause the expulsion of the Cuverian tubules from the body cavity, through a weak place in the gut wall, into the cloacal lumen and thence to the outside through the anus. As the tubules are extruded, contractions of their muscular walls against the enclosed fluid forces them to elongate. When this happens the external glandular layer is disrupted, converting the tubules into long sticky collagenous threads which entrap the predator.

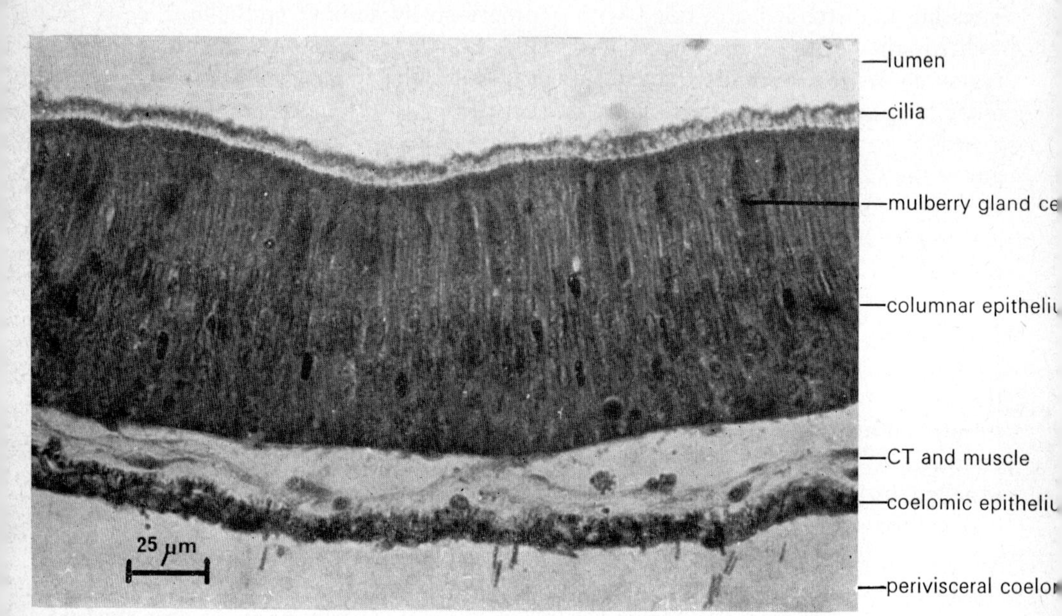

FIG. 8.29. Section through the small intestine wall of *H. forskäli*.

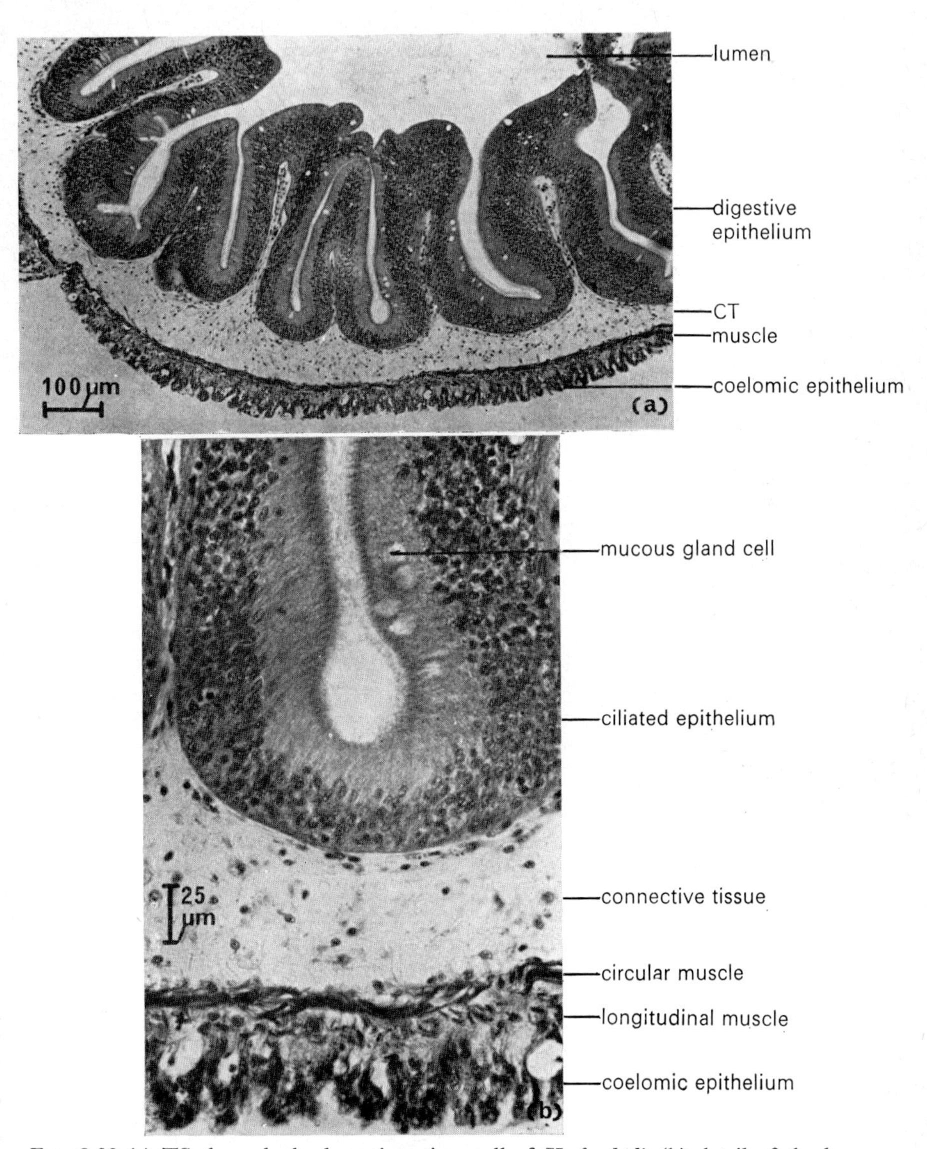

FIG. 8.30 (a) TS through the large intestine wall of *H. forskäli*; (b) detail of the large intestine wall.

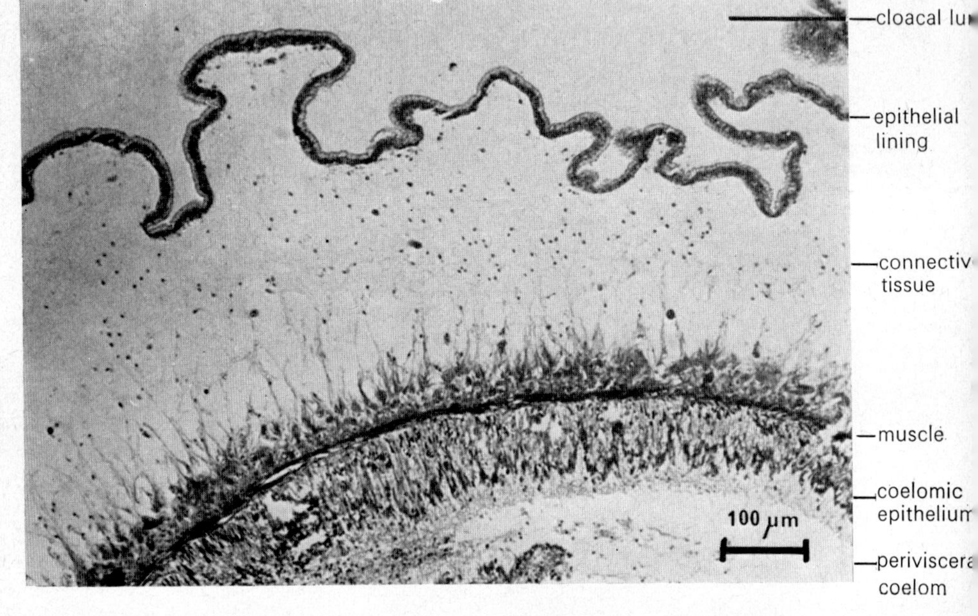

FIG. 8.31. Section through the cloacal wall of *H. forskäli*.

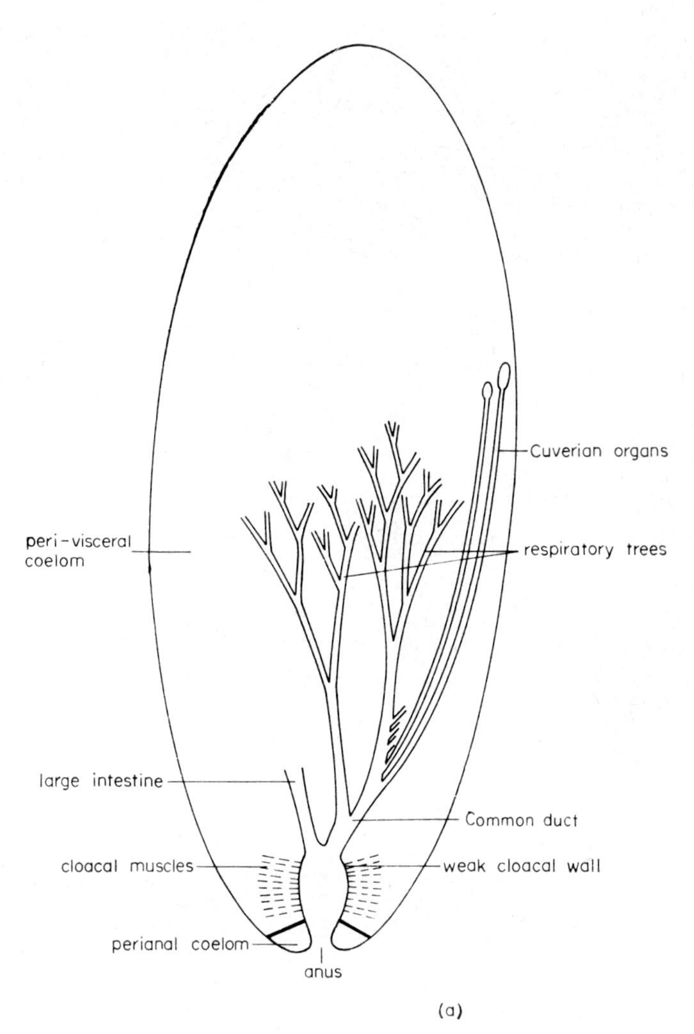
Cuverian organs
peri-visceral coelom
respiratory trees
large intestine
Common duct
cloacal muscles
weak cloacal wall
perianal coelom
anus
(a)

FIG. 8.32(a)

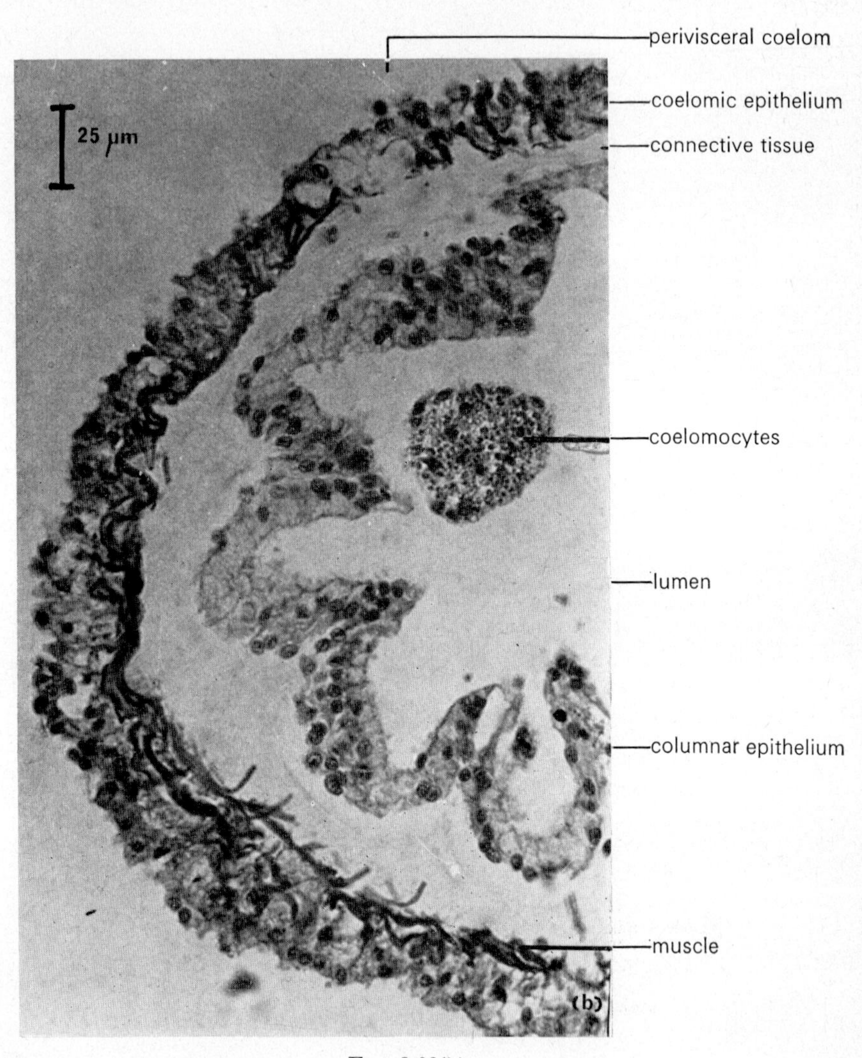
25 μm
perivisceral coelom
coelomic epithelium
connective tissue
coelomocytes
lumen
columnar epithelium
muscle
(b)

Fig. 8.32(b)

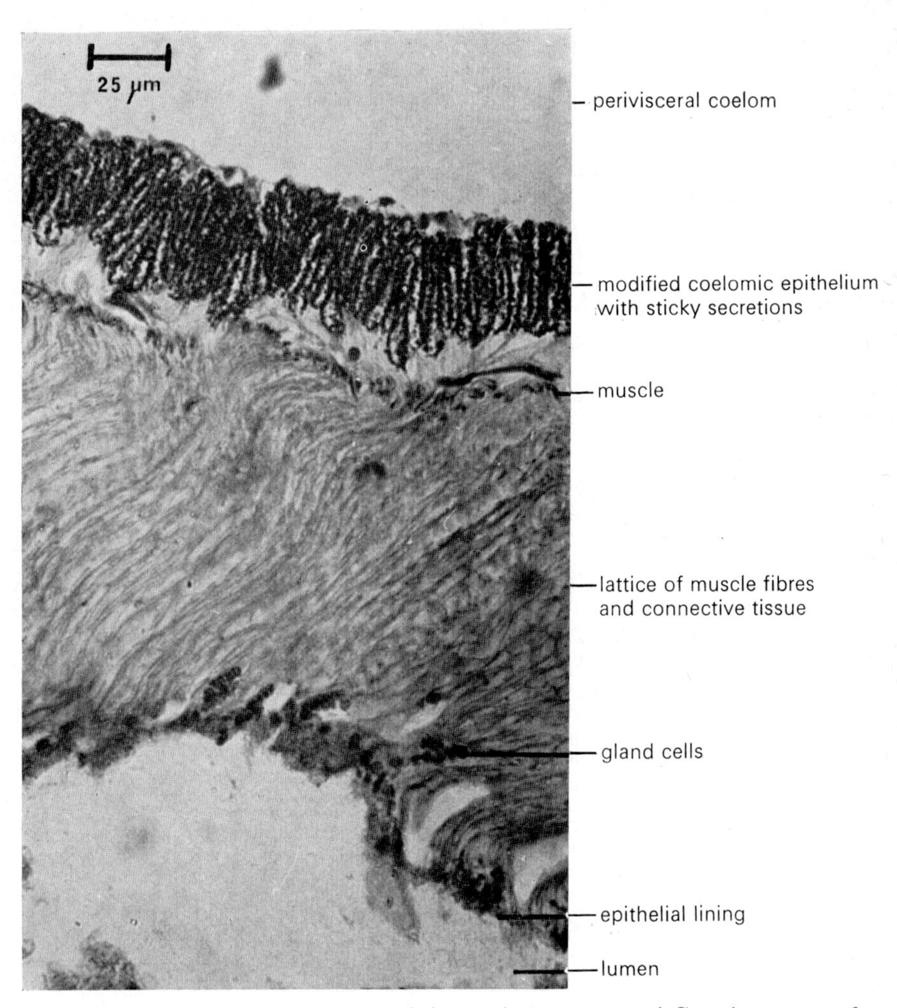

FIG. 8.32 (a) Plan of the arrangement of the respiratory trees and Cuverian organs of *H. forskäli*; (b) TS through a tubule from the respiratory tree of *H. forskäli*; (c) LS through the wall of a Cuverian organ from *H. forskäli*.

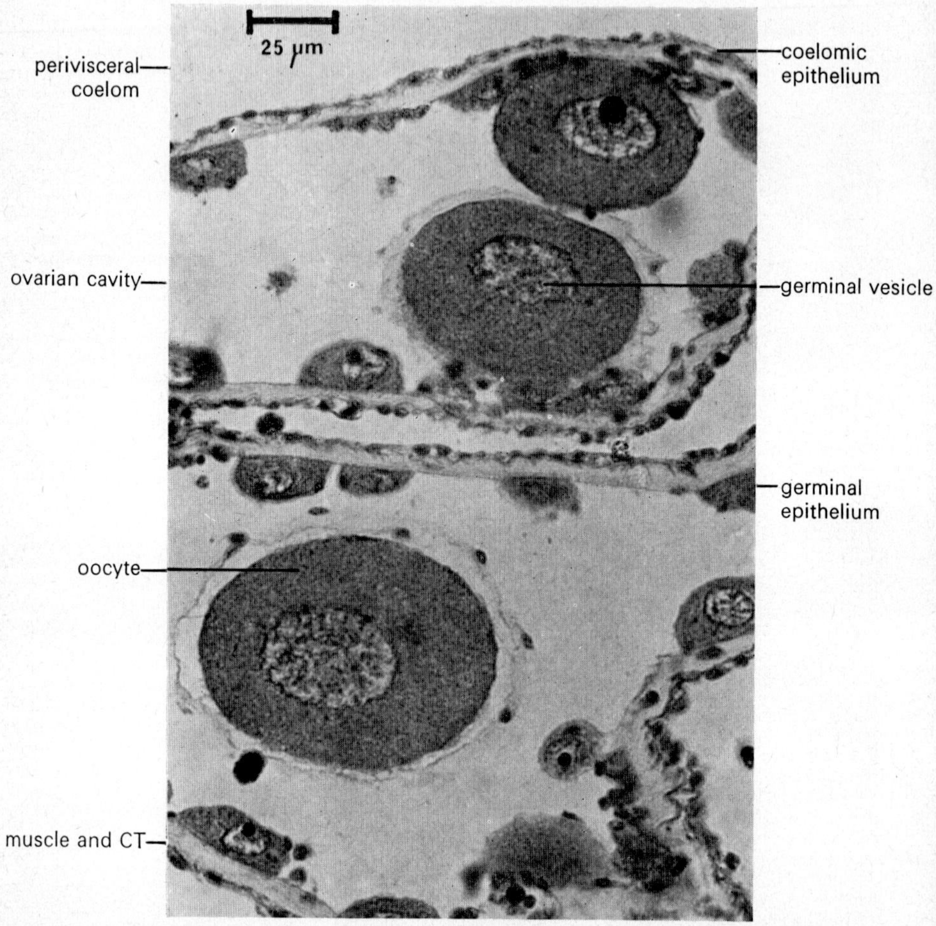

FIG. 8.33. Section through two tubules of the branched ovary of a female *H. forskäli*.

The tubules break free and remain entwined round the predator, while new tubules regenerate quickly in the holothuroid.

Reproduction

Most holothuroids are dioecious, but, unlike other echinoderms, there is only a single gonad which is suspended in the coelom near the oral end of the mid-dorsal interambulacrum. Both male and female gonads are composed of a large tuft of finely-branched tubules covered with thin layers of coelomic epithelium and muscle and lined with germinal epithelium (Fig. 8.33). A single long gonadal duct lined with ciliated epithelium opens through a gonopore located just behind the mouth at the base of the buccal tentacles. As in other echinoderms, fertilization is external and development is through several pelagic larval stages.

Chapter 9

Phylum Hemichordata

For many years the hemichordates were classified as protochordates on the misunderstanding that their small pre-oral gut diverticulum was a skeletal notochord. More recently its true nature has become apparent, and the hemichordates have been divided off as a separate phylum. They are characterized by the possession of a coelom, pharyngeal gill clefts and a dorsal hollow nerve cord in the neck region only. There is no tail, bony tissue or endostyle (see p. 386). Hemichordates are marine, and can be either free-living (Enteropneusta) or sessile (Pterobranchiata); while several fossil forms are known. Hollow arms containing extensions of the coelom are found in some Pterobranchs, resembling the lophophore or food catching organ of certain minor invertebrate phyla and recalling the tube-feet of echinoderms.

Class Enteropneusta

Enteropneusts, or acorn worms, are free-living vermiform creatures normally found in shallow water. They have a large number of gill slits opening to the outside by gill pores, and are further distinguished by a straight gut with a terminal anus.

Example: *Balanoglossus* (sp.)

Balanoglossus is a sluggish acorn worm which lives in a burrow. Its body is divided into three parts: short conical proboscis, wide collar and long cylindrical trunk (Fig. 9.1). The proboscis and trunk are mobile and used for burrowing. Each region contains a segment of coelom, which is split into two lateral halves in the post-oral part of the body.

The body wall is composed of muscles, but a dermis is lacking (Fig. 9.2). The epidermis is a tall columnar epithelial layer bearing cilia which form tracts in some regions, and is well supplied with goblet, granular and mulberry gland cells. The muscle layers are not well represented in the proboscis and collar but the trunk has thick layers of longitudinal muscle, particularly down the ventral surface [Fig. 9.2(c)].

Nervous system

Balanoglossus has a primitive nervous system consisting of a sub-epidermal nerve net (Fig. 9.2), thickened longitudinally to form several nerve cords, represented diagrammatically in Fig. 9.3. Two thick dorsal and ventral cords are joined immediately behind the collar by a thickened pre-branchial or circum-enteric ring. The dorsal cord passes forwards into the collar where it becomes even thicker and invaginates to form a hollow neurotube, a forerunner of the vertebrate hollow nerve cord. There is, however, no evidence that the neurotube possesses any of the integrative properties of the vertebrate spinal cord. Giant nerve cells are present in the neurotube.

There are no special sense organs in hemichordates, the only sensory receptors being cells in the epidermis which send distal processes directly to the nerve net.

Respiration

The respiratory organs of *Balanoglossus* are ciliated gills. The walls of the dorsal part of the pharynx are perforated by a series of gill slits, each leading into a gill pouch which opens to the outside by a small gill pore (Fig. 9.4). Each U-shaped gill slit is separated from the next by a septum, and is supported down the centre by a tongue bar, in a similar fashion to those of *Amphioxus* (see Fig. 10.23). The septa and tongue bars are well vascularized, strengthened by skeletal supports, and covered by ciliated epithelium on their pharyngeal and lateral sides (Fig. 9.5). These cilia draw water currents in through the mouth, and out through the gill slits into the gill pouches and then to the outside.

Circulation

The circulatory system of hemichordates is partially open. Blood from small lacunae round the gut and gills drains forwards into a contractile dorsal vessel lying in a dorsal mesentery. This vessel expands in the posterior part of the proboscis into a venous sinus, which pumps blood into the non-contractile central sinus or "heart" (Fig. 9.6). This is a thin-walled structure situated between the buccal diverticulum and heart vesicle (pericardium) as can be seen in Fig. 9.1. The heart vesicle is composed of connective tissue and muscle fibres, particularly in its ventral wall which is in very close contact with the "heart". The buccal diverticulum, central sinus and heart vesicle are covered with peritoneum which is thrown into a complex of outpocketings constituting the glomerulus (see Figs 9.8 and 9.9), the lumen of which is continuous with the "heart". Contractions of the vesicle wall drive blood forwards from the "heart" to a plexus of cavities in the glomerulus. Two vessels carry blood from the glomerulus to the proboscis wall and two others join to form a single contractile ventral vessel situated just below the gut which supplies the plexus of blood lacunae round the intestine wall and the gill pouches.

Digestion

Balanoglossus is a particulate feeder, food being wafted into the ventral mouth by the action of cilia on the body surface and lining the gill slits.

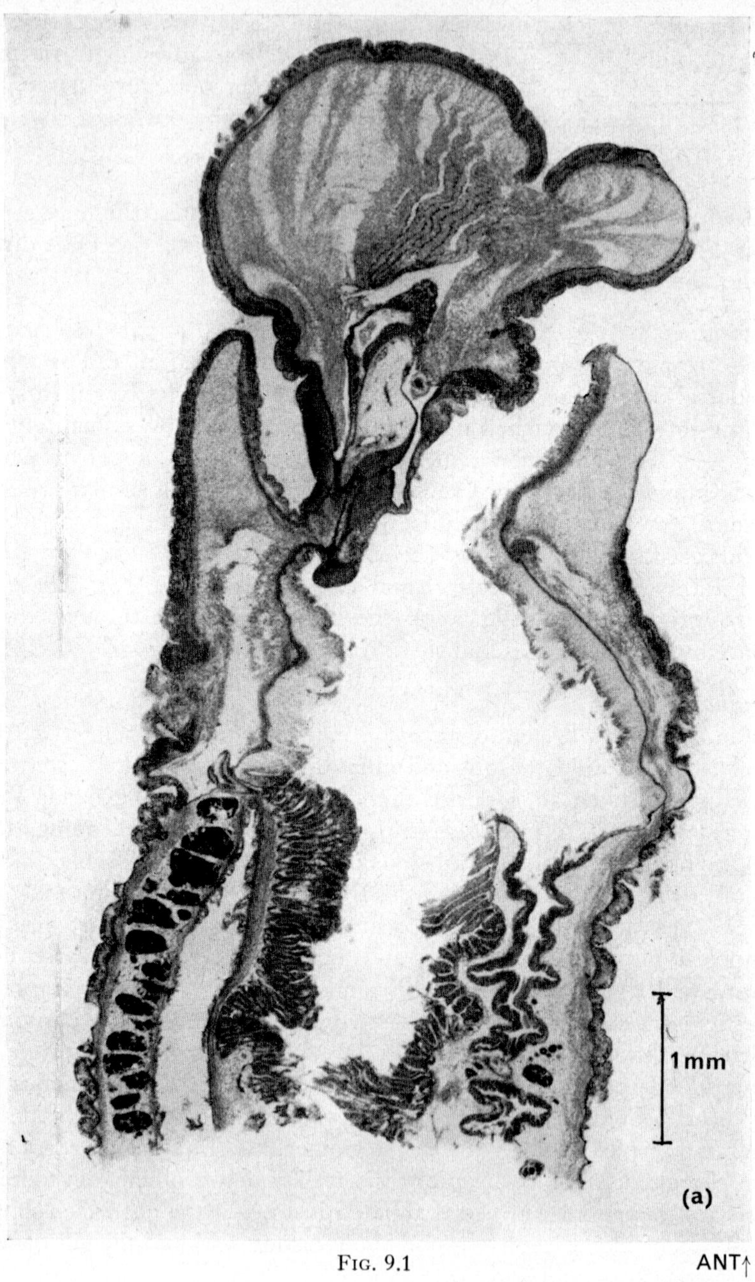

FIG. 9.1

ANT↑

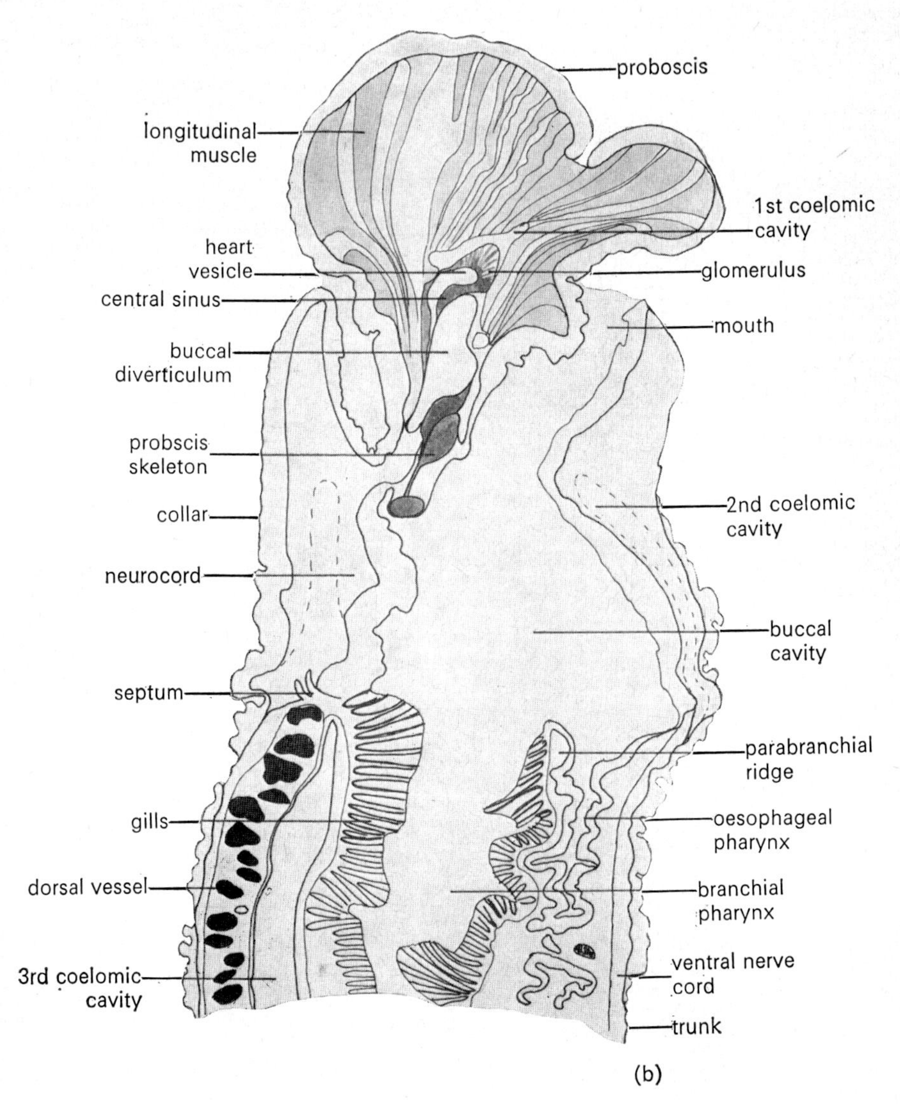

FIG. 9.1 (a) VLS and (b) explanatory drawing of the anterior end of the hemichordate *Balanoglossus sp.*

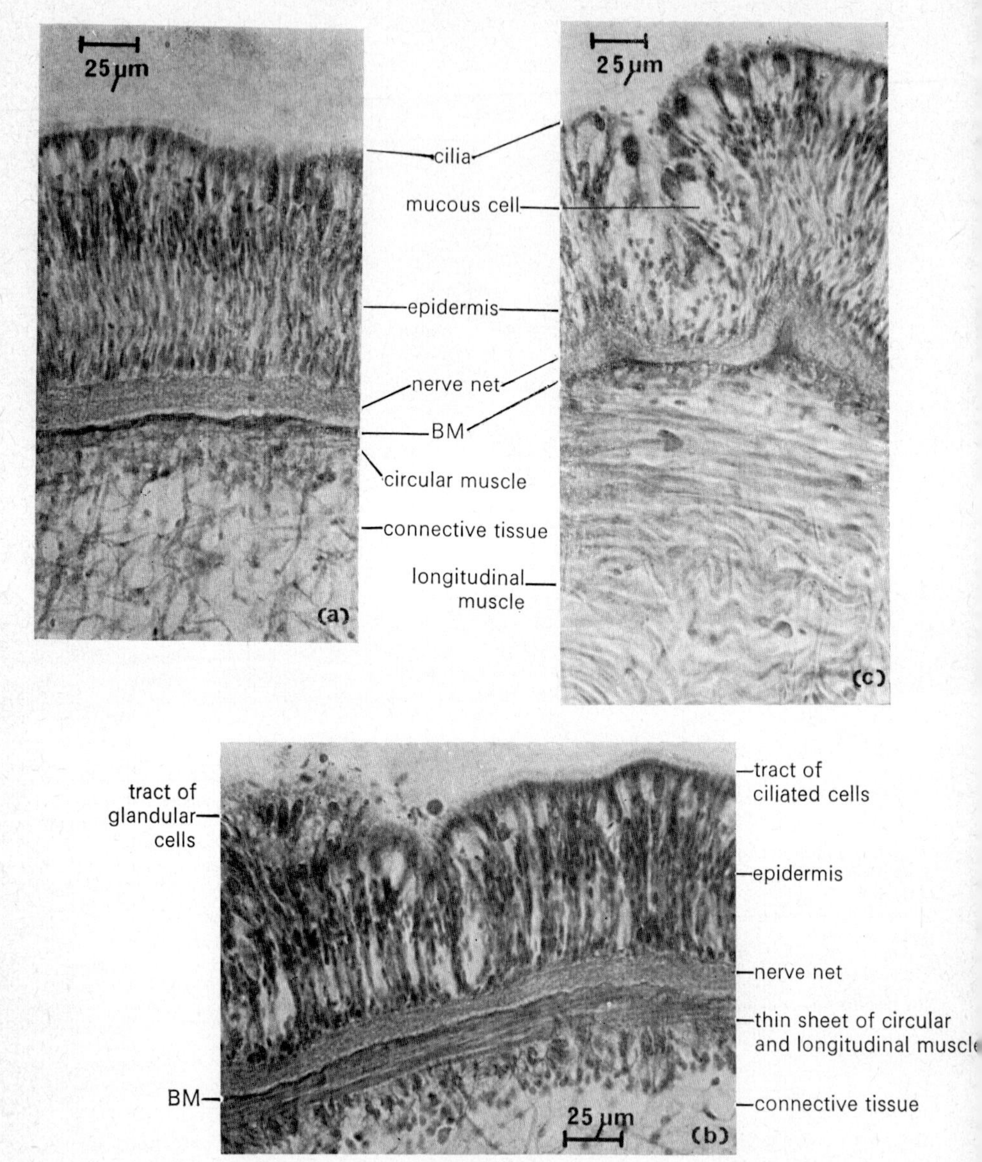

FIG. 9.2. LS through the body wall of *Balanoglossus* in (a) the proboscis region; (b) the collar region; and (c) the trunk region. Note the thick nerve net at the base of the epidermis.

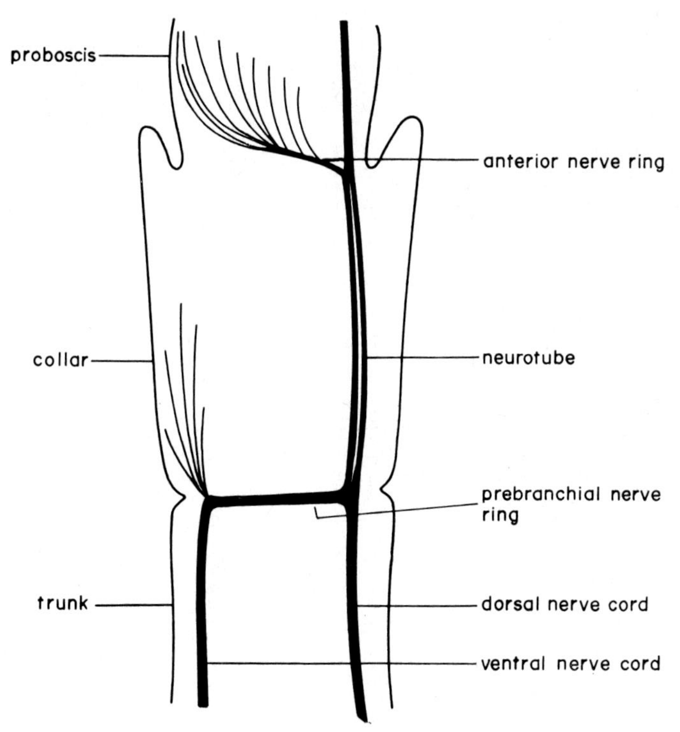

FIG. 9.3. Plan of the anterior end of the nervous system of hemichordates.

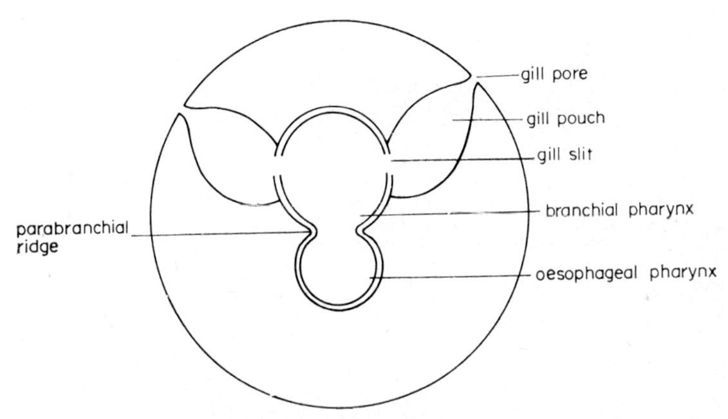

FIG. 9.4. Diagram of a TS through the branchial region of *Balanoglossus*.

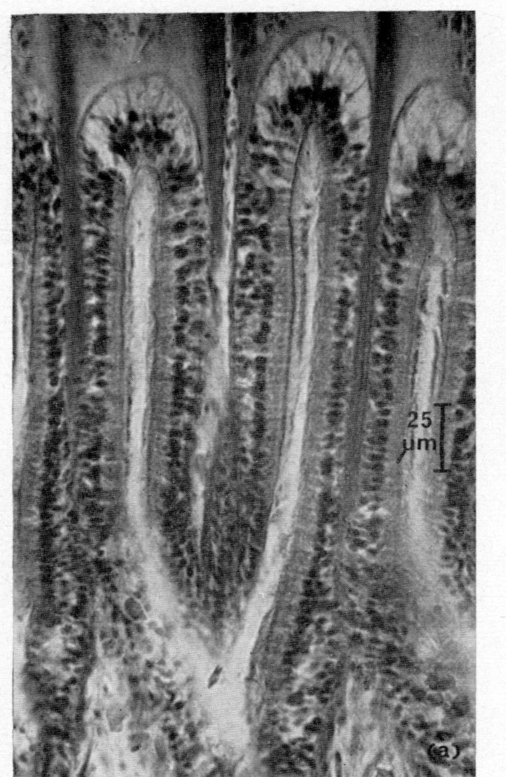

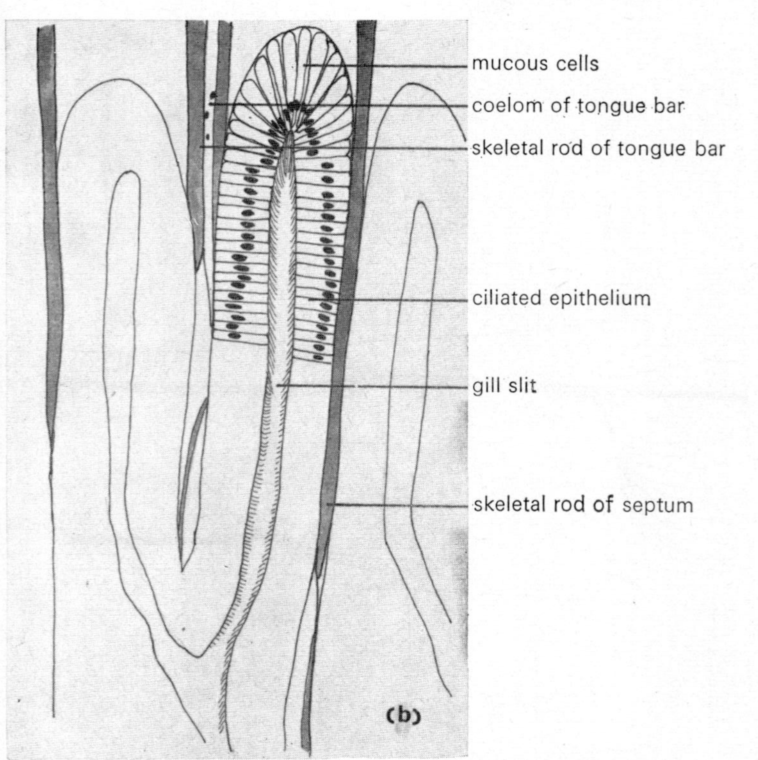

FIG. 9.5 (a) LS and (b) drawing of the pharynx wall of *Balanoglossus*. Note that each septum has a single skeletal supporting rod while each tongue bar bears two skeletal rods.

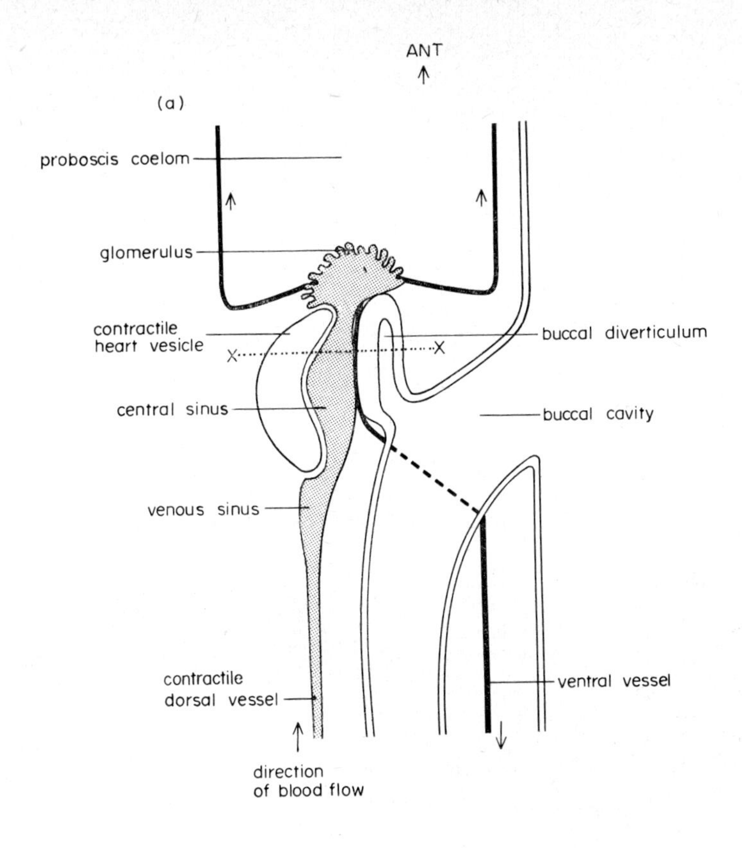

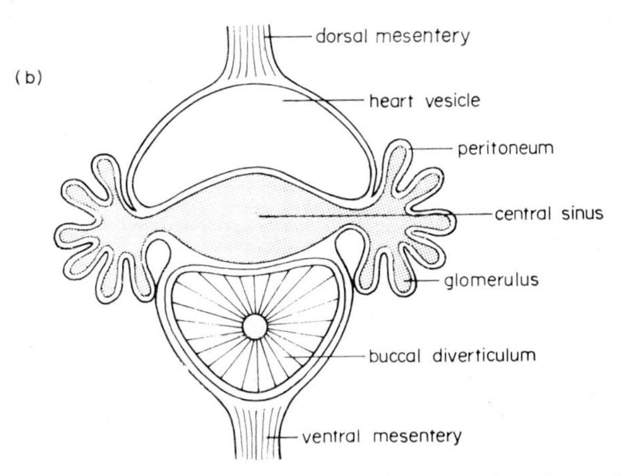

FIG. 9.6 (a) Plan of the anterior end of the circulatory system of *Balanoglossus*, showing the direction of blood flow through the heart and glomerulus [X-X is the section shown in (b)]; (b) plan of the arrangement of heart vesicle, glomerulus and buccal diverticulum in transverse section [X-X in (a)].

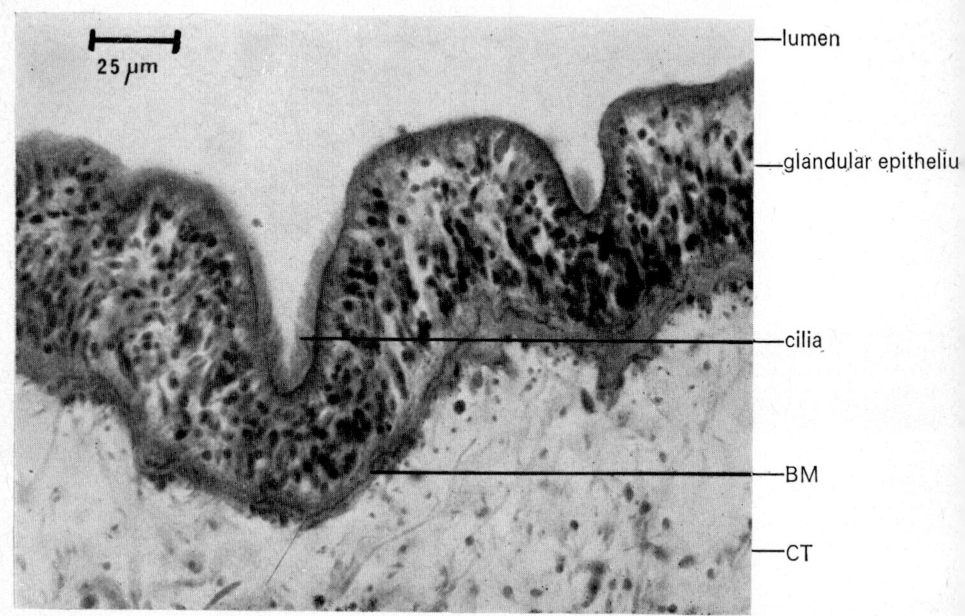

FIG. 9.7. LS through the intestine wall of *Balanoglossus*.

It passes from the wide buccal cavity into the ventral part of the pharynx (oesophageal pharynx) below the gill slits (see Fig. 9.4). The pharynx leads into an oesophagus which is provided with circular muscles that help to move the food onwards. All other regions of the gut have poorly developed musculature, and food is passed chiefly by ciliary action. Stout cilia can be seen in the intestine (Fig. 9.7), part of which bears sacculations and is known as the hepatic region where most digestion occurs. The intestine is followed by a short rectum leading to the outside by an anus guarded by a sphincter muscle.

A long narrow buccal diverticulum, once thought to be a notochord, runs from the buccal cavity into the proboscis where it expands slightly (see Figs. 9.1 and 9.6). It has a lining of tall vacuolated columnar cells (Fig. 9.8), very similar in structure to the wall of the buccal cavity. Its function is still unknown.

Excretion

The glomerulus is accepted as the chief excretory organ in hemichordates. It is composed of a mass of blind-ended evaginations of peritoneum containing blood (Fig. 9.9), and acquired its name because of its resemblance to vertebrate kidney glomeruli. Blood passes into the glomerulus from the central sinus and it is assumed that waste material is filtered off into the coelom.

Reproduction

Hemichordates are dioecious but there is no sexual dimorphism. The gonads are sac-like structures developed from the coelom wall and covered in a membrane continuous with the basement membrane of the epidermis (Fig. 9.10). They open to the outside by a narrow neck and genital pore. Both gonads are situated laterally in the anterior trunk region. Fertilization is external, and development is through a pelagic larva.

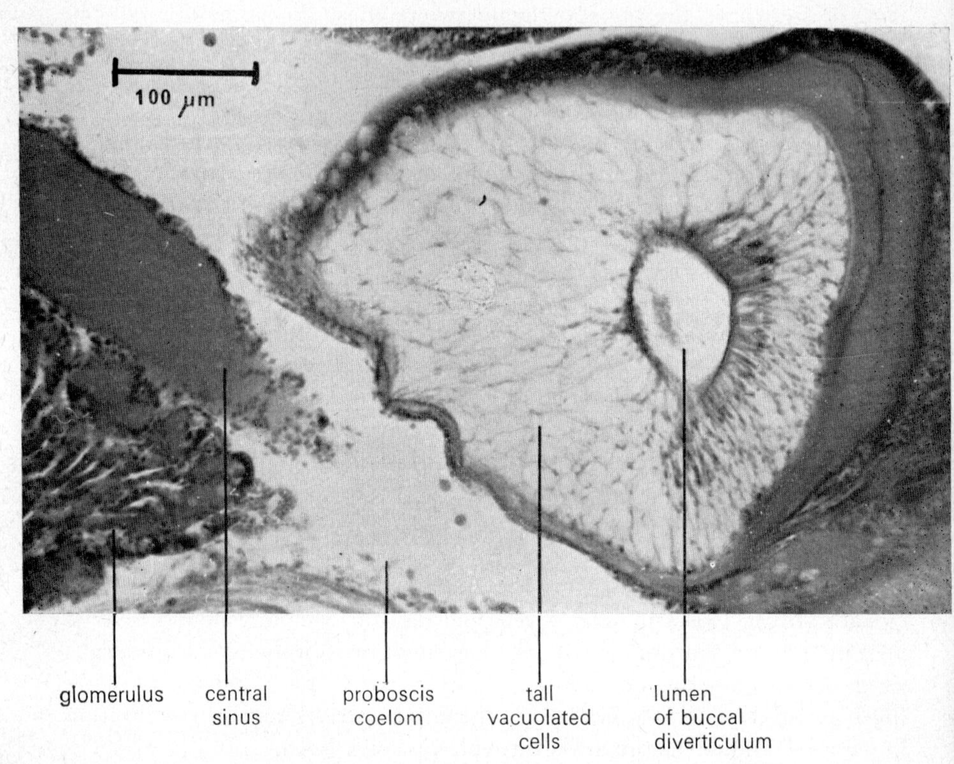

FIG. 9.8. Oblique TS through the buccal diverticulum of *Balanoglossus*. Note its close proximity to the glomerulus.

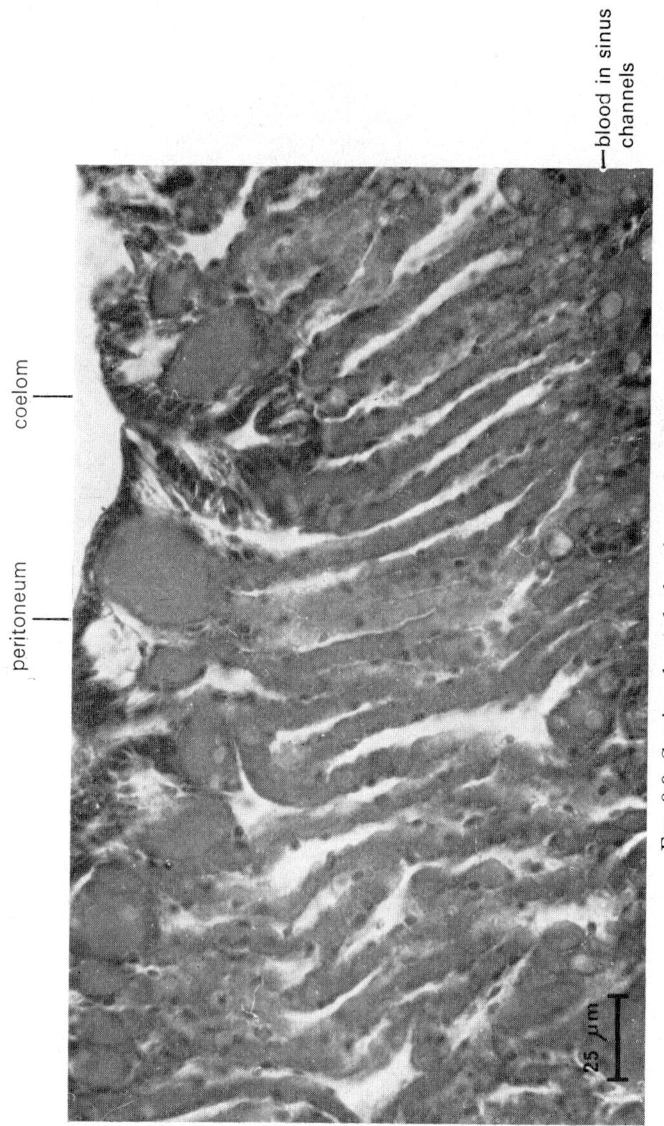

FIG. 9.9. Section through the glomerulus of *Balanoglossus*.

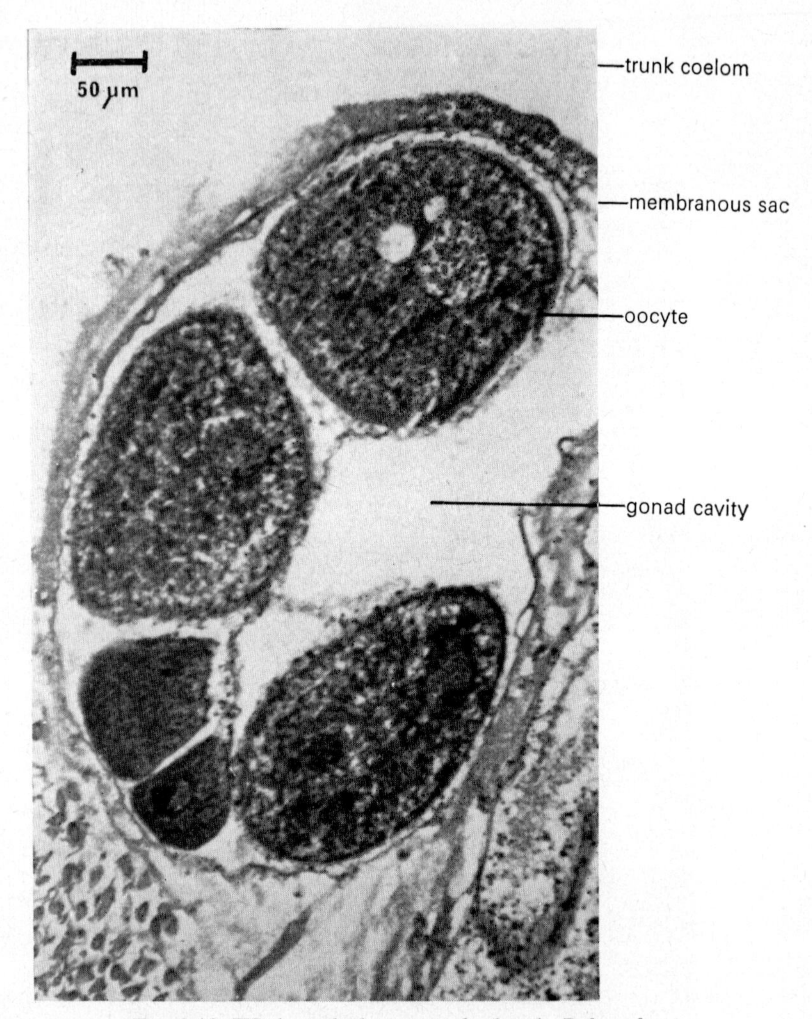

FIG. 9,10. TS through the ovary of a female *Balanoglossus*.

PHYLUM CHORDATA

The chordates take their name from the *chorda dorsalis* or notochord, present at some stage in their life cycle, which acts as a skeletal support and is the forerunner of the vertebral column in vertebrates. They are also characterized by a hollow dorsal nerve cord, and a pharynx that is perforated by slits or pouches which are often used for respiration. They possess a coelom, which may be secondarily divided or reduced. At some stage in their life, all chordates possess a muscular post-anal tail, and an endostyle or ciliated pharyngeal groove which can fix iodine from the surrounding medium and which gives rise to the thyroid glands in higher vertebrates.

Chapter 10

Protochordates

Protochordates possess all the chordate characters, at least at some stage of their life. Thus, they have a dorsal "skeletal" rod (notochord), a dorsal tubular nerve cord, numerous pharyngeal slits, a post-anal tail and an endostyle. Their coelom develops as three chambers (anterior, median and posterior) although in the adult these are often obscured or reduced.

The protochordates comprise two sub-phyla of the Chordata. These are the Urochordata and the Cephalochordata, which are distinguishable by the development and positioning of their notochords.

Sub-phylum Urochordata

The urochordates, or tunicates, are common marine animals. The larval stage is free-living while the adults are generally sessile, although there are a few pelagic species. The hollow dorsal nerve cord and notochord are present only in the larva, and then the notochord is restricted to the tail region. Both structures degenerate in the adult. There are no signs of bony tissue or segmentation such as are seen in higher chordate groups. The coelom is either missing or may be represented by the small epicardium, divided into two sacs, which surrounds the gonads and gut. The main body cavity is an atrium, formed by invagination of the dorsal side of the body and lined with ectoderm. All tunicates are characterized by their external tunic or test composed of tunicin, a material related to cellulose.

Example: *Ciona* (*intestinalis*)

Ciona intestinalis is a simple marine ascidian (sea-squirt) which is distributed widely in the littoral zone around the coasts of Britain. It has a short-lived, symmetrical, tadpole-like larva which does not feed but which allows for dispersion of the species. The adult is sessile and lives anchored to the underside of rocks and breakwaters by a stolon. Although there are colonial sea-squirts, *Ciona* is usually solitary. The adult bears little resemblance to a "typical chordate" as it has the form of a transparent cylindrical sac with no tail. The sac has two openings, both guarded by muscular closing mechanisms. These are the anterior buccal siphon (mouth) and the dorsally opening atrial siphon.

Figure 10.1 shows a vertical longitudinal section through the posterior end of a small *C. intestinalis*. The transparent tunic and small coelom

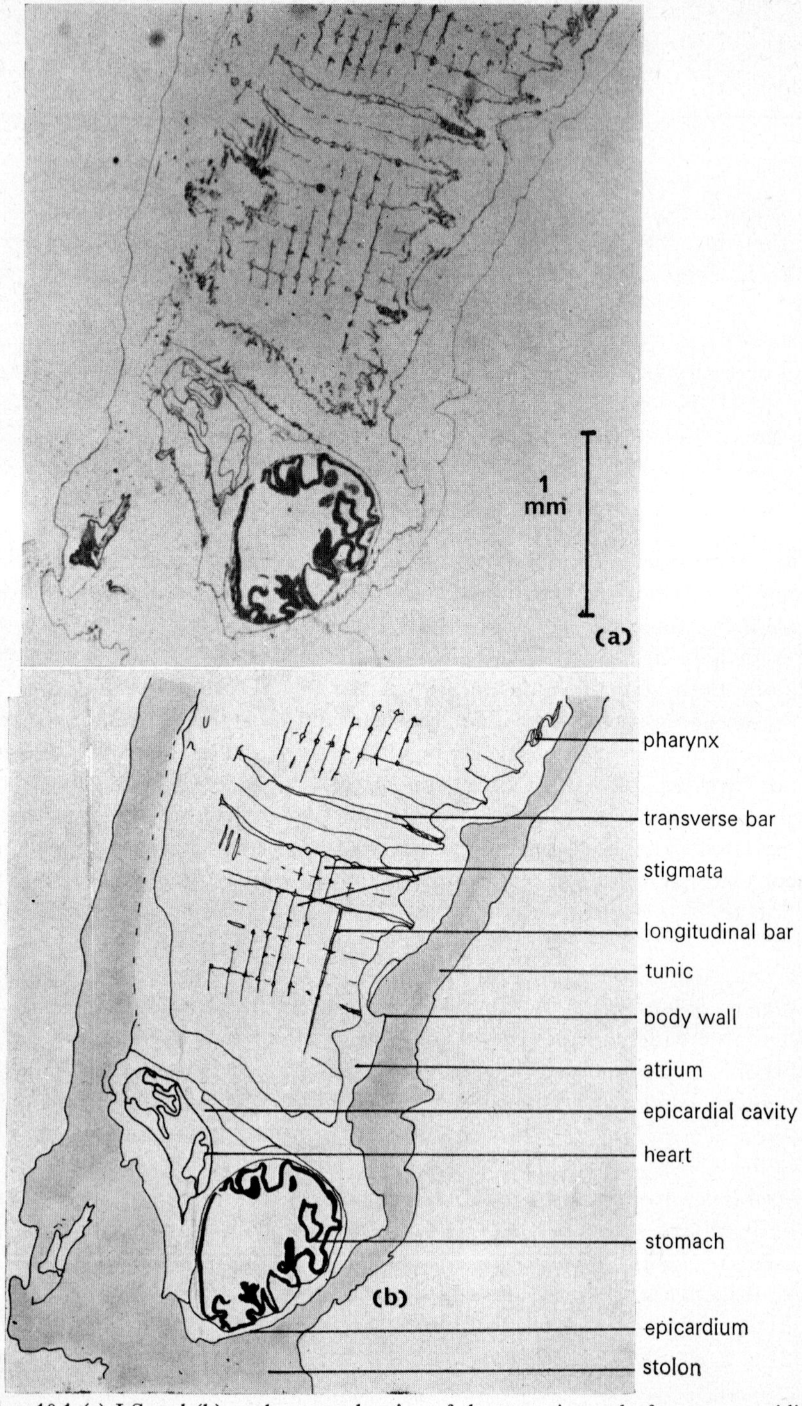

FIG. 10.1 (a) LS and (b) explanatory drawing of the posterior end of a young ascidian (*Ciona intestinalis*).

(epicardial cavity) are visible, while the bulk of the animal is occupied by the extensive pharynx and the atrium. Ascidians possess numerous gill clefts in the pharyngeal wall. The clefts are formed by sub-division of the larval pharyngeal slits, and are separated by external longitudinal and transverse bars (seen in Fig. 10.1). The gill clefts lead from the pharyngeal cavity into the atrial cavity (Fig. 10.2), so that water currents taken in through the buccal siphon pass through perforations in the pharynx wall and are expelled through the atrial siphon.

The body wall consists of a single-layered epidermis (characteristic of invertebrates) overlying connective tissue in which are embedded bands of longitudinal and circular muscle (Fig. 10.3). It is lined on the inside by atrial epithelium. The epidermis secretes a tunic on the outside as a protective device. The tunic is a translucent, elastic coat containing a variety of wandering mesodermal cells and it acts as a barrier to infection as well as to mechanical damage. The tunic is extended at the posterior end to form the stolon, and at the anterior end to line both siphons.

Nervous system

Much of the nervous system of adult ascidians has degenerated, leaving only the cerebral ganglion and its associated nerves. The cerebral ganglion is a small spindle-shaped structure situated in the body wall between the atrial and buccal siphons (Fig. 10.4). It consists of a mass of interlacing nerve fibres with a few peripheral neurones, a typically invertebrate arrangement [Fig. 10.5(a, b)]. Paired anterior and posterior nerves innervate the body wall while a single visceral nerve connects the posterior end of the ganglion with a visceral plexus surrounding the body organs. A narrow rod of cells (dorsal strand) extends from the hind end of the ganglion, close to the visceral nerve, but its function is not yet known.

Closely associated with the ganglion is the neural or subneural gland [Fig. 10.5(c)], a spongy glandular structure thought to be derived from the larval nerve cord. The neural gland is composed of cuboidal epithelium thrown into folds, and tubules which are intermingled with blood lacunae. It opens into the pharynx via a short duct and horse-shoe shaped ciliated funnel (Fig. 10.6). The lumina of gland and duct are often filled with phagocytes which has led to the assumption that the gland is involved in excretion. Because of its glandular nature and its method of attachment to the pharynx, the neural gland has also been suggested as a possible precursor of the vertebrate pituitary, but there is no real evidence for this.

Sense organs

There are no complex sense organs in *Ciona* but it is well supplied with epidermal sensory cells, particularly on the outer and inner surfaces of the siphons. These usually take the form of pairs of pear-shaped cells, often vacuolated, each with a sensory nerve fibre from the narrow end, and

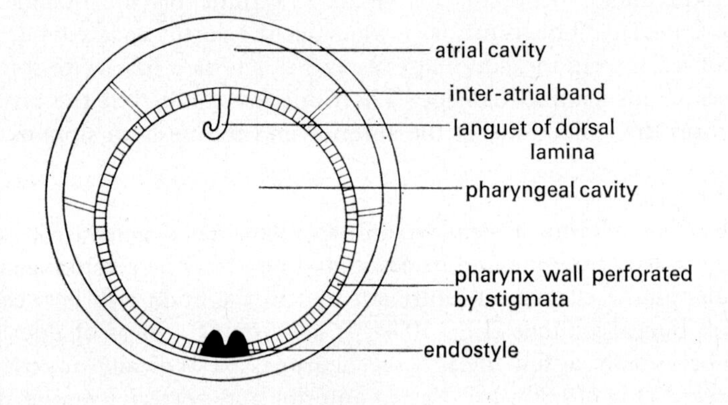

FIG. 10.2. Diagram of a TS through the mid-pharyngeal region of *Ciona.*

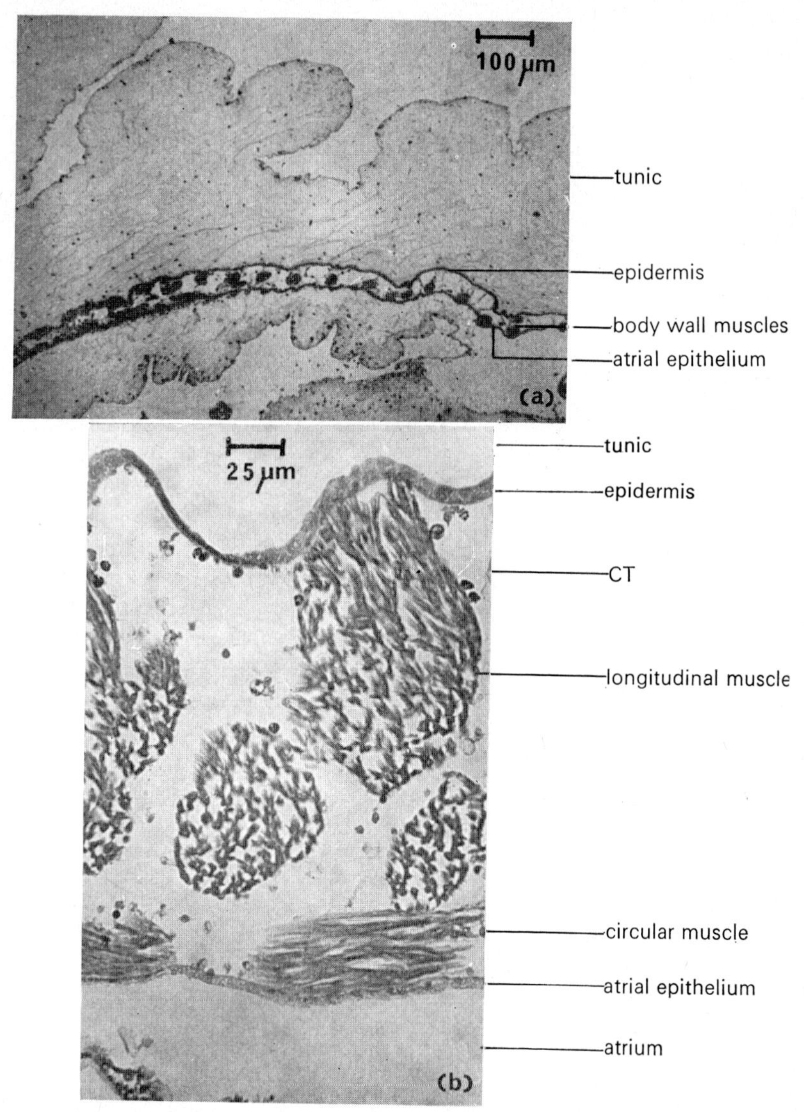

FIG. 10.3 (a) TS through the test and body wall of *C. intestinalis*; (b) detail of the body wall of *C. intestinalis*.

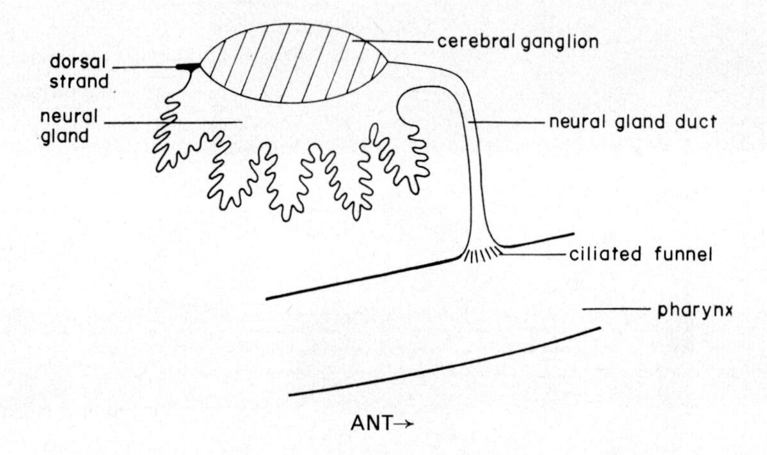

FIG. 10.4. Diagram of the cerebral ganglion and associated structures in *Ciona.*

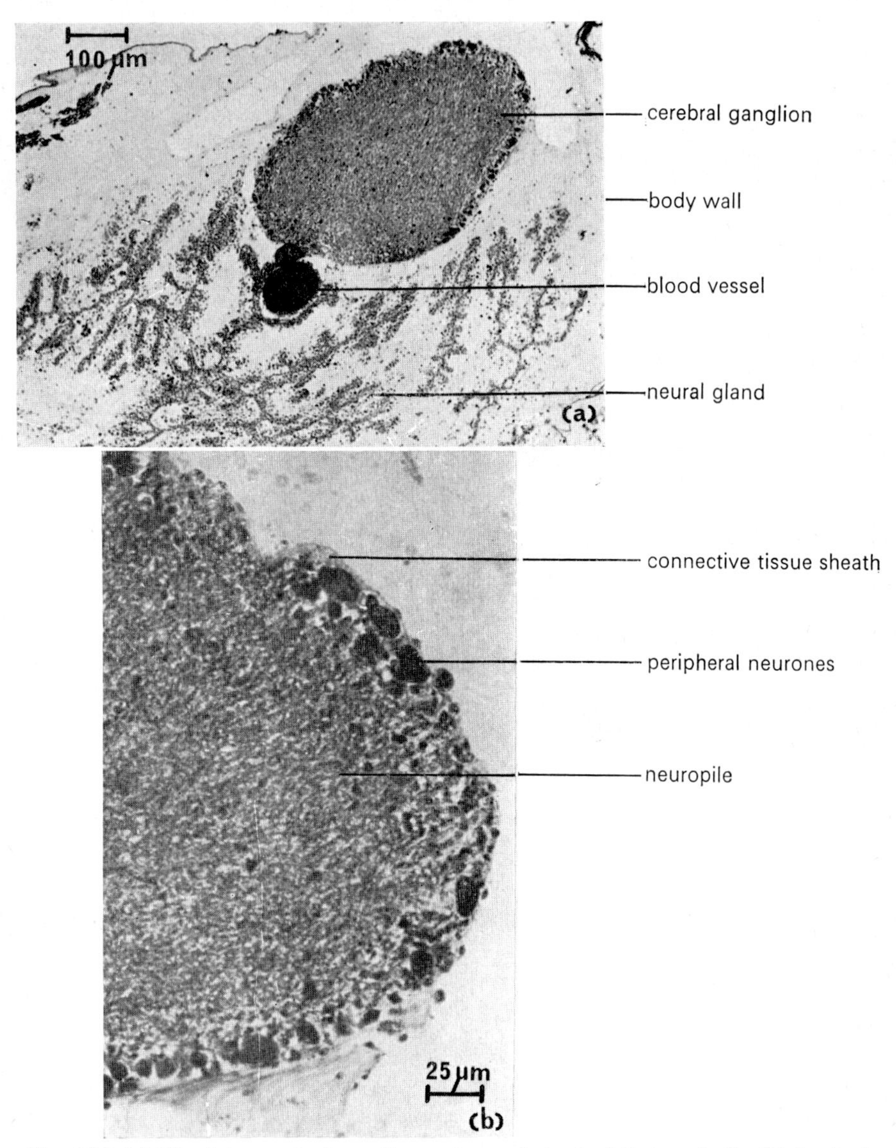

FIG. 10.5 (*a*) TS of the cerebral ganglion and neural gland of *C. intestinalis*; (*b*) detail of the cerebral ganglion.

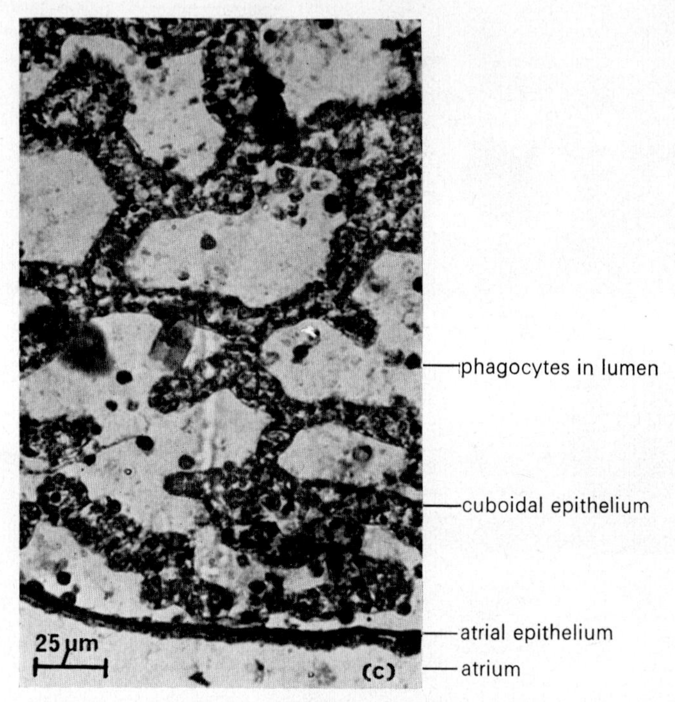

FIG. 10.5 (*c*) Detail of the neural gland of *C. intestinalis*.

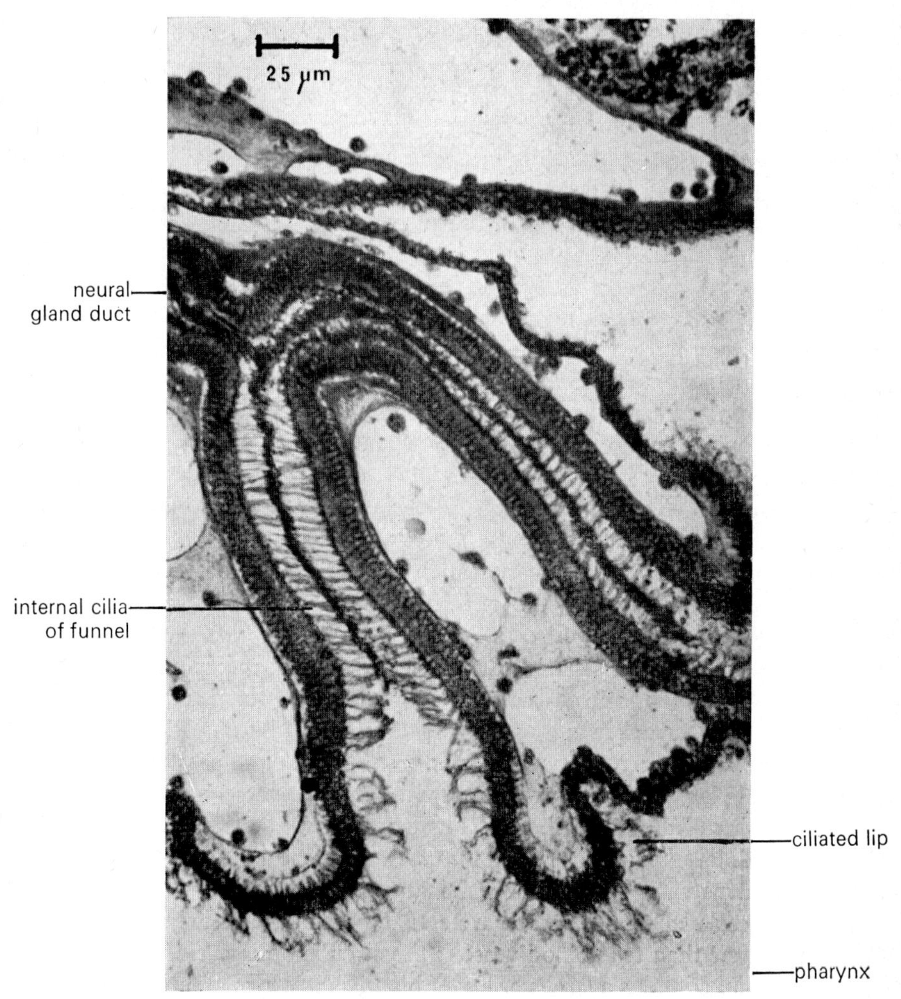

FIG. 10.6. Section through the ciliated funnel of *C. intestinalis*.

surrounded by a group of supporting cells. They appear to be tactile receptors, but some of them may be chemoreceptive. On the epithelium lining the atrial siphon are scattered the cupula organs. These are small groups of sense cells covered by a flag or cupula of tunic, and are thought to sense water currents in the atrial cavity.

Respiration

Unlike the hemichordates, where the gills are primarily used for respiratory purposes, the gills in *Ciona* are only incidentally involved in respiration and will be described under *Digestion*. Exchange of respiratory gases occurs over the whole body surface.

Circulation

There is no evidence of a coelom in adult ascidians but it has been suggested that the epicardium represents the remains of a coelom. The epicardium is formed by two sacs evaginating from the hind end of the pharnyx. The epicardial cavity thus formed surrounds the viscera in *Ciona* with the epicardial walls closely applied to the heart and gut (see Figs 10.1 and 10.8).

Enclosed by epicardial walls is the pericardium, a large bag composed of a sheet of hexagonal cells and a few muscle strands. An invagination of the pericardium forms the heart, which is a short curved tube lying in the pericardial cavity between the gut and the gonads, and attached to the pericardium by a raphe (Fig. 10.7). The heart is composed of a sheet of primitive transversely-striated muscle cells which spiral round the heart. Circulation of blood round the body is tidal since either end of the heart can act as a pacemaker and the direction of beat alternates. The ventral end of the heart leads into a large sub-endostylar vessel passing along the ventral side of the pharynx which gives off transverse branches to the vascular bars of the pharynx wall. This network of channels connects with a longitudinal median dorsal sinus with branches to the viscera, and finally returns via a dorsal abdominal sinus to the dorsal end of the heart. All the vessels lack walls and so are actually sinus channels in the mesenchyme rather than true blood vessels: the circulation is therefore open.

The blood contains a large variety of blood cells which are manufactured in haematogenic tissue, particularly in the tunic and round the intestine (see Figs 10.11 and 10.12).

Digestion

Sea-squirts are ciliary feeders. Water is taken in by the mouth (buccal siphon): large pieces of material are rejected by the tentacular ring guarding the entrance to the pharynx (Fig. 10.8) and small particles are filtered off at the pharynx wall. The pharynx is perforated by a large number of small vertical slits (stigmata) arranged in horizontal rows (see

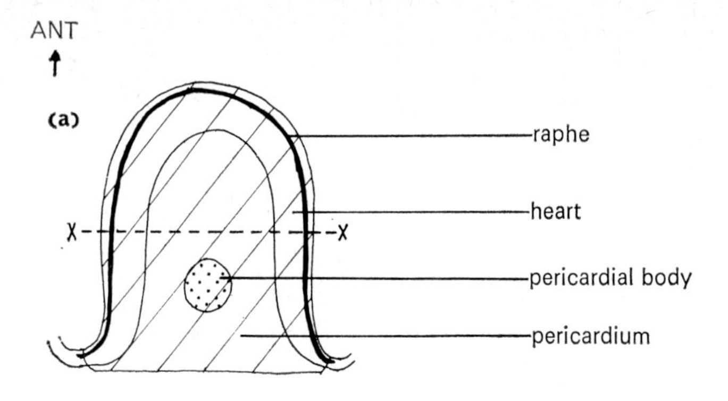

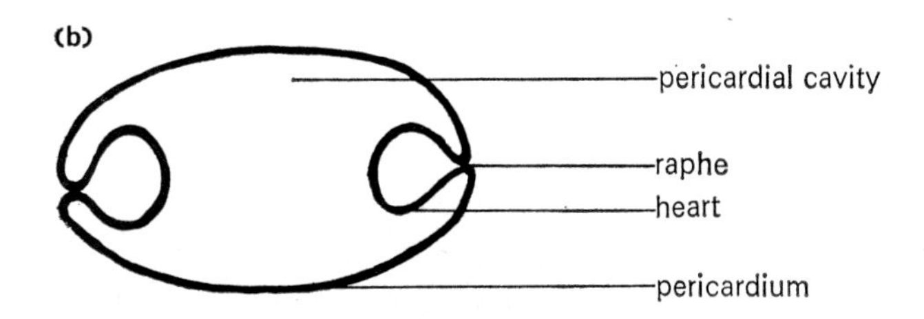

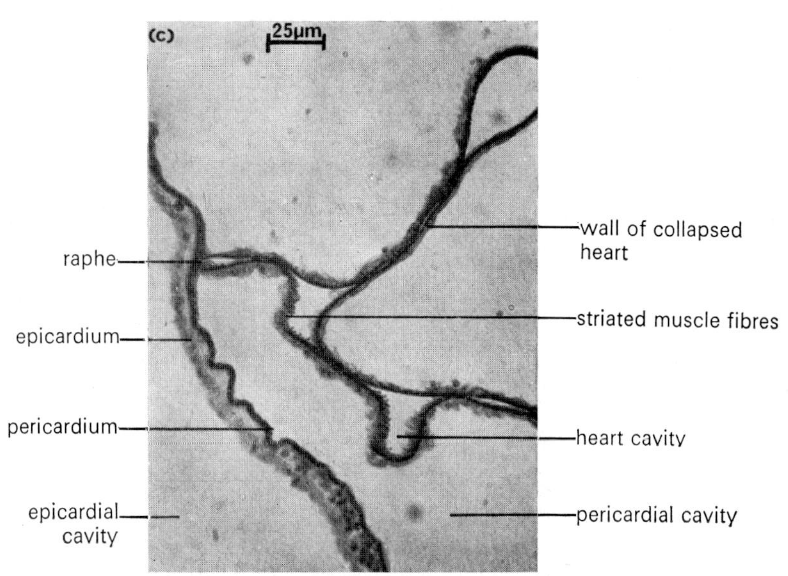

FIG. 10.7 (a) Diagram of the heart and pericardium of *Ciona* [X-X is the section shown in (b)]; (b) diagram of a cross-section of the heart [X-X in (a)]; (c) detail of part of a TS through the heart of *C. intestinalis*. Note that the heart is collapsed and the heart cavity appears to be almost occluded.

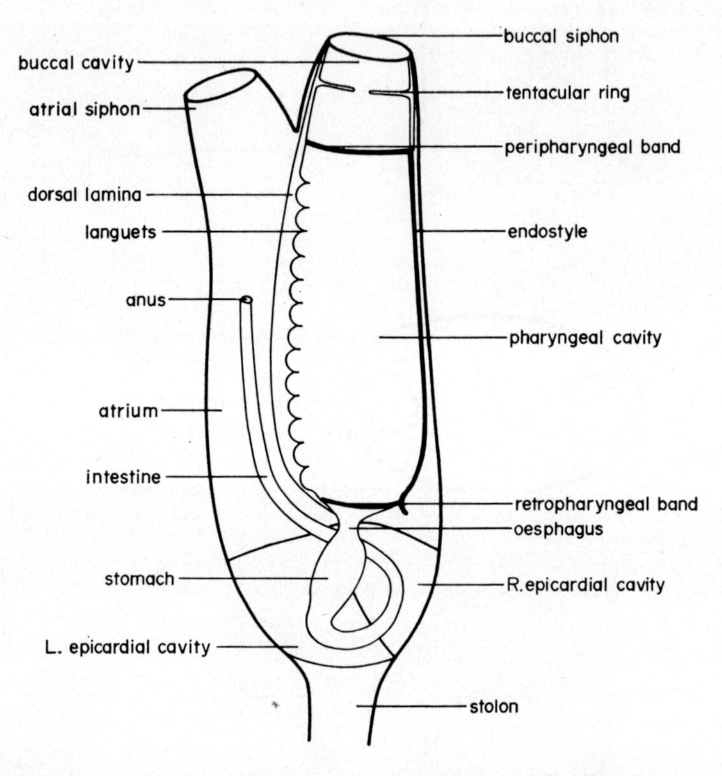

FIG. 10.8. Diagram of an LS of *Ciona* to show the positions of the digestive tract, atrium and epicardial cavities.

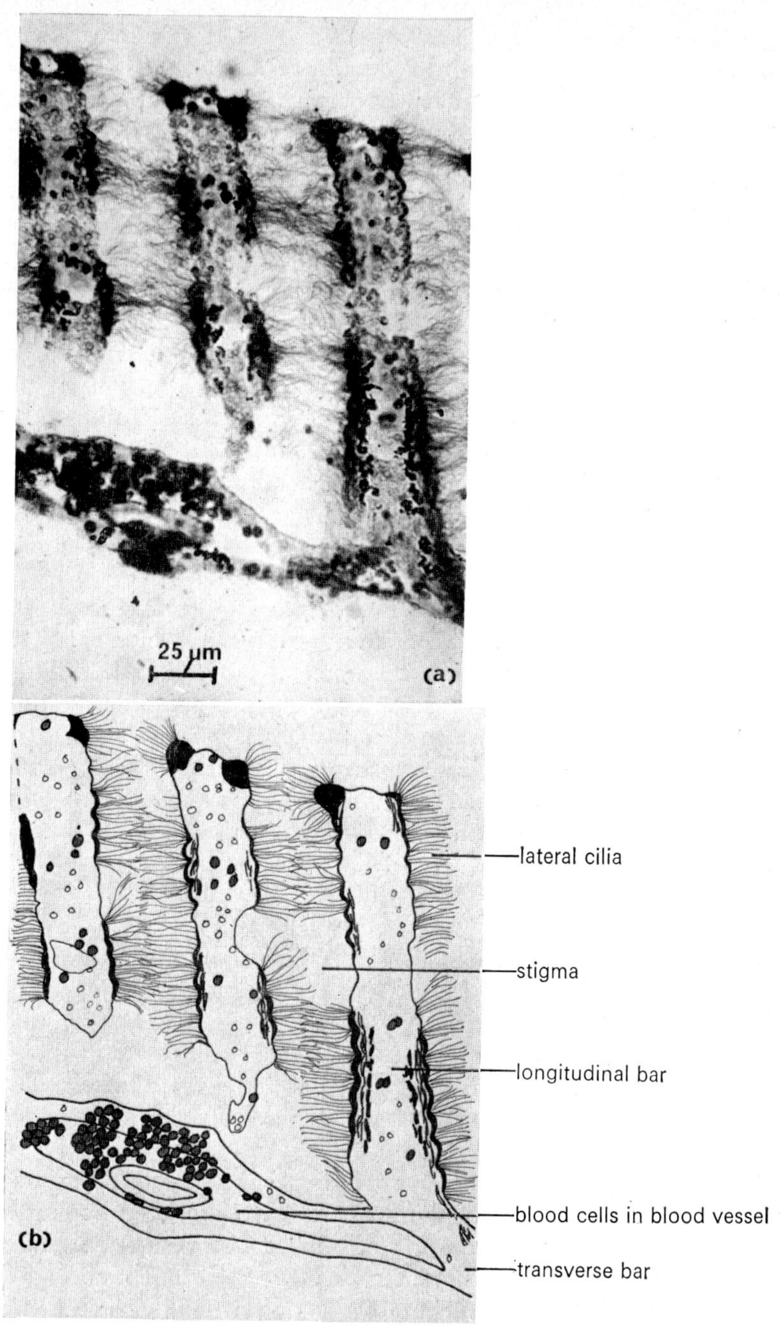

FIG. 10.9 (a) LS and (b) drawing of the pharynx of *C. intestinalis* showing the stigmata. The section has been cut parallel to the pharynx wall.

Fig. 10.1). The stigmata are separated from one another by longitudinal and transverse bars carrying blood vessels and are lined with cells bearing long cilia (Fig. 10.9). The cilia beat from the pharyngeal cavity towards the atrium, thus creating a strong inhalent current which forces water through the stigmata into the atrial cavity and out through the atrial siphon. Food is trapped in mucus produced by cells of the pharynx wall and of the endostyle. The endostyle is a groove running down the ventral side of the pharynx, and is composed of ciliated bands interspersed with areas of gland cells secreting mucus, proteins and mucopolysaccharides (Fig. 10.10). The long cilia at the base of the groove appear to be almost inactive, but the short cilia on the sides of the groove and dorsal lip are responsible for spreading secretions from the gland cells onto the lateral walls of the pharynx. The resulting mucus sheet traps food particles as it moves towards the languets, a row of curved ciliated structures running up the mid-dorsal line of the pharynx (see Fig. 10.2). These languets roll the mucus film into a food cord and pass it back to the oesophagus where its entry is assisted by a ciliated lip. The mucus sheet is delimited by the peripharyngeal and retropharyngeal ciliated bands which mark the anterior and posterior limits of the pharynx (see Fig. 10.8).

The oesophagus is lined with ciliated columnar cells [Fig. 10.11(a)], but bears two ciliated grooves containing gland cells which spiral round the oesophagus, thus channelling the mucus food cord back towards the stomach and adding yet more secretions to it as it goes. The rest of the gut shows a similar structure to that of the oesophagus, with a single-layered epithelium and no muscle fibres [Fig. 10.11(b)]; the stomach having very folded walls and the intestine possessing a typhlosole to increase surface area. The anus opens into the atrium.

Excretion

Urochordates do not possess glomerular excretory organs or nephridia such as are found in hemichordates and cephalochordates. Excretion can occur by several possible means. Wandering amoebocytes and nephrocytes are thought to accumulate waste material as they circulate in the blood stream, and later become fixed in the tissues, particularly round the gonads and digestive loop where they develop vesicles filled with uric acid [see Fig. 10.13(c)].

The pyloric glands formed as gut diverticula and situated on the outside walls of the posterior part of the intestine have also been implicated in excretion. They consist of networks of tubules with vesicular swellings at the ends of the branches and at intervals down the tubules, and are lined with a layer of flattened epithelium bearing a few long cilia (Fig. 10.12). The lumina of the tubules contain crystals and ultimately connect with the anterior end of the stomach. Although it has been suggested that the

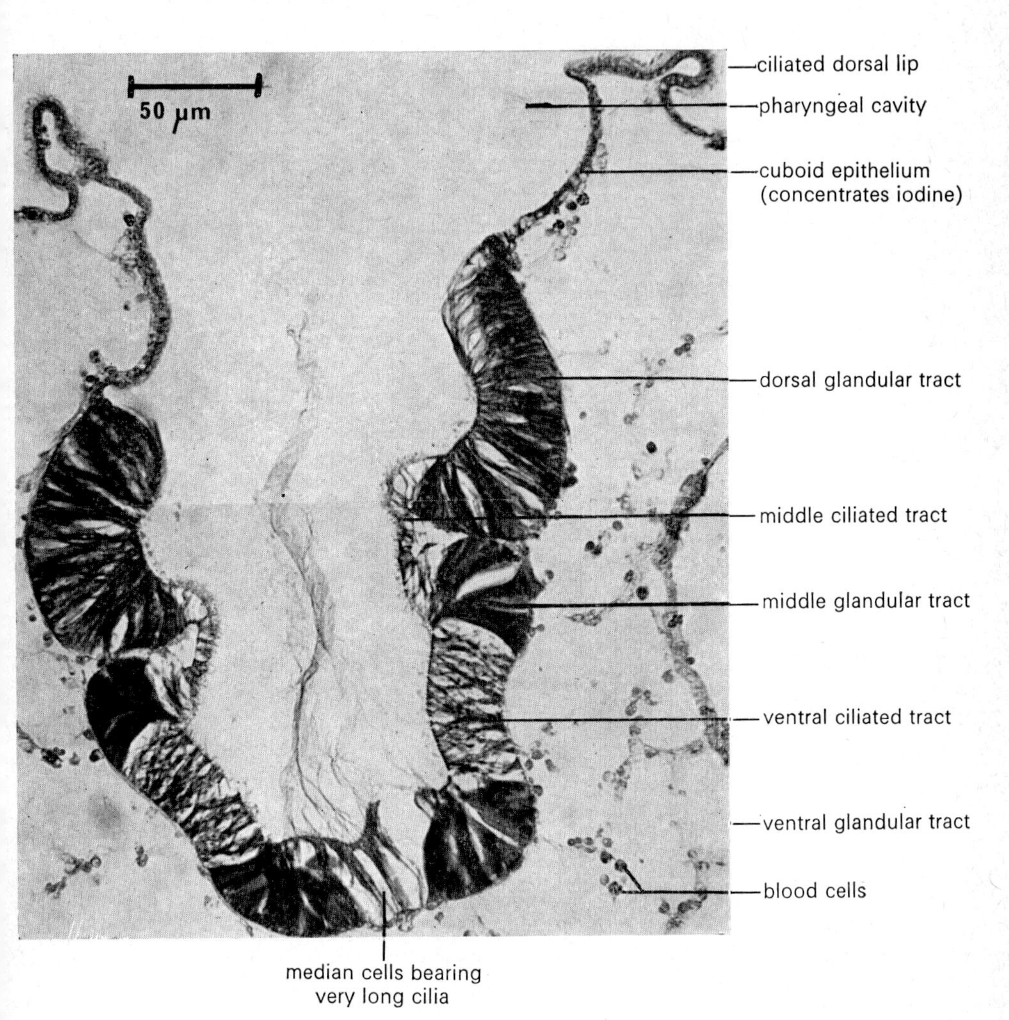

FIG. 10.10. TS of the endostyle which runs down the ventral pharyngeal wall of *C. intestinalis*.

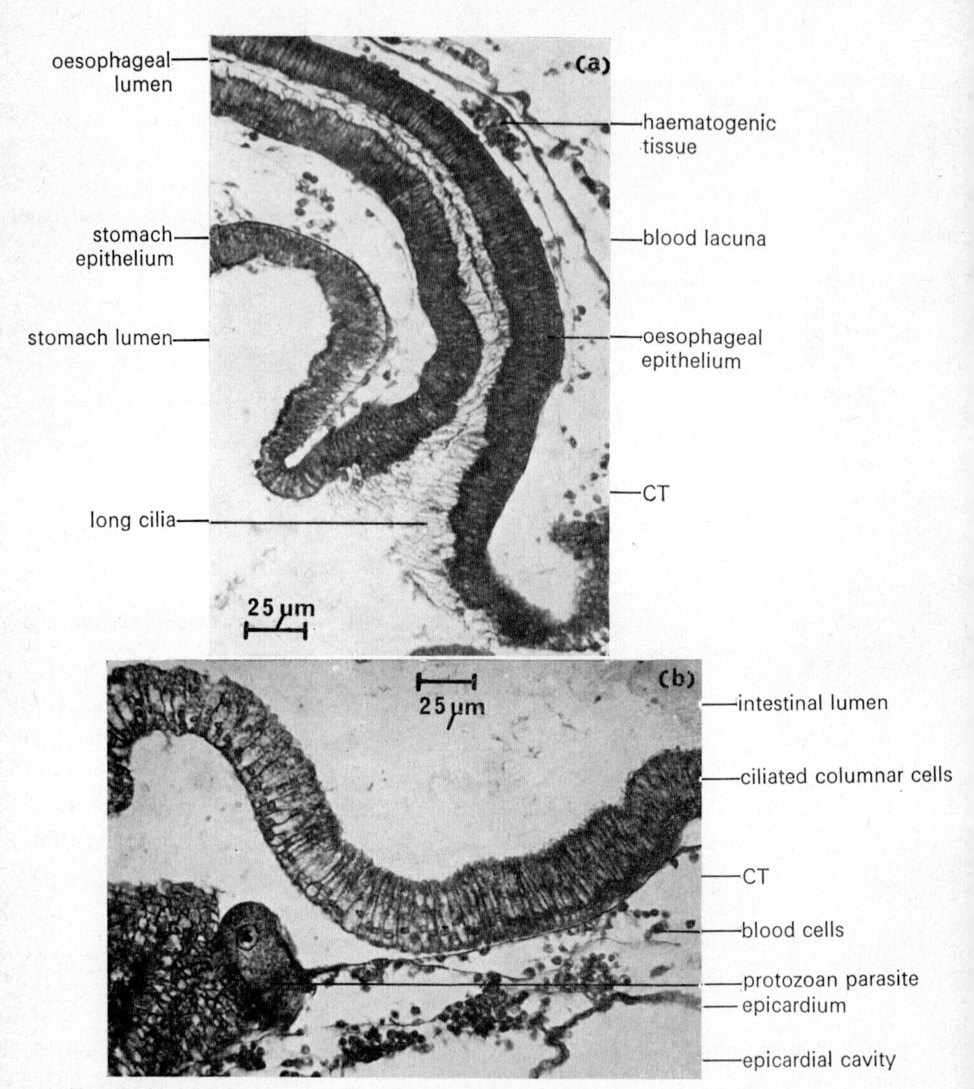

FIG. 10.11 (a) LS through the oesophagus of *C. intestinalis*, showing the long cilia; (b) section through the intestine wall of *C. intestinalis*. Note the lack of gut musculature, and the large protozoan (Gregarian) parasite.

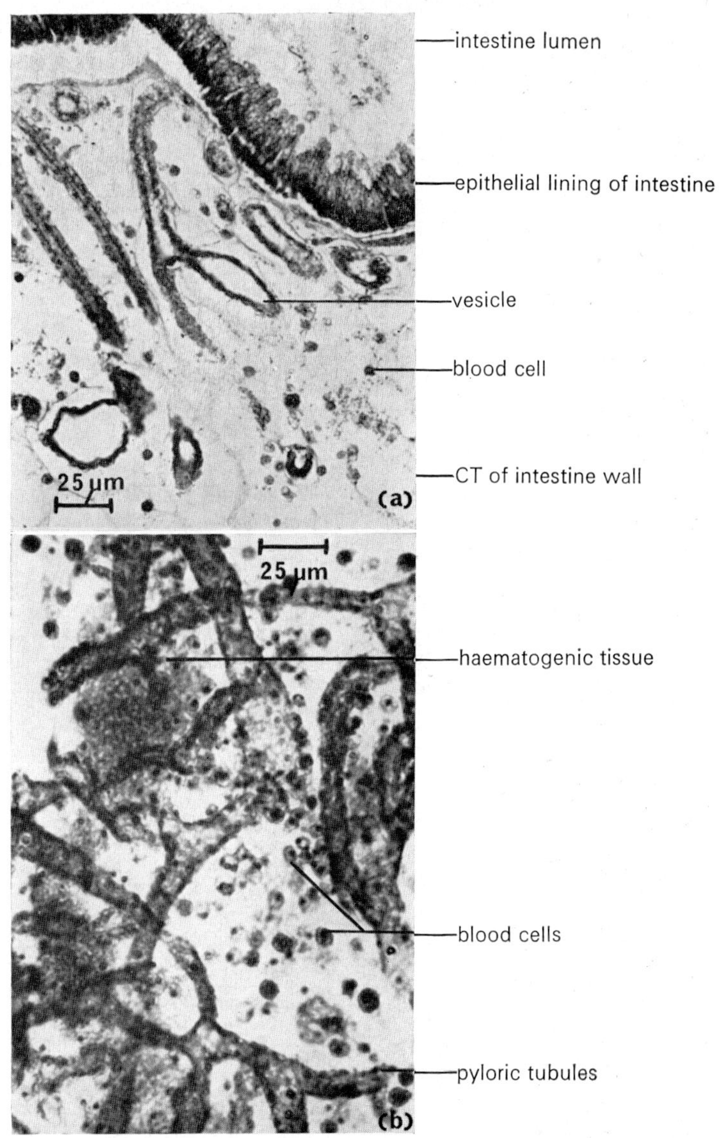

Fig. 10.12 (a) Section through the intestine wall of *C. intestinalis*, showing vesicles and tubules of the pyloric glands; (b) interlacing network of pyloric gland tubules.

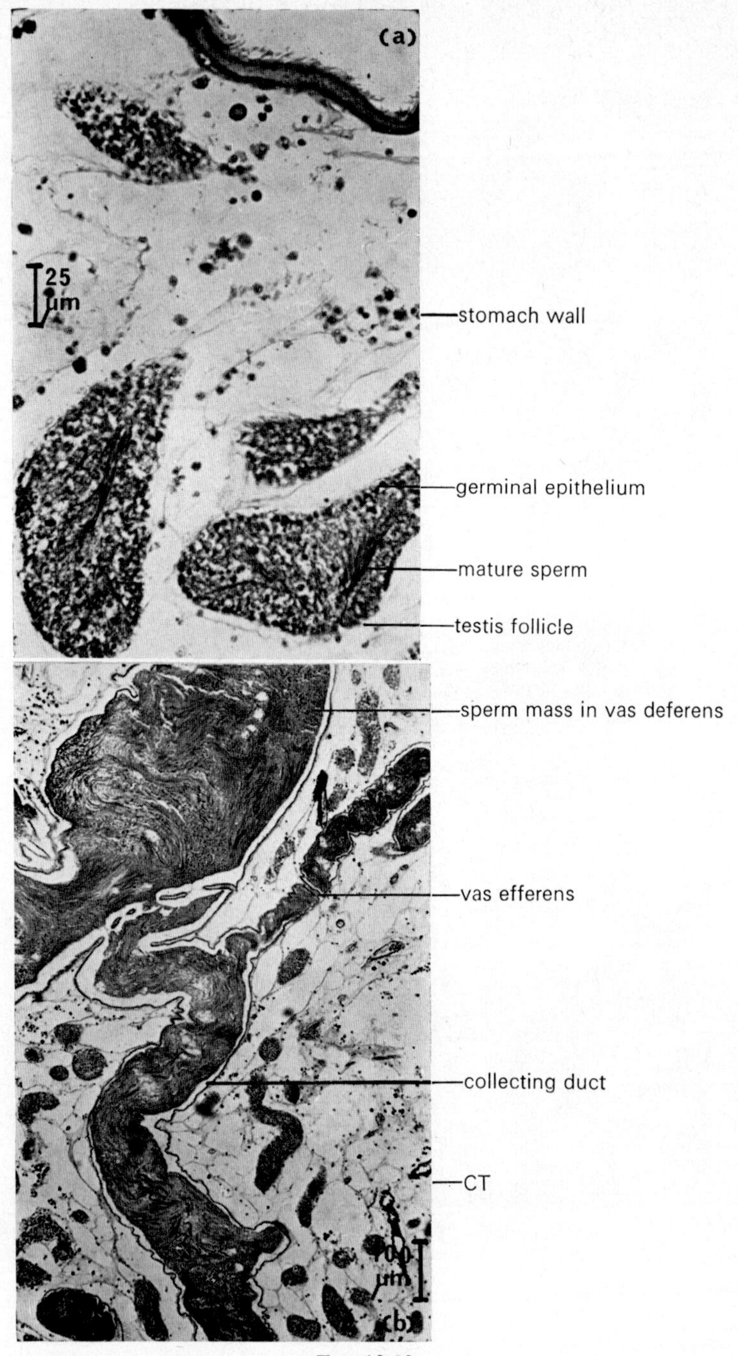
(a)
25 µm
stomach wall
germinal epithelium
mature sperm
testis follicle
sperm mass in vas deferens
vas efferens
collecting duct
CT
100 µm
(b)

Fig. 10.13

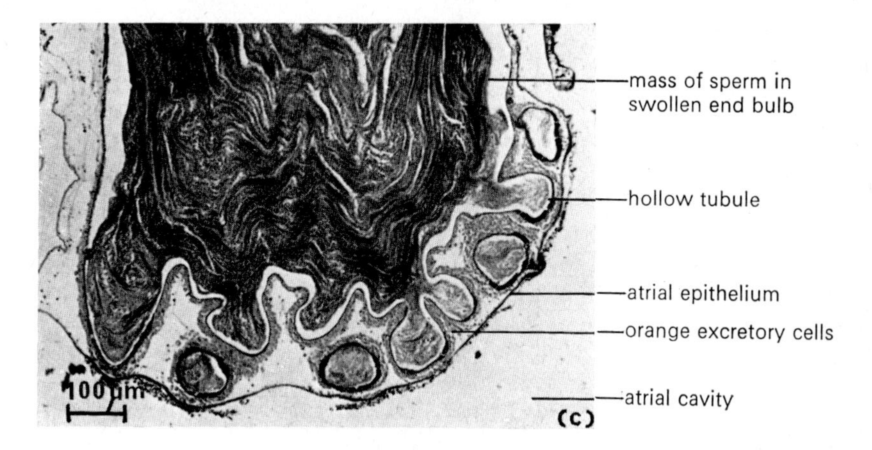

FIG. 10.13 (a) Testis follicles embedded in the stomach wall of *C. intestinalis*; (b) vasa efferentia, collecting duct and vas deferens of *C. intestinalis*. The ducts are packed with ripe sperm; (c) LS through the swollen end-bulb of the vas deferens from *C. intestinalis* showing the crown of finger-like tubules through which ripe sperm are extruded. Note the dense aggregations of orange excretory cells round the fine tubules.

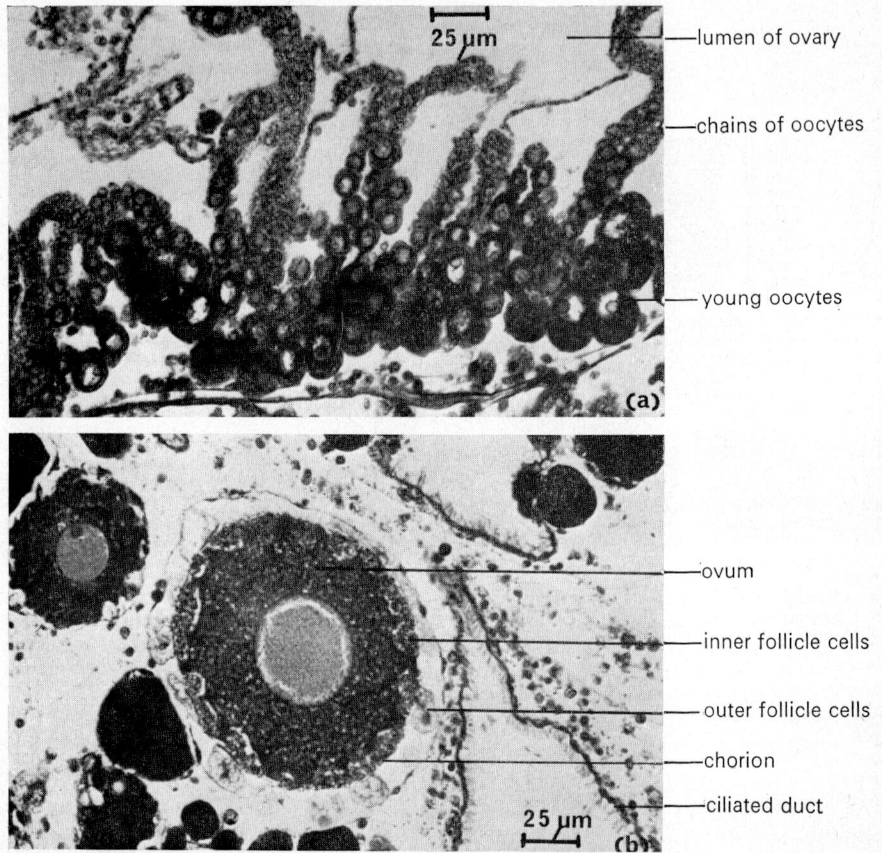

FIG. 10.14 (a) Section through the immature ovary of a young *C. intestinalis*; (b) section through a mature ovary, showing a developing ovum and the ciliated duct.

pyloric glands are excretory in function, it may well be that they have a purely secretory function and secrete digestive enzymes.

Reproduction

Ascidians are generally hermaphrodite and the gonads are situated between the heart and the digestive loop.

The testis is very diffuse and is closely applied to the walls of the stomach and intestine. It is composed of numerous thin branching tubules ending in club-shaped sacs or follicles filled with germinal epithelium [Fig. 10.13(a)]. Sperm with long tails are released into the follicle lumen which leads into a narrow ciliated vas efferens. Large numbers of vasa efferentia empty into two collecting ducts which unite to form a single vas deferens [Fig. 10.13(b)]. The vas deferens terminates in a swollen end-bulb which carries a crown of fine pointed tubules, each of which is hollow and possesses a narrow opening at the tip. [Fig 10.13(c)]. These tubules are characteristically surrounded by orange excretory cells. When accumulated sperm fill the end-bulb, they are squeezed out through the fine holes into the atrial cavity at a point between the anus and atrial siphon.

The ovary is a pear-shaped hollow organ lying above the testis and stomach, and is composed of ciliated epithelium with ventrolateral bands of germinal epithelium (Fig. 10.14). An ovary containing mature eggs has a follicular appearance due to distension of the thin ovarian wall. Each mature ovum has a large spherical nucleus and is surrounded by two layers of follicle cells separated by a membranous "chorion". Several functions have been ascribed to these cells including yolk formation, secretion of an enzyme which will facilitate hatching of the larva, and a flotation mechanism. Ova are shed into the cavity of the ovary, pass down the ciliated oviduct and are discharged into the atrial siphon. The entrance to the oviduct is guarded by a sphincter muscle which may prevent self-fertilization although there is also evidence that ova and sperm from one individual are incompatible. Cross-fertilization, therefore, occurs in the water, and the development of both egg and larva are rapid events.

Sub-phylum Cephalochordata

Cephalochordates consist of a few, closely related forms, often collectively called *Amphioxus*. They are small, elongated, laterally-compressed creatures with pointed ends. They are often regarded as survivors of an ancestral type that gave rise to vertebrates and show a number of basic vertebrate features, but also exhibit some totally different ones. Thus they have a notochord and hollow dorsal nerve cord running the length of the animal (both of which are retained in the adult), but they have no vertebral column. They also possess a muscular post-anal tail and numerous pharyngeal slits, but there are no paired limbs or jaws typical of most vertebrates.

Example: *Amphioxus* (*lanceolatus*)

A. lanceolatus is a small, translucent marine cephalochordate living in shallow water. The larval stage is asymmetrical and swims using cilia. The adult spends much of its life buried in the sand with only the head projecting, but it can swim freely using undulatory movements of the body assisted by poorly developed dorsal, ventral and caudal fins. These can be seen in Fig. 10.15 which shows a young *Amphioxus*. The large pharynx perforated by vertical slits is particularly striking. The coelom is reduced and its tripartite nature obscured. Much of the body space is taken up with the metamerically segmented mesodermal muscles (myotomes) which line the body wall. These muscles, arranged in Vs on either side of the body, surround the nerve cord and skeletal structures. These features are shown in Fig. 10.16, a TS of a young *Amphioxus* in the posterior pharyngeal region.

During development, two lateral folds of the body wall (metapleural folds) grow ventrally, and from them transverse epipleural folds grow across to meet one another. This creates an ectoderm-lined cavity (atrium) around the pharynx and viscera. Anteriorly the metapleural folds form an oral hood round the buccal cavity. The atrium extends upwards into the coelom and completely surrounds the "liver" and gonads. and partly encloses the gut. The dorso-lateral pharynx wall, however, is bordered by two dorso-pharyngeal coelomic canals (see Fig. 10.16). The only other portions of coelom which remain are narrow cavities surrounding the organs, and found beneath the endostyle and in the primary gill bars. The ventral atrium, unlike that of the urochordates, opens ventrally by an atriopore situated in front of the anus (see Fig. 10.15). Behind the atriopore the atrium extends backwards, on the right side only, between the myotomal muscles and the coelomic epithelium (Fig. 10.17).

The myotomes are covered by a thin skin, used in respiration. It consists of a fibrous dermis supplied with nerves and blood-vessels but no glands, and overlaid by a single layer of unpigmented epidermis which is ciliated in the larvae but becomes cuticularized in adults (Fig. 10.18). Compare this skin with that of the vertebrates (e.g. Fig. 11.3).

The main skeletal structure present in *Amphioxus* is the notochord which runs the length of the body underneath the nerve cord and extends forwards beyond the front end of the cord. It is a cylindrical rod composed of a tough collagenous sheath surrounding a longitudinal stack of muscular lamelli (Fig. 10.19, see also Fig. 10.22). These muscles are thought to contain paramyosin, a contractile protein also found in the catch muscles of molluscs, and contractions of the muscle plates appear to produce increased stiffness of the rod during fast swimming. The notochord muscles are under nervous control.

Nervous system

The nervous system of *Amphioxus* looks much more like that of the vertebrates than does that of the urochordates (see Fig. 10.5). The hollow dorsal nerve cord shows the typical vertebrate plan of a small central canal surrounded by a zone of nerve cell bodies and bordered by a rim of axons (Fig. 10.19). Anteriorly the central canal is enlarged to form a ventricle surrounded by nerve cells. This region (cerebral vesicle) may be the forerunner of the vertebrate diencephalon (see Fig. 11.7) but there is no proper brain. A patch of neurosecretory cells in the floor of the vesicle ("infundibular organ") probably secretes a viscous acellular Reissner's fibre, a structure found in the central canal of all vertebrate spinal cords. The nerve cord in *Amphioxus* contains several median multipolar giant nerve cells (Rohde fibres) which appear to act as through-conduction pathways co-ordinating responses involving the whole animal. Segmental nerves arise from the nerve cord, consisting of alternating dorsal and ventral roots. The dorsal roots bear sensory fibres and run between the myotomes, while the ventral roots carry motor nerves to the myotomes. Since the myotomes are staggered on opposite sides of the body, the segmental nerves also alternate in position. The cell bodies of the sensory nerves lie close to the epidermis or scattered along the segmental nerves, and are not collected into dorsal root ganglia as in vertebrates. All peripheral nerves are unmyelinated as in other invertebrates.

Sense organs

Amphioxus is a sluggish creature and is not well supplied with sense organs. Most of the receptors are free nerve endings or sensory hairs connected to nerves in the dermis. There are, however, some photoreceptive cells in the spinal cord, which are protected by asymmetrical pigment cups (Fig. 10.20). Their apparently random arrangement is thought to be related to the spiralling movements executed by *Amphioxus* when it swims. At the anterior end of the nervous system a group of pigment cells form a terminal pigment spot (see Fig. 10.15) which is probably not photoreceptive. Just in front of this spot, a small pit (Kolliker's pit) is lined with sensory epithelium and thought to be chemoreceptive. Pressure- and photo-receptive functions have also been attributed to the "infundibular organ" in the floor of the cerebral vesicle.

Respiration

The epidermis and atrial walls are permeable and appear to be the chief sites of respiratory exchange. The pharyngeal clefts, in *Amphioxus*, as in the urochordates, serve mainly to channel food into the digestive tract,

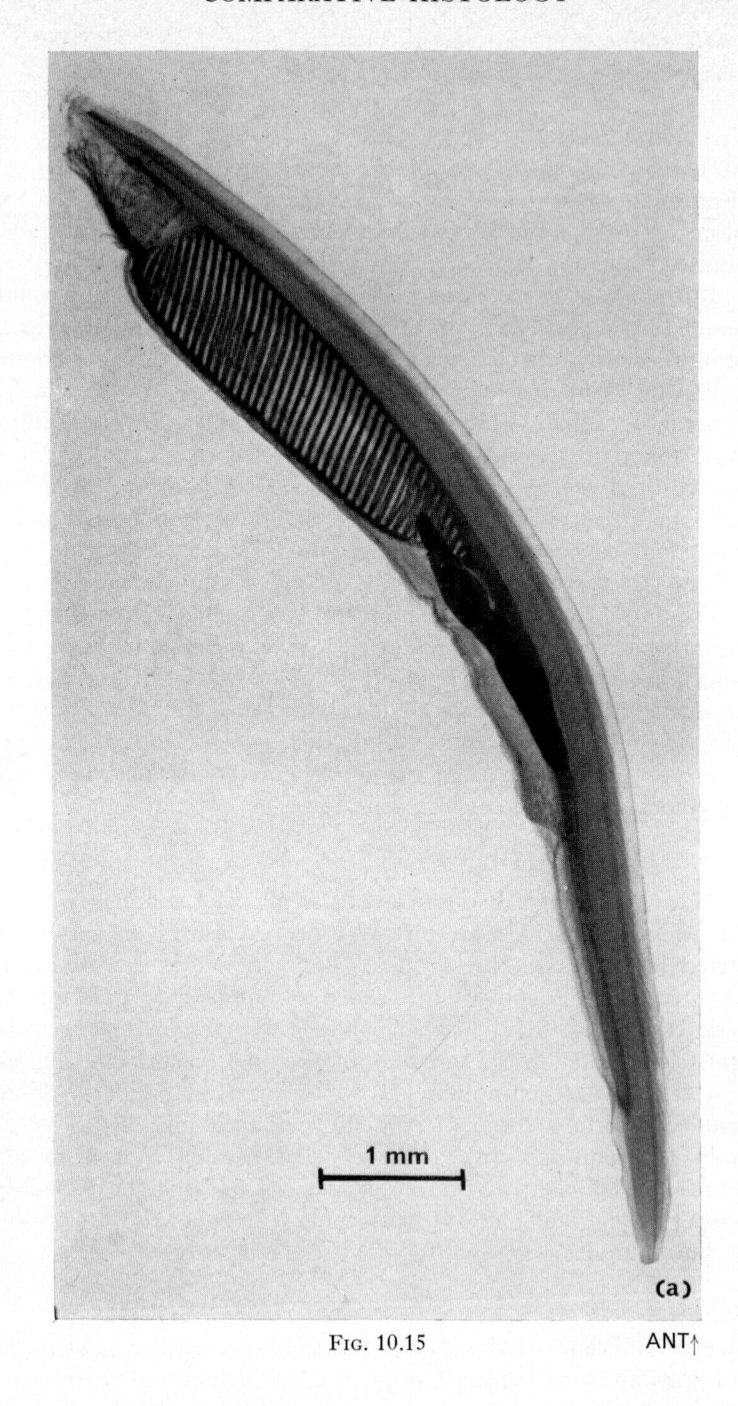

Fig. 10.15 ANT↑

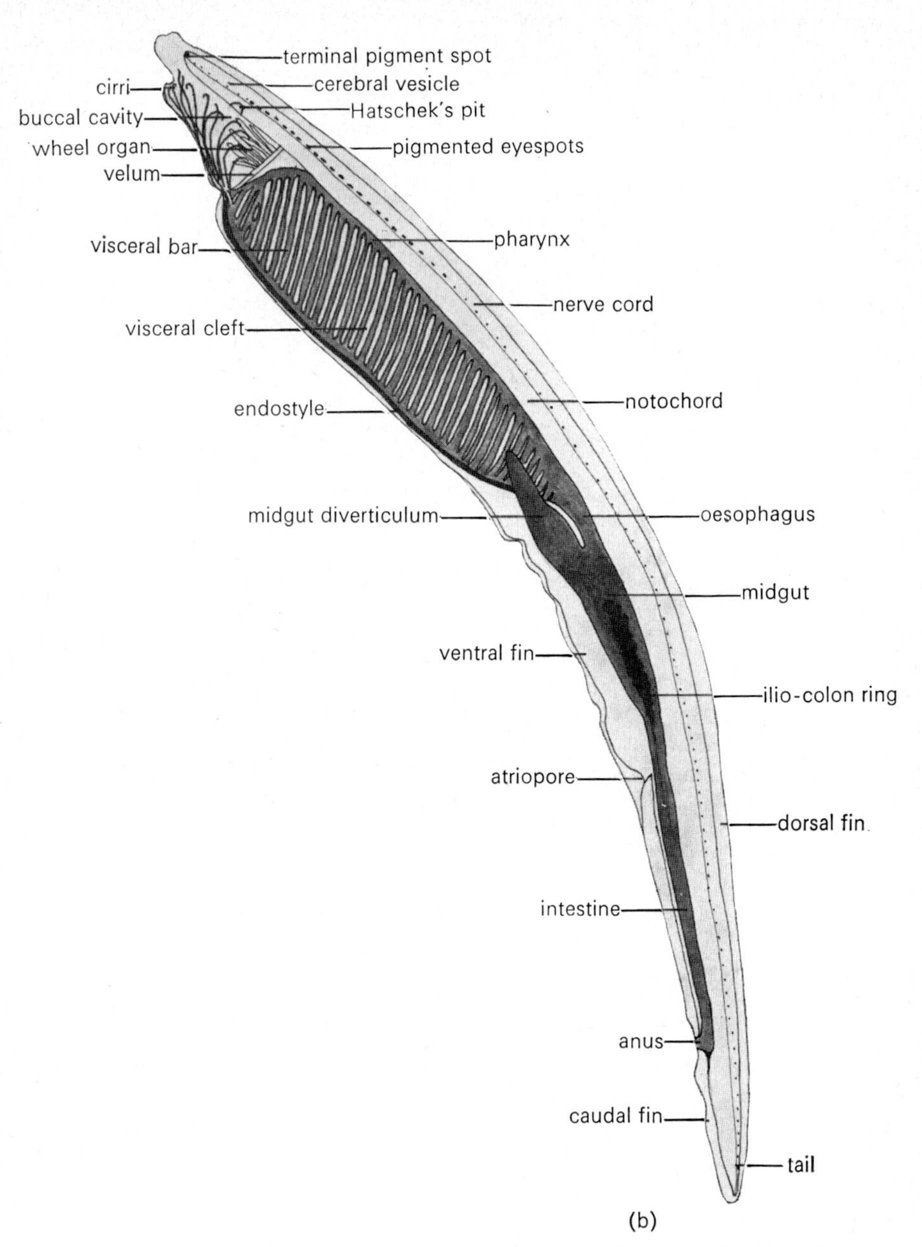

FIG. 10.15 (a) Whole mount; and (b) explanatory drawing of a young *Amphioxus lanceolatus*.

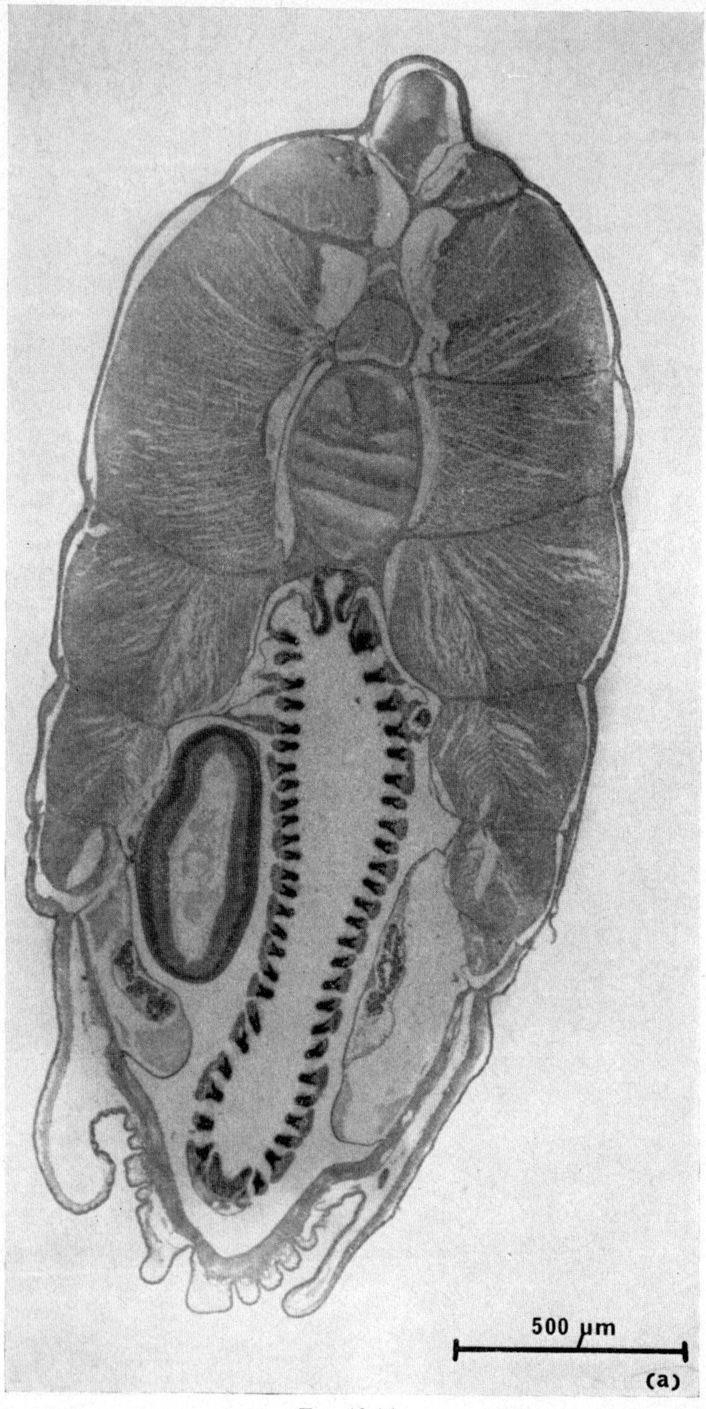

FIG. 10.16

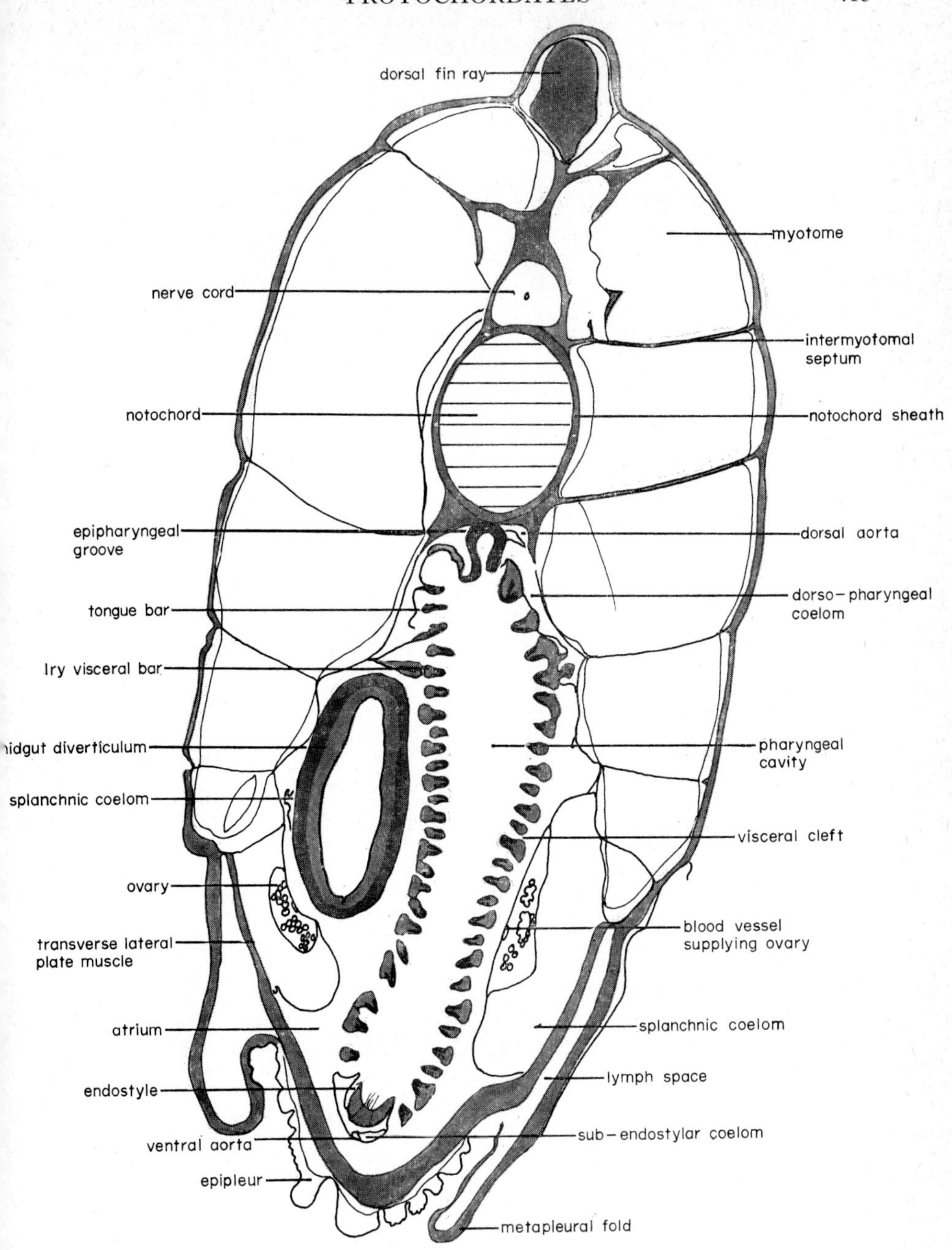

(b)

FIG. 10.16 (a) TS and (b) explanatory drawing of a young female *A. lanceolatus*. The section was cut in the posterior pharyngeal region.

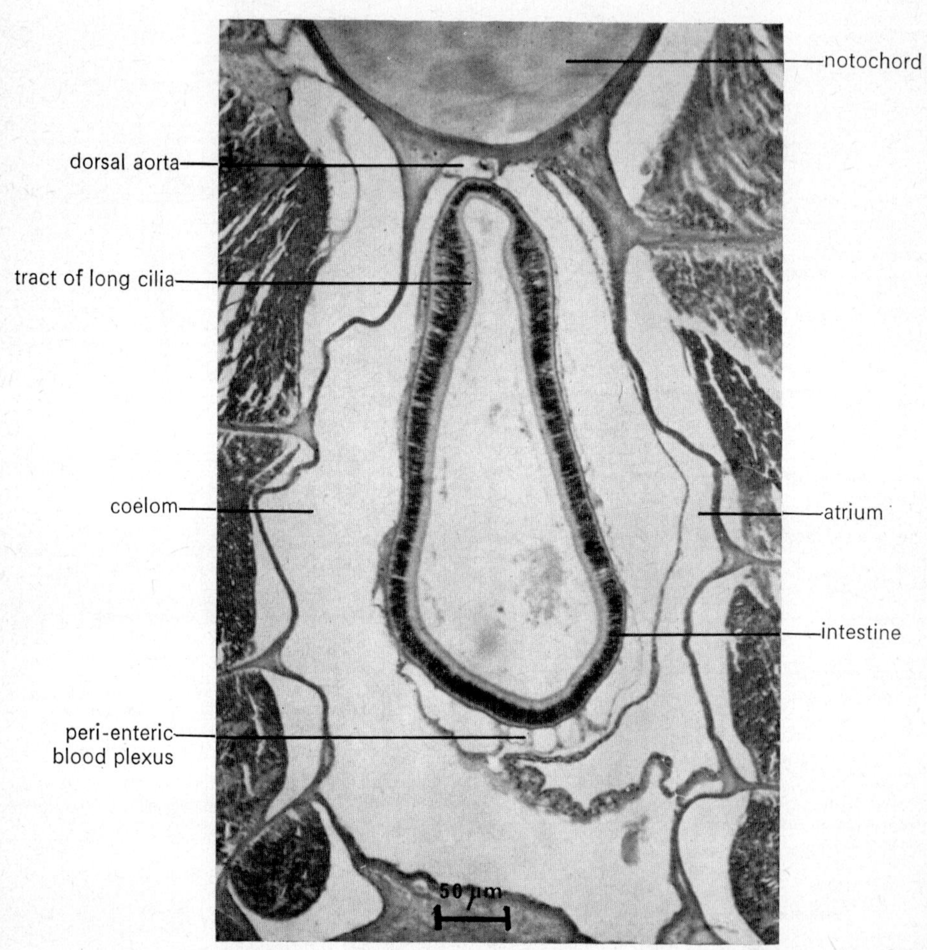

FIG. 10.17. Part of a TS of *A. lanceolatus* in the region behind the atriopore. Note the asymmetrical atrium (on the right side only).

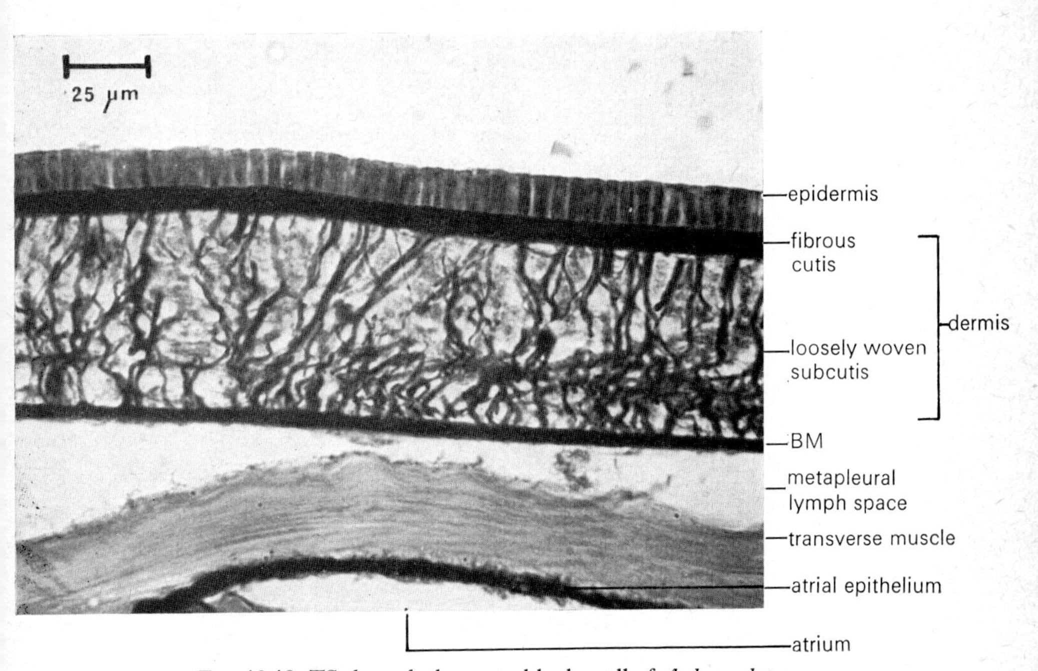

FIG. 10.18. TS through the ventral body wall of *A. lanceolatus*.

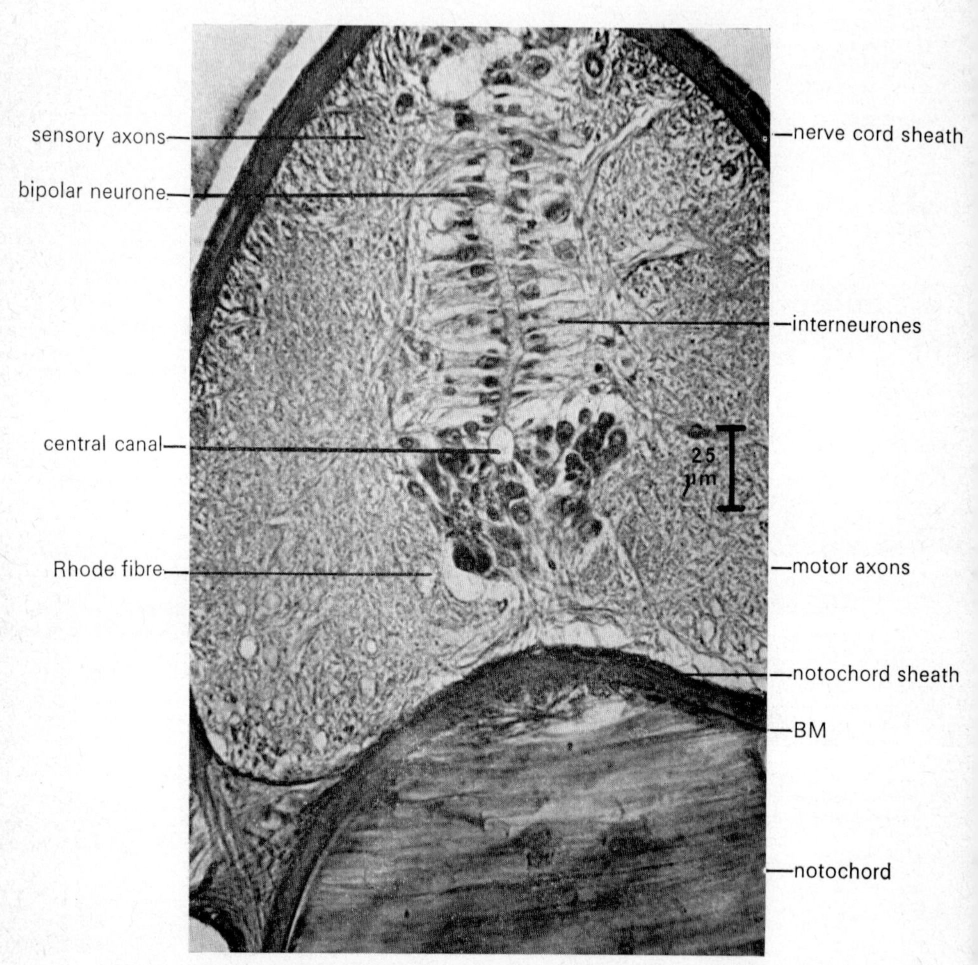

FIG. 10.19. TS through the dorsal nerve cord and notochord of *A. lanceolatus*.

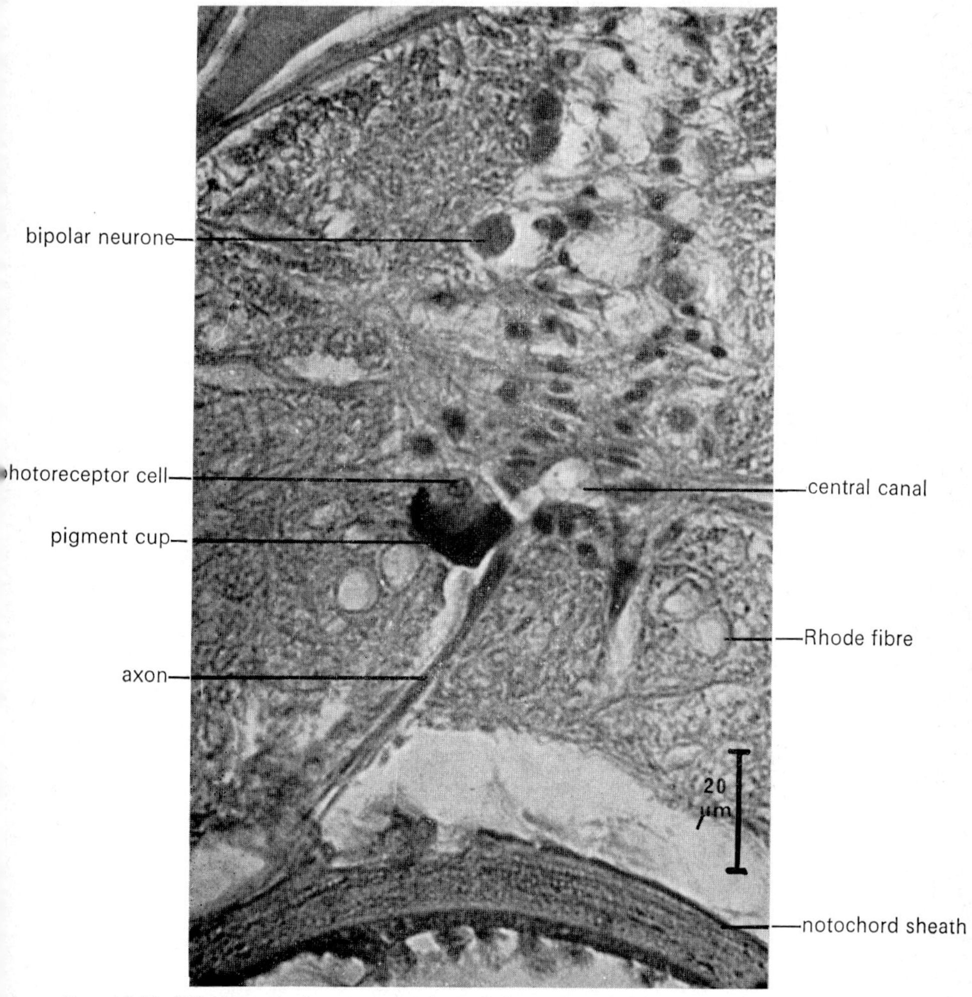

FIG. 10.20. TS through the nerve cord of *A. lanceolatus* in the anal region, showing an asymmetrically sited photoreceptor.

although they may incidentally create water currents which can be utilized in respiratory exchange.

Circulation

The blood flow in *Amphioxus* is typical of a vertebrate in that it flows forwards ventrally and backwards dorsally; but, unlike most vertebrates, there are no capillaries, no respiratory pigments or red blood cells, and no true heart. Blood flows from lacunae in the tissue spaces, through veins into paired Cuverian ducts which open into a large sinus venosus. It passes forwards into a contractile ventral aorta which pumps it into a series of afferent branchial arteries. These have small contractile bulbs (accessory or branchial hearts) at their bases, which boost the circulation. Blood passes through efferent branchial arteries into paired dorsal aortae, and thence to lacunae supplying the tissues. All blood from the intestine goes through a lacunar network round the midgut diverticulum ("liver"), which thus forms a type of "hepatic-portal" system characteristic of vertebrates. This blood passes on to an "hepatic" vein, on the dorsal surface of the mid-gut diverticulum, which runs into the sinus venosus.

Digestion

Amphioxus is a ciliary feeder, ingesting detritus particles swept in by water currents. The form of the digestive tract can be seen in Fig. 10.15. The buccal cavity is surrounded by a circle of stiffened cirri (Fig. 10.21) which assist in food collection. Within the buccal cavity a pre-oral pit (Hatschek's pit), lined with cilia and producing mucus (Fig. 10.22), is linked to a series of branched ciliated grooves (wheel organ) which run towards the mouth. These two structures sort out food particles of the required size, entrap them in mucus and pass them into the pharynx.

The pharynx is perforated by a large number of slits (visceral clefts) which strain off food from the water as it passes through. Each primary cleft is separated from the adjacent one early in development by a primary visceral bar, and later becomes divided into two secondary clefts by the downwards growth of the dorsal wall to form a tongue bar (Fig. 10.23). Both visceral and tongue bars contain skeletal supporting rods and are connected by vascular cross-bars (synapticulae). They can be distinguished in cross-section by the fact that the primary bar contains a small portion of coelom while the tongue bar has no connection with the coelom (Fig. 10.24). The coelomic canals on the outside of the primary bars connect dorsally with the dorso-pharyngeal coelom and ventrally with the sub-endostylar coelom (see Fig. 10.16). The bars are covered by ciliated epithelium: lateral cilia create water currents and frontal cilia waft sheets of mucus and food towards the epipharyngeal groove, where cilia convey it backwards towards the oesophagus. A shallow endostyle lined with mucous cells and long cilia (Fig. 10.25) adds more mucus to the food and

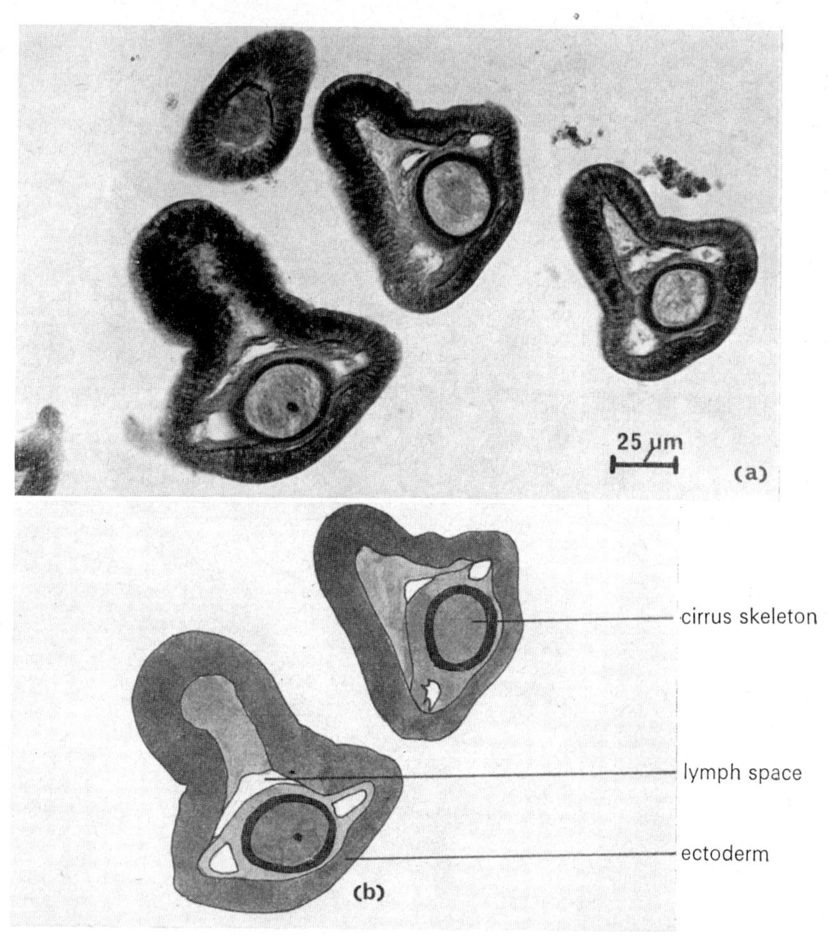

Fig. 10.21 (a) Section through stiff cirri in the buccal cavity of *A. lanceolatus*; (b) drawing of two cirri seen in (a).

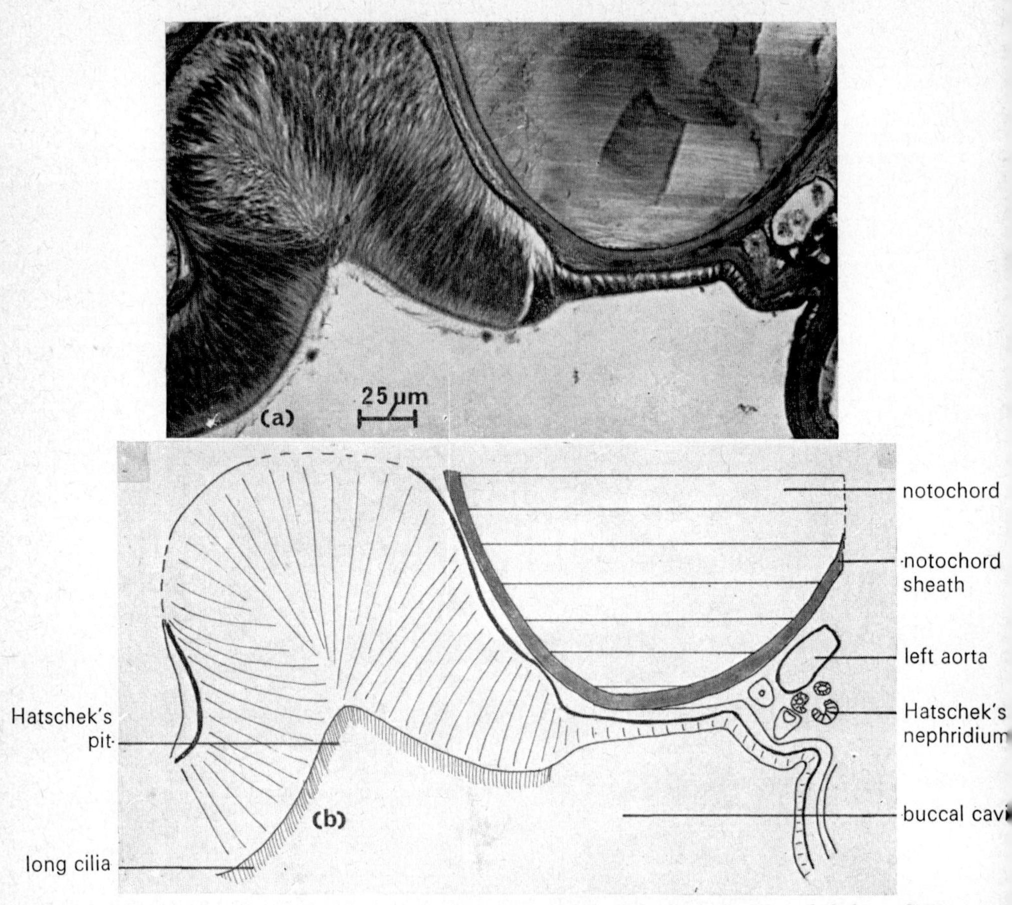

FIG. 10.22 (a) TS and (b) drawing of the dorsal wall of the buccal cavity of *A. lanceolatus*, showing part of the notochord and the ciliated Hatschek's pit.

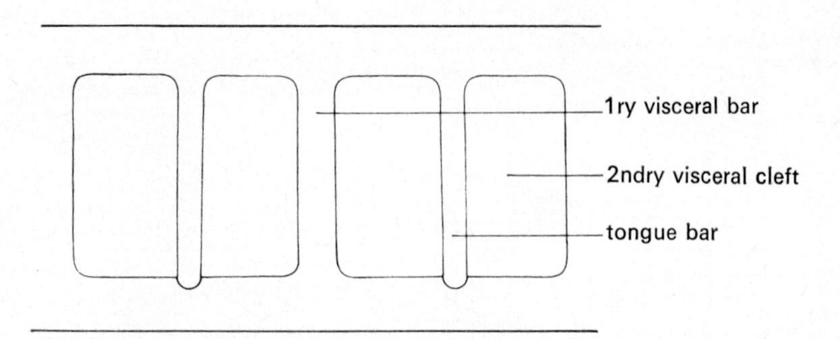

FIG. 10.23. Diagram showing a side view of the visceral clefts of *Amphioxus*.

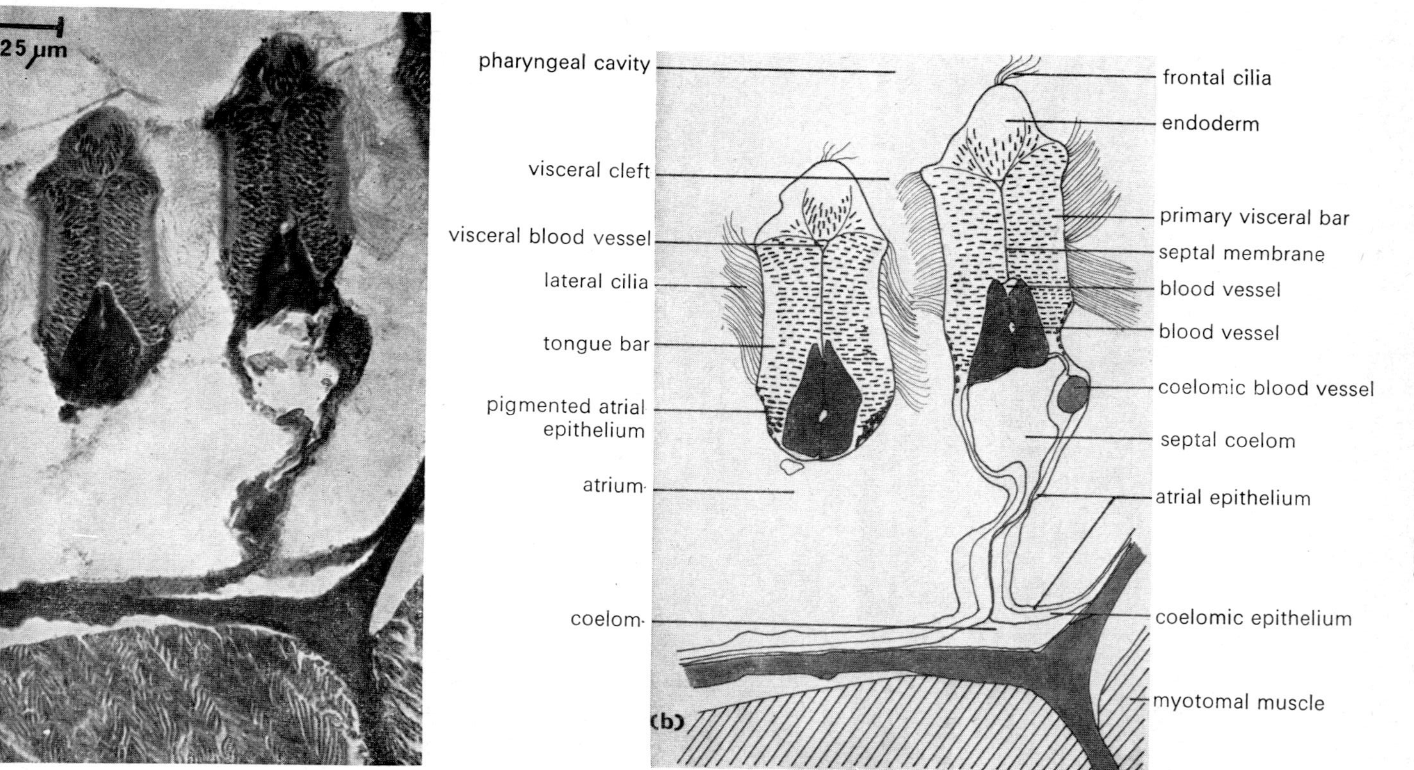

FIG. 10.24 (a) TS and (b) drawing of two pharyngeal bars of *A. lanceolatus*.

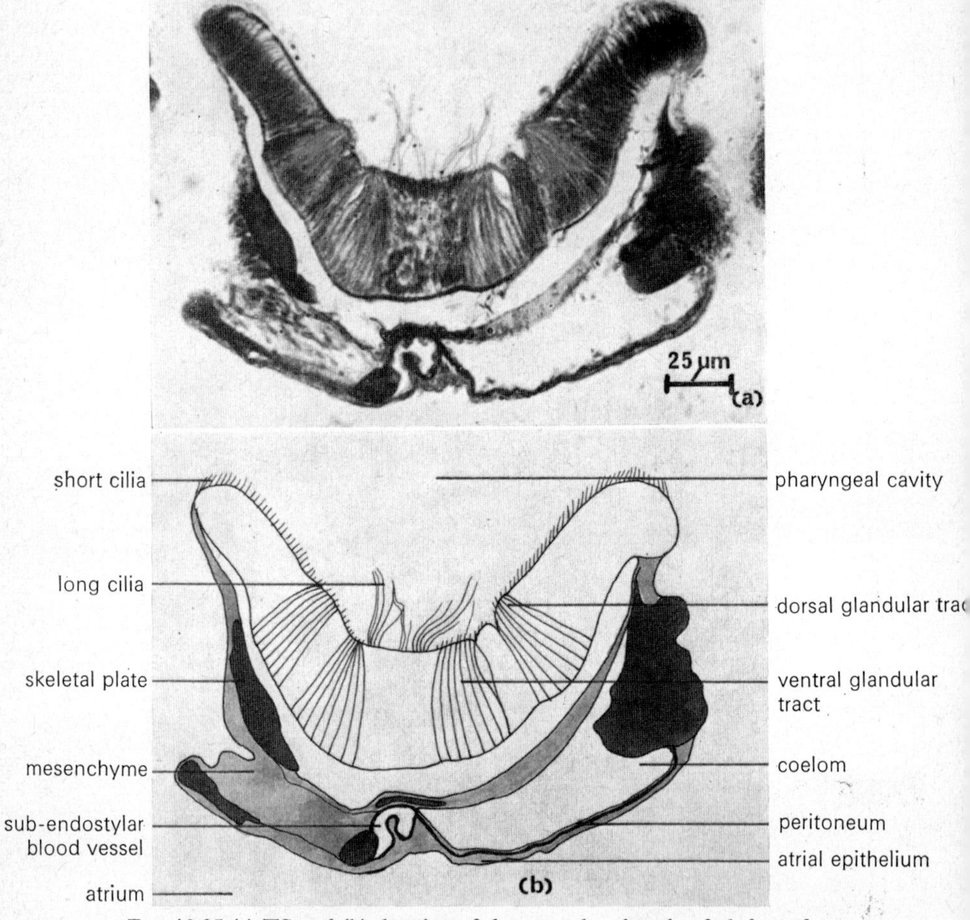

FIG. 10.25 (a) TS and (b) drawing of the ventral endostyle of *A. lanceolatus*.

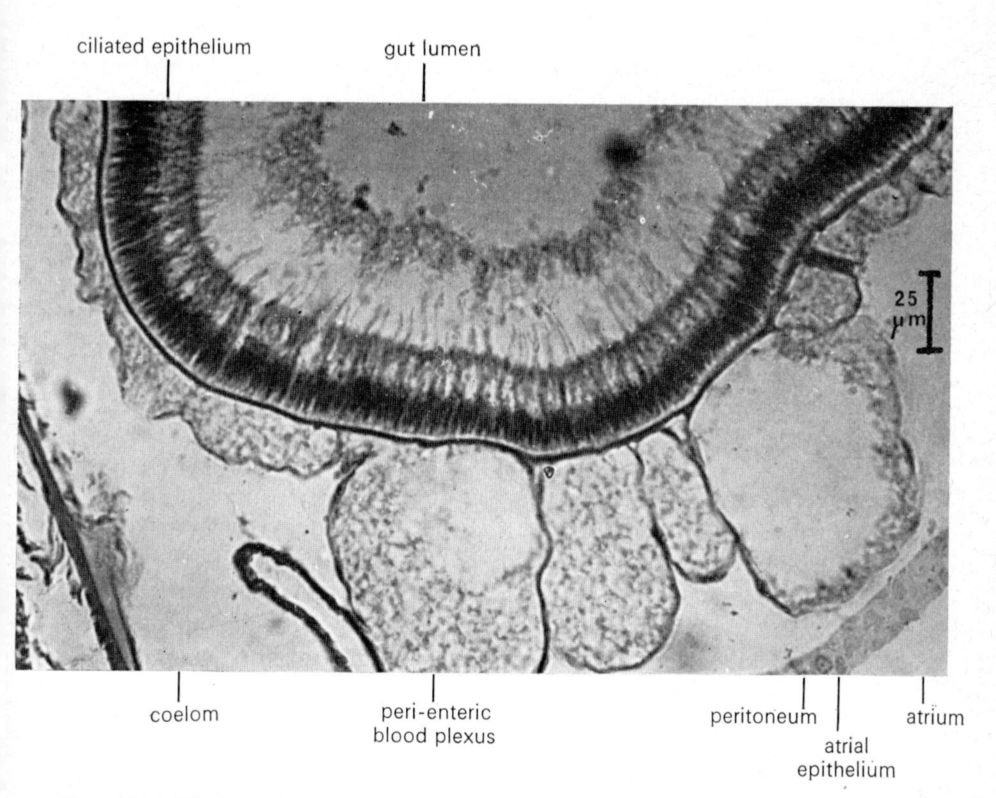

FIG. 10.26 TS through the intestine wall of *A. lanceolatus* showing the extensive peri-enteric blood plexus.

helps to sweep it from the pharynx, through a short oesophagus into the mid-gut (Fig. 10.26). The mid-gut is a straight tube lined with a single layer of ciliated epithelium. Again, there are no muscles present. Digestion takes place throughout the gut, but the enzymes responsible are secreted in the laterally-compressed mid-gut diverticulum (variously called the "liver" or caecum). In fact, this diverticulum from the anterior end of the mid-gut corresponds to the pancreas of other chordates rather than the liver as its tall columnar secretory cells (Fig. 10.27) produce digestive enzymes which are then swept out into the mid-gut in a mucus cord by the long cilia. The mid-gut is separated from the intestine by an ilio-colon ring (see Fig. 10.15), a ring of tightly packed ciliated cells which are specialized for rotating the cord of mixed food and mucus together with the cord of enzymes from the mid-gut diverticulum, and thus pulling the food through the gut. The smallest particles to break off are swept by cilia into the mid-gut diverticulum and ingested, while the residue is passed through the intestine to the anus.

Excretion

The excretory organs in *Amphioxus* are segmentally arranged protonephridia derived from ectoderm, and are thus very similar to those of annelids and completely different from the mesodermal vertebrate kidneys. The nephridia lie in the upper part of the coelomic canals of the primary visceral bars, above the pharynx. Each consists of many flame-cells (solenocytes) opening into a nephridial sac which voids excretory material into the atrium by a nephridiopore situated opposite the dorsal end of a tongue bar (Fig. 10.28). The solenocytes project into the dorso-pharyngeal coelom but have no connection with it. They are closely applied to blood-vessels (glomeruli) which issue from the dorsal ends of the pharyngeal vessels. Fluid is driven through the solenocytes by long flagella. Because of their small size, the nephridia are difficult to see in light microscope sections, but Fig. 10.29 shows their position in the pharynx wall and parts of their structure.

Reproduction

The sexes are separate in *Amphioxus* but there is no sexual dimorphism. The gonads are typically invertebrate in being numerous. Each gonad takes the form of a hollow sac bounded by a single layer of coelomic epithelium and lined with germinal epithelium (Fig. 10.30). There are no ducts: the gametes develop in the gonad cavity and are shed through a temporary pore into the atrium. The sperm have long tails and the eggs are small and yolky. Fertilization is external. Development is by a long-lived pelagic larva.

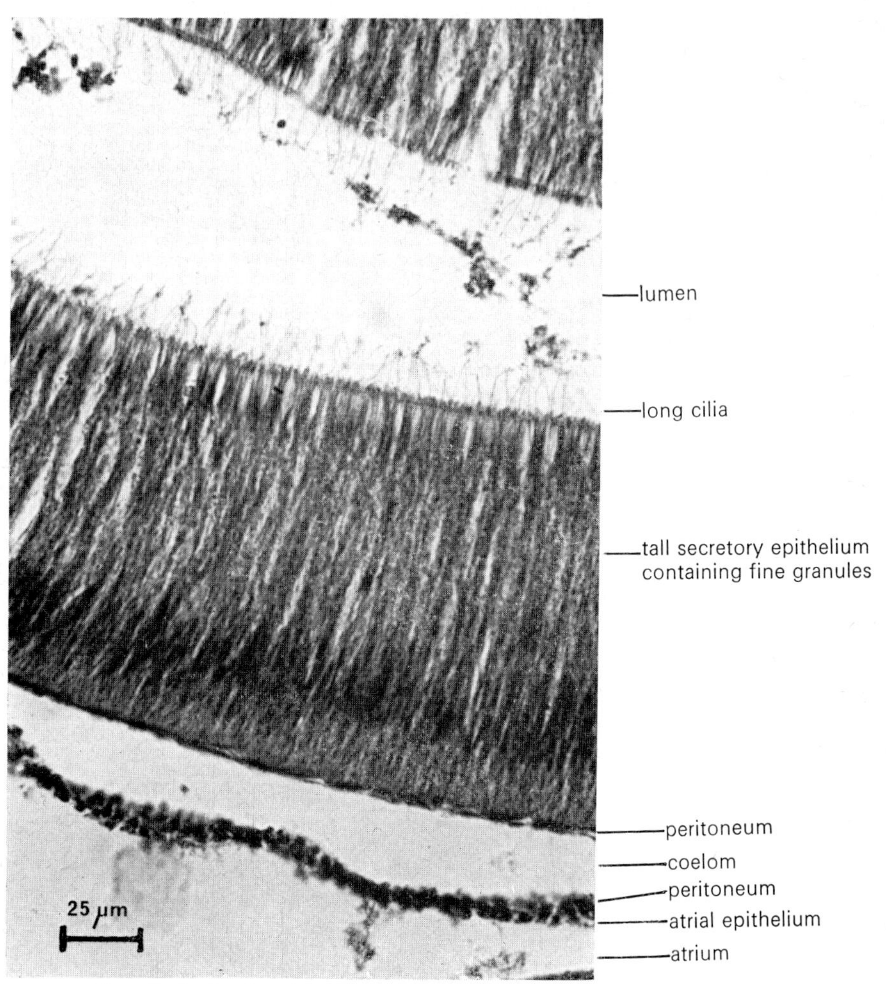

FIG. 10.27. Section through the midgut diverticulum of *A. lanceolatus*.

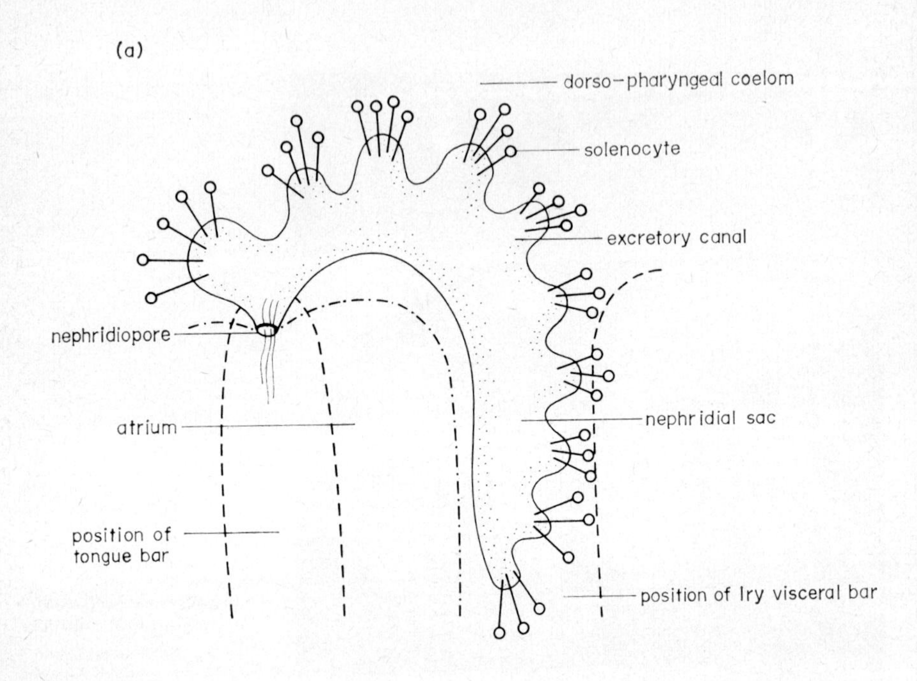

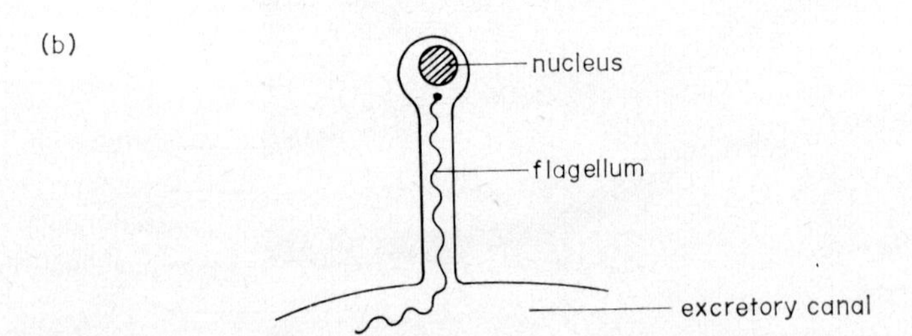

FIG. 10.28 (a) Plan of nephridium structure and position in *Amphioxus*; (b) diagram of a single solenocyte from the nephridium.

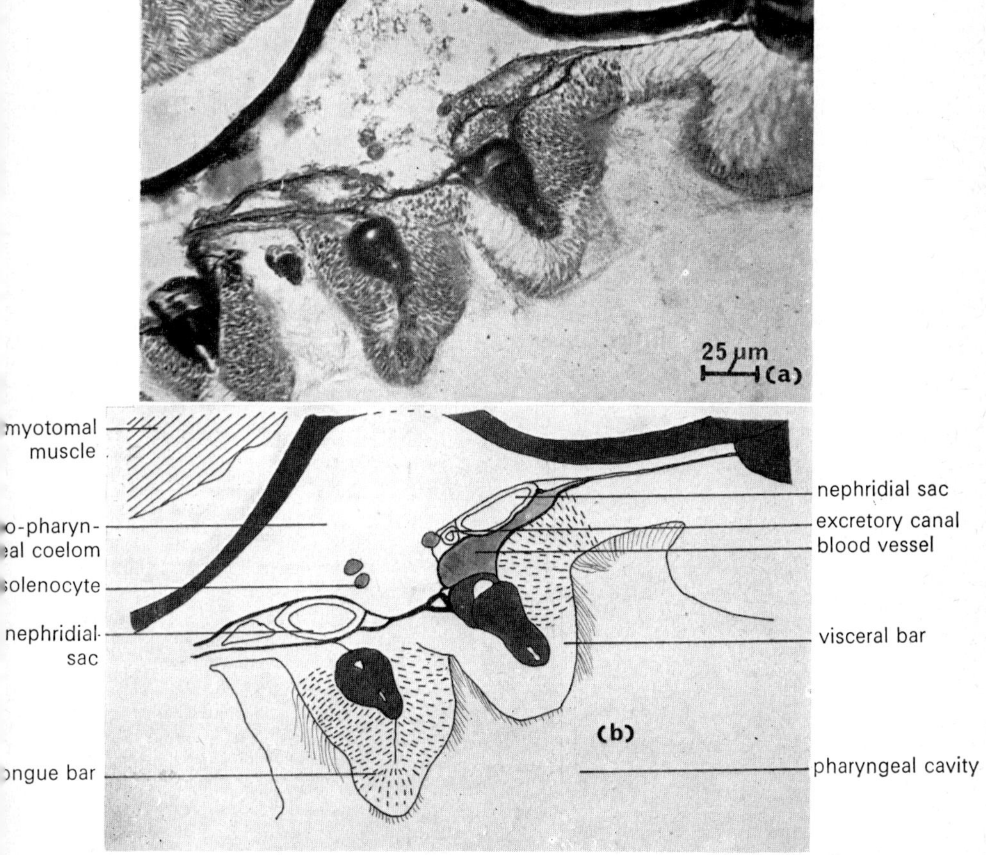

FIG. 10.29 (a) TS and (b) drawing of part of the dorsal pharyngeal wall of *A. lanceolatus* showing portions of a nephridium.

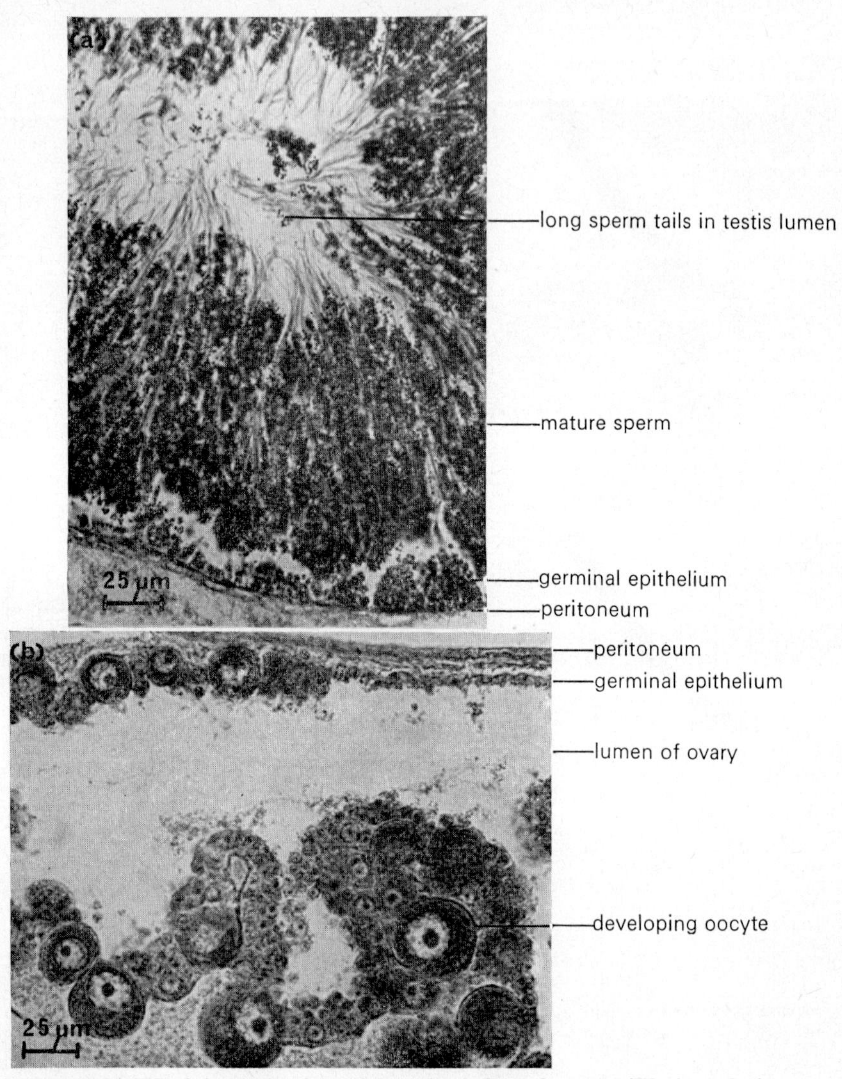

FIG. 10.30 (a) TS of the testis of an adult male *A. lanceolatus*; (b) TS through the ovary of a young female *A. lanceolatus*.

Sub-Phylum Vertebrata (Craniata)

The vertebrates are so called because, in the adult, the notochord is more or less completely replaced by a segmented vertebral column composed of hard material (cartilage or bone). The hollow dorsal nerve cord, situated above the notochord, is expanded anteriorly to form a brain which is also protected by a hard box (cranium). The brain becomes progressively more complicated as the associated anterior receptive systems (e.g. vision, hearing, smell) develop. The vertebrate body is bilaterally symmetrical about an anterior-posterior axis, and is normally divided into a head, trunk and post-anal tail. The tail, in most aquatic and a few terrestrial forms, is large and used for locomotion. Paired fins or limbs assist movement, or in the higher vertebrates are the chief means of locomotion. The striated somatic mesodermal muscles form segmental blocks (myotomes) which contract in sequence during locomotion of the lower vertebrates. In tetrapods myotomes are retained only in the embryo: in the adult there are specialized longitudinal muscles associated with particular skeletal components. The coelom is restricted to the trunk and may be sub-divided. There is a closed circulatory system (i.e. blood is confined to vessels) with a heart responsible for circulating the blood. Vertebrate blood characteristically possesses erythrocytes or red blood cells which carry the respiratory pigment (haemoglobin). As in all chordates, the vertebrates have pharyngeal slits or pouches at some stage during their development. In fish and some Amphibia these give rise to gills which are used in respiration, while in other tetrapods the respiratory structures are lungs which are thought to be derived from a posterior pair of gill pouches. All vertebrates are covered by a relatively thick skin consisting of a multi-layered epidermis overlying a vascularized dermis, and often bearing keratinized structures (e.g. scales, hairs and claws) for protective purposes.

Vertebrates may be subdivided into two main groups or superclasses: Pisces and Tetrapoda. The Pisces includes all the fish, many of which are extinct. Modern fish comprise the classes Agnatha, Chondricthyes and Osteichthyes. The Tetrapoda includes all those vertebrates the adults of which have four legs (although they may be secondarily modified or absent), i.e. the classes Amphibia, Reptilia, Aves and Mammalia. Of these, the Amphibia do not possess an amnion surrounding the embryo (see p. 11) and are therefore referred to as anamniotes, while the other three classes do have an amnion and are called amniotes.

Chapter 11

Class Agnatha

The most primitive vertebrates were jawless. Most of the Agnatha are known only as fossils; the sole surviving ones belonging to the Cyclostomata or round-mouthed agnathans. These are free-swimming eel-like creatures with a round suctorial mouth, but with no jaws. They swim using lateral undulations of the body assisted by movements of the dorsal and caudal fins. There are no paired limbs characteristic of higher vertebrates. Cyclostomes are soft-bodied and are protected by a thick glandular skin. They lack bones but have a cartilaginous skeleton as a means of support and protection. This skeleton is reduced and internal in present-day cyclostomes although many fossil agnathans had a heavy exoskeleton. There is a notochord, and signs of a vertebral column are present consisting of a fibrous neural tube with rudimentary neural arches in the form of scattered cartilaginous plates. Larger plates form an incomplete cranium round the brain. The central nervous system shows few signs of the enormous development that occurs in many higher vertebrates. This feature is associated with slow movements and poorly represented sense organs. Cyclostomes are further characterized by the possession of numerous gill pouches which are used as respiratory structures.

There are two groups of modern cyclostomes. The hagfish (Order Myxinoidea) are exclusively marine and are modified for a parasitic mode of life. They have suctorial mouthparts fringed with tentacles, and they burrow into the bodies of dead or dying fish.

The best known cyclostomes are the lampreys (Order Petromyzontia) which have a very widespread distribution. They possess a freshwater ammocoete larva which filter-feeds and is very like *Amphioxus*. At the end of larval life metamorphosis occurs to an adult which can be marine or freshwater and which generally lives ectoparasitically off other fish.

Example: *Lampetra* (*planeri*)

L.planeri, the lesser freshwater or brook lamprey, is a smooth-skinned cylindrical creature with the tail region flattened laterally. The head end, shown in VLS in Fig. 11.1, bears a concave ventral suctorial mouth, while dorsally there is a single median nostril. Lampreys have two dorsal fins far back on the body, the posterior dorsal fin being continuous with the caudal fin. In females there is a small anal fin behind the cloaca. There are no lateral fins. *L.planeri* is somewhat atypical in that, as an adult, the anterior end of the gut closes off, so, although the animal uses its mouth for attachment, it is not an ectoparasite. The ammocoete larva of *L.planeri*, shown in TS in Fig. 11.2, is typical, and is important in its resemblance to the invertebrates.

Lampreys are covered by a smooth skin, which, unlike the invertebrate integument, is formed of many layers of cells. The outer ectodermal layer (epidermis) consists of small cuboidal epithelial cells interspersed with larger granular or club-shaped mucous cells. It is covered by a cuticle which is not always seen in adults but is easily visible in the larva (Fig. 11.3). The inner layer of skin (dermis) is mesodermal in origin, and is differentiated into a fibrous layer (thick in the adult, thin in the larva) overlying a thin coat of pigment and a subcutaneous layer of loose texture often containing adipose (fat) tissue. Lamprey skin possesses chromatophores but there are no scales such as are found in higher fish.

The endoskeleton is composed of cartilage rather than bone. Cartilage has a uniform, avascular protein matrix which is secreted by rounded mesodermal cartilage cells (chondroblasts). This cartilage varies from one part of the animal to another. Generally in the larva it has a rather loose fibrous structure [Fig. 11.4(a)] while the main part of the adult skeleton is composed of typical hyaline cartilage, with chondrocytes (inactive chondroblasts) packed closely into a little dense matrix [Fig. 11.4(b)].

The notochord of lampreys also helps to provide rigidity (Fig. 11.5). It is surrounded by a tough collagenous notochord sheath, and has the appearance of a stack of plates, consisting of large vacuolated cells. It is clear that the notochord of lampreys is not composed of the same stiff muscle as in *Amphioxus* (see Fig. 10.19), but it undoubtedly plays a similar role in providing an incompressible rod to assist swimming movements.

Nervous system

The nervous system of lampreys is considerably more advanced than that of *Amphioxus*, and shows the characteristic plan present in all vertebrates (Fig. 11.6). The dorsal nerve cord extends nearly the whole length of the vertebral column, is flattened in TS (see Fig. 11.5), and has a central canal lined with ciliated ependymal cells and surrounded by nerve cell bodies with a rim of axons and glial cells round the outside (Fig. 11.7). The central canal is expanded at the anterior end to form a series of interconnecting ventricles, and is filled with cerebrospinal fluid (c.s.f.) which brings nutrient and oxygen to the CNS, removes metabolites and provides a physical buffer for the soft nervous tissue. The brain is a thickened area of cells round the ventricles, and consists of fore-brain (prosencephalon), mid-brain (mesencephalon) and hind-brain (rhombencephalon); primitively associated with the senses of smell, sight and hearing respectively (Figs 11.6 and 11.8). Since the brain in lower vertebrates is chiefly concerned with integrating sensory information, there is considerable hypertrophy of some of the special sense areas, particularly those controlling smell and vision (olfactory lobes and optic tectum). The

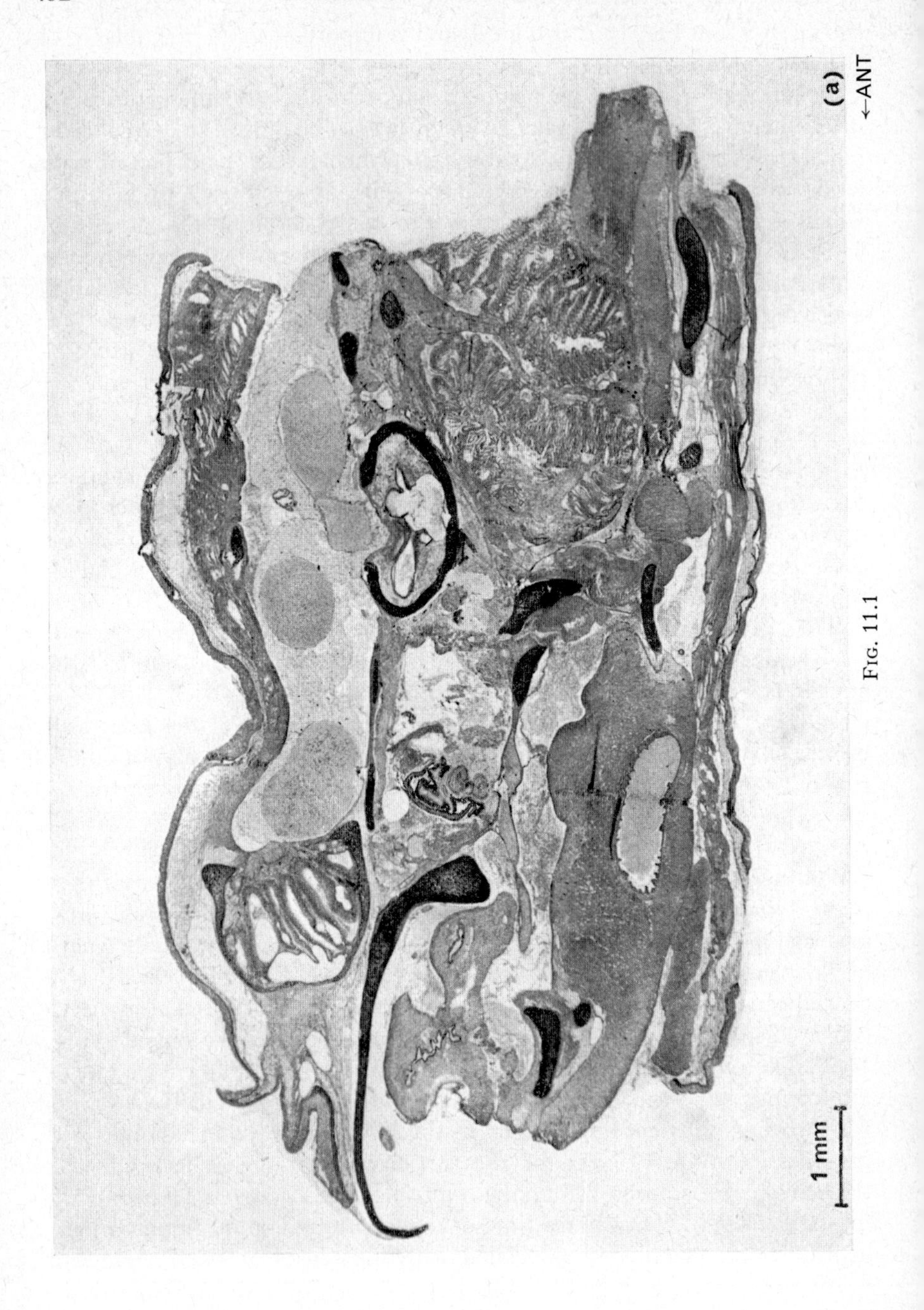

Fig. 11.1

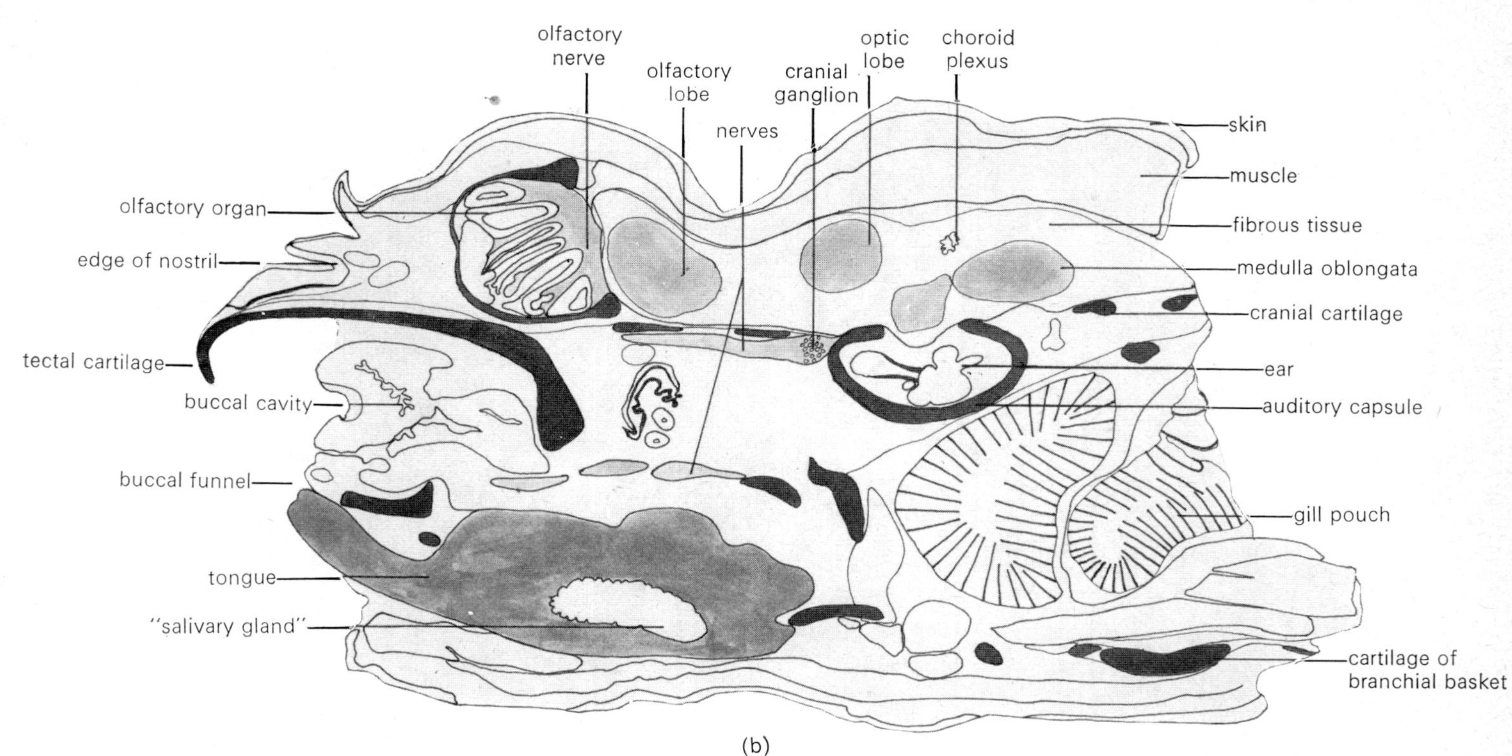

FIG. 11.1 (a) VLS and (b) explanatory drawing of the anterior end of an adult brook lamprey (*Lampetra planeri*).

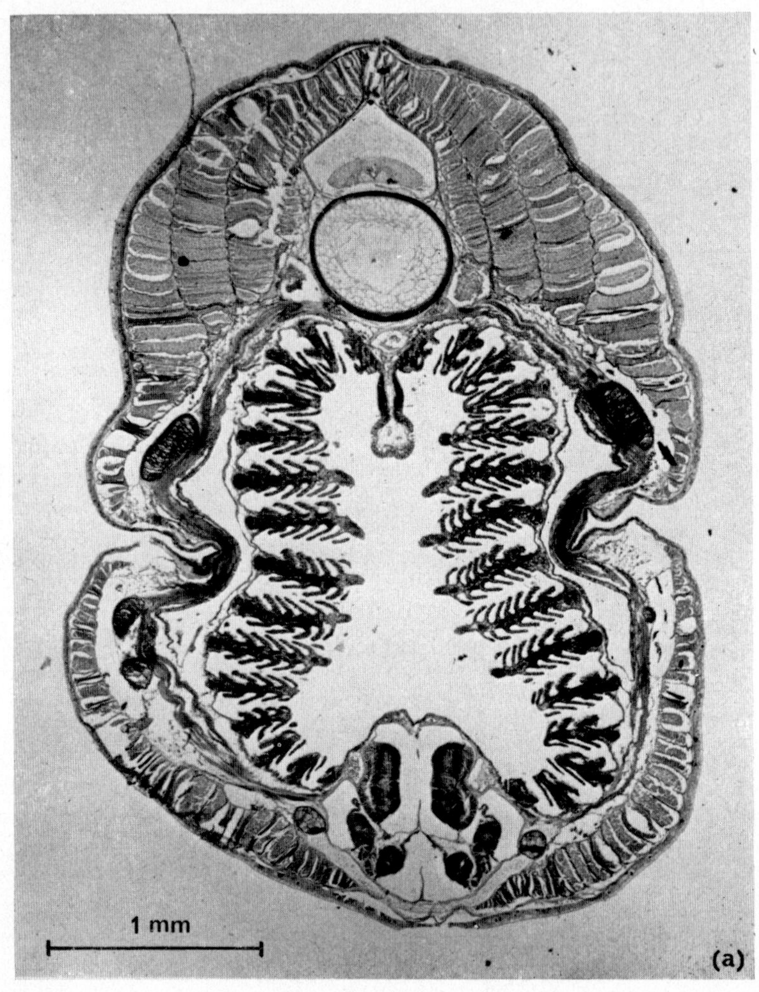

Fig. 11.2

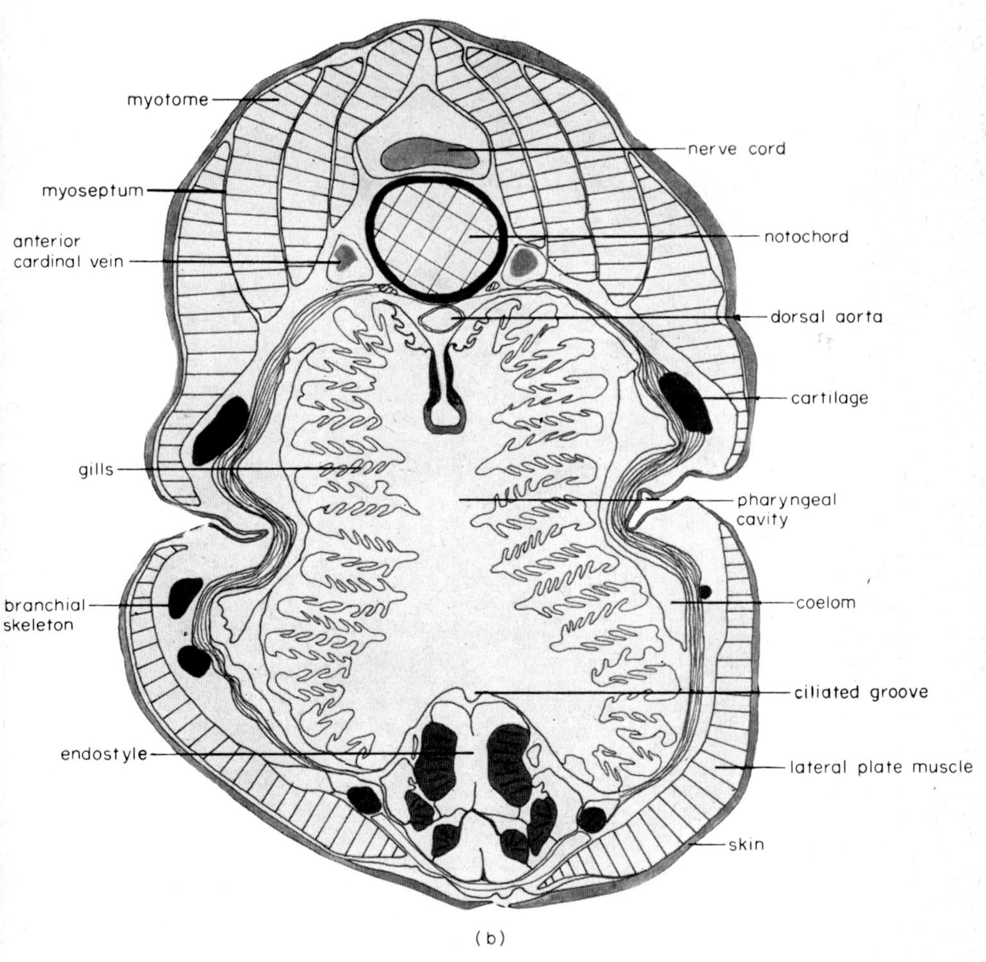

FIG. 11.2 (a) TS and (b) explanatory drawing of the branchial region of an ammocoete larva of *L. planeri*.

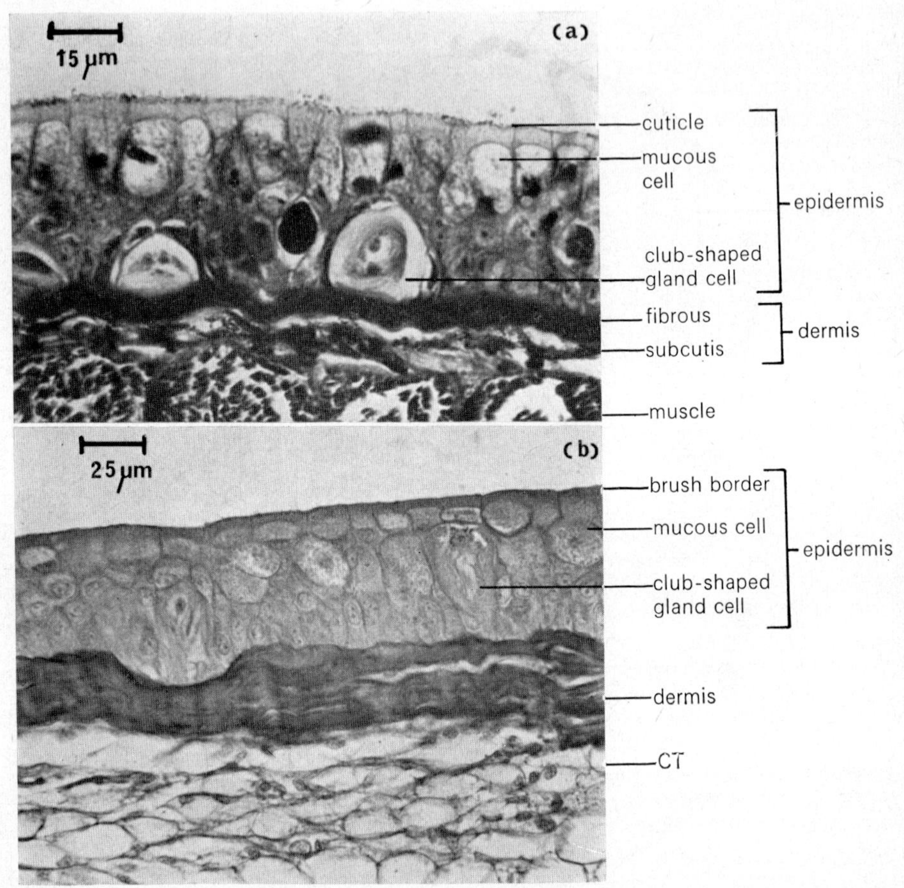

FIG. 11.3 (a) Section through the skin of an ammocoete larva; (b) section through the skin of an adult *L. planeri*.

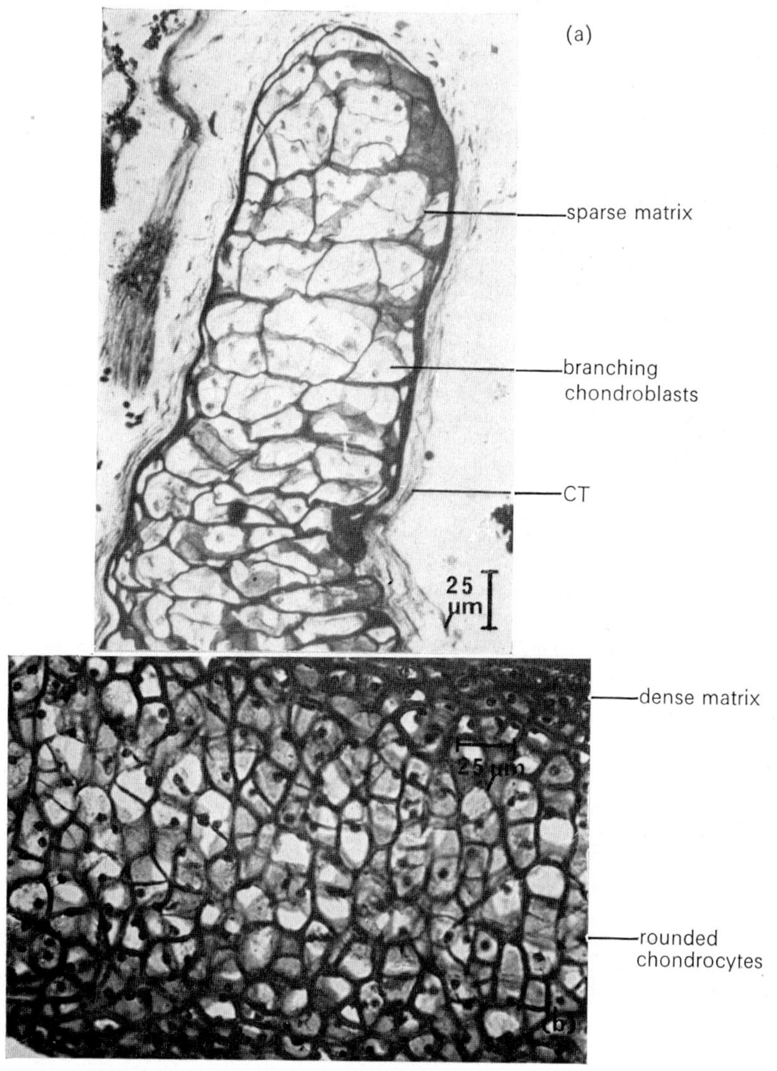

FIG. 11.4 (a) Mucous cartilage from the branchial basket of an ammocoete larva; (b) hyaline cartilage from the cranium of an adult *L. planeri*.

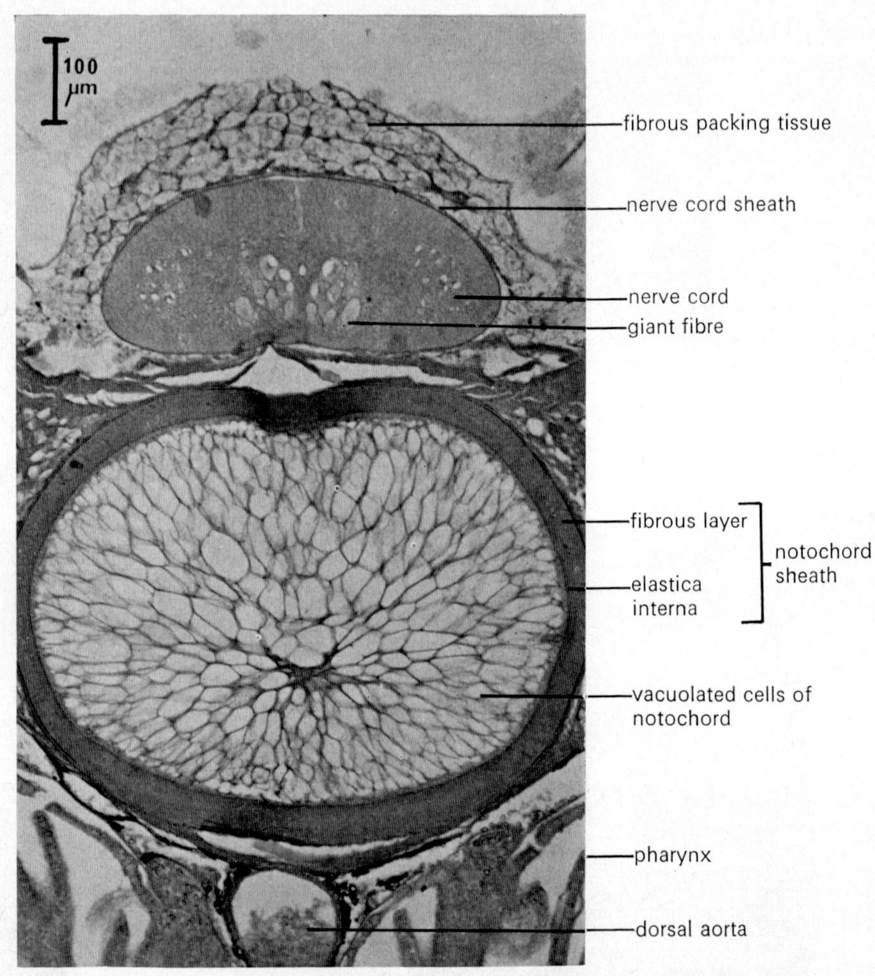

FIG. 11.5. TS through the spinal cord and notochord of an adult *L. planeri*.

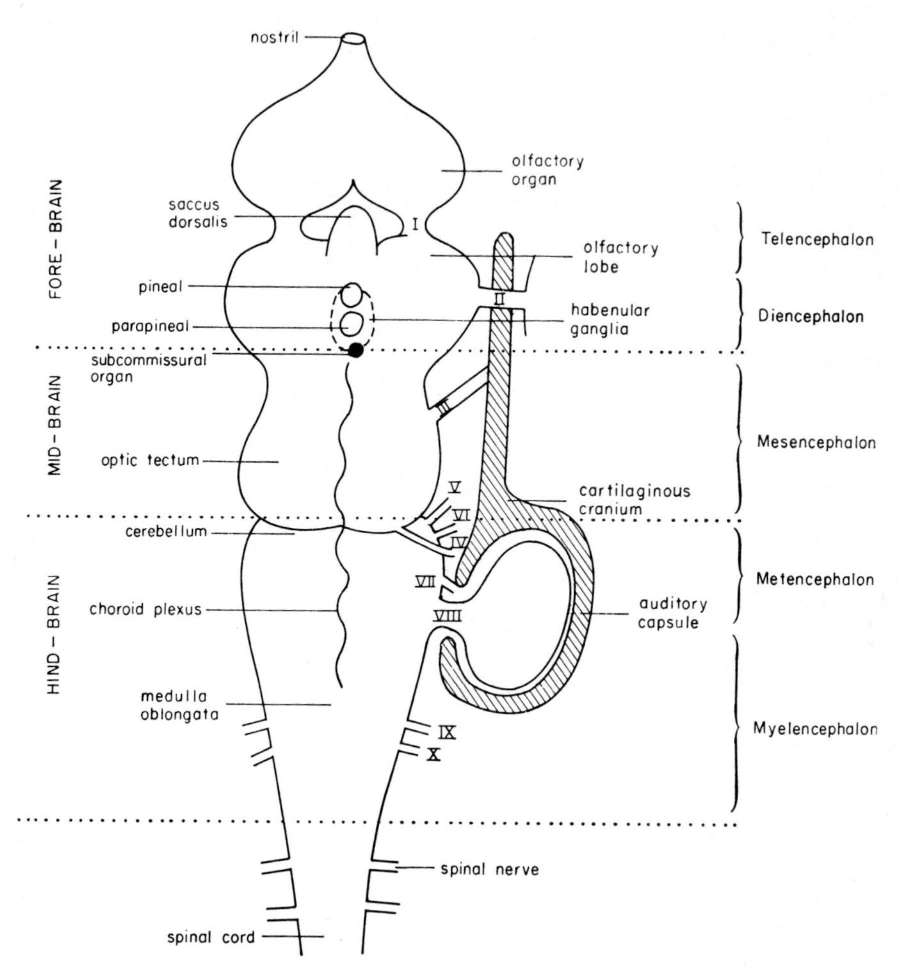

FIG. 11.6. Plan of the central nervous system of a lamprey (as viewed from the dorsal side), showing the major divisions characteristic of all vertebrate brains. Roman numerals denote cranial nerves.

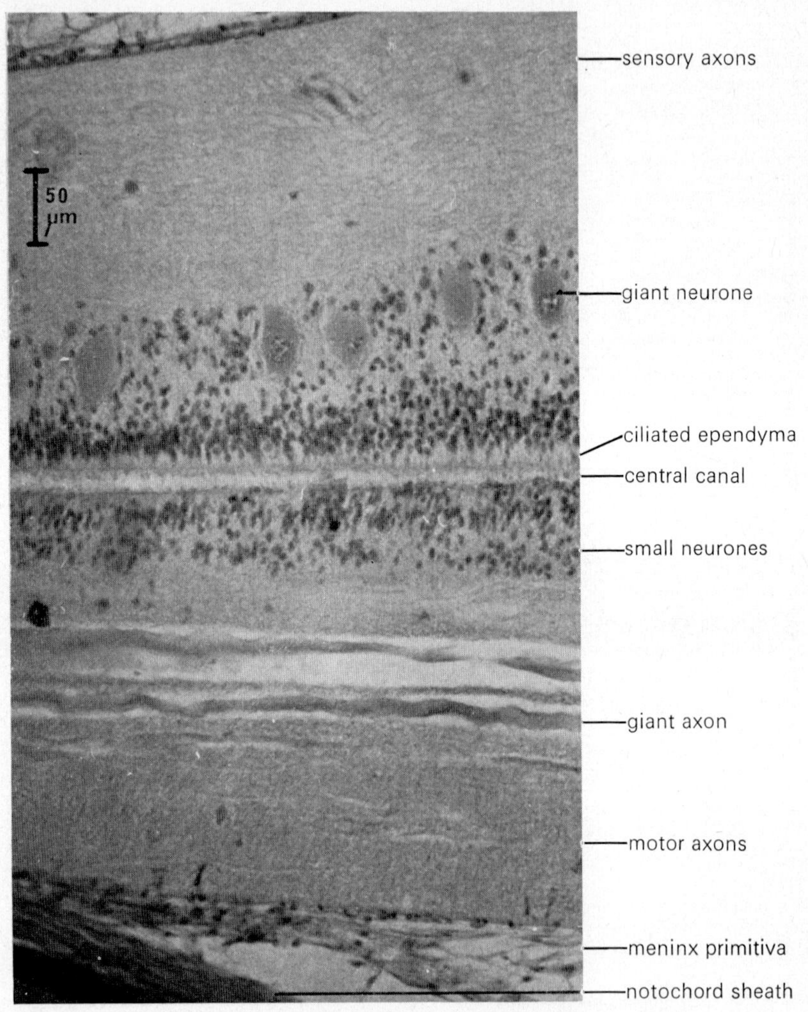

Fig. 11.7. VLS through the spinal cord of an adult *L. planeri*.

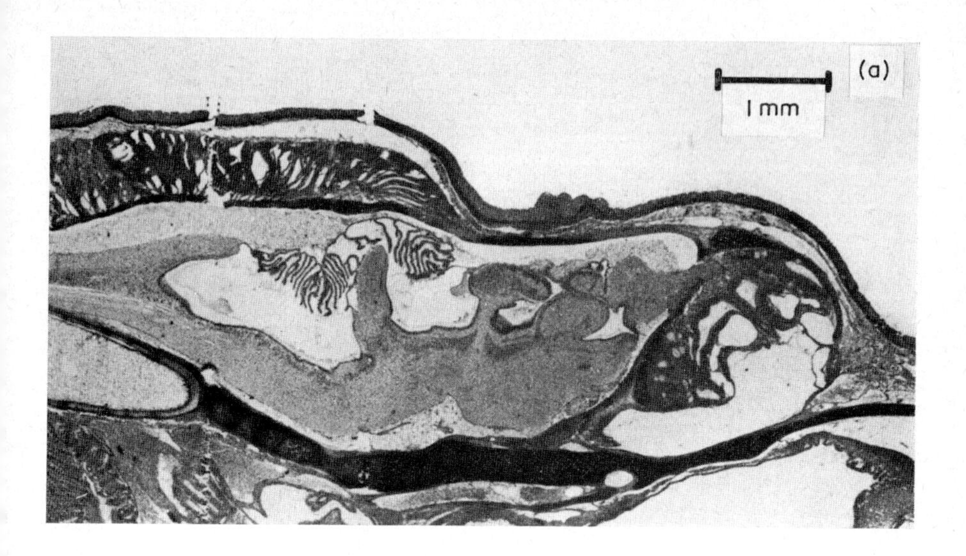

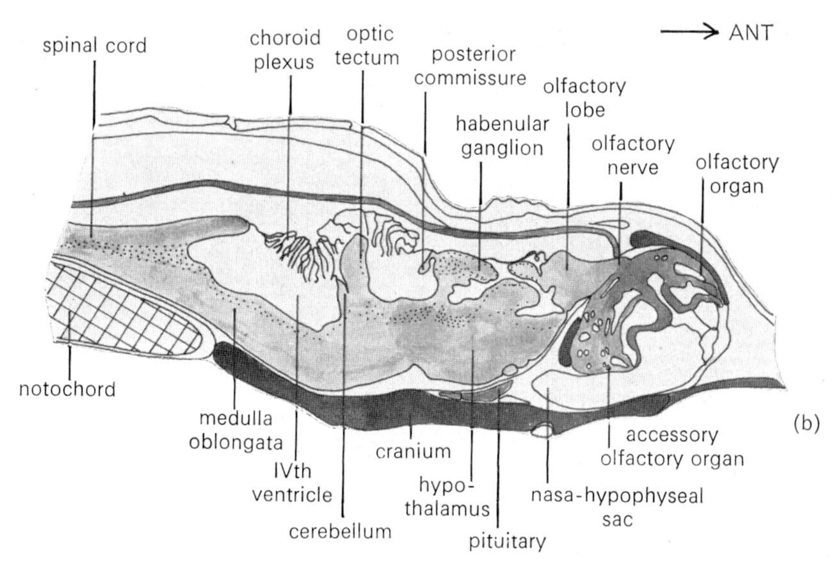

FIG. 11.8 (a) VLS and (b) drawing of the brain of an adult *L. planeri*.

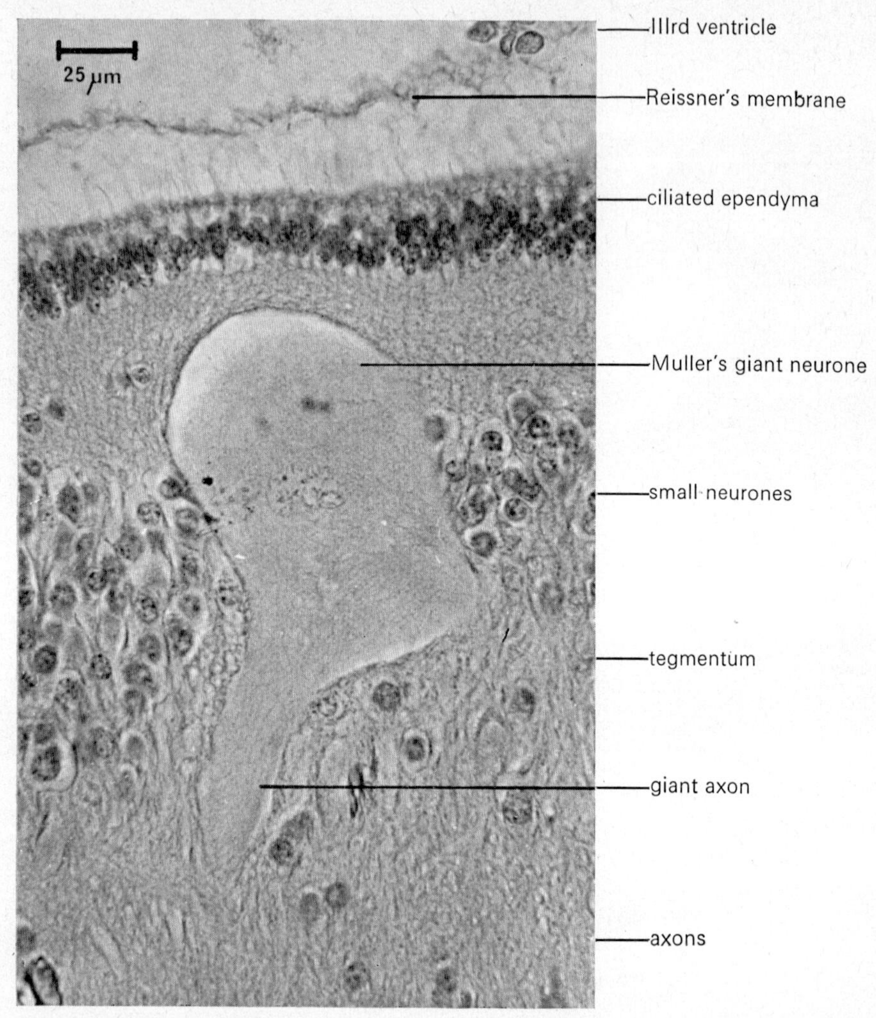

FIG. 11.9. Section through part of the floor of the midbrain (tegmentum) of an adult *L. planeri*, showing a giant Muller's neurone.

hind end of the tectum is evaginated to form optic lobes. In accordance with the lamprey's sluggish, semi-parasitic mode of life, the areas controlling hearing, lateral line, taste and proprioception are poorly developed. In the mid- and hind-brain regions several giant Muller's neurones originate (Fig. 11.9). They are formed by the fusion of many neurones during development (note several nuclei) and run the length of the spinal cord without crossing contralaterally. A pair of large Mauthner neurones occur in the floor of the medulla oblongata: their axons cross over (decussate) before they run down the spinal cord. Both sets of giant cells are thought to be concerned with co-ordinating locomotory movements. The medulla [Fig. 11.10(a)] is the chief centre of co-ordination in lampreys and contains an abundance of neurones, although their arrangement is similar to that of the spinal cord. The roof of the medulla is a thin, convoluted, membranous sheet of ependymal epithelium forming the choroid plexus, which extends over the whole of the IIIrd and IVth ventricles and forms the site for exchange of materials between the blood and c.s.f. [Fig. 11.10(b), see also Figs 11.8 and 11.13]. In vertebrates the CNS is enclosed in vascularised connective tissue membranes (meninges) which provide mechanical protection. In lampreys there is only one layer, the meninx primitiva.

Nerves connecting the brain to some of the more anterior segments have become modified to form special cranial nerves with associated ganglia. Those of the lamprey are similar to those of other vertebrates except that the cranial ganglia are separated from the brain and lie outside the cranium (Fig. 11.11). They are surrounded by connective tissue capsules and contain bipolar and unipolar cells: ganglia of most other vertebrates contain only unipolar and pseudo-unipolar cells. The cranial nerves from the eyes (optic nerves) cross over to the optic lobes on the opposite side of the brain. The cross-over point is called the optic chiasma (Fig. 11.12). The spinal nerves show the same plan as in *Amphioxus* in that the dorsal and ventral roots remain separate. The dorsal roots emerge in the intersegmental region and are mainly sensory with bipolar cells collected into dorsal root ganglia. The ventral roots are primarily motor and emerge in the middle of each segment. Nerve cell bodies occur in ganglia, but are also found scattered along some peripheral nerves. There are no myelinated fibres. Widespread neurosecretory neurones occur throughout the central nervous system. There are two neurohaemal organs—the pituitary and the sub-commissural organ (Fig. 11.13) which secretes the Reissner's fibre (see p. 409).

Sense organs

The lamprey is not well supplied with sense organs although there are many free nerve endings in the skin, but a few of its special senses are important in the adult.

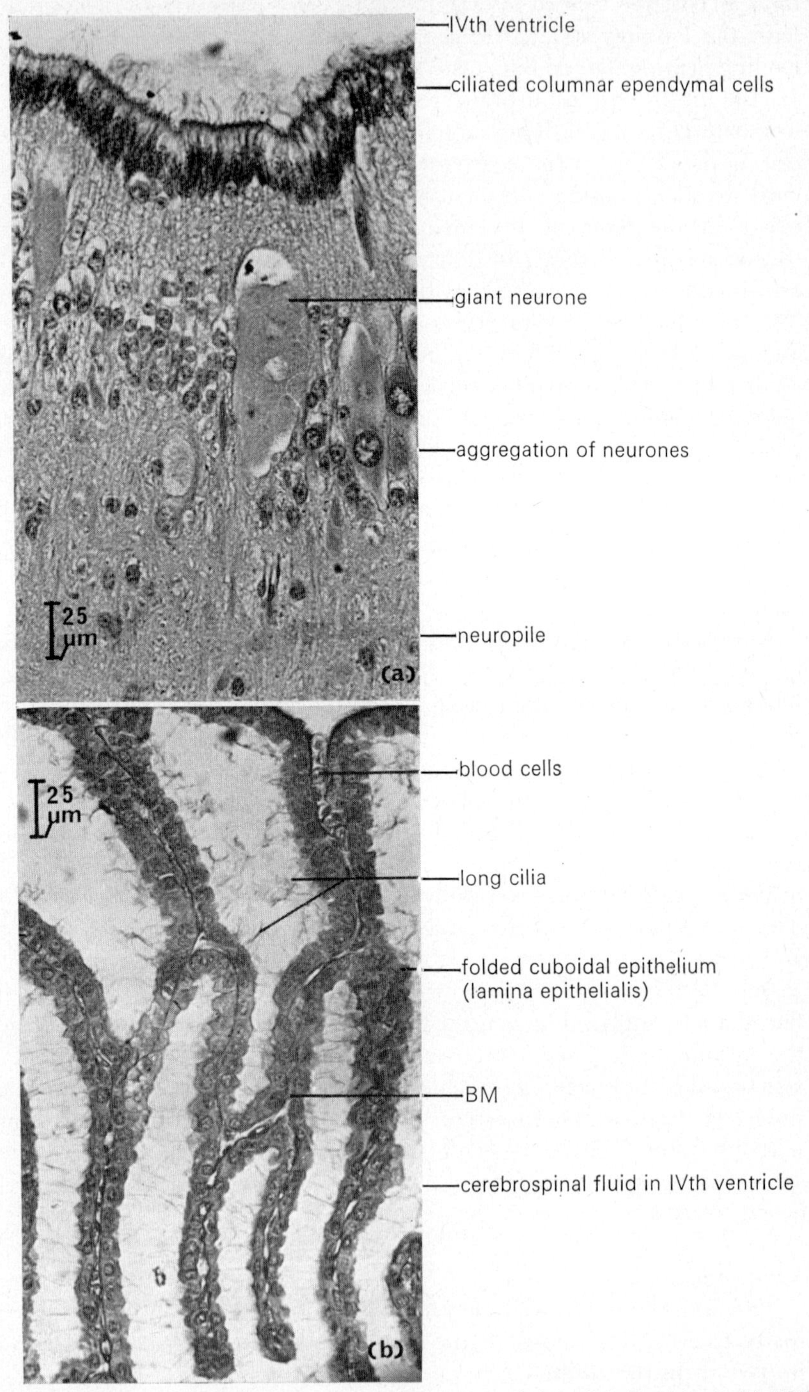

FIG. 11.10 (a) Section through the medulla oblongata of the adult *L. planeri*; (b) VLS through the folded epithelial roof of the hindbrain which forms an extensive choroid plexus.

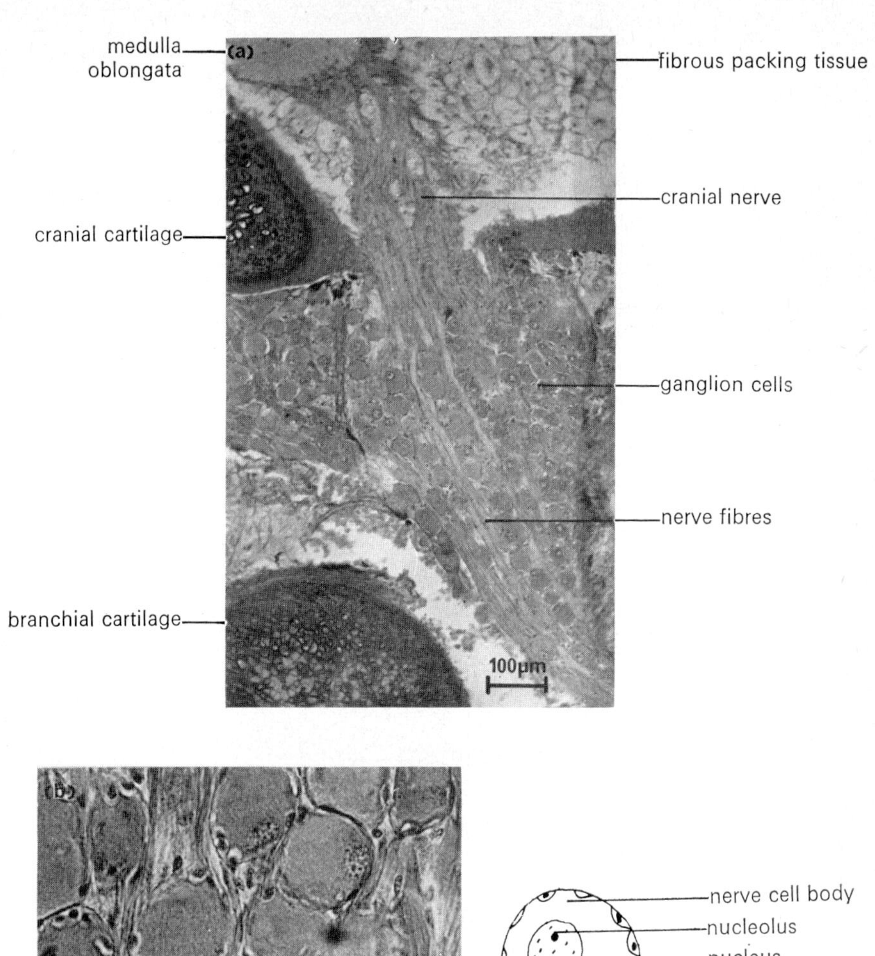

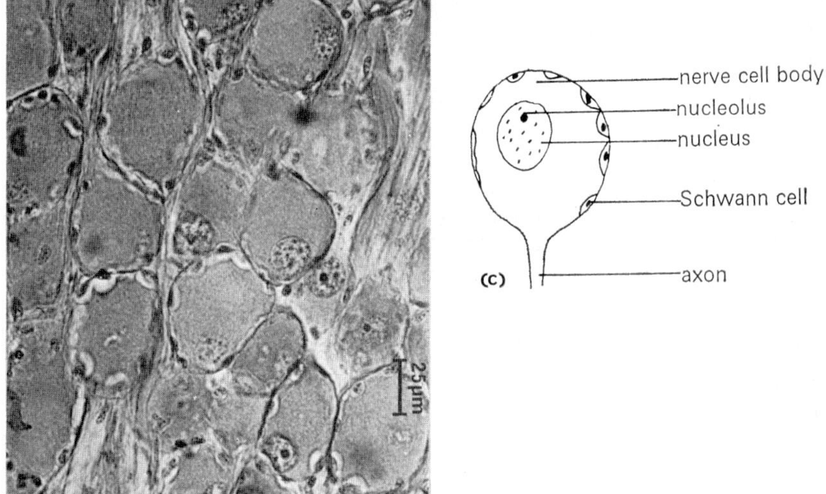

FIG. 11.11 (a) Section through a cranial ganglion in an adult *L. planeri*; (b) detail of unipolar and bipolar neurones in the cranial ganglion; (c) drawing of one unipolar neurone from a cranial ganglion showing the investing Schwann cells.

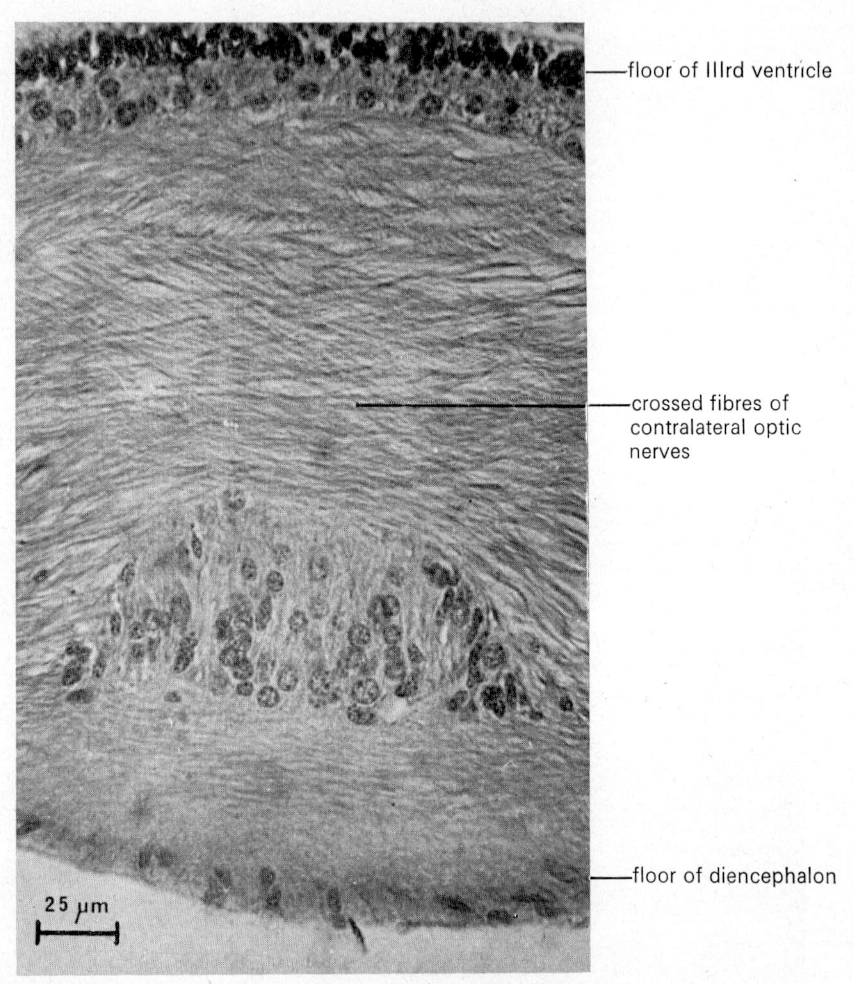

FIG. 11.12. TS through the diencephalon of the brain of an adult *L. planeri*. Note the crossed optic nerve fibres forming an optic chiasma.

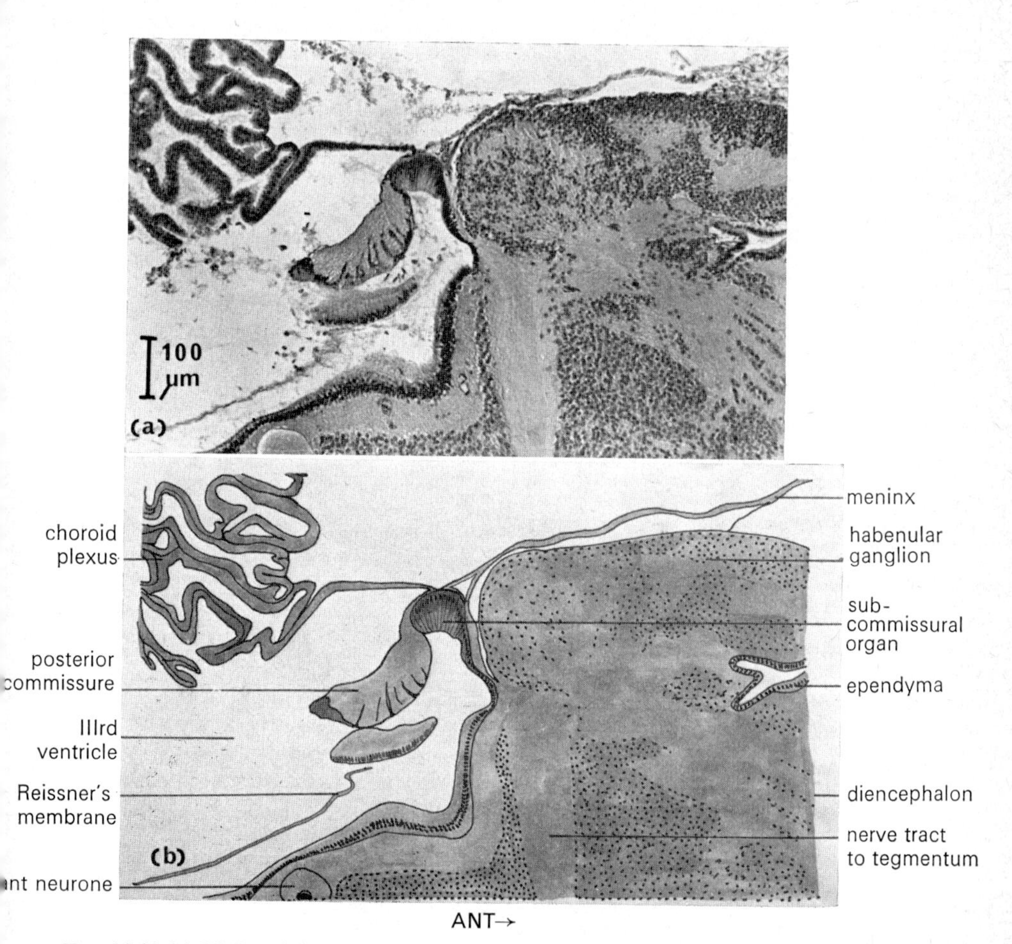

FIG. 11.13 (a) VLS and (b) drawing of the brain of an adult *L. planeri* at the junction of the forebrain and midbrain. Note the large habenular ganglion (which receives nerve fibres from the pineal), the posterior commissure and the subcommissural organ secreting a Reissner's fibre.

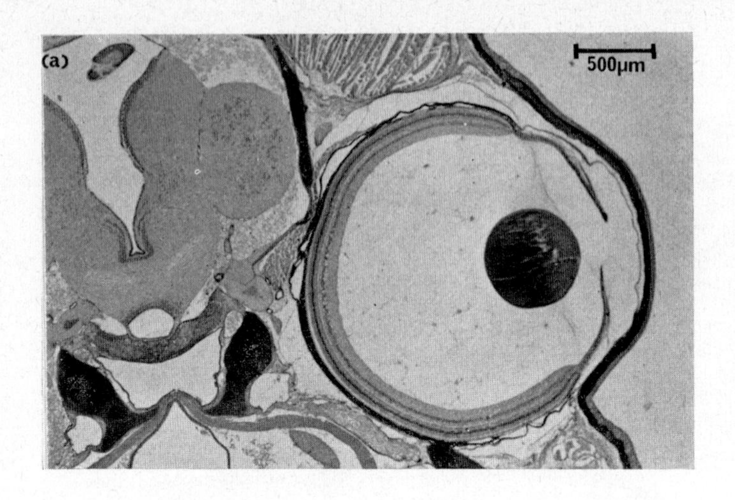
(a)
500μm

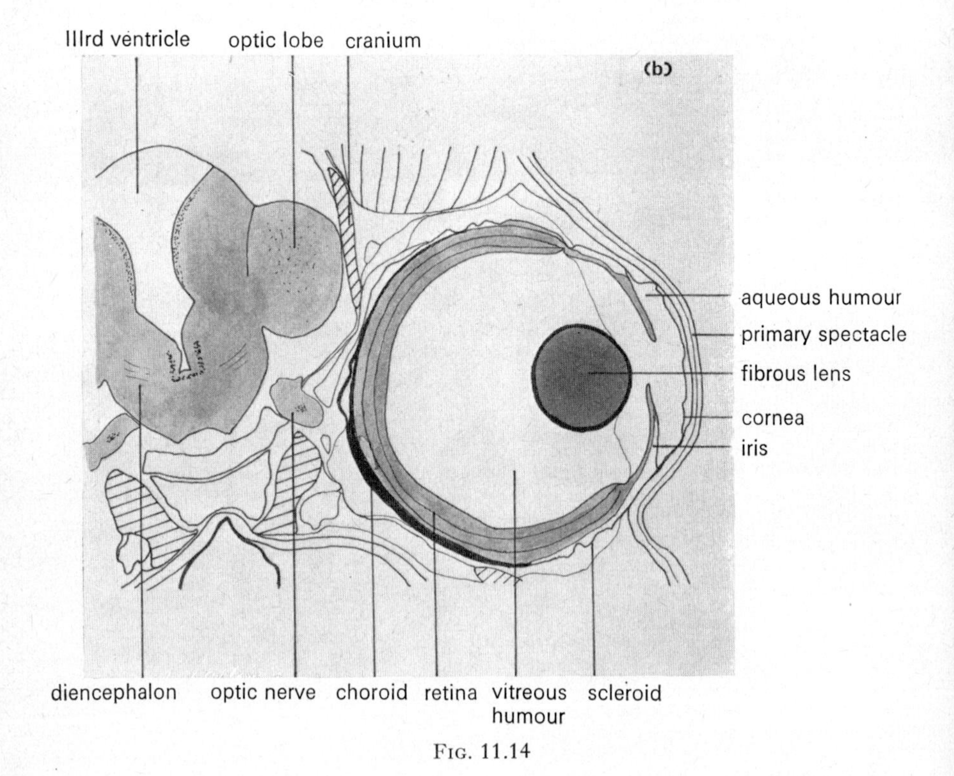
IIIrd ventricle
optic lobe
cranium
(b)
aqueous humour
primary spectacle
fibrous lens
cornea
iris
diencephalon
optic nerve
choroid
retina
vitreous humour
scleroid

FIG. 11.14

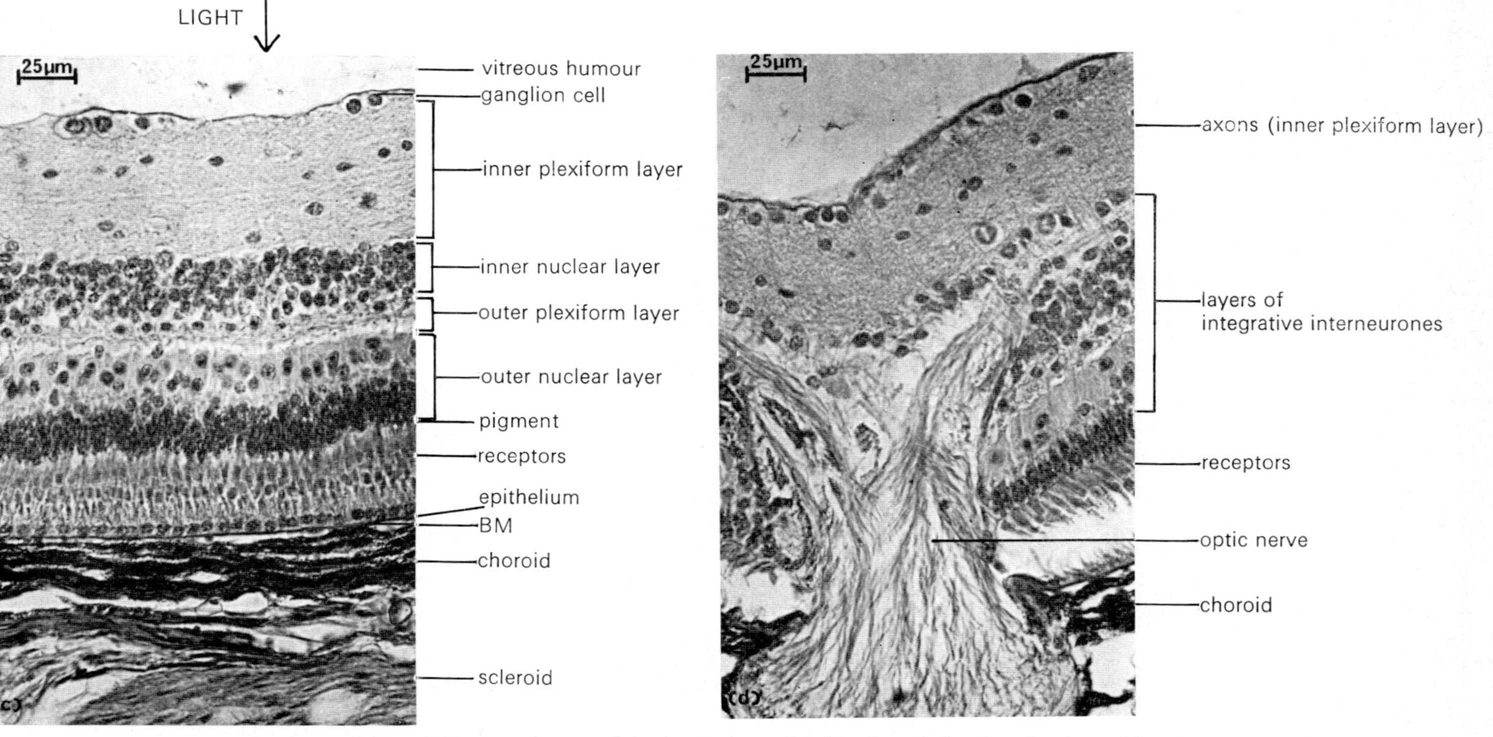

FIG. 11.14 (a) TS through part of the head of an adult *L. planeri*, showing the eye, optic nerve and optic lobe of the midbrain; (b) explanatory drawing of (a); (c) detail of the retina in the eye of an adult *L. planeri*; (d) detail of the optic nerve fibres leaving the retina.

The ammocoete larva lies buried in mud with only the tail end projecting. Its eyes are covered by opaque skin, but the general body surface, particularly that of the hind end, is sensitive to light. After metamorphosis, however, the paired eyes are capable of image formation and are covered by clear areas of skin. Like most of the lower vertebrates the lamprey does not have binocular vision as the eyes are placed laterally on the head. The eyes are very similar to those of other vertebrates, except that they are attached to the rim of the eye orbit and not to the overlying skin [Fig. 11.14(a, b)]. The transparent area of skin above the eye (primary spectacle) is composed of two layers of stratified squamous epithelium and merely acts as a protective device. A flexible cornea covers the surface of the eye and is in contact with the large spherical yellow lens. Body wall muscles inserted on to the rim of the cornea (extrinsic corneal muscles) contract, causing flattening of the cornea which pushes the lens towards the centre of the eye. Since the lens is the main refractive structure in the eye, such movement will alter the focus point, and is thus a device for allowing adjustment of the eye to focus on points at different distances (accommodation). The retina [Fig. 11.14(c)] is avascular and is composed of two types of photoreceptors (although it is not yet possible to classify these, as with the rods and cones of other vertebrates). At least one visual pigment has been identified. The retina is inverted, with the optic nerve fibres (cranial nerve II, formed by evagination of the brain diencephalon) overlying the receptors [Fig. 11.14(d)]. It is backed by a highly vascular, pigmented choroid coat which nourishes the retina and acts as a screen to prevent light scatter outside the eye. The shape of the eye is maintained by an outer thin but strong scleroid coat.

There are also two light-sensitive structures formed by evagination of the diencephalon roof, comparable to paired "median eyes" (see Fig. 11.6). They are both covered by skin, and are small flattened sacs lined with ciliated cells. The right member of the complex, the pineal (Fig. 11.15), has its lining differentiated into a *pellucida* and a "retina" which connects by nerve fibres with the habenular ganglion of the mid-brain. There is some evidence that in some animals the pineal is secretory, but in the lamprey it functions primarily as a non-image forming photoreceptor. It may influence activity of the sub-commissural organ of the mid-brain (a known neurosecretory organ), and almost certainly is involved in colour change. There is no evidence that the left member of the complex, the parapineal (which lies underneath the pineal), acts as a photoreceptor.

All aquatic vertebrates possess a complex acoustico-lateralis system with which they detect disturbances in the water. There are two acoustic sense organs: the ear and lateral line. Both are mechanoreceptors with sensitive hair cells (neuromasts) which are deformed by sound or shock waves.

The vertebrate ear is adapted to receive and amplify relatively high

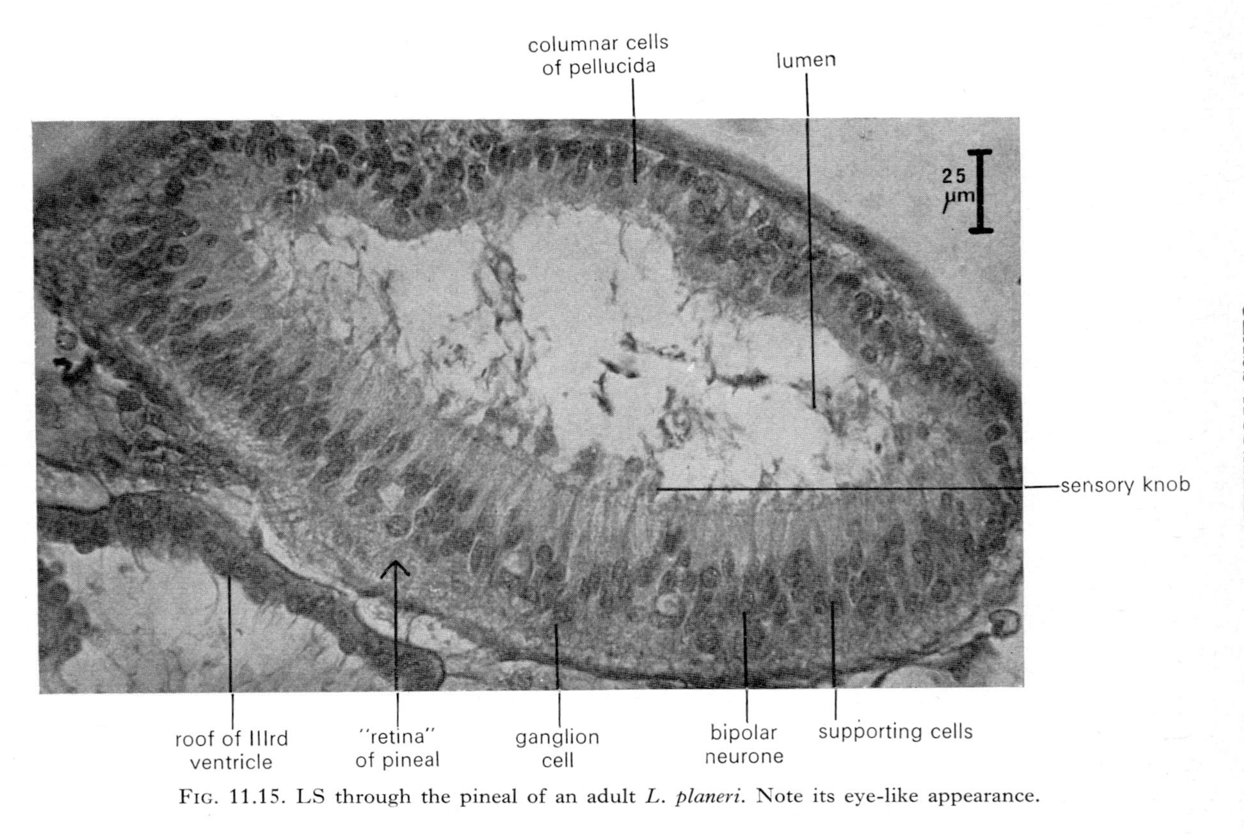

Fig. 11.15. LS through the pineal of an adult *L. planeri*. Note its eye-like appearance.

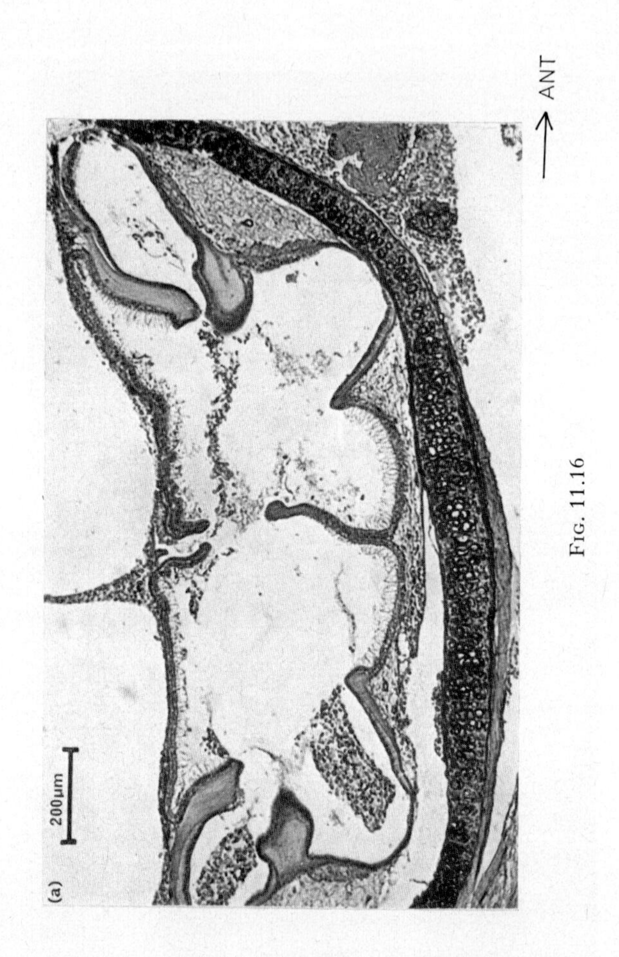

Fig. 11.16

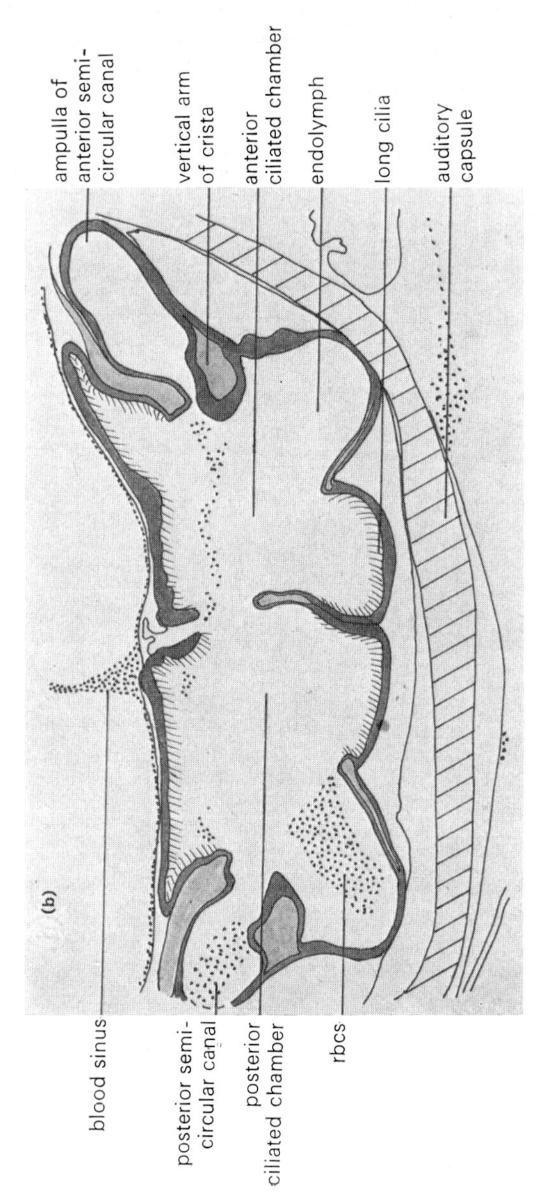

FIG. 11.16 (*a*) VLS and (*b*) drawing of the middle region of the ear of an adult *L. planeri*. Note the long cilia lining the ciliated chambers.

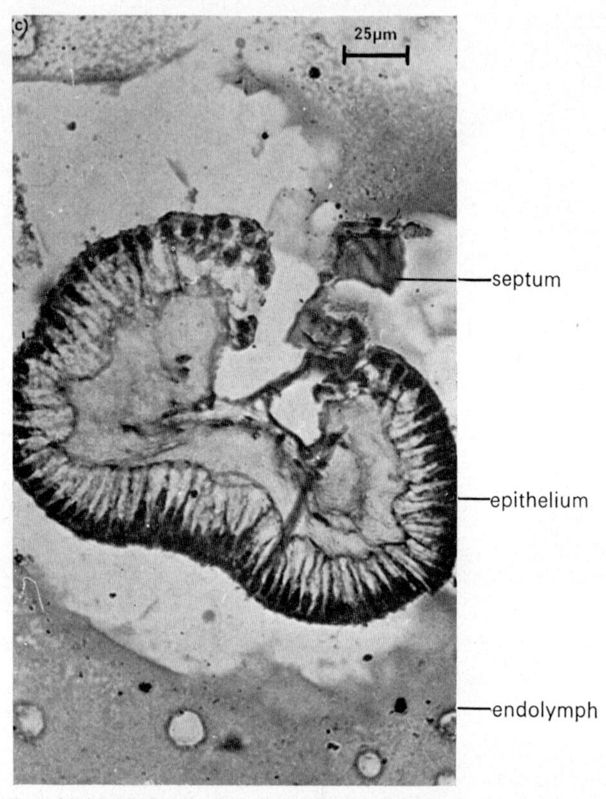

FIG. 11.16 (*c*) detail of the spindle organ from the lamprey ear.

frequency sounds. These are detected primarily in a labyrinth of three otolith organs (a dorsal utriculus, a ventral sacculus and the lagena which is a ventromedial pocket of the sacculus). These chambers have thin epithelial walls strengthened by collagen fibres, and bear raised patches of sensory epithelium (maculae) composed of neuromast cells. These are groups of sensory cells bearing long cilia surrounded by supporting (sustentacular) cells (see Fig. 14.12). The chambers are filled with fluid (endolymph) in which are suspended large calcareous masses (otoliths). These otoliths have a specific gravity slightly greater than that of the endolymph so that they tend to rest on or just above the neuromasts, and, when they are moved about by shock waves in the endolymph, they catch and stimulate the sensitive hairs. Semi-circular canals develop as dorsal folds of the utriculus wall. They are also fluid-filled and are expanded at one end to form ampullae which bear sensory epithelial patches (cristae). Sensitive hairs from these neuromasts project into a large gelatinous cupula (see Fig. 13.9) which is moved about by waves in the endolymph. The utriculus and semi-circular canals are concerned chiefly with maintenance of equilibrium while the sacculus and lagena are concerned mainly with hearing.

The lamprey ear is relatively simple. It is unique among vertebrates in that the central portion of the labyrinth (vestibule) is divided by an incomplete septum into two utricular chambers [Fig. 11.16(a, b)]. These chambers are lined with cuboid epithelial cells bearing long cilia, which create currents in the endolymph. A large otolith chamber lying beneath the ciliated sacs is partially divided into anterior and posterior chambers corresponding to the sacculus and lagena of higher vertebrates, but bears a single *macula communis* and encloses a single large calcareous otolith mass. The lamprey labyrinth bears only two semi-circular canals which join together to open into the vestibule. At the point where they meet there is an epithelial fold which is reinforced by a rigid spindle-shaped rod (spindle organ) shown in Fig. 11.16(c). It has been suggested that the function of the spindle organ is to act as a vane, preventing endolymph movements in the canals from interfering with one another and with movements in the vestibule. The ears are enclosed by cartilaginous auditory capsules (see Fig. 11.1).

The lateral line system acts as a "near-field acoustic detector", that is a displacement detector for shock waves, current and low frequency sounds produced close to the animal. In cyclostomes the lateral line is represented by small pigmented pits on the head and along several lines down the trunk. The pits contain neuromast cells connected by cranial nerves to a poorly developed acoustic area in the hindbrain.

Lampreys have a keen sense of smell. A single median olfactory organ, formed by fusion of paired nasal sacs, leads to the outside by a single nostril (see Fig. 11.6). The posterior walls of this organ are ridged and folded into lamellae which are lined with typical vertebrate olfactory epithelium [Fig. 11.17(a)]. This consists of ciliated bipolar neurones whose axons form the olfactory nerves (cranial nerve I), interspersed with sustentacular supporting cells and small irregular basal cells [Fig. 11.17(b)]. Situated just behind and ventral to the olfactory organ is a glandular accessory olfactory organ [Fig. 11.17(c)] surrounded by a blood sinus. It consists of follicles, lined with cubical epithelium and filled with colloidal material, which open directly into the posterior end of the olfactory organ. It probably secretes mucus to help keep the olfactory epithelium moist.

Endocrine organs

A long nasal-hypophyseal sac (Rathke's pouch) extends backwards from the nostril and olfactory organ, underneath the brain to contact the evaginated floor (infundibulum) of the diencephalon (Fig. 11.18). Eventually this fused structure forms a compact mass of tissue which becomes the pituitary gland. It remains in contact with the nasal pouch by strands of tissue, and differentiates into a glandular portion derived from Rathke's pouch and a nervous portion from the brain. They are known in the lamprey as the adenohypophysis (divided into pro-, meso- and meta-adenohypophyses) and neurohypophysis (Fig. 11.19). These regions are not directly comparable with the three or four lobes of other vertebrates. There is very little histological differentiation between the two parts, and there is probably no vascular connection between them. The lamprey pituitary is characterized by the presence of large cysts, connected by small canals to the nasohypophyseal sac. The pituitary of vertebrates secretes numerous hormones, some of which are neurosecretions while others are glandular secretions. It has often been called the "master-gland" or the "leader of the endocrine orchestra" because of its overall controlling function of other endocrine organs.

In addition to the pituitary, vertebrates possess a large number of other ductless glands, or endocrine organs producing hormones. In the adult lamprey a follicular thyroid gland develops from the larval endostyle (see Fig. 11.22). Its cells concentrate iodine and bind it to amino acids, thus forming the thyroid hormones which are concerned with controlling metabolism. The thyroids of all vertebrates have a similar structure (see Fig. 14.15) and store these hormones inside the follicles attached to a colloidal protein. In the lamprey thyroid follicles lie scattered along the jugular veins below the pharynx, in a similar position to the endostyle.

Vertebrate "adrenals" consist of two tissues which may be separated or closely applied, and are usually associated spatially with the kidneys and gonads. Inter-renal tissue, of mesodermal origin, generally consists of interlacing clusters or cords of cells interspersed with blood-vessels. It secretes steroid hormones active in the control of carbohydrate metabolism, water and electrolyte balance, reproduction and stress. Chromaffin tissue

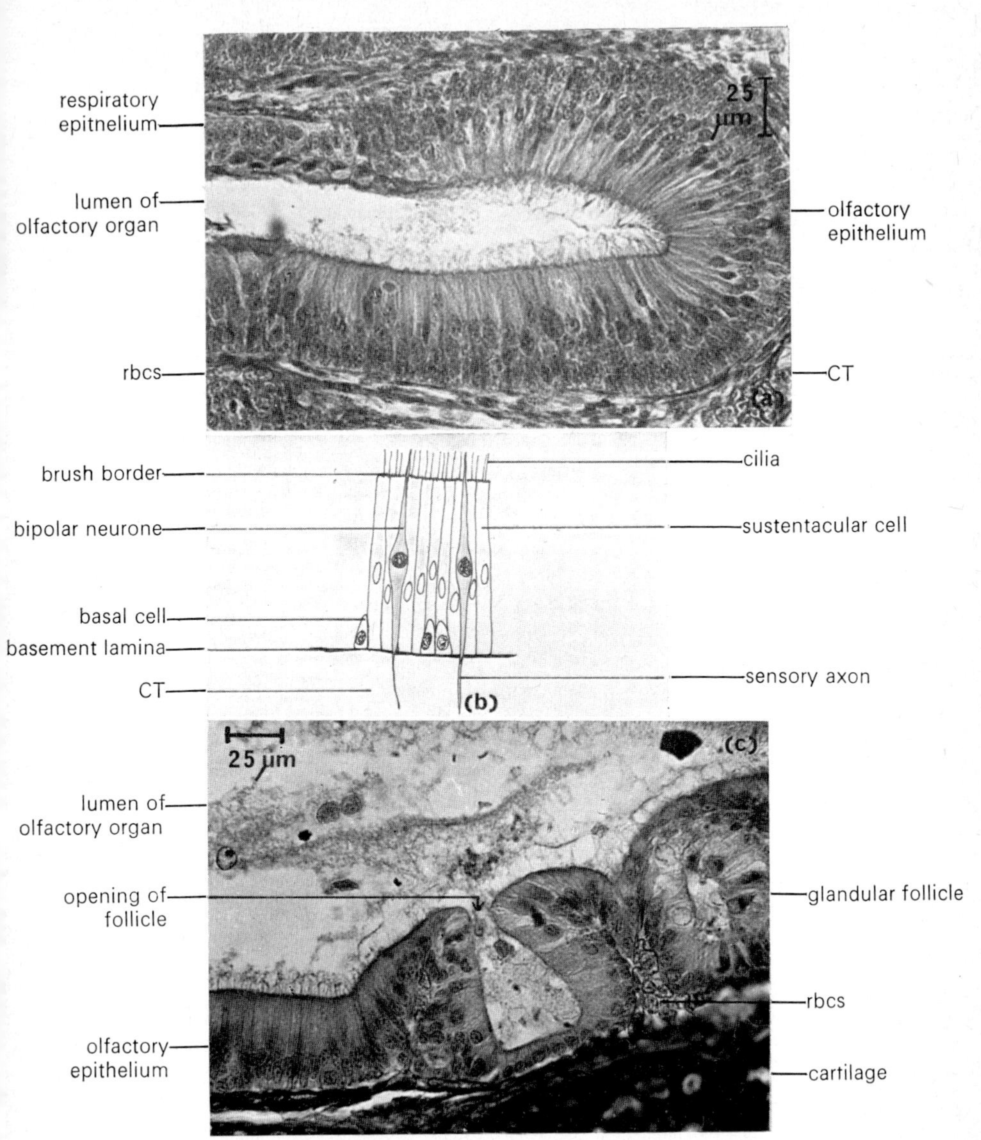

FIG. 11.17 (a) Sensory epithelium of the olfactory organ from an adult *L. planeri*; (b) drawing of a typical vertebrate olfactory epithelium. (c) section through glandular follicles of the accessory olfactory organ of *L. planeri*. Note that the follicles open by short ducts into the olfactory organ.

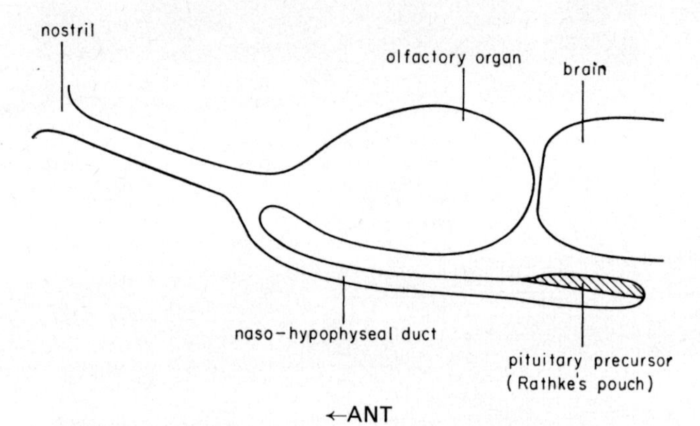

FIG. 11.18. Diagram showing the position of the nasohypophyseal sac in the lamprey.

(forming the supra-renals of some lower vertebrates) is derived from neuroectoderm and consists of large irregular granular cells which produce neurosecretions involved in the control of metabolism and fright reactions. The term "adrenal" is used only for those higher vertebrates in which the two tissues form a single gland. In cyclostomes the two tissues are separate; inter-renal tissue occurring as cell clusters along the major abdominal veins, particularly in the region of the kidney, while the chromaffin cells are grouped round the dorsal aorta.

Other endocrine organs present in the lamprey include the pancreas and gonads which will be dealt with in the sections on digestion and reproduction respectively.

Respiration

Lampreys respire using gills (see Figs 11.1 and 11.2). In the larva the pharynx leads from the oesophagus to the gut, and is perforated by small round openings with gill plates between them. During metamorphosis the pharynx becomes divided to form, in the adult, a blind-ended respiratory pharynx lying below a pharyngo-oesophagus (see Fig. 11.24). This is highly atypical of vertebrates. The pharynx is constricted internally into seven pairs of pouches, each separated from the next by a wide inter-branchial septum bearing blood-vessels. The pouches have muscular walls which draw water in through the mouth to the respiratory pharynx, and pump it out through valves guarding the openings of the gill pouches. The respiratory epithelium lining the pouches is folded into gill filaments bearing secondary lamellae (platelets), with large cells between them which are involved in salt regulation (Fig. 11.20). The platelets are well-vascularized and are the site of respiratory exchange.

Circulation

The circulatory system of lampreys follows the same general plan as in *Amphioxus*, except that in lampreys, as in higher vertebrates, there is a heart to pump blood round the circulatory system and the blood system is completely closed. Blood passes from arteries to veins via small, thin-walled capillaries which are normally the site for exchange of nutrients, metabolites and respiratory gases between the blood and the tissues.

The heart consists of four chambers: a small thin-walled sinus venosus, two muscular chambers (atrium and ventricle) and a small contractile conus arteriosus, formed by a slight thickening of the proximal end of the ventral aorta. (The atrium is often called the auricle, but this term should be reserved for the auricle of the mammalian heart which has a different origin). In the ammocoete larva the heart is straight and thin-walled, while in the adult it is folded into a compact S-shaped structure and the walls of the atrium and ventricle are very muscular (Fig. 11.21). This change at metamorphosis is associated with increased resistance to blood

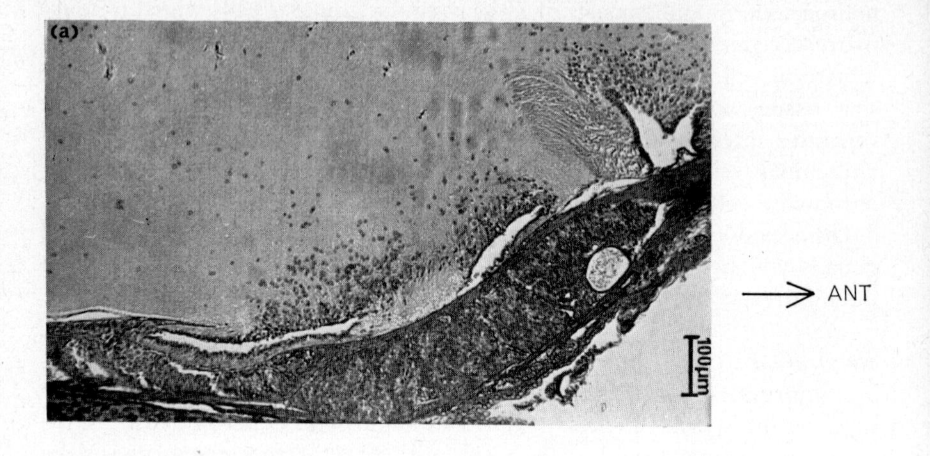
(a)
ANT
100µm

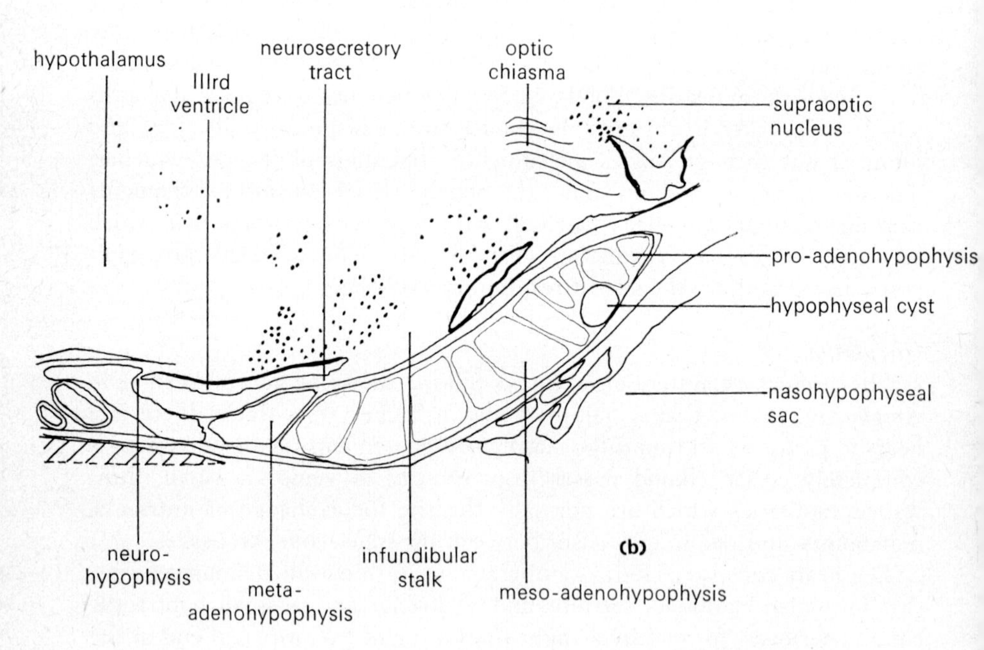
hypothalamus
IIIrd ventricle
neurosecretory tract
optic chiasma
supraoptic nucleus
pro-adenohypophysis
hypophyseal cyst
nasohypophyseal sac
(b)
neuro-hypophysis
meta-adenohypophysis
infundibular stalk
meso-adenohypophysis

FIG. 11.19

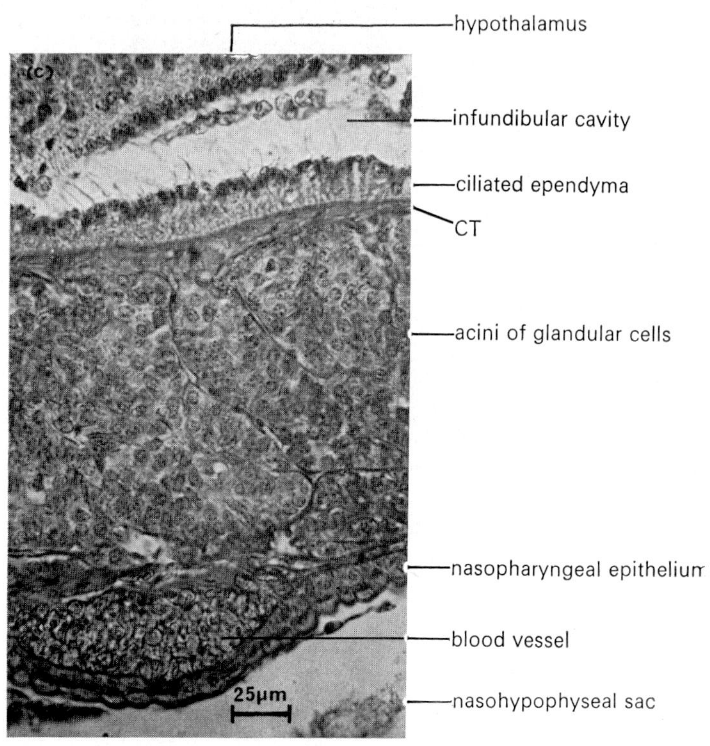

FIG. 11.19 (a) VLS and (b) drawing of the pituitary and hypothalamus of the adult *L. planeri*; (c) detail of the secretory meso-adenohypophysis of the pituitary of *L. planeri*.

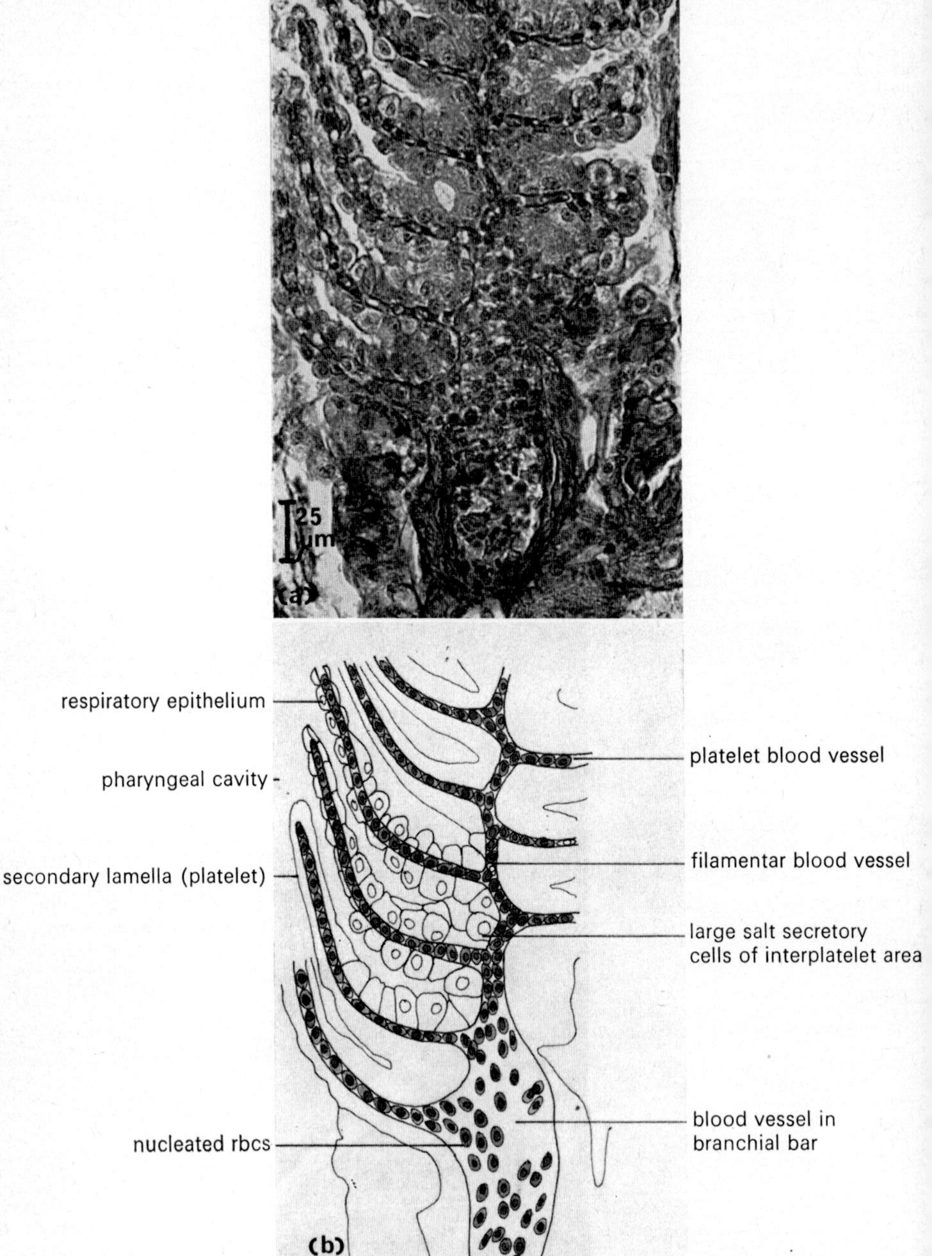

FIG. 11.20 (a) LS and (b) drawing of a gill filament from an adult *L. planeri*.

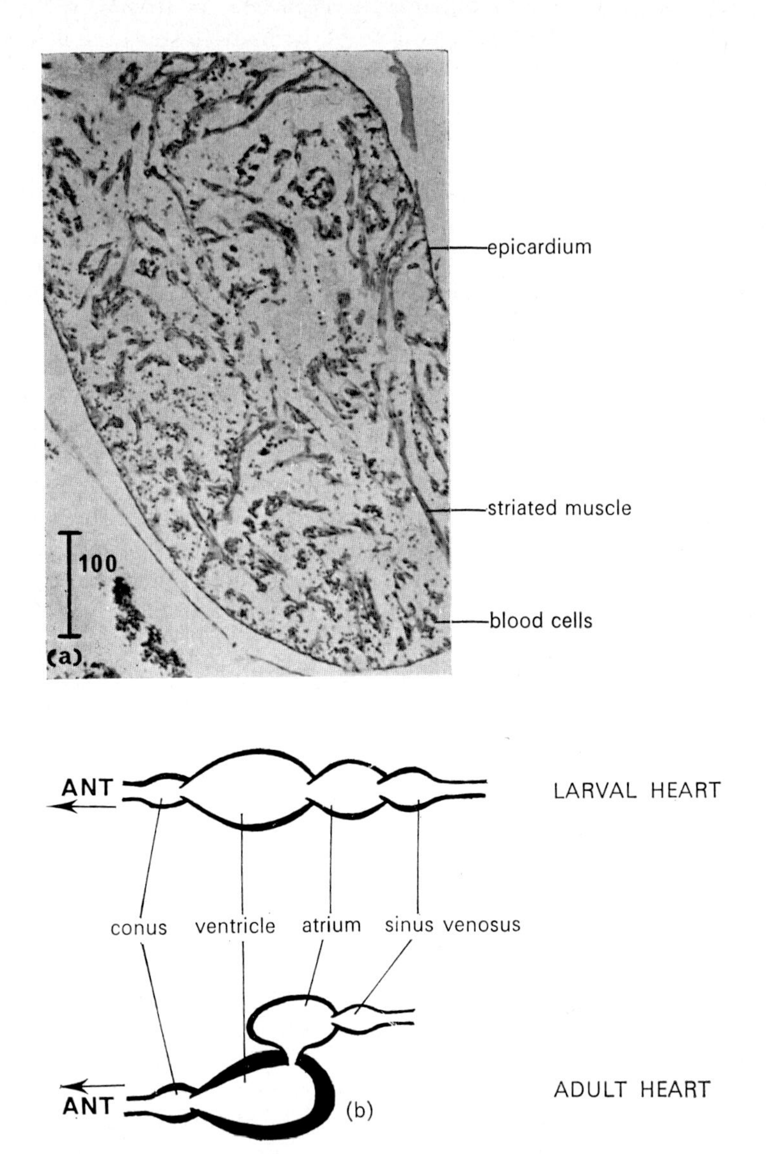

FIG. 11.21 (a) Oblique TS through the heart ventricle of a larval *L. planeri*. Note the thin muscular walls; (b) diagram showing the arrangement of the larval and adult hearts in the lamprey in VLS.

flow by the small-bore capillaries, particularly at the finely divided respiratory surface. Blood returning from the body in veins is collected in the sinus: the four heart chambers then contract in sequence, forcing the blood in an anterior direction into the arteries. Valves present within the heart and the blood-vessels prevent backflow of blood. The heart is suspended in a division of the coelom (pericardium) which has a thick wall, is supported by a cartilaginous skeleton and retains a connection with the body cavity.

All vertebrates including the lamprey possess an hepatic-portal circulation, whereby blood is taken directly from the intestine to the liver (where digestive products can be used or stored) before being returned to the heart. A second portal system is found in many vertebrates (the renal-portal system) which conveys venous blood to the kidneys before returning it to the heart. Such a system is, however, not present in lampreys (or mammals); the kidneys in these animals receive an arterial blood supply. The blood of cyclostomes contains a respiratory pigment which is housed within large (25–30μ) nucleated red blood corpuscles (erythrocytes), and which increases the efficiency of oxygen carriage by the blood. The erythrocytes are manufactured in the kidney or the wall of the intestine: there are no special organs set aside for their production. White blood cells (leucocytes) also occur, but their site of origin is not known. They are involved in combating infection, inflammation and injury.

Digestion

Ammocoete larvae are filter-feeders, relying on muscular action of the velar flap and branchial basket to create feeding currents; unlike *Amphioxus* which uses cilia. A complex saccular endostyle (or sub-pharyngeal gland), lying at the base of the pharynx, consists of a series of clumps of mucous cells (glandular tracts) opening by short ducts into a branched lumen (Fig. 11.22, see also Fig. 11.2). The lumen is divided by a septum into two anterior chambers and four posterior chambers. Copious supplies of mucus are squeezed from the endostyle lumen, through a median duct, into the pharyngeal cavity. Here they entangle small food particles which are then passed on to the intestine by the action of pharyngeal muscles and by the cilia of a peripharyngeal groove which runs round the anterior end of the pharynx. The larval intestine is lined throughout by a single layer of epithelium composed of longitudinal tracts of ciliated mucous cells and enzyme secreting cells (Fig. 11.23). It is the principal site of digestion and absorption.

In general, adult lampreys are ectoparasites and have a relatively simple gut, shown in plan in Fig. 11.24. The anterior mouth is a circular opening surrounded by a muscular sucker and fringed with fleshy papillae armed with rows of horny epidermal teeth which are used for rasping. Lying in the floor of the buccal cavity is a large tongue, provided with complex

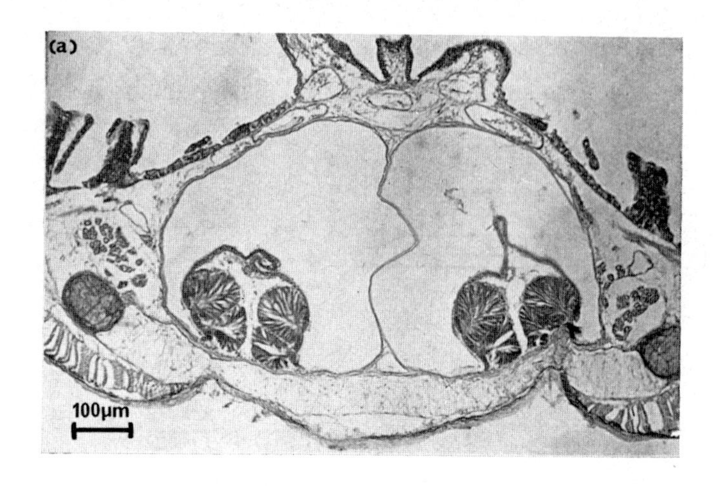

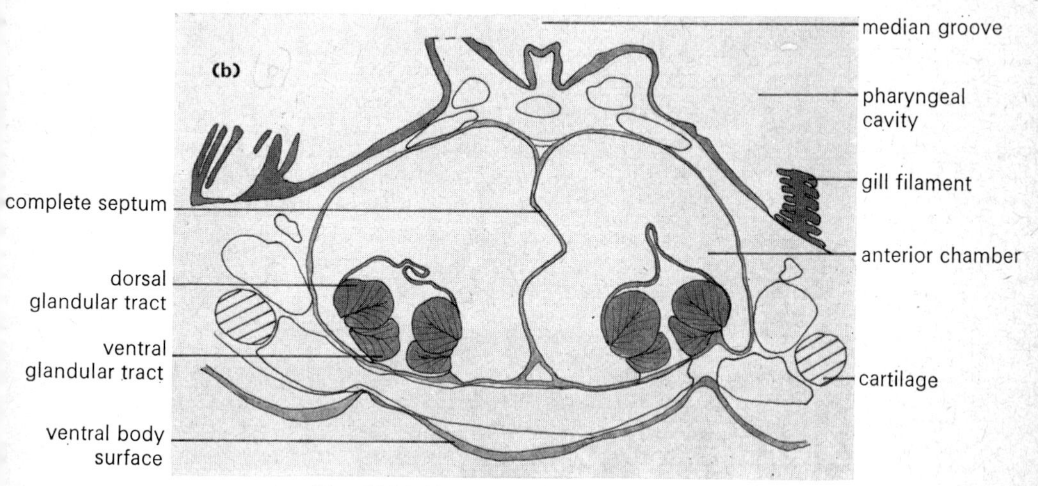

FIG. 11.22 (*a*) and (*b*) see p. 467 for legend.

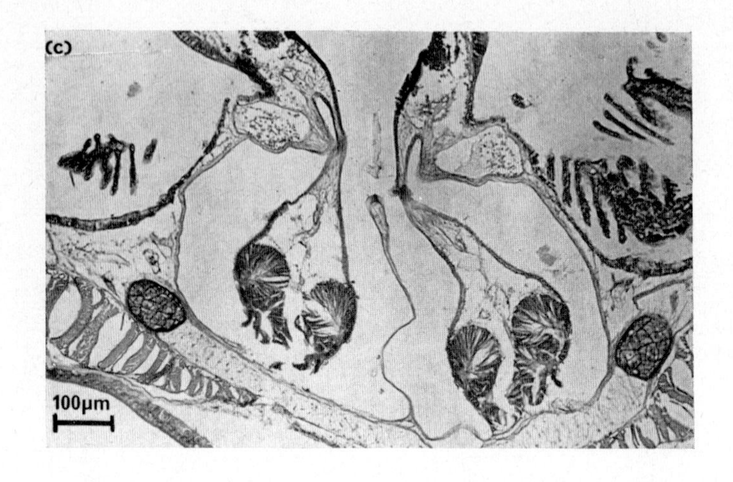
(c)
100μm

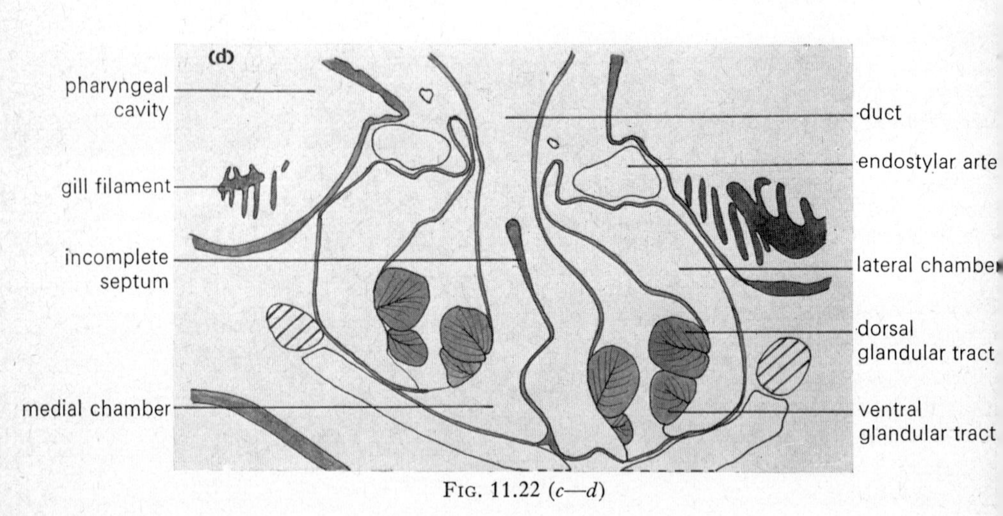
(d)
pharyngeal cavity
gill filament
incomplete septum
medial chamber
duct
endostylar arte
lateral chambe
dorsal glandular tract
ventral glandular tract

FIG. 11.22 (*c—d*)

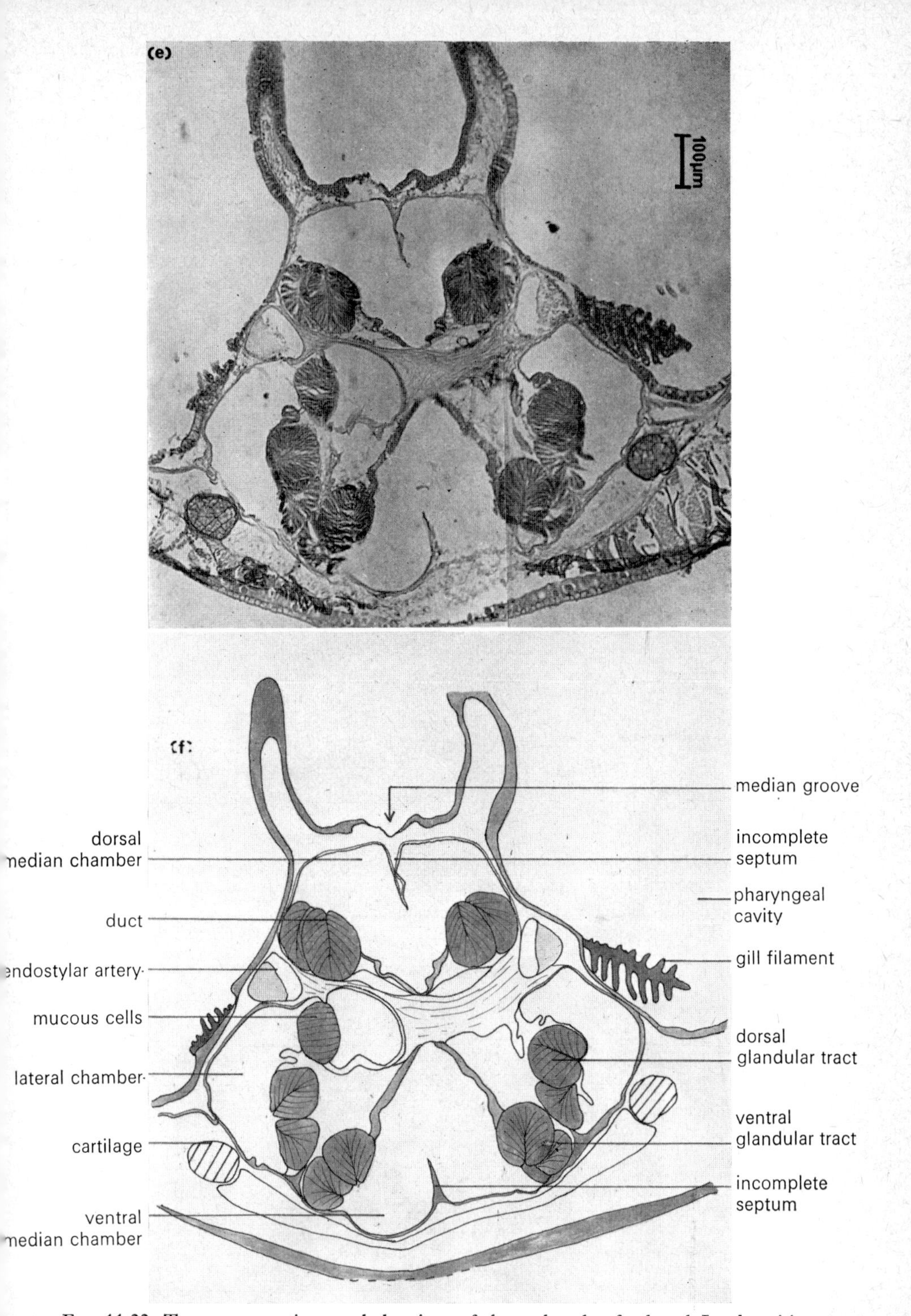

FIG. 11.22. Transverse sections and drawings of the endostyle of a larval *L. planeri* in (a, b) the anterior region, (c, d) the middle region (passing through the main duct), and (e, f) the posterior region.

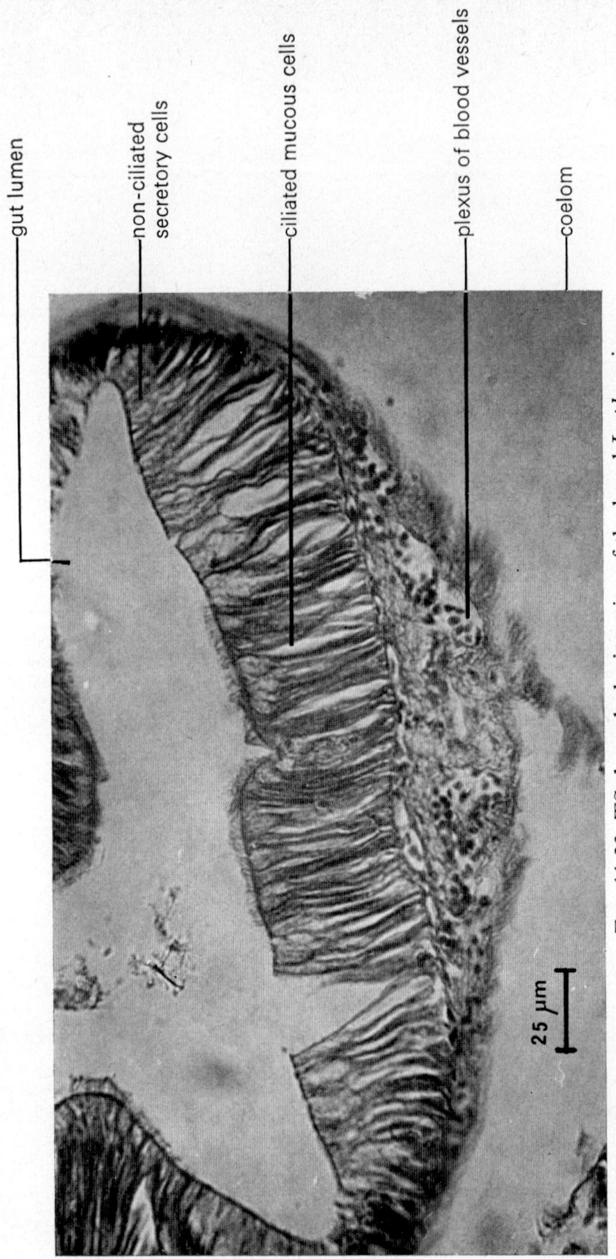

FIG. 11.23. TS through the intestine of the larval *L. planeri*.

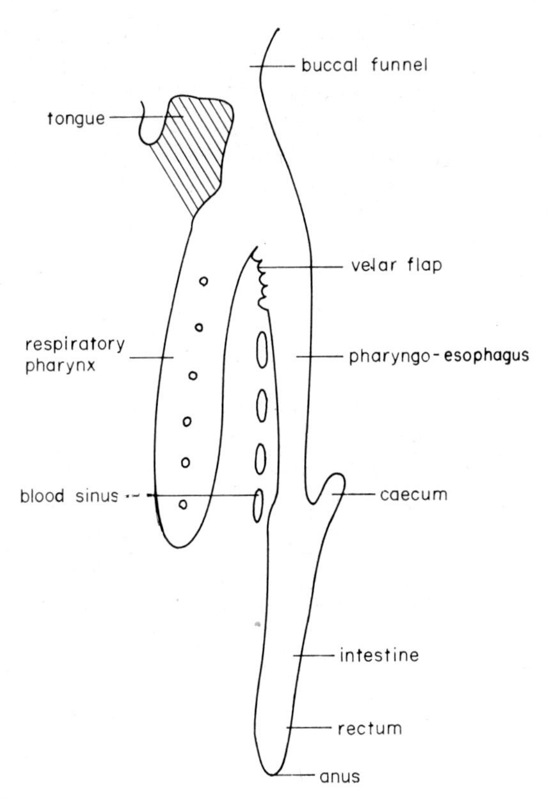

Fig. 11.24. Plan of a typical adult lamprey digestive tract.

musculature and mucous cells [Fig. 11.25(a)]. In some species it bears teeth, but not in *L.planeri* which is not an ectoparasite although it has been seen attached to other lampreys or fish, particularly during spawning. Embedded in the tongue are paired saccular "salivary glands" with folded epithelial walls [Fig. 11.25(b)] which open by narrow ducts to the base of the tongue. They are said to produce an anticoagulant, although this again is unlikely in *L.planeri*. Behind the buccal cavity the ventral respiratory pharynx is guarded by a fleshy velar flap with cartilaginous supports, which is probably sensory in function in the adult. The dorsal pharyngo-oesophagus, a straight tube lined with simple cubical epithelium leads by a narrow slit into the intestine. There is no true stomach in cyclostomes, but the intestine is normally expanded slightly at its anterior end, and is also widened at the hind end to form a rectum. The gut lies freely in the body cavity, attached to the body wall by ligaments at each end. The intestine of the adult is similar to that of the larva, with a fold (typhlosole) running down it in a spiral fashion, forming the "spiral valve". As in the larvae, food is moved through the gut by ciliary action. In *L.planeri*, the adult lives for only a short time and does not feed: its intestine is occluded by tissues at the anterior end.

In both the larva and adult lamprey there is a conical liver with a gall bladder embedded in it; both of which open by ducts into the intestine. The liver is composed of cords of liver cells, enclosing a branched network of minute bile channels (canaliculi), interspersed with blood sinusoids (Fig. 11.26). It probably plays a key role in metabolism. Small masses of pancreatic tissue lie closely applied to the wall of the anterior part of the intestine, particularly round the caecum, where they are compacted to form a "cranial pancreas". This pancreatic tissue consists of separate patches of acinar cells, possibly with a digestive function, and endocrine "islet" tissue which probably secretes hormones involved in carbohydrate metabolism.

Excretion

The excretory organs of vertebrates are kidneys, derived from coelomoducts. Kidneys are composed of renal tubules (nephrons) which develop from segmental blocks of mesoderm (nephrotomes). Unlike most other body organs they have to function at a very early stage of development, and therefore have to be able to operate in a variety of different environments during growth. The earliest tubules to appear during development are the most anterior (pronephric) ones, which retain their open connection with the coelom. This can be seen in Fig. 11.27 which shows the convoluted pronephric tubules and ciliated funnel (nephrostome) of an ammocoete larva (see also Fig. 11.28). At metamorphosis the pronephric tubules degenerate and their function is taken over by the more posterior mesonephric tubules, forming a "back kidney" (opisthonephros). The dorsal

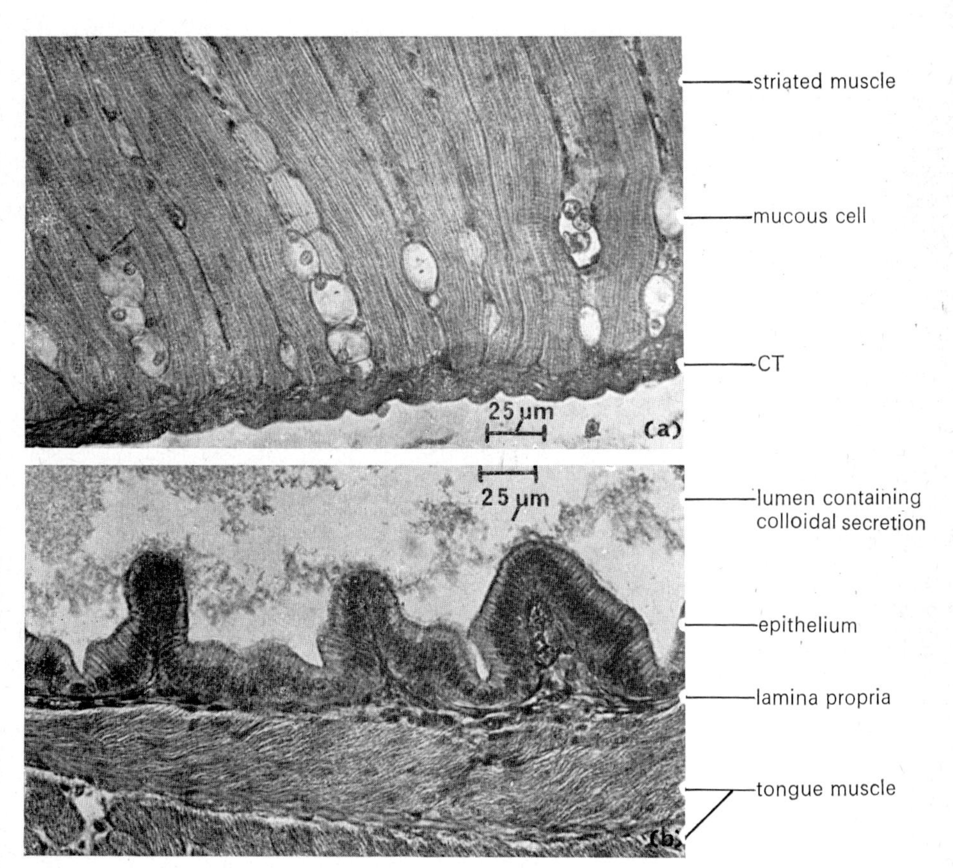

FIG. 11.25 (a) LS through the underside of the tongue of an adult *L. planeri*; (b) section through part of the tongue of an adult *L. planeri* showing the wall of the "salivary gland".

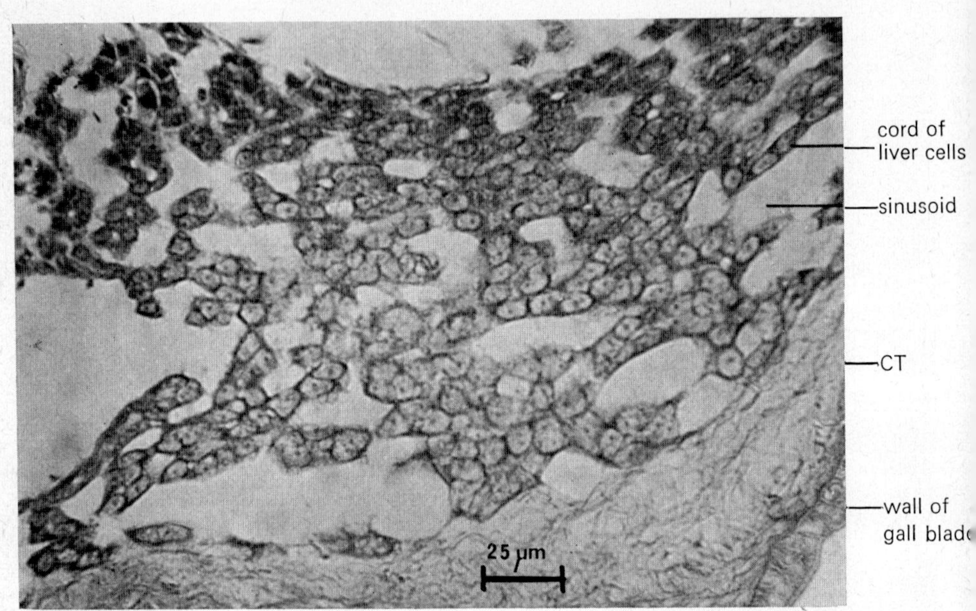

FIG. 11.26. Section through the liver of an ammocoete larva of *L. planeri*.

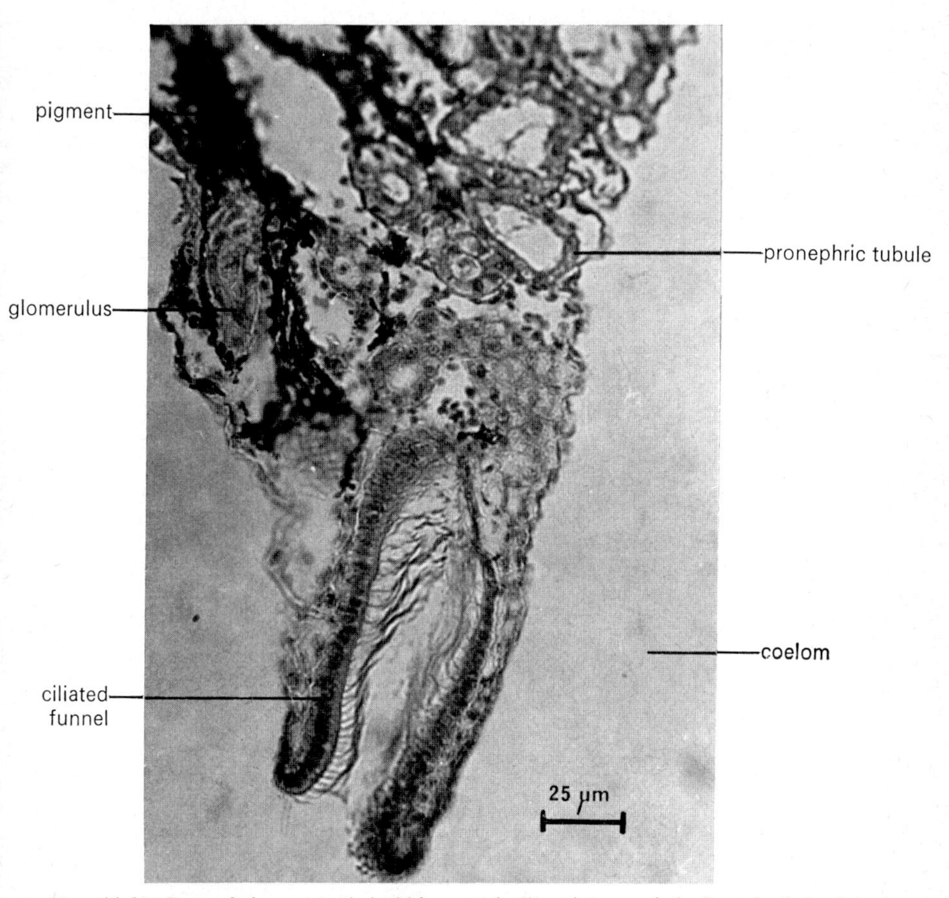

FIG. 11.27. Part of the pronephric kidney and ciliated pronephric funnel of the larval *L. planeri*.

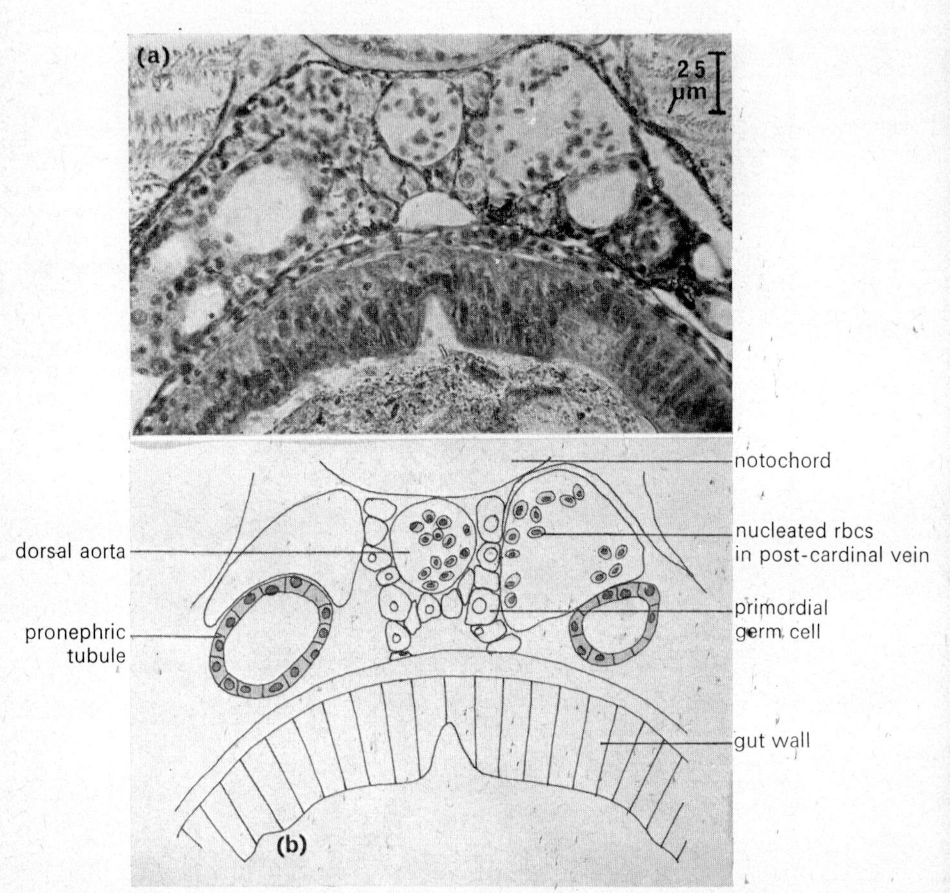

FIG. 11.28 (a) TS and (b) drawing of a young larval *L. planeri* in the dorsal gut region, showing pronephric tubules and the developing gonads.

part is compact and appears to be composed of lymphoid tissue: it is probably the region where blood cells are produced. The ventral part consists of numerous tubules closely associated with a few large tufts of capillaries (glomeruli). These tubules have a similar structure to the pronephric tubules, but they have lost their connection with the coelom and terminate as blind-ended renal capsules, which filter off fluid from the blood in the glomeruli by ultrafiltration. This fluid, containing ammonia, is discharged into paired mesodermal archinephric ducts which run along the free edge of the kidneys and eventually open into a small urinogenital sinus behind the rectum. The sinus opens, by a small papilla, into a cloacal pit which also contains the anus.

Salt and water regulation occurs in the kidneys and across the gills (see Fig. 11.20). In the marine lamprey this function is assisted by salt secretory glands similar to those occurring in reptiles.

Reproduction

Cyclostomes are dioecious, each individual possessing a single large medial gonad suspended in a sheet of peritoneum (mesentery). The gonads develop as invaginations from median genital ridges of mesoderm into the abdominal cavity (Fig. 11.28). They are very simple in form, with the germinal epithelium covering the surface of the gonad.

In the male, the testis consists of clumps of germ cells forming follicles. They are unique among vertebrate seminiferous tubules in having no ducts but merely rupture and release ripe sperm into the body cavity. In the female the germ cells are surrounded by a single layer of follicular epithelial cells with a nutritive function. There is no division of the ovary into cortex and medulla as seen in higher vertebrates, and there are no ducts to the outside.

Both sets of gametes are shed into the coelom, and travel via genital pores into the urinogenital sinus, and thence to the cloacal pit, and outside. Fertilization is external. In *L.planeri* there is an unusually long larval stage, followed by late differentiation and development of the gonads, and a short adult stage.

Chapter 12

Class Chondrichthyes

All vertebrates other than agnathans have jaws, and are known as gnathostomes. They are further distinguished from other chordates by their paired limbs, increased cephalization, and by the possession of three semi-circular canals in the ear.

Chondrichthyes are cartilaginous fish with jaws. Living members belong either to the elasmobranchs (i.e. the sharks, rays and related forms) or to the holocephalans (i.e. chimaeras). Elasmobranchs are almost all marine, except for a few species which have secondarily invaded freshwater. They are either fast-moving predators or sluggish bottom-living scavengers, the latter being dorso-ventrally flattened.

Example: *Scyliorhinus* (*canicula*)

The common dogfish is an elasmobranch belonging to the order Selachii, or the typical present-day sharks. Selachians are characterized by their narrow-based fins, claspers formed from the pelvic fins, and the possession of yolk-laden eggs which are often encased in horny shells (although in several cases the young are hatched inside the mother's body and are born alive). The dogfish is a carnivore and has a long flexible muscular body. The flattened head and upturned (heterocercal) tail act against one another to allow the fish to swim horizontally against the effects of gravity. Stability in the water is given by the paired lateral fins (pectoral and pelvic) and the two triangular dorsal fins and single ventral fin. The anterior edges of the pectoral fins are free, so that they are relatively mobile and can make fine adjustments to movement. The dogfish has paired eyes, and two nostrils which are situated in front of the mouth. Just behind the eyes lie the spiracles; each representing a first gill slit containing a rudimentary gill (pseudobranch). Behind the spiracles lie five pairs of laterally placed gill slits which open separately to the surface (while in the Holocephali all the gill openings are covered by a flap). Figure 12.1 is a TS of a young dogfish in the branchial region, and shows that the gills are extensive and of laminar type, with filaments fixed to the interbranchial septae. The most anterior branchial arch is modified to form the jaws. The hind end of the upper jaw is slung from the cranium by a stout hyomandibular muscle (rather than fused to it as in holocephalans), which results in powerful mobile jaws that are used for catching prey, defence and other activities. The cranium is a stout, cartilaginous box with nasal and auditory capsules fused onto it. The body musculature is well developed and arranged in characteristic W-shaped myotomes, shown in section in Fig. 12.2 which is a TS through the intestinal region of a young dogfish.

Dogfish are protected by a tough skin covered by dermal denticles or placoid scales with backwards-directed points (Fig. 12.3). These denticles bear a resemblance to teeth, and are larger on the dorsal surface of the body. They are continuously replaced from underneath. Colour patterns on the skin are provided by dermal chromatophores.

Cartilage is widespread as a skeletal tissue in elasmobranchs. It shows the typical form of hyaline cartilage, with chondrocytes in lacunae embedded in an extensive dense matrix (Fig. 12.4). This produces a tough yet light-weight and flexible structure. A notochord occurs in the embryo, but it is gradually replaced by a vertebral column. The outermost coat of the notochord sheath is invaded by spindle-shaped mesenchyme cells, which are arranged in a circular fashion to form the perichordal tube [Fig. 12.5(a)]. This later becomes chondrified and converted into a rigid, vertebral rod (spinal column). This rod is composed of non-articulated, biconcave (amphicoelous) vertebral bodies (centra) linked together by the remains of the notochord and by intervertebral cartilages [Fig. 12.5(b)]. Each vertebra bears processes above and below the centrum. These are the neural and haemal arches which protect the nerve cord and the dorsal aorta respectively. The arches can be seen in Fig. 12.6, which is a TS through the posterior region of the body. Note the heavy musculature in this region, which provides the main locomotor thrust for the dogfish. Also visible is the supporting skeleton for a dorsal fin, which takes the form of a central cartilaginous supporting rod and horny dermal rods (ceratotrichia).

Nervous system

The central nervous system of elasmobranchs is much more advanced than that of cyclostomes, with a large brain (Fig. 12.7). Development of the forebrain (cerebrum) pushes the olfactory lobes to a more anterior position, and partially takes over their function. A keen sense of smell is associated with large olfactory organs and lobes. Sight is less important, and the optic tectum is fairly small. The dogfish depends a great deal on the acoustico-lateralis system and on internal proprioceptive information. The area receiving nerves from these two sources, the cerebellum, is very large and an important integration centre. It is the only area of the dogfish brain which shows a cortical arrangement of neurones; a notable feature of co-ordination centres (see also Figs 14.7, 15.5 and 17.6). The medulla oblongata contains columns of neurones (such as are seen in the spinal cord), representing control centres for an autonomic nervous system (ANS). The ANS is concerned with control of visceral functions (e.g.

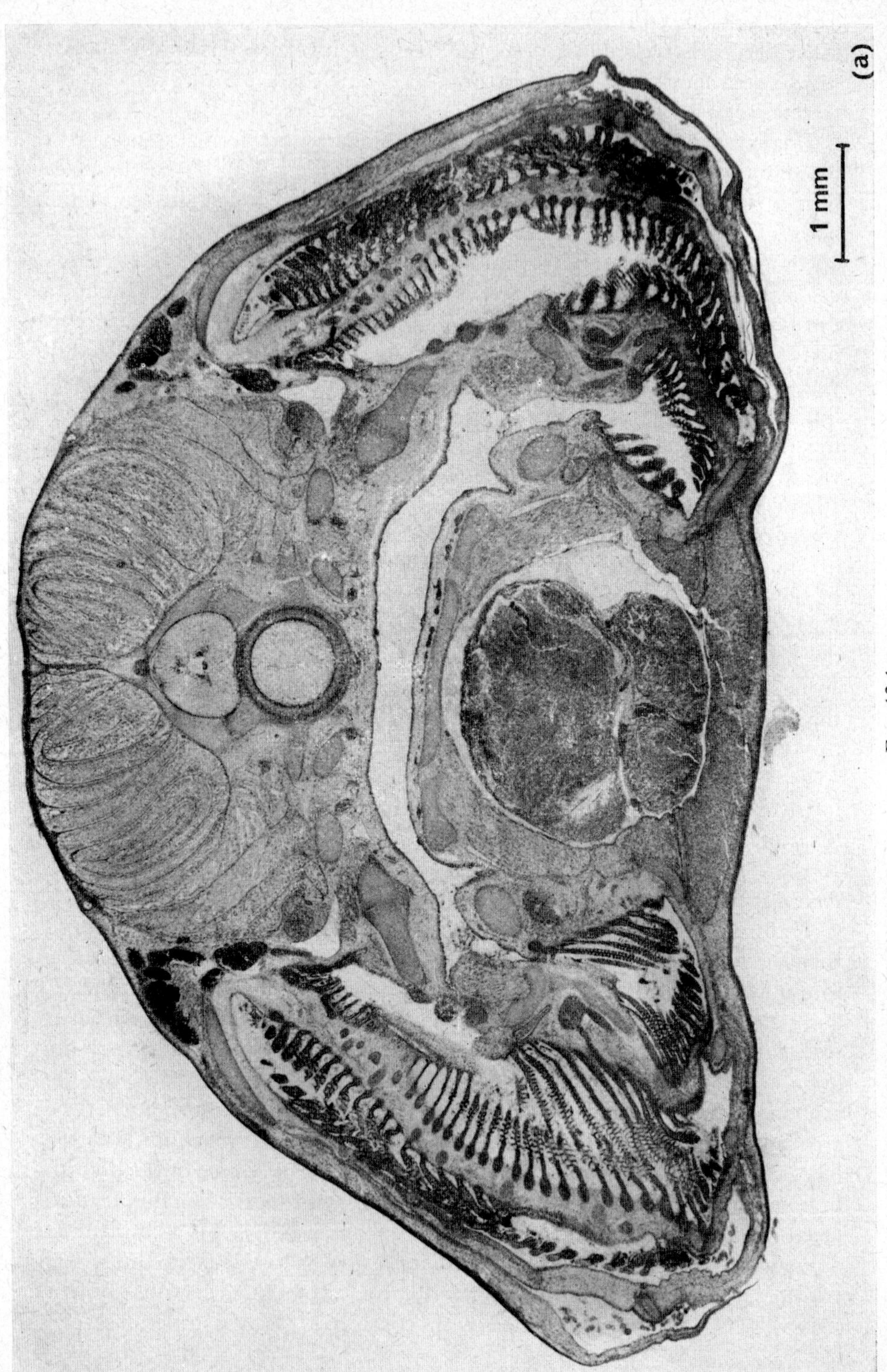

Fig. 12.1

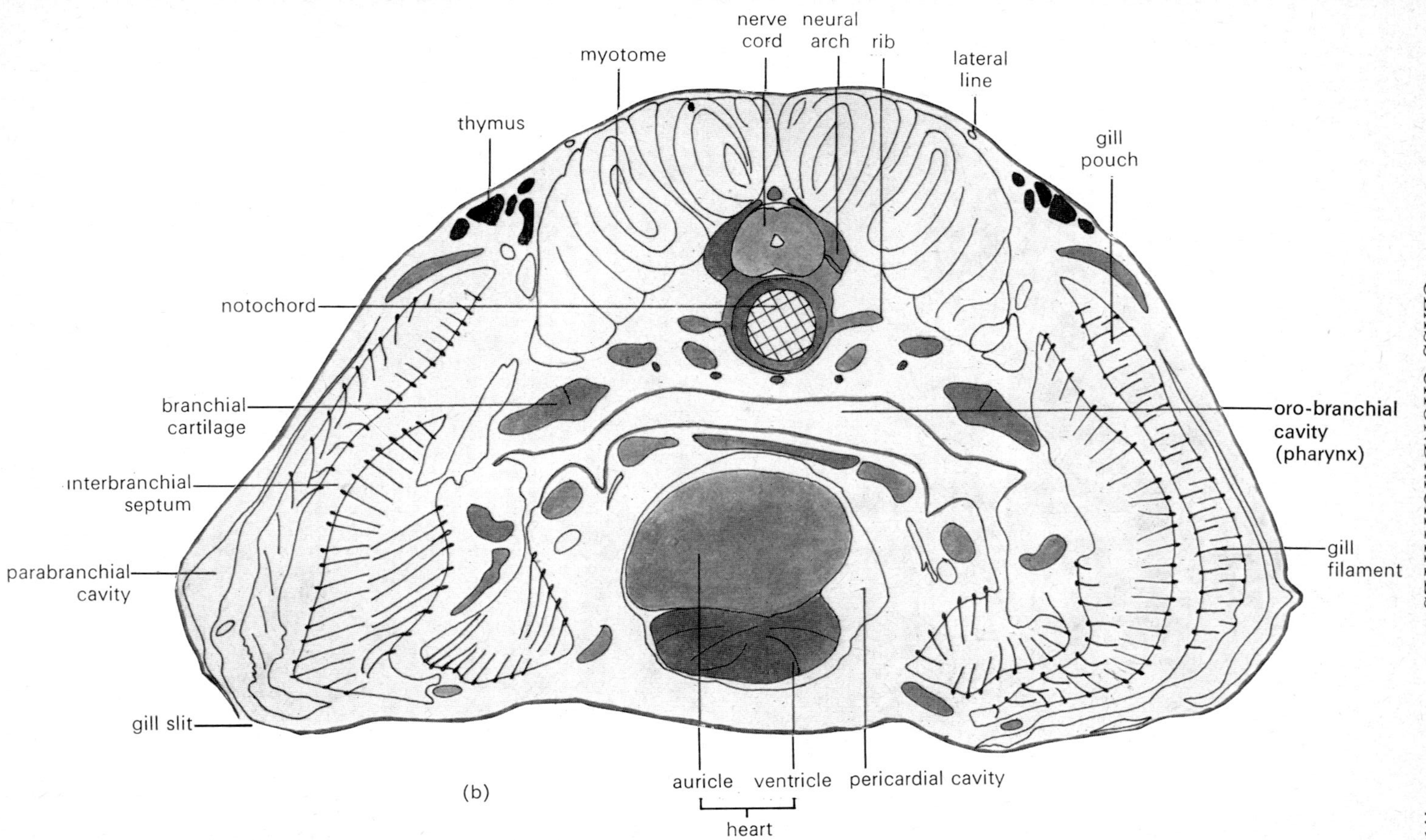

FIG. 12.1 (a) TS and (b) explanatory drawing of a young dogfish (*Scyliorhinus canicula*) in the branchial region.

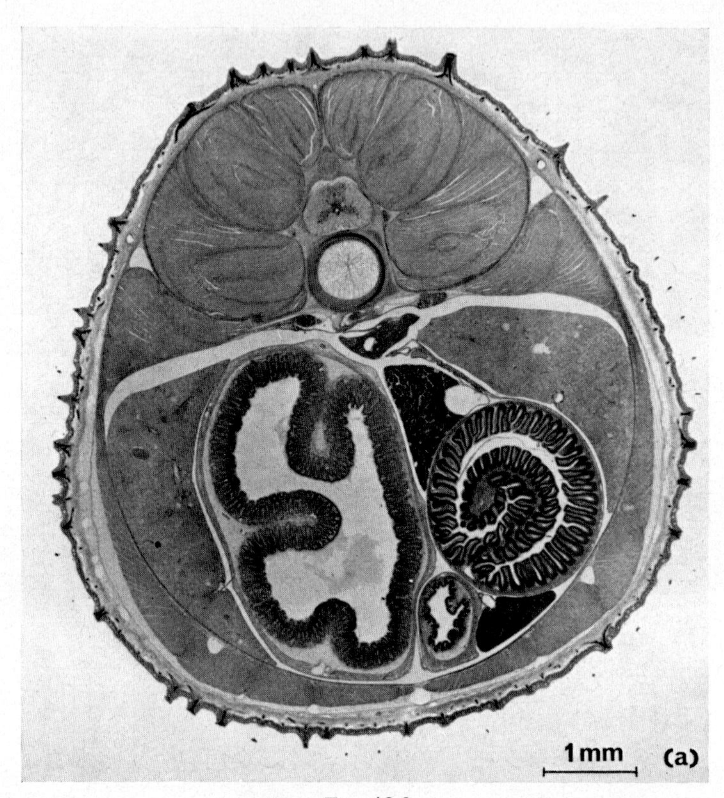

Fig. 12.2

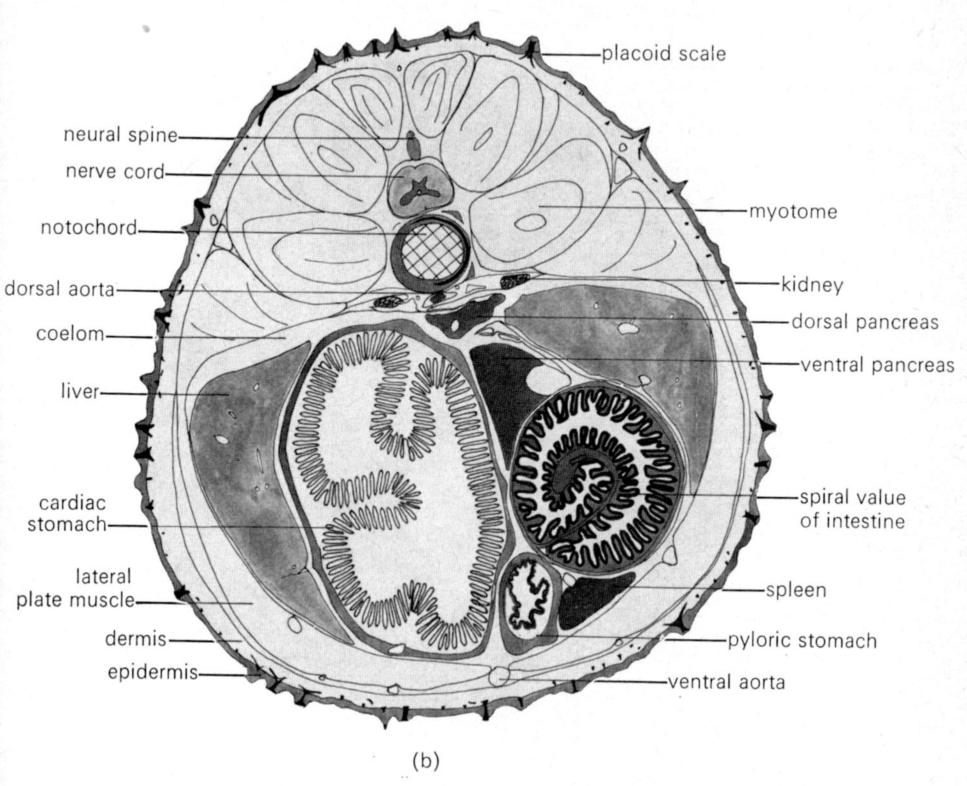

(b)

FIG. 12.2 (a) TS and (b) drawing of a young dogfish in the intestinal region.

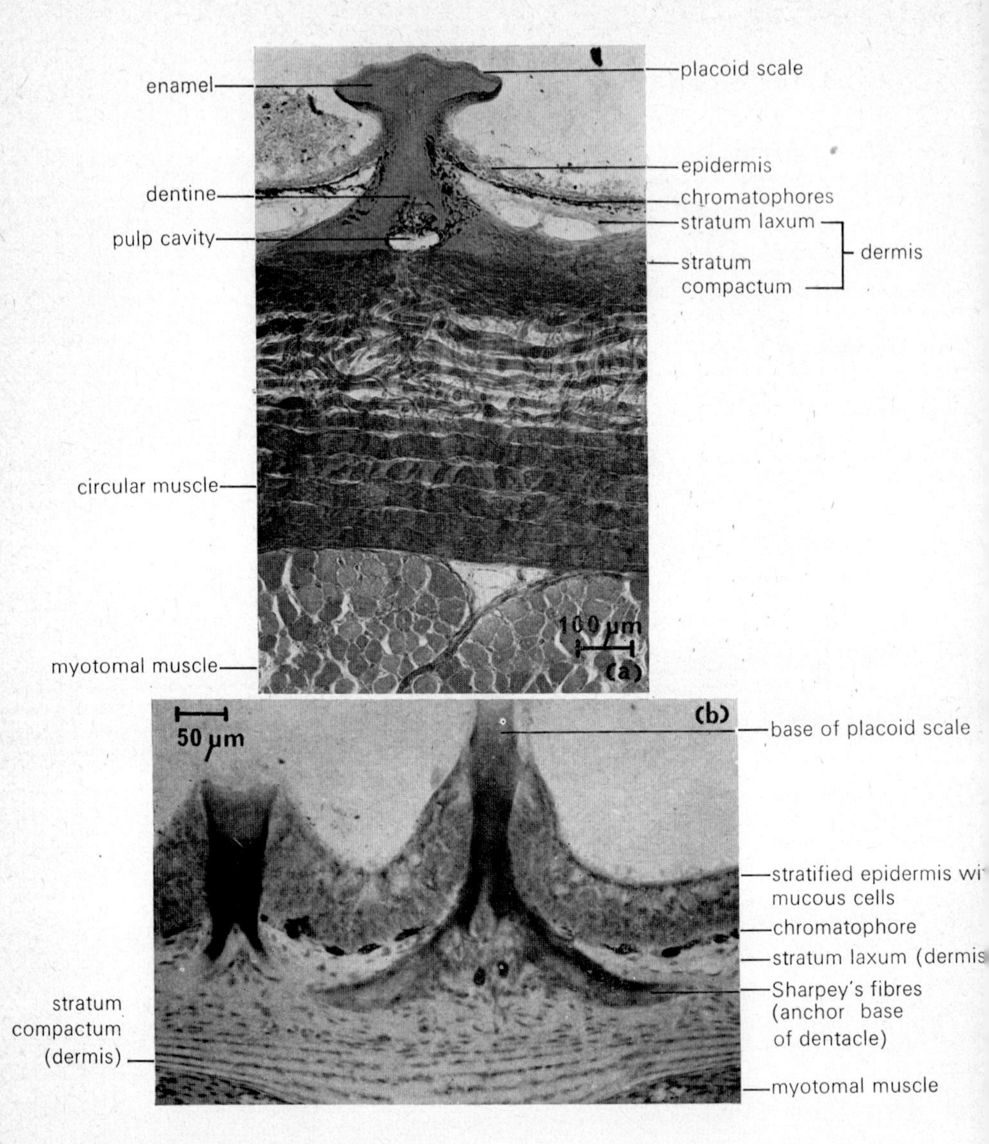

FIG. 12.3 (a) TS through the body wall of a dogfish showing a placoid scale; (b) details of dogfish skin, showing the bases of two placoid scales.

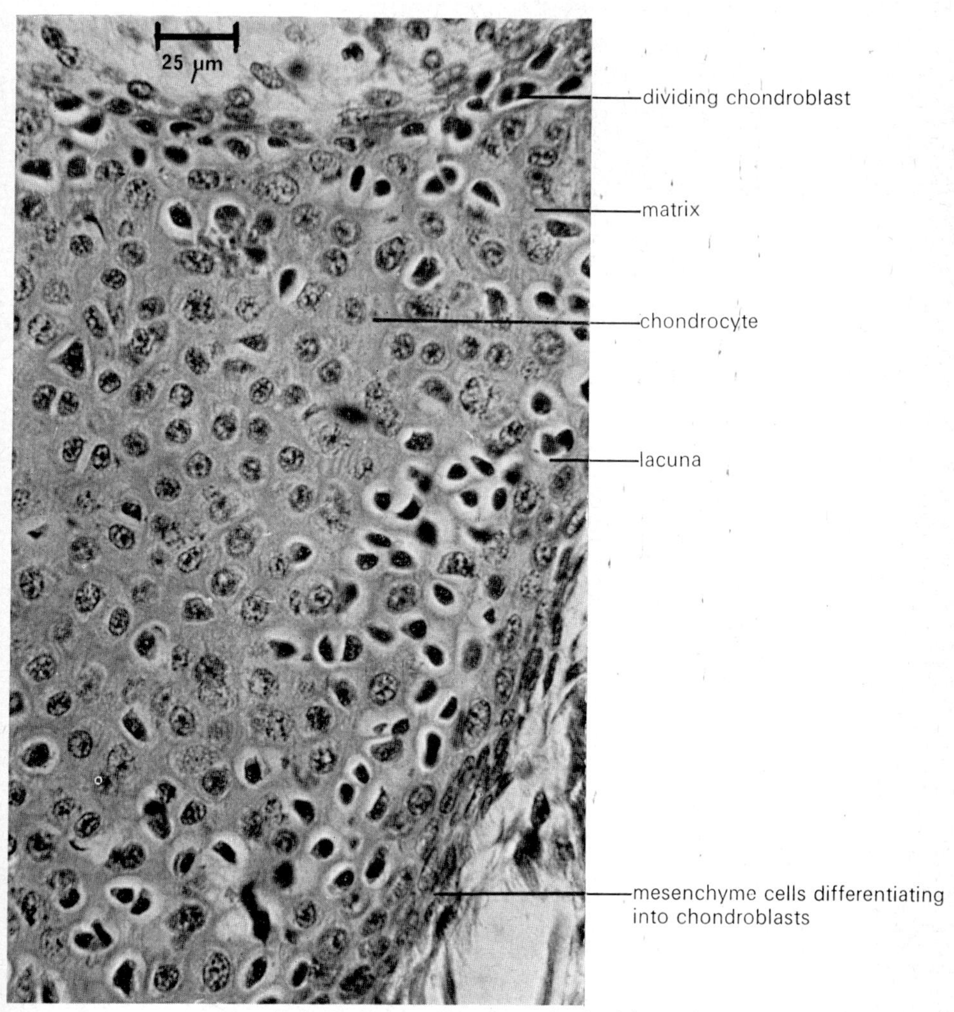

FIG. 12.4. Hyaline cartilage from the dogfish cranium.

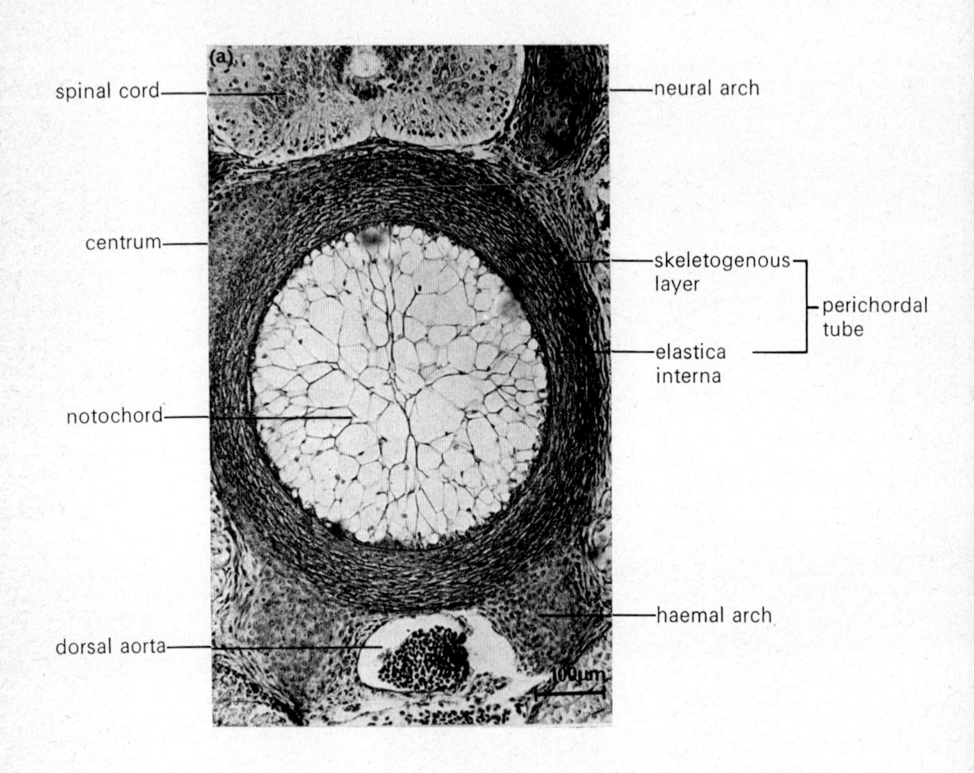

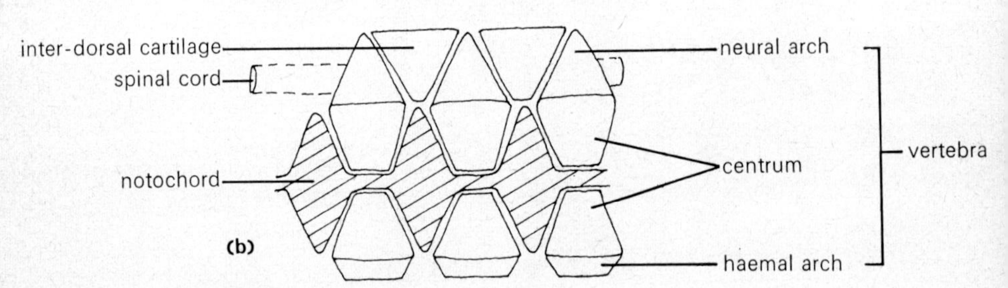

FIG. 12.5 (a) TS through the notochord and perichordal tube of a young dogfish; (b) diagram of a VLS through the dogfish spinal column.

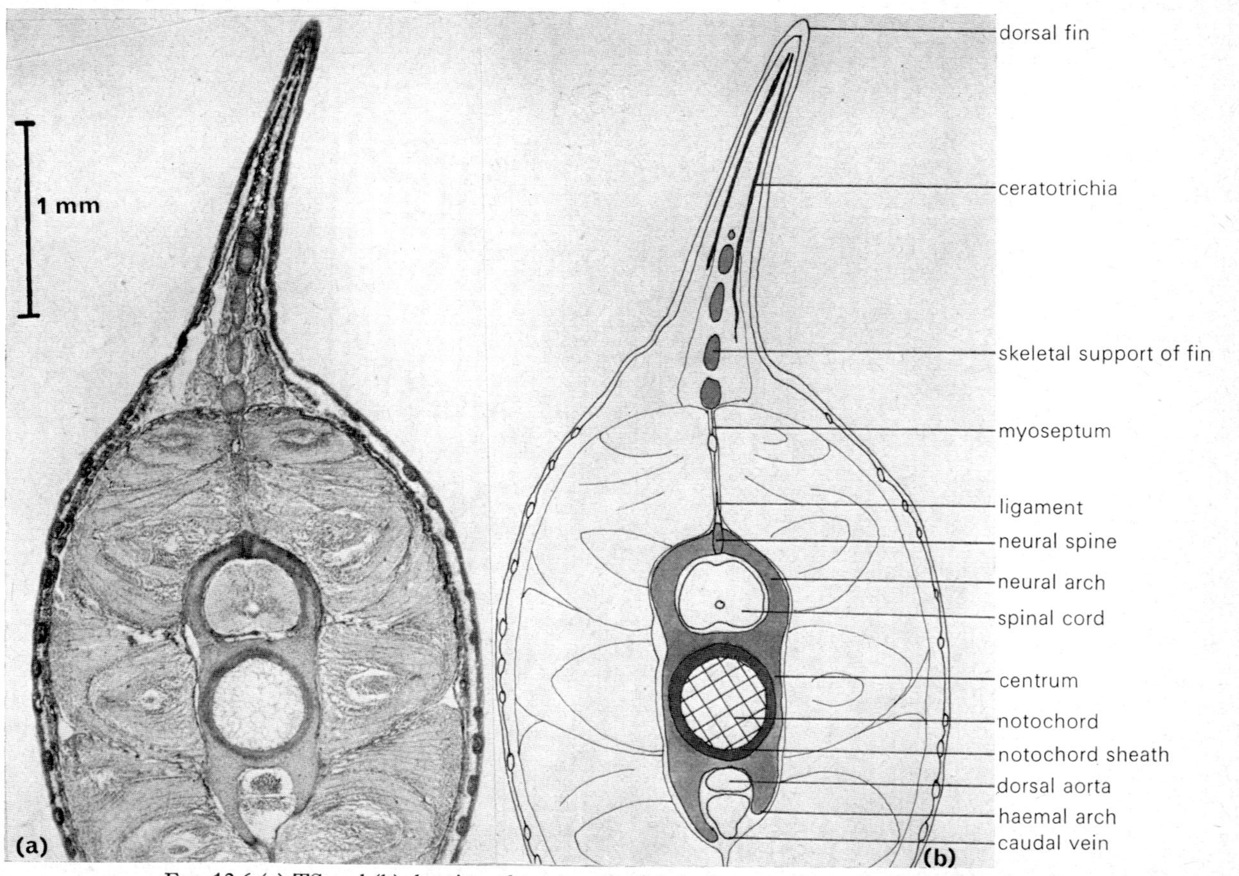

FIG. 12.6 (a) TS and (b) drawing of a young dogfish in the posterior region, showing the skeletal components of the vertebral column and the structure of the posterior dorsal fin.

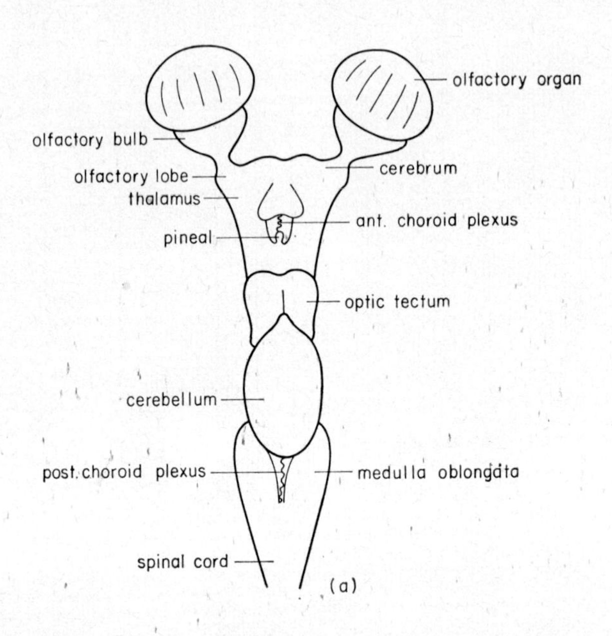

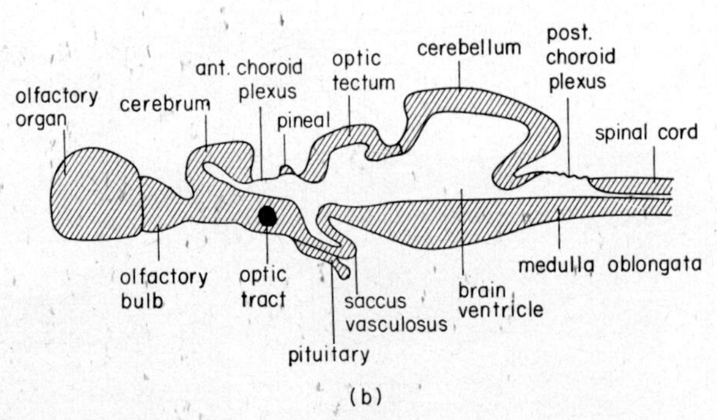

FIG. 12.7 (a) Plan of a dogfish brain viewed from the dorsal side; (b) diagram of a VLS of the dogfish brain.

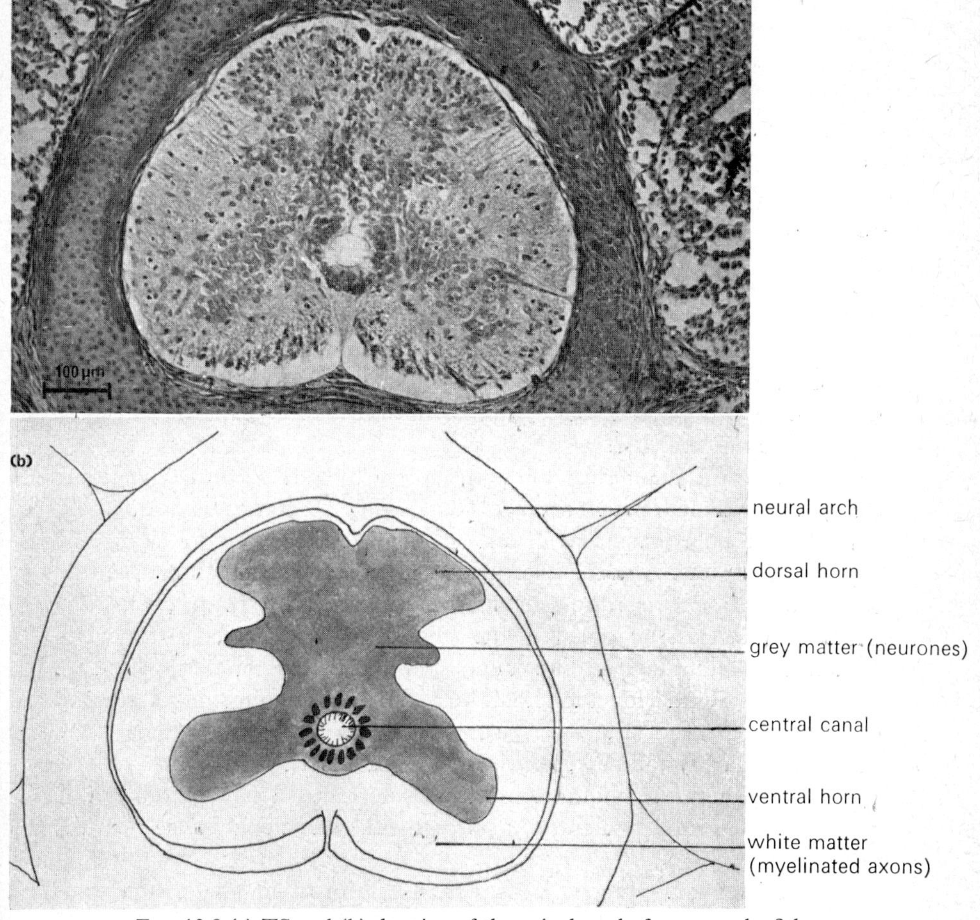

FIG. 12.8 (a) TS and (b) drawing of the spinal cord of a young dogfish.

respiration, circulation, digestion and excretion) and has peripheral ganglia. It is not, at this stage of evolution, divided into sympathetic and parasympathetic portions such as are found in higher vertebrates. The spinal cord (Fig. 12.8) is more rounded in cross-section than that of cyclostomes, and clearly shows a pattern of horns of grey matter (neurones) surrounded by white matter (myelinated axons). The dorsal and ventral roots of the spinal cord join to form segmental spinal nerves, each with an associated spinal ganglion. In all gnathostomes the nerve fibres (except the very small ones) are myelinated, i.e. covered with a white fatty membranous sheath (myelin). This arrangement allows fast conduction down the nerves; a function allotted to the giant fibres of invertebrates. Vertebrates (other than cyclostomes) do not, in general, possess giant fibres although there is a pair of giant Mauthner neurones in most aquatic vertebrates (see Fig. 13.4).

Sense organs

Dogfish have large numbers of sense organs, including many free nerve endings in the skin and viscera, proprioreceptors, mechanoreceptors, chemoreceptors and photoreceptors. All the special senses are much better developed than in the Agnatha, in accordance with the dogfish's more active mode of life.

The eye is nearly spherical and is supported by a thick connective tissue sclera, strengthened with cartilage (Fig. 12.9). A cornea is formed by fusion of the outermost coat of the sclera with the overlying skin. The large spherical lens is normally adjusted for distant vision, but it can be focused on near objects: the lens is swung forwards by contractions of a ventral *protractor lentis* muscle. The amount of light entering the eye is controlled by a mobile iris. The retina is composed only of rods, which are photoreceptors with a low threshold, for vision in dim light such as is found below the surface of the sea. Visual acuity in elasmobranchs is often increased by a crystalline *tapetum lucidum* in the choroid which reflects light back through the photoreceptors a second time. This tapetum can be occluded by choroid pigment, but no pigment movements are seen in the retina.

The ear of elasmobranchs is similar to that of other gnathostomatous fish. There are three otolith chambers and three semi-circular canals (anterior vertical, posterior vertical and horizontal), which thus allow appreciation of movement in three different planes at once. Elasmobranchs are the only group in which the original connection (*ductus endolymphaticus*) remains between the labyrinth and the outside of the head.

The lateral line system is in the form of rows of neuromast organs lying in canals which connect with the surface by pores. The dogfish bears one canal on each side of the anterior end of the trunk (see Fig. 12.1). They run forwards onto the head and branch into smaller cephalic canals.

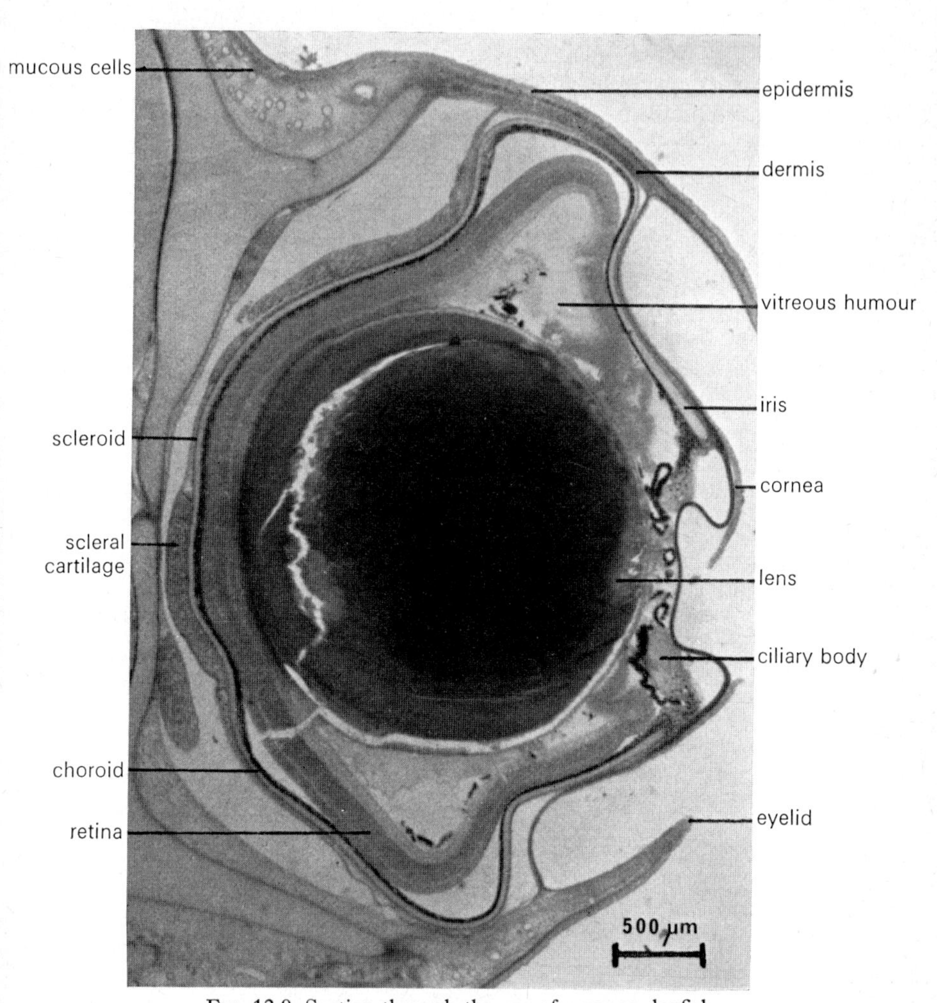

FIG. 12.9. Section through the eye of a young dogfish.

Each neuromast organ consists of many sensory hair cells covered by a large gelatinous cupula (Fig. 12.10). Isolated neuromasts also occur scattered over the head region. These are the large ampullae of Lorenzini, which have been variously implicated in electroreception, thermoreception and pressure reception.

Chemosensory reception in fish covers a wide range of sensitivities, and it is therefore chiefly for convenience that here, as in other vertebrates, external chemoreception is divided into smell or olfaction (distance chemoreception) and taste or gustation (contact chemoreception). The terms have more relevance when one is dealing with terrestrial animals.

The olfactory organs in fish are paired pits in the head, lined with sensory epithelium and connected to the forebrain by the first cranial nerve. In elasmobranchs these pits open by two nostrils, situated towards the ventral side of the head, which are partially divided by a fold of skin into an anterior inlet and a posterior outlet. This arrangement ensures that water circulates past the elaborately folded sensory epithelial lining of the olfactory organs. This olfactory epithelium differs from that of cyclostomes only in the presence of large mucous cells (Fig. 12.11): there is no accessory glandular organ for mucus production. The gustatory organs (taste buds) are restricted to the mouth and pharynx of elasmobranchs. Typically they consist of small clusters of sensory cells in a pit in the epidermis.

Elasmobranchs and teleosts both possess a saccus vasculosus lying behind the pituitary. It is a saccular organ with thin, well-vascularized walls lined by an epithelium consisting of ciliated sensory cells and supporting cells. It connects by nerves to the hypothalamus, and is supposedly sensory in function, although its exact role does not appear to be known.

Endocrine organs

The pituitary of elasmobranchs shows the characteristic vertebrate pattern of development: a pocket (infundibulum or median eminence) in the floor of the hypothalamus grows down towards, and eventually fuses with, an invagination of the stomodaeum (Rathke's pouch). The pituitary (hypophysis) lies in a depression of the skull and consists of an anterior glandular portion (of stomodaeal origin) which occupies most of the gland, and a small posterior part of nervous origin. This is not histologically different from the anterior part and so does not form the true pars nervosa of higher vertebrates. Elasmobranchs are peculiar in that a ventral lobe extends from the pituitary on a stalk.

Associated with the ventral side of the spinal cord in the posterior region is a caudal neurosecretory organ (urophysis). In elasmobranchs this takes the form of a slight swelling of the spinal cord containing a number of large Dahlgren neurosecretory cells with short axons. Their

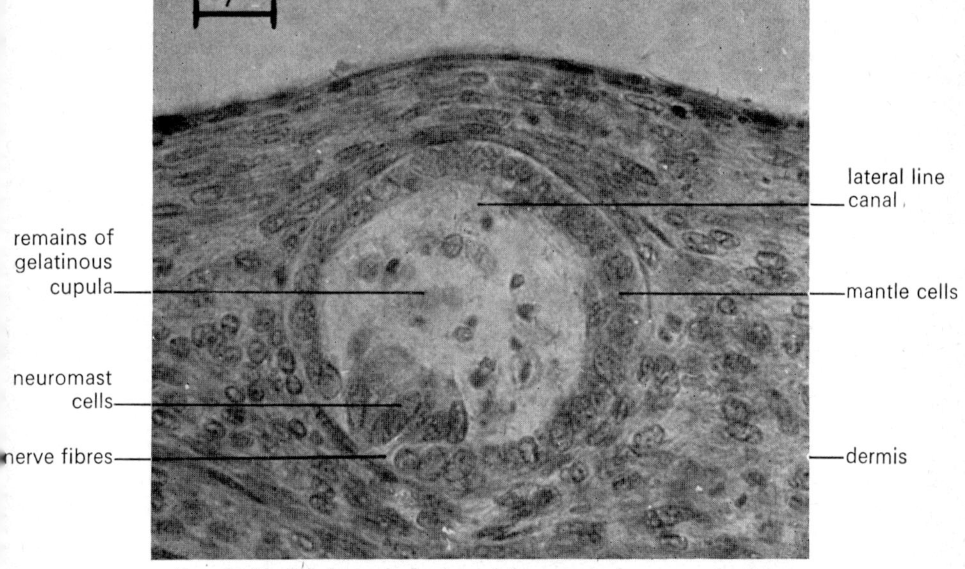

FIG. 12.10. TS through the lateral line canal of a young dogfish.

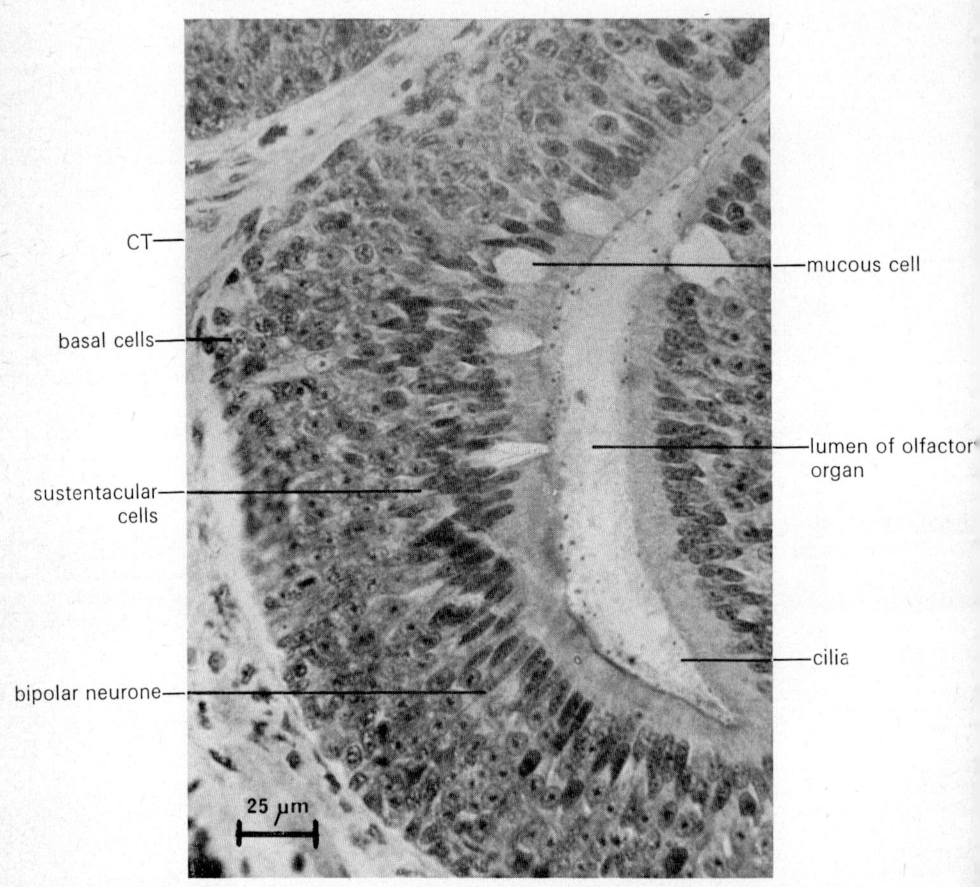

FIG. 12.11. Olfactory epithelium from the olfactory organ of the dogfish.

secretions are released into the renal-portal circulation, and are thought to play a part in control of salt and water balance.

The thyroid gland arises as an evagination of the midventral pharynx floor. In a few species the stalk attaching it to the pharynx may persist, but normally it is lost and the gland is therefore isolated. The follicular structure of thyroid glands is remarkably constant throughout the vertebrates (see Fig. 14.15) but the arrangement of thyroid tissue varies considerably. In elasmobranchs the thyroid is a single compact gland covered by a thick connective tissue coat, and lying in the anterior bifurcation of the ventral aorta.

Chromaffin tissue in elasmobranchs is closely associated with the sympathetic ganglia from which it arises (Fig. 12.12). The more anterior nests of chromaffin tissue are fused together (suprarenals), while the most posterior ones are embedded in the kidneys. The inter-renal tissue occurs as a discrete gland between the kidneys.

A new endocrine organ arises in sharks in the form of an ultimobranchial gland which lies between the ventral wall of the oesophagus and the sinus venosus of the heart. It is a small follicular gland in the dogfish, thought to have arisen from evaginations of the most posterior branchial pouches. It appears to play a part in control of calcium and phosphate levels in the body (Fig. 12.13).

Respiration

In the dogfish five pairs of gills are situated between the pharynx (orobranchial cavity) and the outside. Gill slits are separated by muscular interbranchial septa, each of which bears one afferent blood-vessel, paired efferent blood-vessels and a skeleton (branchial arch and its extensions, the gill rays) which supports the gill filaments [Fig. 12.14(a), see also Fig. 12.1]. The gill filaments have highly vascular walls folded into secondary lamellae [Fig. 12.14(b)]. These lamellae are lined by a single layer of squamous epithelium and composed of narrow supporting pillar cells which enclose capillary blood channels between them (Fig. 12.15). There is thus a very small amount of tissue separating blood and seawater, which allows for efficient exchange of oxygen. Branchial muscles pump water in through the mouth and the spiracles, which represent a modified first pair of gill slits and which possess a few ridges on their anterior walls, of the same histological structures as the gill. Water is then pumped through the sieve formed by the secondary gill lamellae, in the opposite direction to the blood flow in the capillary channels; thus forming an efficient counter-current system for the optimal exchange of oxygen.

Circulation

The finely divided gill surface of elasmobranchs presents an even higher resistance to blood flow than in the lamprey; thus the heart is larger and

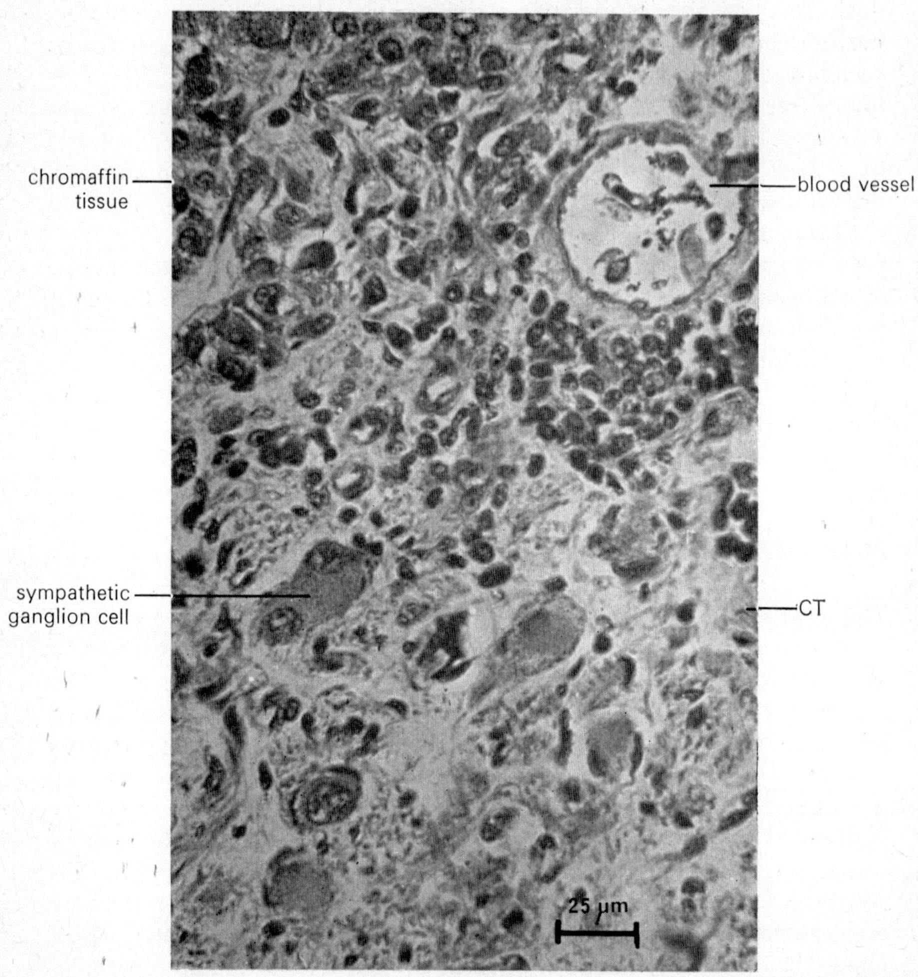

FIG. 12.12. Section through a posterior supra-renal body (chromaffin tissue) of the dogfish. Note the close proximity of the chromaffin tissue to several large sympathetic ganglion cells.

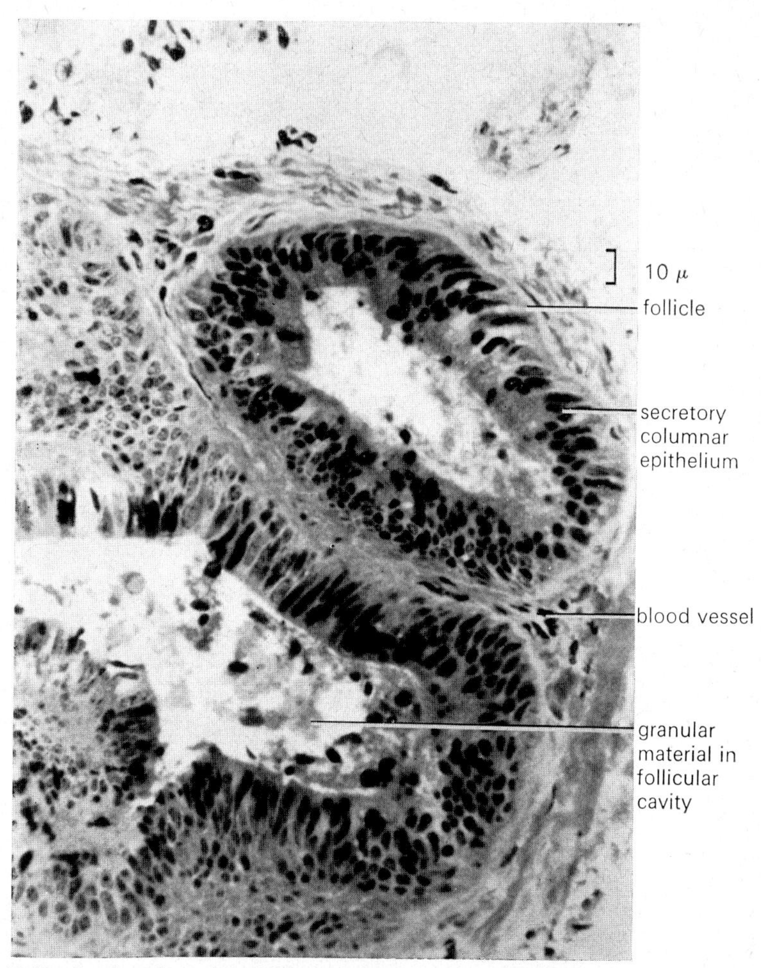

FIG. 12.13. Section through a follicular ultimobranchial gland from an elasmobranch (*Squalus suckleyi*). Photograph by kind permission of D. Harold Copp & W. A. Webber FIG. 2 Chapter 7 *in*: "Fish Physiology" Vol II (eds. W. S. Hoar and D. J. Randall) Academic Press, 1969.

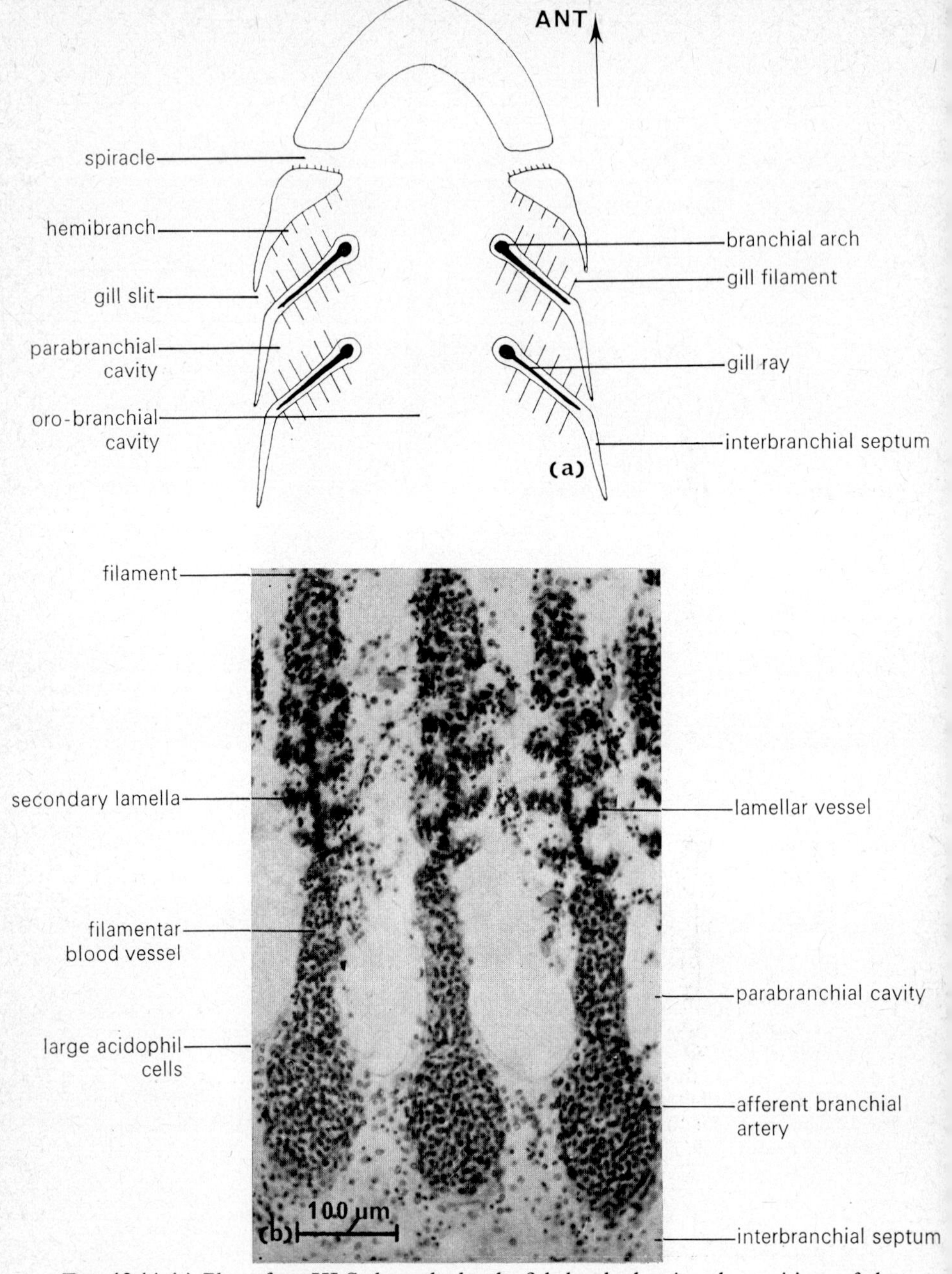

FIG. 12.14 (a) Plan of an HLS through the dogfish head, showing the positions of the spiracle and most anterior gill slits. Note the long interbranchial septae fringing the gill slits, and the gill filaments attached to them; (b) LS through part of a dogfish gill lamella, showing the filaments and secondary lamellae.

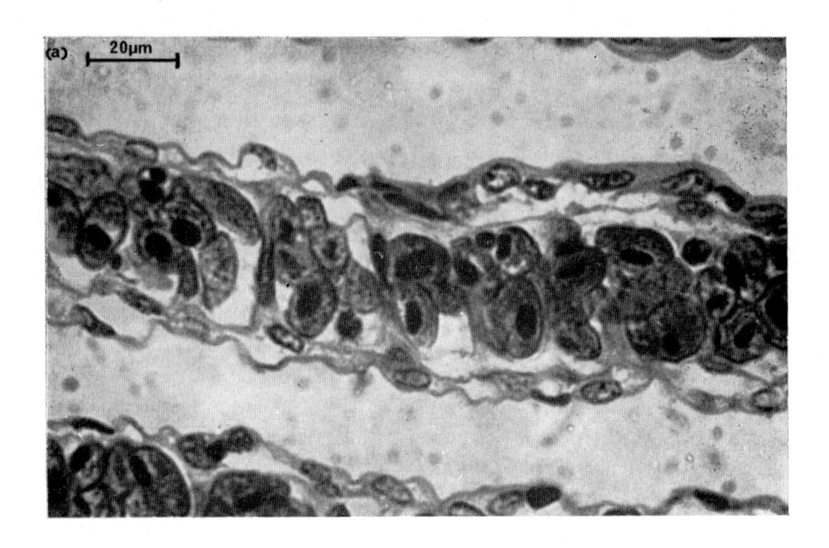

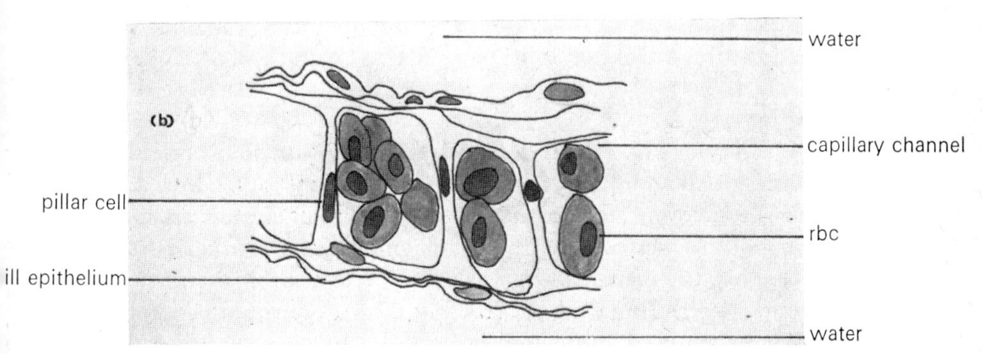

Fig. 12.15 (a) LS and (b) drawing of a secondary lamella of the dogfish gill. Note the small distance between the outside medium and the blood inside the capillary channels.

even more muscular (Fig. 12.16). Blood collects in a large sinus venosus, and is pumped forwards by the three contractile chambers: the triangular atrium, extremely muscular ventricle and the conus arteriosus. Two sets of valves in the conus prevent backflow of blood into the ventricle. The heart is enclosed in an inelastic pericardium which retains a connection with the body cavity.

A diagnostic feature of elasmobranchs is that the venous system is expanded in places to form sinuses. As in most vertebrates there are two portal systems: the hepatic-portal system carrying venous blood from the gut to the liver, and a renal-portal system conveying venous blood to the kidneys.

A small spleen is present which produces large (20 μ) oval, nucleated erythrocytes (Fig. 12.17). It lies close to the digestive tract (see Fig. 12.2) and is drained by the hepatic portal vein. White cells are mainly produced in lymphoid tissue, such as Leydig's organ in the oesophageal mucosa (an organ peculiar to elasmobranchs) and the thymus. All vertebrates except cyclostomes possess lymphoid thymus tissue arising from evaginations of the gill pouches. In dogfish this takes the form of irregular masses of tissue lying above the gill slits (see Fig. 12.1). They consist of dense whorls of epithelial cells surrounded by connective tissue (Fig. 12.18), and do not show any division into cortex and medulla as in higher vertebrates. The thymus appears also to play a role in immunogenesis, and has had an endocrine function attributed to it, particularly during early development.

Digestion

The dogfish gut (Fig. 12.19) demonstrates the basic plan present in all gnathostomes, although there may be modifications according to diet. Dogfish are carnivorous, feeding mainly on small molluscs and crustaceans. The large mouth is surrounded by cartilaginous jaws bearing backwards pointing teeth which are continuously replaced and are similar to the dermal denticles covering the body (see Fig. 12.3). The mouth and pharynx are lined with stratified epithelium, unicellular mucous glands and small taste papillae. The gills and spiracle open into the pharynx (see Fig. 12.1). Ciliated epithelium is found in the anterior regions of the gut of all the lower vertebrates, as shown in the dogfish oesophagus (Fig. 12.20), while the more posterior regions of the gut are provided with smooth muscles which undergo peristalsis and move food in a distal direction. The long U-shaped stomach is divided into cardiac and pyloric portions [Fig. 12.21 (a,b)]. Its lining is thrown into rugae, and consists of columnar epithelium containing many mucous cells. In the cardiac portion, large pyramidal secretory cells line the gastric glands and produce both enzymes and hydrochloric acid [Fig. 12.21(c)]. In most vertebrates the stomach is the region where food is stored while the first stages of digestion occur. The acid produced in the stomach assists hydrolysis of food and prevents

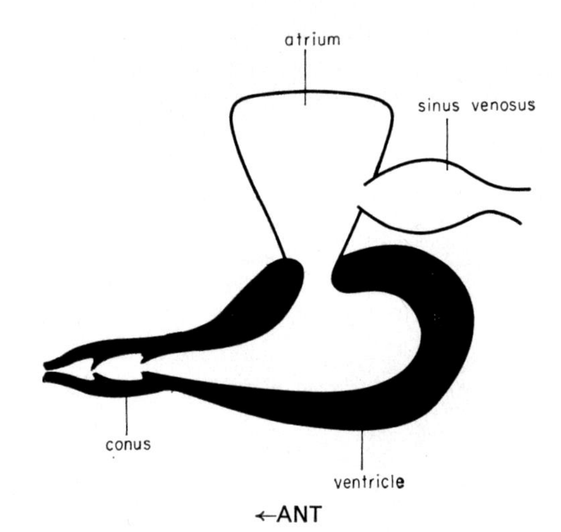

FIG. 12.16. Diagram of a VLS through the dogfish heart.

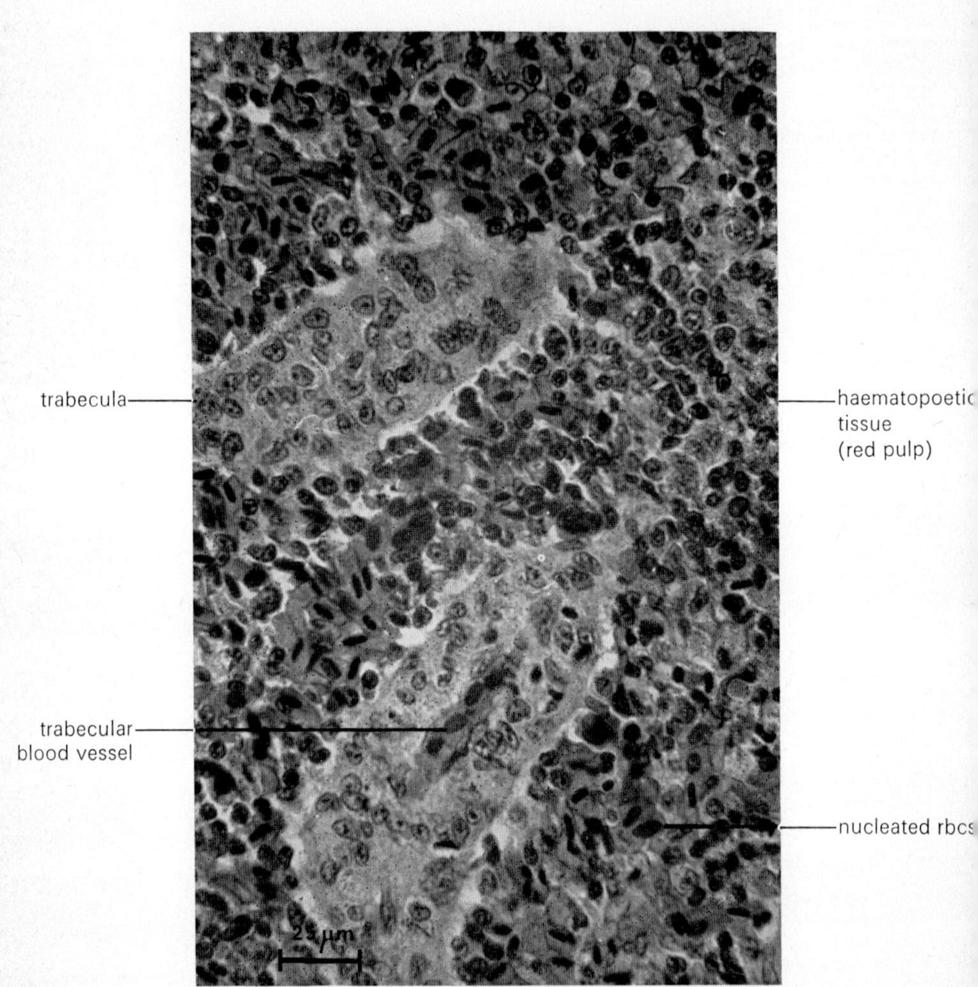

FIG. 12.17. Section through part of the dogfish spleen.

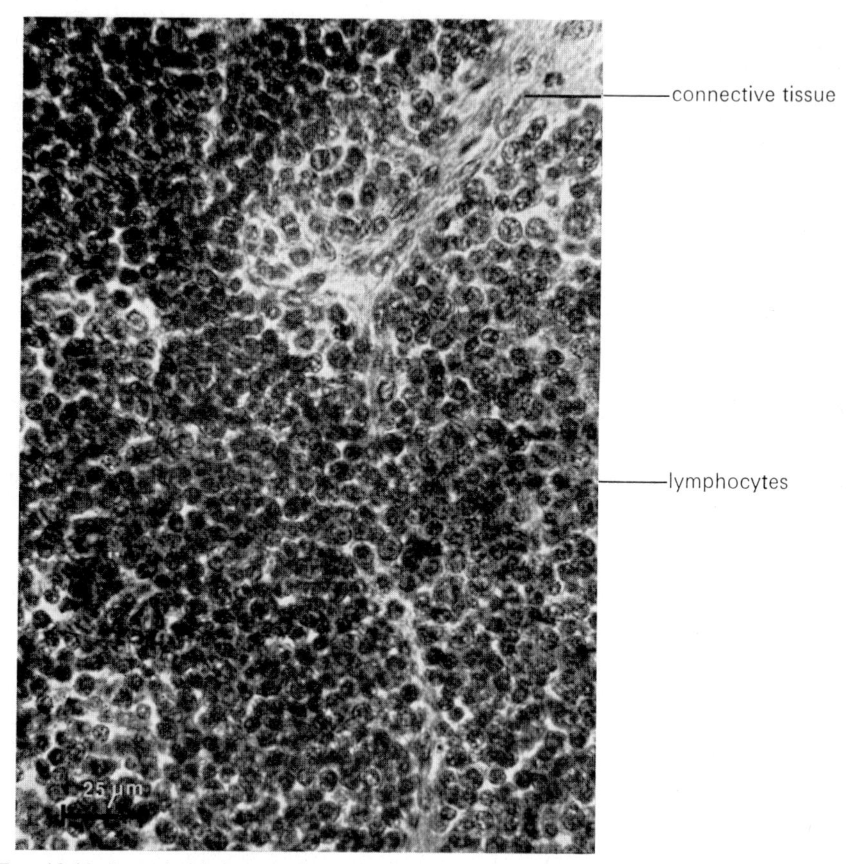

FIG. 12.18. Lymphoid thymus tissue situated in the dorsal wall of the dogfish gill pouch.

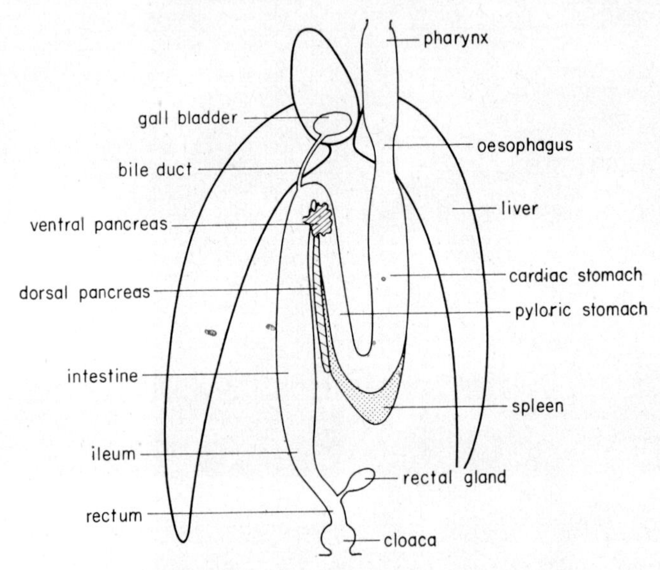

FIG. 12.19. Plan of the dogfish digestive tract, as viewed from the ventral side.

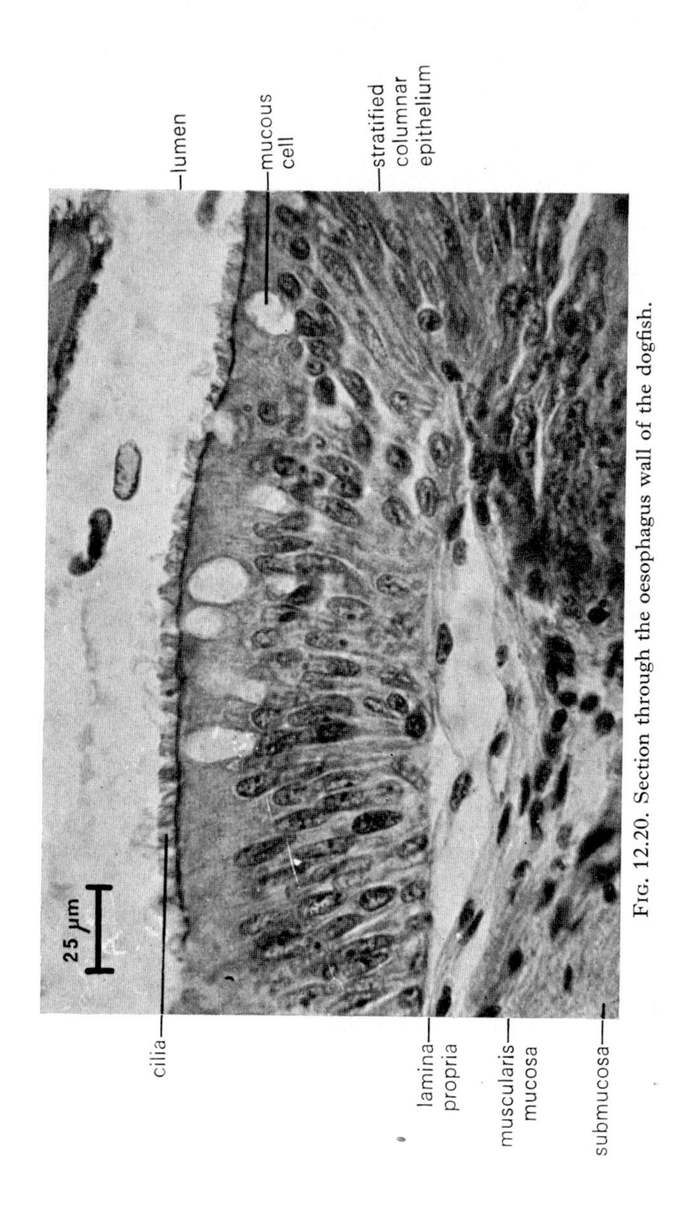

Fig. 12.20. Section through the oesophagus wall of the dogfish.

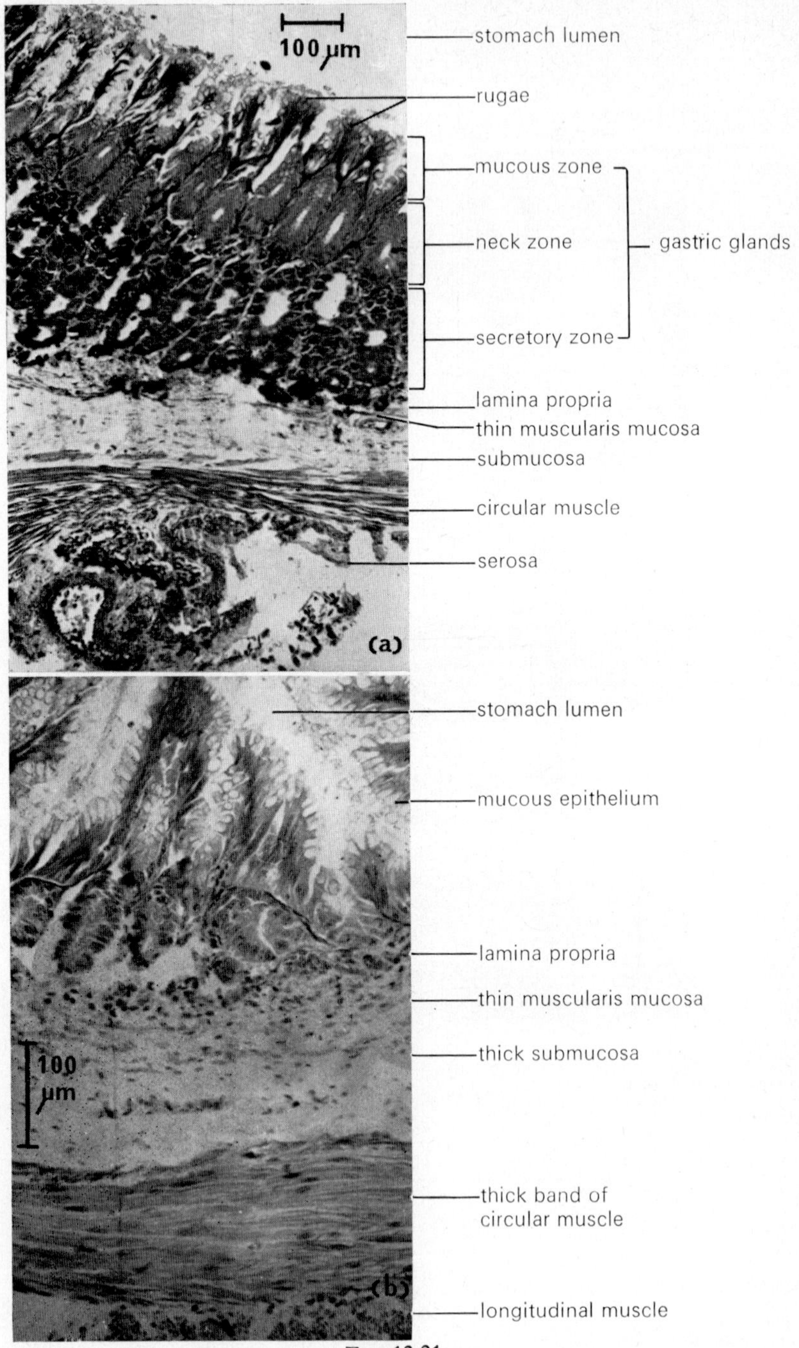

100 μm
stomach lumen
rugae
mucous zone
neck zone
secretory zone
gastric glands
lamina propria
thin muscularis mucosa
submucosa
circular muscle
serosa
(a)
stomach lumen
mucous epithelium
lamina propria
thin muscularis mucosa
thick submucosa
100 μm
thick band of circular muscle
(b)
longitudinal muscle

FIG. 12.21

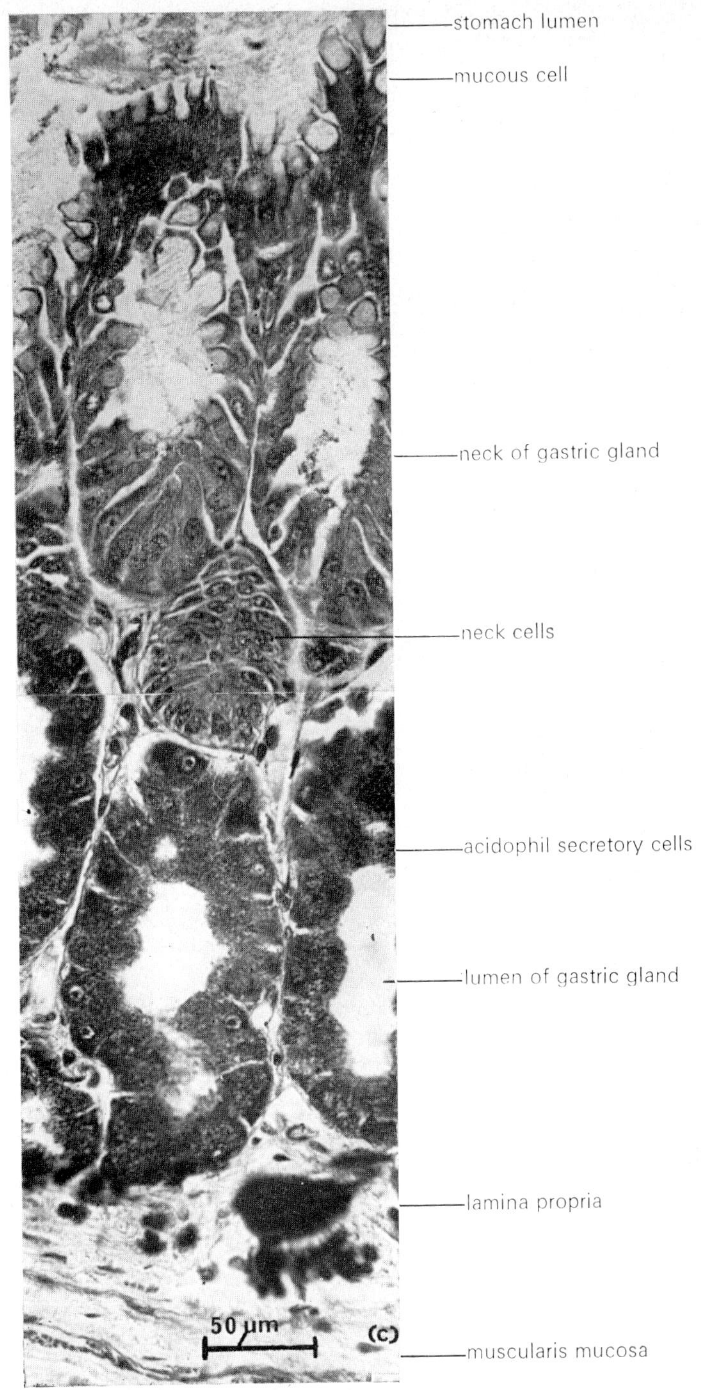

FIG. 12.21. TS through (a) the cardiac stomach; and (b) the pyloric stomach of the dogfish; (c) detail of the gastric glands of the dogfish cardiac stomach.

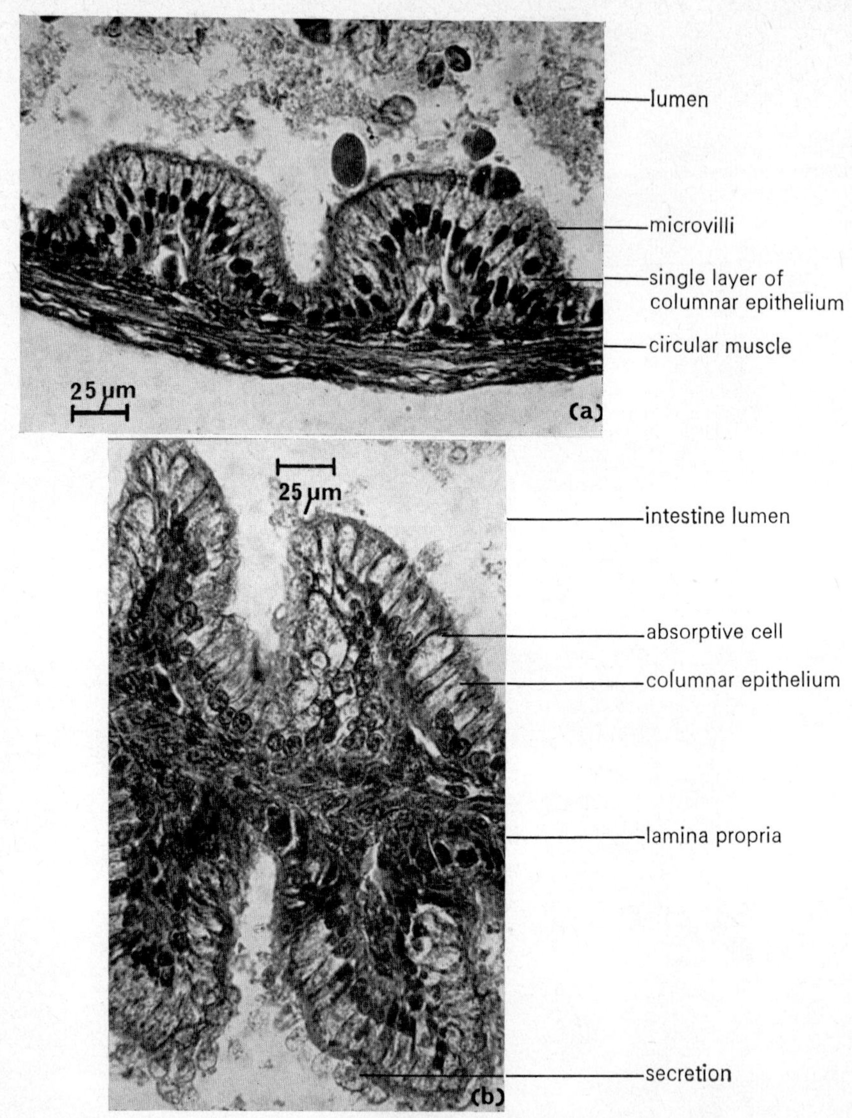

FIG. 12.22 (a) TS through the intestine wall of the dogfish. Note the lack of muscularis mucosa and submucosa; (b) section through the spiral valve of the dogfish intestine.

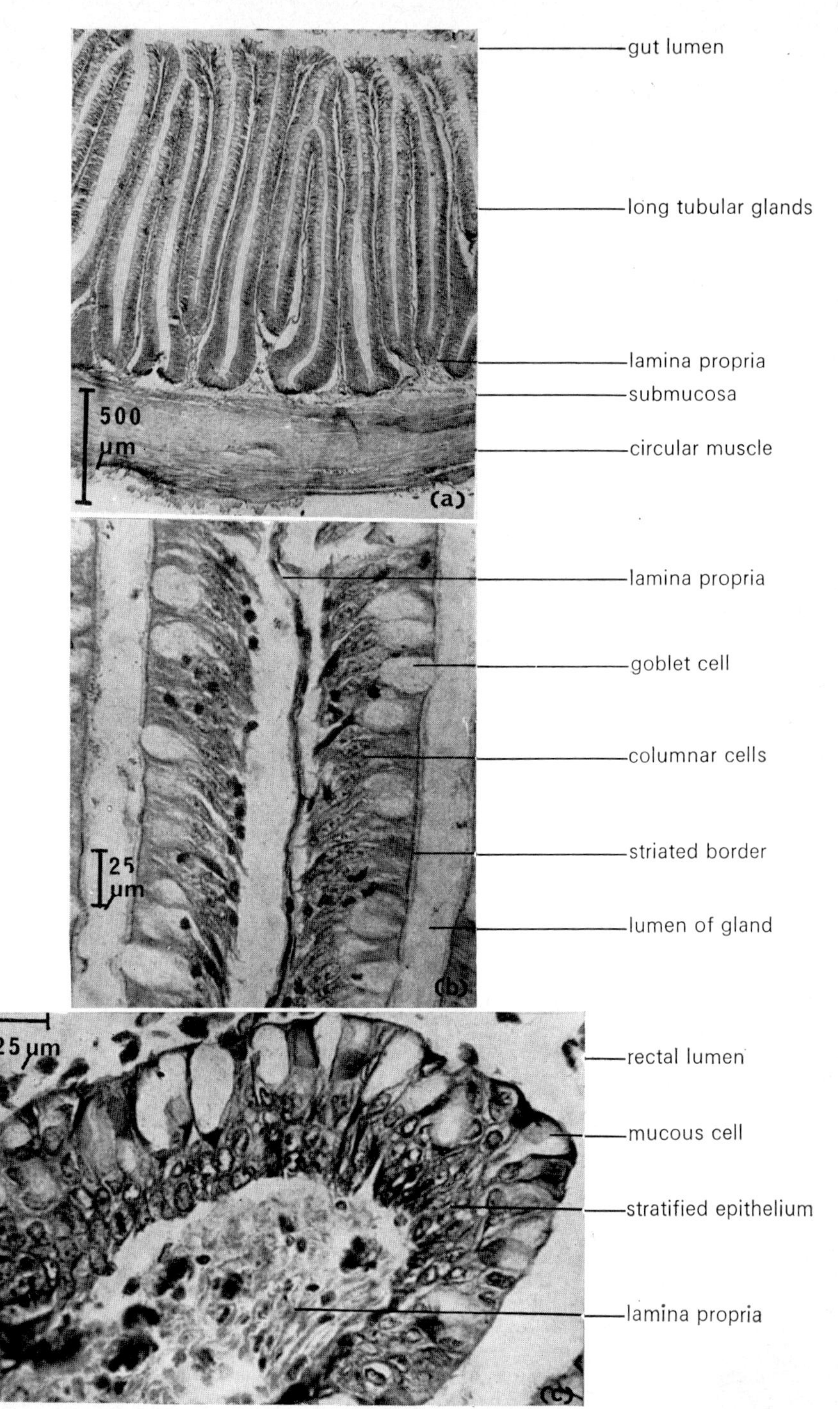

FIG. 12.23 (a) TS through the ileum wall of the dogfish; (b) detail of the tubular glands of the ileum; (c) detail of the rectal wall of the dogfish.

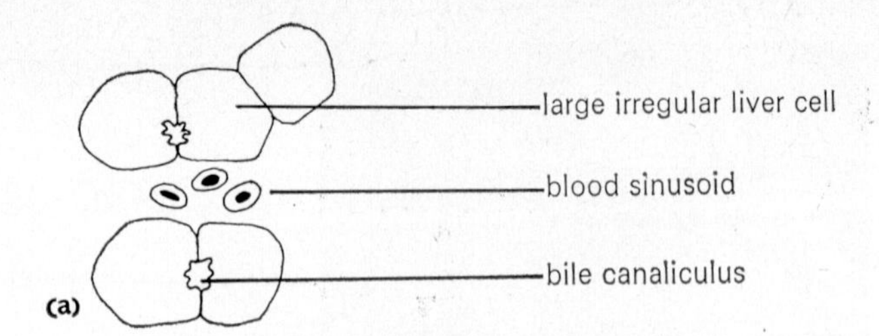

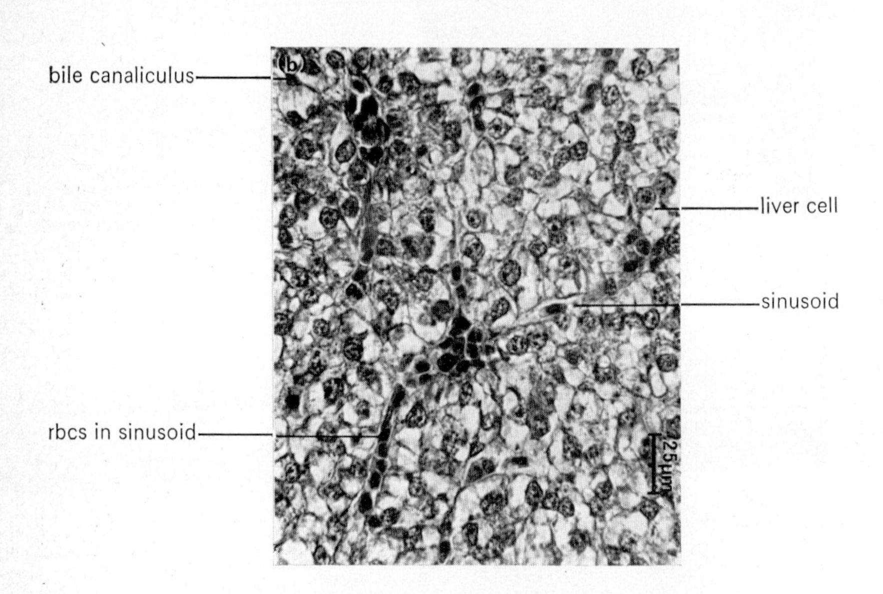

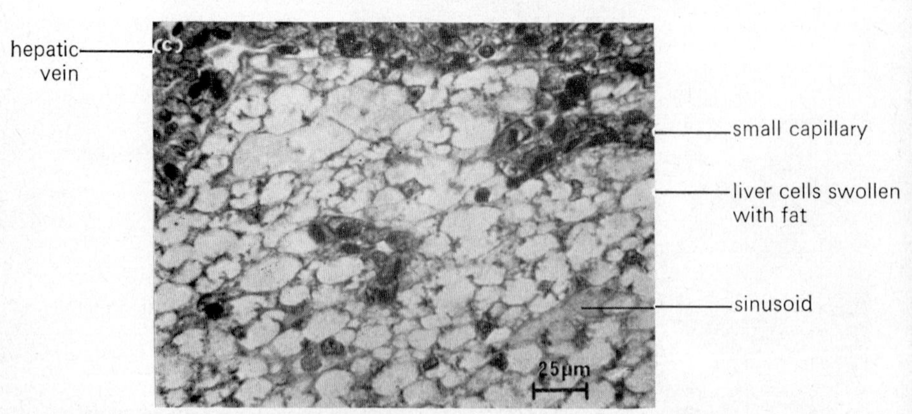

FIG. 12.24 (a) Diagram of the arrangement of cellular components of the dogfish liver. This arrangement is typical of most lower vertebrates; (b) section of the dogfish liver in a secretory phase; (c) section of the dogfish liver in a storage phase. Note the liver cells swollen with unilocular fat globules.

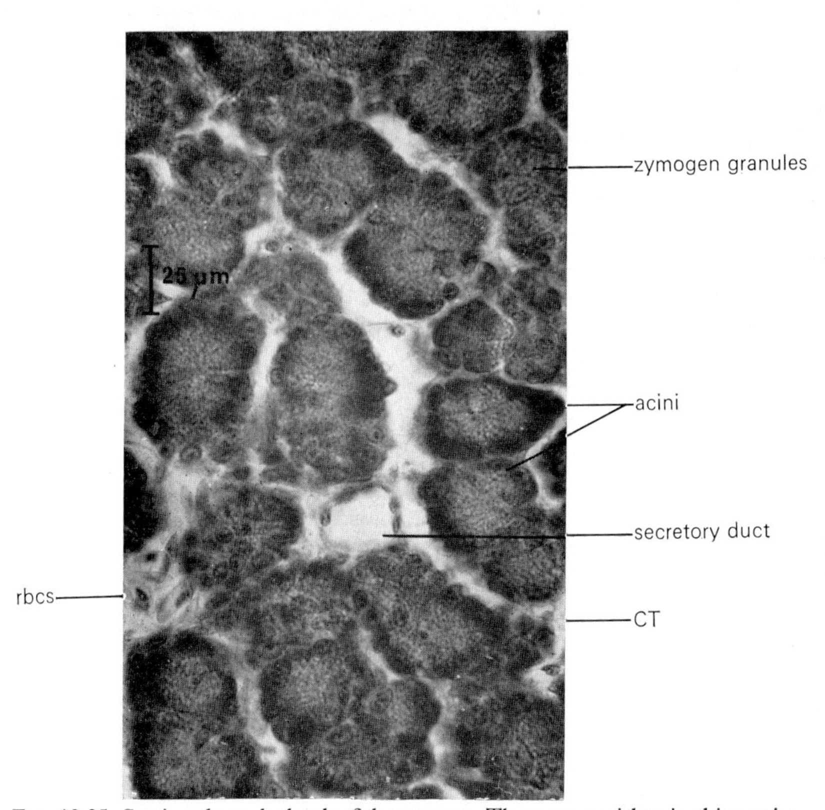

FIG. 12.25. Section through the dogfish pancreas. There are no islets in this section.

bacterial decay. The pyloric stomach has a thick layer of circular muscle which forms a pyloric sphincter at the distal end of the stomach, separating it from the intestine. The intestine is a short straight tube lined with a single layer of folded epithelium [Fig. 12.22(a)] and provided with a spiral valve, or fold of epithelium which runs a spiral course down its length [Fig. 12.22(b), see also Fig. 12.2]. This serves to increase surface area for digestion. The hind end of the intestine, the ileum, possesses long tubular mucous glands [Fig. 12.23(a, b)] and runs into a short rectum lined with stratified epithelium and mucous cells [Fig. 12.23(c)], before opening into the cloaca.

Digestion in gnathostomes is aided by two "digestive glands", the liver and pancreas. The liver arises as a diverticulum from the anterior end of the intestine, and its functions are to assist digestion, receive (via the hepatic-portal vein) food absorbed from the gut and store it. The liver structure of all vertebrates is very similar. Large polyhedral liver cells [Fig. 12.24(a)] absorb food materials from the blood sinusoids surrounding them, store fat, protein and carbohydrates, and secrete bile. This bile, which is involved in emulsification of fats in the gut, drains into a network of minute canaliculi between the liver cells and thence by an hepatic duct to the gall bladder to be stored. Figure 12.24(b, c) shows a dogfish liver during phases of secretion and fat storage. The gall bladder lies underneath the median liver lobe in dogfish and drains by a bile duct into the anterior end of the intestine. The pancreas is a long bilobed structure lying between the stomach and intestine. It is composed largely of masses of exocrine acinar cells which produce digestive enzymes (Fig. 12.25). Small groups of endocrine islet cells occur, in the dogfish, as an outer layer along the epithelium of small pancreatic ducts, but they are not apparent at all times of the year. As in higher vertebrates the islet cells probably secrete hormones involved in carbohydrate metabolism. A pancreatic duct also drains into the intestine.

Excretion

Many of the problems of osmoregulation and excretion are solved in elasmobranchs by the presence of large amounts of urea in the blood, which thus maintains the blood at a high osmotic concentration. This is achieved by having the gills impermeable to urea, and by possessing the facility to reabsorb urea in the kidney. The kidney of the adult dogfish is opisthonephric, and has lost the segmental arrangement of the cyclostomes (Fig. 12.26). In the female, the anterior part of the kidney forms a narrow strip of soft tissue composed of small tubules, which join up with nephric tubules from the posterior portion and dilate to form a pair of long urinary sinuses. These sinuses unite posteriorly into a median sinus which opens by a papilla into the cloaca. In the male, the anterior portion of the kidney is degenerate and no longer functional as a renal organ; it is invaded by the

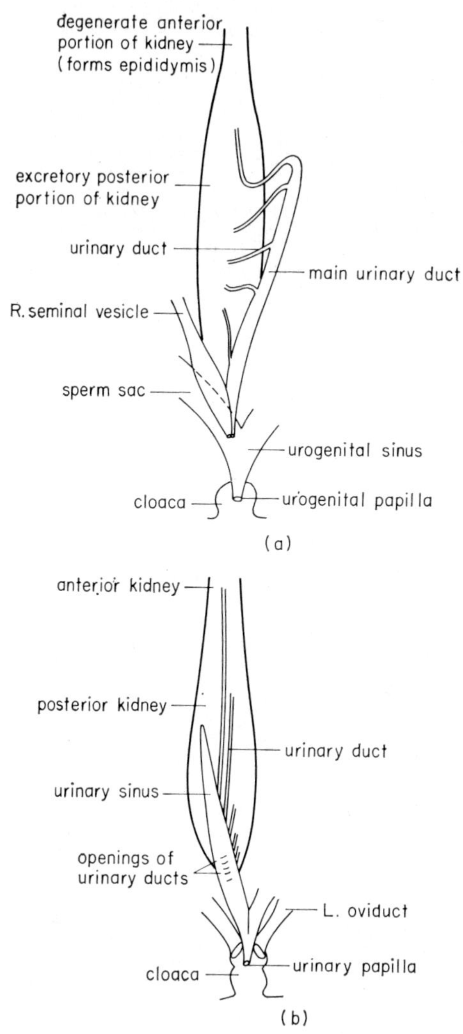

FIG. 12.26 (a) Plan of the urinary system of the male dogfish (R. side only) as viewed from the ventral side; (b) plan of the urinary system of the female dogfish (R. side only) as viewed from the ventral side.

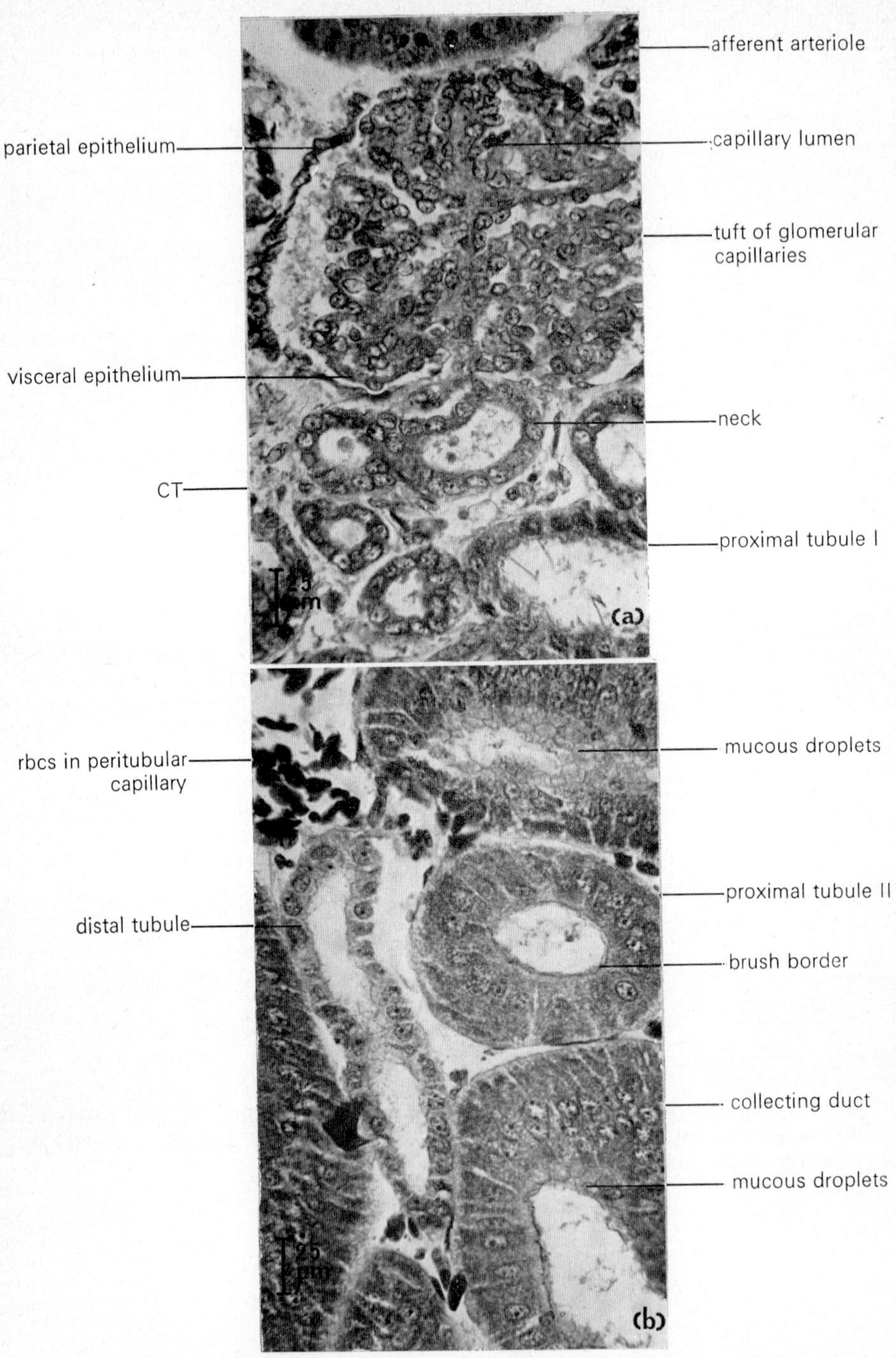

FIG. 12.27 (*a*) Section through the kidney of an adult dogfish, showing parts of the nephric tubule and a large renal capsule; (b) section through parts of the nephric tubule of the dogfish kidney.

testis and becomes the epididymis. The posterior part of the kidney is a compact lobular body composed of numerous long coiled nephric tubules (Fig. 12.27). Each tubule ends in a large renal corpuscle which surrounds a tuft of glomerular capillaries. This is the site of ultra-filtration of dissolved substances from the blood into the nephron. Each tubule consists of a narrow neck region, a long proximal tubule divided into two distinct regions (I and II), and a short distal tubule. Reabsorption of water, some dissolved substances such as sugars and amino acids, and urea occurs across the walls of the nephric tubule. The paucity of glomeruli and the poor degree of vascularization combined with long reabsorbing tubules leads to a low urine flow in the dogfish. Urine formed in the nephrons drains into a few large collecting ducts and out via the urinary ducts into the urinary sinuses.

The rectal gland is also involved in excretion. It is a compound tubular gland (Fig. 12.28) composed of secretory tubules lined with cuboidal epithelium, radiating outwards from a central canal lined with stratified epithelium, which in turn joins by a short duct to the rectum (see Fig. 12.19). The rectal gland secretes excess salt out of the body.

Reproduction

The sexes are separate in dogfish: plans of the male and female reproductive systems are shown in Fig. 12.29.

The testes are long paired organs composed of numerous seminiferous tubules or ampullae. As in most fish and amphibia, each of these ampullae shows synchronous development (i.e. all the cells within it are at the same stage of development at any one time). Each testis is divided into a ventral zone of tubulogenesis [Fig. 12.30(a)], various intermediate zones showing different stages of division, through to a dorsal zone of ripe ampullae [Fig. 12.30(b)]. Sperm are nourished by large Sertoli cells which regress after sperm discharge. Clusters of Leydig cells in between the ampullae secrete gonadial hormones. Mature sperm are released into the testis lumen and thence through small vasa efferentia into the epididymal tubules [Fig. 12.31(a)]. These tubules run directly into paired vasa deferentia whose thick walls expand posteriorly to form two seminal vesicles [Fig. 12.31(b)]. Sperm are stored in the seminal vesicles and two thin-walled sperm sacs which open off them. The seminal vesicles finally join to form the urinogenital sinus which opens into the cloaca in common with the excretory ducts and the intestine (see also Fig. 12.26).

In the female, the left ovary remains undeveloped in the adult while the single large right ovary is found attached to a peritoneal fold of the body wall. The ovary is lobed and divided into medulla and cortex. The medulla or centre of the ovary is composed of a mass of connective tissue, nutritive cells and blood vessels while the ova and their surrounding

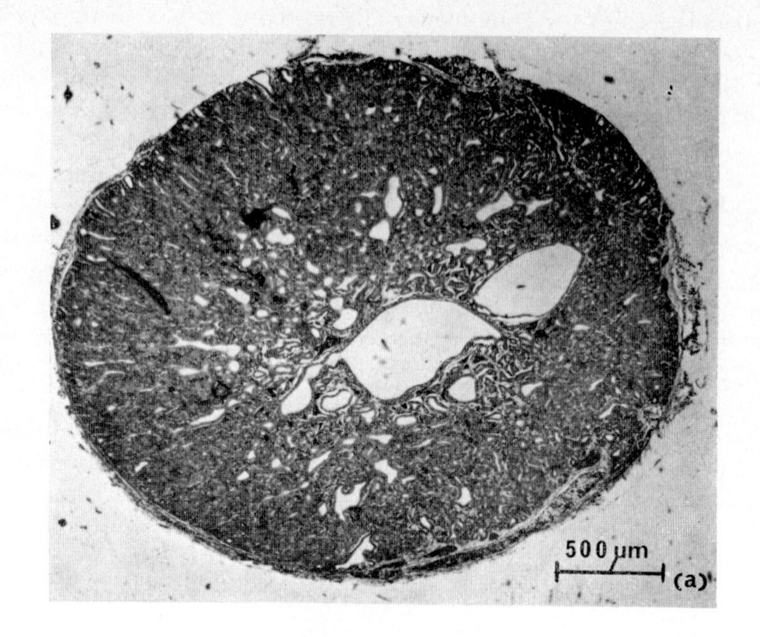
500 µm
(a)

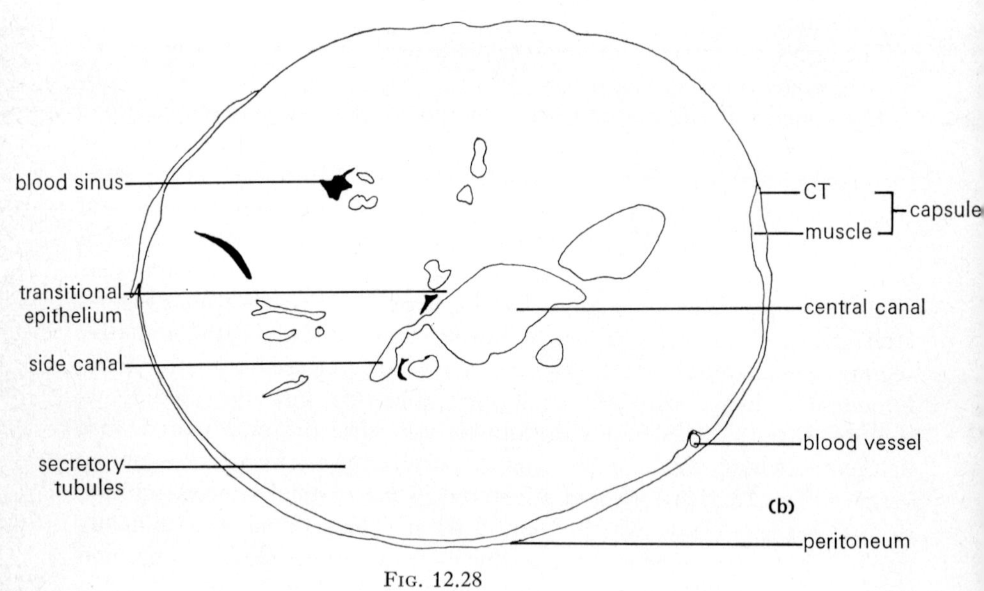
blood sinus
transitional epithelium
side canal
secretory tubules
CT
muscle
capsule
central canal
blood vessel
(b)
peritoneum

FIG. 12.28

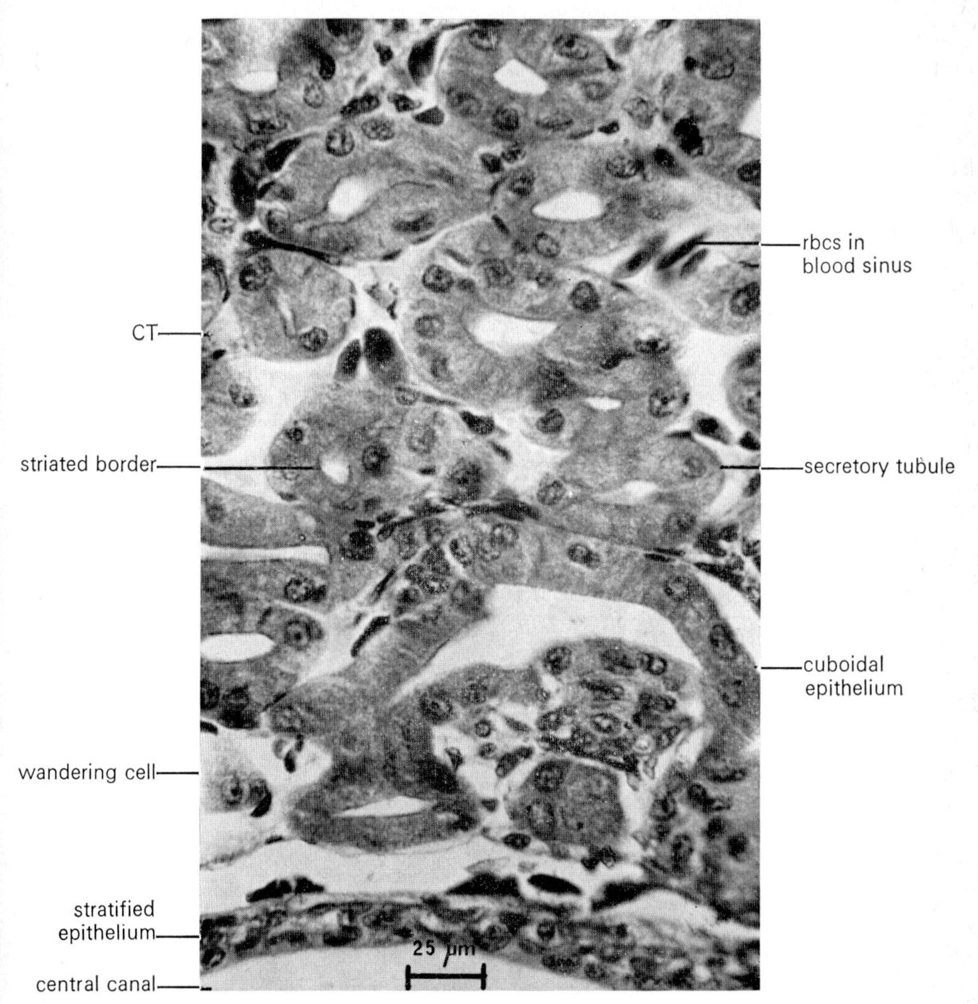

FIG. 12.28 (a) TS and (b) drawing of a TS of the dogfish rectal gland; (c) detail of the rectal gland structure in the region adjoining the central canal.

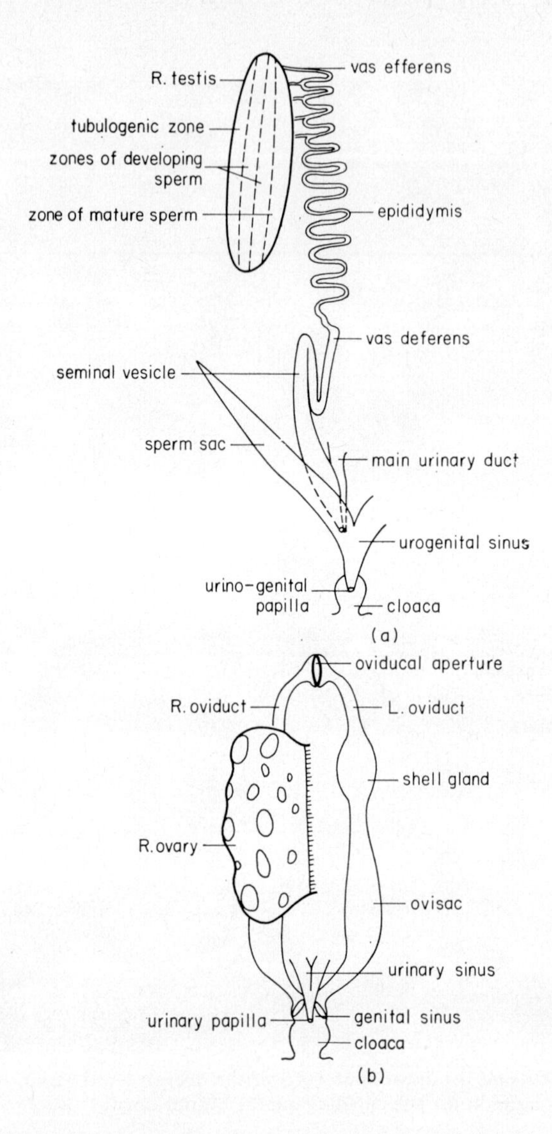

FIG. 12.29 (a) Plan of the reproductive system of the male dogfish (R. side only) as viewed from the ventral side; (b) plan of the reproductive system of the female dogfish as viewed from the ventral side. Note that there is only one (R) ovary.

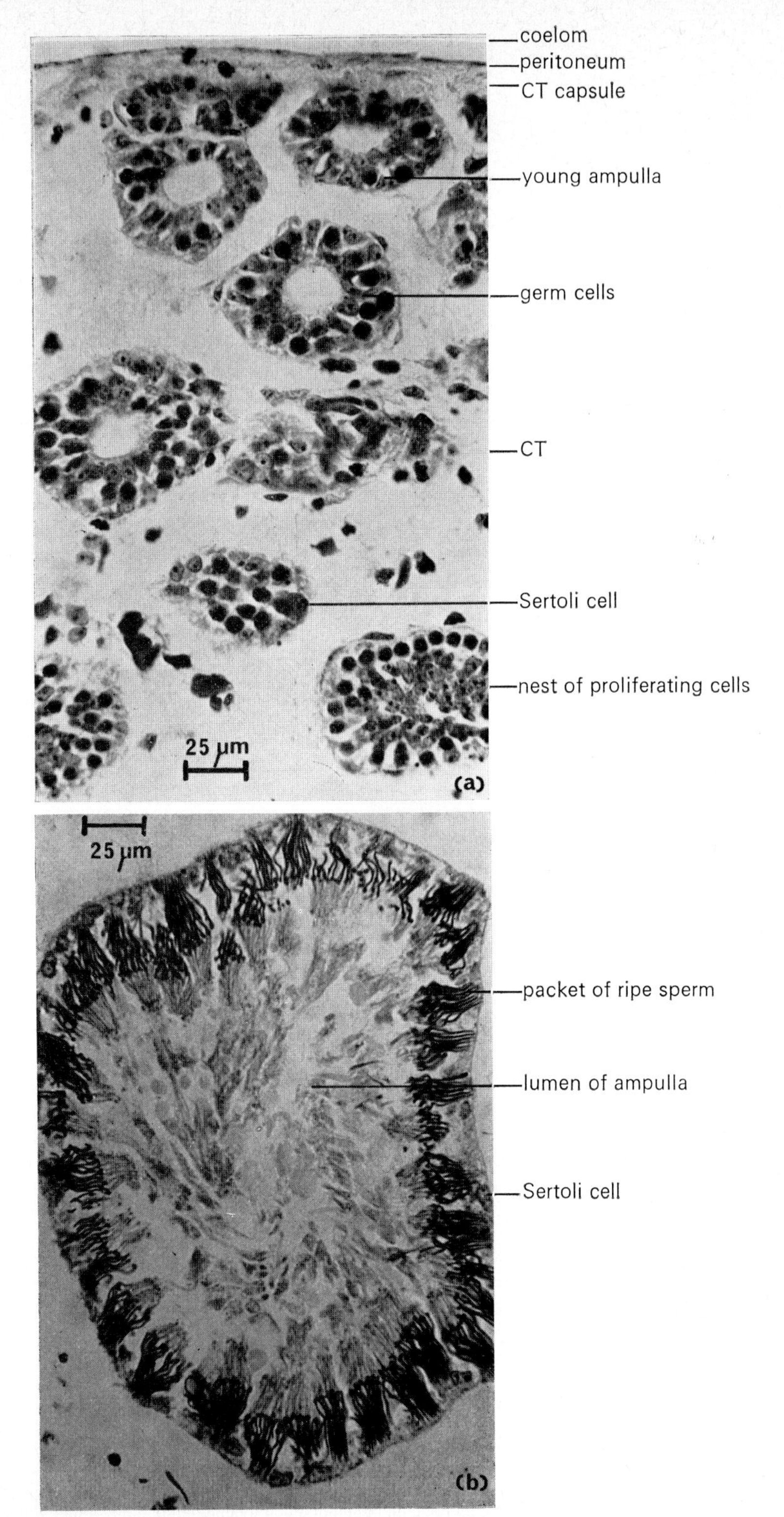

FIG. 12.30 (a) Section through the tubulogenic zone of the testis from an adult male dogfish; (b) TS through an ampulla containing ripe sperm, from the testis of an adult male dogfish.

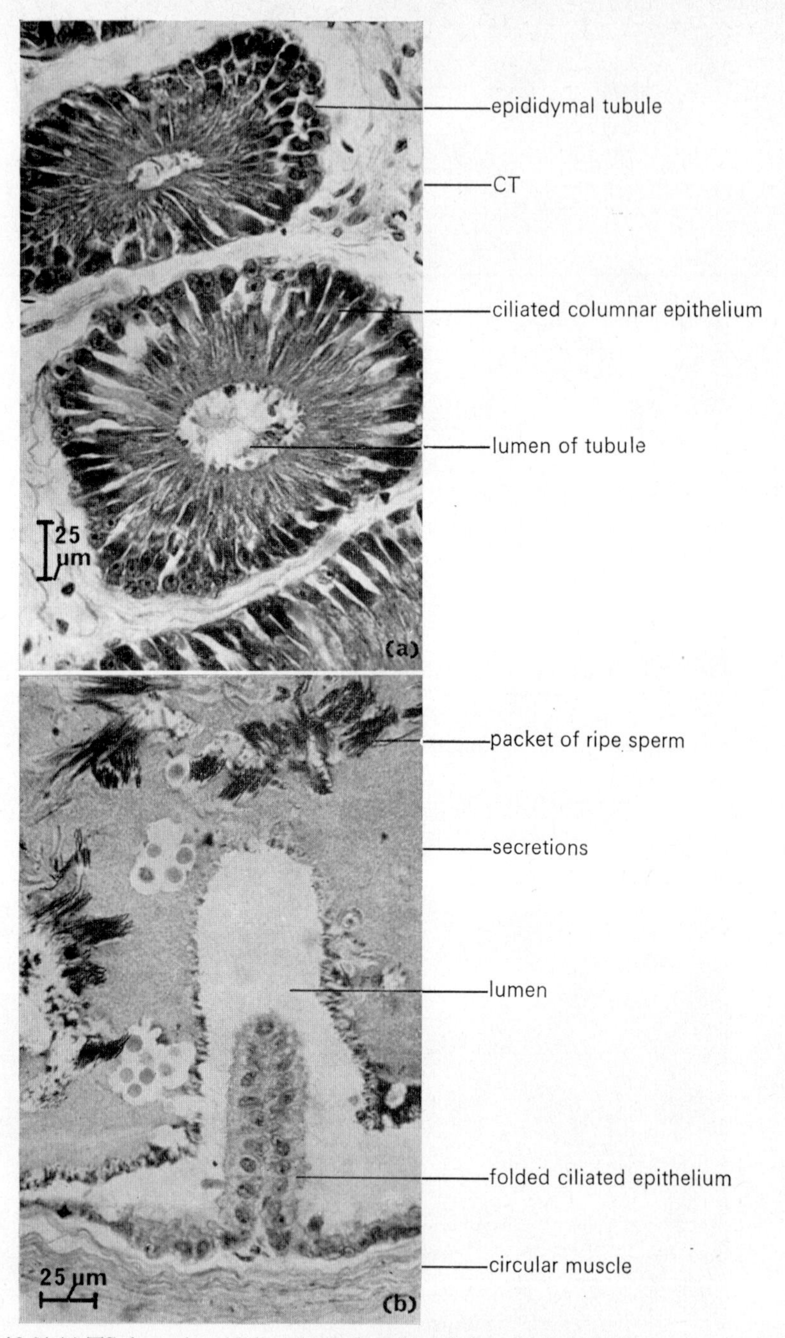

FIG. 12.31 (a) TS through epididymal tubules of an adult male dogfish; (b) section through the seminal vesicle wall of an adult male dogfish. Note the packets of ripe sperm in the lumen of the seminal vesicle.

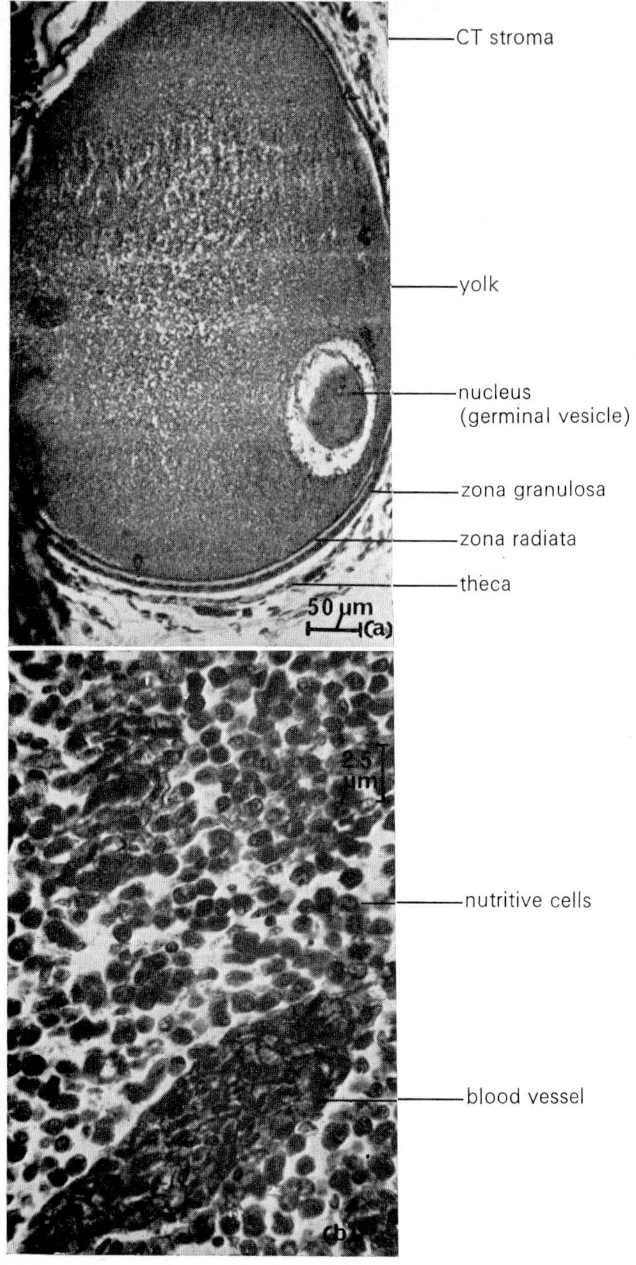

FIG. 12.32 (a) Developing ovum from the cortex of the ovary of an adult dogfish; (b) nutritive cells from the medulla of the ovary of an adult female dogfish.

follicular cells constitute the cortex (Fig. 12.32). The follicle cells produce yolk and female hormones. Ripe ova project up from the surface of the ovary and are released directly into the body cavity: here they are captured by a pair of ciliated funnels lying on either side of the liver. These funnels are derived from pronephric funnels and lead into two long thick-walled oviducts lined with folded ciliated epithelium and a thick muscle layer. About half-way down the oviducts are found the oviducal or shell glands. The anterior portion of each gland is a mass of branched tubules lined with epithelial cells containing mucoprotein (albumen). The posterior region has a very folded mucosa forming large villi with simple branched tubular glands between them. These glands secrete ovokeratin from which the horny egg-case is formed. The two oviducts are thickened at their posterior ends to form ovisacs. These unite to become the uterus which opens into the common cloaca.

Fertilization is internal, and occurs when the eggs are in the ovisac. The male has a pair of claspers that are modified pelvic fins. They are scroll-like organs containing erectile tissue and are used to introduce sperm into the female. Dogfish are oviparous and the young develop inside the egg-case.

Chapter 13

Class Osteichthyes

The Osteichthyes, or bony fish, possess a skeleton that is partly or completely made of bone. They have not more than five gill slits covered by a bony operculum, and are clothed by flat, overlapping dermal scales. They usually possess an anterior gut diverticulum which forms an air-bladder or lungs, or may be modified to form a hydrostatic swim-bladder. They show increased complexity of anterior receptors and an associated brain development. The Osteichthyes are a large class which is divided into the Choanichthyes (Sarcopterygii) and the Actinopterygii. The Choanichthyes (including the lungfish) have fins with fleshy lobes at the base and an axial skeleton of the type thought to have given rise to the tetrapod limb. The Actinopterygii, or ray-finned fishes, have membranous fins each of which is supported by reduced bony fin-rays radiating from the base. They possess external nostrils and are covered by cycloid (evenly curved) or ctenoid (comb-shaped) bony scales. Actinopterygians occur in marine or freshwater habitats, and are often partially adapted for life out of water. They tend to have complex reproductive mechanisms.

Example: *Lebistes* (*reticulatus*)

The guppy belongs to the largest group of living actinopterygians, the Teleostei or modern bony fish; a group showing great adaptive radiation. With the development of an air-bladder, teleosts have attained neutral buoyancy, and so have lost the flat head and upturned tail of elasmobranchs. In general, their bodies are shortened, particularly between the pectoral and pelvic girdles, and are flattened laterally, while the tail or caudal fin is symmetrical (homocercal). This arrangement leads to more efficient swimming movements. Teleosts are further distinguished from elasmobranchs by the possession of distinct vertebrae, and pectinate (toothed) gills which open into a single cavity guarded by a muscular operculum. They have no spiracle, spiral valve in the intestine, optic chiasma in the brain or conus in the heart.

The guppy, known also as the millions fish, is small and comes in a large number of colour and form varieties. A freshwater teleost, it originates from South America and lives on mosquito larvae and detritus. It is a member of the Cyprinodontiformes, or toothed carps, which possess a large mouth fringed by small conical teeth. Guppies have paired fins but there is no adipose (posterior dorsal) fin. The first dorsal fin ray is often

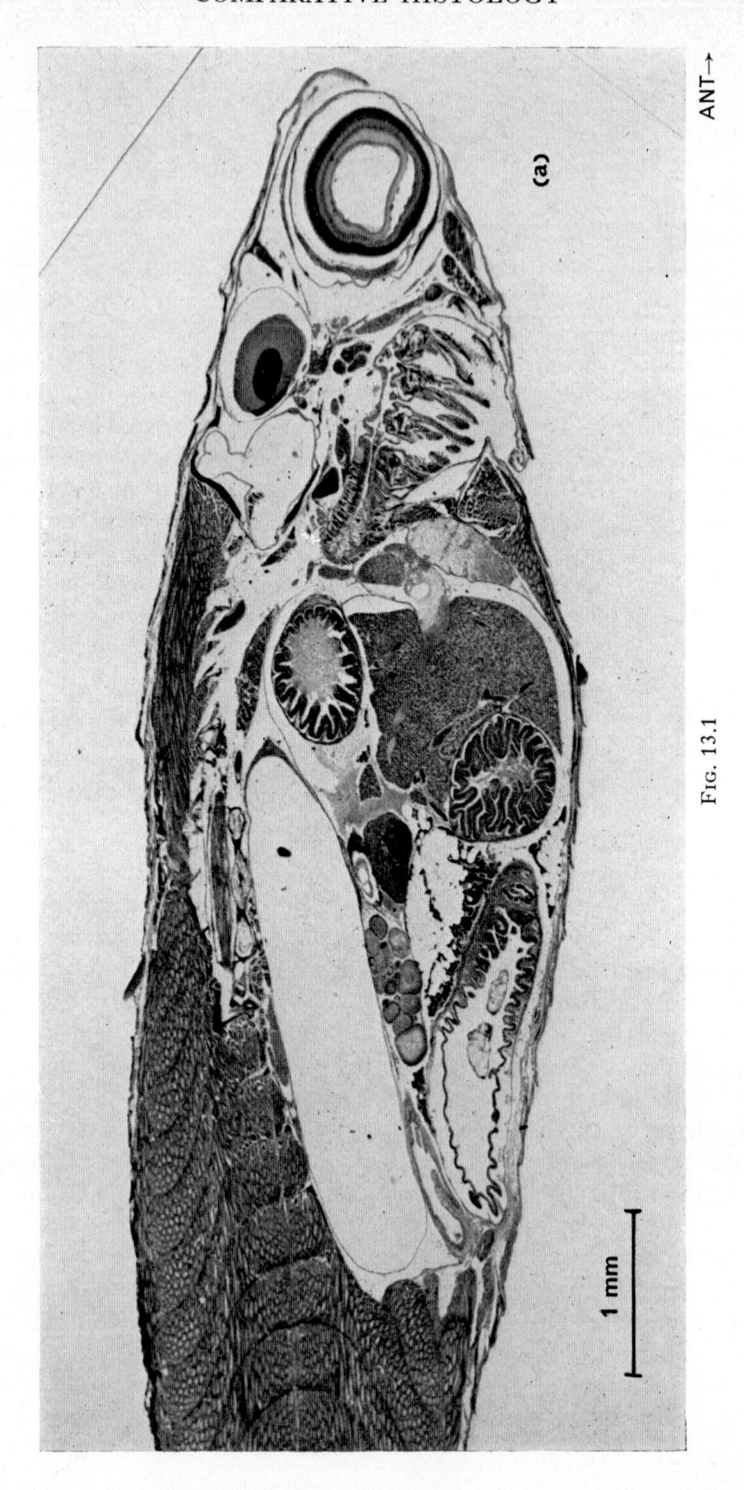

Fig. 13.1

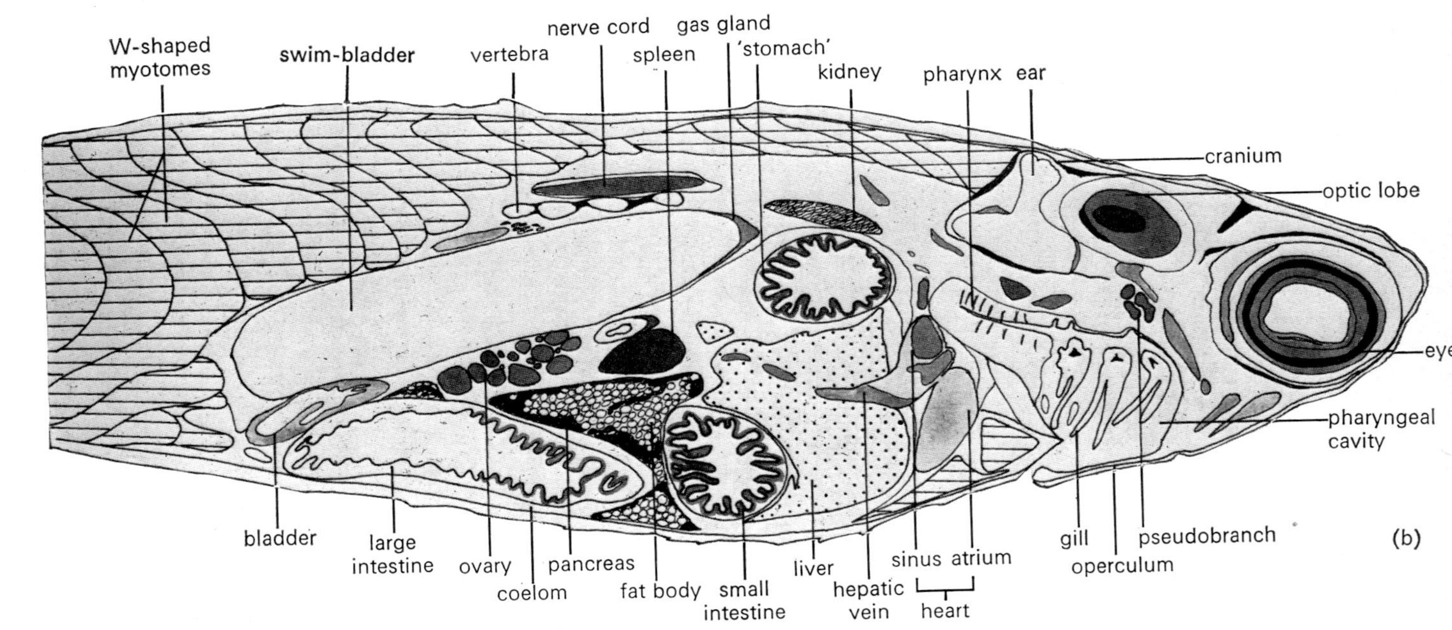

FIG. 13.1 (a) VLS and (b) explanatory drawing of the anterior end of a young female guppy (*Lebistes reticulatus*).

stiffened. Guppies are viviparous and the sexes show dimorphism, the male being brightly coloured. Figure 13.1 is a VLS of a young female guppy, showing the positions of the viscera and swim-bladder, and the W-shaped myotomes for fast swimming.

The body wall is covered by a skin consisting of a thin, stratified epidermis overlying a fibrous dermis (Fig. 13.2). The epidermis bears mucous cells and covers the dermal scales that project onto the surface. These scales are thin, almost circular plates composed of fibrous lamellae covered by a bony surface layer and often bearing spines. They overlap, and thus form a flexible yet tough and relatively impermeable coat over the fish. The dermis is poorly vascularized and bears many chromatophores.

A much reduced and constricted notochord is present, but the functions of support and protection of the spinal cord have been taken over by the vertebral column. During development the notochord sheath or perichordal tube [see Fig. 12.5(a)] becomes completely invaded by bone and is built into a complex jointed rod which allows for rigidity with flexibility. Each unit (vertebra) consists of a biconcave (amphicoelous) central disc (centrum) with an associated neural arch and spine, plus other attachments (such as ribs or haemal arches). Adjacent vertebrae articulate on one another by fibrous ligaments produced from the remains of the notochord. Each vertebra corresponds to a myotome, but it is placed intersegmentally so that each myotomal muscle block pulls on two adjacent vertebrae. In the young teleost a series of cartilaginous plates fuse with the auditory (otic), optic and olfactory capsules to form a complete, protective box (chondrocranium) round the brain. This structure later becomes partially replaced and surrounded by bone, forming the skull.

Nervous system

The nervous system of teleosts is similar to that of elasmobranchs, but the brain (Fig. 13.3) shows increased development, particularly of the cerebellum and medulla. These regions receive inputs from the lateral line, which becomes one of the chief sense organs. In some fish the lateral line is sensitive to electric currents generated by modified trunk muscles which form an electric organ or series of electroplates. Such an electric sense is normally used by fish which live in shallow murky waters.

The spinal cord has undivided dorsal horns of grey matter, and separated ventral horns. Dorsal and ventral roots join to form spinal nerves. The medulla possesses two giant Mauthner neurones (Fig. 13.4) lying near the mid-line: their axons decussate and descend in the spinal cord just below the central canal (see Fig. 13.19). They are present in all teleosts and function in the co-ordination of fast swimming escape reactions. The medulla roof forms a choroid plexus. The cerebellum is very large and its thickened roof bulges out backwards, and anteriorly under the optic tectum as *valvuli cerebelli*. The paired hollow optic lobes are much larger

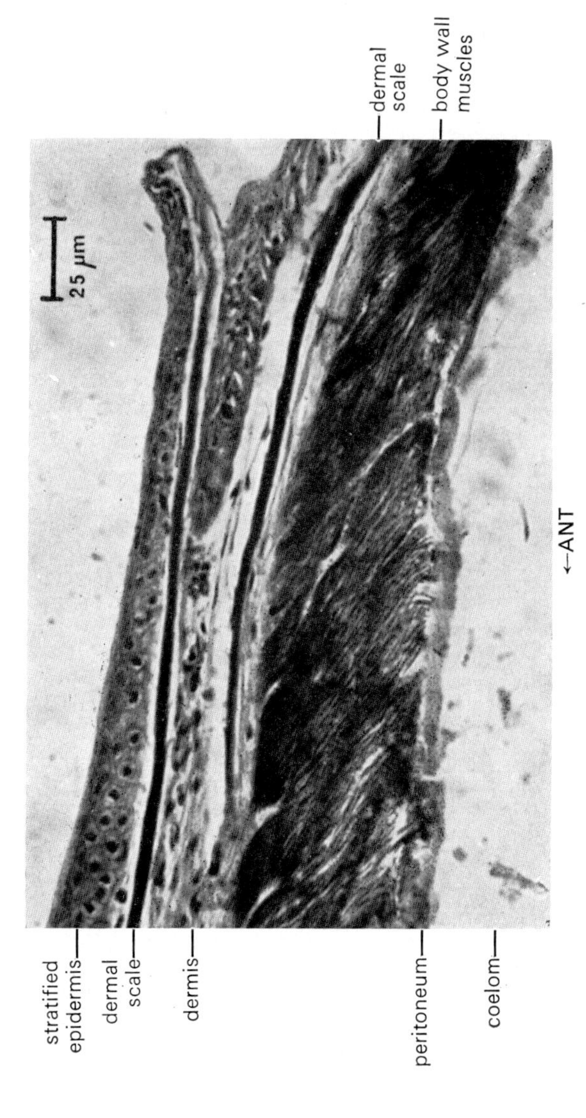

FIG. 13.2. Section through the body wall of the guppy, showing the overlapping dermal scales.

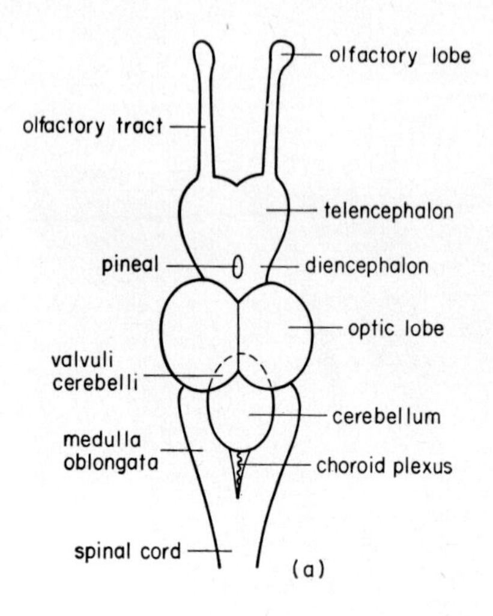

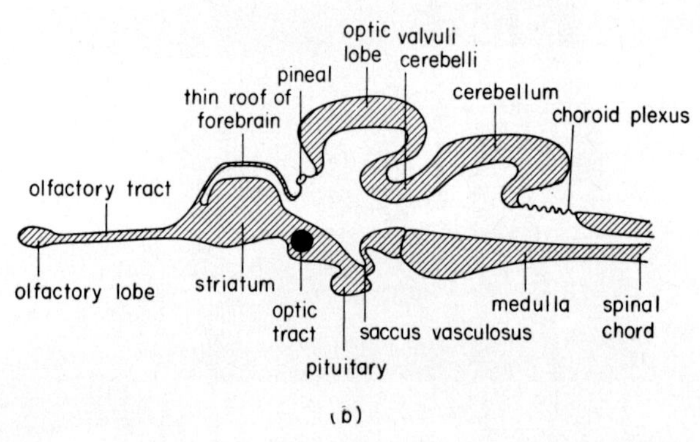

FIG. 13.3 (a) Plan of a teleost brain viewed from the dorsal side; (b) diagram of a VLS of the teleost brain.

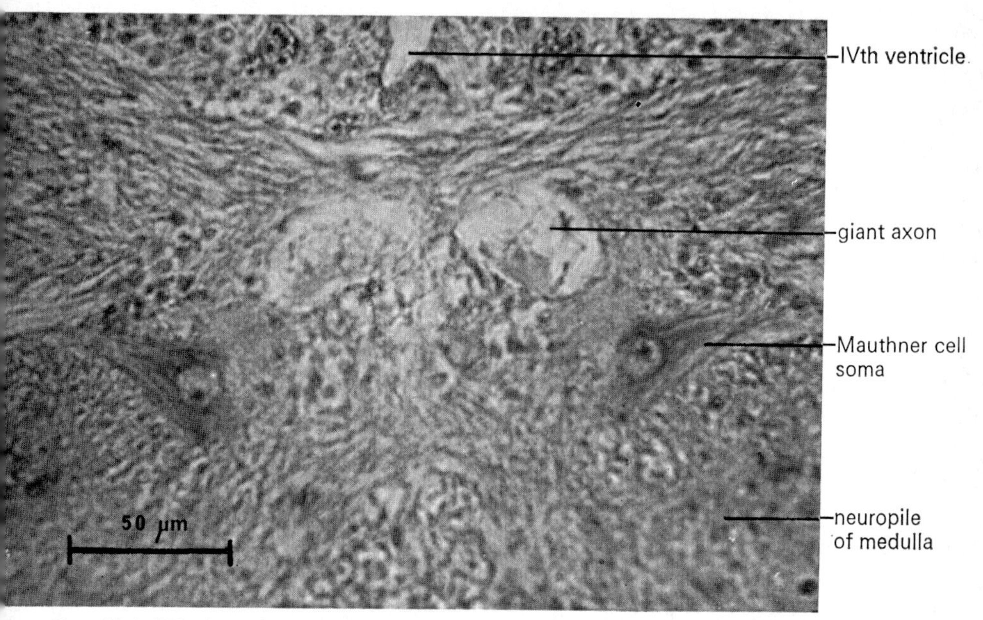

FIG. 13.4. TS through part of the medulla oblongata of the guppy brain, showing the giant axons and cell bodies of the paired Mauthner neurones. (Both axons and cell bodies are visible in the same section as the axons run forwards for a short distance before turning backwards and decussating.)

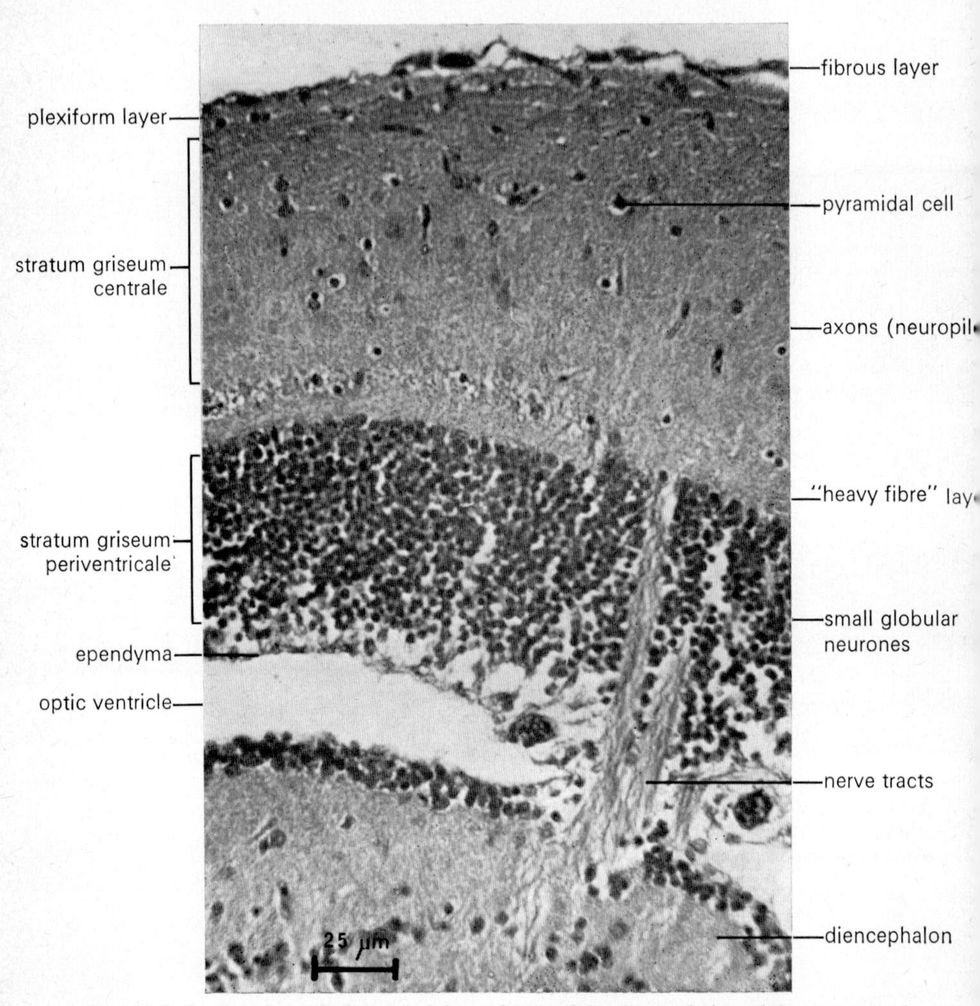

FIG. 13.5. Section through the optic lobe and part of the diencephalon of the guppy brain.

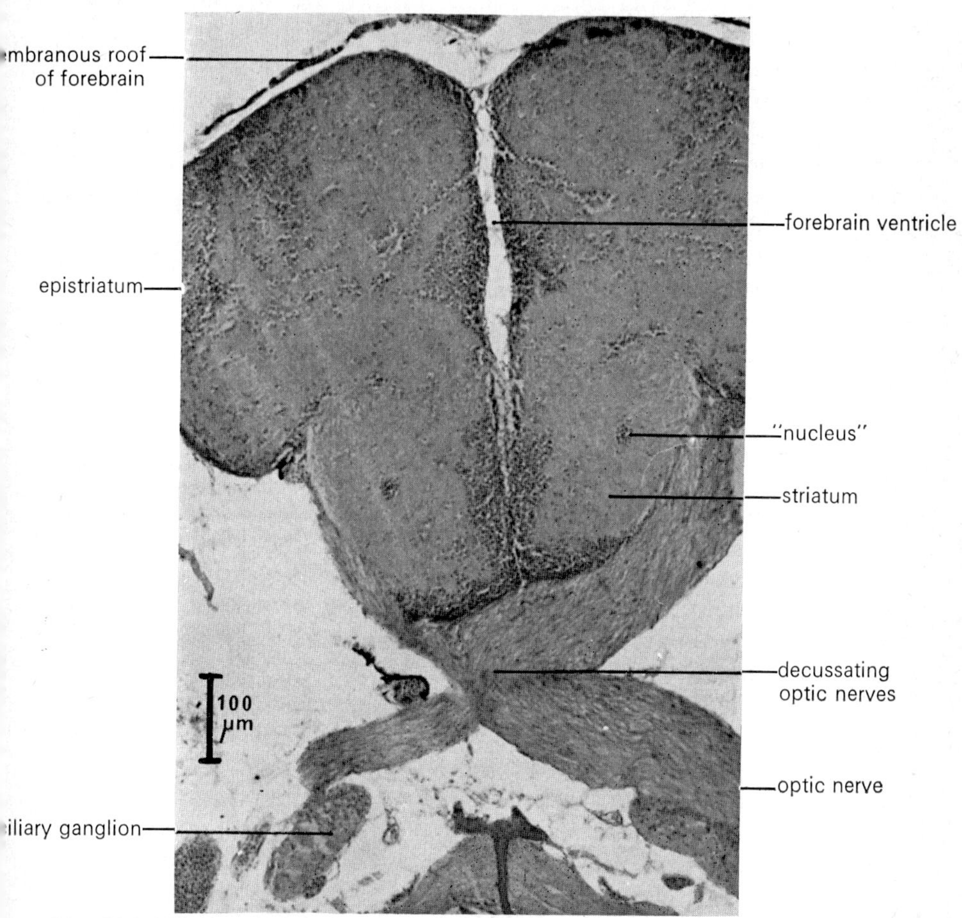

FIG. 13.6. TS through the guppy brain in the posterior forebrain region. Note the thin membranous roof of the forebrain, and the decussating optic tracts.

than in elasmobranchs, and show a well-developed laminar pattern (Fig. 13.5). There is no optic chiasma, the optic nerves decussating outside the brain (Fig. 13.6, compare with Fig. 11.12). The diencephalon is small with a large dorsal pineal of receptor-like structure, and a ventral infundibular evagination which forms a large saccus vasculosus just behind the pituitary. The saccus vasculosus is lined with epithelium composed of sustentacular cells and ciliated sensory cells. Because the cilia are stimulated by movements of the c.s.f. it has been suggested that the saccus vasculosus acts as a depth (pressure) receptor, but there is also evidence for its secretory function. A secretory subcommissural organ is located at the hind end of the diencephalon. The cerebrum is small and is characteristically everted in teleosts, i.e. appears as a solid mass of neurones (basal nuclei) covered dorsally and laterally by a thin, non-nervous roof (Fig. 13.6).

Teleosts are the first group to show chains of autonomic sympathetic ganglia. These look similar to cerebrospinal ganglia except that they consist of small, irregular, multipolar neurones which do not all have Schwann cells (see Fig. 11.11).

Sense organs

Teleost fish are generally active creatures well supplied with sense organs. Numerous simple nerve endings, chemoreceptors, temperature receptors, osmoreceptors and proprioreceptors occur both on the outside and inside of the body, while the organs of special sense show an increased development and specialization.

The eye (Fig. 13.7), which conforms to the general vertebrate plan, not only protrudes from the body but also possesses a spherical lens which bulges through the iris, and thus ensures a wide field of vision. The section shown has only passed through the edge of the lens, so the large size of the latter is not apparent. The lens, normally adjusted for near vision, can be swung backwards by a *retractor lentis* muscle attached to its posterior edge, and thus focused on distant objects. The retina is composed of two types of photoreceptors: long, thin rods with a low threshold to light, and short, stout cones with a high threshold. Teleosts often have twin cones in the retina, although the significance of these is uncertain. Unlike the retina of the elasmobranch, the teleost retina (Fig. 13.8) can show adaptation to either light or dark. This is achieved by movements of both pigment and receptors. On exposure to light the external pigment moves inwards towards the centre of the eye, as do the cones, while the rods are withdrawn into the pigment layer [Fig. 13.8(a)]. The visual cells have contractile portions next to the receptor which allows considerable movement [see also Fig. 14.11(b)]. In the dark, conversely, rods are moved inwards towards the light, while cones and pigment move outwards [Fig. 13.8(b)]. The retina is nourished either by a layer of blood-vessels on its

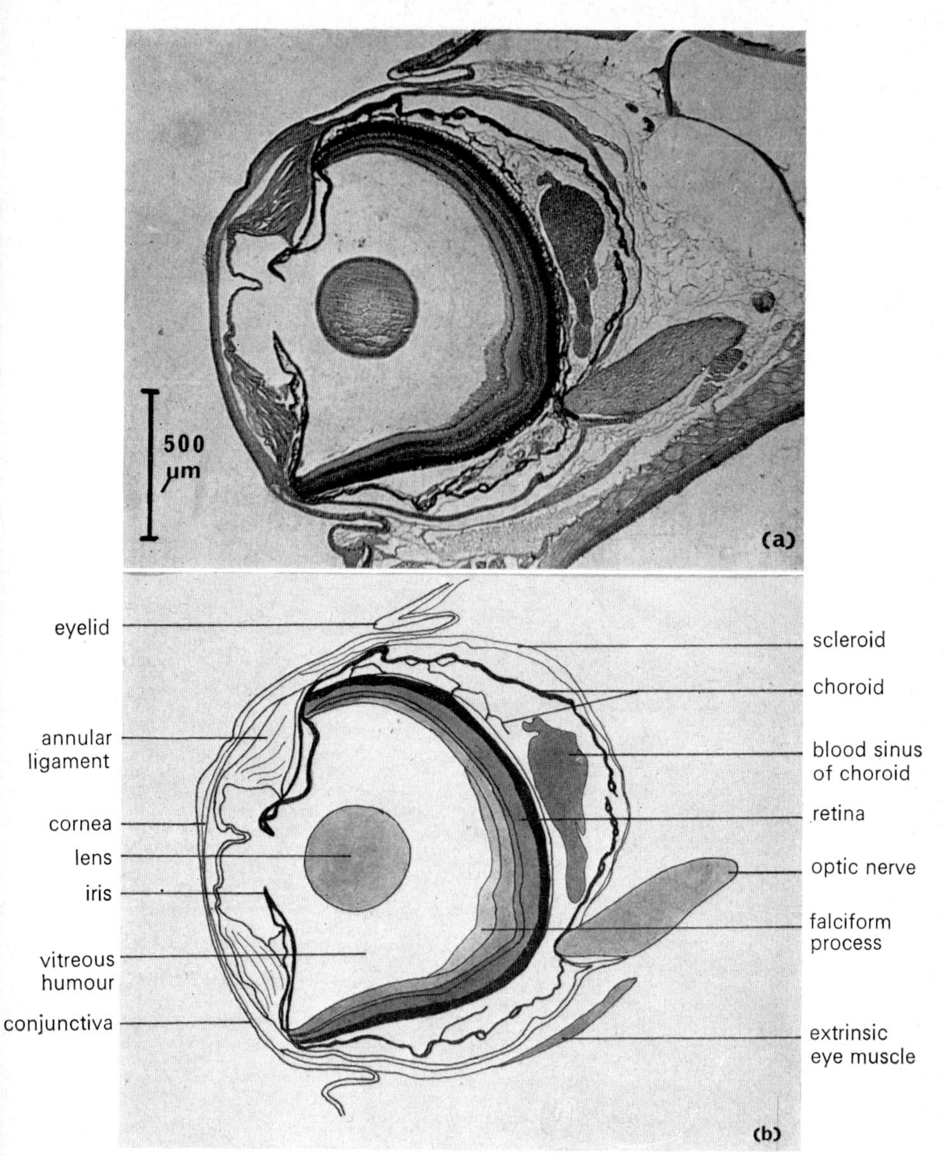

FIG. 13.7 (a) Section and (b) drawing of the eye of the guppy. Note the two layers of the choroid coat enclosing a large blood sinus between them.

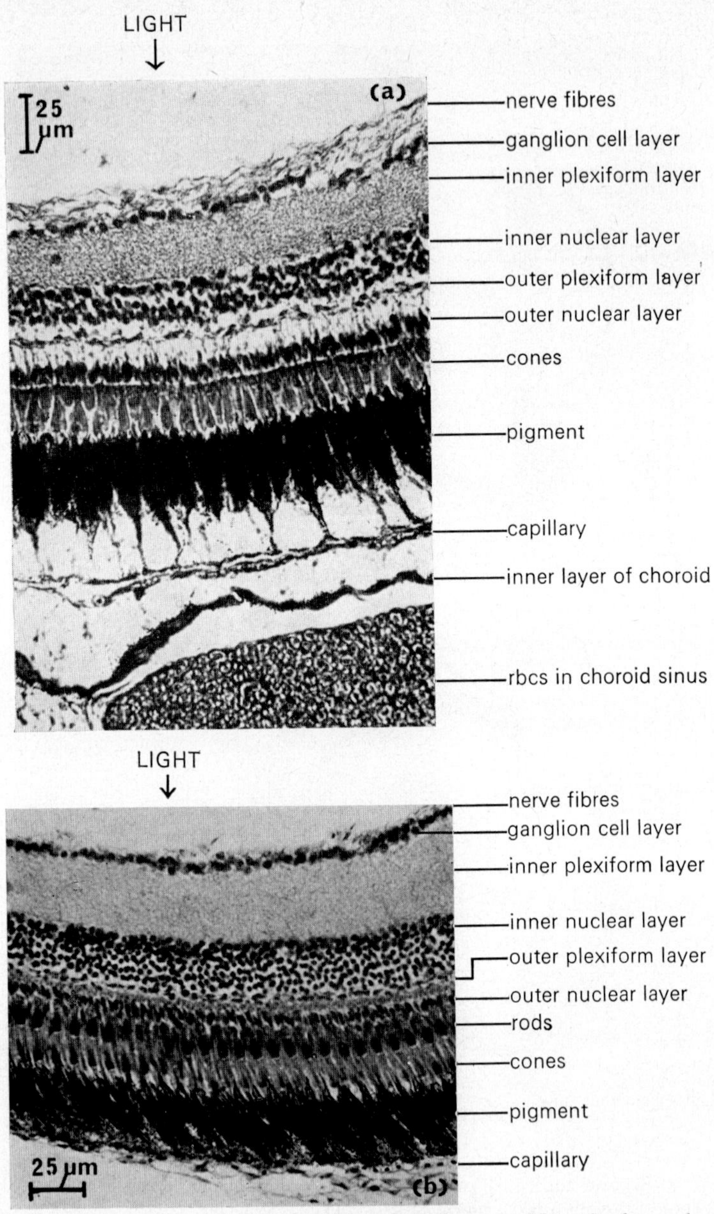

FIG. 13.8 (a) Details of the retina from a light-adapted guppy. Note the forward position of the pigment covering the rods, while the cones are exposed; (b) details of the retina from a dark-adapted guppy. Note the retracted position of the pigment, leaving the rods and cones exposed.

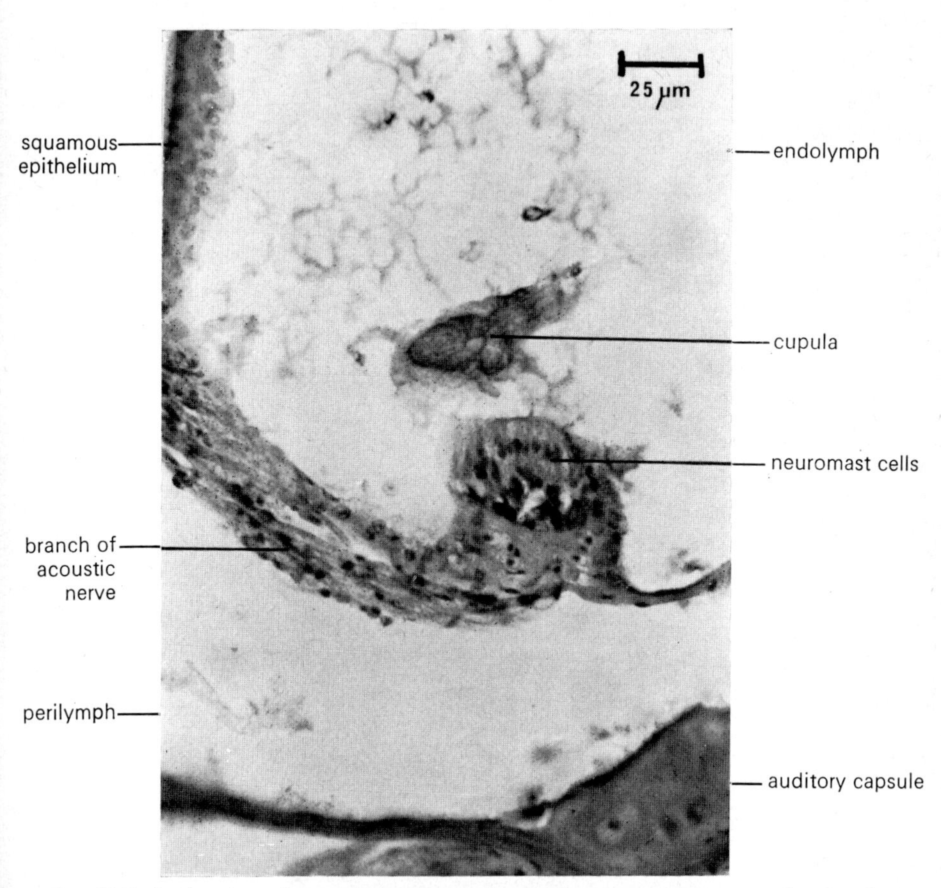

FIG. 13.9. Section through a semi-circular canal in the inner ear of the guppy, showing the gelatinous cupula and the neuromast cells of the crista.

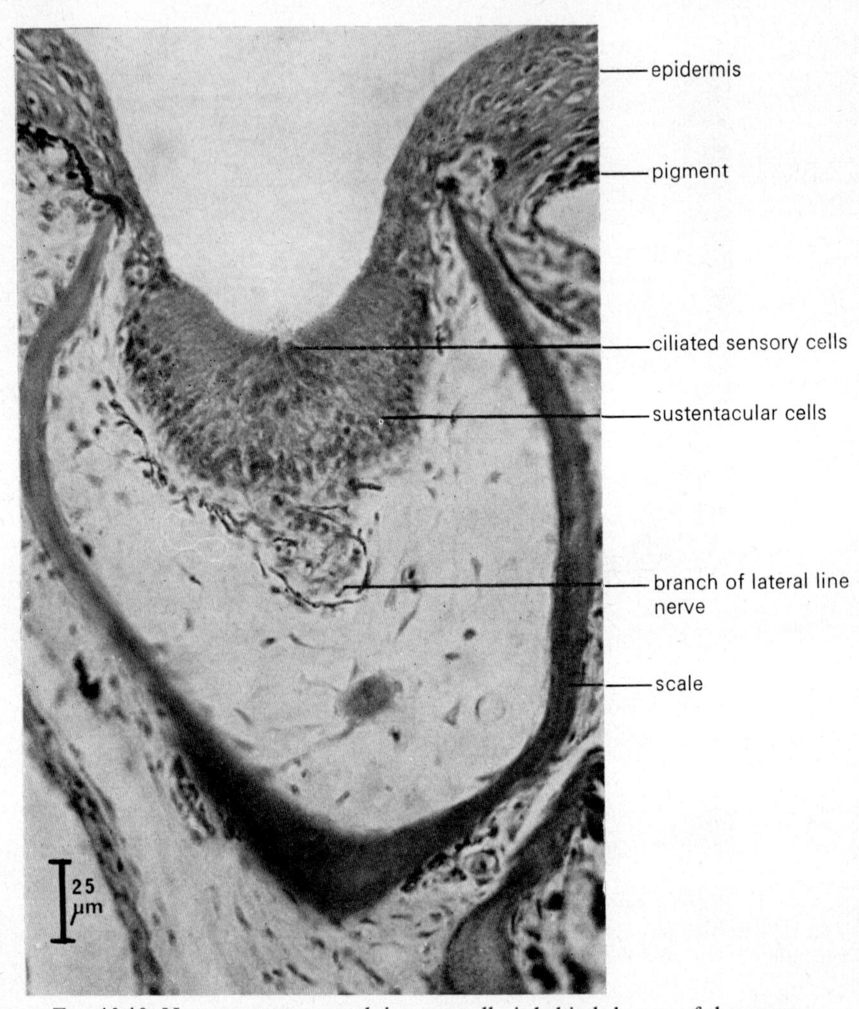

FIG. 13.10. Neuromast organ sunk into a small pit behind the eye of the guppy.

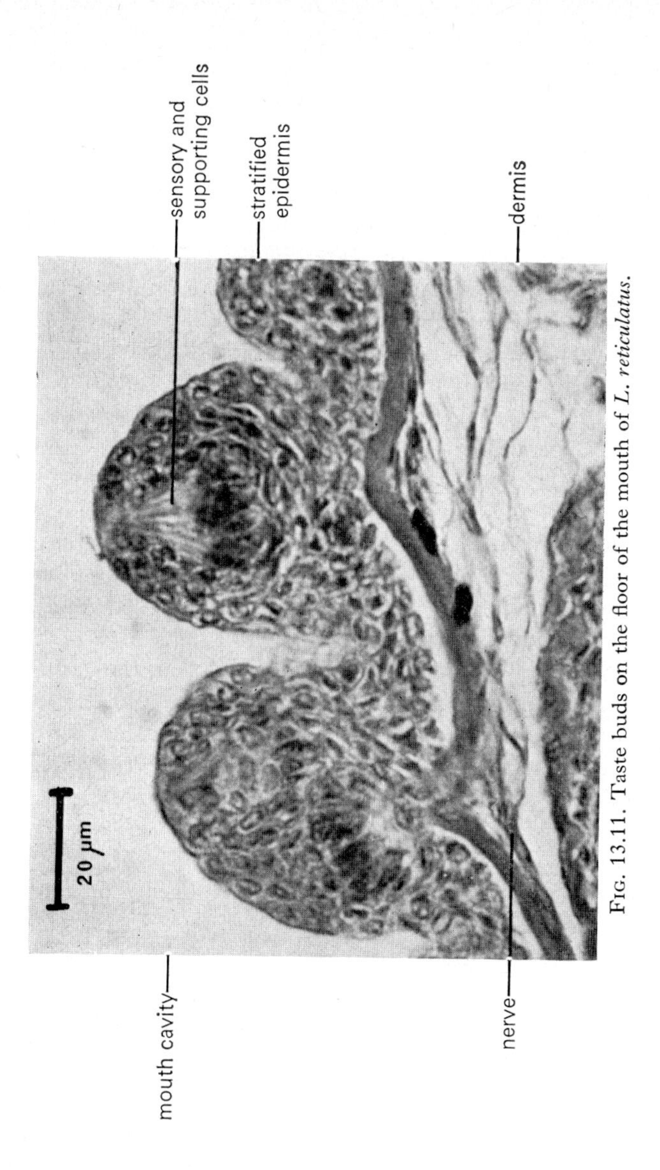

Fig. 13.11. Taste buds on the floor of the mouth of *L. reticulatus*.

inner surface, or by a falciform process, a pigmented vascular fold coming from the choroid and lying inside the retina. The retina is backed by a heavily vascularized choroid, and a thin scleroid reinforced with cartilage. The cornea is of mesodermal origin and overlaid with epidermis. In some species, particularly those which come out of water, there is fusion of the epidermal and dermal layers to form a secondary spectacle for extra protection.

The ear of teleosts is very similar to that of elasmobranchs except that the *ductus endolymphaticus* is lost. The shapes and positions of the various components are somewhat different, in that the otolith organs are more separated from each other and from the semi-circular canals, and the otoliths are very big. Part of a semi-circular canal is shown in Fig. 13.9, demonstrating the cupula overlying the neuromast cells.

Similarly the lateral line system has the same components, although in teleosts the canals are usually restricted to the head, and are sunk into or below the scales. Single scattered neuromasts are also found (Fig. 13.10).

The olfactory organs lie in paired pits on the dorsal side of the head. There are two nostrils, each one having a separate anterior inlet and posterior outlet. The epithelial lining is like that of elasmobranchs except that in teleosts the areas of sensory epithelium are separated by areas of ciliated columnar epithelium.

Taste buds occur widely in teleosts, embedded in the epithelium of the mouth, pharynx, and gill pouches (Fig. 13.11, see also Fig. 13.21). They are densely packed in the roof of the mouth where they form the palatal organ. Taste buds are thought to be more conspicuous in freshwater than marine fish.

Endocrine organs

The pituitary of teleosts is much more advanced than that of elasmobranchs, with two histologically distinct portions (Fig. 13.12). The major part derives from the stomodaeum and is glandular, often separable into pars glandularis and pars intermedia; while the posterior part forms a pars nervosa consisting of neurosecretory nerve endings of the hypothalamo-hypophyseal tract and blood sinuses. There is considerable interdigitation of the two regions. In a few teleosts the stalk once connecting the stomodaeum and pars glandularis is retained as a ciliated duct opening into the buccal cavity.

A second neurosecretory organ is associated with the caudal end of the spinal cord in teleosts. This is the urophysis which lies laterally and ventrally to the spinal cord and consists of numerous Dahlgren neurosecretory cells which are quite small in the guppy but which often reach a large size. In some fish (Fig. 13.13) the urophysis forms a distinct organ with the neurosecretory tracts making a stalk analogous to the pituitary stalk of higher vertebrates.

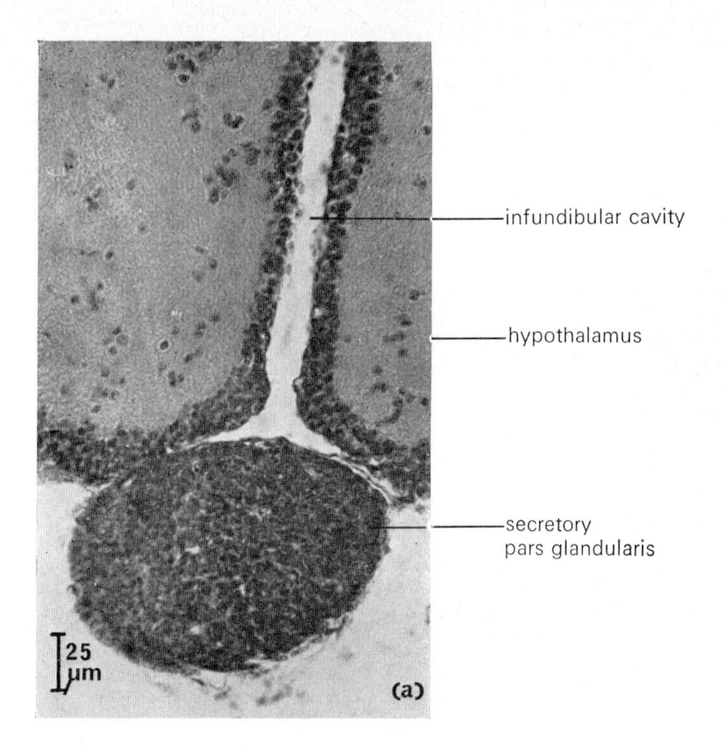

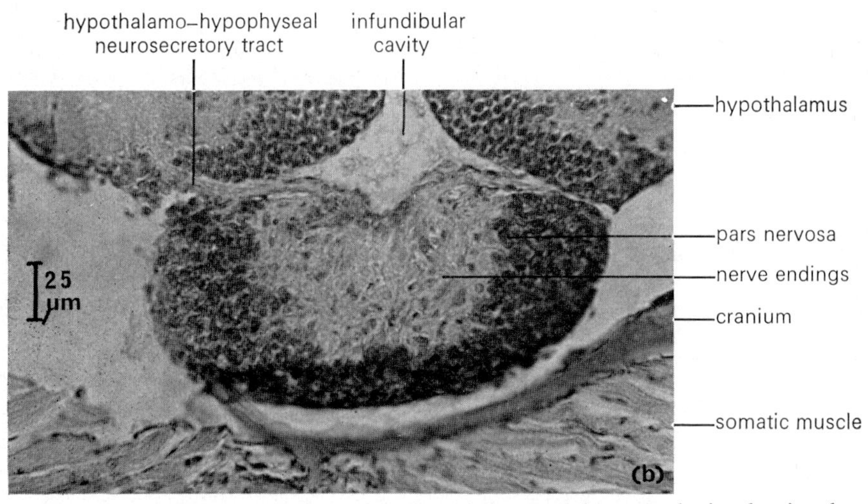

FIG. 13.12 (a) TS through the base of the hypothalamus of the guppy brain, showing the anterior lobe of the pituitary; (b) TS through the posterior lobe of the guppy pituitary, showing the hypothalamo-hypophyseal tract connecting the pars nervosa to the hypothalamus of the brain.

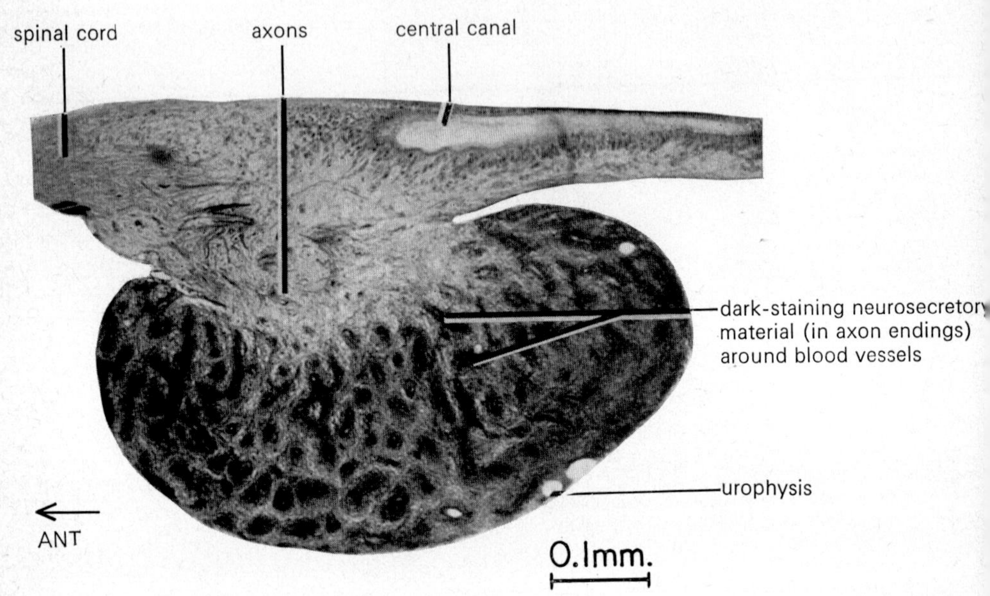

FIG. 13.13. VLS through the caudal neurohaemal organ (urophysis) of a teleost (*Gillichthys mirabilis*). This section was stained to show concentrations of neurosecretory material (in axon endings) around blood vessels. Photograph by kind permission of Howard A. Bern Fig. 1c Chapter 8 *in*: "Fish Physiology" Vol II (eds W. S. Hoar and D. J. Randall) Academic Press, 1969.

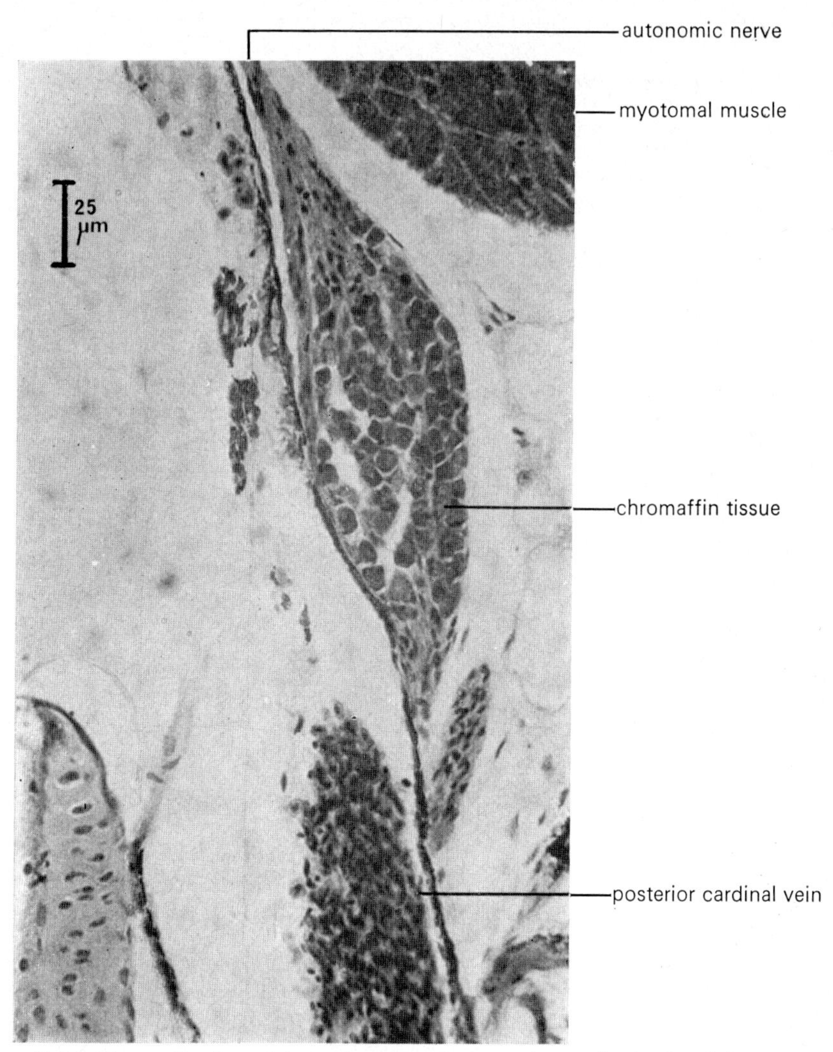

FIG. 13.14. Chromaffin tissue lying adjacent to the posterior cardinal vein of the guppy.

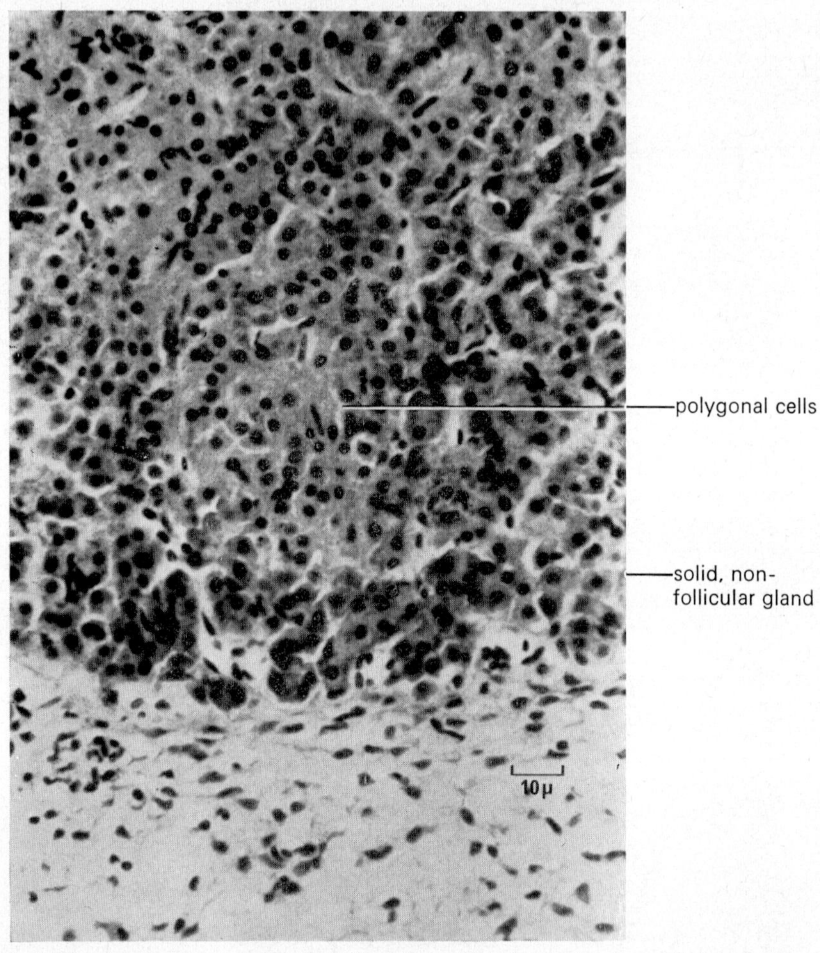

FIG. 13.15. Section through the solid ultimobranchial gland of a teleost (*Salmo gairdneri*). Note the cords and clumps of polygonal cells. Photograph by kind permission of D. Harold Copp & W. A. Webber, Fig. 5 Chapter 7 *in*: "Fish Physiology" Vol II (eds. W. S. Hoar and D. J. Randall). Academic Press, 1969.

The thyroid consists of numerous follicles scattered along the ventral aorta and afferent branchial vessels, while in a few species they form compact masses between the bases of the first gill pouches. The follicles have a typical structure and manufacture thyroid hormones.

Inter-renal and chromaffin tissue are normally found close to one another in teleosts. They occur in the region of the anterior end of the kidney and round the posterior cardinal veins (Fig. 13.14).

Ultimobranchials also develop at the bases of the last branchial arches and lie between the oesophagus and sinus venosus as in elasmobranchs, but in teleosts they are solid glands composed of cords and clumps of polygonal cells (Fig. 13.15). However, they also function in controlling calcium and phosphate levels.

Respiration

Most teleosts use gills as the main respiratory surface, although accessory respiratory structures also occur (e.g. skin, pharyngeal and buccal diverticula), particularly in those species adapted to spend periods of their life out of water. The guppy, however, is entirely aquatic and uses only gills.

There are five gills, although the first half-gill (hemibranch), which is modified to form a pseudobranch, is often covered by epithelium. In the guppy there are no visible pseudobranchs as they have sunk inwards from the branchial chamber and are separated from it by stratified epithelium and connective tissue (see Fig. 13.1). Each pseudobranch, however, still takes the form of leaf-like lamellae composed of large rounded granular cells clustered round branching capillaries and surrounded by a pigmented sheath (Fig. 13.16). It has been suggested that such a glandular pseudobranch, typical of many teleosts, may have an endocrine function as yet unspecified.

Compared with the elasmobranchs, the interbranchial septae of teleosts are much reduced, and thus the lamellae project freely into the opercular cavity [Fig. 13.17(a), see also Fig. 13.1]. This arrangement provides a larger respiratory surface in contact with the water. The gill filaments also bear secondary lamellae, and are well vascularized with single afferent and efferent vessels [Fig. 13.17(b)]. The lamellae have the same structure as those of elasmobranchs (see Fig. 12.15). Water is taken in through the mouth, and a nearly-continuous flow of water over the gill lamellae is ensured by the co-operation of buccal and opercular muscular pumps. Water is expelled through a single opercular opening on each side, the openings being guarded by opercular flaps. Occasional mucous cells and large salt-secreting cells occur at the bases of the gill lamellae, and assist in osmotic and ionic regulation.

Almost all teleosts possess an air-bladder or swim-bladder which lies below the kidneys and above the gut, extending the length of the abdominal cavity (see Fig. 13.1). In some fish a pneumatic duct connects

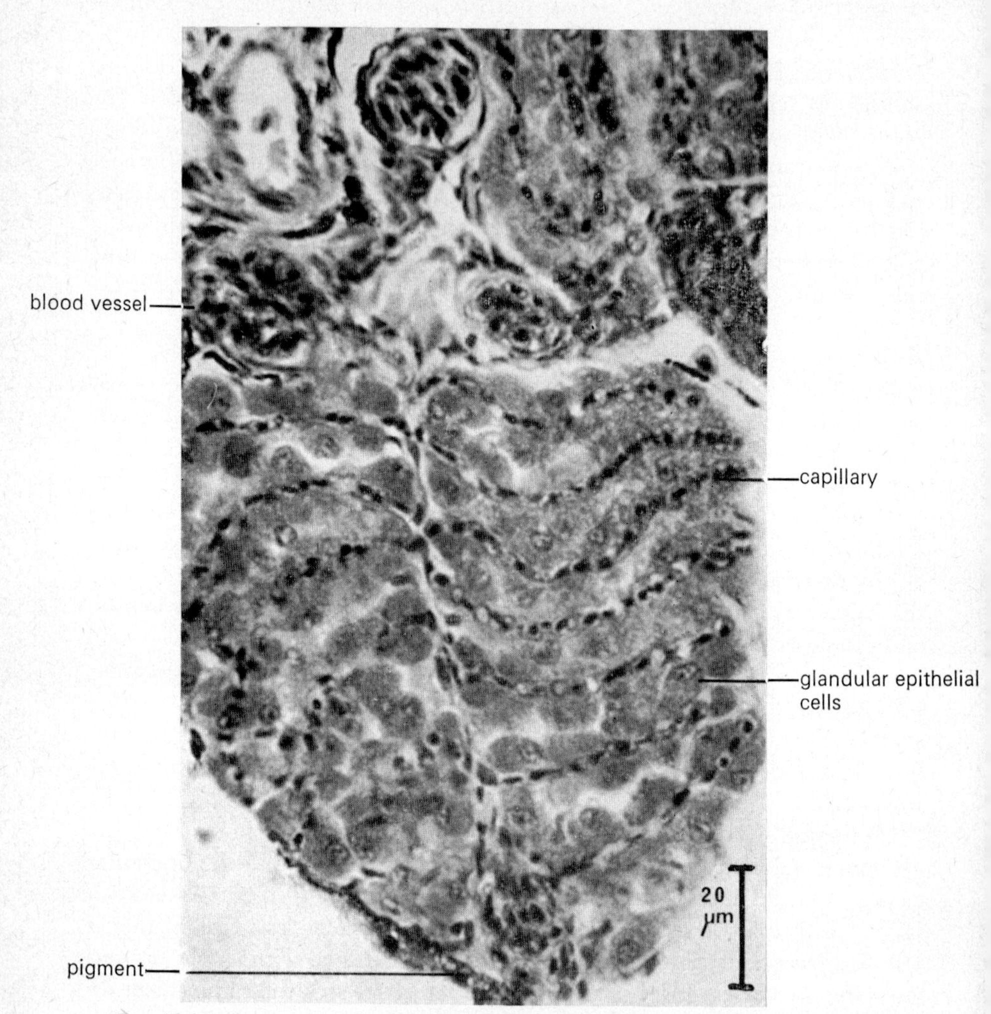

FIG. 13.16. Section through the glandular pseudobranch in the neck of the guppy.

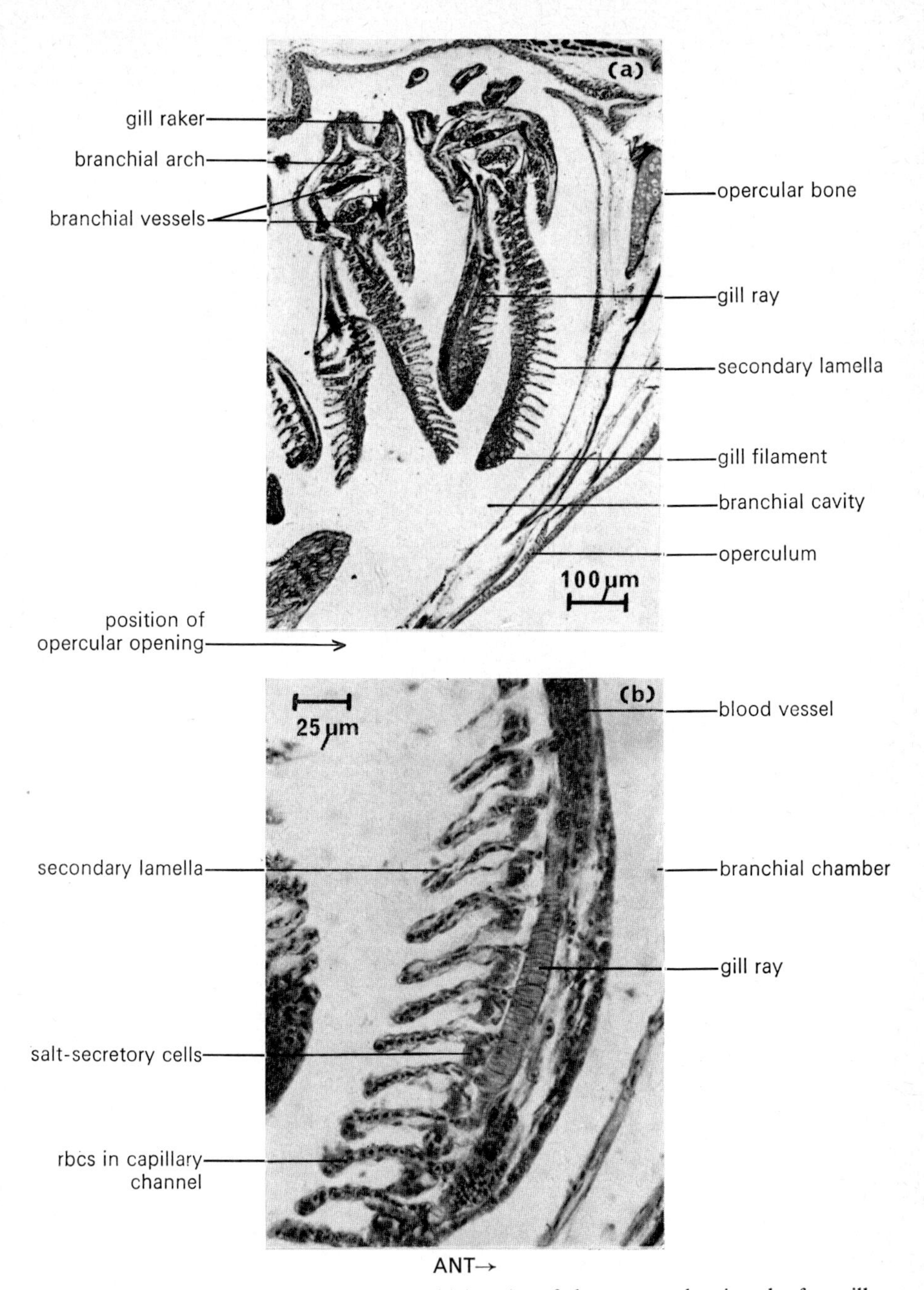

FIG. 13.17 (a) Section through the branchial cavity of the guppy, showing the free gill lamellae, the operculum and the single opercular opening; (b) detail of one gill filament from the guppy gill.

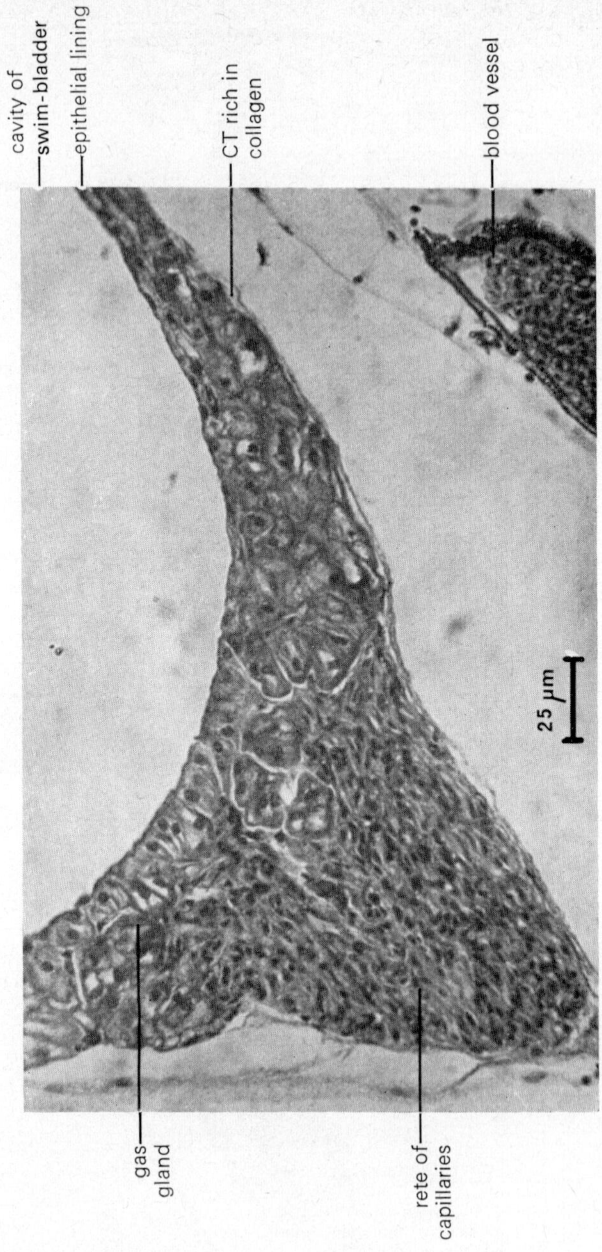

FIG. 13.18. VLS through the anterior end of the guppy swim-bladder, showing the gas gland and rete mirabile.

the bladder to the gut, but in others, including the guppy, the bladder is completely closed. There is a simple saccular swim-bladder in the guppy, with an anterior gas gland supplied by a rete mirabile of blood capillaries (Fig. 13.18). Gas can be secreted into the bladder by metabolic activity of the gas gland; and can be released across an area of thin epithelium on the general surface of the bladder (or via the pneumatic duct in those species which possess one). In some fish this area, known as the oval, is restricted and can be shut off from the rest of the bladder by a sphincter muscle.

Circulation

As in other fish, the teleost circulation is single and possesses one pump. This is the four-chambered heart, composed of typical striated cardiac muscle and enclosed in a thin elastic pericardium. There is no connection between pericardial and peritoneal chambers as in lower fish. Large paired Cuverian veins empty venous blood into a thin-walled sinus venosus and thence into a single atrium and thick-walled ventricle (Fig. 13.19). There is no contractile conus, but instead an elastic bulbus arteriosus at the base of the ventral aorta which is separated from the heart by a pair of valves and which serves to smooth the pressure pulse created by the ventricular beat.

The circulatory system is similar to that of elasmobranchs except that there is a circulation associated with the swim-bladder. The major change in circulatory pattern, however, comes in the Choanichthyes where an important circulation develops to the air-bladder or lungs and is returned directly to the heart. This feature is retained and developed further in the tetrapods.

The erythrocytes of teleosts are small (12 μ), oval and nucleated, and produced in the spleen (Fig. 13.20) and kidney [see Fig. 13.24(a)]. White blood cells are formed in the thymus, which is a large distinct organ in the medial wall of the branchial cavity in teleosts with a structure similar to that of elasmobranchs (see Fig. 12.18).

Digestion

Guppies eat detritus and small invertebrates, and have an unspecialized gut lined throughout with simple columnar epithelium. Fish are unusual in possessing striated muscles round the buccal cavity, pharynx and oesophagus, but the rest of the gut musculature is smooth. There are no specialized glands.

The wide mouth of the guppy is fringed with small teeth for grasping food. The buccal cavity and pharynx are regions for the reception of food, and are lined with stratified epithelium bearing mucous cells, taste buds and more pointed teeth (Fig. 13.21). A short oesophagus leads into the intestine which is long and coiled to provide an increased surface area for

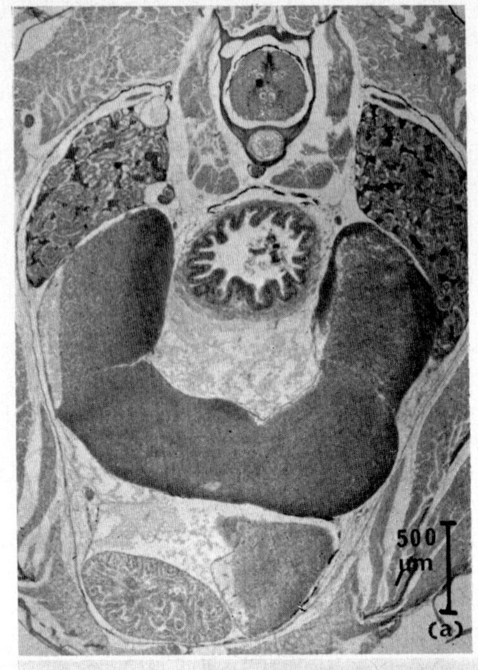

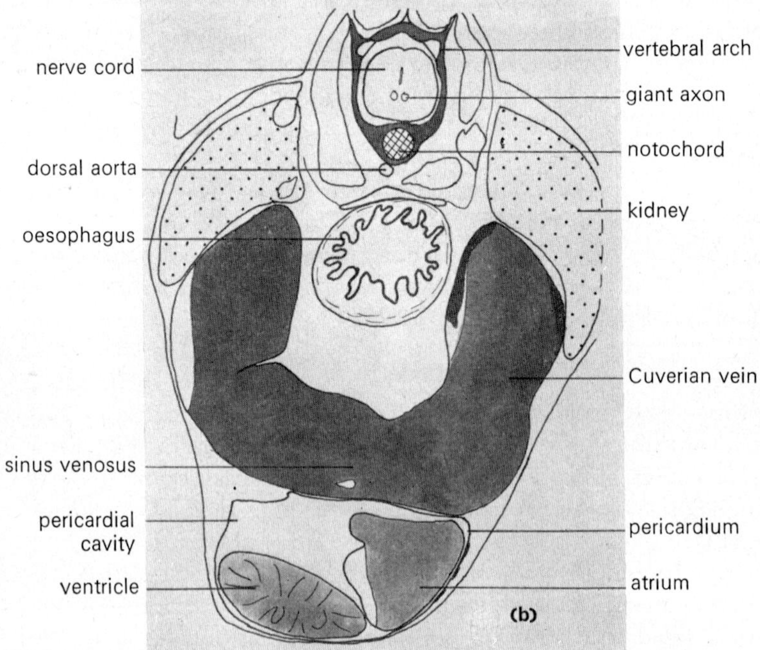

FIG. 13.19 (a) TS and (b) drawing of the guppy to show the large Cuverian veins. Note the thin pericardium round the heart.

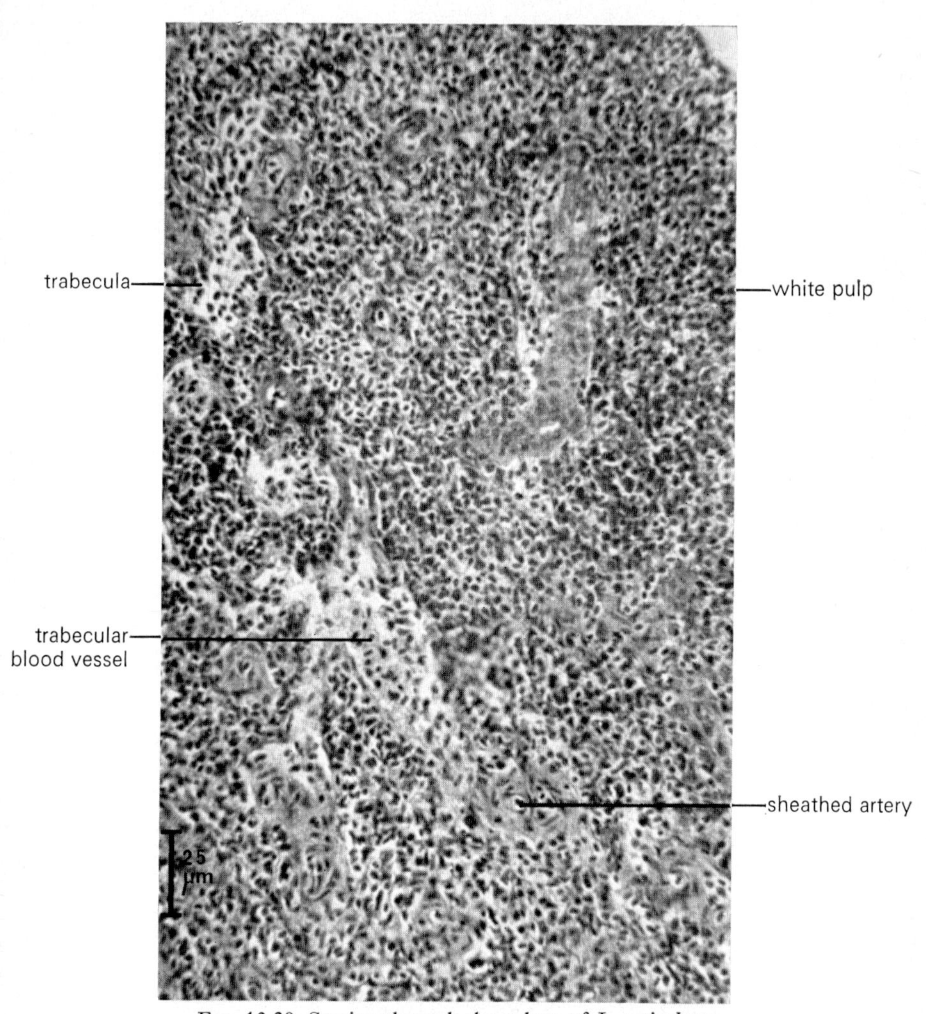

FIG. 13.20. Section through the spleen of *L. reticulatus*.

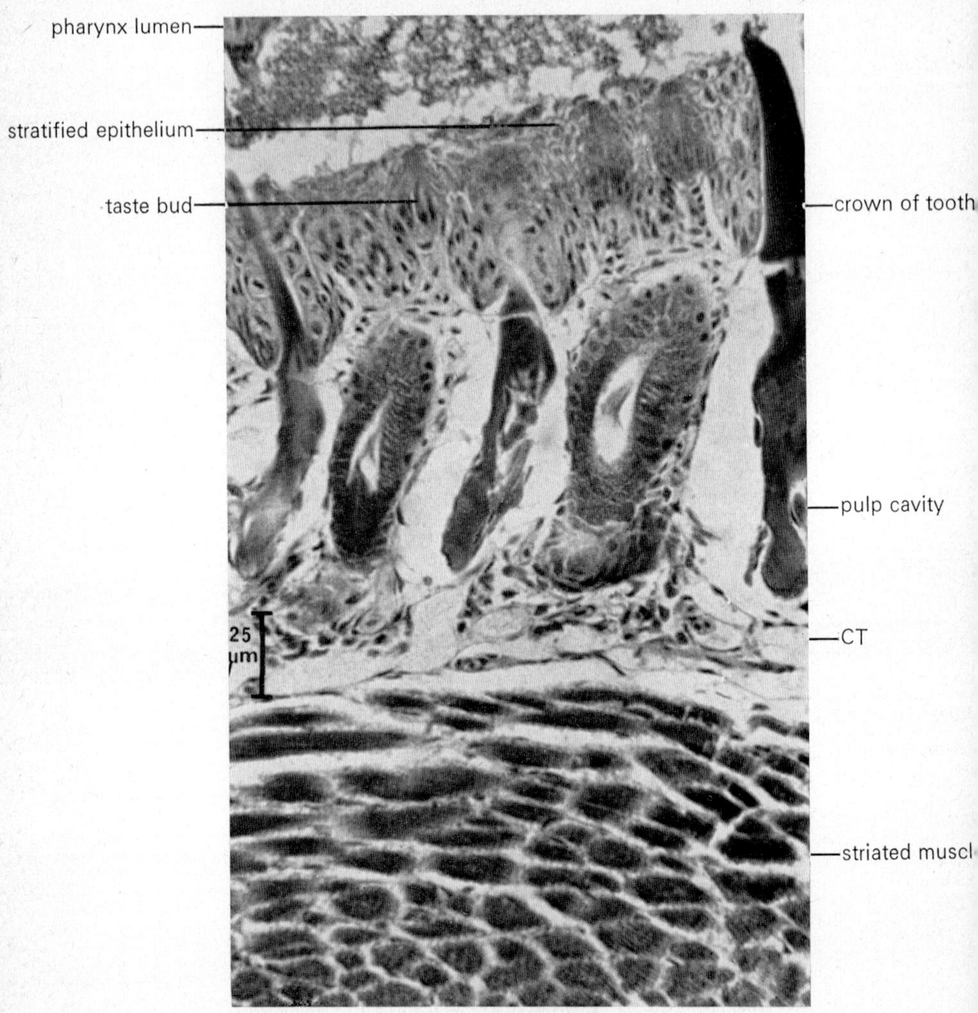

Fig. 13.21. LS through the pharynx wall of the guppy, showing taste buds and the pointed pharyngeal teeth. Note the extensive striated musculature.

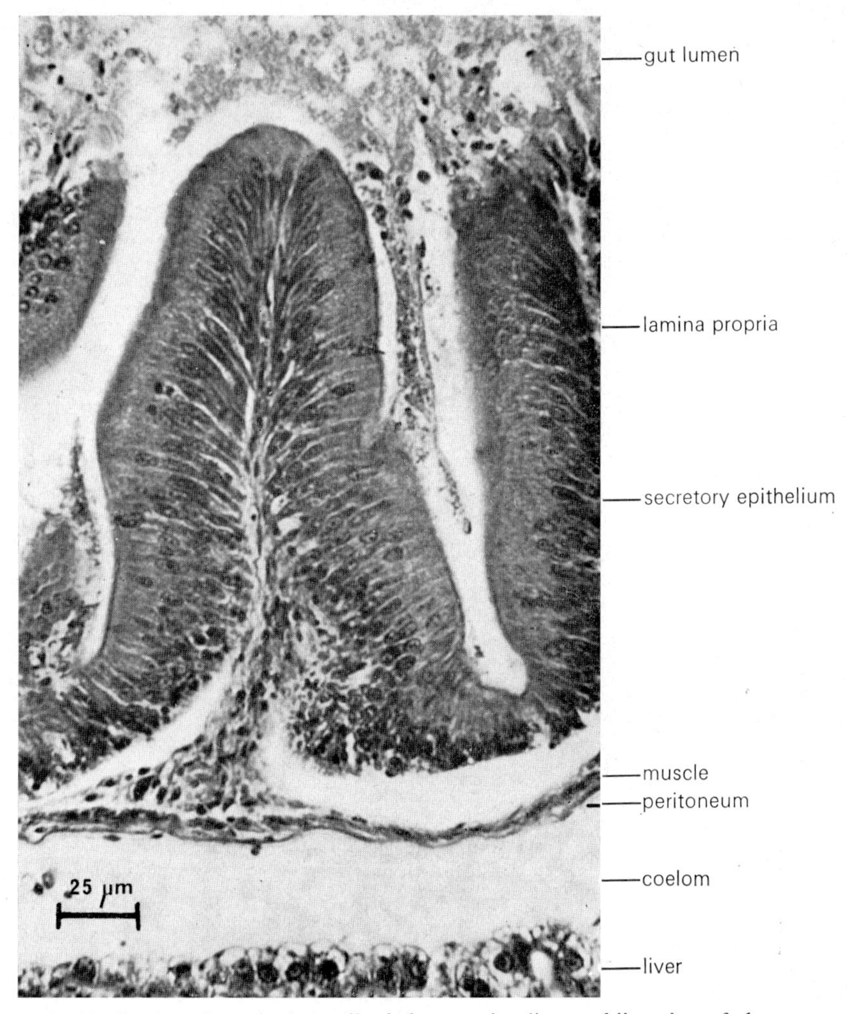

FIG. 13.22. Section through the wall of the anterior "stomach" region of the guppy intestine.

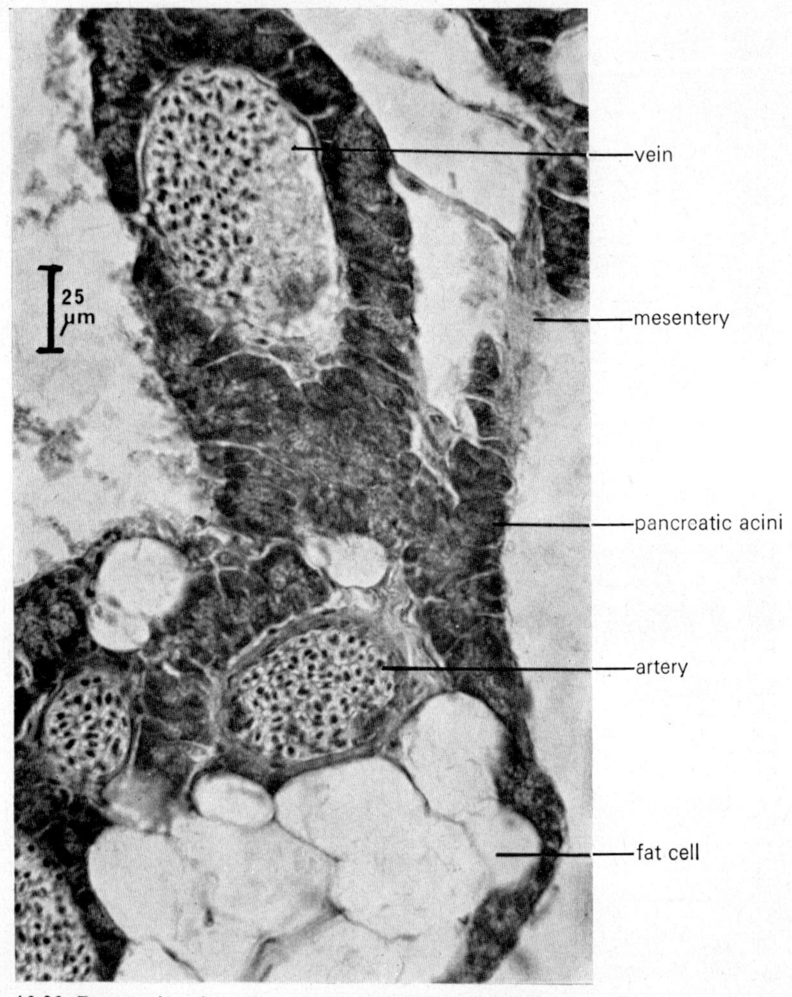

FIG. 13.23. Pancreatic acinar tissue scattered among unilocular fat cells in the abdominal mesentery of the guppy.

digestion. It is lined with a folded epithelium, which is taller and thrown into bigger folds in the anterior region, which thus approximates to a "stomach" (Fig. 13.22). The anterior portion is mainly secretory: the posterior portion, which has a lower epithelium, is chiefly absorptive. The hind-gut (rectum) is often separated from the intestine by an ilio-caecal ring, but it is similar in structure to the intestine although generally it has thicker muscle layers and more mucous cells.

Of the glands associated with digestion, the liver (and gall bladder) has the same characteristic structure as in elasmobranchs. The pancreas occurs as diffuse masses of glandular acinar cells suspended in mesentery and covering most of the intestine surface (Fig. 13.23). A few small isolated groups of endocrine islets of Langerhans occur as separate structures from the exocrine pancreas.

Excretion

Most teleost fish excrete ammonia as their main nitrogenous waste product. Some ionic regulation and disposal of waste substances occurs across the gill surface, but the chief excretory organs in the adult are paired opisthonephric kidneys. These are diffuse elongated structures lying above the swim-bladder (see Fig. 13.1). The degree of development of the nephric tubules can be correlated with environment. Thus in many marine teleosts, which are not subjected to osmotic inflow of water but rather have to counteract outflow, the glomeruli are small or secondarily absent, the proximal tubules constricted and the distal tubules often missing. As guppies are freshwater fish, their kidneys possess fairly large, well-vascularized renal corpuscles and long coiled proximal and distal tubules, as well as characteristic nests of haematopoietic tissue [Fig. 13.24(a)] and isolated encapsulated Stannius corpuscles which may play a role in the manufacture or storage of steroid hormones. Urine passes backwards from the kidneys in paired nephric ducts [Fig. 13.24(b)], which originate from the archinephric ducts and which unite posteriorly to form a mesodermal bladder derived from part of the embryonic cloaca. There is no cloaca in adult teleosts, and the urinary bladder opens directly to the outside via the urinary papilla.

Reproduction

All teleosts are dioecious, and the gonads are generally paired although they tend to be single in species where the young are born live (such as the guppy).

In the male guppy the testis is composed of strings of cysts (also called ampullae or acini) bound together by connective tissue into compact lobules (Fig. 13.25). Each cyst is lined with epithelium bearing germ cells and supporting Sertoli cells, and contains sperm at the same stage of development. All the sperm of a number of cysts ripen at the same time and

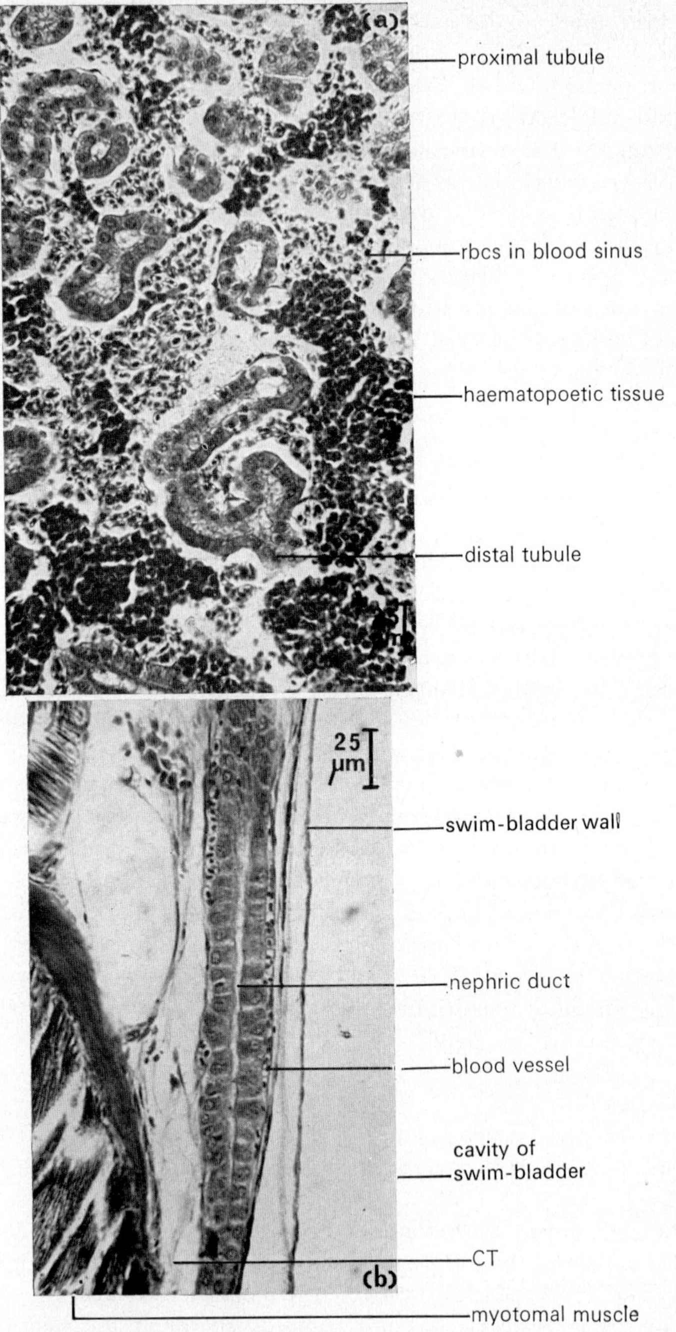

FIG. 13.24 (a) Section through the kidney of a young adult guppy. Note the dark nests of haematopoietic tissue; (b) LS through the nephric duct in the posterior abdominal region of the guppy.

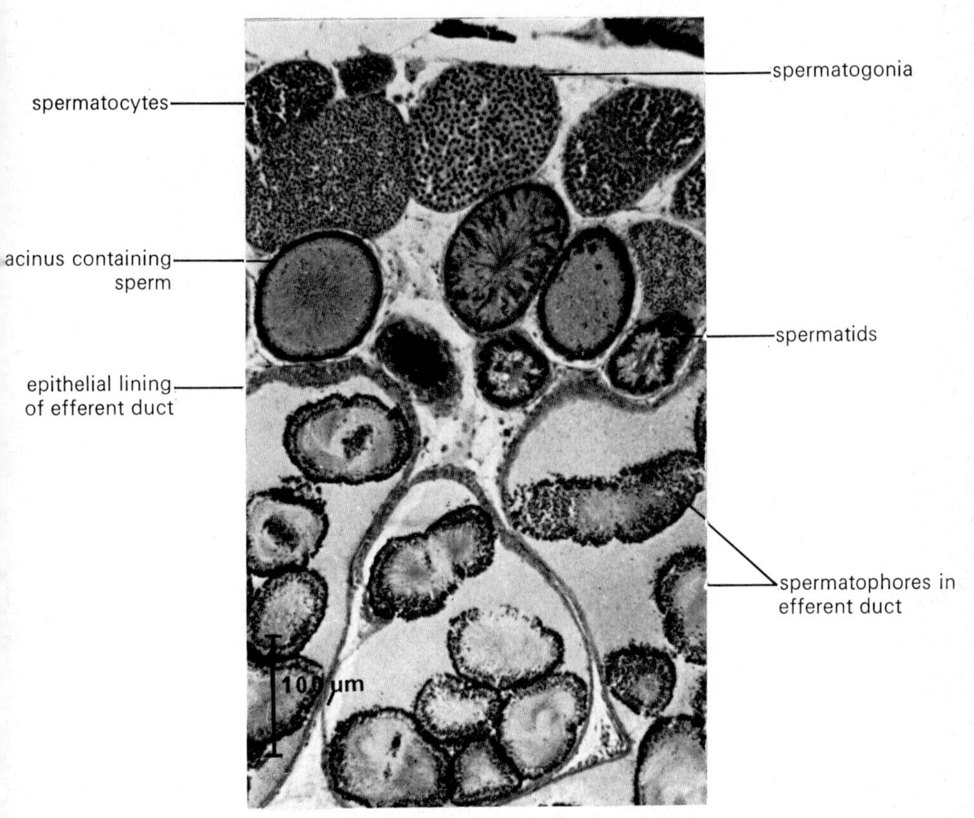

FIG. 13.25. Section through the testis of an adult male guppy.

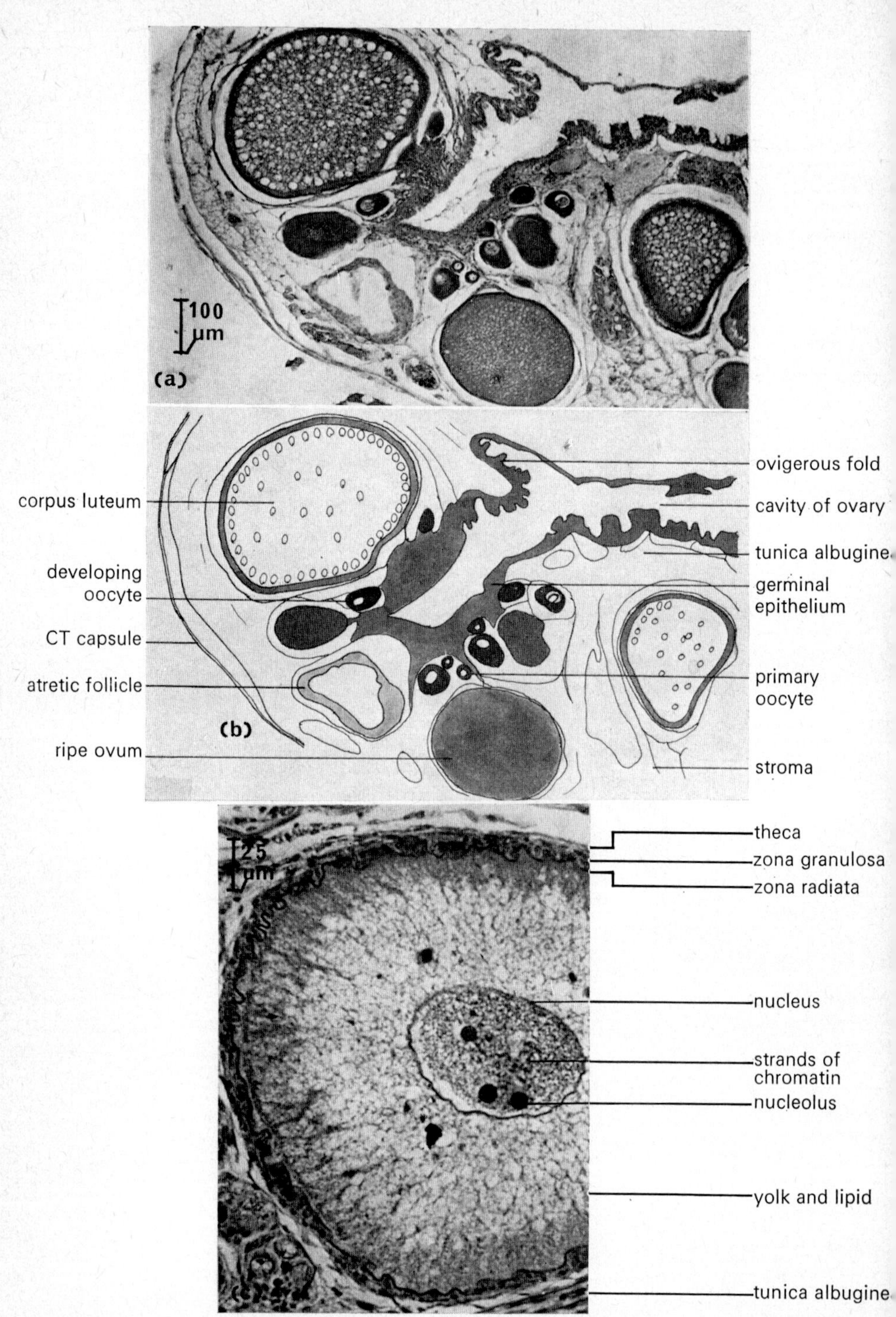

FIG. 13.26 (a) TS and (b) drawing of the ovary of an adult female guppy; (c) detail of ovum structure.

are discharged as spermatophores into small efferent ducts and thence into a long winding spermatic duct continuous with the vas deferens. The male genital duct remains separate from the excretory ducts along its whole length, and terminates in a long tubular copulatory organ (gonopodium) formed by modification of the anal fin rays.

Female teleosts are unique among vertebrates in having hollow ovaries lined with germinal epithelium. This lining of the guppy ovary [Fig. 13.26 (a, b)] is thrown into ovigerous folds, and numerous oocytes surrounded by a vascular stroma develop at the edges of these folds. Each oocyte [Fig. 13.26(c)] is surrounded by layers of follicular cells, which deposit yolk and secrete hormones. A corpus luteum is formed in the ovary after shedding of the ovum; and is probably concerned with hormone production as well as with phagocytosis of yolk and blood cells. Ripe ova are released into the cavity of the ovary which is continuous with the simple ciliated oviduct. This is atypical of vertebrates generally, but is an adaptation found in many teleosts to the release of numerous eggs which might otherwise get lost in the body cavity.

Fertilization is internal, and is effected by the gonopodium of the male assisted by the pelvic fins. Guppies are one of the few groups of teleosts which are ovoviviparous. i.e. the eggs continue to develop into young inside the mother but receive no assistance or nourishment from her.

Chapter 14

Class Amphibia

The Amphibia are anamniote tetrapods (see p. 429) many of which have become adapted for life on land, although they usually return to water for breeding. Most amphibians have free-living aquatic larvae which respire using gills. The gills, pisciform body and fins (without fin-rays) of the larvae are retained in some adult urodeles while other amphibians show metamorphosis from an aquatic larva into an adult which is normally partly terrestrial. Adult amphibians have two sets of paired limbs which are used for locomotion (although these are secondarily lost in the Apoda). The skeleton is reduced in weight and the skull is moveable on the vertebral column due to formation of special cervical vertebrae. Larval gills are replaced as respiratory organs in the adult by simple lungs and well-vascularized skin. The skin does not bear scales (except in the Apoda and some fossil species) but its outer layer is keratinized to resist abrasion and desiccation. The sense organs are also modified for life on land. The nostrils open into the oral cavity: while the ear shows the first development of a middle ear cavity for the reception of vibratory stimuli in air, although the system for receiving low frequency vibrations in water is usually retained.

Example: *Rana* (*temporaria*)

R. temporaria is a commonly occurring frog belonging to the Anura, one of the three living orders of Amphibia. The Anura are particularly well adapted for an amphibious mode of life, and undergo a dramatic metamorphosis. *Rana* has an aquatic larval stage which is omnivorous (chiefly herbivorous), and metamorphoses into a primarily terrestrial insectivore. Figure 14.1 is a VLS through a young frog (*R. temporaria*) and shows that the trunk is shortened and dorso-ventrally flattened. The tail (found in lower vertebrates) is lost and the caudal vertebrae are fused into an elongated urostyle. The limbs are highly specialized for both swimming and jumping.

In the larva, all respiratory exchange occurs across the gills and body wall. The larval skin is extremely thin and consists of ciliated epithelium resting on a basement membrane [Fig. 14.2(a)]. Dermis and muscle only develop at the end of metamorphosis and during the juvenile stage. In the adult, although remaining as a site for respiration and osmoregulation, the skin has a complex structure of stratified epidermis and thick fibrous dermis containing blood vessels, nerves and chromatophores [Fig. 14.2(b)]. The epidermis gives rise to multicellular glands which sink down into the dermis. They are mainly mucus-producing, which helps to keep the skin moist for respiration. On the dorsal surface larger serous "poison" glands occur which are used for protective purposes. The outermost layer of the epidermis is keratinized and replaced continuously from underneath [Fig. 14.2(c)]. It is often thickened to form horny warts or papillae. In the mating male, a pad appears on the hand of thickened dermis containing modified serous glands, and with a papillated surface covered by cornified epidermis [Fig. 14.2(d)]. Such mating pads are used for gripping the females.

The larval skeleton is cartilaginous (Fig. 14.3), and a notochord is present. Ossification occurs at metamorphosis, beginning with invasion of the cartilage by blood-vessels and modified mesenchyme cells (osteoblasts) which induce calcification and destroy the chondrocytes. Later this calcified cartilage is replaced by bony trabeculae or lamellae, and the bone surface is covered by a dense connective tissue periosteum (Fig. 14.4). In the spinal column, vertebrae formed in this way have procoelous centra, i.e. with the anterior concave face articulating on the posterior convex face of the adjacent one. This gives a much more flexible jointed rod needed for terrestrial locomotion and support of the body weight. The notochord remains only as inter-vertebral pads. The skeleton in adult frogs is reduced and, therefore, light. The vertebral column consists of only nine vertebrae and a tubular unsegmented urostyle containing an ossified remnant of the notochord.

Metamorphosis in anurans is accompanied by numerous striking changes (e.g. in skin structure, gut structure, chemical nature of blood pigments and enzymes, and development of limbs). One of the most intriguing changes is the degeneration and reabsorption of the long mobile tadpole tail. Figure 14.5(a, b) show a VLS of the hind end of a tadpole at the climax of metamorphosis, and demonstrate the typical pattern of degeneration, i.e. shortening of the tail, blackening due to accumulation of melanophores, and necrosis of certain tissues (e.g. epidermis, muscle, nerve cord, notochord and notochord sheath) [Fig. 14.5(c)]. The epidermis sloughs off, but autolysis of other degenerate tissues is followed by phagocytosis by invading mesenchyme cells.

Nervous system

The amphibian brain is similar to that of teleost fish, except that the small fore-brain (cerebrum) is inverted, i.e. has a thick roof (Fig. 14.6). The two most important senses to the frog, olfaction and sight, are well represented by sizeable olfactory lobes and large optic lobes. The brain ventricle expands in the tectal region to form optic ventricles, and the nerve cells have moved away from their position lining the central canal to form a distinct cortical arrangement of layers nearer the surface (Fig. 14.7). The tectal region is the chief co-ordination centre in amphibians, while the cerebellum is very small. The pineal is present in amphibians as a simple

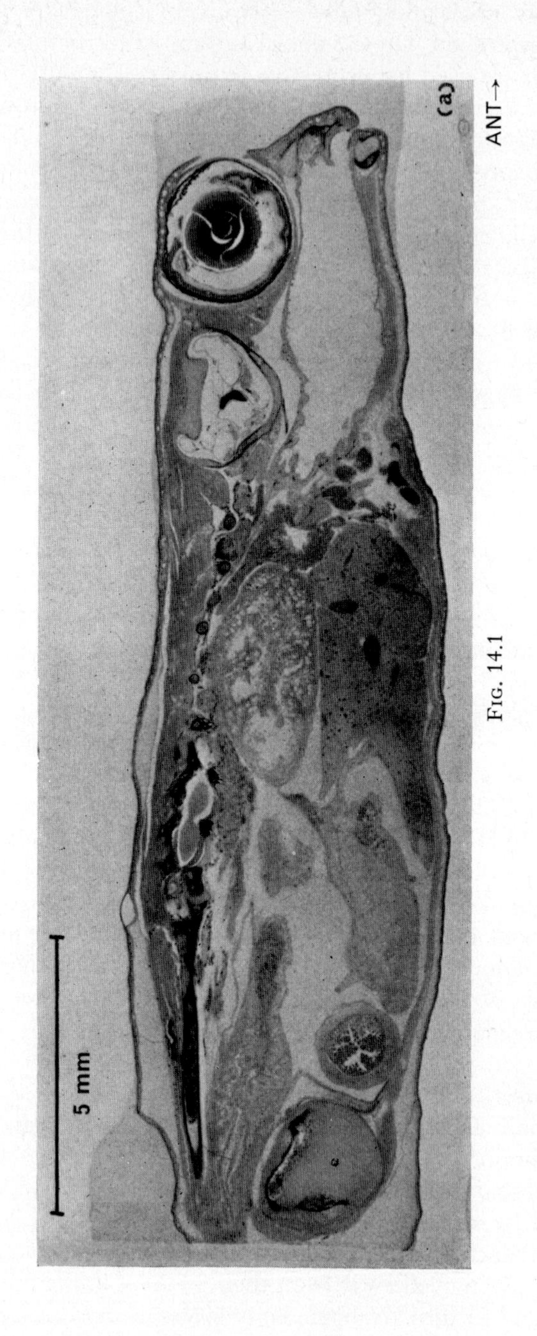

Fig. 14.1

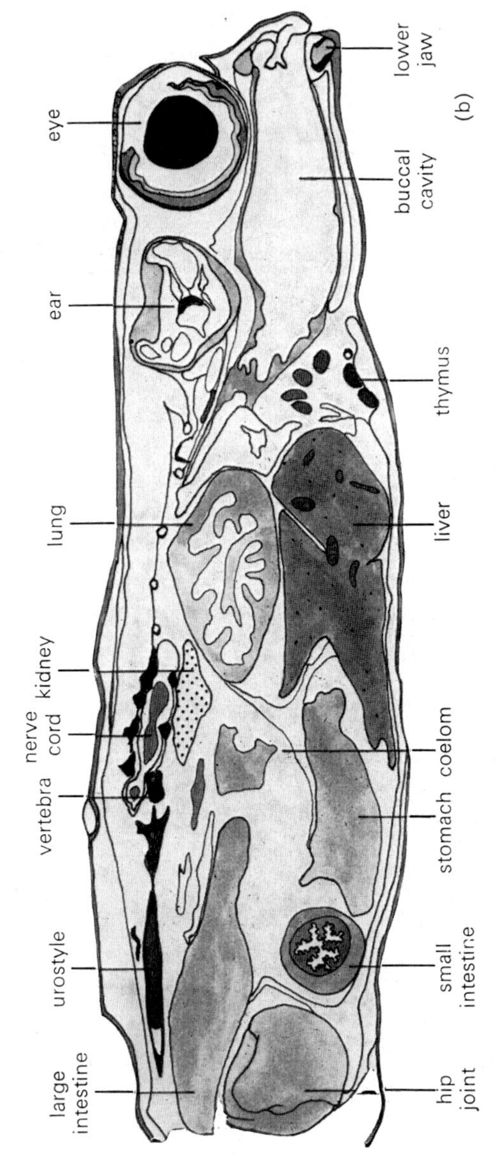

FIG. 14.1 (a) VLS and (b) explanatory drawing of a young frog (*Rana temporaria*).

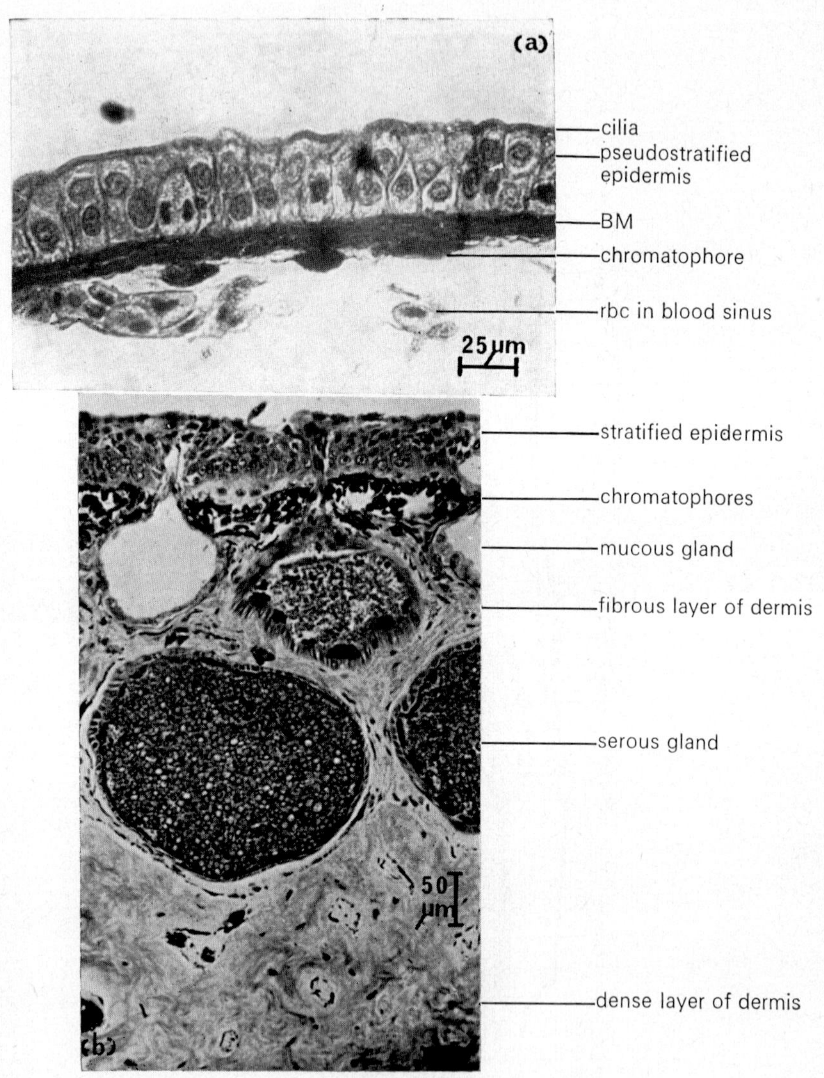
(a)
cilia
pseudostratified epidermis
BM
chromatophore
rbc in blood sinus
25 μm
stratified epidermis
chromatophores
mucous gland
fibrous layer of dermis
serous gland
50 μm
dense layer of dermis
(b)

FIG. 14.2

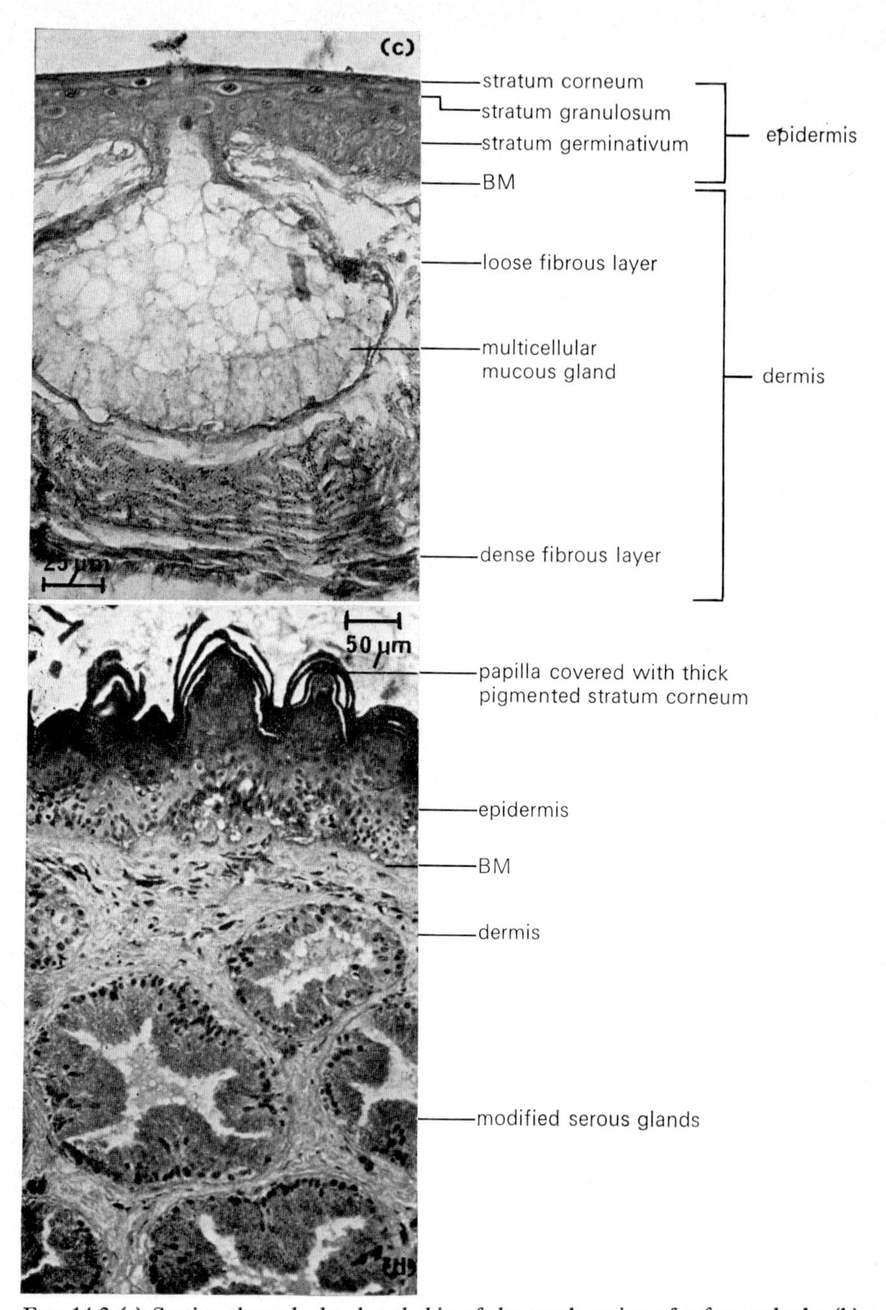

FIG. 14.2 (a) Section through the dorsal skin of the trunk region of a frog tadpole; (b) section through the dorsal skin of the trunk region of an adult frog; (c) detail of the epidermis and a multicellular gland from the skin on the head of an adult frog; (d) section through the nuptual pad on the forelimb of an adult male frog in the mating season.

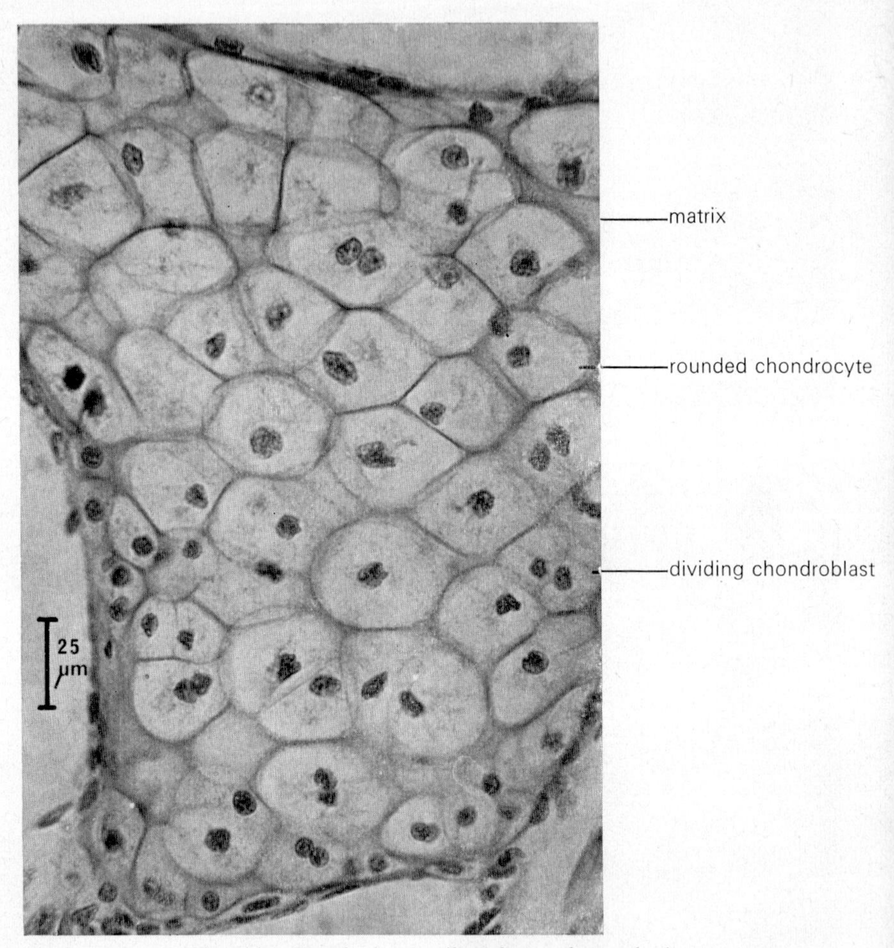

Fig. 14.3. Developing cartilage from a frog tadpole.

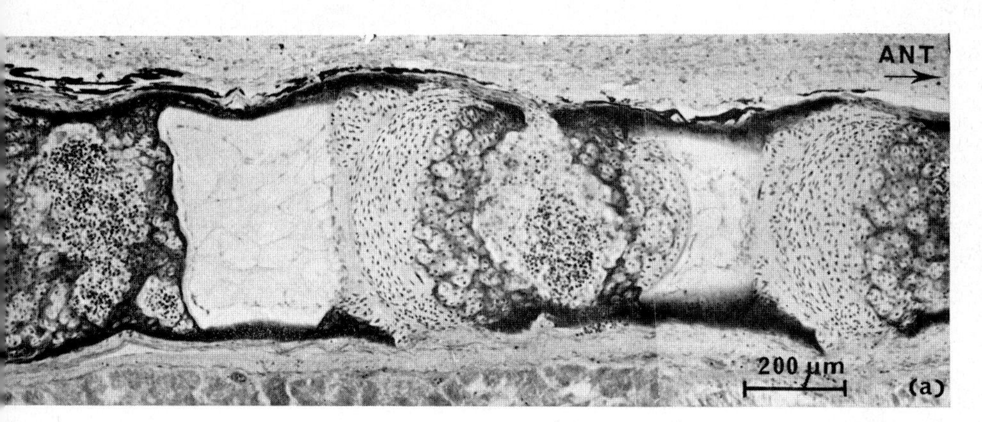

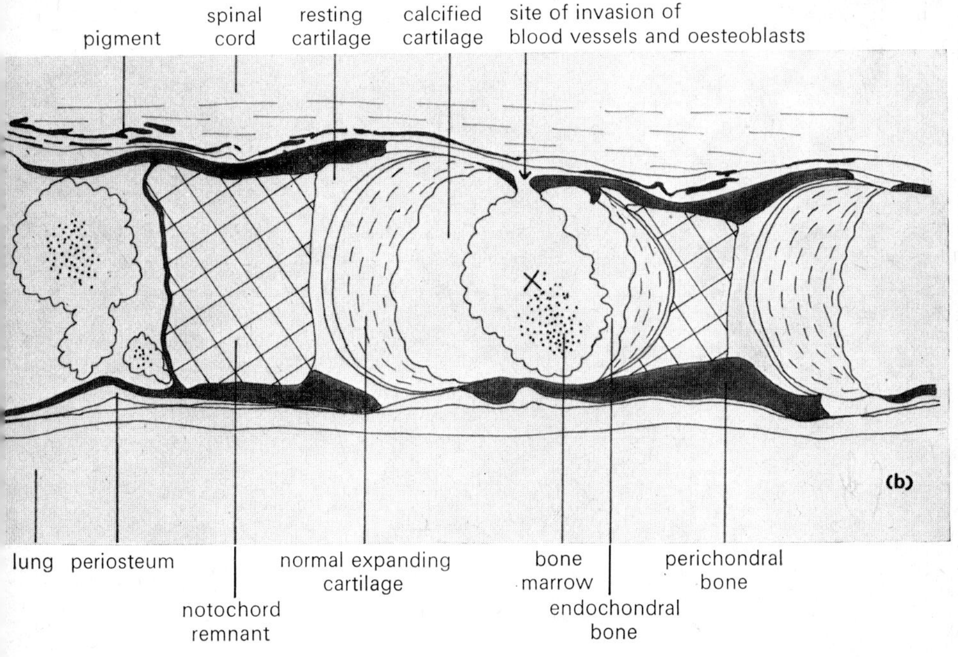

IG. 14.4 (a) VLS and (b) drawing of developing vertebrae in the spinal column of a young frog. 'he larval cartilage is being replaced by adult bone. Endochondral bone formation occurs at the entre (marked X) and spreads in both directions. Perichondral ossification occurs superficially.

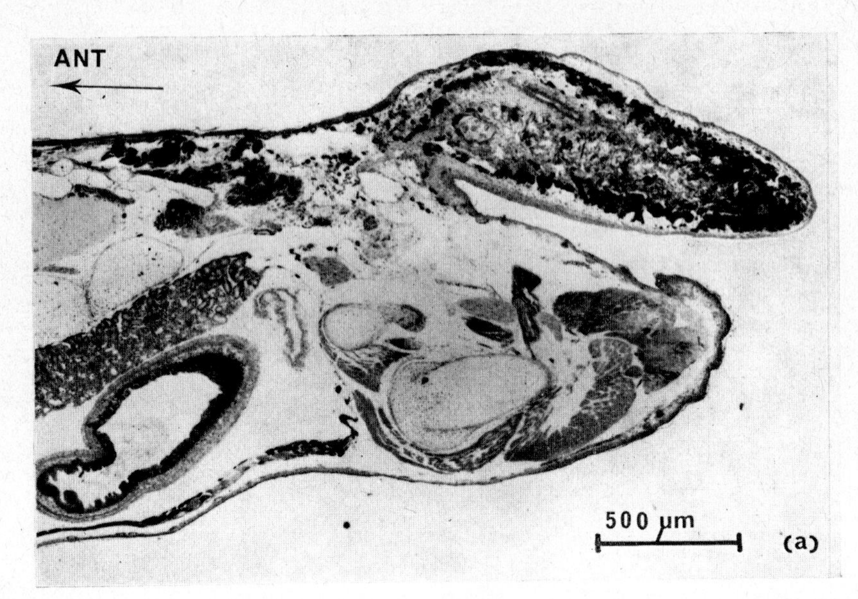
ANT
500 µm
(a)

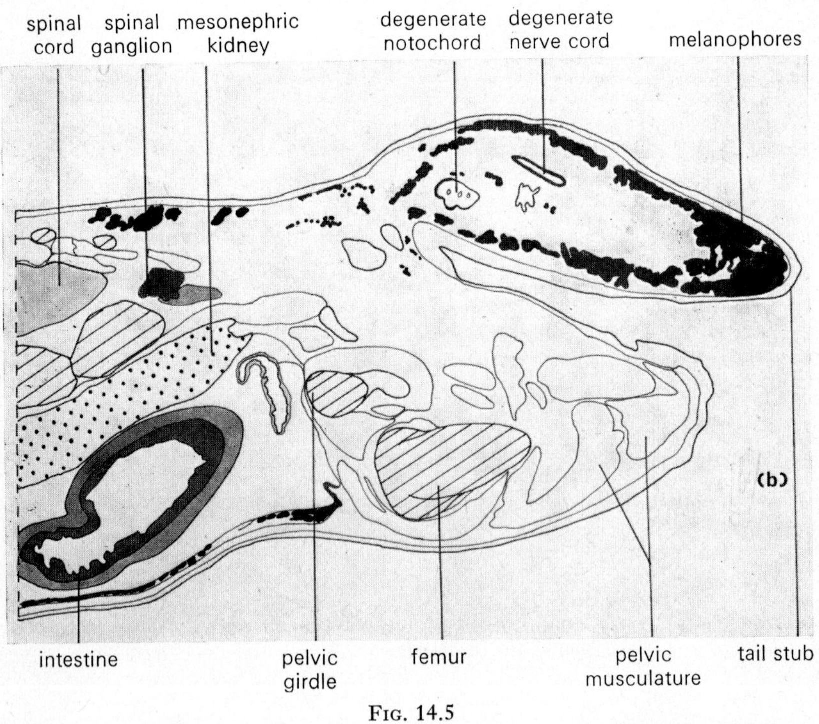
spinal cord
spinal ganglion
mesonephric kidney
degenerate notochord
degenerate nerve cord
melanophores
(b)
intestine
pelvic girdle
femur
pelvic musculature
tail stub

Fig. 14.5

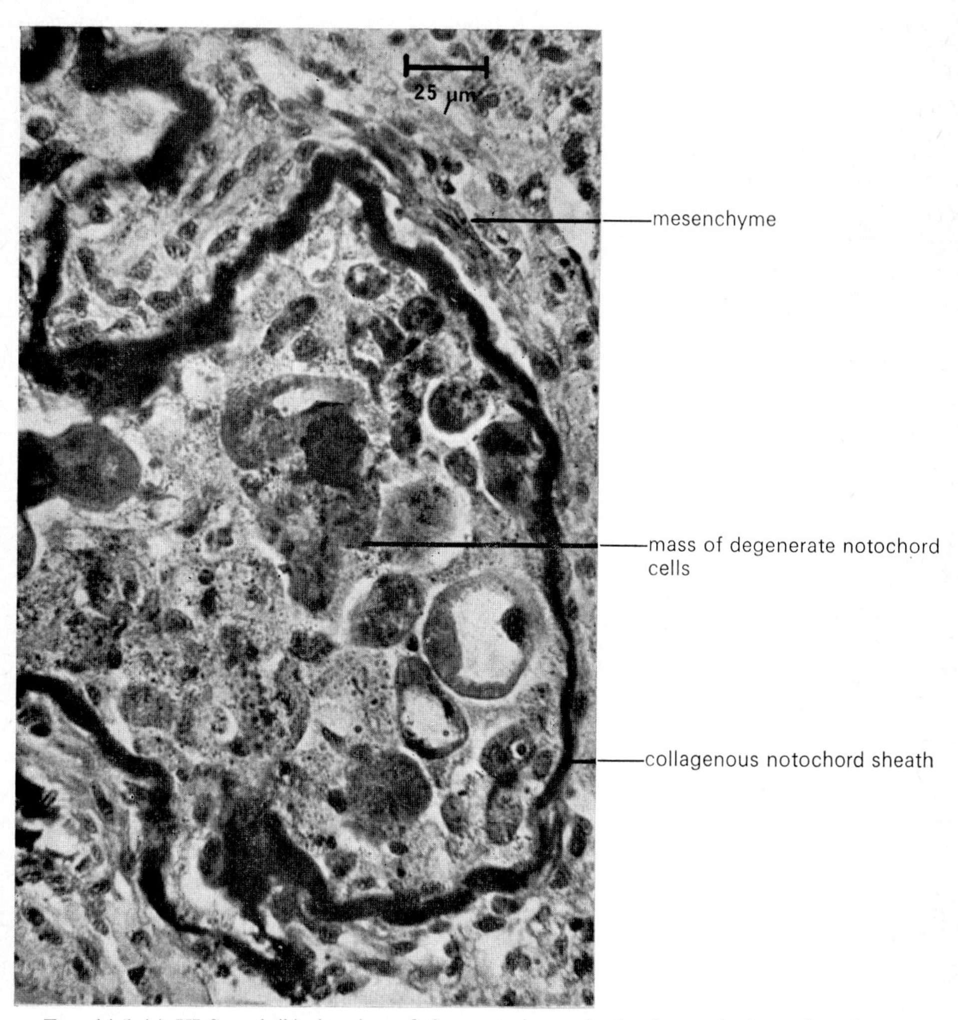

FIG. 14.5 (a) VLS and (b) drawing of the posterior end of a frog tadpole undergoing metamorphosis; (c) detail of the degenerating notochord and notochord sheath in the tail region of a metamorphosing tadpole.

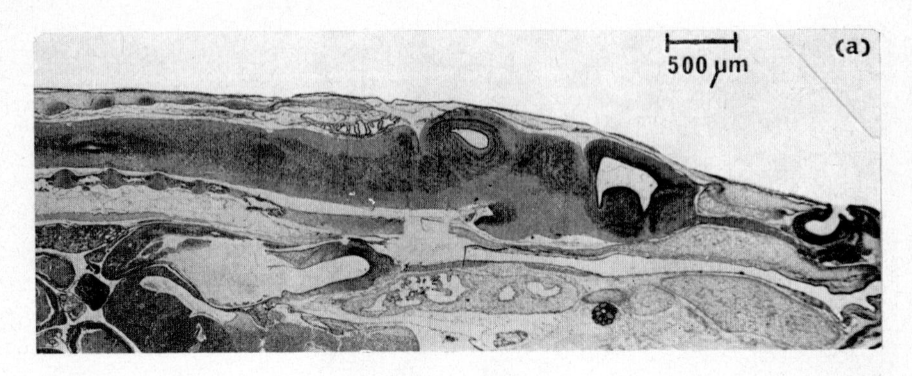

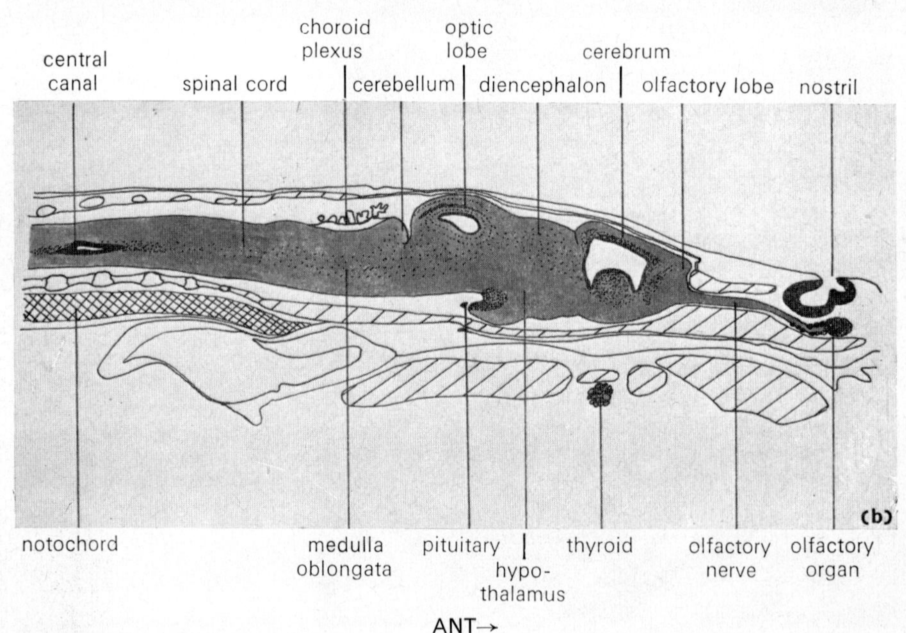

FIG. 14.6 (a) VLS and (b) drawing of a tadpole brain, showing the relative positions of the main regions.

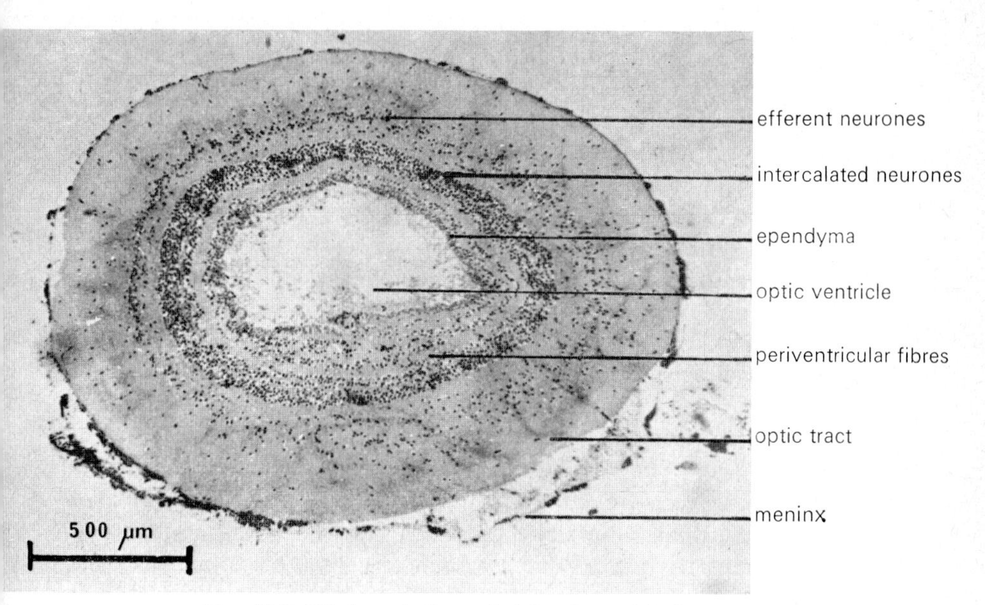

FIG. 14.7. TS through the optic lobe of an adult frog.

sac, or it has disappeared. Large Mauthner cells (see Fig. 13.4) occur in the medulla oblongata of larvae and aquatic adults, but are lost in the adults of those species that undergo complete metamorphosis.

There are cranial and spinal nerves, but the spinal nerve arrangement is somewhat altered by the development of limbs and by the shortening of the tail. The spinal cord possesses marked dorsal and ventral horns. The CNS is enclosed in two meningeal membranes in amphibians.

Sense organs

Amphibians are well supplied with free nerve endings and simple sense organs, but the most important senses of the frog are sight and olfaction, although hearing is also quite good.

The spherical eyeball houses a lens (Fig. 14.8) that is flatter and farther from the cornea than in fish. As in elasmobranchs vision is normally fixed on distant objects, and the lens is focused on near objects by being swung forwards by contractions of dorsal and ventral *protractor lentis* muscles. The rounded cornea (Fig. 14.9), which forms the main refractive structure in terrestrial vertebrates, is kept moist in air by oily secretions from the orbital Harderian glands (Fig. 14.10). The upper eyelids of frogs are fixed, but the lower lids are transparent and are capable of flicking over the eye, thus spreading the lubricant. The amount of light falling on the retina is controlled by an iris of circular and radial muscles. The retina [Fig. 14.11 (a)] is characterized by the presence of few ganglion cells, and consists of mixed photoreceptors: rods, used for vision in dim light, and cones for bright light vision [Fig. 14.11(b)]. Both types of photoreceptor have an outer photosensitive segment, and an inner region consisting of metabolically-active ellipsoid, contractile myoid, a nucleus (lying in the outer nuclear layer of the retina) and a conducting "footpiece" which makes synaptic contact with relay neurones (in the inner nuclear layer) and the ganglion cells. Light or dark adaptation of the amphibian retina is effected by movement of peripheral pigment. During light adaptation this pigment moves towards the light source and thus shields the photoreceptors, while during dark adaptation [as in Fig. 14.11(a)] the pigment moves distally leaving the photoreceptors uncovered so that they can receive all available light.

The inner ear of amphibians is similar to that of fish. The otolith chambers bear discrete maculae and the otoliths are large (Fig. 14.12). However, the inner ear is linked to a flexible tympanic membrane on the outside of the head by a bony columella. This structure transmits air-borne vibrations to the inner ear, and is enclosed in a cavity forming the middle ear, which connects with the pharynx.

Large paired olfactory organs open to the outside by nostrils and into the oral cavity by choanae. They are divided into chambers lined with olfactory epithelium bearing large multicellular flask-shaped Bowman's glands

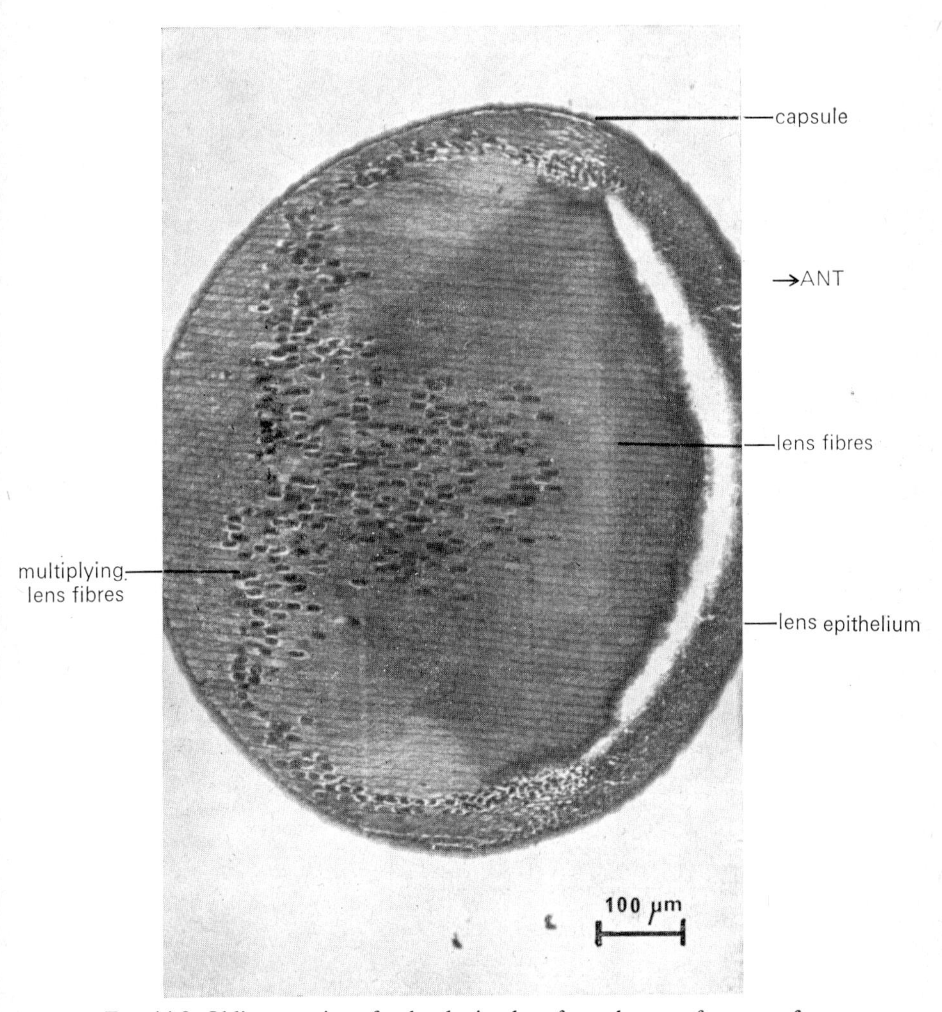

FIG. 14.8. Oblique section of a developing lens from the eye of a young frog.

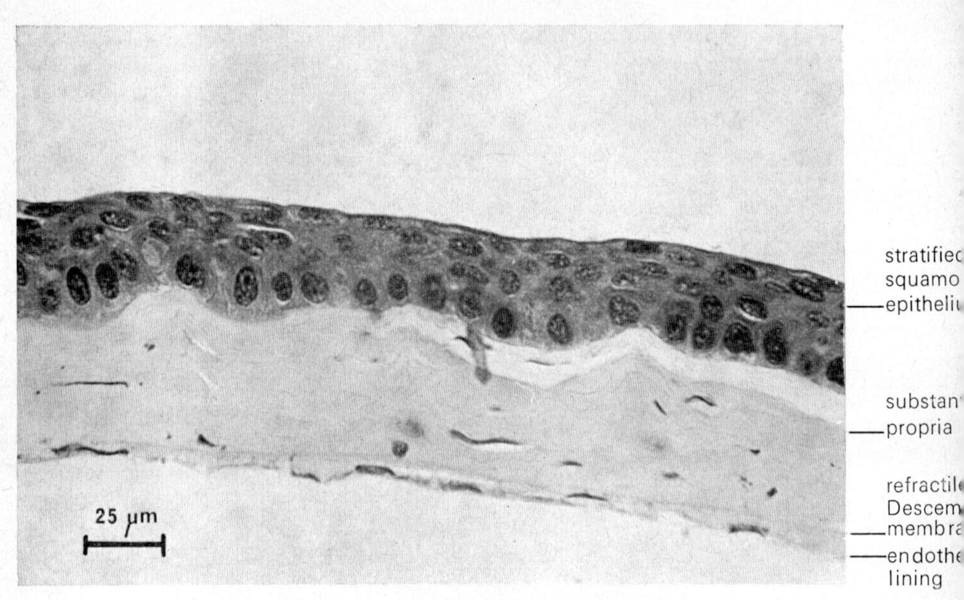

FIG. 14.9. Section through the cornea from the eye of a young frog.

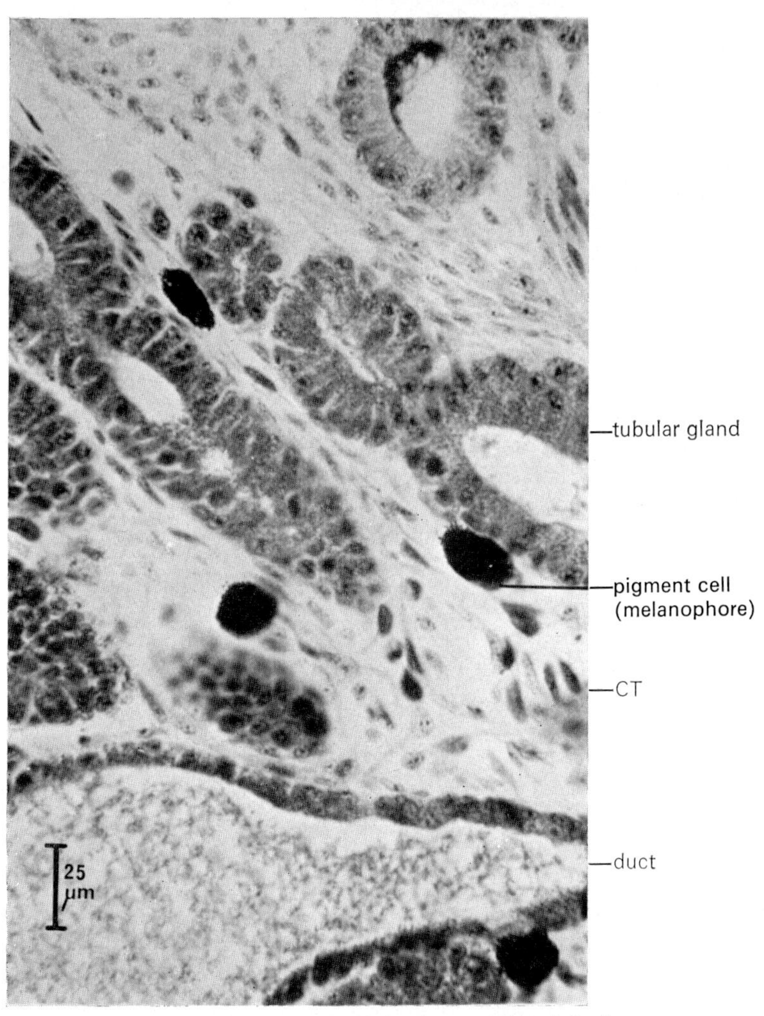

FIG. 14.10. Harderian glands in the eye orbit of the frog.

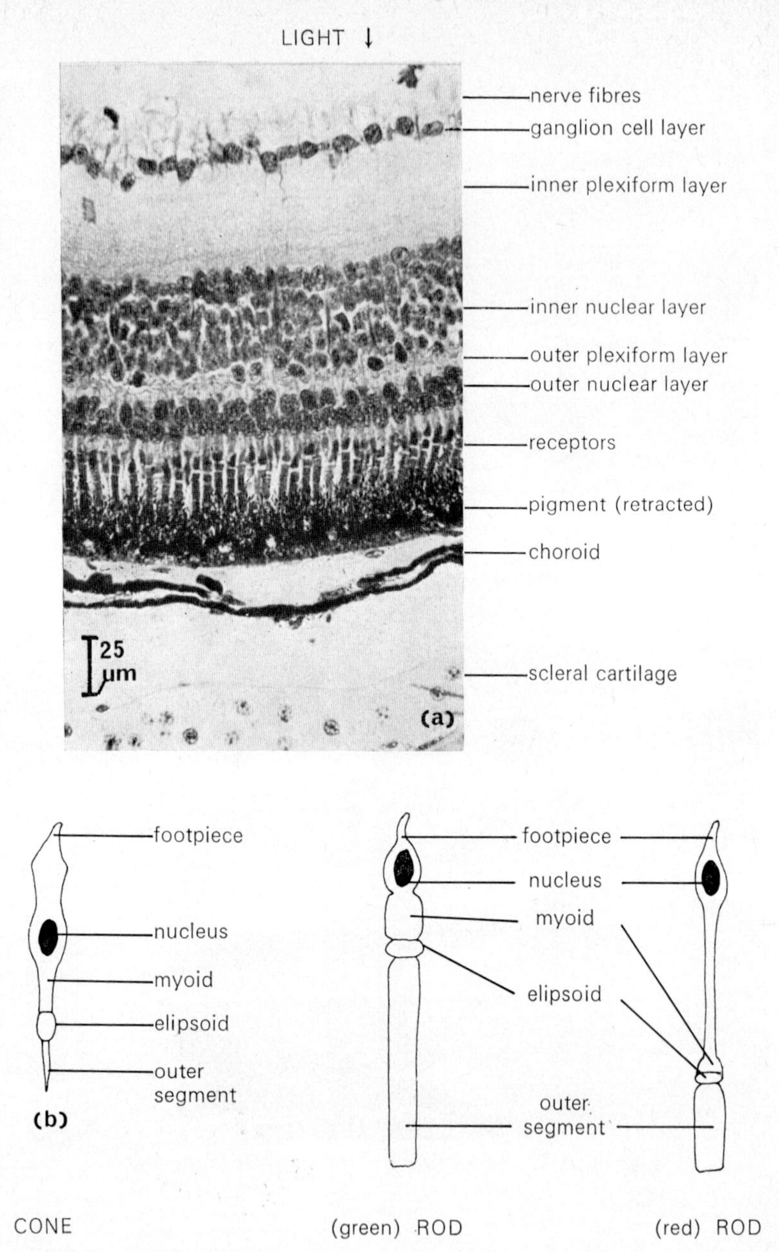

FIG. 14.11 (a) Section through the dark-adapted retina of a frog; (b) diagram of the different types of photoreceptors found in frogs. Some of the rods are more sensitive to red light, and some to green.

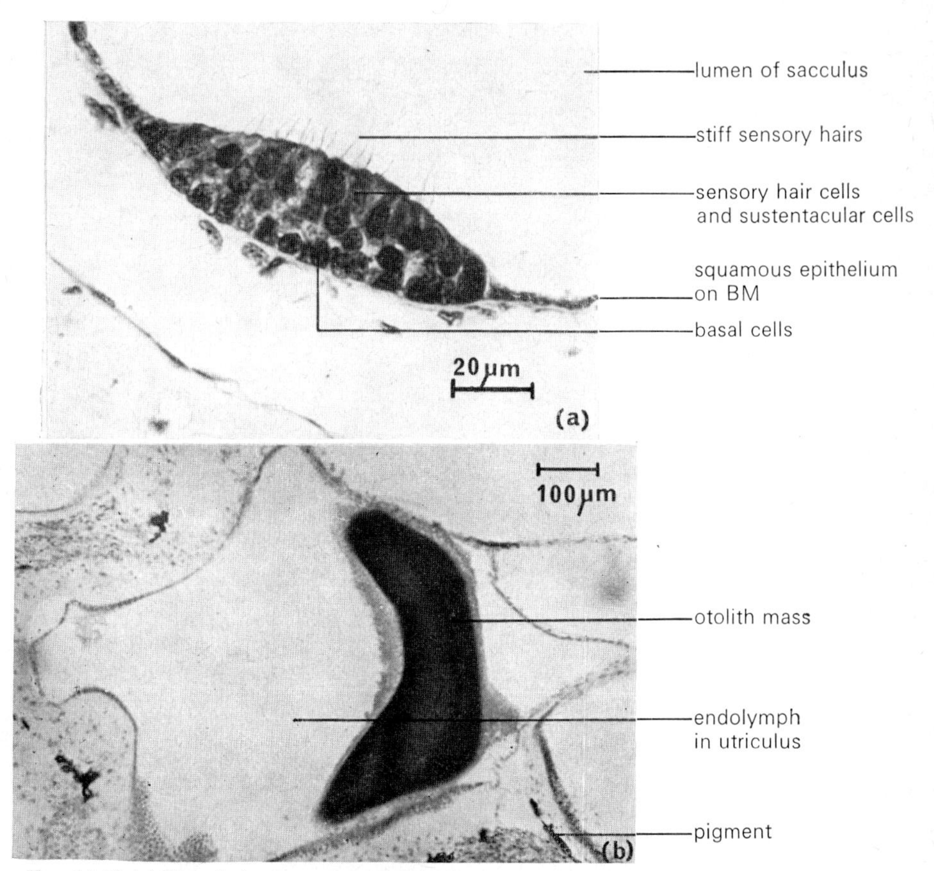

FIG. 14.12 (a) Macula in the sacculus of the inner ear of the frog; (b) Large otolith mass in the utriculus of the frog inner ear.

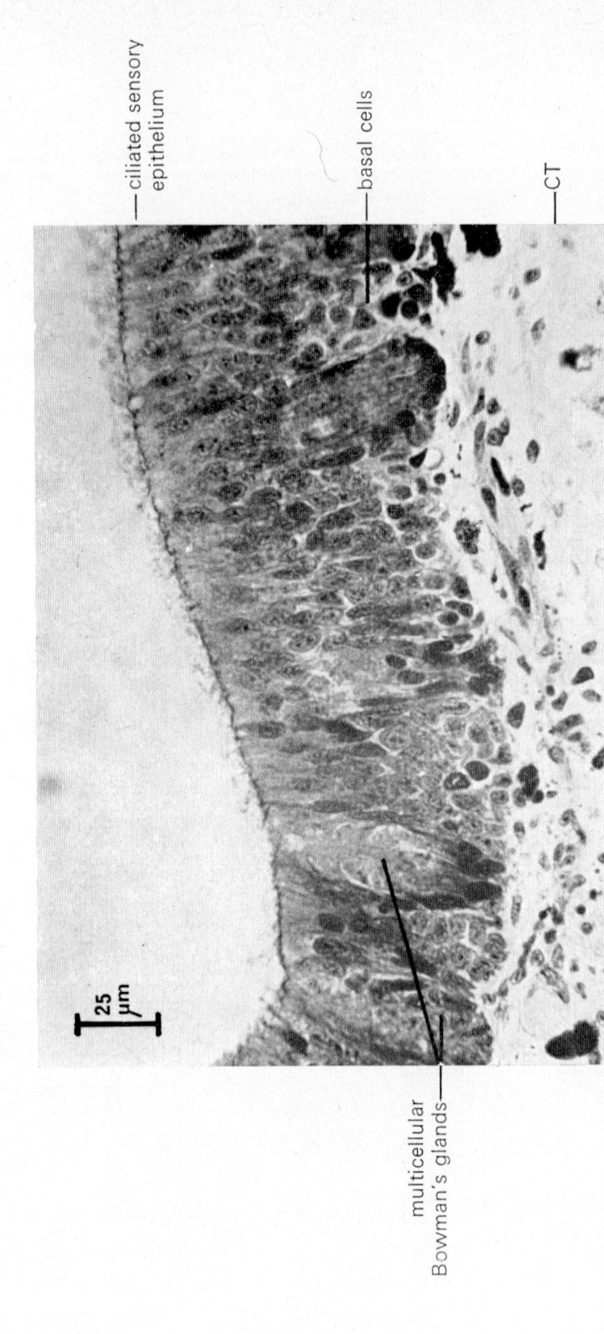

Fig. 14.13. Olfactory epithelium containing Bowman's mucous glands lining the olfactory organ of the frog.

which produce mucus to keep the olfactory epithelium moist in air (Fig. 14.13). In the medial wall of the middle cavity of the olfactory organ is a small recess lined with vomeronasal epithelium (similar to the olfactory epithelium except that there are no Bowman's glands) connected by nerves to a small accessory olfactory bulb in the brain. This is the forerunner of Jacobson's organ in reptiles, but in the Anura it has no direct connection with the oral cavity and no obvious function.

Endocrine organs

The pituitary of frogs is clearly differentiated into glandular and nervous portions, and is well developed. The glandular portion shows considerable hypertrophy during metamorphosis (Fig. 14.14) correlated with increased output of many of its hormones at this time. A pars intermedia, involved in colour change, is distinguishable while a pars tuberalis is thought to be present but is not identifiable on the section shown. The pars nervosa is small.

Amphibia have compact paired thyroid glands connected by an isthmus in the ventral neck region (see Fig. 14.6). The thyroids show typical follicular structure (Fig. 14.15) in larvae and adults, but are particularly active just prior to metamorphosis when they undergo explosive production of thyroid hormones. Accessory follicles sometimes occur in the neck region.

Chromaffin tissue and inter-renal tissue are normally associated in amphibians into a distinct adrenal gland close to the anterior end of the kidney [Fig. 14.16(a)]. Small nests of chromaffin tissue, however, also occur along the edges of the cardinal veins in the abdominal region [Fig. 14.16(b)].

Parathyroid glands make their first appearance in the Amphibia. They are four small glands originating from ventral outgrowths of the pharyngeal pouches. They lie close to the thyroids and have a typical structure of cords of small cells arranged in whorls (see Fig. 16.14). They tend to degenerate in winter and are most active in spring. They are involved in control of calcium and phosphate levels in the body, and thus take over the functions of the ultimobranchials found in lower vertebrates. The ultimobranchials may persist but are gradually submerged by the thyroid.

Respiration

The frog, which lives in water or in damp terrestrial conditions, still uses its moist, well-vascularized skin for respiration. Indeed in some purely aquatic amphibians there are no other respiratory structures. The lining of the buccal cavity is also used by the frog, and shows evagination of capillaries into the buccal epithelium. Accessory respiratory structures are present in the tadpole in the form of, first, external and then internal gills [Fig. 14.17(a, b)]. In the adult there are simple paired saccular lungs opening directly into the pharynx. The vascularized epithelial lining of the

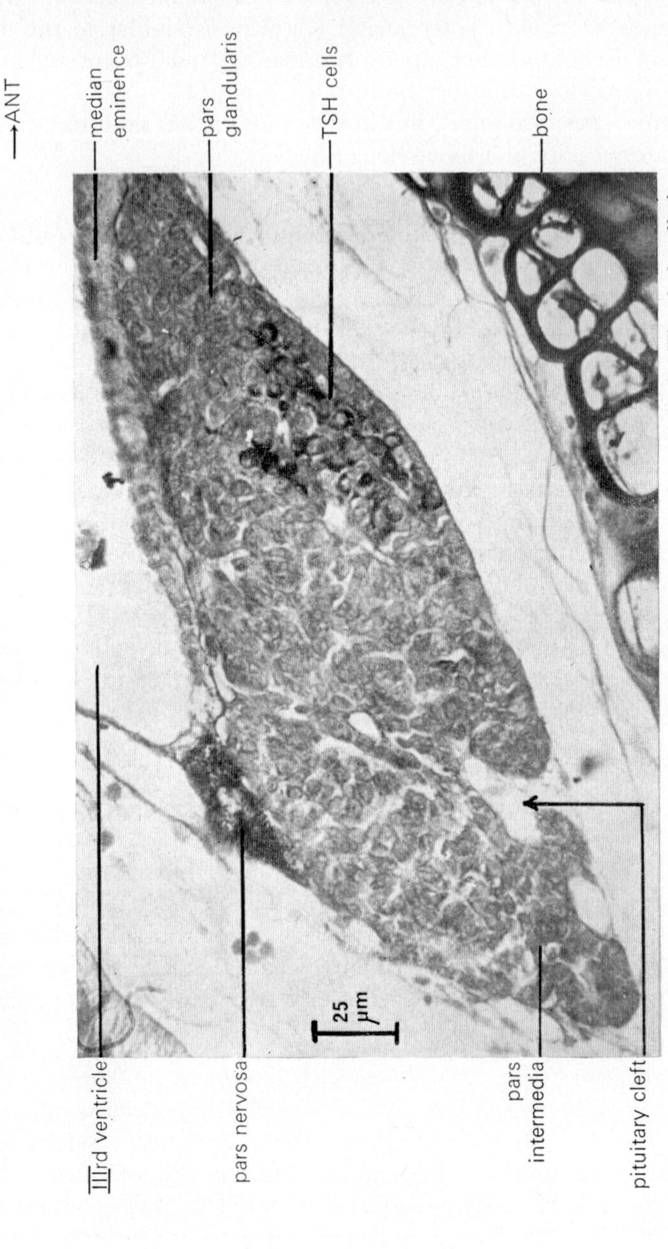

FIG. 14.14. VLS through the pituitary of a metamorphosing tadpole. The dark cells in the pars glandularis have been stained for one of the pituitary hormones (TSH) involved in metamorphosis.

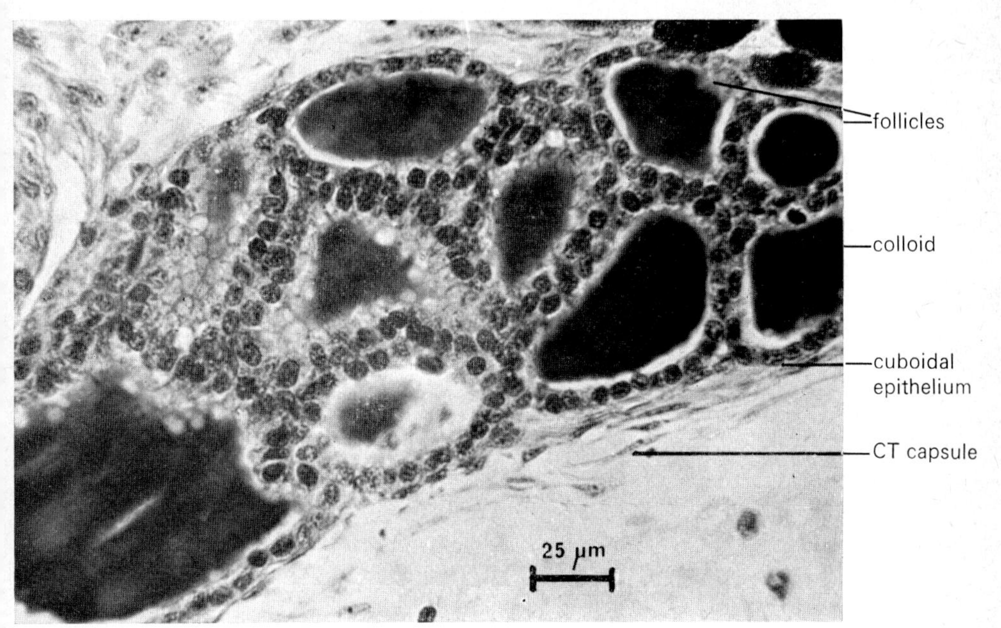

FIG. 14.15. Section through the thyroid gland of a tadpole undergoing metamorphosis. Thyroid hormones are stored within the colloid.

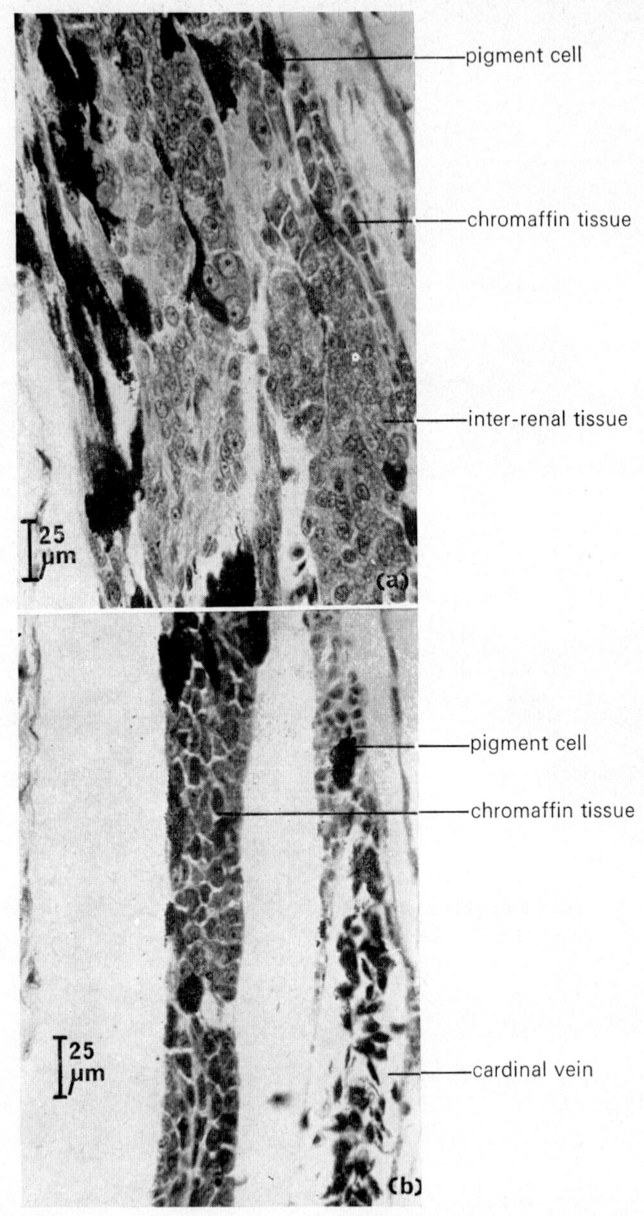

FIG. 14.16 (a) Section through the frog adrenal gland, showing the juxtaposition of inter-renal and chromaffin tissue; (b) an isolated strip of chromaffin tissue lying adjacent to the cardinal vein of the frog.

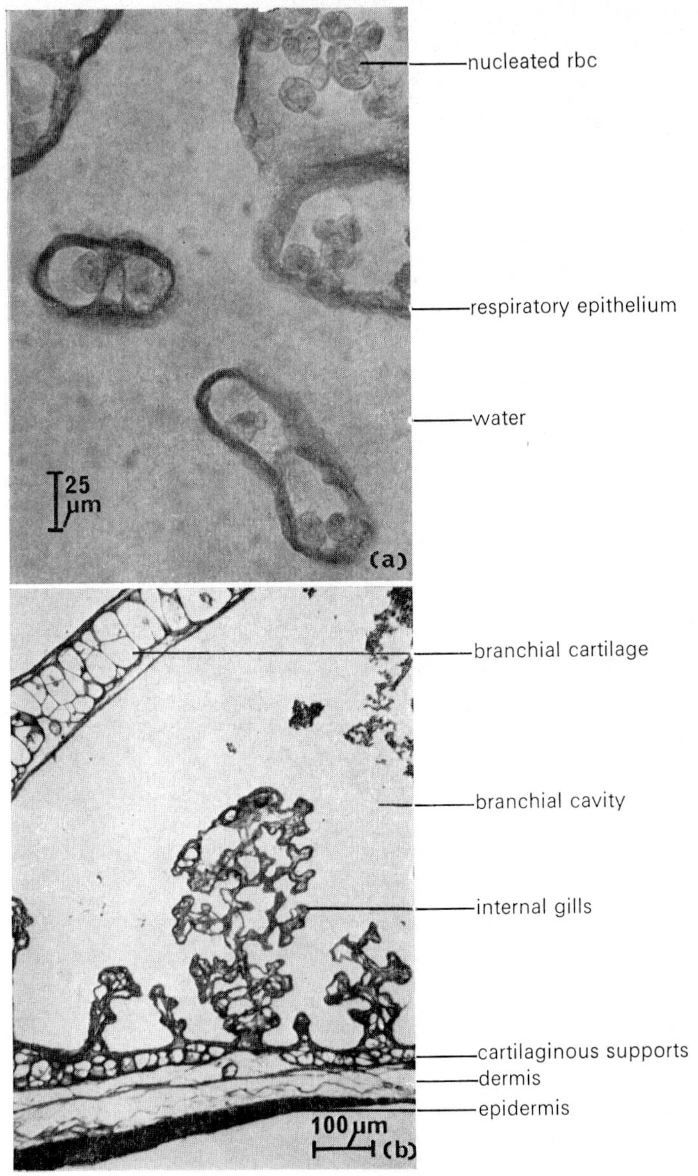

FIG. 14.17 (a) Section through the external gills of a young tadpole; (b) section through the branchial cavity and internal gills of a tadpole shortly before metamorphosis.

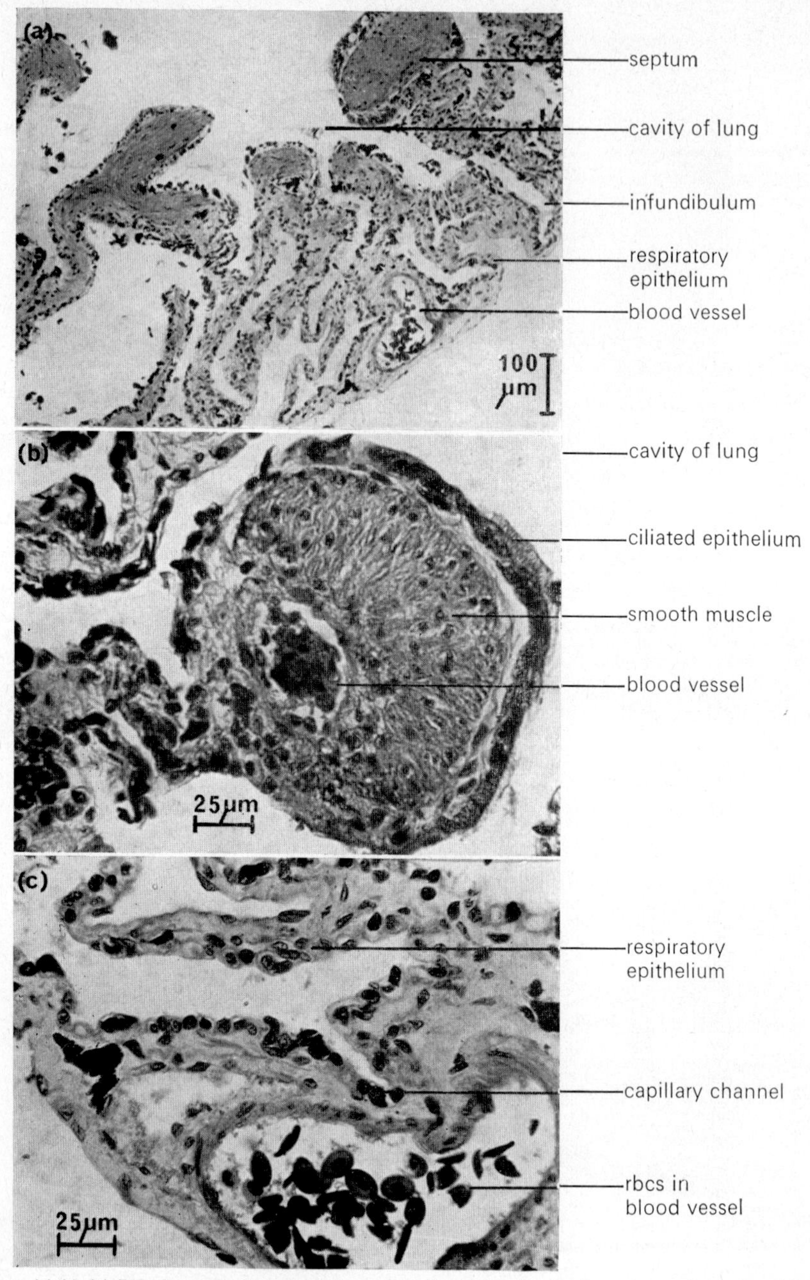

FIG. 14.18 (a) TS through the lung of an adult frog; (b) detail of a septum from the adult frog lung; (c) detail of the infundibular region of the frog lung.

lungs is smooth in very young frogs, but as the animal grows older the lining is thrown into folds (septae) separating infundibula lined with thin squamous respiratory epithelium, which thus form blind-ended alveoli (Fig. 14.18).

Circulation

The circulatory system of amphibians is very different from that of fish. This is associated with the presence of lungs in the adult. Even in the larva [Fig. 14.19(a)] where the lungs are functionless and there is only one atrium in the heart, the afferent gill arteries (such as are found in fish) are replaced by three pairs of arterial arches (carotid, systemic and pulmonary), although another pair sometimes persists. In the adult [Fig. 14.19(b)], where the lungs are used, there is a functional separation of blood flowing from the heart to the lungs (via the pulmonary arch) and that circulating round the rest of the body (via the carotid and systemic arches). Such a "double" circulation is correlated with the presence of an inter-atrial septum (i.e. effectively there are two atria) and a spiral valve in the truncus arteriosus: features not found in lungless amphibians. The adult frog heart [Fig. 14.19(c)] possesses a thin-walled sinus venosus, paired atria (one receiving blood from the lungs and the other from the body) and a thick-walled ventricle. The ventricle wall has numerous trabeculae on its inner surface which probably help to keep separate streams of blood returning from the pulmonary circuit and the body. The first part of the truncus (bulbus cordis) is composed of cardiac muscle and is contractile; the second part comprises the short ventral aorta. The heart, including the bulbus but not the ventral aorta, is enclosed in a thin pericardium.

Erythrocytes are generally large in amphibians (20μ in frogs), especially in the urodeles where they may reach 50μ in diameter. They are formed in the kidney of the larva, and primarily in the spleen of the adult. White blood cells are produced in small paired thymus glands lying in the neck region (see Fig. 14.1). They are very similar to those of reptiles (see Fig. 15.18). Amphibia also possess primitive "tonsils" in the form of masses of lymphoid tissue in the roof of the mouth.

Digestion

The frog tadpole is omnivorous, and feeds by rasping off bits of plant material and other food with its horny jaws and trapping the resulting particles in mucus. Sheets of mucus and food are then wafted by cilia along pharyngeal grooves lining the outer edges of the branchial arches and thence into the gut. The branchial sieve consists of a series of plates extending from the branchial arches, each of which has its epithelial surface thrown into complex secondary and tertiary folds. Water currents are created by masticatory muscles in the mouth. The larval gut is long, coiled, and lined with a simple unciliated epithelium (Fig. 14.20).

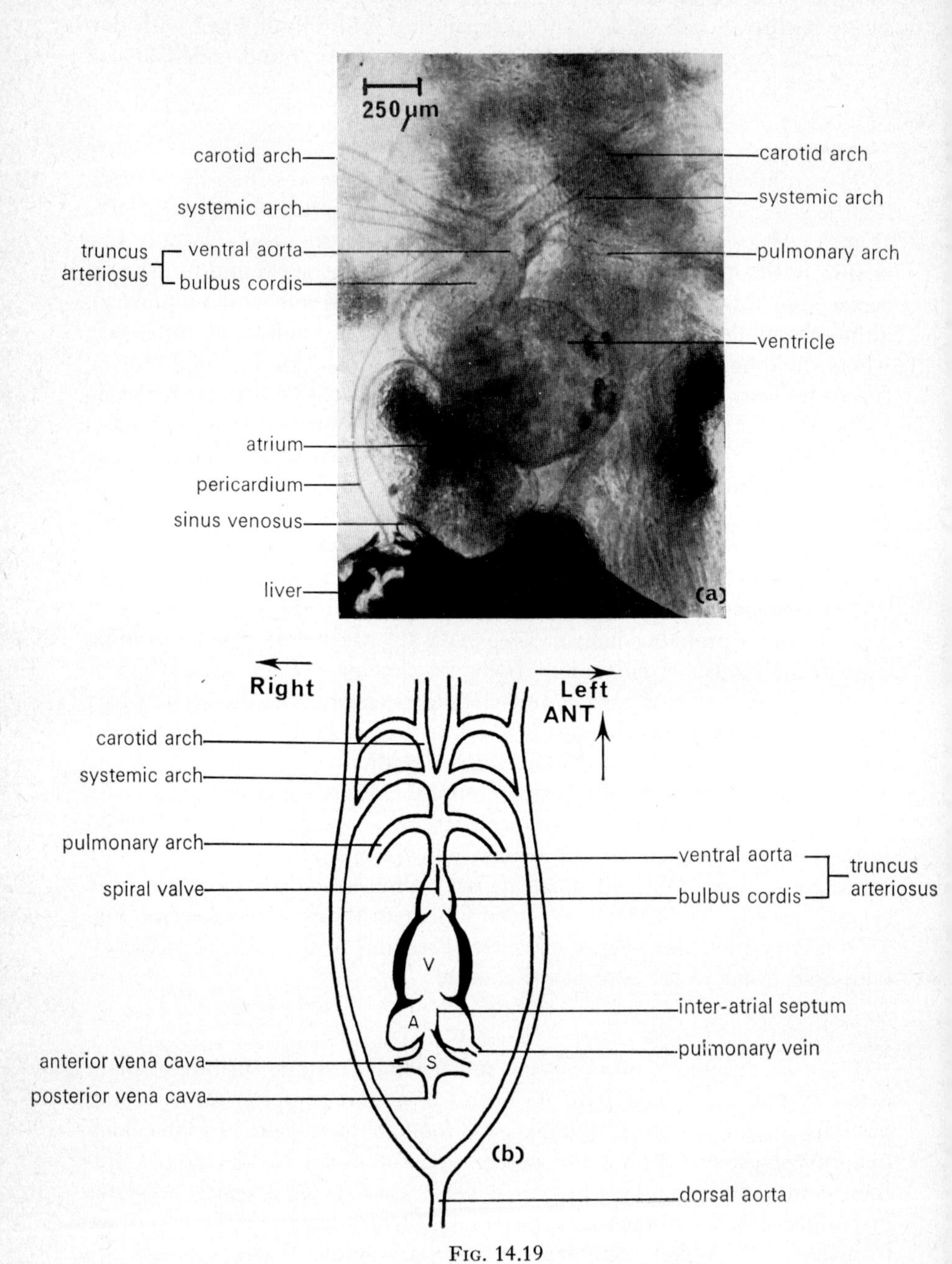

250 μm
carotid arch
systemic arch
truncus arteriosus
ventral aorta
bulbus cordis
atrium
pericardium
sinus venosus
liver
carotid arch
systemic arch
pulmonary arch
ventricle
(a)
Right
Left
ANT
carotid arch
systemic arch
pulmonary arch
spiral valve
anterior vena cava
posterior vena cava
ventral aorta
bulbus cordis
truncus arteriosus
inter-atrial septum
pulmonary vein
V
A
S
(b)
dorsal aorta

Fig. 14.19

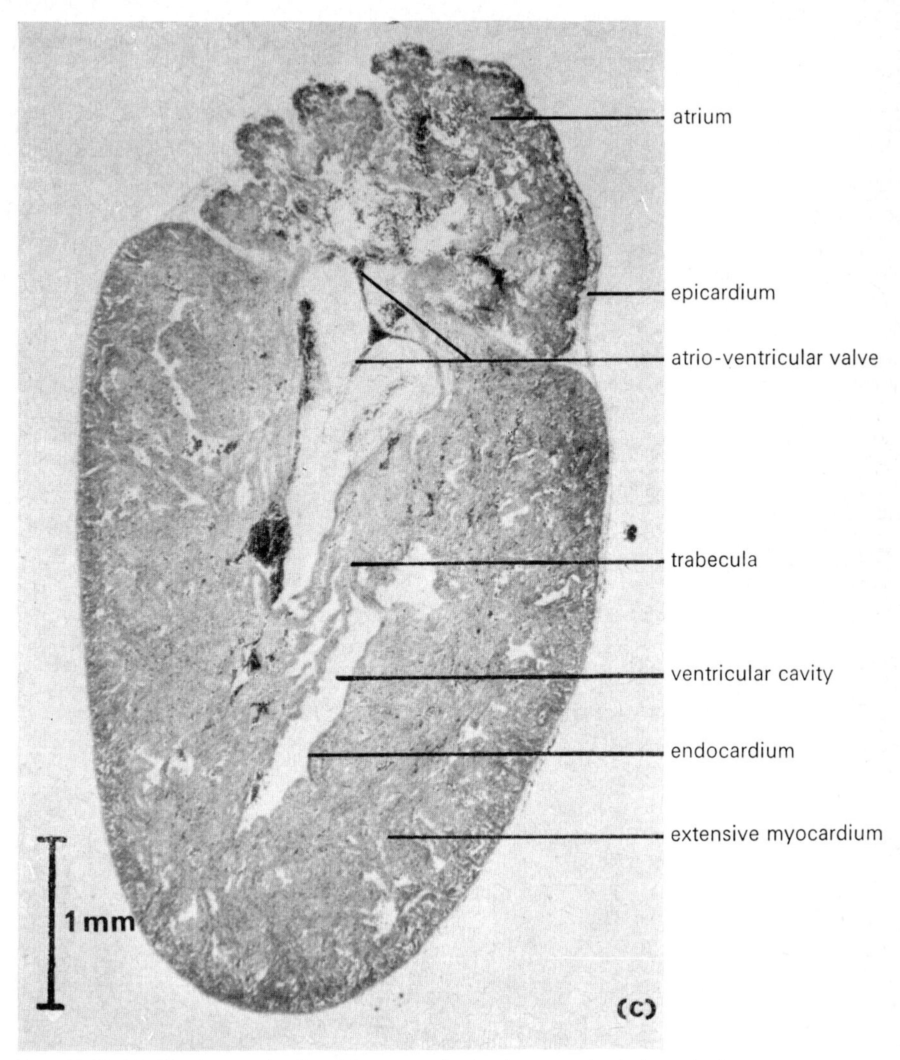

FIG. 14.19 (a) The heart of a live *Xenopus* tadpole photographed through the transparent body wall. Note the arrangement of the arterial arches leaving the heart; (b) plan of the heart and aortic arches of the adult frog. (A = atrium, S = sinus venosus, V = ventricle); (c) LS through the heart of an adult frog (*R. temporaria*).

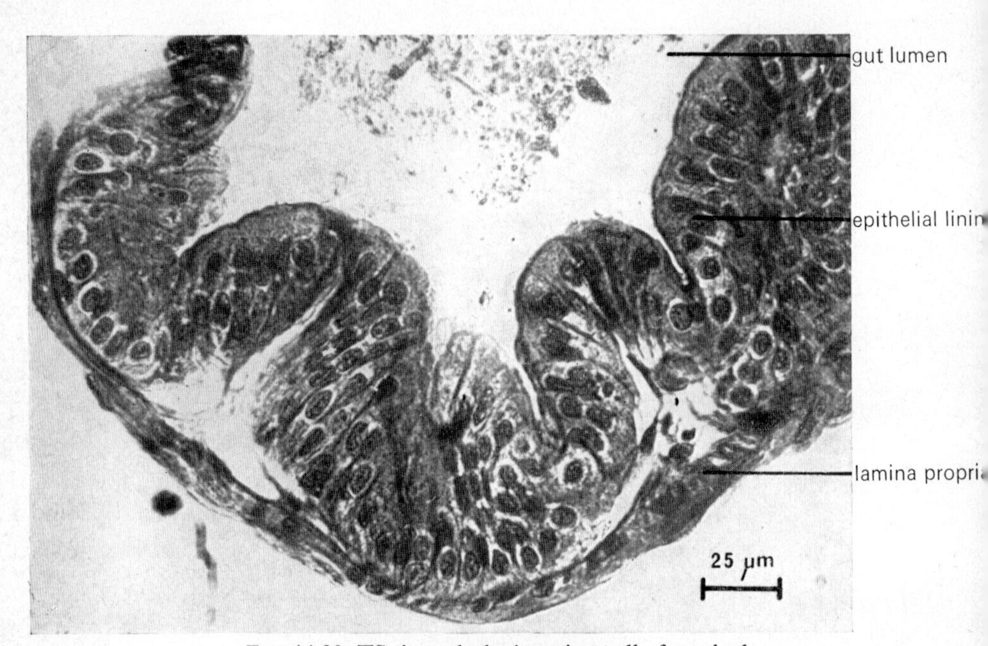

FIG. 14.20. TS through the intestine wall of a tadpole.

The adult frog is carnivorous, and its digestive tract shows a much greater degree of specialization and localization of function than in teleosts. Movement of food through the gut is by cilia assisted by contractions of muscles that appear at metamorphosis. Prey (chiefly in the form of insects) is snapped up by the wide mouth under water, or, on land, is caught up on the sticky tongue. The tongue is a muscular structure well supplied with cilia, mucous cells and branched tubular "salivary" glands containing mucous and serous cells (Fig. 14.21). It is attached to the anterior floor of the mouth and can be flicked out very rapidly to entrap prey. The upper jaw is lined by small conical bony teeth which are only used to assist food capture. The buccal cavity is large (see Fig. 14.1) and is provided with taste buds and capillaries and lined with ciliated epithelium interspersed with unicellular and multicellular glands producing mucus and adhesive secretions.

The short oesophagus (Fig. 14.22) has a ciliated epithelial lining thrown into folds. The stomach (Fig. 14.23) is a large U-shaped sac lined with tubular gastric glands producing enzymes and mucus. The principal site of digestion is the small intestine. This is differentiated into a proximal duodenum which has thick muscle layers, submucosal glands and receives bile and pancreatic ducts; and a distal ileum with a folded epithelium of columnar and mucous cells bearing a striated border but no cilia. In *Rana* a flap-like valve separates the small intestine from the large intestine, which is lined with columnar epithelium and plentiful goblet cells. The large intestine opens into the cloaca, a muscular pocket lined with cloacal glands. The large intestine and cloaca are both sites of salt and water reabsorption in amphibians.

The liver is bilobed, of similar structure to that of the fish, and bears on its undersurface a large spherical gall-bladder to store bile produced in the liver. The anuran pancreas is a compact structure, forming two separate lobes in the adult. The dorsal lobe is embedded in the dorsal mesentery and the ventral lobe lies between the gut and liver. The pancreas has the same acinar structure as in fish, but there is less connective tissue separating the lobules, so it appears to be a more compact organ. Large islets of Langerhans are present (Fig. 14.24) which secrete hormones involved in metabolism.

Excretion

Adult frogs excrete chiefly urea, while the larvae excrete ammonia. Much excretion and osmoregulation occurs across the permeable surfaces of the skin and respiratory organs, particularly in the larva; but in the adult the chief excretory organs are paired opisthonephric kidneys. These are more compact than in fish and possess more nephric tubules. The large number of glomeruli (about a thousand in each frog kidney) can be seen in Fig. 14.25, but the renal corpuscles are fairly small particularly in fully

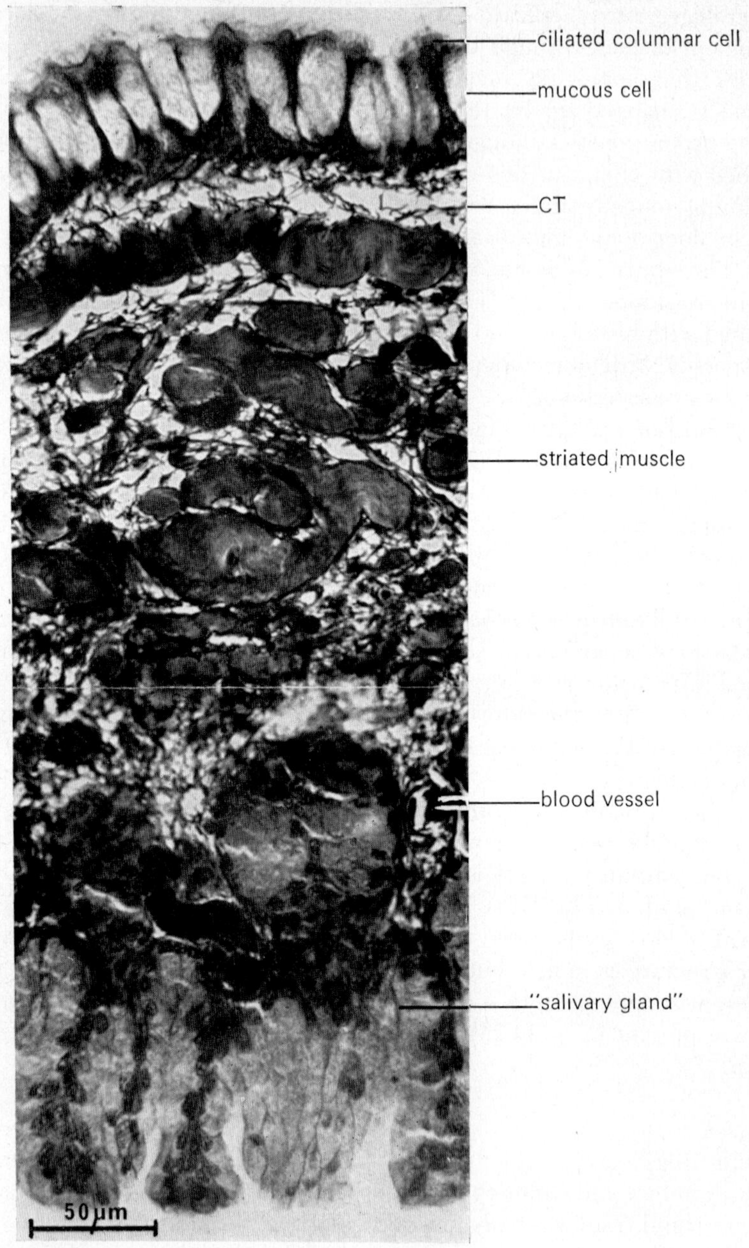

FIG. 14.21. VLS through the tongue of an adult frog. Note the extensive striated musculature and the lining of ciliated and mucous cells.

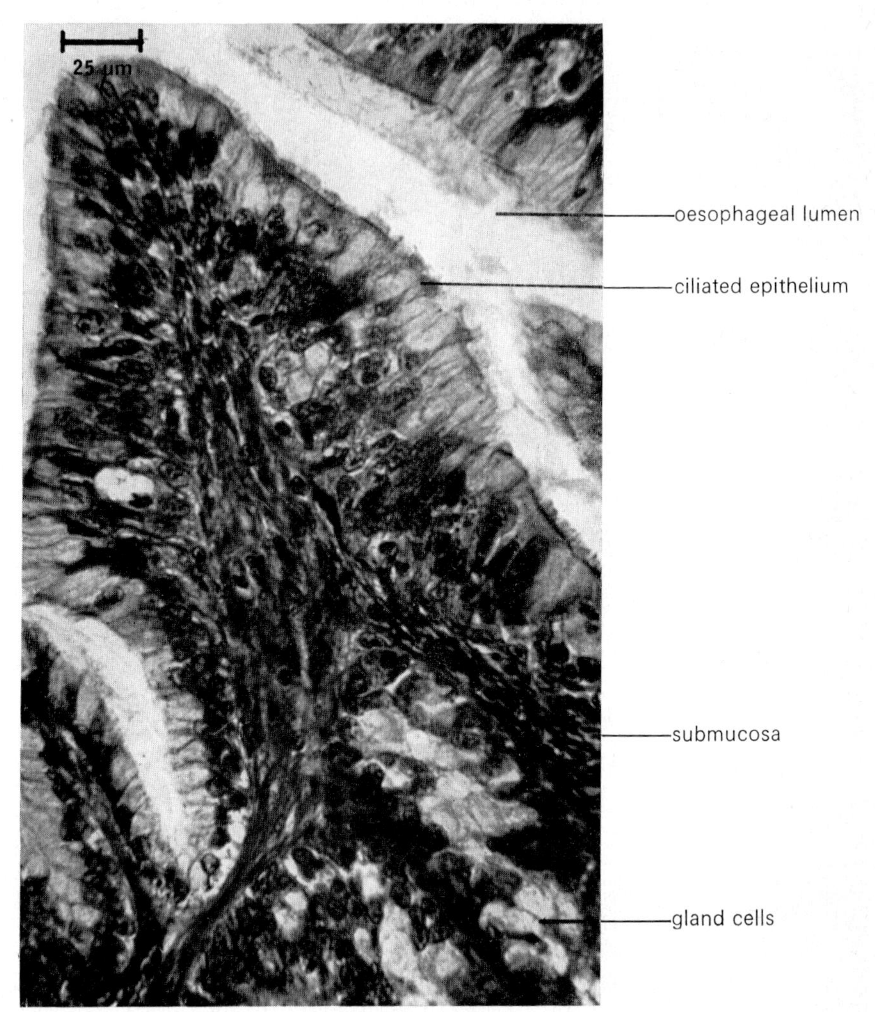

FIG. 14.22. Section through the oesophageal wall of an adult frog. Note the folded lining and ciliated epithelium.

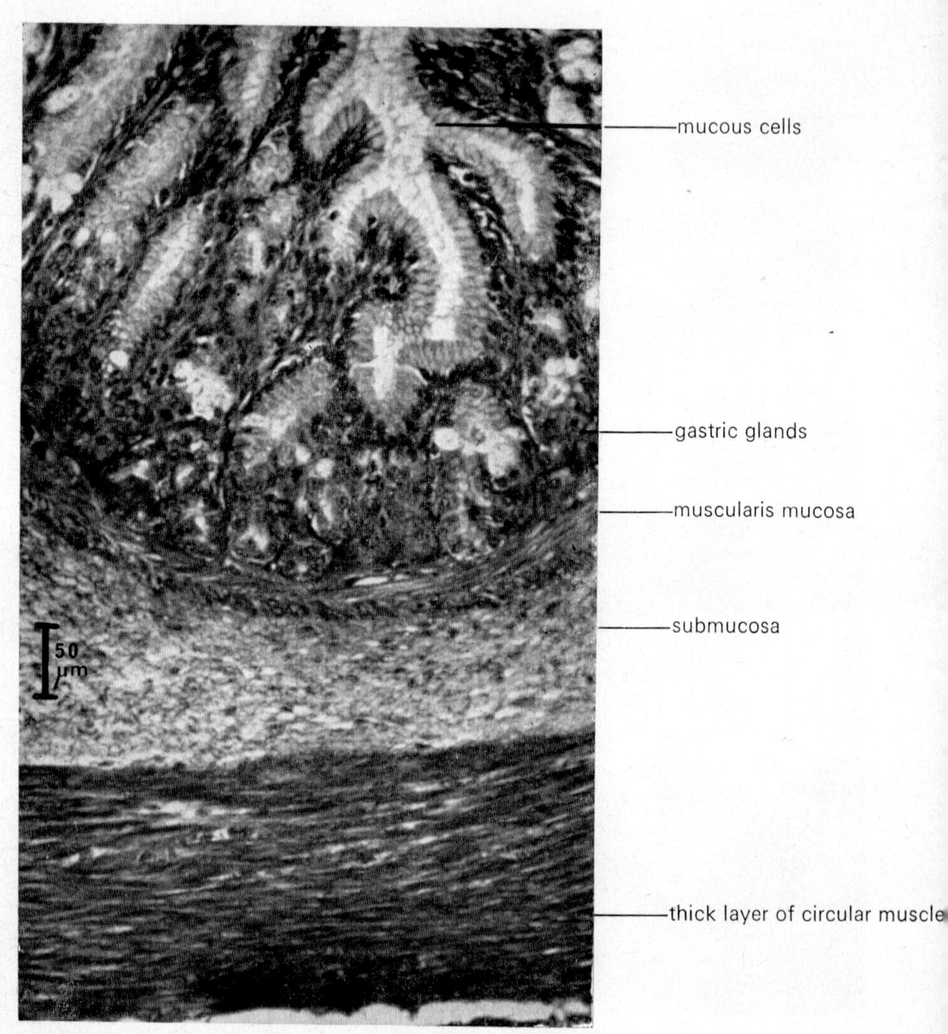

FIG. 14.23. TS through the stomach wall of an adult frog.

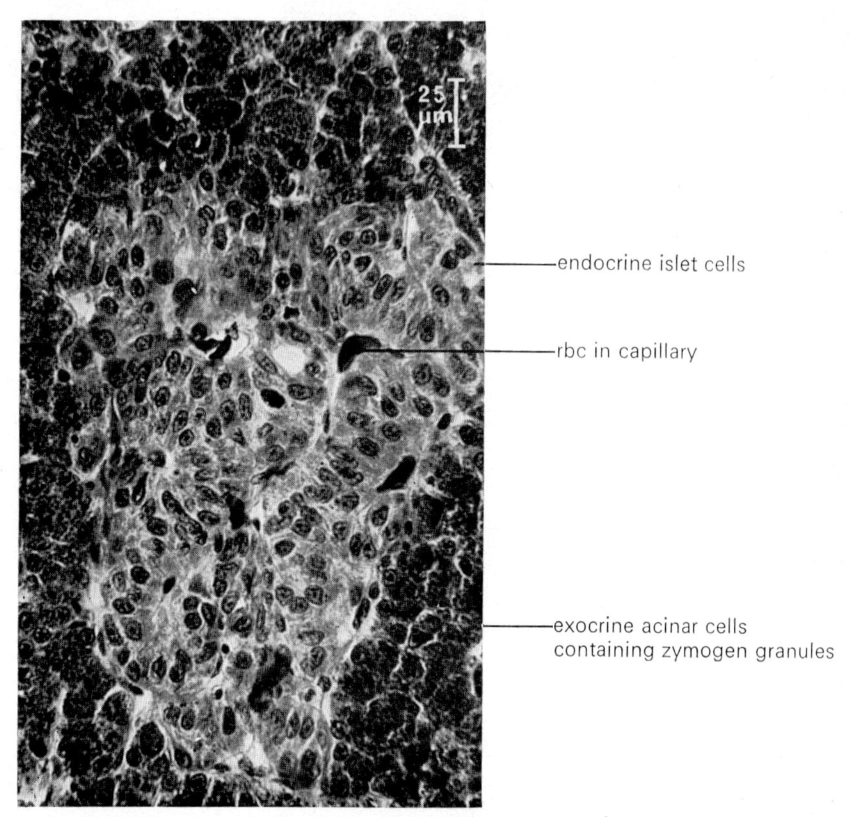

FIG. 14.24. Section through the pancreas of an adult frog, showing a large (lightly stained) islet of Langerhans. The various types of islet cell can be differentiated histochemically.

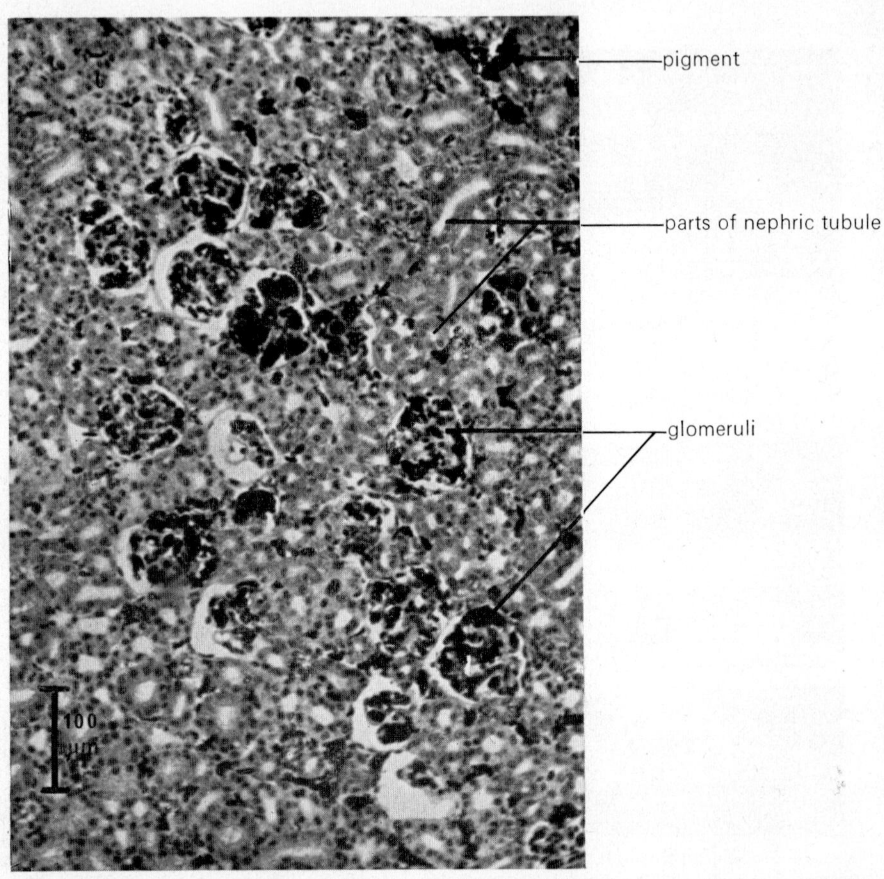

Fig. 14.25. Section through the kidney of an adult frog. Note the scattered glomeruli and dense accumulations of pigment.

terrestrial species which must conserve water. Each nephron has a narrow ciliated neck, a proximal tubule with a thick brush border, an intermediate segment of ciliated cuboidal epithelium and a distal tubule of low, unciliated columnar epithelium. The collecting ducts drain by paired nephric ducts into a urinary bladder which develops as a pocket in the cloacal wall. A copious urine flow is maintained to prevent hydration of the body, but mechanisms for reabsorption of water in the kidney tubules, cloaca and large intestine allow adjustment of the amounts of fluid excreted. Indeed, in amphibians living in dry conditions, most of the water filtered off at the kidneys is later reabsorbed.

Reproduction

Almost all amphibians return to water to breed, as the eggs require water for their development. The sexes are separate and fertilization is external, but elaborate mating and clasping behaviour ensures efficient fertilization.

The paired round testes of the male are capped by fat bodies. Each testis consists of short, blind-ended seminiferous tubules. A large number of cells in a given area of a seminiferous tubule mature synchronously, thus giving the appearance of cysts, each at a different stage of development (Fig. 14.26). Sperm are released into the tubule lumen, and travel, via a series of small vasa efferentia formed by outgrowths of the kidney into the gonad, into the nephric ducts. Thus, in the frog, as in the dogfish the nephric ducts serve for the conduction of both sperm and urine. At their lower ends they possess thin-walled diverticula (seminal vesicles) which are used to store sperm.

The ovaries (Fig. 14.27) of the female frog form irregular masses of germinal tissue surrounding a number of cavities. There is no solid stroma; the medulla consisting of a thin vascularized connective tissue meshwork. The cortex remains as narrow layer of germinal epithelium during the winter, but in spring it develops into folds of germinal tissue containing clumps of oocytes. Each developing ovum is surrounded by a theca of smooth muscle and follicle cells secreting hormones. Lipid and yolk accumulate in the ova for its nourishment. Atretic follicles can sometimes be seen, but corpora lutea do not develop. As the ova are released, by rupture, into the body cavity they are captured by ciliated funnels at the ends of two long convoluted oviducts. The oviduct walls [Fig. 14.28(a)] are ciliated, and the distal portion of each is expanded to form an ovisac [Fig. 14.28(b)]. Here, albumen and jelly are secreted round the eggs which may be stored for a time before passing via a short vagina to the cloaca and thence to the outside.

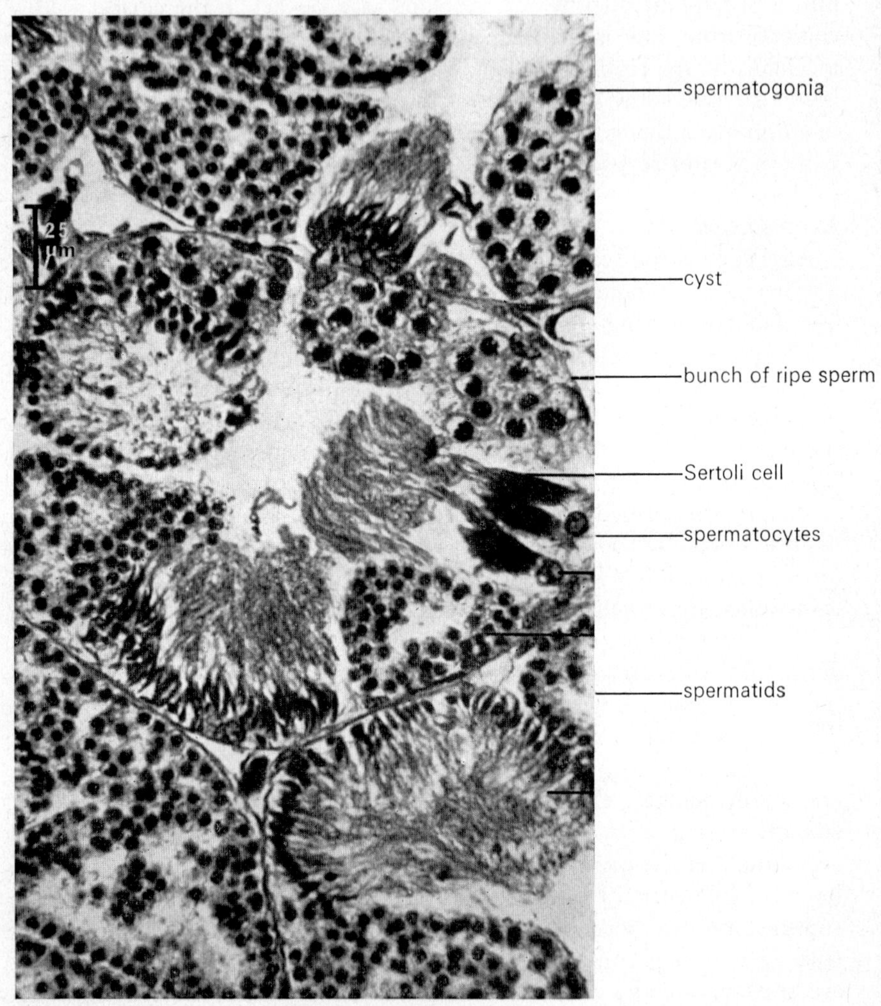

FIG. 14.26. TS through seminiferous tubules of the testis from an adult male frog. Note synchronous development of germ cells in each "cyst".

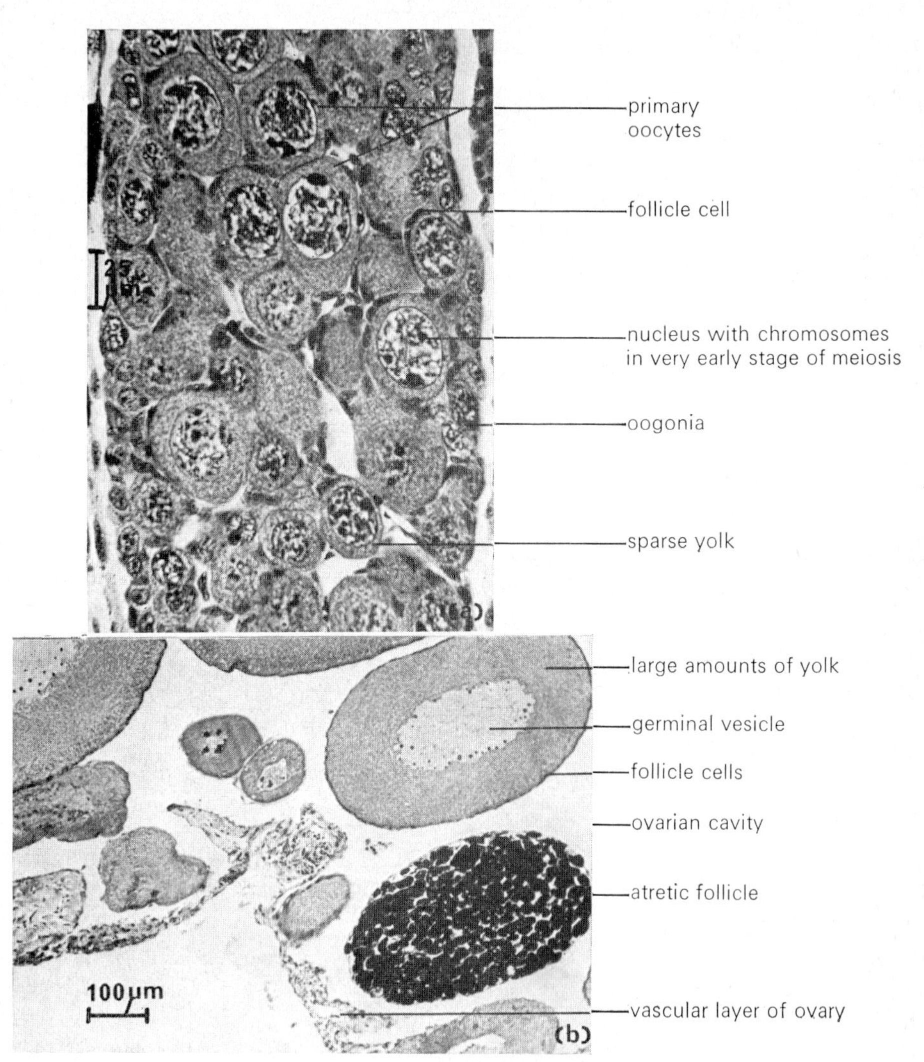

FIG. 14.27 (a) Section through the ovary of a young female frog just after metamorphosis; (b) section through the ovary of an adult female frog.

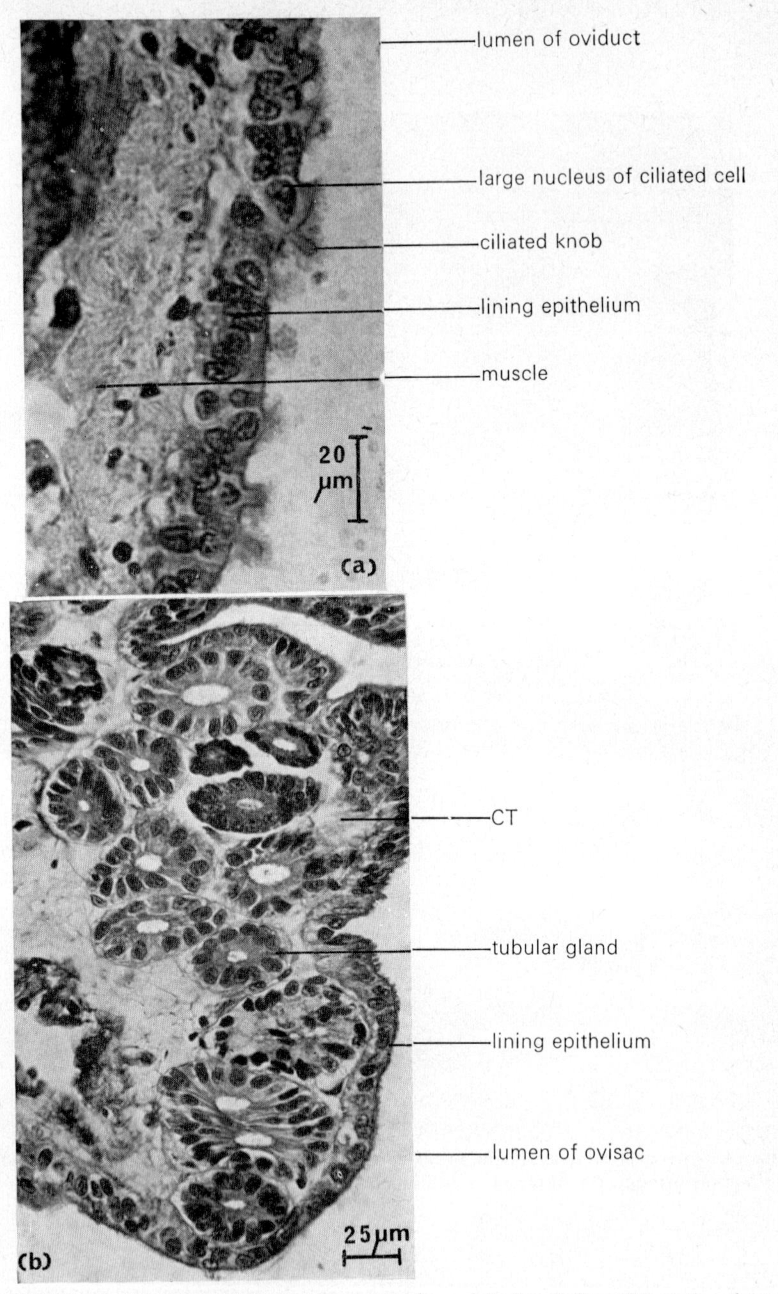

FIG. 14.28 (a) Section through the oviduct wall of an adult female frog; (b) section through the wall of the ovisac from an adult female frog, showing the tubular secretory glands.

Chapter 15

Class Reptilia

Reptiles are fully terrestrial vertebrates and can live independently of water, even in the early stages of their life cycle. They are assisted in this by the development of a cleidoic egg, with a large yolk supply and extra-embryonic membranes (amnion, allantois and chorion) (see Fig. v) which ensure protection and a supply of food, albumen, water and oxygen for the embryo when the egg is laid on land. There are also some viviparous reptiles, and in a few of these the extra-embryonic membranes form a sort of placenta.

Most reptilian orders are now extinct, modern reptiles being represented by only four living orders. They are cold-blooded amniote tetrapods, occurring mainly in tropical environments. The body is elongate and is generally clearly divided into a small head, neck, rounded trunk and (usually) long cylindrical tapering tail. The pentadactyl limbs, secondarily absent in some species, are placed ventrally and are capable of supporting the body weight on land. Reptiles are covered by a layer of horny epidermal scales (sometimes fused completely) which helps resist desiccation and abrasion. Functional gills are not developed at any stage. All reptiles use lungs as their respiratory organs, and the paired nasal apertures are used for breathing as well as olfaction. Reptiles are further characterized by the structure of their ear and by having a transverse anal opening, paired copulatory organs and a cloacal bladder. There is no larval stage, and, therefore, no metamorphosis.

Example: *Lacerta* (*vivipara*)

The common lizard is widespread in Europe, and lives in dry terrestrial conditions.

It belongs to the most modern order of reptiles, the Squamata (lizards and snakes); an order of diapsid reptiles in which the temporal region of the skull is reduced to leave only one temporal arch (lizards) or none (snakes). The vertebral column is very flexible, with the sacral region (supporting the hindlimbs) reduced. Lizards show autotomy, whereby the tail can be detached at pre-determined planes of weakness in the caudal vertebrae. This feature is thought to be used as a means of escape, as the tail can later be regenerated. Squamates possess a well-developed Jacobson's (vomeronasal) organ for the reception of taste and smell.

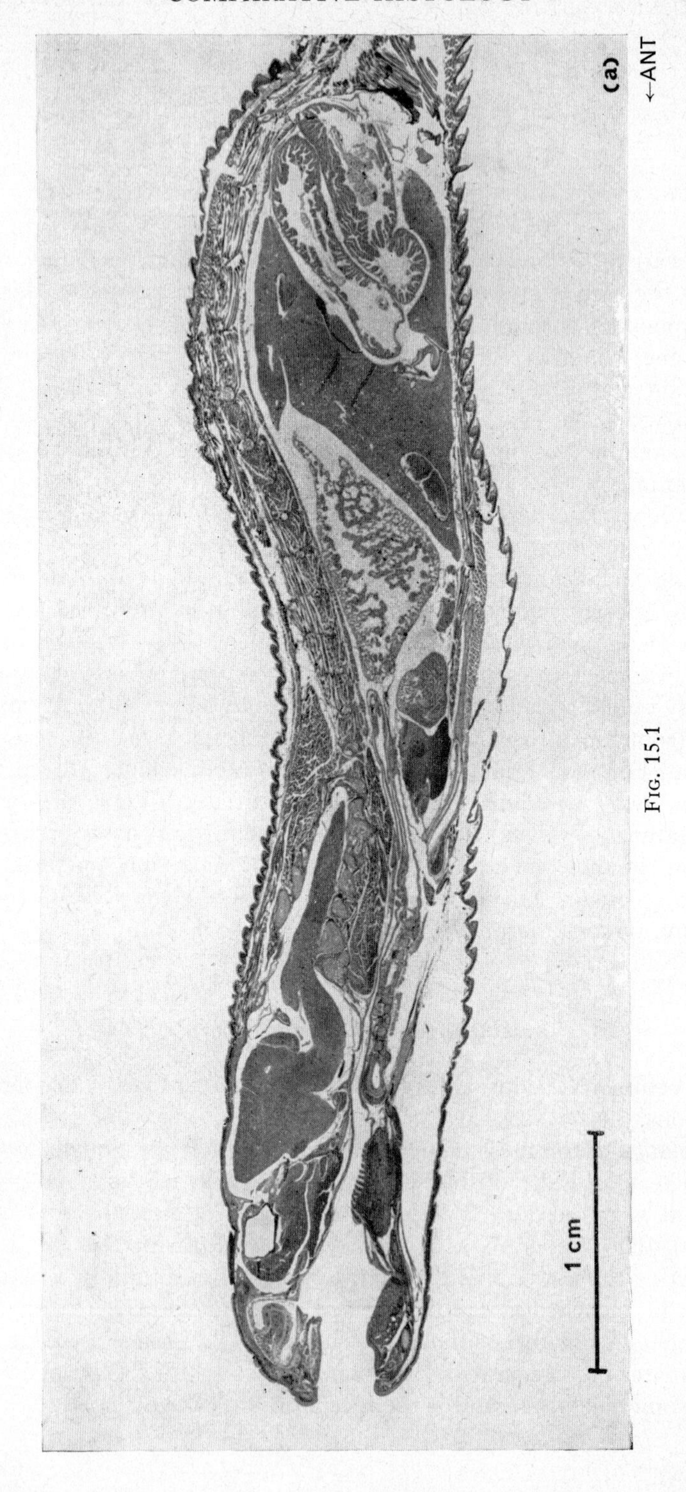

FIG. 15.1

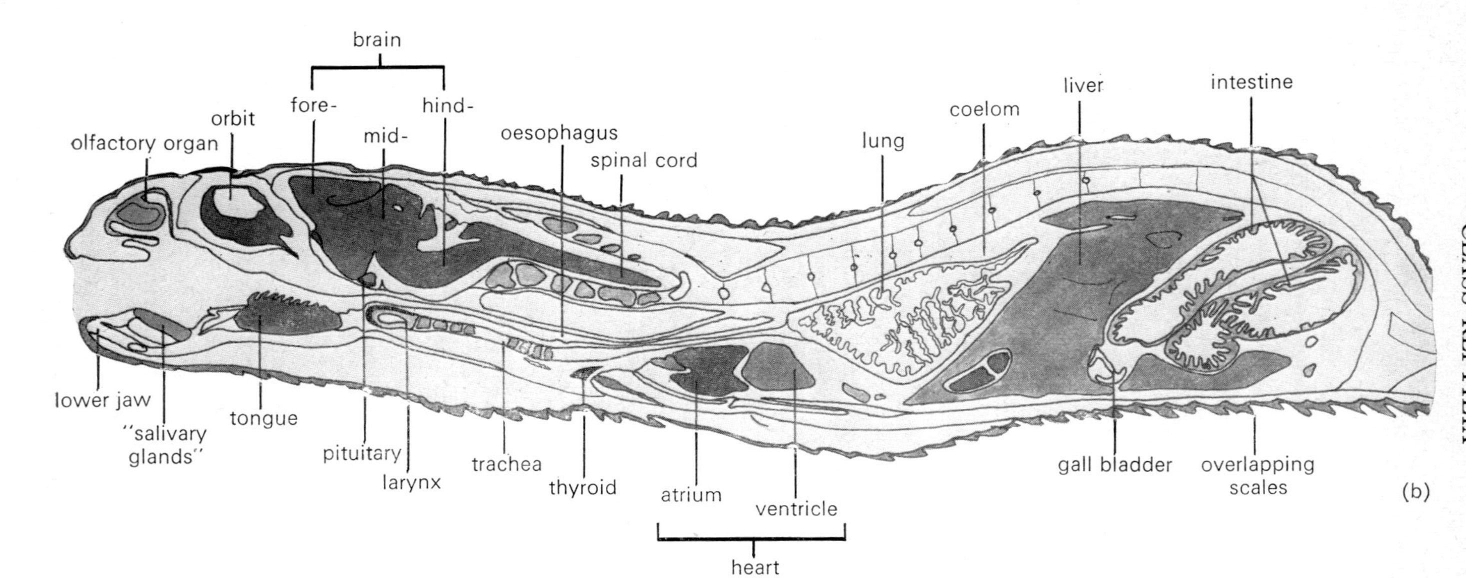

Fig. 15.1 (a) VLS and (b) explanatory drawing of the body of a young lizard (*Lacerta vivipara*). The tail of this lizard has not been included in the photograph.

Figure 15.1 is a VLS of a young *L. vivipara*, showing the arrangement of viscera, the long flexible vertebral column, the mobile neck and the overlapping scales. The skin (Fig. 15.2) is relatively impermeable and heavily pigmented. It is devoid of glands, except in the region of the mouth, anus and thighs where short tubular slime-producing glands occur. The dermis is thrown into regular folds, each covered by horny epidermal scales, the outer part of which is continuously being worn away and renewed from underneath. These scales are not homologous with the dermal bony scales. of fish, although some lizards do also possess dermal osteoscutes which become fused to the skull bones.

Cartilage persists in the skull, limb girdles, sternum and articulating ends of bones, but the remainder of the adult skeleton is ossified. Bone is a well-vascularized tissue formed by osteoblasts which are irregular branched cells, secreting a proteinaceous matrix that becomes hard, opaque and calcified. It can take the form either of hard lamellar bone (see Fig. 17.5) or of spongy (cancellous) bone which is an open network of bony bars (trabeculae) interspersed with marrow consisting of fat, haematopoetic tissue and blood-vessels (Fig. 15.3). Spongy bone always is encased in hard bone. It is lighter than lamellar bone and occurs during bone formation, although it may persist in the adult. Bone may be formed directly within the mesenchyme, as in the dermal scales of fish (see Fig. 13.2) or by replacing cartilage (see Figs 14.4 and 15.3).

The skull consists of two boxes: an inner chondrocranium of cartilage surrounding the nasal organ, inner ear and brain; and an outer bony cranium. The vertebral column is divided into five regions (cervical, thoracic, lumbar, sacral and caudal), each with characteristic vertebrae. In *Lacerta* there are two sacral vertebrae. The column is composed of numerous articulating, fully-ossified procoelous centra bearing spines and single-headed ribs (except in the abdominal region). The notochord is retained only as fibrous inter-vertebral discs between the centra.

Nervous system

The brain (Fig. 15.4) is similar to that of the frog although in lizards there is a ventral flexure at the junction of the brain and spinal cord (see also Fig. 15.1), which is associated with greater manoeuvrability of the head. Two small cerebral hemispheres bear dorso-lateral bands of cortex (Fig. 15.5) which act as an association centre and are the fore-runners of the mammalian neopallium (cerebral cortex). The fore-brain ventricle splits into two lateral ventricles which are continued forwards into the paired olfactory bulbs. The roof of the diencephalon takes the form anteriorly of a choroid plexus and posteriorly of the pineal complex. This complex includes the "median eye" [Fig. 15.6(a)], a probably photoreceptive structure thought to be of parapineal or parietal origin. It is situated close under the dorsal skin, which is translucent at this point

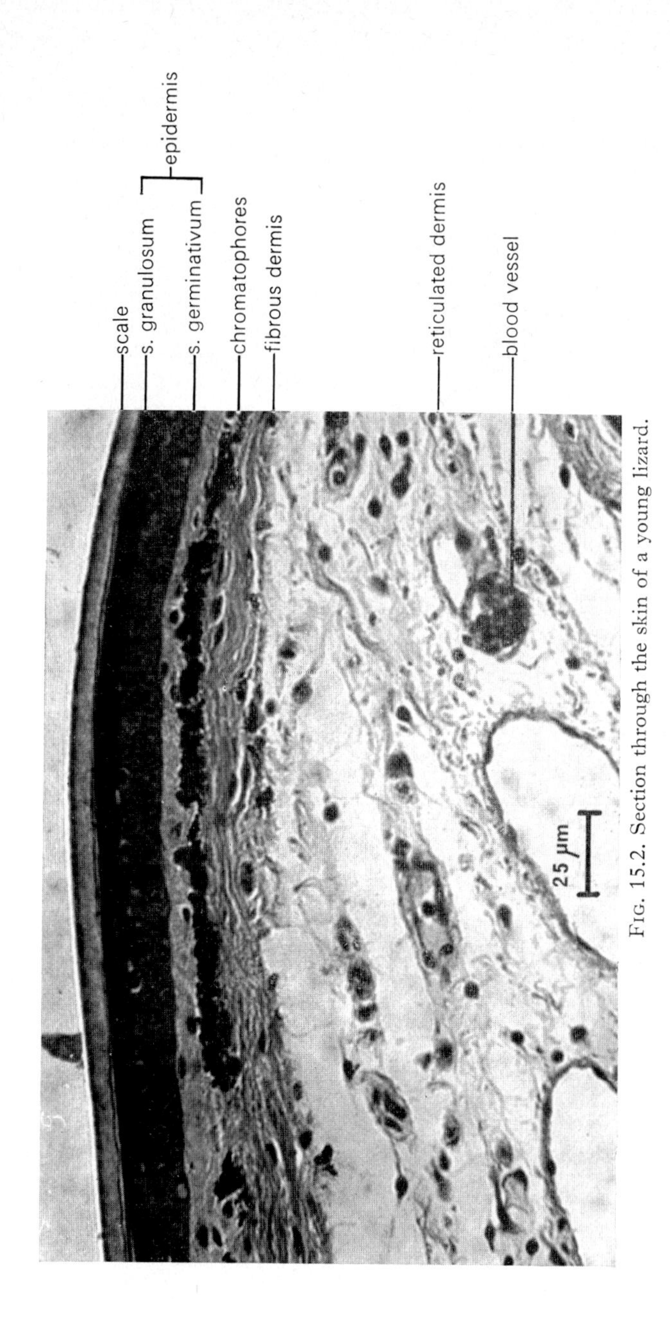

Fig. 15.2. Section through the skin of a young lizard.

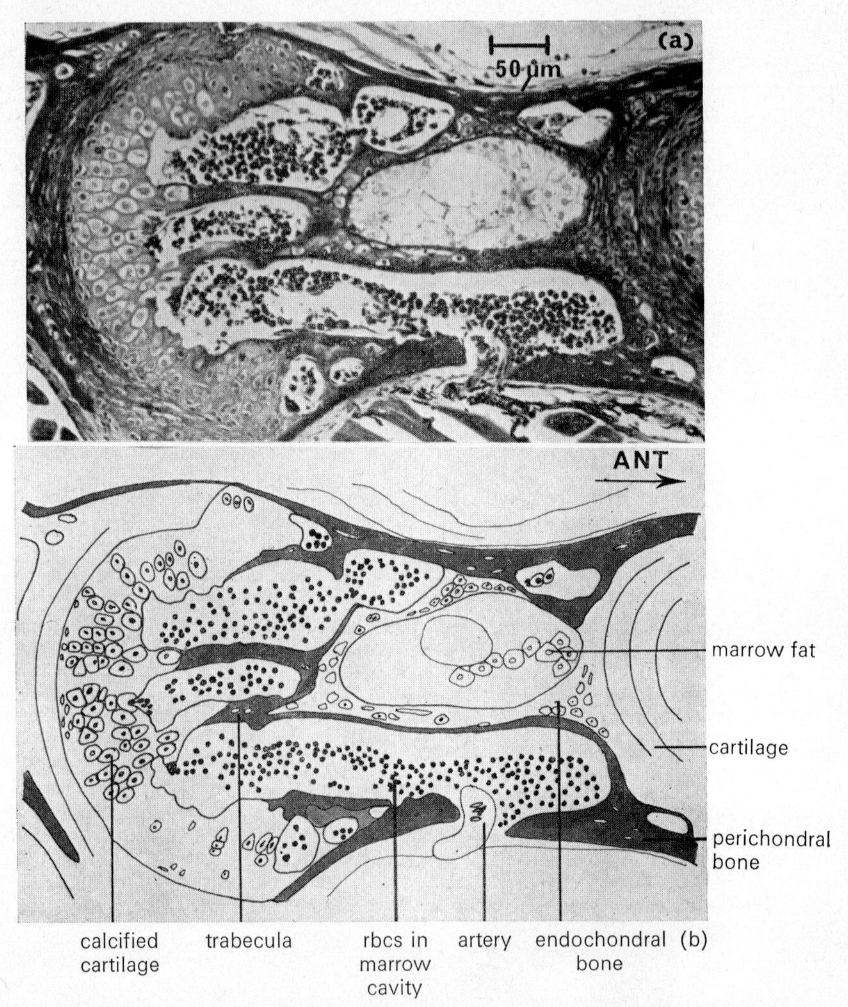

FIG. 15.3 (a) VLS and (b) drawing of a developing vertebra in a young lizard. Note that the main body of the cartilaginous centrum has been replaced by spongy (cancellous) bone.

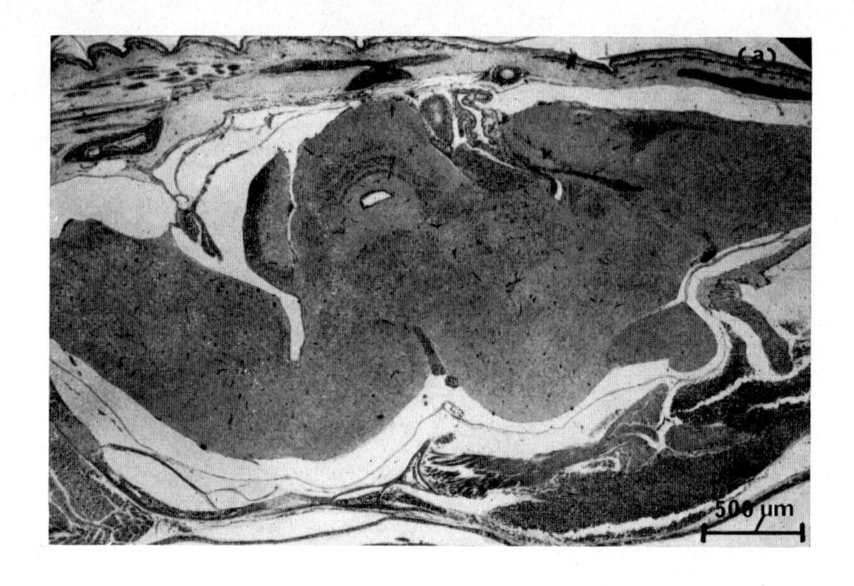

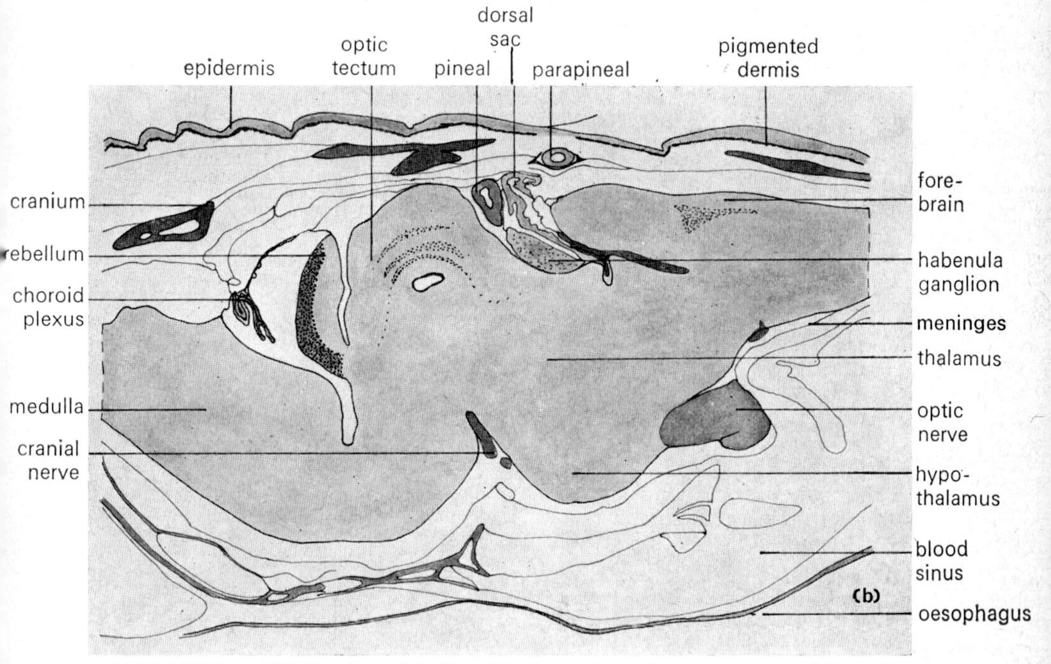

FIG. 15.4 (a) VLS and (b) drawing of the lizard brain, to show the arrangement of the main regions. Note the position of the pineal and parapineal.

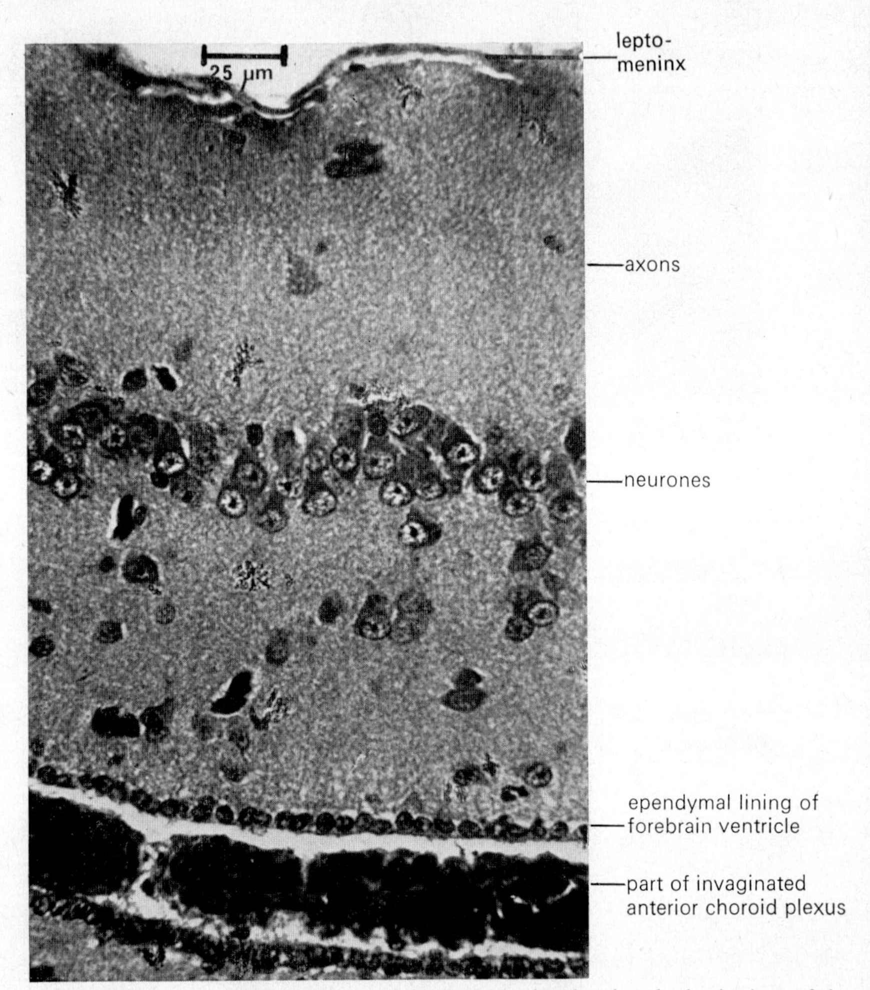

FIG. 15.5. VLS through the cerebrum of the lizard brain, showing the beginnings of the cortical arrangement of neurones.

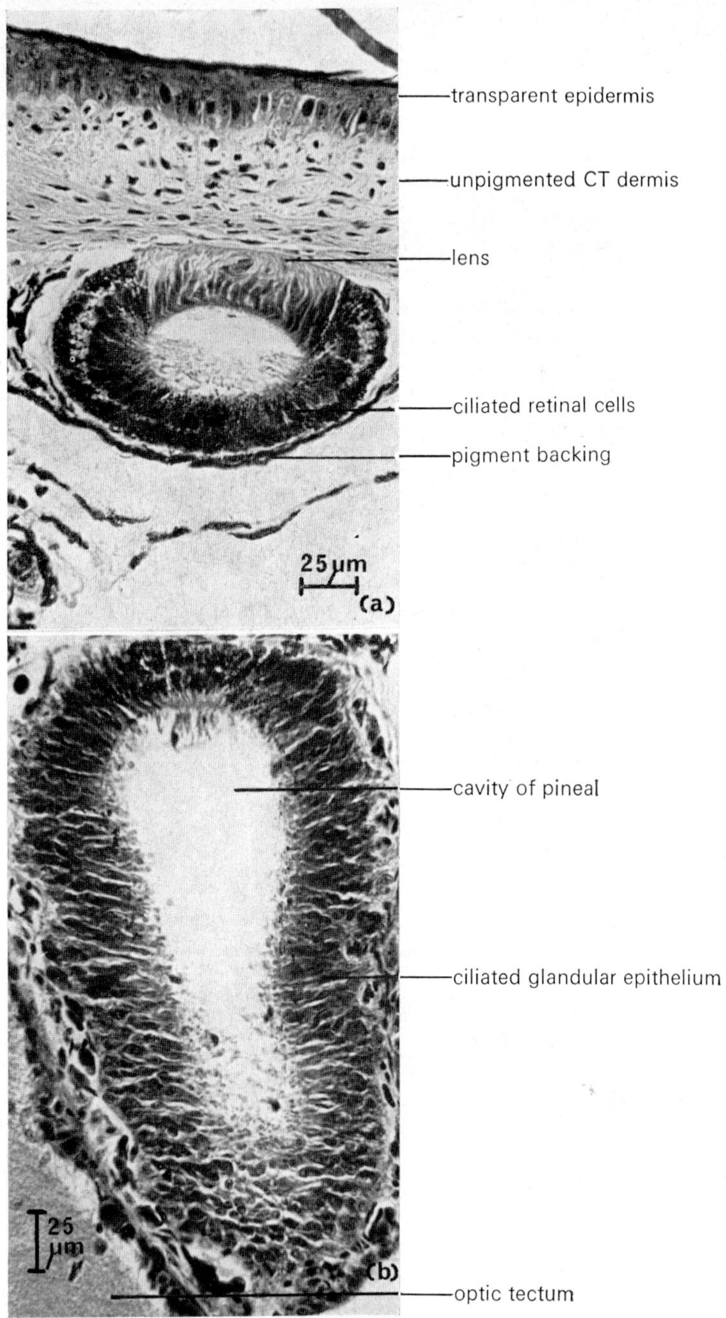

FIG. 15.6 (a) LS through the parapineal (parietal body) of the lizard brain. Note the "eye-like" appearance, and the lack of pigment in the overlying skin; (b) LS through the glandular pineal of the diencephalon region of the lizard brain.

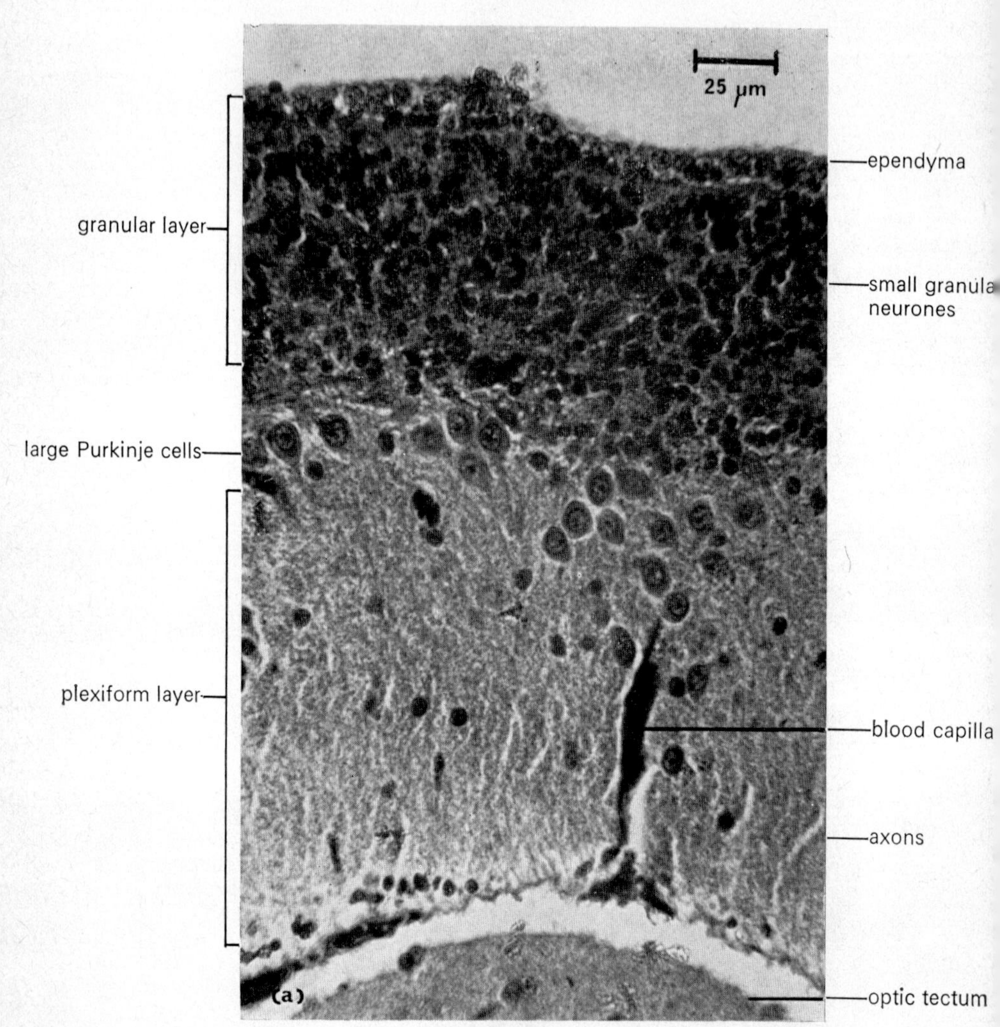
25 μm
granular layer
large Purkinje cells
plexiform layer
ependyma
small granula
neurones
blood capilla
axons
optic tectum
(a)

Fig. 15.7

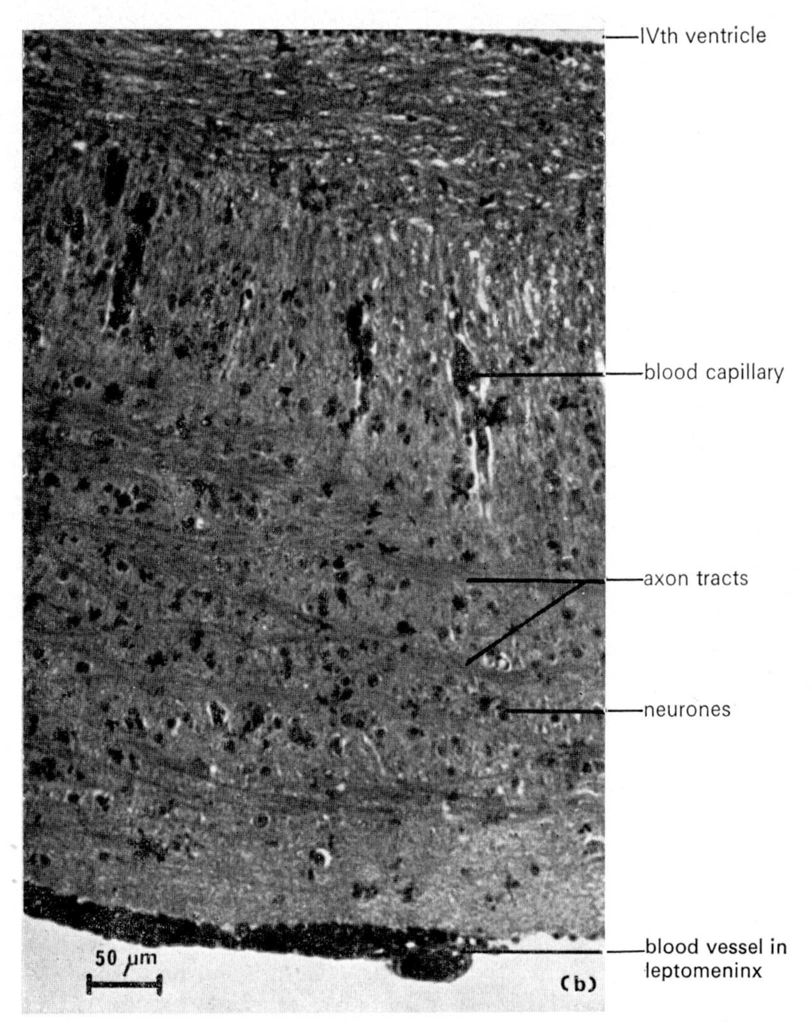

FIG. 15.7 (a) VLS through the cerebellum of the lizard brain. Note the large Purkinje neurones which are the main output of the cerebellum; (b) VLS through the medulla oblongata of the lizard brain.

(see Fig. 15.4), and is connected to the brain by a nerve. Lying close behind it is a glandular epiphyseal body (usually called the pineal) which has been implicated in hormonal control of colour change [Fig. 15.6(b)]. The functions of both these organs are, however, not yet fully established.

The paired optic lobes are well developed as the mid-brain forms the chief co-ordination centre in lizards (in snakes the tectum is similar in form to that of mammals and does not give rise to optic lobes; but this is atypical of reptiles). The cerebellum [Fig. 15.7(a)] is small but shows the usual arrangement of peripheral neurones. In the medulla oblongata [Fig. 15.7(b)] motor and sensory nerve tracts break up into groups ("nuclei") of nerve cells controlling various autonomic functions. The spinal cord is slightly enlarged opposite the two pairs of limbs and tapers into the tail region, but in cross-section has the normal appearance of dorsal and ventral horns of grey matter. Twelve pairs of cranial nerves are found, and paired chains of sympathetic ganglia. There are two meningeal membranes covering the CNS; an inner leptomeninx and an outer dura mater.

Sense organs

Sight, hearing and temperature reception are normally the most important senses in squamates, although this is not true for some burrowing forms. In lizards, small myelinated nerve fibres with expanded tips situated in the epidermis probably act as temperature receptors. In snakes pit-organs are used for thermoreception.

The eyes are large, and are fringed by eyelids. In some species (e.g. snakes) the eyelids fuse and form a transparent tertiary spectacle, while in others they are lined by an epidermal conjunctiva which forms a fold (nictitating membrane) that can be drawn over the surface of the eye and spreads lubricating fluid. This fluid is secreted by Harderian glands (Fig. 15.8) on the inner and outer surfaces of the orbit. These glands are modified in terrestrial reptiles for salt secretion. The scleroid coat of the eye is strengthened by a ring of bony ossicles, and the choroid bears a vascular pigmented process (conus) which projects into the vitreous humour and is thought to be a nutritive structure (Fig. 15.9). Accommodation is achieved in all amniotes by altering the shape of the lens rather than its position. The amniote lens is a relatively plastic structure and its focal length can be changed by making it more or less spherical. In lizards this is brought about by striated ciliary muscles contracting against an annular pad round the lens periphery. Deformation of the eyeball is prevented by the scleral ossicles combined with a viscous vitreous humour within the eye. The lizard retina consists solely of cones for daylight vision, including double cones which may serve as detectors of polarized light. The pupil is round, and often has small notches in it to let in minute amounts of light.

The reptile ear is divided into outer, middle and inner parts. The outer ear in lizards consists of folds of skin above a circular smooth tympanic

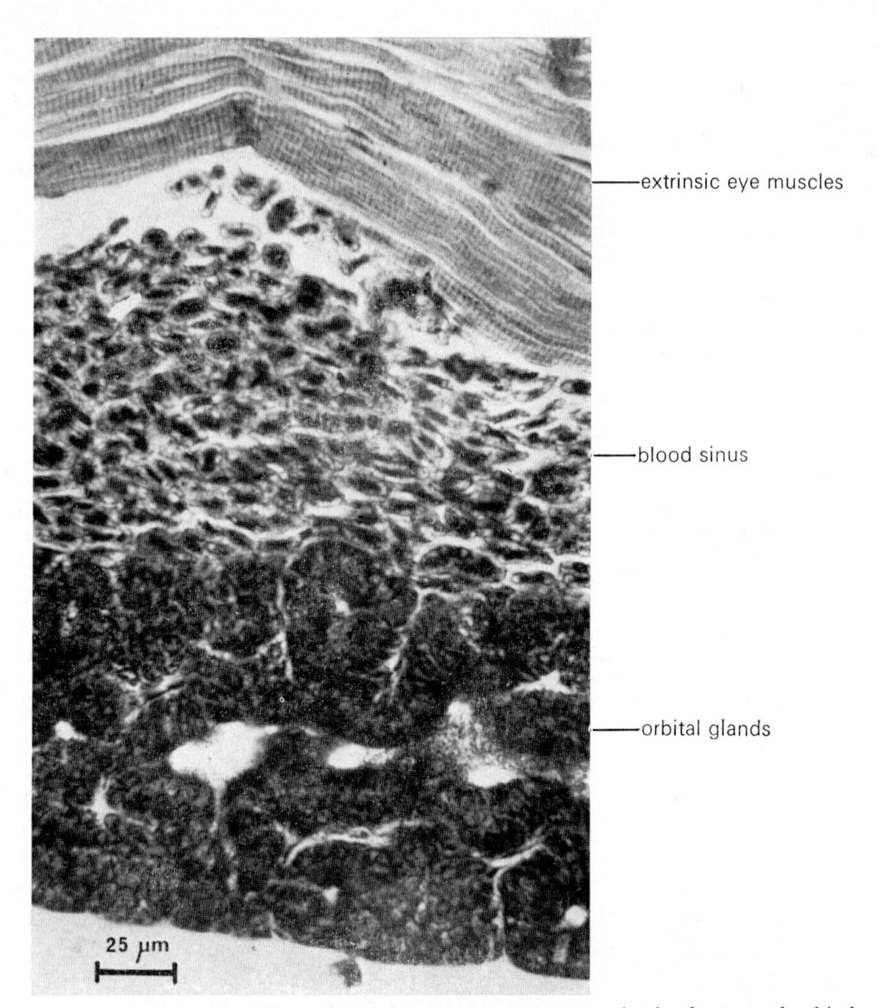

Fig. 15.8. Harderian glands underlying the extrinsic eye muscles in the ventral orbital region of the lizard.

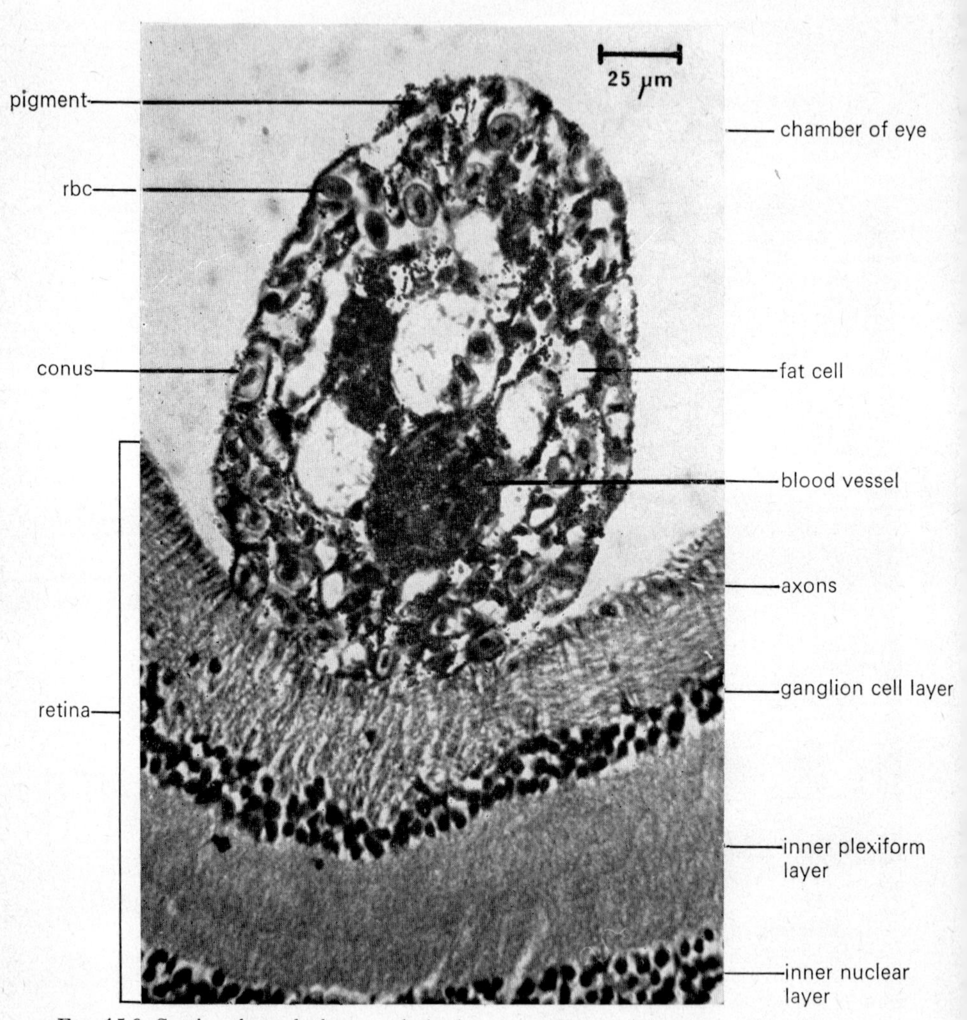

FIG. 15.9. Section through the vascularised conus and part of the retina of the lizard eye.

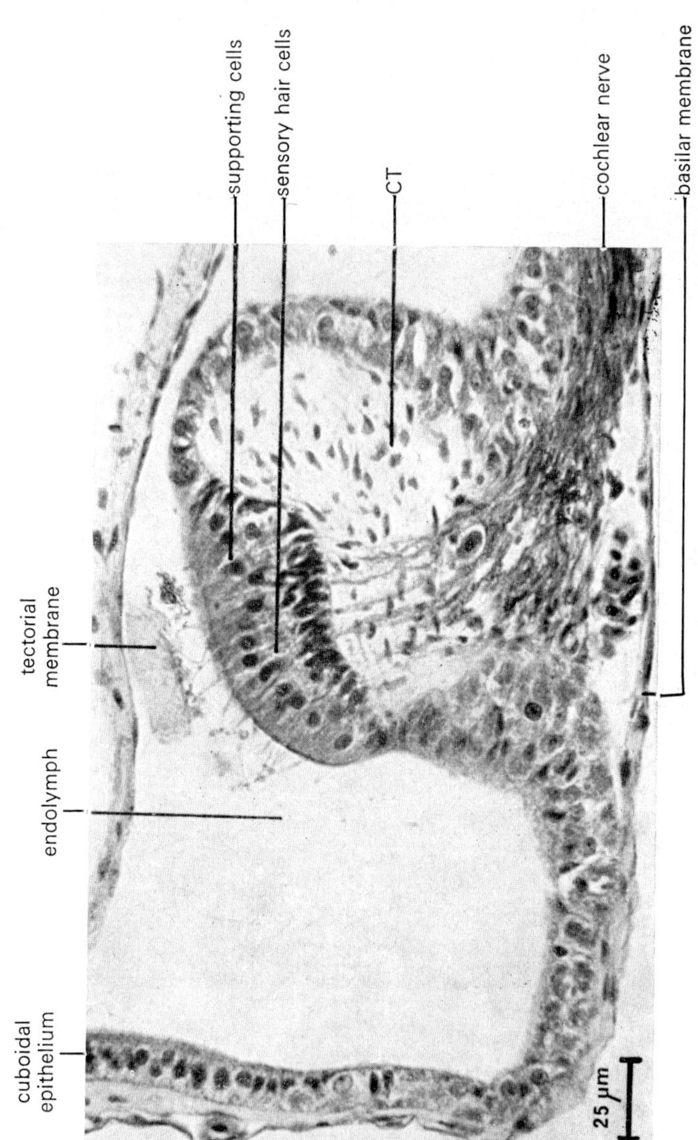

FIG. 15.10. Details of the basilar papilla in the cochlear duct of the inner ear of the lizard.

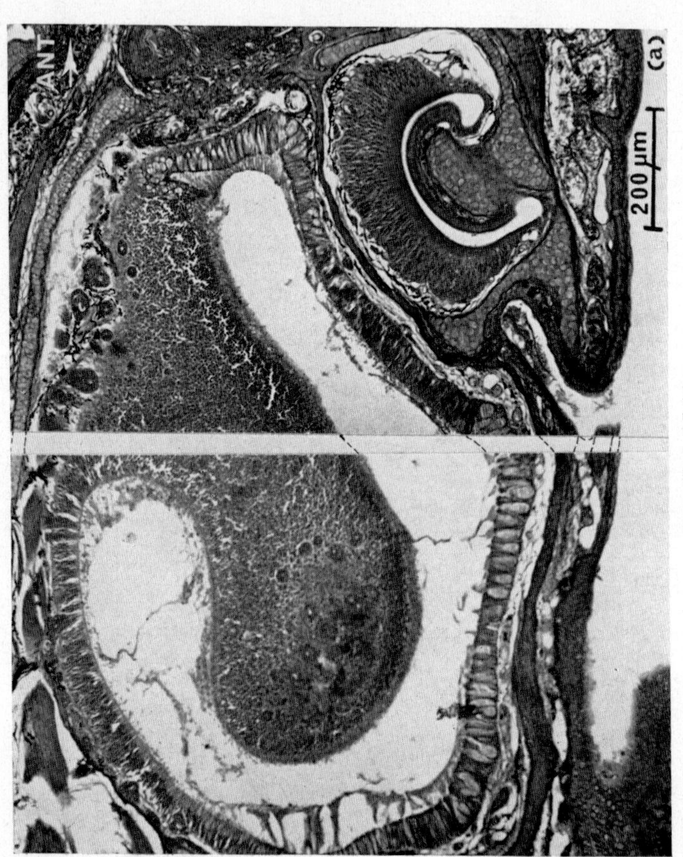

Fig. 15.11

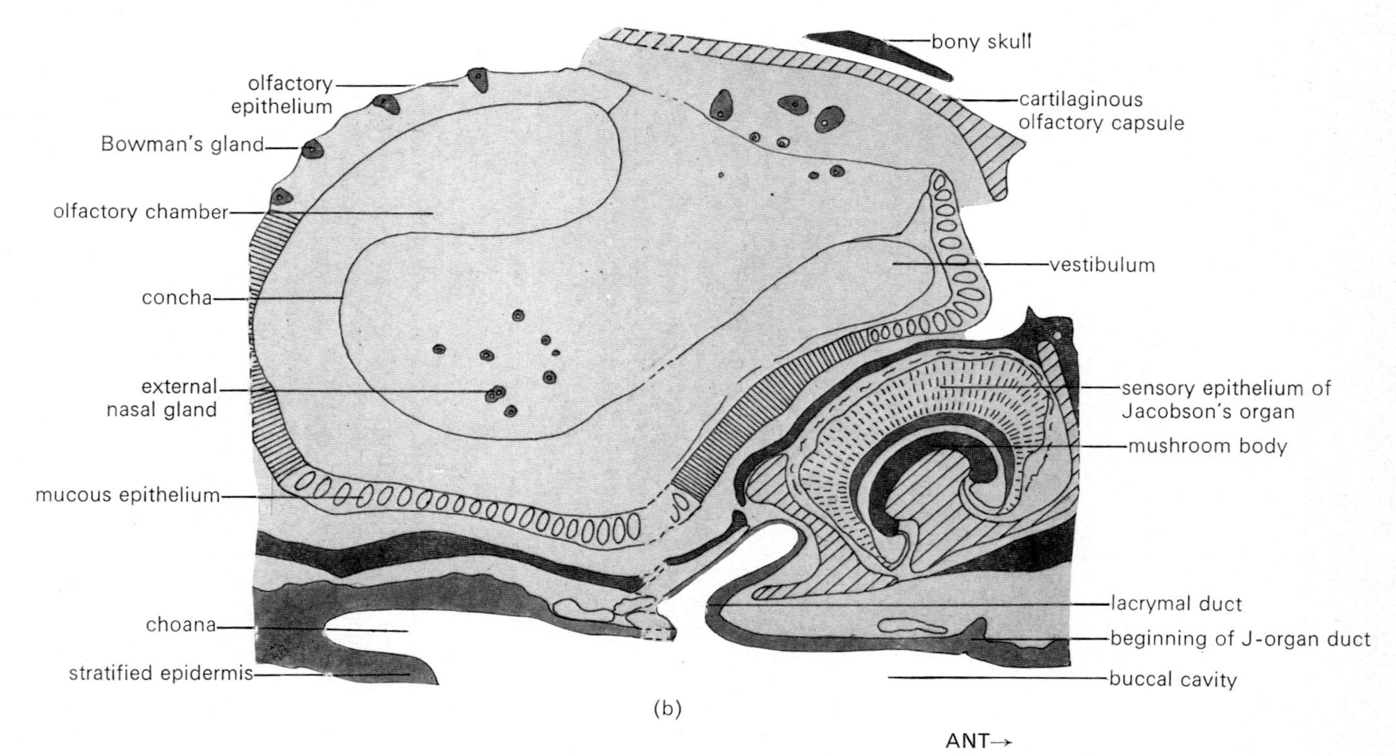

Fig. 15.11 (a) Composite VLS and (b) drawing of one side of the head of *L. vivipara*, showing an olfactory organ and one Jacobson's vomeronasal organ.

membrane situated just behind the eye. The middle ear is a cavity between the typanum and the skull, which is perforated at this point by two holes (fenestra ovalis and fenestra rotunda). The air-filled cavity is traversed by a columellar bone (stapes) and connects with the pharynx by a long auditory (Eustachian) tube. The stapes is fixed internally to the f. ovalis, and externally to an extracolumellar cartilage (extrastapes) which rests on the tympanum. Sound waves deform the membranous tympanum, are transmitted by the extrastapes and stapes to the f. ovalis where they set up waves in the perilymph surrounding the membranous labyrinth of the inner ear. The inner ear consists of three semi-circular canals and three otolith chambers bearing maculae. A small tubular utriculus and large sacculus are lined with squamous epithelium. Only the tip of the third chamber forms the lagena while the central part forms the cochlear duct. This part is lined with cuboidal epithelium and has stretched across it a basilar membrane, the middle part of which is modified to form a basilar papilla bearing sensory hair cells (Fig. 15.10). Waves set up in the perilymph are transmitted to the cochlea duct where they produce shearing movements of the tectorial membrane against the sensory hair cells on the basilar membrane. This structure is simpler than the complex cochleas of birds and mammals.

The olfactory organs of reptiles are large paired sacs, each one opening externally to a nostril by a short vestibule, and internally almost directly into the roof of the mouth by a slit-like choana (Fig. 15.11). In each main sac there is a median projection (concha) from the lateral wall. Conchae are not found in Amphibia but occur in all higher tetrapods and correspond to the turbinal bones of mammals. The nasal sacs are lined with mucous epithelium with patches of sensory epithelium containing Bowman's mucous glands [Fig. 15.12(a)].

Accessory chemosensitive organs, responsible for gustation (taste) rather than smell, occur in squamates. These are the paired Jacobson's organs which lie just anterior to the olfactory organs (see Fig. 15.11). They open by narrow ducts into the roof of the mouth in front of the choanae. It seems probable that lacrymal fluid containing odorous particles is carried by cilia from the choanal groove up the duct to the Jacobson's organ. In snakes the tip of the tongue may be inserted into the organ, but this is not likely in lizards as the duct is too narrow. Jacobson's organs [Fig. 15.12(b)] possess a ventrally placed "mushroom body" covered with non-sensory ciliated epithelium while the dorsal surface is lined with a stratified sensory epithelium bearing a striated border of microvilli but no cilia. Jacobson's organs, also known as vomeronasal organs, are better developed in squamates than in any other animal. They are innervated from accessory olfactory bulbs in the brain.

Branched tubular serous (nasal) glands lie dorsolaterally to the nasal cavity outside the cartilaginous capsule (Fig. 15.13). They probably secrete the lacrymal fluid which dissolves the scent particles.

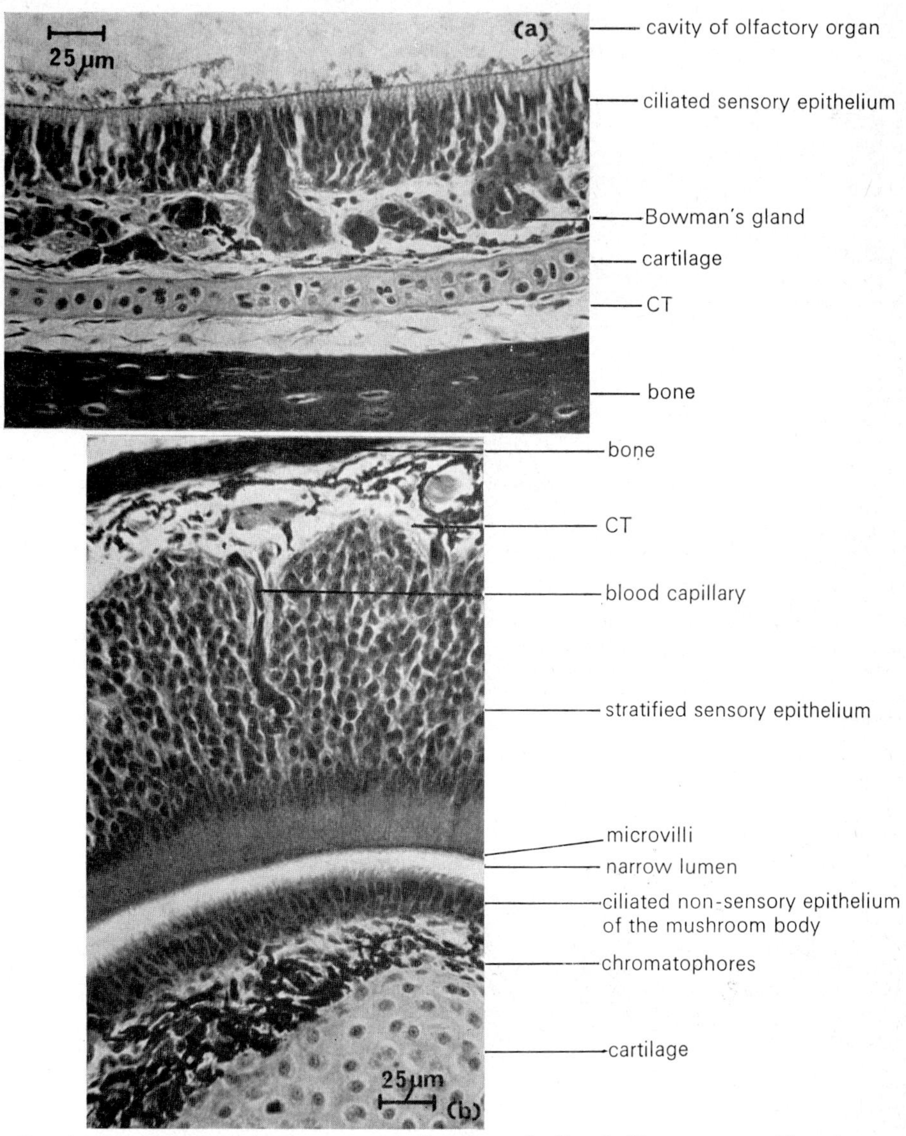

FIG. 15.12 (a) Detail of the sensory epithelium from the lizard olfactory organ. Note the multicellular Bowman's mucous glands; (b) details of the sensory epithelium from Jacobson's organ and of the mushroom body of the lizard.

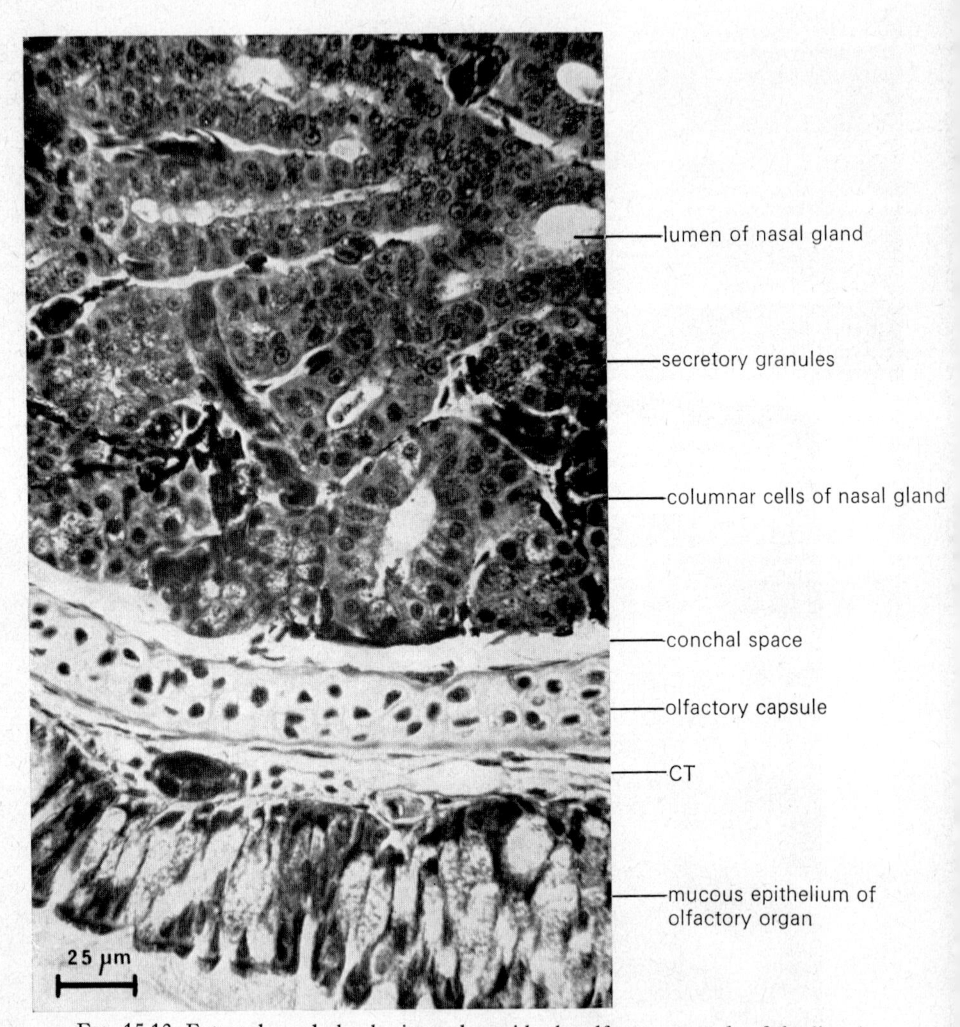

FIG. 15.13. External nasal glands situated outside the olfactory capsule of the lizard.

Endocrine organs

Lizards have a well-developed pituitary (Fig. 15.14) attached to the floor of the midbrain region by an infundibular stalk. It is clearly divided into three parts: a large pars glandularis separated from the pars nervosa by a narrow strip of pars intermedia. No pars tuberalis is present.

There is a single large thyroid gland situated ventrally above the heart (see Fig. 15.1). It has a typical follicular structure.

The paired adrenals are discrete glands located near the gonads. Each possesses a dorsal layer of chromaffin tissue covering a mass of inter-renal cell cords which are arranged very irregularly (Fig. 15.15). A few isolated nests of chromaffin tissue occur scattered among the inter-renal cells. Little work has been done on the physiology of the adrenals in lizards.

Ultimobranchials persist in reptiles, but their original function of controlling calcium and phosphate balance has been taken over by the parathyroids. The parathyroids of all amniotes are very similar (see Fig. 16.14). Lizards have two small parathyroids situated near the thymus. They are composed of cords of chief cells interspersed with sinusoids.

Respiration

Although gill slits are present in the pharynx wall of embryo lizards, they are not used in respiration and are rapidly replaced by saccular lungs. In the adult the lungs consist of large paired cavities divided into alveoli by spongy septa [Fig. 15.16(a)]. They are similar to frog lungs but slightly more sacculated. In geckos and chamaeleons the posterior lung surface is extended into thin air-sacs, but in *Lacerta* it is merely less deeply divided. The trachea leading from the pharynx to the lungs (see Fig. 15.1) is strengthened by incomplete cartilaginous rings [Fig. 15.16(b)] and divides into two bronchi before entering the lungs.

Circulation

The circulatory system of reptiles is almost completely divided into two functional streams: one to the lungs and the other round the body. The lizard heart has a small sinus venosus, a complete atrial septum (i.e. two atria), and a thick-walled ventricle which is almost completely divided by a muscular septum. Figure 15.17(a) illustrates the three-dimensional nature of the cardiac muscle network in the ventricle wall. A short truncus arteriosus, separated into three parts by folds, channels deoxygenated blood from the right side of the ventricle through the pulmonary arch to the lungs and through the left systemic arch [Fig. 15.17(b)]. Oxygenated blood returning from the lungs is channelled from the left side of the ventricle through the right systemic arch to the body. All venous blood returning from the posterior end of the body returns to the heart via the kidneys (renal-portal system) or liver (hepatic-portal system).

The site of formation of the large (20-25 μ) erythrocytes in adult lizards

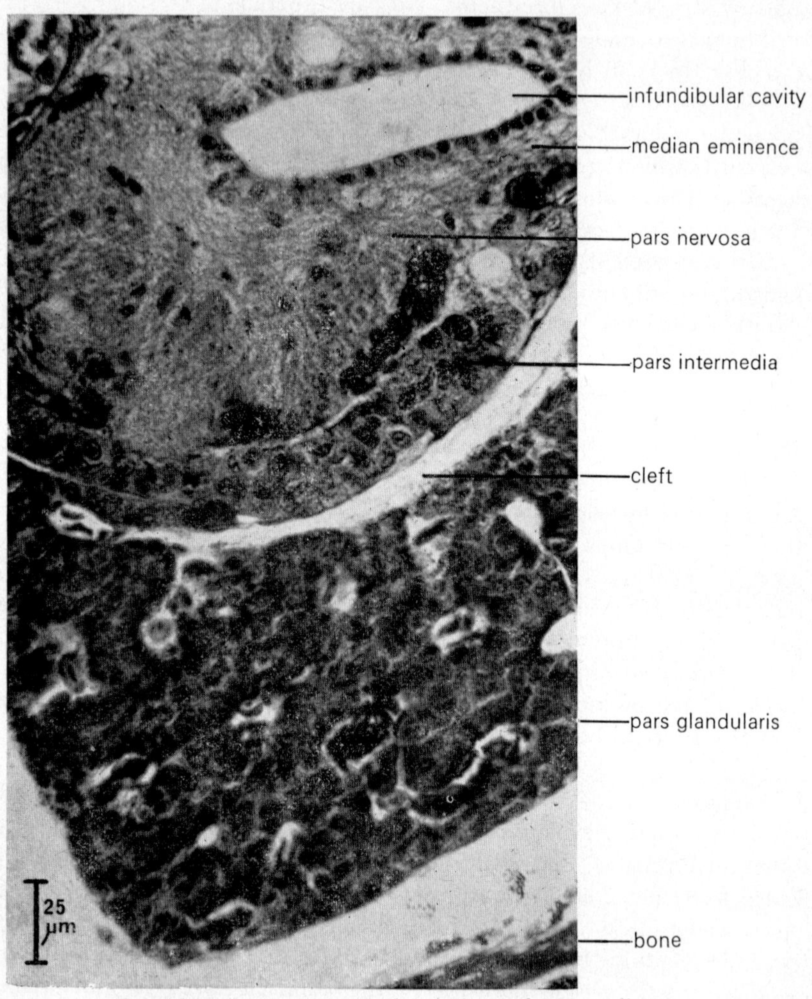

FIG. 15.14. Section through the pituitary of a young lizard.

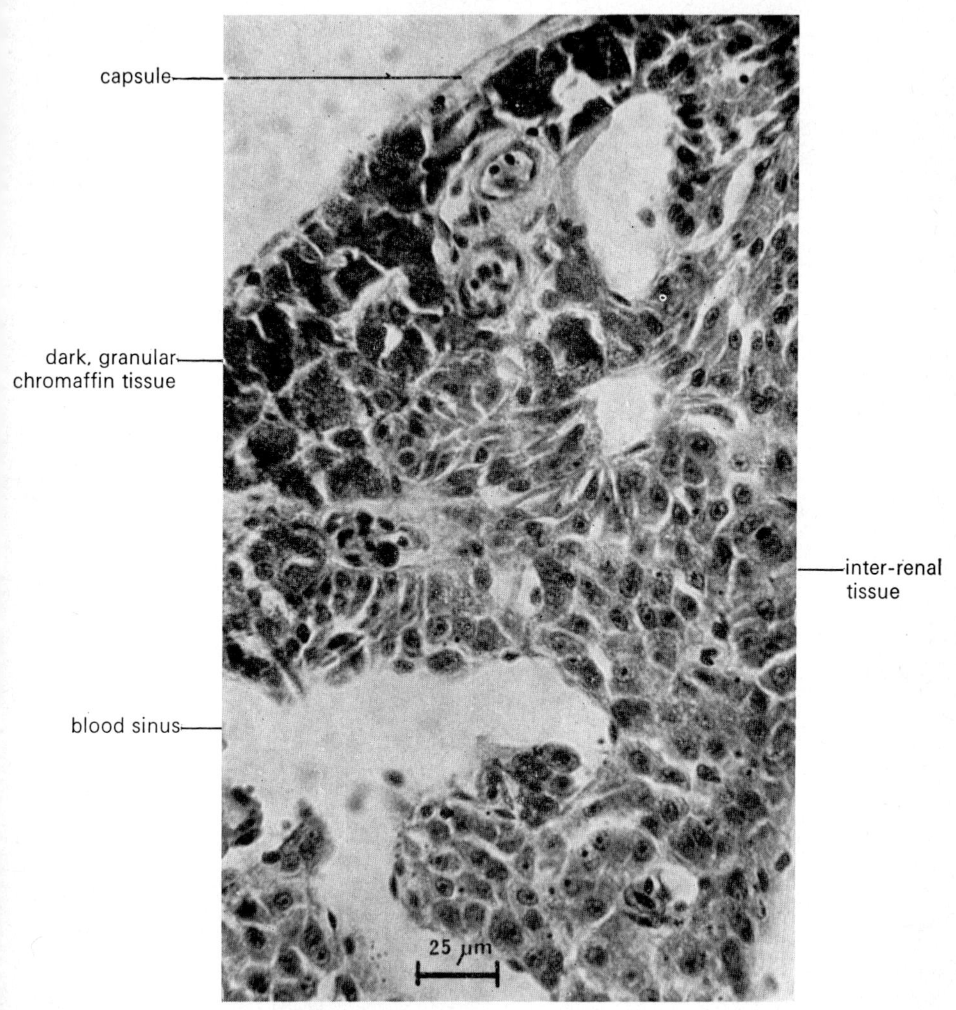

Fig. 15.15. Section through the adrenal gland of a young lizard. Note the dorsal layer of dark-staining chromaffin tissue, and the numerous blood sinuses.

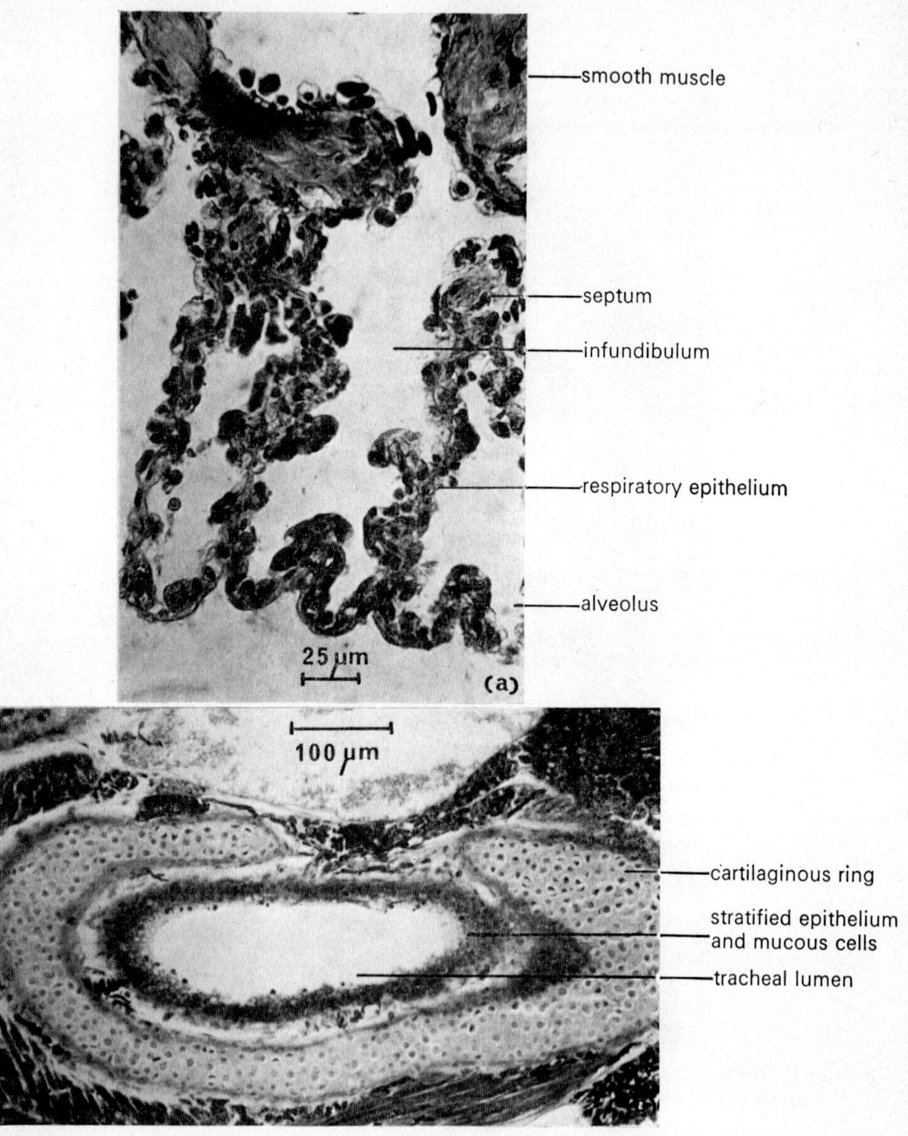

FIG. 15.16 (a) Section through the lung of an adult lizard; (b) TS of the lizard trachea, showing an incomplete cartilaginous supporting ring.

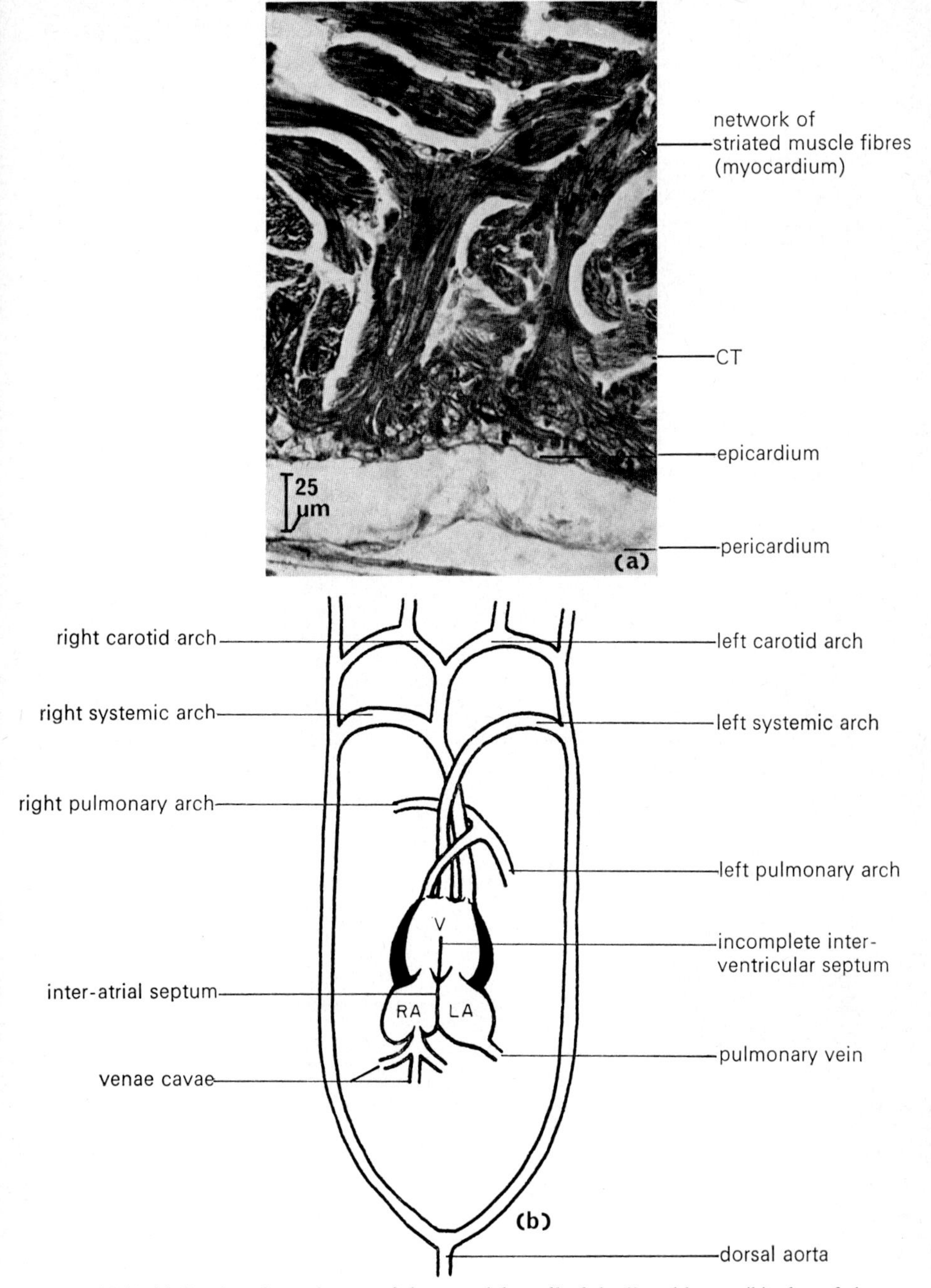

FIG. 15.17 (a) Section through part of the ventricle wall of the lizard heart; (b) plan of the heart and aortic arches in the lizard. (LA = left atrium, RA = right atrium, V = ventricle).

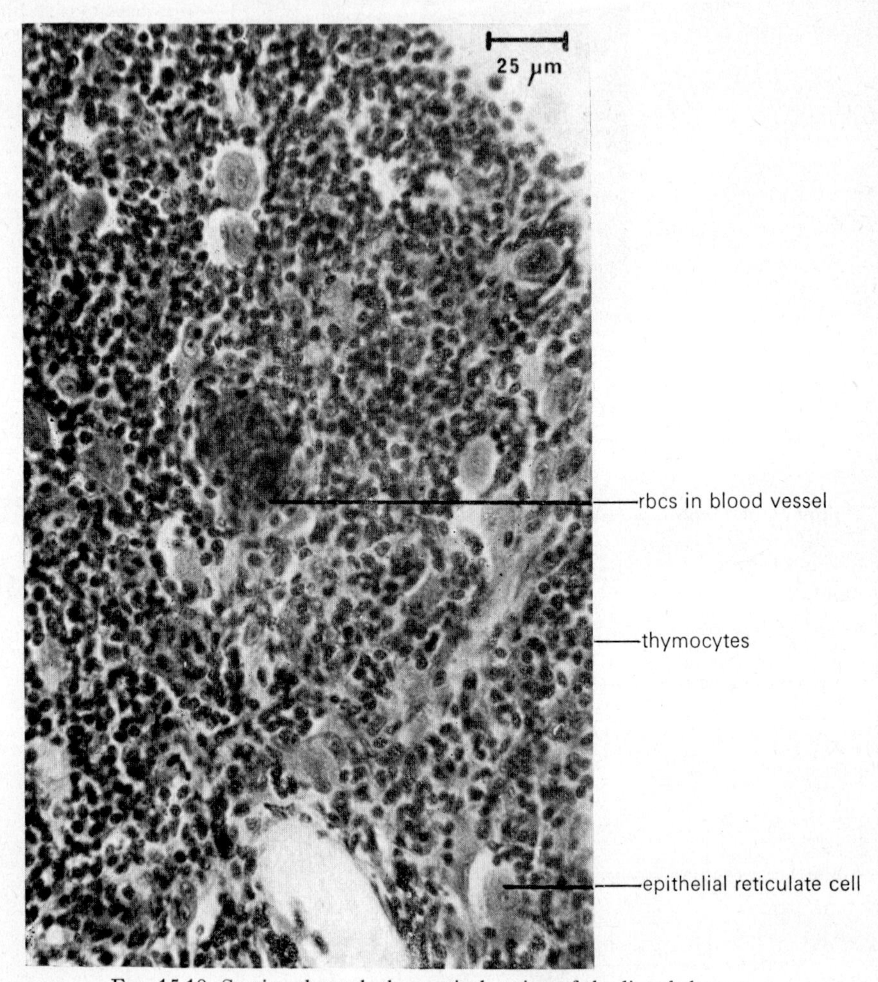

Fig. 15.18. Section through the cortical region of the lizard thymus.

is the bone marrow (see Fig. 15.3), while the small round spleen lying in the dorsal mesentery takes over the function of their destruction. Lymphocytes are formed in two small bilobed thymus glands in the neck (Fig. 15.18). The thymus glands of higher vertebrates are differentiated into cortical and medullary regions. The cortex contains blood-vessels and dense aggregations of thymocytes while the medulla consists of a loose meshwork of connective tissue, reticular cells and small whorls of cells which form thymic corpuscles.

Digestion

Lizards are carnivorous, eating chiefly insects, spiders and worms. Both upper and lower jaws bear small conical teeth along their margins (Fig. 15.19). The teeth are used for gripping and retaining food rather than for chewing it. The mouth is lined with stratified epithelium, mucous cells and taste buds; while masses of tubular mucous glands and a few serous glands (Fig. 15.20) take the place of true salivary glands which are not present in reptiles. The poison glands of snakes are formed from modified "salivary glands". The tongue (Fig. 15.21) is long, narrow and muscular. It is notched anteriorly and bears large backwards-pointing papillae which help to direct food towards the oesophagus. The oesophagus is lined with a single non-folded layer of ciliated epithelium bearing mucous goblet cells (Fig. 15.22). The stomach (Fig. 15.23) is lined with gastric glands and has a thick layer of circular muscle. The intestine is differentiated into a coiled small intestine with a deeply-folded mucosa (Fig. 15.24) and a short wider large intestine with smaller folds and more mucous cells. The gut empties, together with ducts from the kidneys and gonads, into a shallow cloaca which opens to the outside by a transverse anus. The cloaca is lined with stratified epidermis and bears tubular glands (Fig. 15.25) which probably play a part in reproduction.

The pancreas and liver are similar to those of the amphibia. Figure 15.26(a) is a photograph of the lizard liver and shows the relationship of liver cells, blood sinusoids and bile canaliculi. Bile is stored in the gall-bladder [Fig. 15.26(b)]. The structure of gall-bladders is uniform throughout the vertebrates; they are bags composed of thin sheets of muscle surrounded by a connective tissue coat and peritoneum, and lined with a single layer of columnar epithelium.

Two large fat bodies (Fig. 15.27) lie in the ventro-posterior region of the abdomen, and probably act as food reserves. They are white fat, typical of that found in all vertebrates. Small fat globules coalesce to form large (unilocular) fat droplets.

Excretion

As in all the amniotes, the kidneys of adult reptiles are metanephric (i.e. formed from the posterior nephrotomes). In the lizard the paired

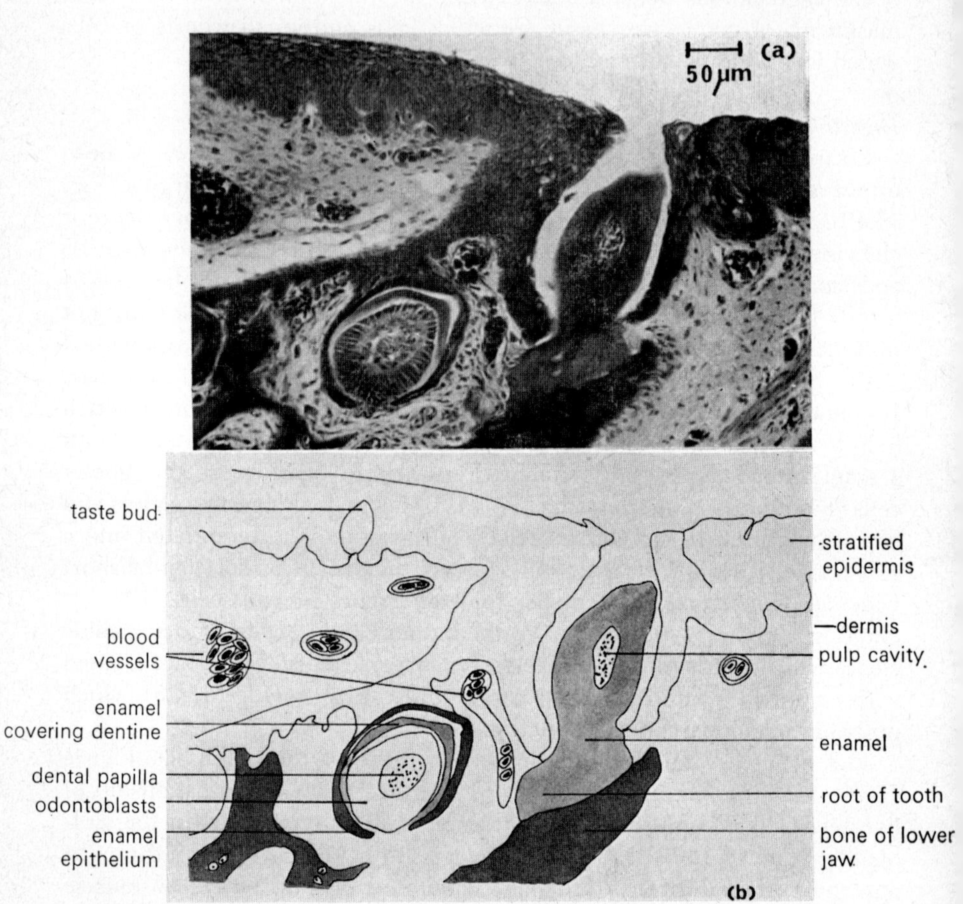

FIG. 15.19 (a) LS and (b) drawing of part of the lower jaw of a young lizard, showing developing teeth.

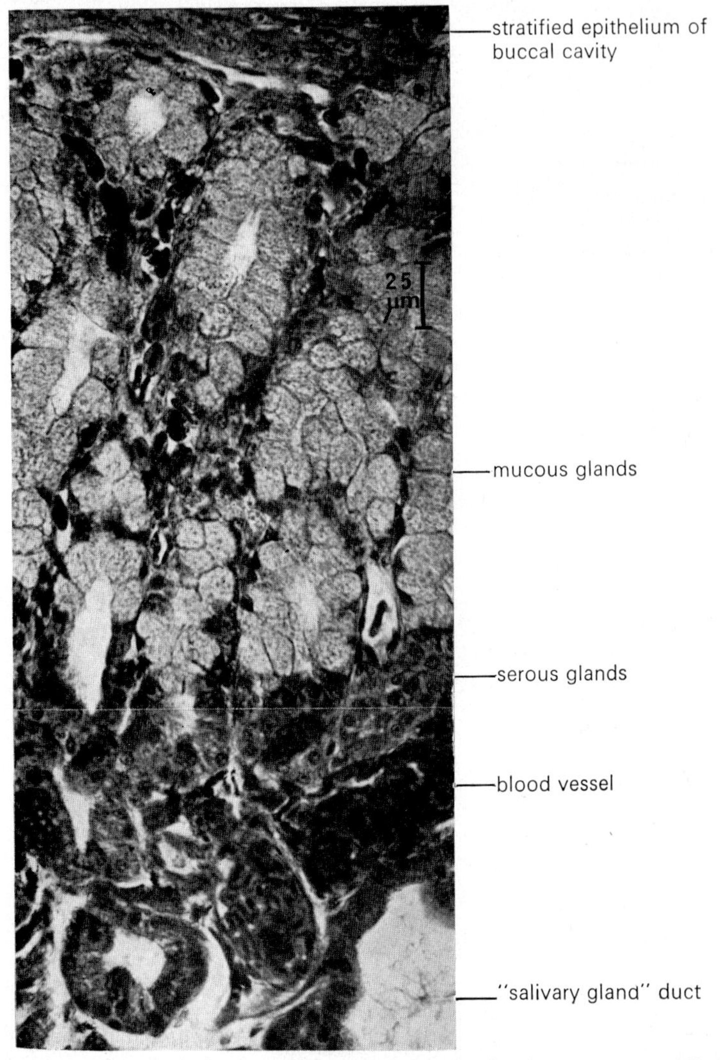

FIG. 15.20. Section through the floor of the lizard mouth, showing two types of "salivary glands".

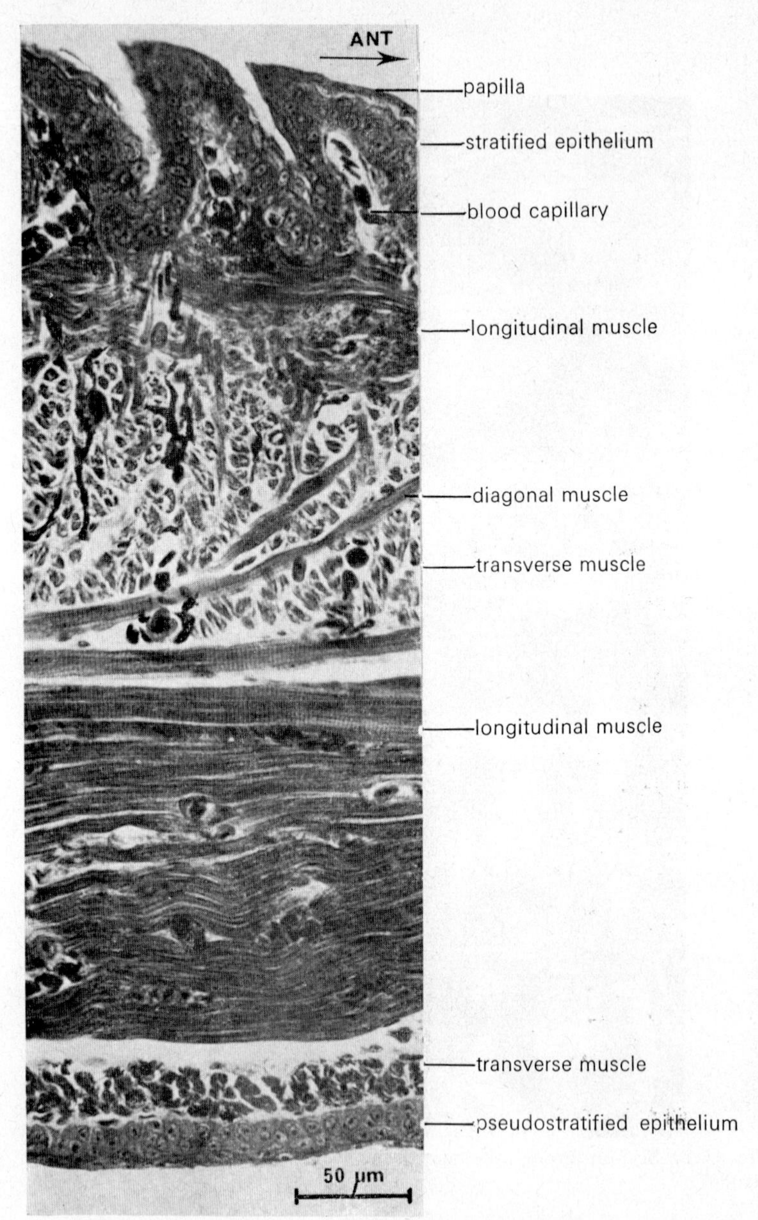

FIG. 15.21. VLS through the tongue of a lizard, showing the extensive striated muscles and the backwards pointing papillae.

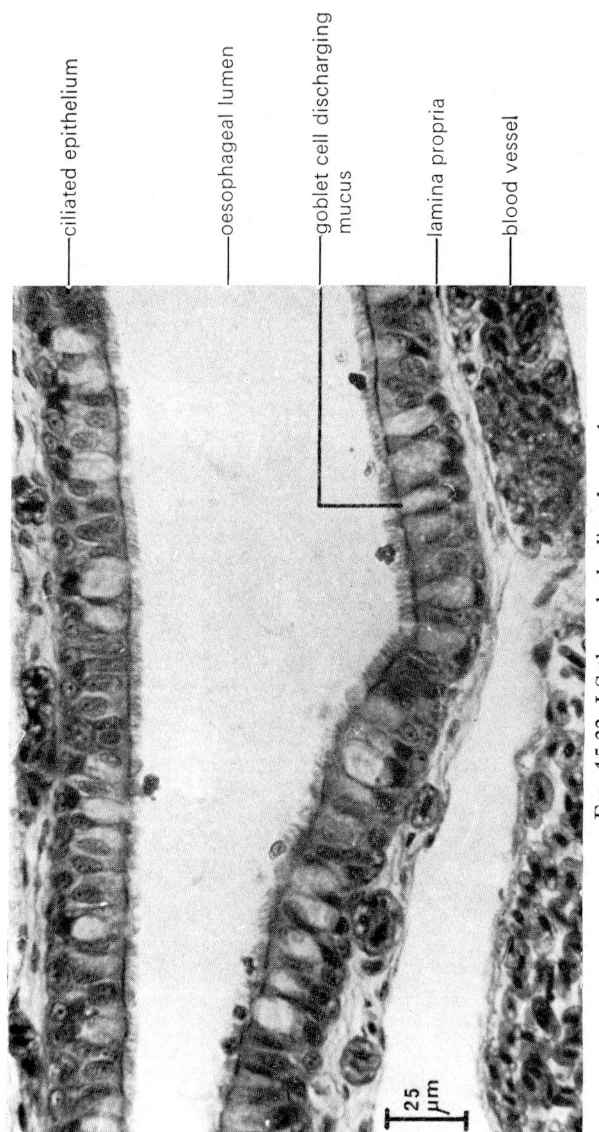

FIG. 15.22. LS through the lizard oesophagus.

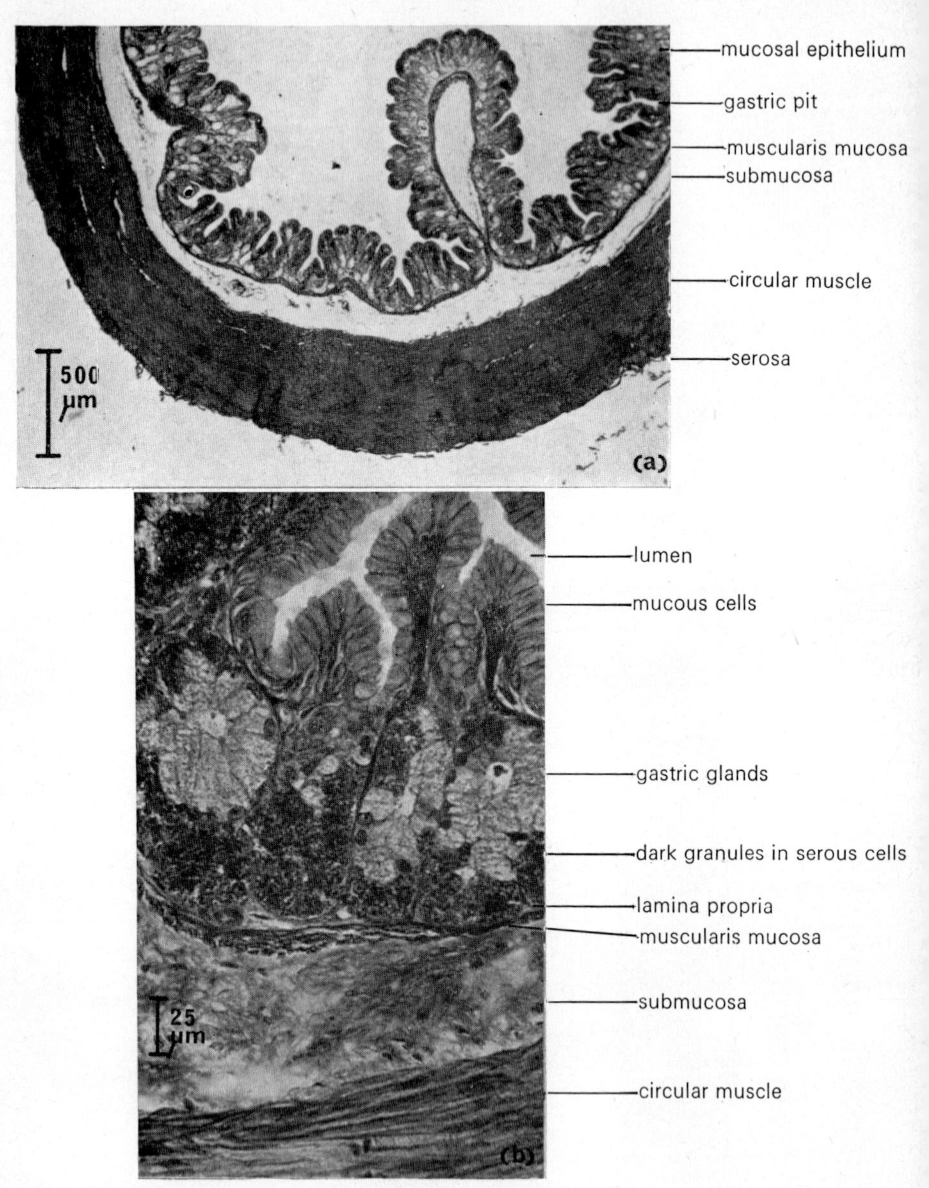

FIG. 15.23 (a) TS through the lizard stomach showing the thick layer of circular muscle; (b) TS through the stomach lining of the lizard, showing the gastric glands.

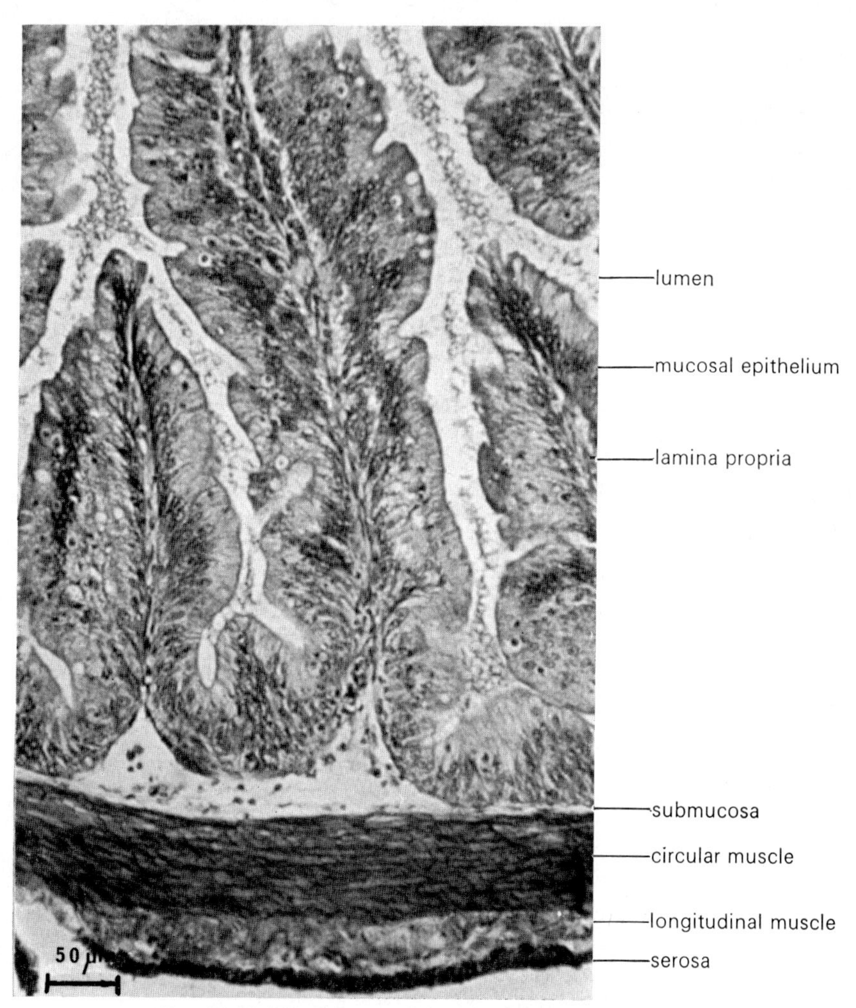

FIG. 15.24. TS of the small intestine wall of the lizard.

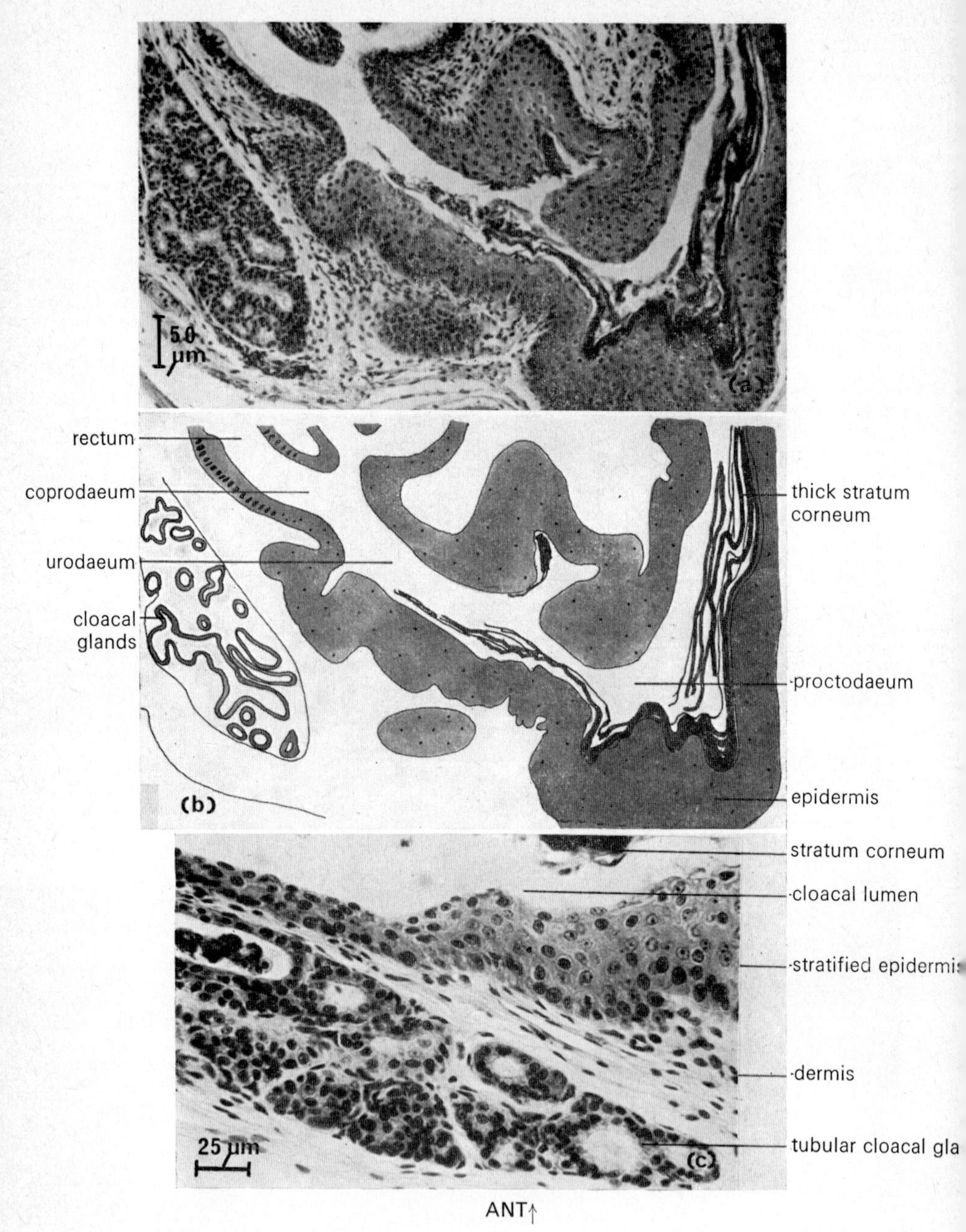

FIG. 15.25 (a) VLS and (b) drawing of the cloaca of the lizard. Note the heavily cornified epidermal lining; (c) detail of the tubular glands in the cloacal wall.

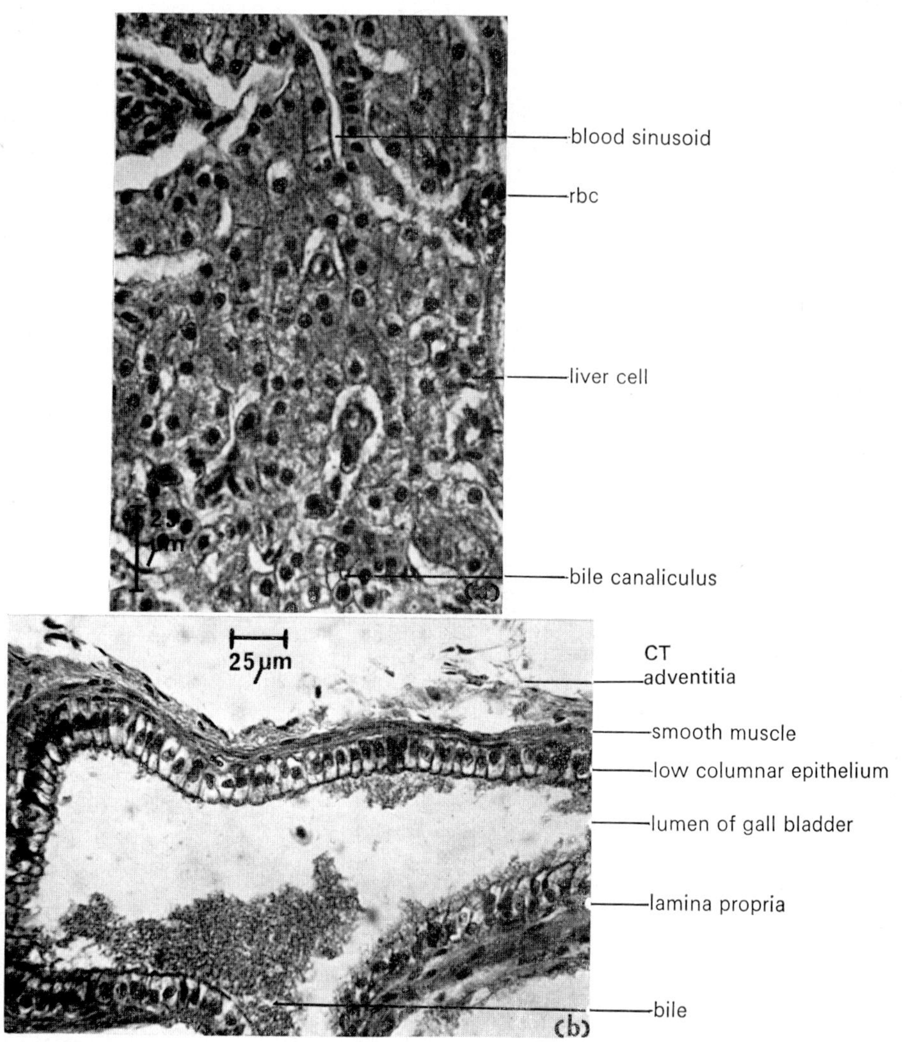

FIG. 15.26 (a) Section through the lizard liver; (b) section through part of the gall bladder of the lizard.

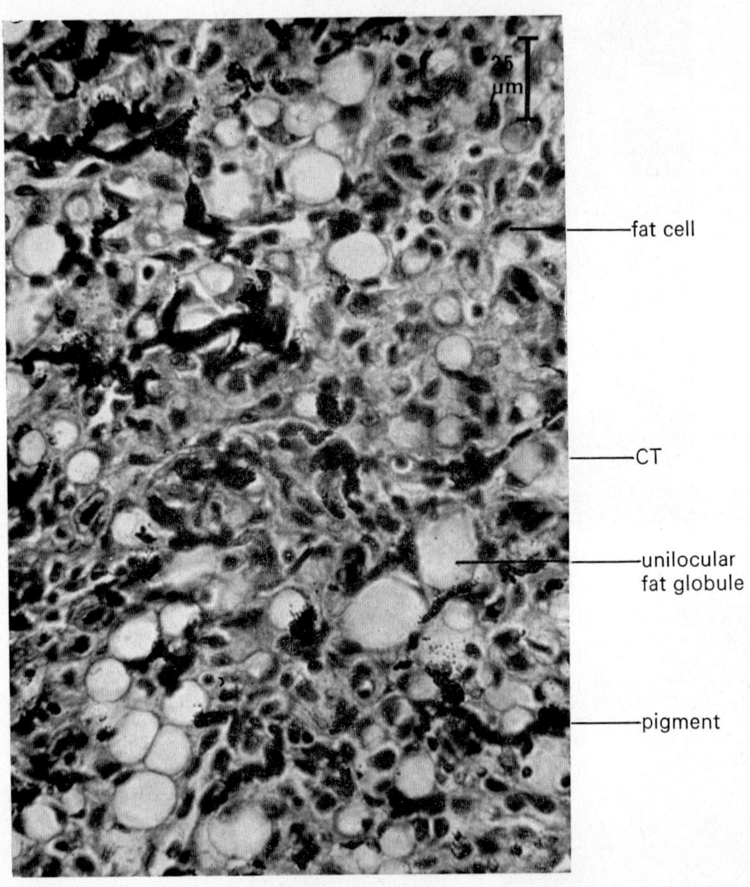

FIG. 15.27. Section through a fat body in the posterior ventral abdominal region of a young lizard. Note the unilocular fat globules.

kidneys lie closely applied to the dorsal abdominal wall, covered on their ventral surface only by peritoneum. Each kidney is divided into numerous small lobules, each composed of a few nephric tubules with small glomeruli (Fig. 15.28). Proximal and distal tubules are separated by small ciliated intermediate segments, and the large collecting ducts are characteristically grouped together. Small amounts of fluid are filtered off from the blood, and much of this is reabsorbed across the kidney tubules, thereby conserving as much body water as possible. The kidneys are drained by ureters, ducts new to amniotes (Fig. 15.29). Ureters are formed as outgrowths from the archinephric ducts, while the latter are modified for the transport of gonad products. The ureters empty into the cloaca, and urine is stored temporarily in a thin-walled (allantoic) bladder (Fig. 15.30) which opens into the ventral side of the cloaca. Further reabsorption of water occurs across the walls of the bladder and cloaca under hormonal control. Excess salts taken in with the food are also eliminated in the cloacal region as well as by the orbital salt glands.

Reproduction

Lizards are dioecious: a plan of the reproductive systems is shown in Fig. 15.29. The males have special eversible copulatory organs (hemipenes) which are pockets of skin containing blood-vessels and erectile tissue and often bearing spines. They are situated next to the cloacal opening, are attached to the caudal vertebrae by retractor muscles, and can be everted into the female cloaca.

The paired oval testes are composed of numerous looped seminiferous tubules opening into a central canal. Unlike the lower vertebrates, the seminiferous tubules of reptiles show continuous development of sperm [Fig. 15.31(a)]; a feature typical of amniotes. Ripe sperm are released into the central canal which connects by a series of small vasa efferentia [Fig. 15.31(b)] with the anterior end of the embryonic archinephric duct, which has become the epididymis and vas deferens. The genital ducts open into the cloaca, and sperm travel down grooves from the ends of the vasa deferentia to the hemipenes.

L. vivipara is ovo-viviparous i.e. the young develop inside the body of the mother for several months, and are fully formed when they emerge. This feature is thought to allow such species as *L. vivipara* to live in colder environments than is possible for oviparous species. In the female, paired ovaries produce a relatively small number of ova which mature in a linear array [Fig. 15.32(a)]. The eggs are large, filled with yolk and surrounded by a thick layer of follicular cells secreting nutrients and hormones [Fig. 15.32 (b)]. When ripe, the eggs are shed into the body cavity and pass through wide ostia into the oviducts. The oviducts are folded and looped at their anterior ends, and the walls are glandular, secreting albumen and (in

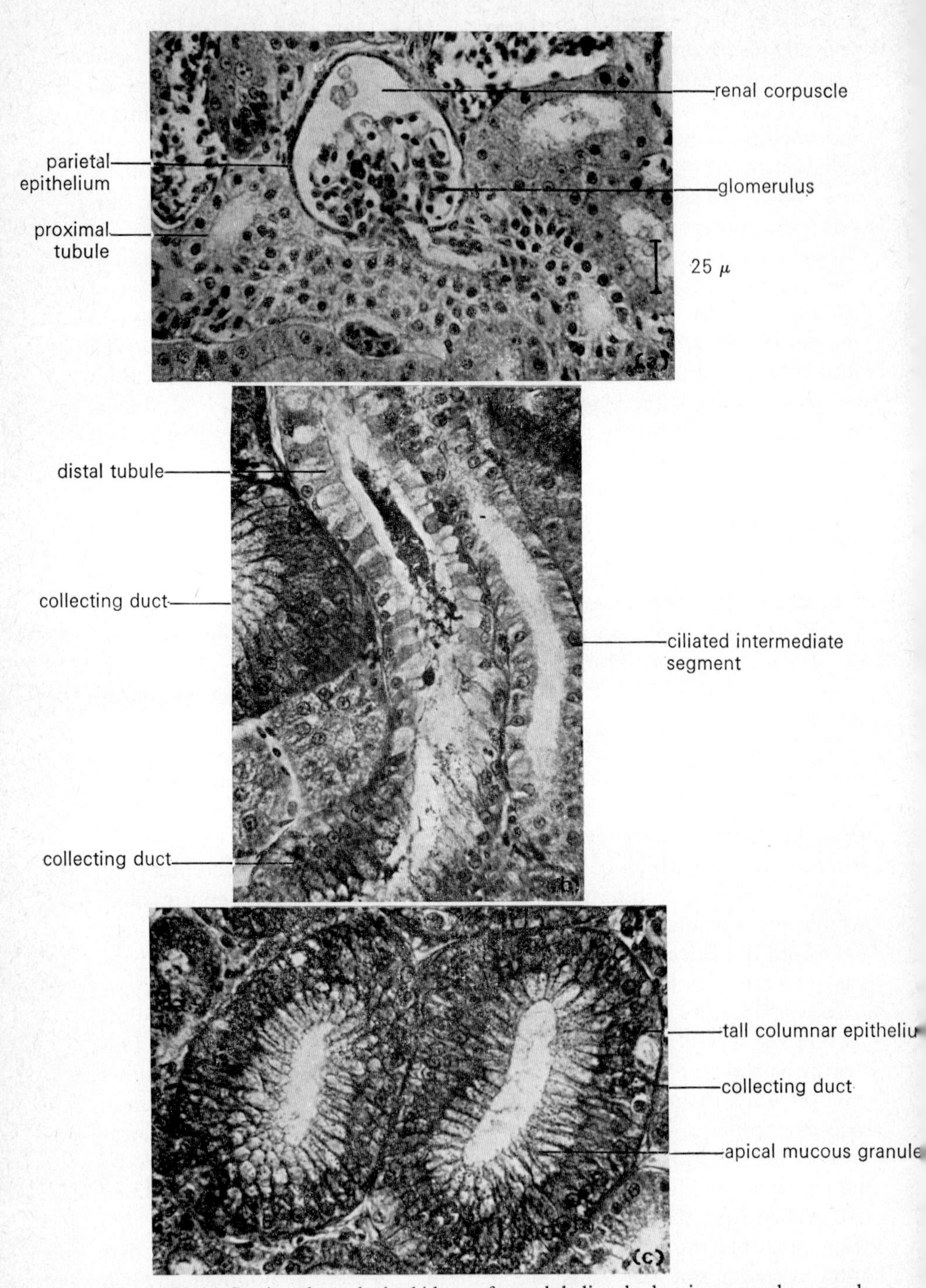

FIG. 15.28 (a) Section through the kidney of an adult lizard, showing a renal corpuscle and parts of a nephric tubule; (b) details of the intermediate and distal tubules of the lizard kidney; (c) TS of two collecting ducts of the lizard kidney.

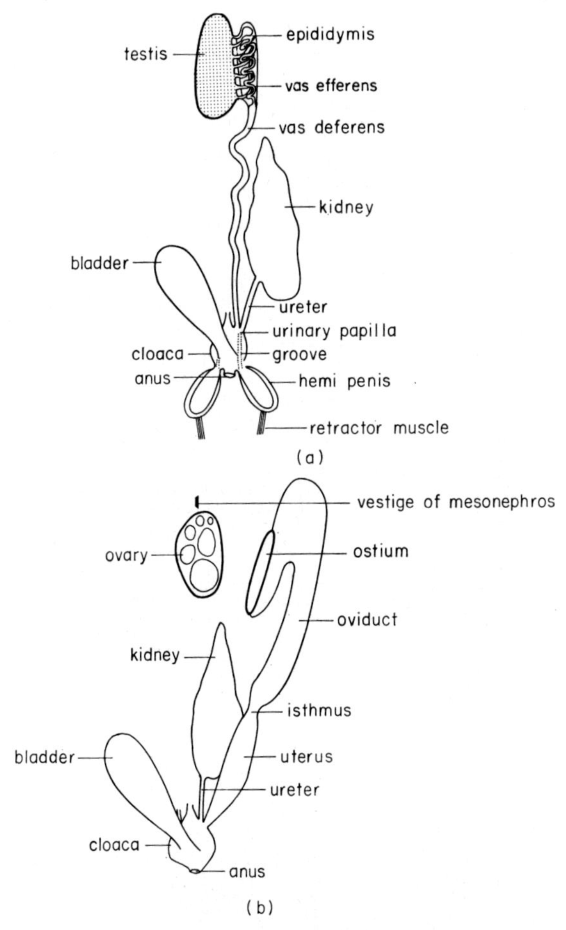

FIG. 15.29 (a) Plan of the urinogenital system of a male lizard (L. side only) as viewed from the ventral side. The bladder has been displaced to the right; (b) plan of the urinogenital system of a female lizard (L. side only) as viewed from the ventral side. The bladder has been displaced to the right.

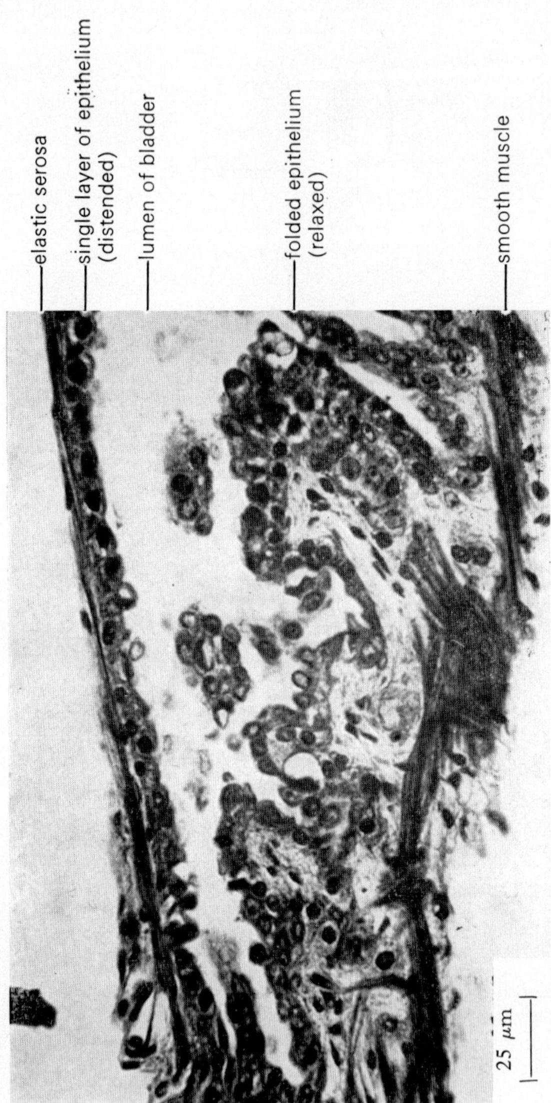

FIG. 15.30. Section through the bladder of the lizard. Note that the dorsal wall, which is stretched, is thin and lined with a single layer of epithelium; while the ventral wall, which is contracted, has a thicker folded appearance.

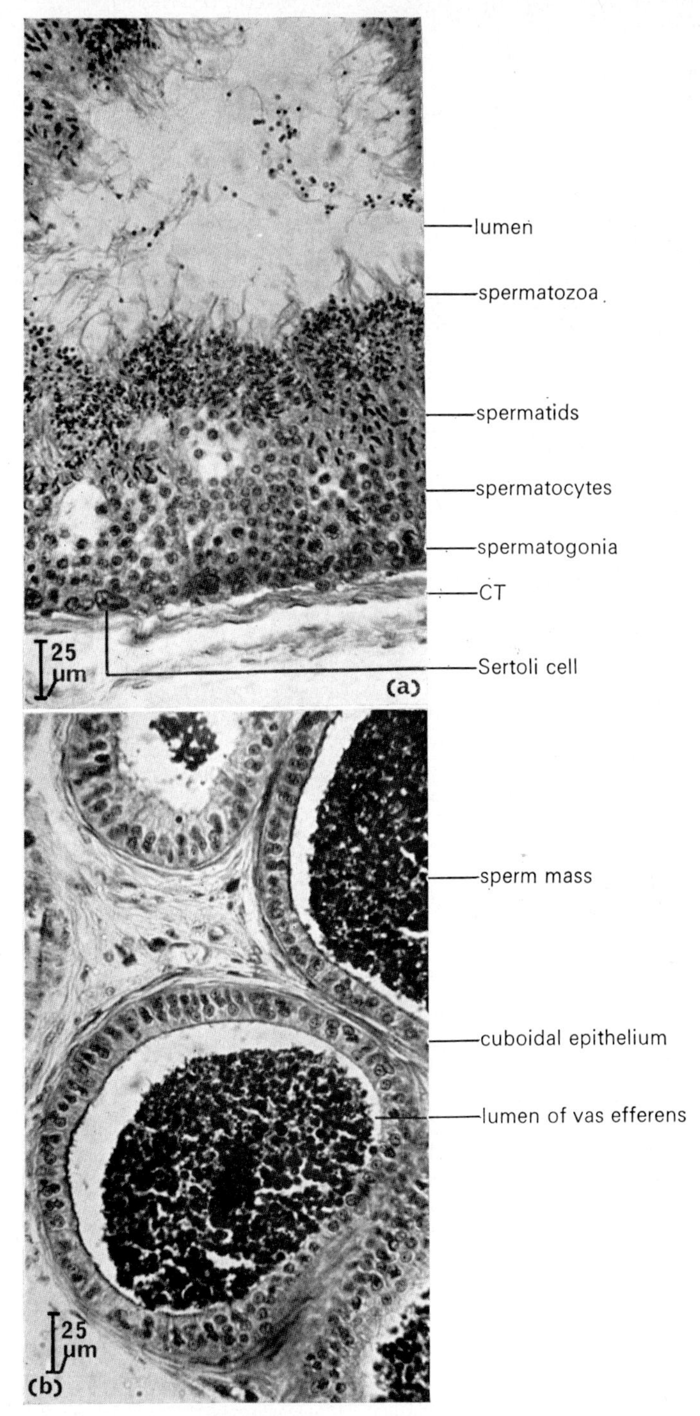

Fig. 15.31 (a) TS through a seminiferous tubule of the testis of an adult male lizard. Note that all stages of spermatogenesis can be seen within one tubule, i.e. development of germ cells is asynchronous: (b) vasa efferentia full of ripe sperm in the adult male lizard.

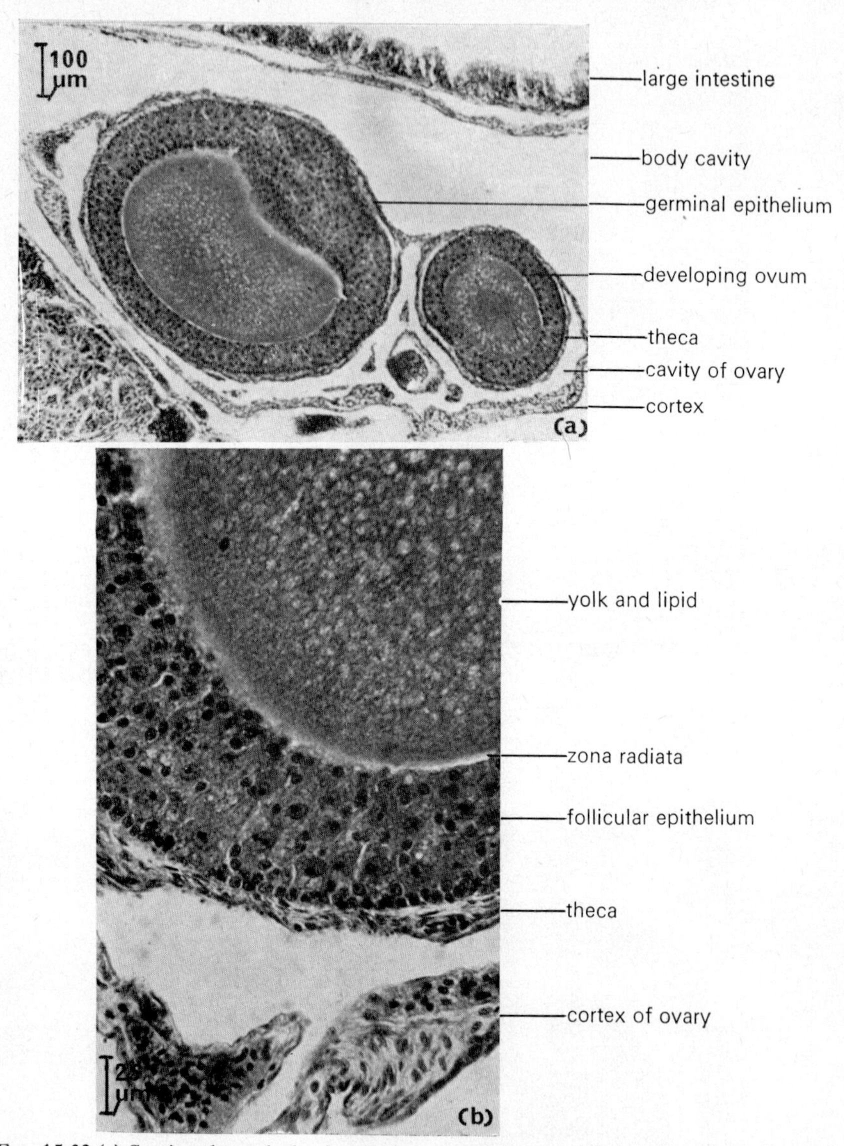

FIG. 15.32 (a) Section through the developing ovary of a young female lizard; (b) detail of ovum structure.

oviparous species) a calcareous shell. At their posterior ends they form thick-walled uteri, where fertilization occurs and where, in ovo-viviparous species such as *L. vivipara*, the eggs develop inside their thin membranous "shells". The oviducts open into the cloaca just in front of the ureters. The thin-shelled eggs are laid, but the fully-developed young inside them rupture the egg membranes immediately and free themselves.

Chapter 16

Class Aves

Birds are warm-blooded amniotes, highly adapted for an aerial mode of life. Modern birds are remarkably uniform in structure and all show modifications for flight, even if the powers of flight have been secondarily lost. Birds are tetrapods: but the forelimbs are normally modified as wings with the sternum and pectoral girdle adapted to take large wing muscles, while the hind limbs and pelvic girdles are adapted for a bipedal mode of life when the bird is not flying. The skeleton is greatly reduced in weight, the teeth are lost, the tail is reduced and the bones are hollow and often filled with thin-walled extensions from the lungs (air-sacs). A constant high body temperature is generated and maintained by very efficient respiratory and metabolic systems, and by a thick insulative layer of overlapping feathers covering the body. Birds have a keen sense of sight: the eyes and orbits are enlarged, and a considerable portion of the brain is given over to visual association. The sense of smell is reduced, but hearing is important and the inner ear is equipped with a cochlea specialized for picking up airborne sound waves. All birds are oviparous, and incubate their cleidoic eggs in a short space of time by making use of their high body temperature. When the young are hatched they are dependent on the parents for a time after birth.

Example: *Gallus* (*domesticus*)

The domestic fowl, despite its history of domestication, shows typical bird characteristics. There is a small head, long mobile neck, thick-set trunk and short tail bearing a set of long "tail" feathers. The mouth is terminal, the nostrils set far back, and the eyes very large and protected by two mobile lids and a nictitating membrane. The fowl belongs to the superorder Neognathae, which includes the majority of living birds. They possess a light horny beak and a sternum bearing a pronounced keel for attachment of wing muscles. These sternal muscles can be seen in Fig. 16.1, which shows a VLS of the trunk region of a young chick shortly before hatching. At this stage the abdomen wall is extremely thin, and the intestine often bursts through it during fixation (as shown in the photograph).

The body wall is covered by a thin and delicate skin which is attached loosely to the underlying musculature and encloses numerous air spaces. It is attached to the skeletal system at the jaws and feet, while extensions of the skin form combs (or webs on the feet of some aquatic species). The epidermis produces keratinized outgrowths which can take the form of scales and claws (on the legs and feet), beaks (modified jaws covered by epidermal sheaths), or feathers which are usually heavily pigmented and arranged in complex tracts. There are two chief kinds of feathers: soft down feathers found in young birds and adults, and larger contour feathers which form the outer covering of adults and are used for flight. Both types of feather develop from conical dermal papillae (Fig. 16.2), which are covered by epidermis and have cores of mesodermal pulp. They are continuously replaced from underneath. Feathers are provided with sensory nerve endings at their bases, and may be raised and lowered by dermal muscles. The dermis of the skin is swollen and heavily vascularized in certain areas, such as the cere round the nostrils or the brood patch on the breast of some birds. The skin is devoid of glands except for the uropygial or preen gland (Fig. 16.3). This is a compound alveolar gland situated on a dorsal papilla of the uropygium (see Fig. 16.1). It secretes an oily material which the bird spreads over its feathers to make them water-proof, and it is particularly well developed in aquatic birds.

The bird skeleton is composed of very light, hard bone. The skull has a characteristically rounded cranium and pointed jaws. The vertebral column is very long, especially in the neck region, and flexible due to the heterocoelous (saddle-shaped) vertebrae (see Fig. 16.1). The centra articulate by lubricated cartilaginous plates, and bear double-headed ribs. The sternum (see Fig. 16.16) is formed of true bone, unlike the calcified sternal cartilage of reptiles. The last thoracic, lumbar and sacral vertebrae are fused to form a rigid syn-sacrum to support the hind limbs (which have moved to an antero-ventral position for a bipedal mode of life). The vertebral column ends in a characteristic pygostyle formed of four fused caudal vertebrae.

Nervous system

The bird brain (Fig. 16.4) is primarily concerned with vision, control of movement, and instinctive stereotyped behaviour. The olfactory bulbs are small and on short stalks. The cerebral hemispheres are large but consist mainly of basal (motor) ganglia or striatum with a poorly developed cortex. The diencephalon is almost obscured by the cerebrum. There is no parapineal (parietal) but the pineal is present as a glandular structure (Fig. 16.5) and may act as an endocrine organ. The optic lobes are large and tend to be pushed outwards to a more ventro-lateral position than in lower vertebrates (see Figs 16.8 and 16.9). The optic tracts show complete decussation. The cerebellum is large, lobed and crenulated (Fig. 16.6) and possesses a distinctive layer of Purkinje neurones, which are its main output (efferent) components.

The spinal cord shows marked enlargements opposite the limb origins, associated with the innervation of the wings and legs. The caudal swelling (intumescentia lumbalis), peculiar to birds, consists of masses of cells rich

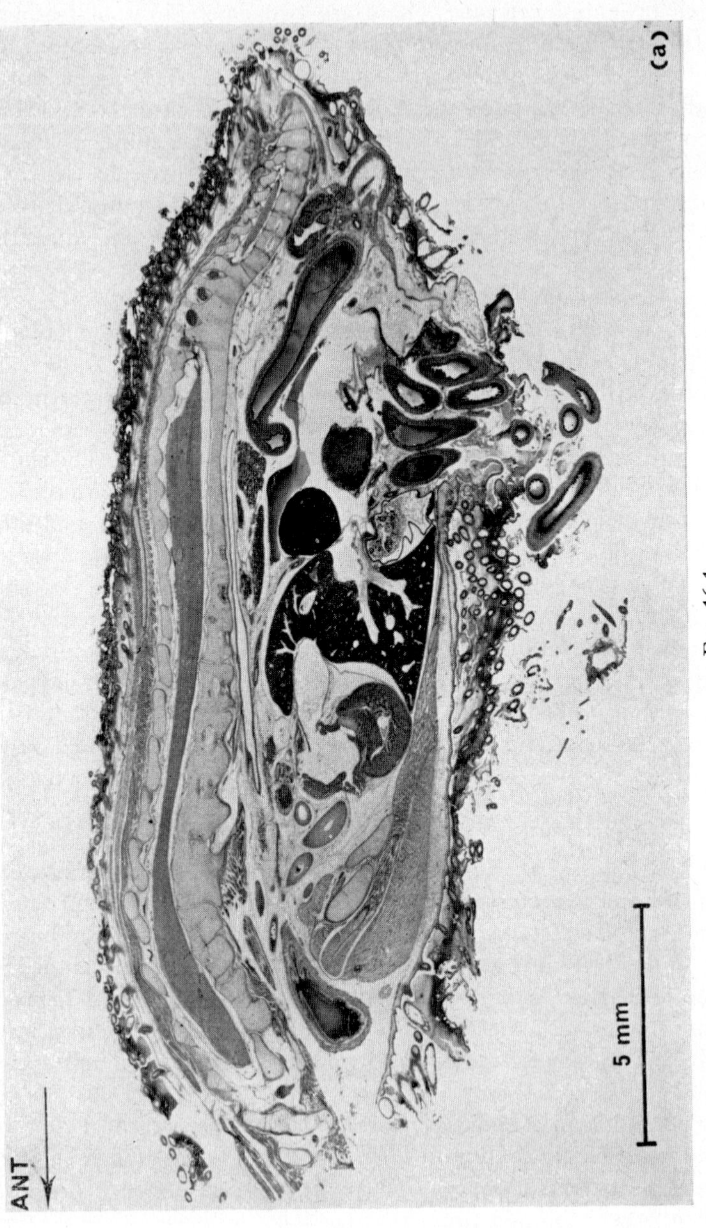

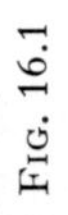

Fig. 16.1

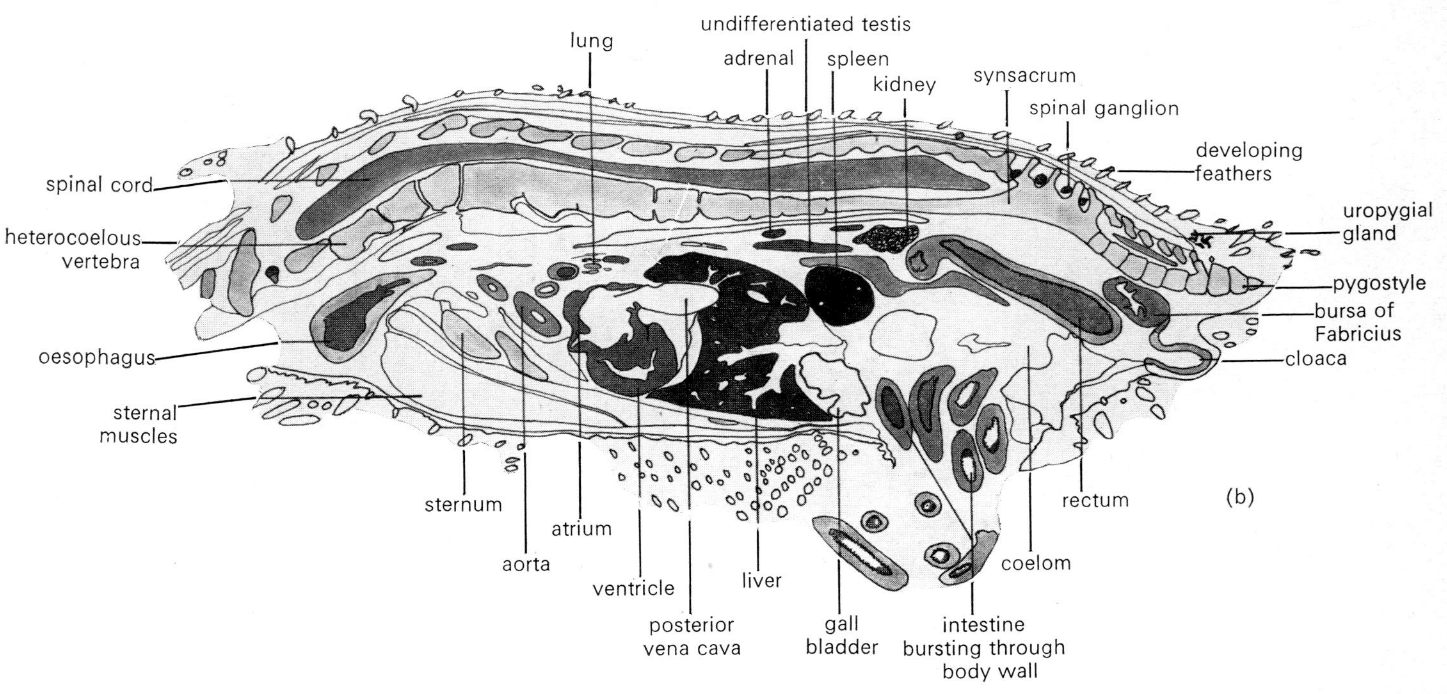

Fig. 16.1 (a) VLS and (b) explanatory drawing of the body of a young chick (*Gallus domesticus*) just prior to hatching. The dark material surrounding the chick and filling the gut and air-passages is amniotic fluid. The extrusion of part of the intestine out through the abdomen wall is an artifact caused by the fixation process.

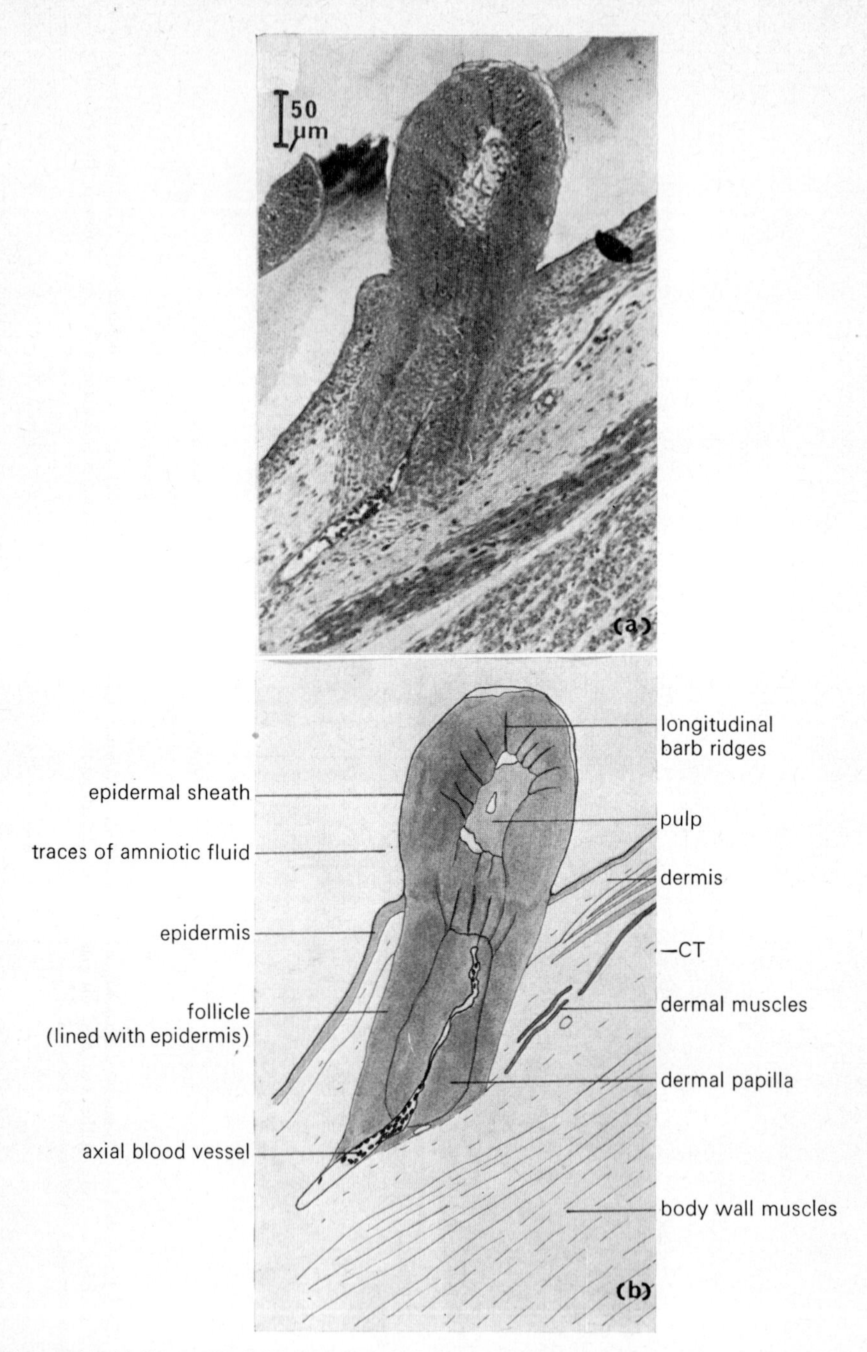

FIG. 16.2 (a) Section and (b) drawing of the skin of a young chick, showing a developing down feather.

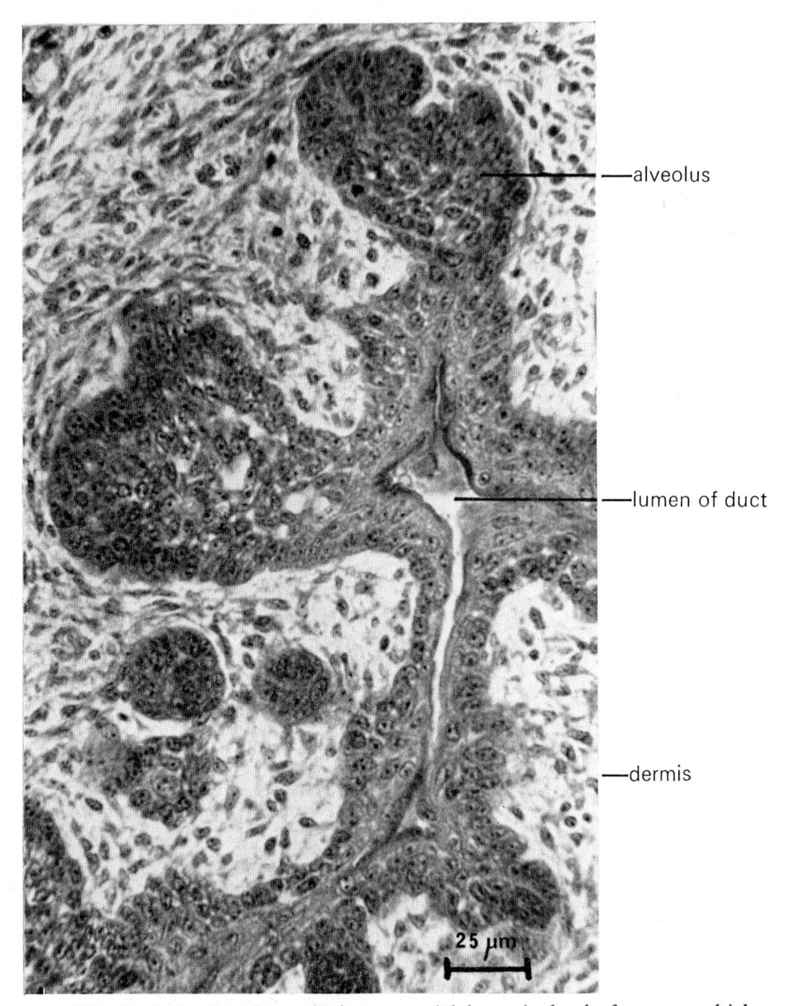

FIG. 16.3. Section through the uropygial (preen) gland of a young chick.

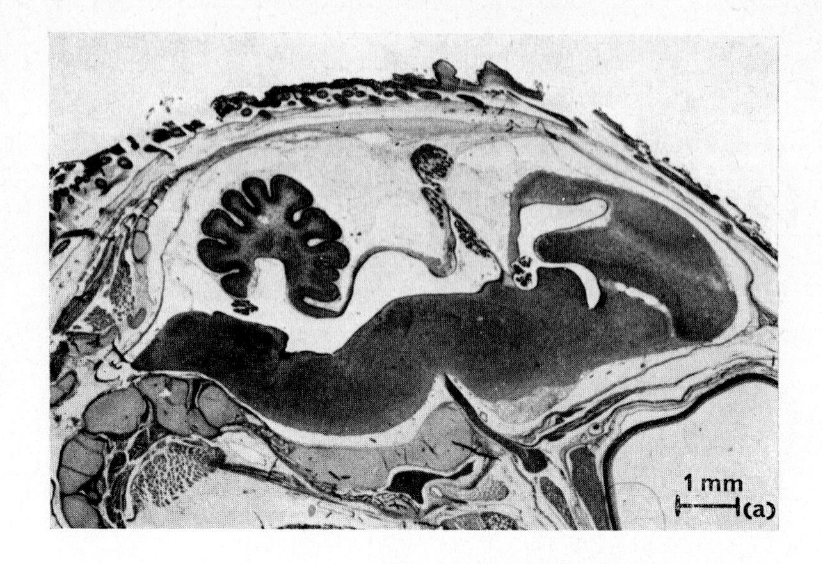

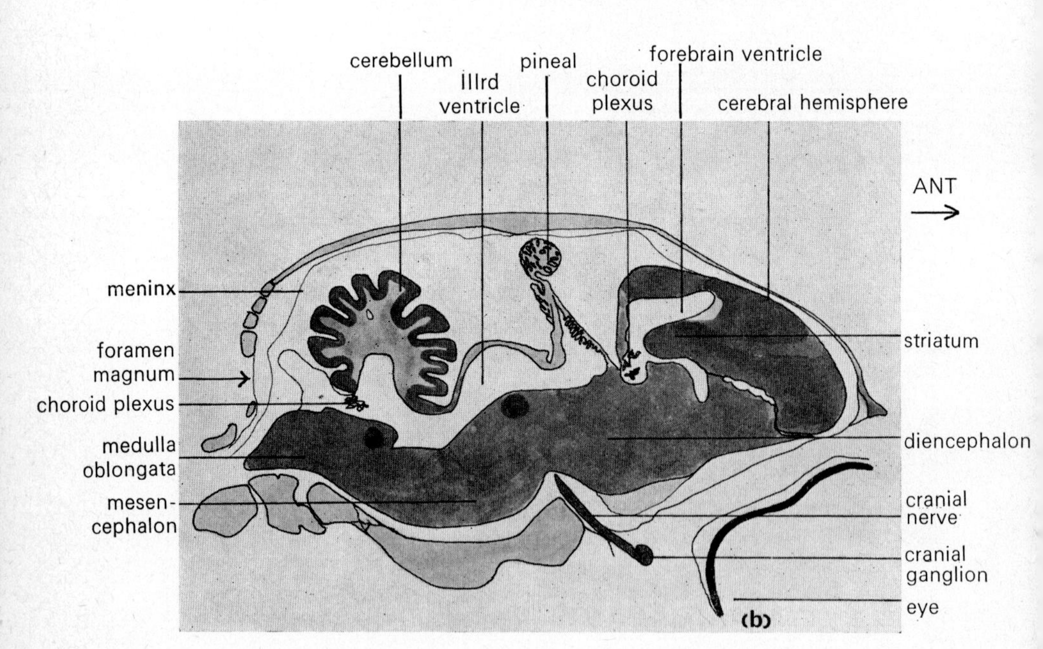

FIG. 16.4 (a) VLS and (b) drawing of the brain of a young chick. Note the large crenulated cerebellum, and the forebrain occupied mainly by striatum.

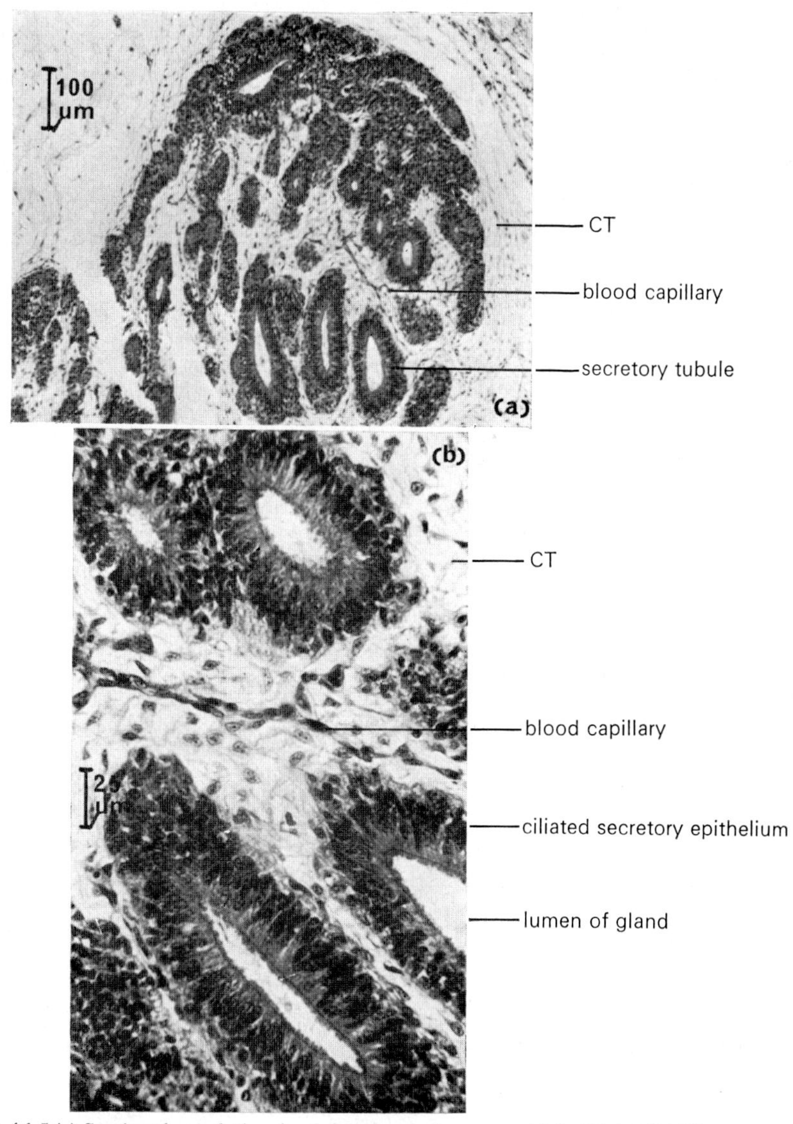

FIG. 16.5 (a) Section through the glandular pineal of a young chick; (b) detail of the secretory tubules of the chick pineal.

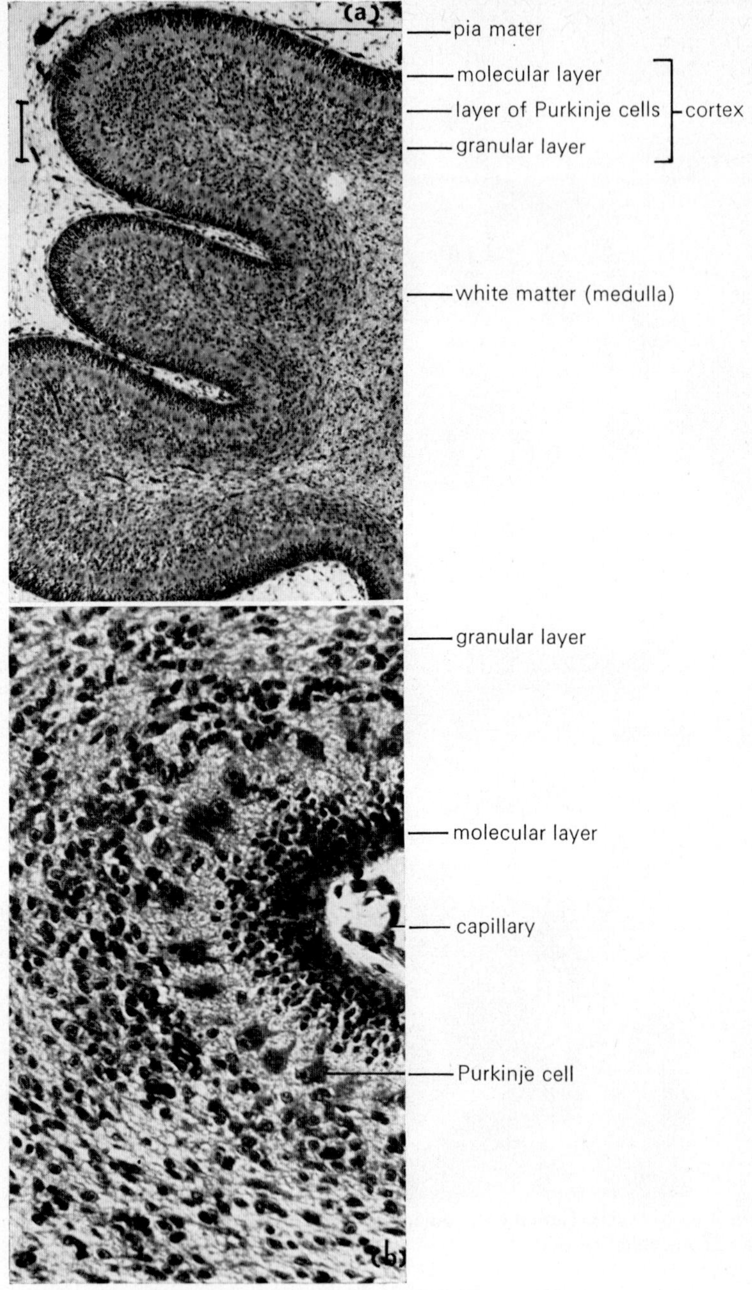

FIG. 16.6 (a) LS through part of the chick cerebellum; (b) detail of the cerebellar cortex. Note the large Purkinje cells.

in glycogen, rather than of grey matter, and possibly plays a role in metabolism of the nervous system. There are twelve pairs of cranial nerves and numerous segmentally arranged spinal nerves with dorsal ganglia.

Sense organs

Birds have a plentiful supply of interoreceptors, but receive most of their information about the outside world through the eye and ear. Taste buds are found in the buccal cavity especially on the base of the tongue, but are not raised on papillae. Possibly only the beak and feet possess a sense of touch although tactile receptors have been reported in the dermis. The olfactory organ is reduced, and there is no Jacobson's organ in birds or any vomeronasal epithelium except during early stages of development.

The most important sense is sight. The eyes are large and, in some nocturnal species, may be very deep. The cornea (Fig. 16.7) which is the main refractive structure, bulges forwards. It is kept moist by copious secretions from orbital Harderian glands. The lens is soft and its shape can be altered by the action of striated ciliary muscles, which thus effect accommodation. There is a thin scleroid bearing ossicles which help to maintain the shape of the eye. The choroid bears a large pleated vascular membrane (pecten) which projects into the vitreous cavity (Fig. 16.8). It is similar in structure to the lizard conus and is thought to assist in nutrition of the eye. However, it has also had ascribed to it functions of adjusting intraocular pressure during accommodation, assisting in focusing, and increasing visual acuity. The retina is composed mainly of cones, for daylight vision, with some areas (foveae) containing only cones. Birds have a wide visual field which can be increased by turning the head, but they do not have binocular vision.

Birds have acute hearing over a wide frequency range. The external ear is a short tube covered with feathers (to reduce air turbulence) and leads to the tympanic membrane. The middle ear (Fig. 16.9), as in reptiles, possesses a bony columella and cartilaginous extrastapes. The inner ear (Fig. 16.10) shows development of a specialized cochlea with the lagena remaining as a patch of sensory cells at its tip. The basilar papilla, seen in reptiles, is elongated and enlarged to form an organ of Corti consisting of sensory hair cells and supporting cells lying on a basilar membrane and overlaid by a tectorial membrane (Fig. 16.11). Airborne vibrations, transmitted via the columella and perilymph to the endolymph within the cochlea, cause movements of the tectorial membrane relative to the sensory cells, thus stimulating them. Different parts of the organ of Corti respond to different frequencies of sound. The membranous labyrinth also possesses a sacculus and utriculus surmounted by large semi-circular canals (Fig. 16.10). Their size is probably associated with the good sense of orientation that birds possess.

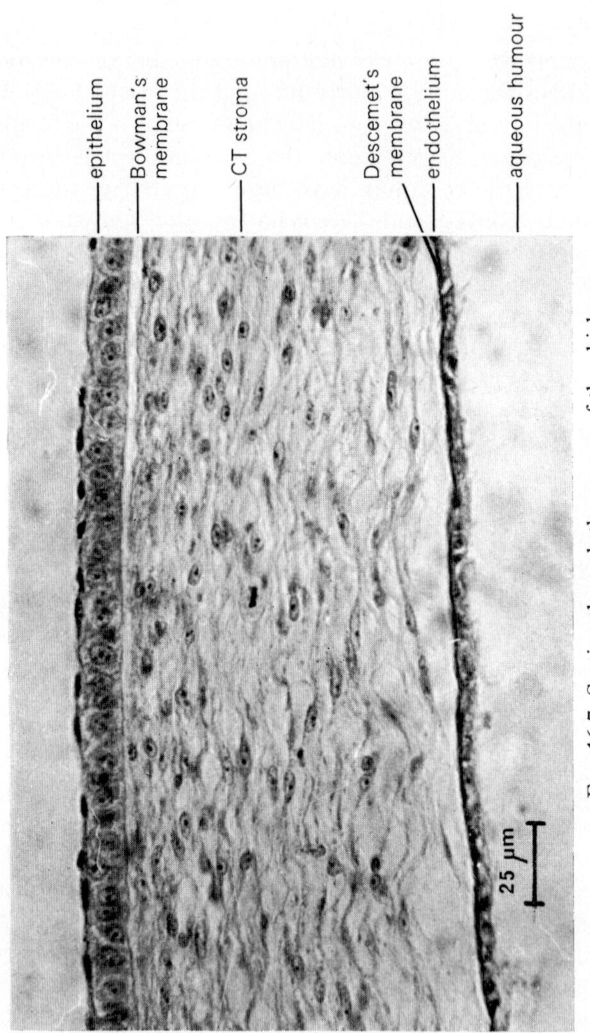

FIG. 16.7. Section through the cornea of the chick eye.

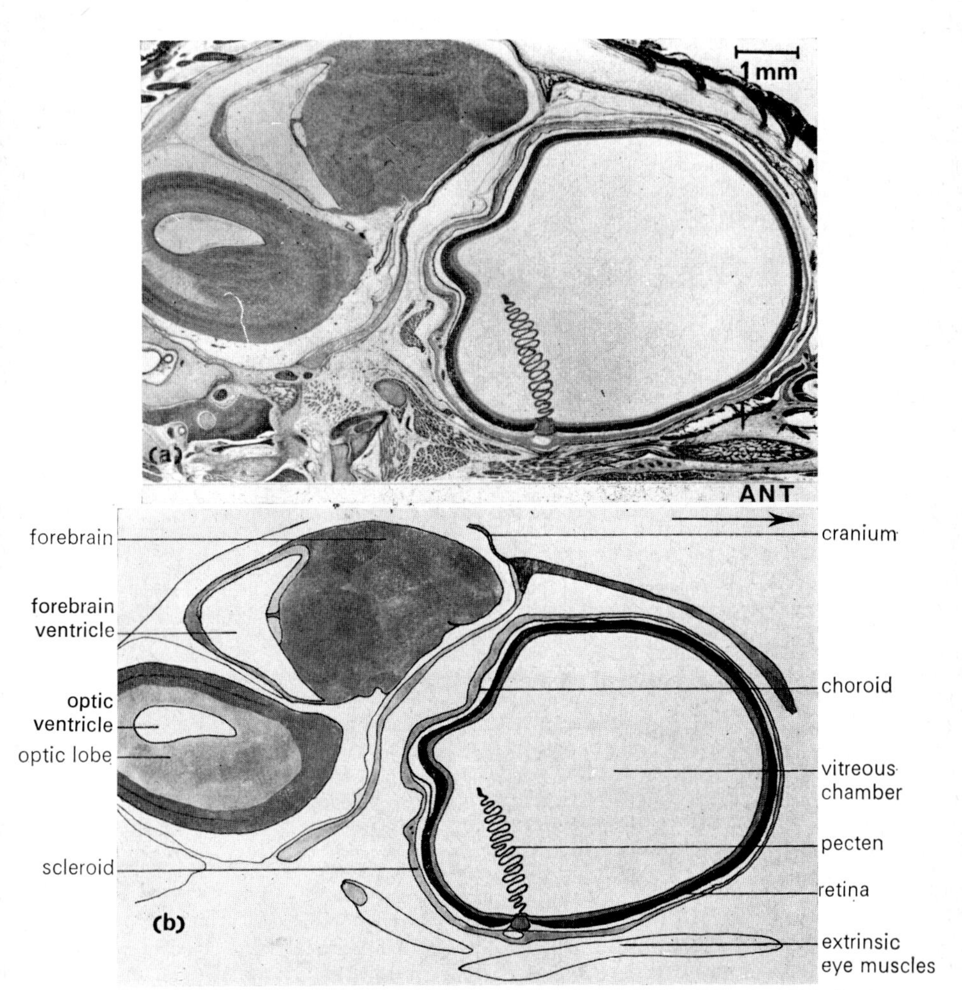

FIG. 16.8 (a) VLS and (b) drawing of one side of the chick head, showing part of the brain and the extensive pecten in the eye.

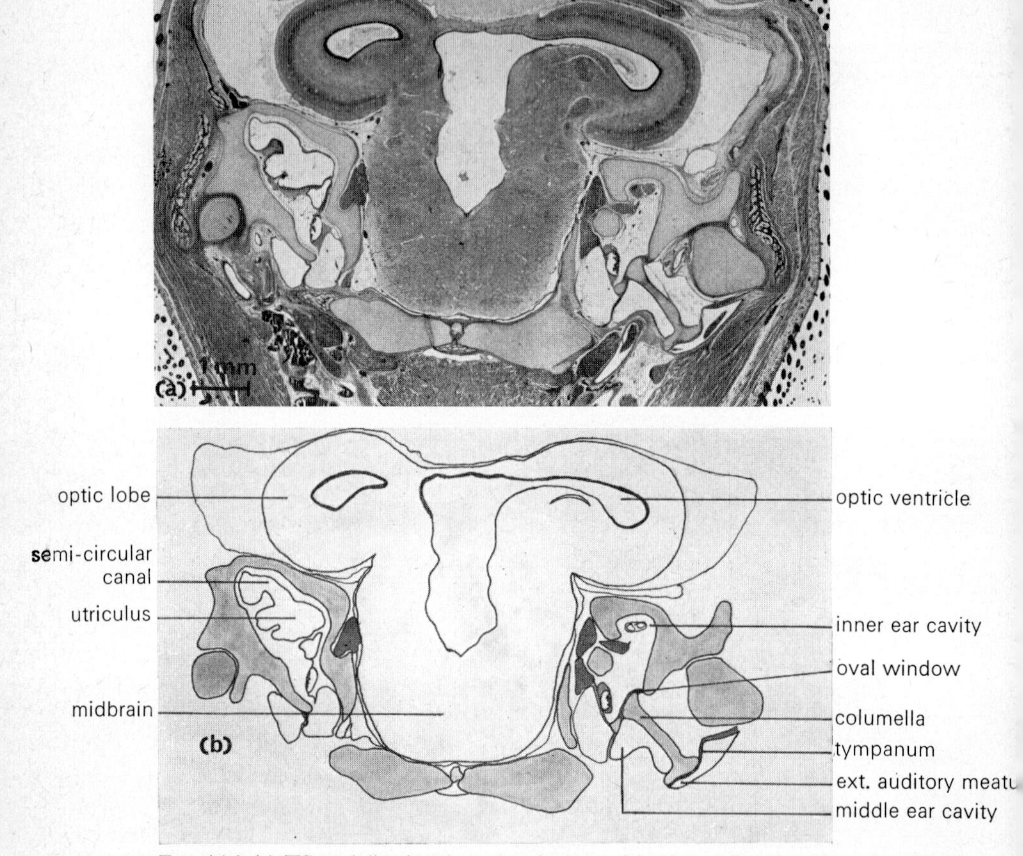

FIG. 16.9 (a) TS and (b) drawing of a chick head in the midbrain region, showing parts of the middle and inner ears.

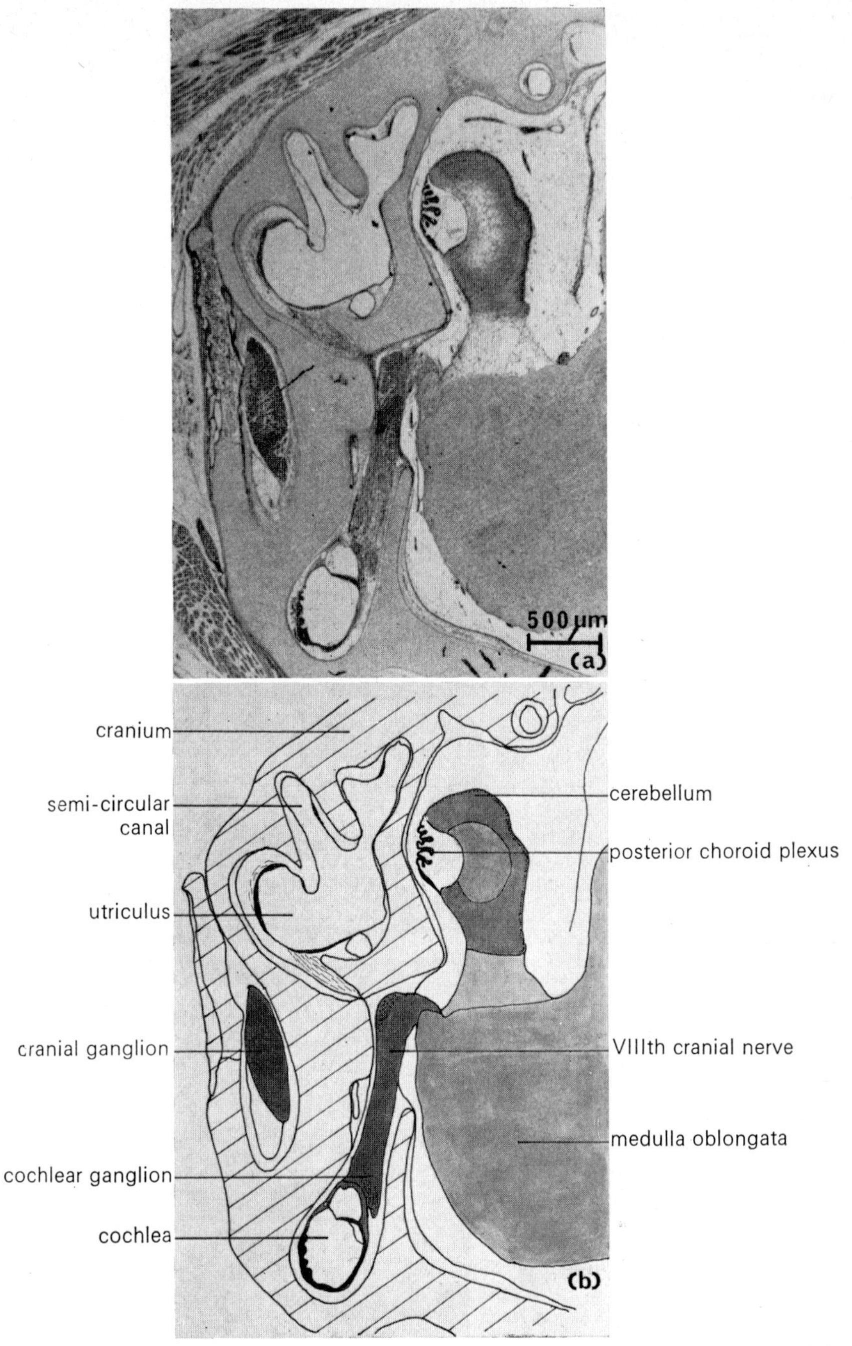

FIG. 16.10 (a) VLS and (b) drawing of a chick head showing parts of the inner ear and the innervation of the cochlea by the VIIIth cranial nerve.

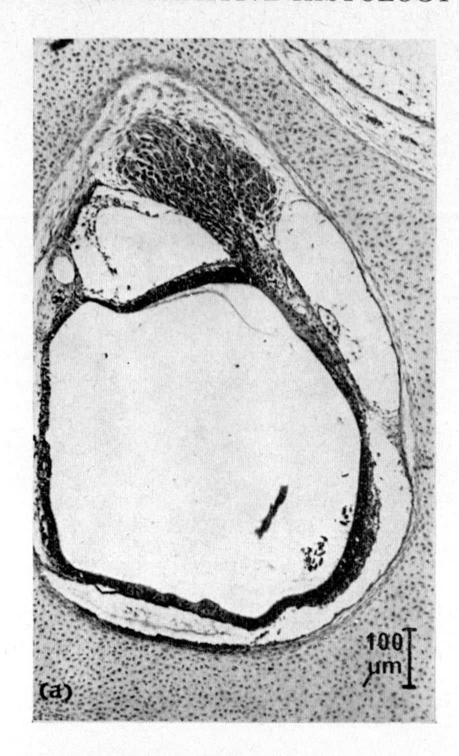

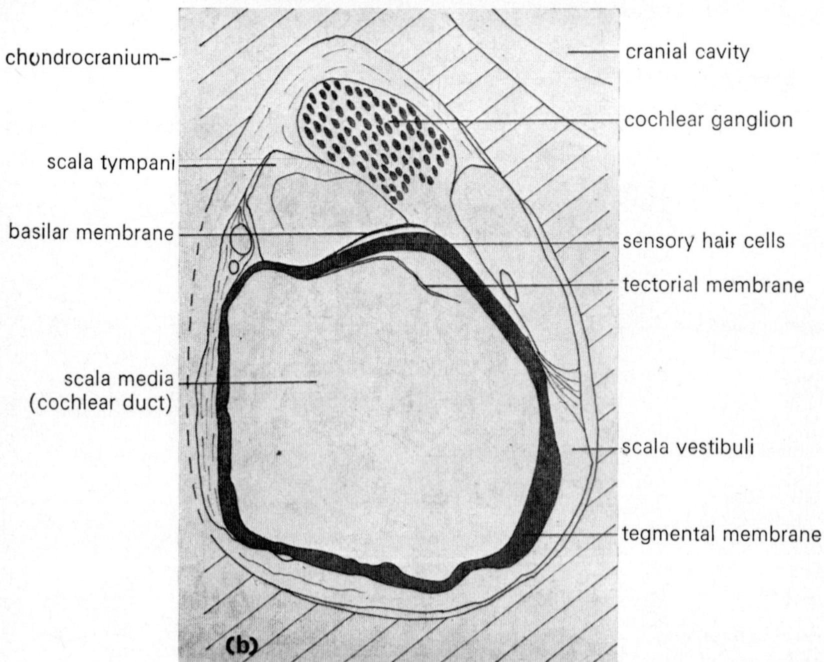

FIG. 16.11 (a) TS and (b) drawing of the cochlea of a young chick.

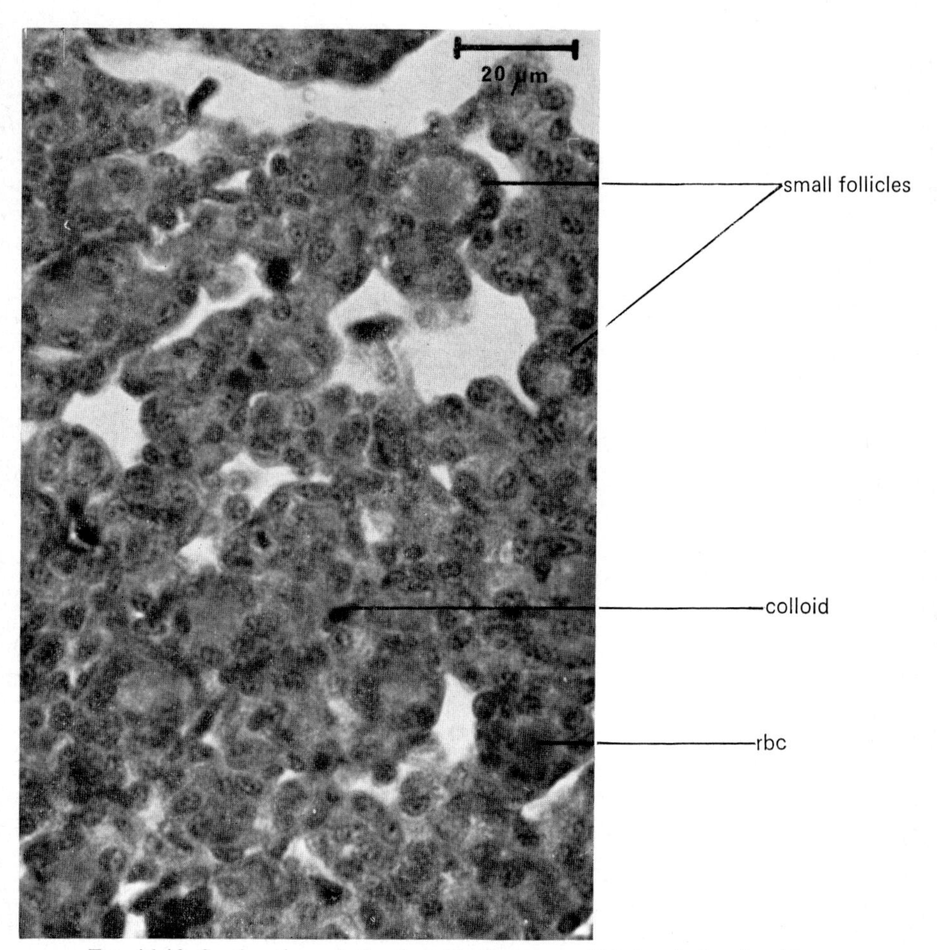

FIG. 16.12. Section through the thyroid of a young chick just before hatching.

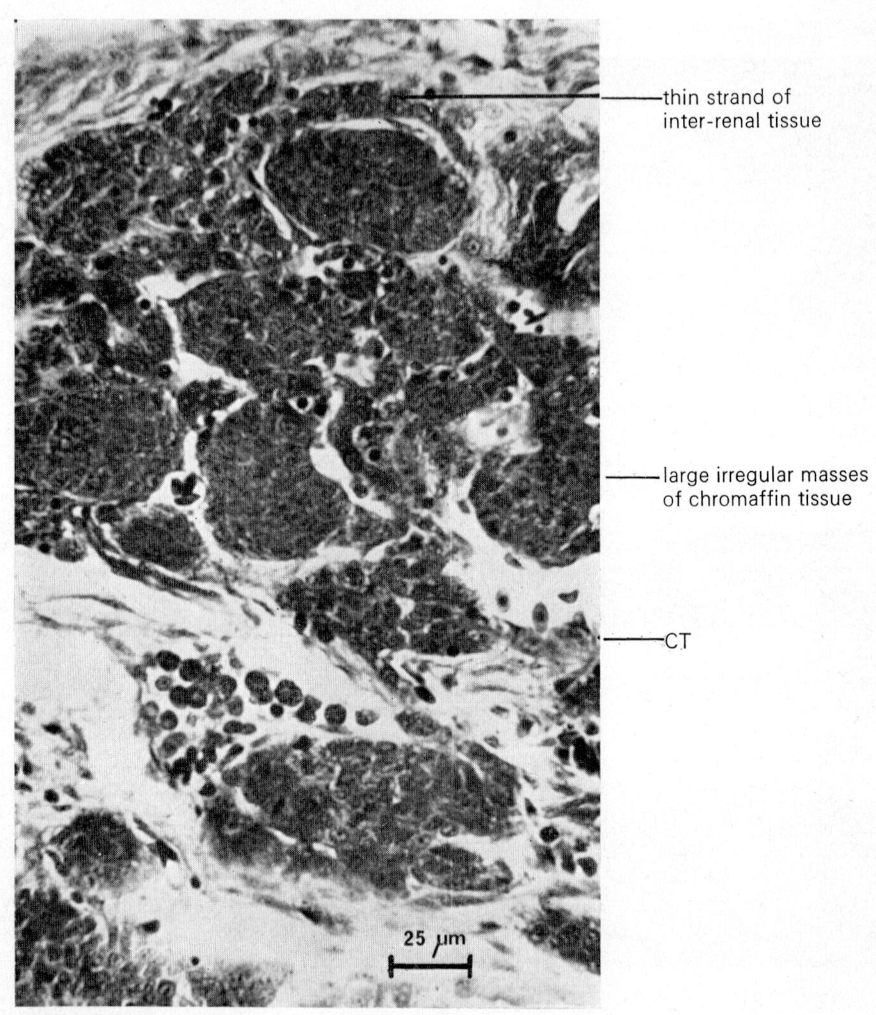

FIG. 16.13. Section through the adrenal gland of a young chick, showing the complete intermingling of inter-renal and chromaffin tissue.

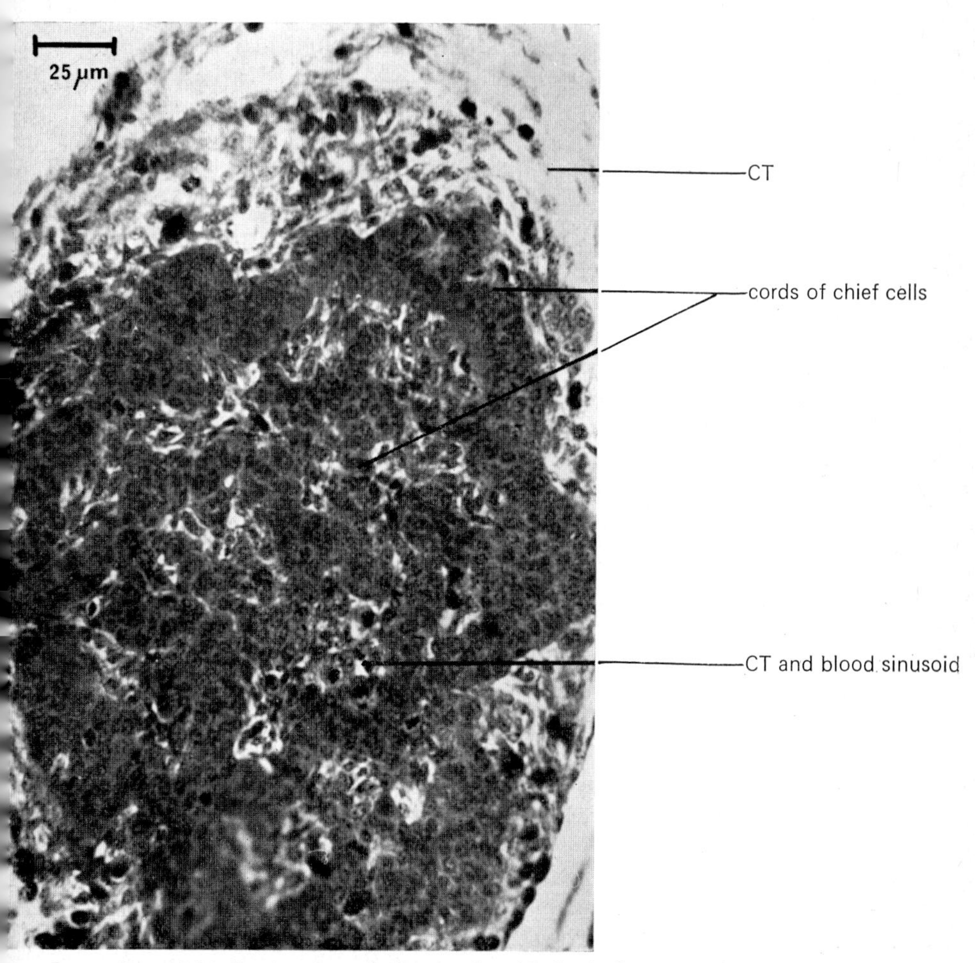

FIG. 16.14. Section through the parathyroid gland of a young chick.

Endocrine organs

As in all amniotes the pituitary occurs on the end of an infundibular stalk. The large pars glandularis and the pars tuberalis are derived from the stomodaeum and are separated by connective tissue from a small pars nervosa. There is apparently no pars intermedia.

As a result of neck elongation the paired thyroid glands lie farther from the head than in lower vertebrates. They are located just anterior to the systemic arch (see Fig. 16.17) and have a typical follicular structure. Figure 16.12 shows the thyroid of a chick just prior to hatching. Note that the follicles are much smaller than in the thyroid of a tadpole just before it metamorphoses into a frog (emphasising the important role of the thyroid in amphibian metamorphosis).

Chromaffin and inter-renal tissue are intermingled in birds to form paired adrenal glands, lying antero-dorsally to the kidneys. Each adrenal (Fig. 16.13) consists of thin strips of inter-renal tissue interspersed with masses of chromaffin tissue.

Two small pairs of parathyroid glands are situated in the neck region just behind the thyroids (see Fig. 16.17). They are composed of cords of chief cells surrounding connective tissue and blood sinusoids (Fig. 16.14). The parathyroids are responsible for control of calcium and phosphate balance.

Respiration

Birds have a very high metabolic rate, and so a most efficient respiratory system is necessary to provide the large supply of oxygen required. The respiratory structures are lungs, assisted in their function by thin-walled, avascular air-sacs which act as reservoirs and which send diverticula into spaces within the bones (Fig. 16.15). The trachea, often very long, is supported by complete cartilaginous rings and lined with a ciliated epithelium containing mucous cells. At its lower end, just as it splits into two bronchi, the trachea is slightly thickened and enlarged to form an organ peculiar to birds, the syrinx or sound-producing organ. Noise is generated by vibrations of semi-lunar membranes within the syrinx, and the pitch and quality of the sound can be altered by a complex of muscles surrounding it. The bronchi are strengthened by incomplete cartilaginous rings and divide, as they enter the compact spongy lungs, into secondary bronchi (Fig. 16.16). These branches communicate with the air-sacs [Fig. 6.17(a)] which are extensions of the ends of the primary and secondary bronchi. Small tertiary bronchi (parabronchi) anastomose with one another, and give off at right angles numerous branching air-capillaries [Fig. 16.17(b)] which radiate in all directions. They are composed of thin respiratory epithelium which is closely applied to the lung capillaries. All the air-passages in the bird lung are open-ended and communicate with one another, so there is little or no dead space, such as is found in other tetrapods whose lungs are composed of blind-ending alveoli.

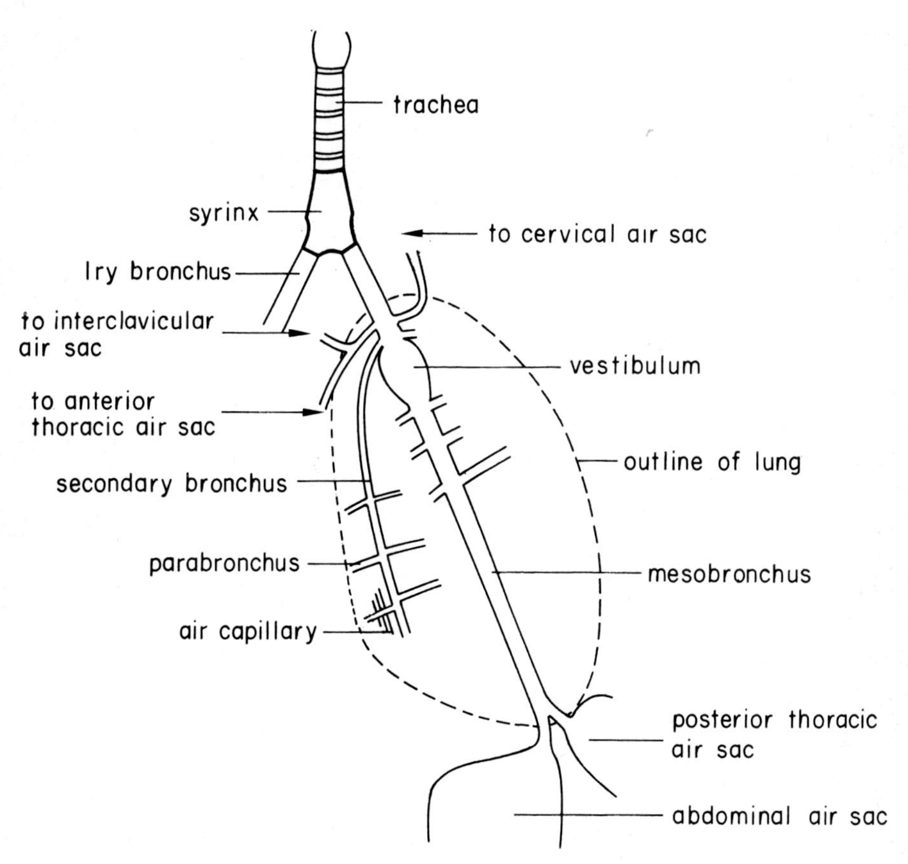

FIG. 16.15. Plan of the bird respiratory system.

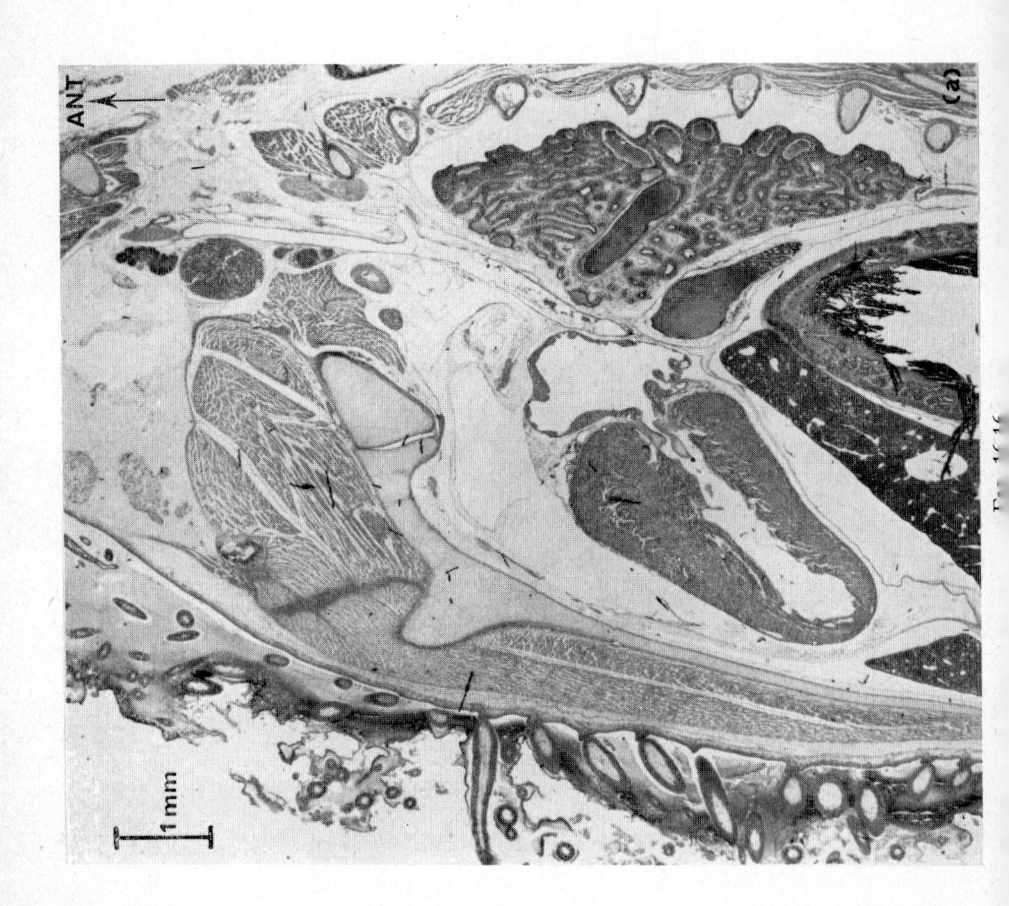
ANT
(a)
1 mm

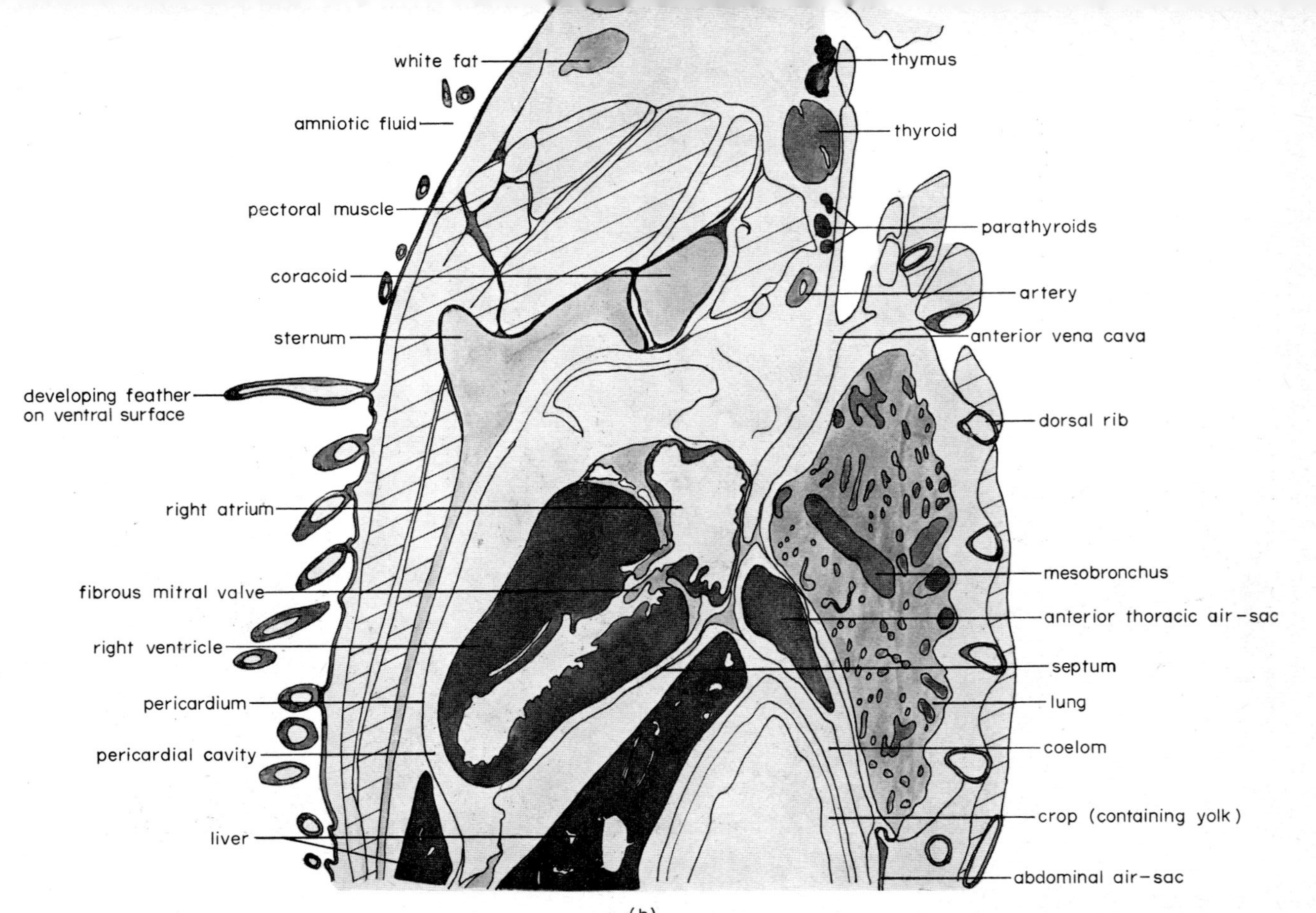

FIG. 16.16 (a) VLS and (b) drawing of the anterior trunk region of a young chick, showing the position of the heart and lung in the thoracic cavity.

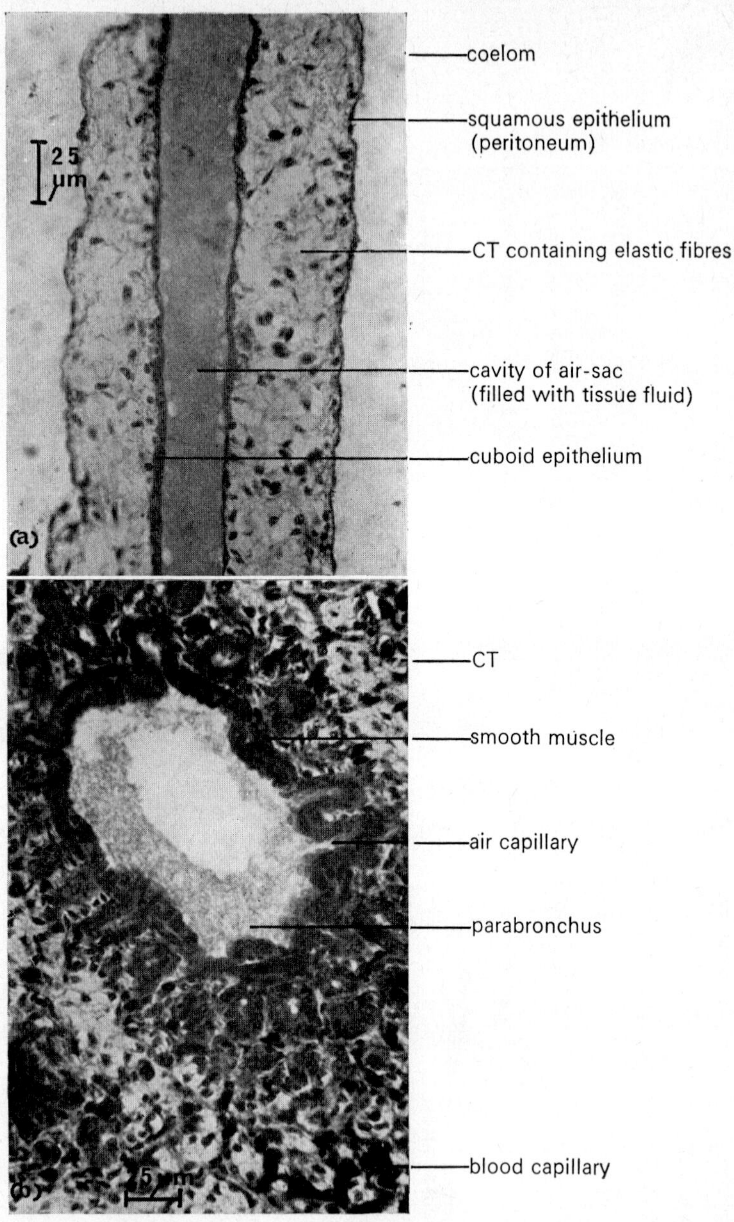

FIG. 16.17 (a) LS through the posterior thoracic air-sac of a young chick just prior to hatching. Note the tough connective tissue walls. At this stage the cavity of the air-sac is filled with tissue fluid; (b) section through the chick lung showing a parabronchus with numerous air capillaries radiating from it into the lung matrix.

The part of the coelom containing the lungs and anterior air-sacs is divided off from the rest of the body cavity containing the posterior air-sacs by an oblique septum composed of fibrous tissue (Fig. 16.16). In some species this septum is split to form two "diaphragms". Ventilation of the lungs is achieved by altering the volume of the thoracic cage. At rest this is effected by sternal and intercostal (rib) muscles, assisted to some extent by the pulmonary diaphragm, while during flight, wing movements play a large part. Air is drawn chiefly into the posterior air-sacs, and appears to flow through the parabronchi and air-capillaries both during inspiration and expiration.

Circulation

The high metabolic rate and high body temperature of birds require a most efficient circulation of blood round the body. The circulatory system is completely divided into a pulmonary circuit and a visceral circuit. The heart is very large (Fig. 16.16), and its division into two atria and two ventricles, presaged in the reptiles, is now complete. Thus deoxygenated blood returning from the body to the right side of the heart is channelled directly to the lungs, and oxygenated blood from the lungs is returned to the left side of the heart to be pumped round the body. The two sides of the heart act as separate but linked pumps. The ventricles are large with thick muscular walls and the atria have thin muscular walls. A sinus venosus generally appears only briefly during early development and becomes incorporated into the right atrium. A muscular valve separates right atrium and ventricle, while all the other heart valves are membranous. In adults the left aortic arch (L. systemic) is lost, while the right arch is retained as the dorsal aorta and receives only oxygenated blood (Fig. 16.18). The bulk of the blood returning from posterior regions of the body is not, as in lower vertebrates, taken through the renal-portal system, but is returned directly to the heart. Only a small renal-portal system remains, but the hepatic-portal system is still well developed.

Erythrocytes are oval and nucleated, tending to be smallest in actively flying birds and largest in flightless birds (15 μ in the fowl). Their chief site of production is the bone marrow, while they are destroyed in the small spleen. It is thought, however, that the spleen has some control over erythrocyte production in the bone marrow, and can take over production in an emergency. Birds are well provided with lymphoid tissue manufacturing white blood cells. Numerous nests of thymus tissue are present in the neck region showing typical cortical and medullary structure [Fig. 16.19(a)] while an accessory lymphoid organ peculiar to birds (bursa of Fabricius) is present as a dorsal diverticulum of the cloaca [Fig. 16.19(b)]. Both thymus and bursa are largest in young birds. Pharyngeal tonsils are also found.

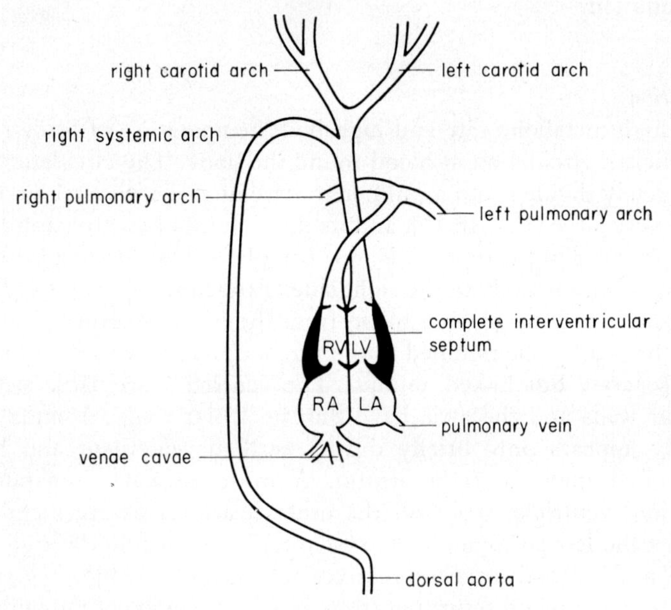

FIG. 16.18. Plan of the heart and aortic arches of the bird. (LA = left atrium, RA = right atrium' LV = left ventricle, RV = right ventricle).

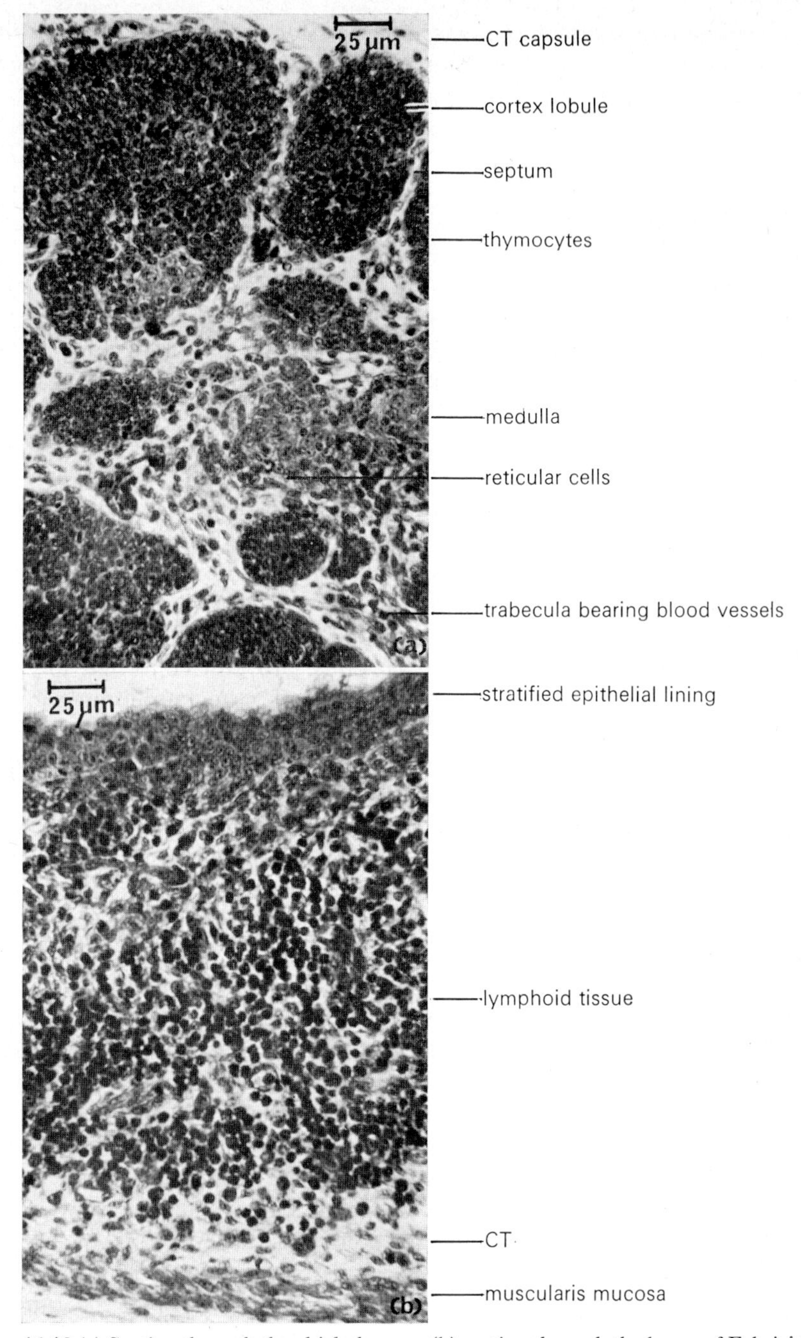

FIG. 16.19 (a) Section through the chick thymus; (b) section through the bursa of Fabricius in the dorsal wall of the chick cloaca.

Digestion

The digestive tract of birds is modified for their mode of life, with considerable reduction of the anterior end and a shifting of the large storage parts (i.e. the centre of gravity) towards the trunk.

Teeth are replaced by a light horny beak which shows great variation according to diet. The tongue [Fig. 16.20(a)] is long and pointed, sometimes strengthened by cartilage but usually with little or no intrinsic musculature. It has a cornified surface and is usually provided with taste buds and "salivary glands". The "salivary glands" [Fig. 16.20(b)] are mucigenic tubular glands, which in some species secrete an adhesive solution to assist with nest-building. The buccal cavity and oesophagus are lined with cornified epithelium to resist damage by hard food (Fig. 16.21). The oesophagus is very long and wide to accommodate large pieces of food, since they cannot be chewed up. Oesophageal glands are normally simple tubular glands secreting lubricative fluid, but in pigeons they are specialized to produce "milk" for feeding the young. The hind end of the oesophagus is modified to form a large storage chamber (crop) which is similar in structure to the rest of the oesophagus. Thus food can be held for long periods and channelled through gradually into the digestive part of the gut. In some birds the crop is used as a resonating organ during singing.

The stomach is divided into two portions: proventriculus and gizzard. The proventriculus, similar to the stomach of other vertebrates, has a thick folded mucous membrane with typical gastric glands. This is the site of initial digestion. The gizzard is the site of trituration and has a very specialized structure (Fig. 16.22). It has thick complex muscular walls and is lined with a dense collection of tubular glands producing a hard material (koilin) related to keratin. This substance, in conjunction with stones and grit taken into the gizzard with the food, forms an effective macerating surface for hard food and is produced in largest quantities in gramnivorous species such as the fowl. It is absent in carnivorous birds. The small intestine is the principal site of digestion. It is characterized by a mucosa which extends mobile finger-like villi out into the gut lumen and dips down towards the submucosa forming crypts of Lieberkühn lined with glands. In this birds are like mammals and unlike other vertebrates. The large intestine has lower folds of mucosa, fewer digestive gland cells and possesses more mucous goblet cells (Fig. 16.23). A pair of caecae occur at the junction between the small and large intestines. The rectum is unfolded, has even more mucous cells and empties into a cloaca. The cloacal pouch (bursa of Fabricus) atrophies in adult birds.

The glands associated with the digestive tract are similar to most other vertebrates. The liver is bilobed with two ducts leading into the anterior end of the small intestine. There is a small gall-bladder in the chick for concentrating bile produced by the liver, but a gall-bladder is missing in

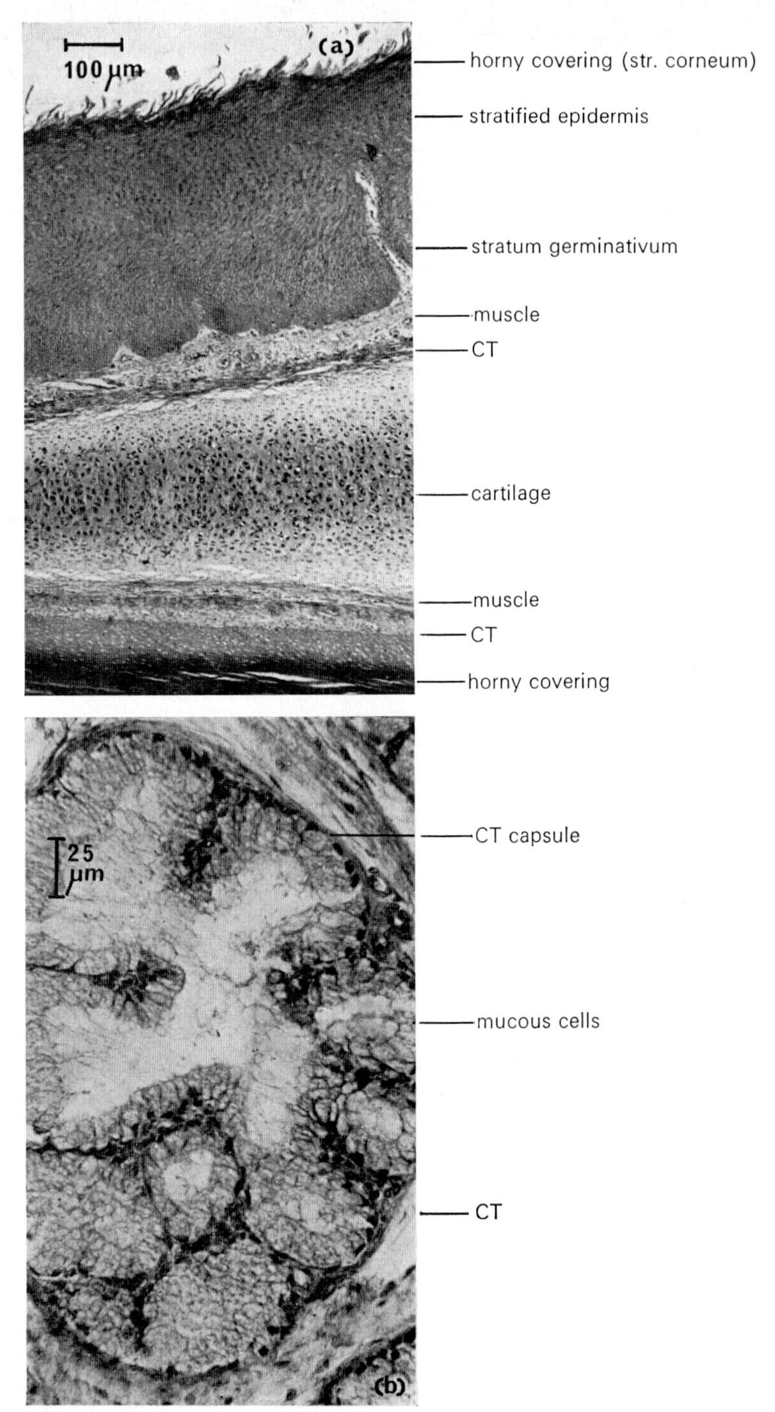

FIG. 16.20 (a) VLS through the tongue of a chick. Note the cornified covering and the skeletal support; (b) section through a mucous "salivary gland" of the chick.

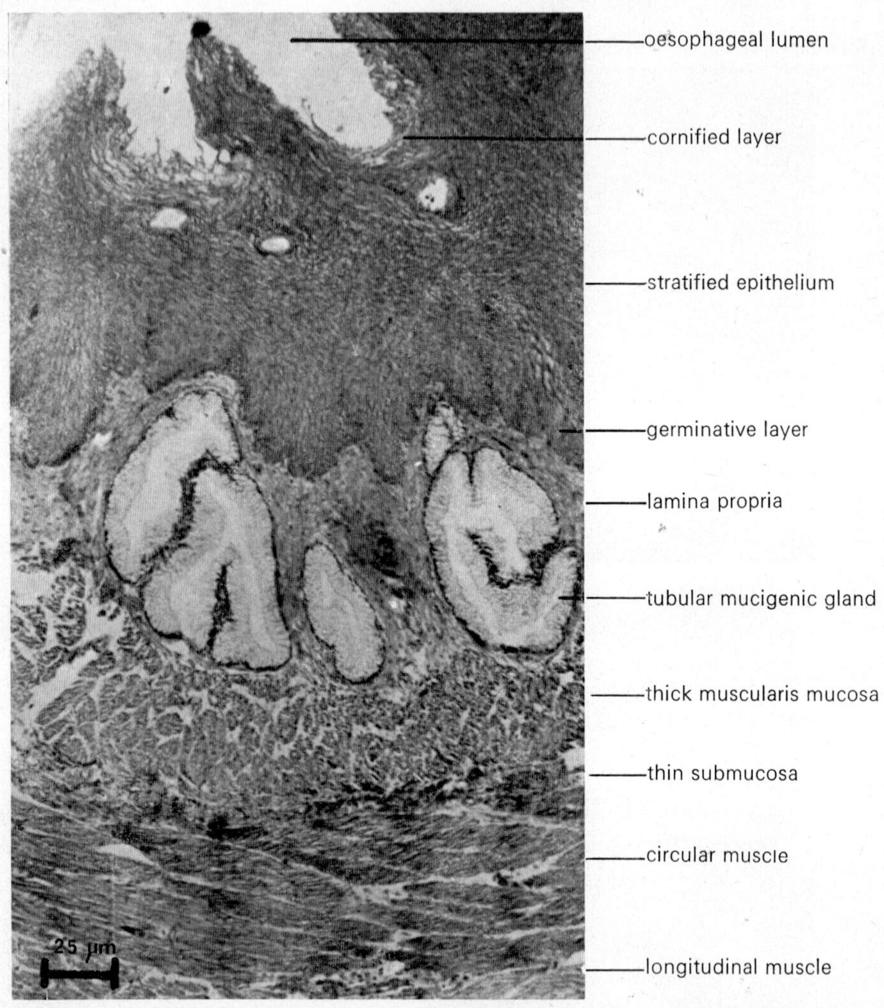

FIG. 16.21. TS through the oesophageal wall of the chick. Note the thick cornified lining.

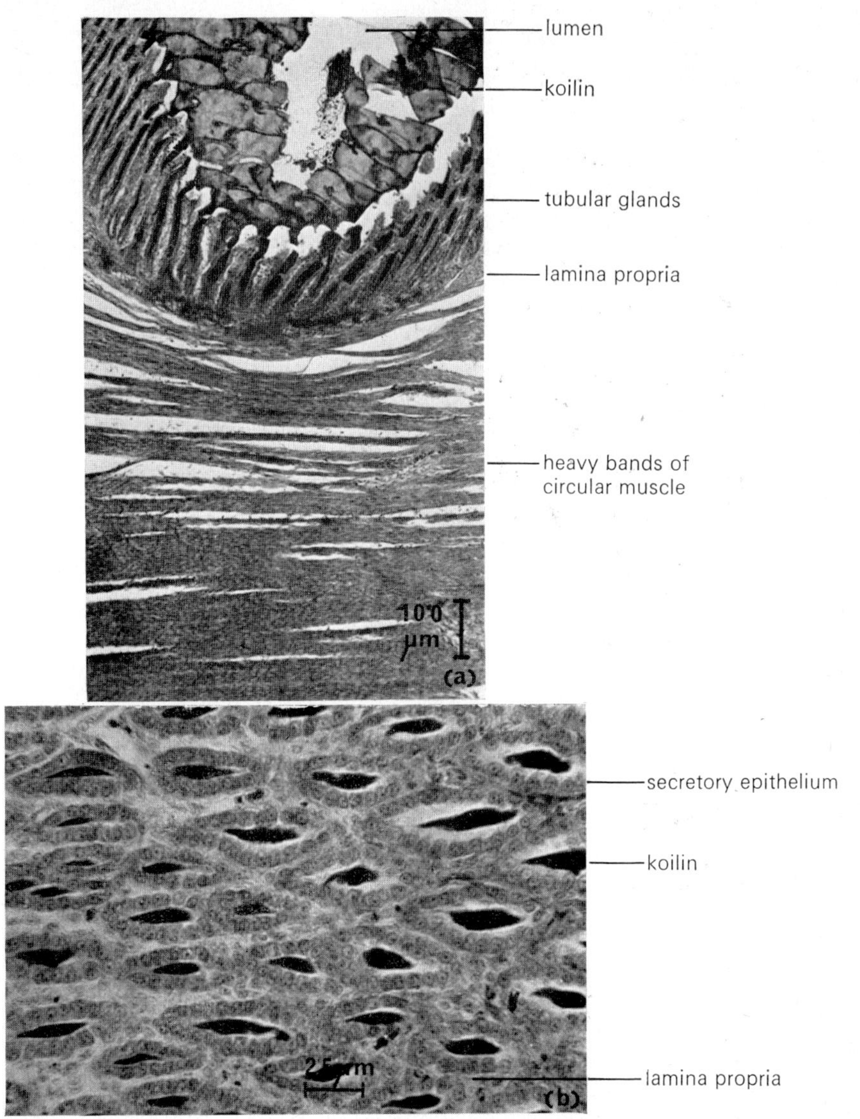

FIG. 16.22 (a) TS through the gizzard wall of a chick; (b) TS of the tubular glands lining the chick gizzard.

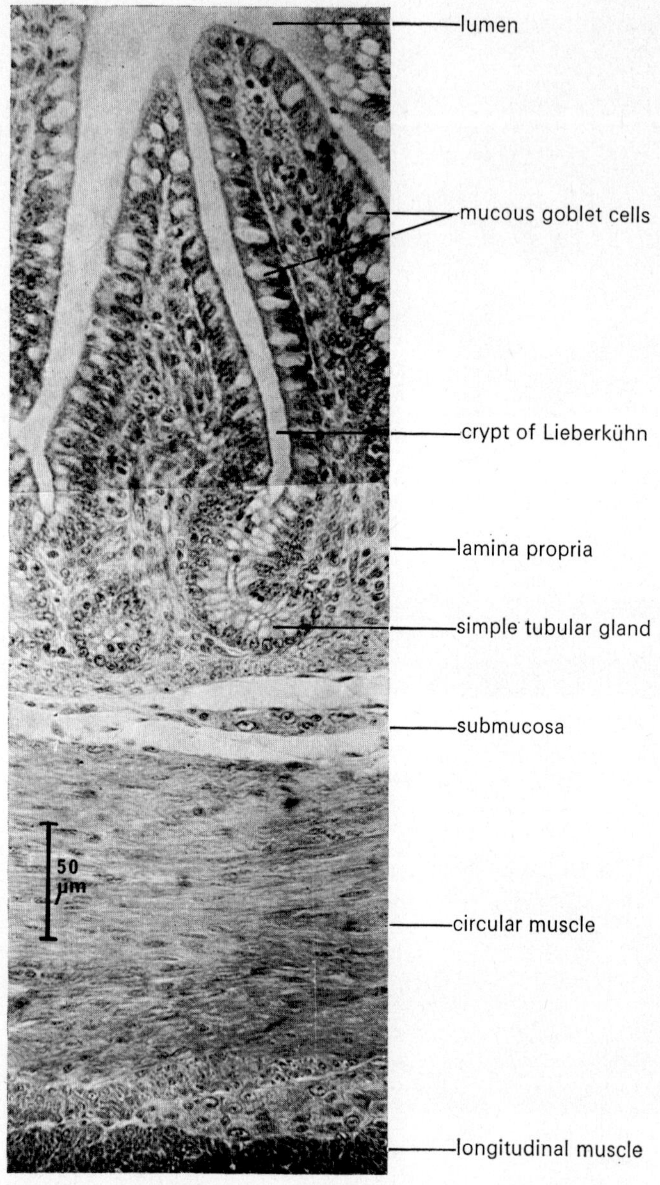

FIG. 16.23. TS through the wall of the large intestine of the chick.

many species except those that have an oily diet. The pancreas is a compact organ composed of typical acini with aggregates of islet tissue scattered amongst them.

Excretion

Due to their high respiratory and metabolic rates, birds tend to lose a lot of water. They compensate for this by having most efficient kidneys, which, unlike those of other vertebrates so far encountered, can produce hypertonic urine. This facility is shown best by small birds living in dry climates, but is present in all species.

The paired kidneys are long, lobed structures (see Fig. 16.25) divided into small numbers of lobules. Each lobule contains a branch of ureter draining the collecting tubules, and is divided into medullary and cortical parts, thus indicating the trend in mammals. The kidneys are metanephric and are composed of a very large number (twice as many as mammals) of small glomeruli and their associated nephric tubules (Fig. 16.24). Some of the nephrons possessing only proximal and distal tubules occur in the cortex, while others have a thin-walled intermediate segment (loop of Henle), part of which dips down into the medulla and is responsible for water reabsorption. Reabsorption also occurs across the rectal and cloacal walls. Urine and faeces are voided together as a semi-solid waste material containing uric acid and related compounds.

Birds, like reptiles, have nasal glands modified for salt secretion, but these are of major importance only in marine birds.

Reproduction

The reproductive system of birds is similar to that of lizards, except that in the female the right ovary and oviduct are usually only vestigial (Fig. 16.25).

The male has a pair of ovoid testes consisting of numerous anastomosing seminiferous tubules opening into a cavity (antrum). Ripe sperm pass from each antrum into small efferent ducts, and via a convoluted epididymis into a vas deferens which in some species is slightly enlarged at its hind end to form a seminal vesicle. In a few species (including ducks and geese) a penis is formed from a thickening of the ventral wall of the cloaca while a dorsal groove acts as sperm channel. In all other birds copulatory organs are missing, and sperm are transferred directly from the genital papillae of the male into the female cloaca. Many parts of the genital tract hypertrophy during the sexual season, while the testes of all birds, except pigeons, regress at the end of the breeding season.

The remaining left ovary of the female is large and irregular in the adult, and shows seasonal regression in most species. This does not happen, however, in the domestic hen which ovulates spontaneously and lays eggs almost all the year round. The ovary consists of a heavily vascularized

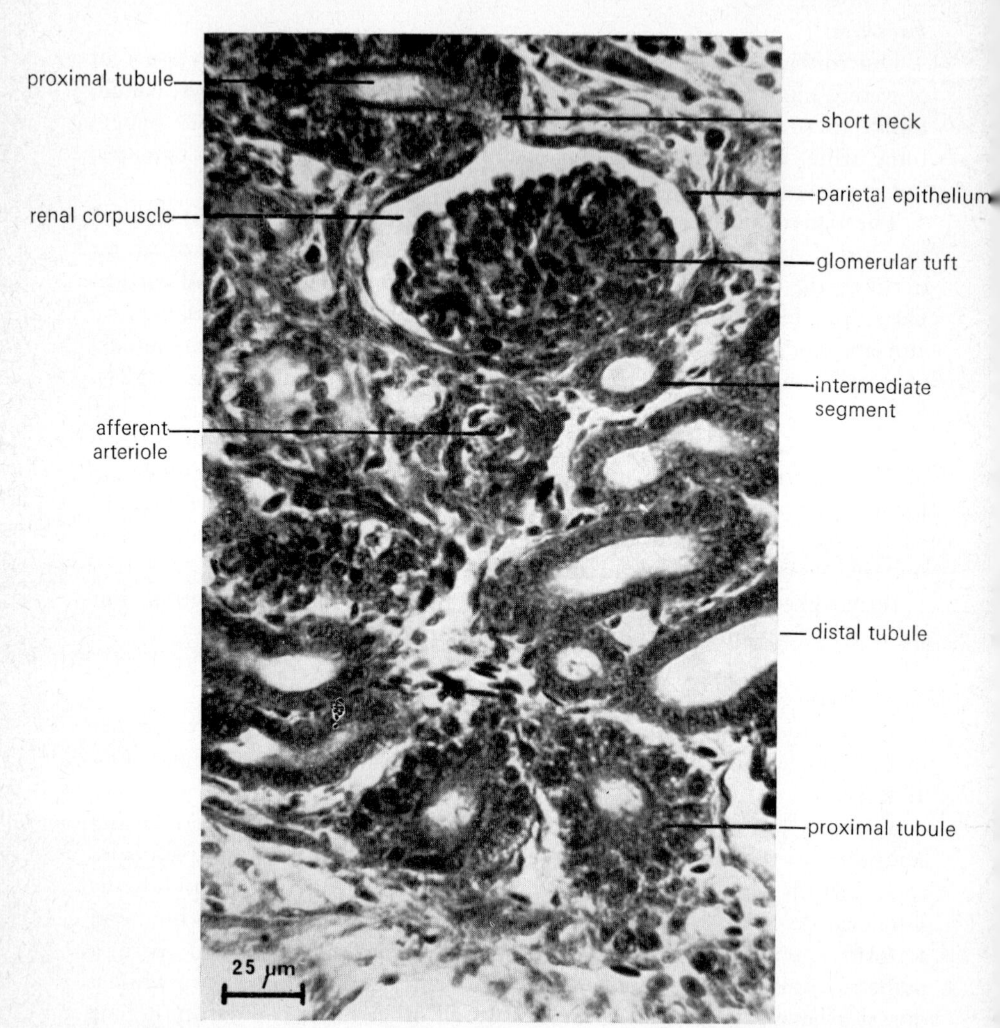

Fig. 16.24. Section through the kidney of the chick, showing a renal corpuscle and parts of the nephric tubule.

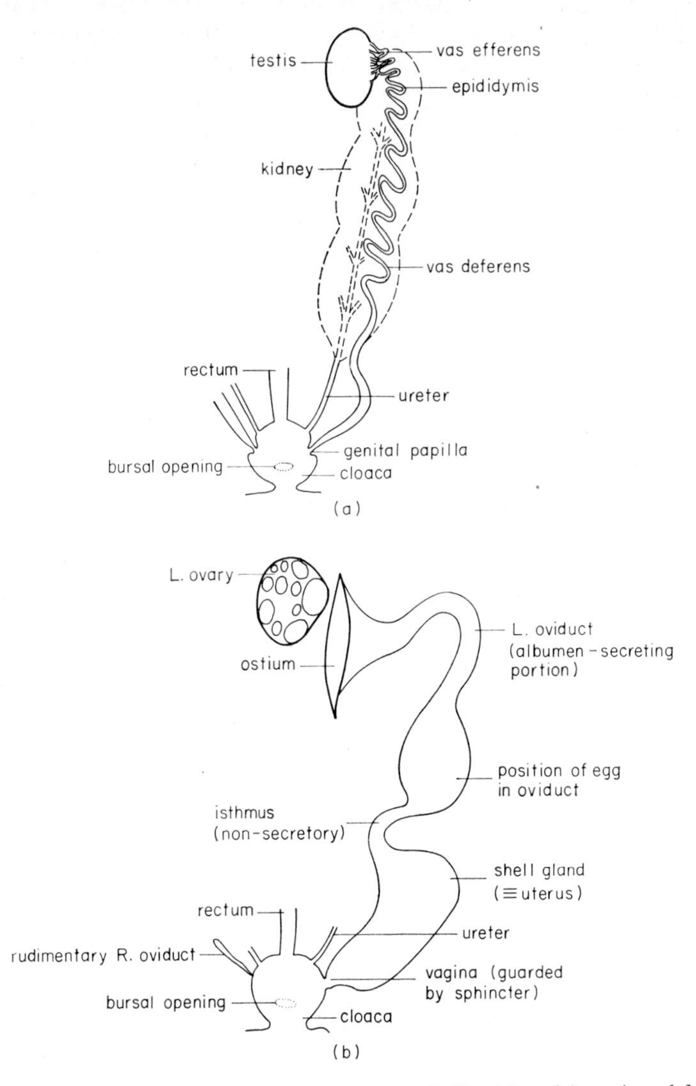

FIG. 16.25 (a) Plan of the urinogenital system of a cock (L. side only) as viewed from the ventral side; (b) plan of the reproductive system of the hen as viewed from the ventral side. Note that the R. ovary is missing and the R. oviduct vestigial.

connective tissue medulla surrounded by a cortex containing developing ova and bordered by germinal epithelium. Ova, when released, are captured by a wide ciliated membranous ostium at the anterior end of the long coiled oviduct. The oviduct walls are glandular and coat the eggs with albumen, shell membranes and a calcareous shell. Fertilization occurs in the oviduct, and the eggs complete with shell are extruded from the cloaca.

Both the testes and the ovary act as endocrine organs, secreting hormones involved in reproduction.

Chapter 17

Class Mammalia

Mammals are the most highly organized vertebrates, and show adaptations for exploiting a wide range of environments including seawater, freshwater and air as well as extreme conditions on land. Therefore, they show perhaps greater diversity of structure than any other vertebrate class and, because of this variation, there are only a few characteristics common to all mammals. However, they are all warm-blooded amniotes which nourish their young for a time after birth on milk produced by special mammary glands.

Typical modern mammals are covered by a coat of epidermal hair (scales often persist as well), and generally have four limbs each bearing five clawed digits, although both these features may be secondarily altered. The body is divided into a large head, a distinct neck, a compact trunk and a long tail (also frequently lost). Each half of the lower jaw is reduced to only one bone, while the remaining jaw bones develop into a chain of small bones in the middle ear. The inner ear contains a curved cochlea. The brain shows increased development particularly in the area (cerebral cortex) concerned with intelligence and associative learning. Mammalian teeth are complex and show differentiation of form and function: they are often used as diagnostic characters of different mammalian orders. As in birds there is a double blood circulation and a four-chambered heart. The thorax is divided from the abdomen by a muscular diaphragm which is used in respiration. In most female mammals there is cyclic activity of the gonads (oestrous cycles), and all but a few species (e.g. monotremes) are viviparous. Parental care of the young is highly developed.

The majority of modern mammals belong to the sub-class Eutheria, or placental mammals, which have a highly organized allantoic placenta (see p. 11). This enables the young to be nourished *in utero* for a relatively long period so that they are in an advanced state when born. Placentals are further characterized by the absence of a cloaca, and by the appearance of a new bridge of nervous tissue (corpus callosum) linking the two cerebral hemispheres in the brain.

Example: *Mus* (*musculus*)

The house-mouse belongs to the order Rodentia, or gnawing mammals. This is a large order of relatively unspecialized mammals characterized by their teeth and associated jaw muscles which are adapted for gnawing. The teeth are reduced in number to a single pair of incisors and three pairs of

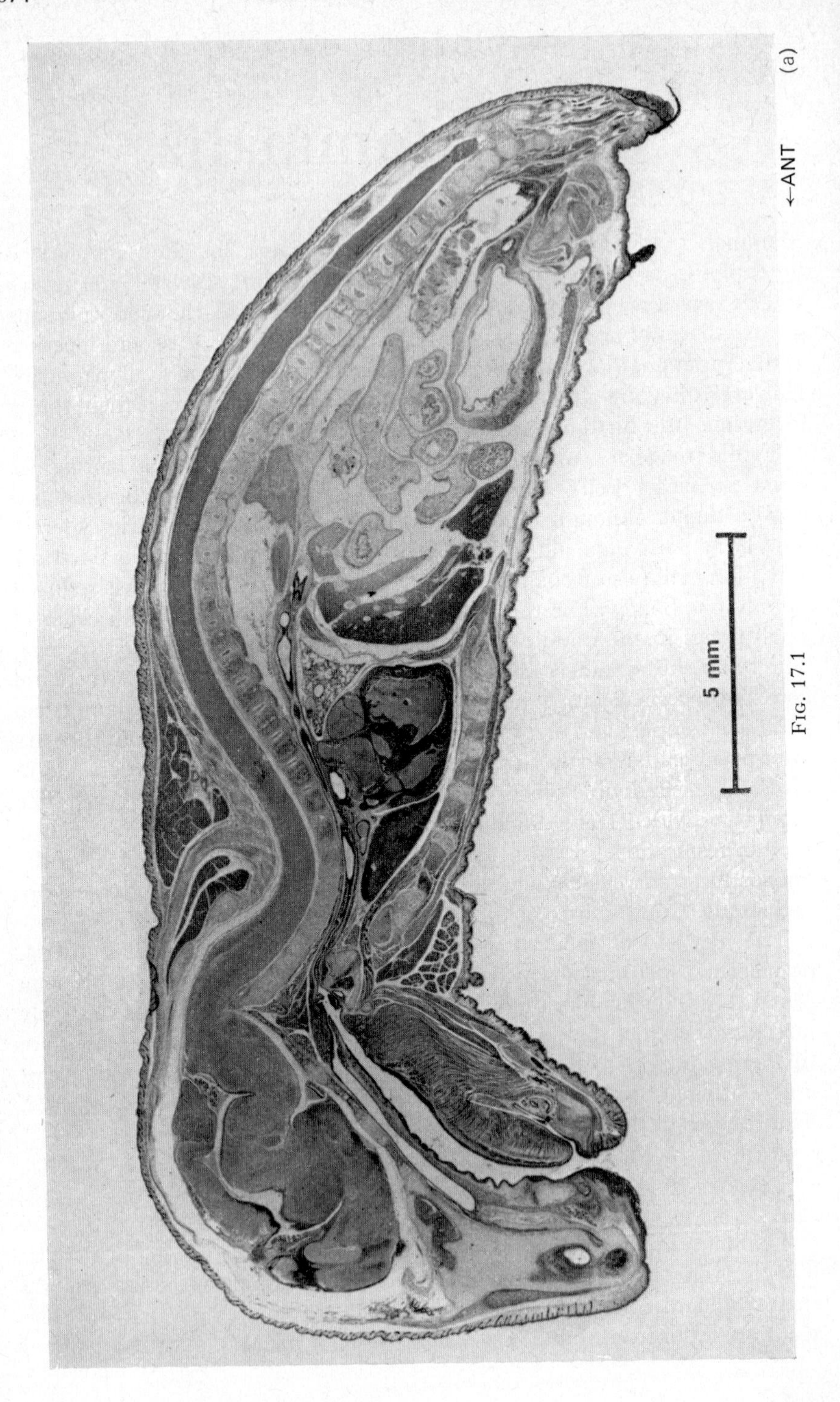

FIG. 17.1

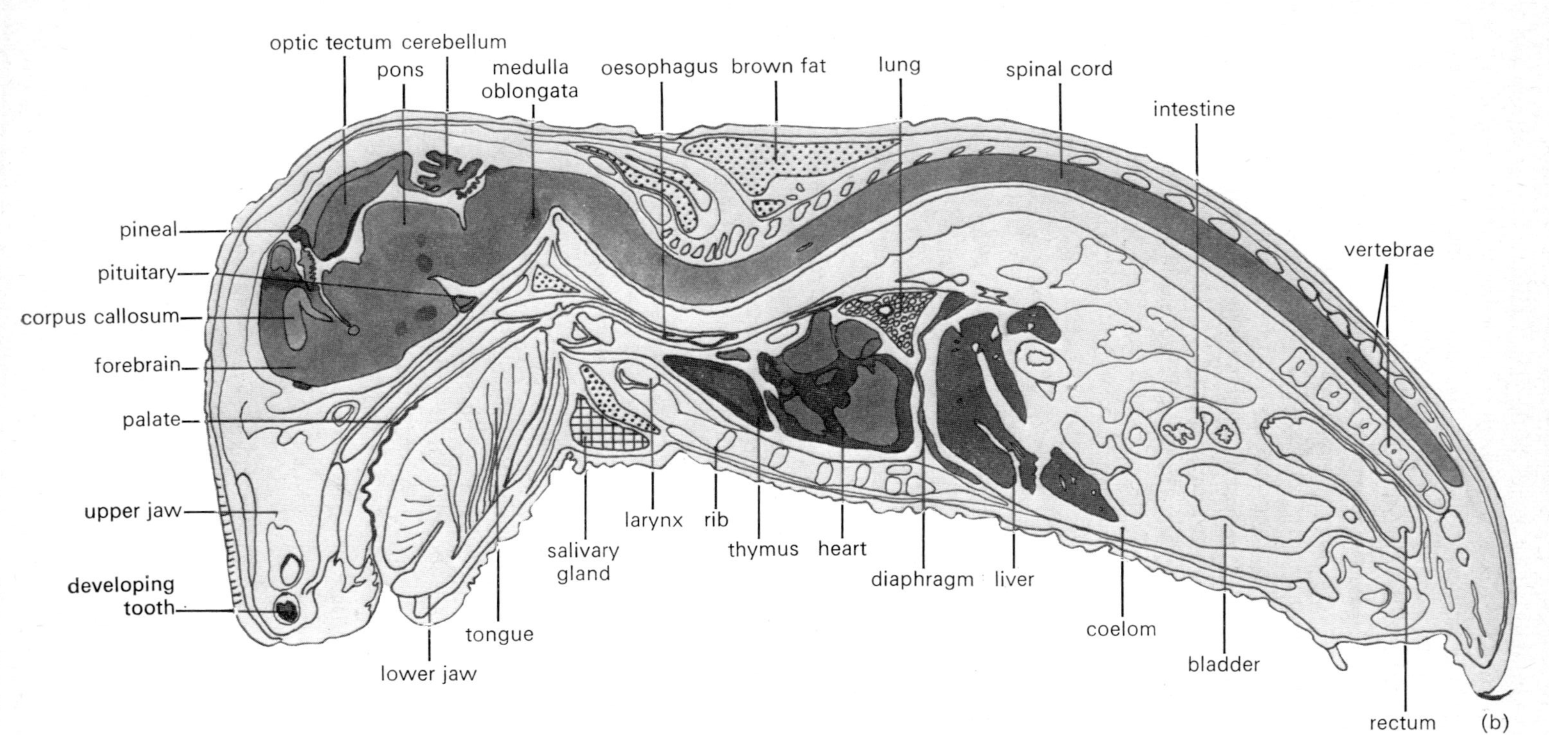

Fig. 17.1 (a) VLS and (b) explanatory drawing of a 1-day old mouse (*Mus musculus*).

cheek teeth (molars) on each side of the jaw. Canine teeth are absent and are replaced by a gap (diastema) which is used for manipulating food. The incisors are large and grow continuously: the outer enamel layer is missing at the back so that continual wear grinds a sharp chisel-like edge to the tooth, perfect for biting. The molars have rows of cusps for grinding, which function is assisted by the loose jaw articulation and massive jaw muscles.

The mouse is very common and has a world-wide distribution. It is mainly nocturnal and is provided with well-developed sense organs. It has a pointed face, long hind legs and short fore legs which are often used for manipulation of food. The feet and tail are hairless but the rest of the body is covered uniformly with hair. Figure 17.1 shows a VLS of a 1-day old mouse. Notable features are the flexible vertebral column and the complexity of the brain, which develops at a faster rate than most other tissues and, therefore, is disproportionately large at this stage. The thorax containing heart and lungs is bounded by vertebral column, rib-cage and domed diaphragm.

Mammalian skin shows the ultimate degree of complexity being tough and flexible, and bearing hair (fur) and a number of other epidermal derivatives (scales, horns, claws, nails, quills and prickles) as well as complicated glands. The glands are basically of two types: simple or branched tubular sweat glands (absent in rodents and various other mammals), and branched alveolar sebaceous glands. The scent glands of carnivores and the mammary glands present in all mammals are modified sebaceous glands. Newborn mice are naked, but even at this stage developing hairs can be seen (Fig. 17.2) which soon cover the animal in a thick pelt. In those regions where there is no fur, the outermost layer of the epidermis (stratum corneum) is thickened particularly in regions of most wear (e.g. the soles of the feet). Some of the hairs (vibrissae), especially round the eyes and mouth, are very long and thick (Fig. 17.3). They are associated with sensory nerve endings and have an important role as tactile receptors.

Two kinds of adipose tissue are found in mammals. White fat (see Fig. 15.27) is widespread as in other vertebrates. During its development the cells accumulate droplets of fat (multilocular fat), which later coalesce to form large rounded globules that obliterate the cytoplasmic contents of the cells. This type of unilocular fat is found in adults. Figure 17.4 shows brown fat which is a tissue peculiar to mammals, in particular to hibernants and many new-born mammals. Brown fat is multilocular, is well-vascularized and has a high content of unsaturated fat. It is distributed around the neck region (see Fig. 17.1) and its presence in an animal is associated with an ability to withstand cold, due to its thermogenic properties.

The endoskeleton of adult mammals is ossified. Most of the bones in tetrapods are composed of hard lamellar bone (Fig. 17.5). Hard, or compact, bone is extremely strong and is composed of units (osteons) each consisting of a central Haversian canal surrounded by concentric layers of

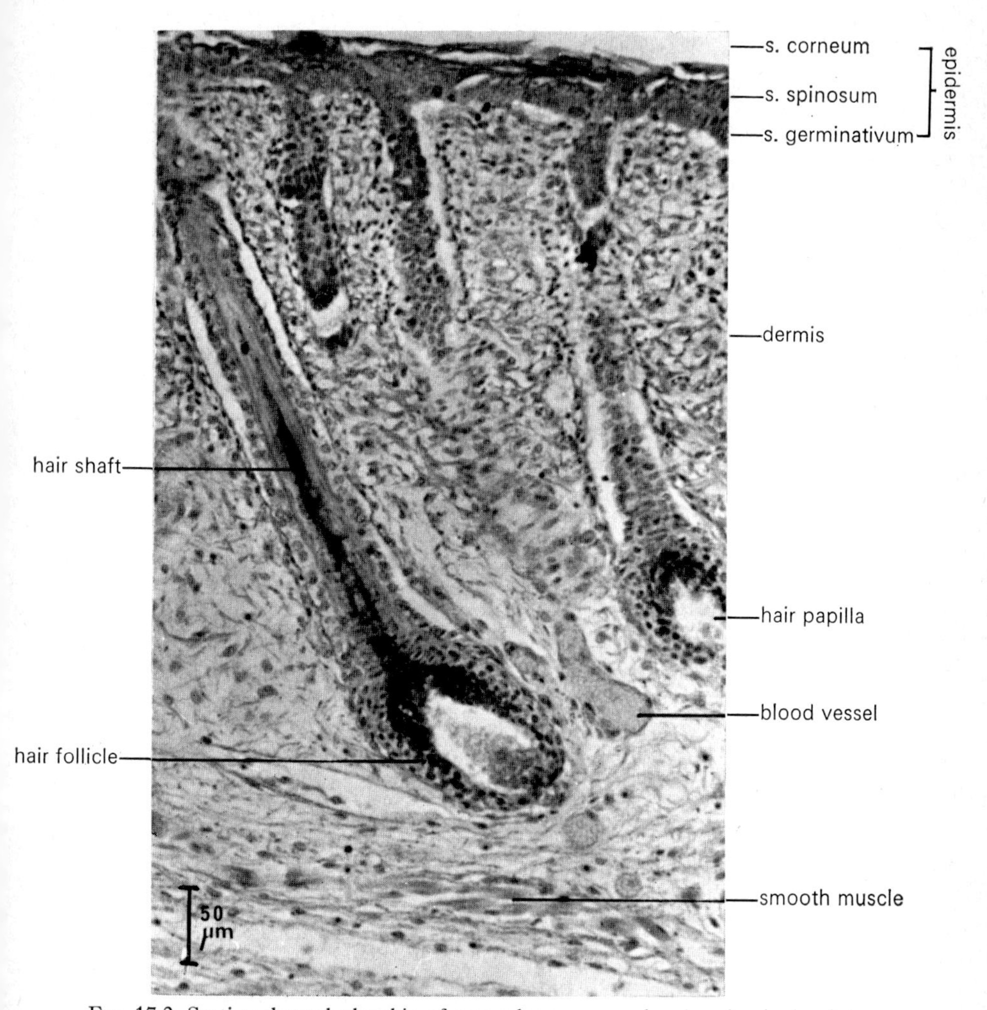

FIG. 17.2. Section through the skin of a new-born mouse showing developing hairs.

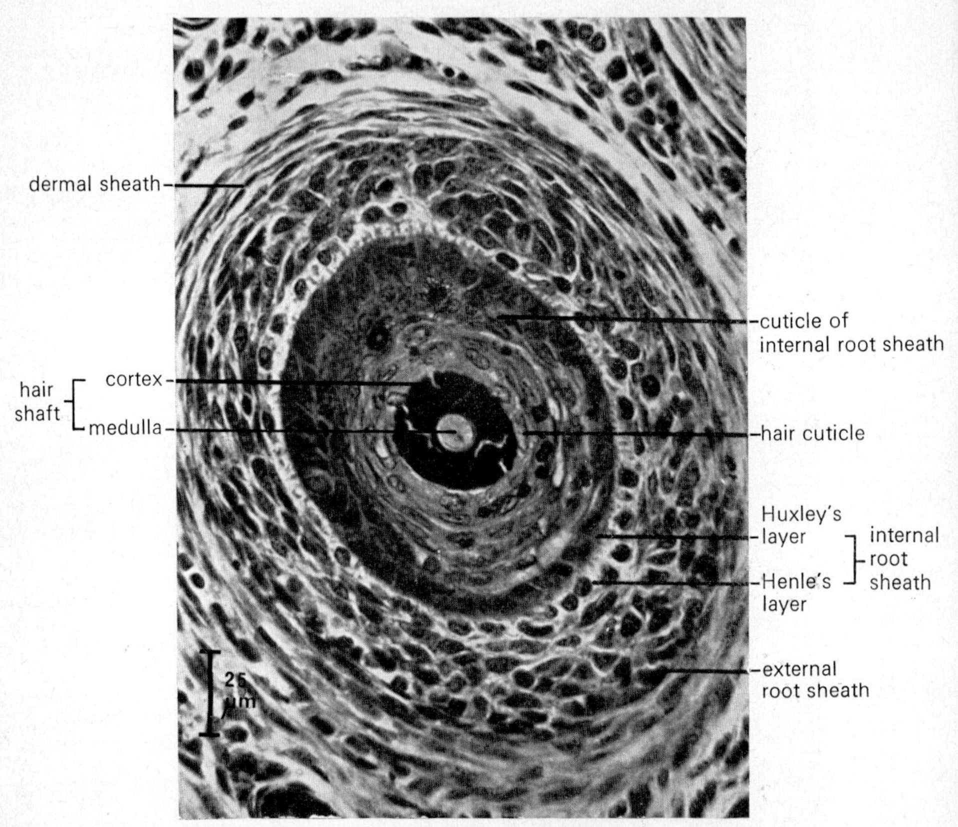

FIG. 17.3. TS of a vibrissa close to the mouth of a young mouse.

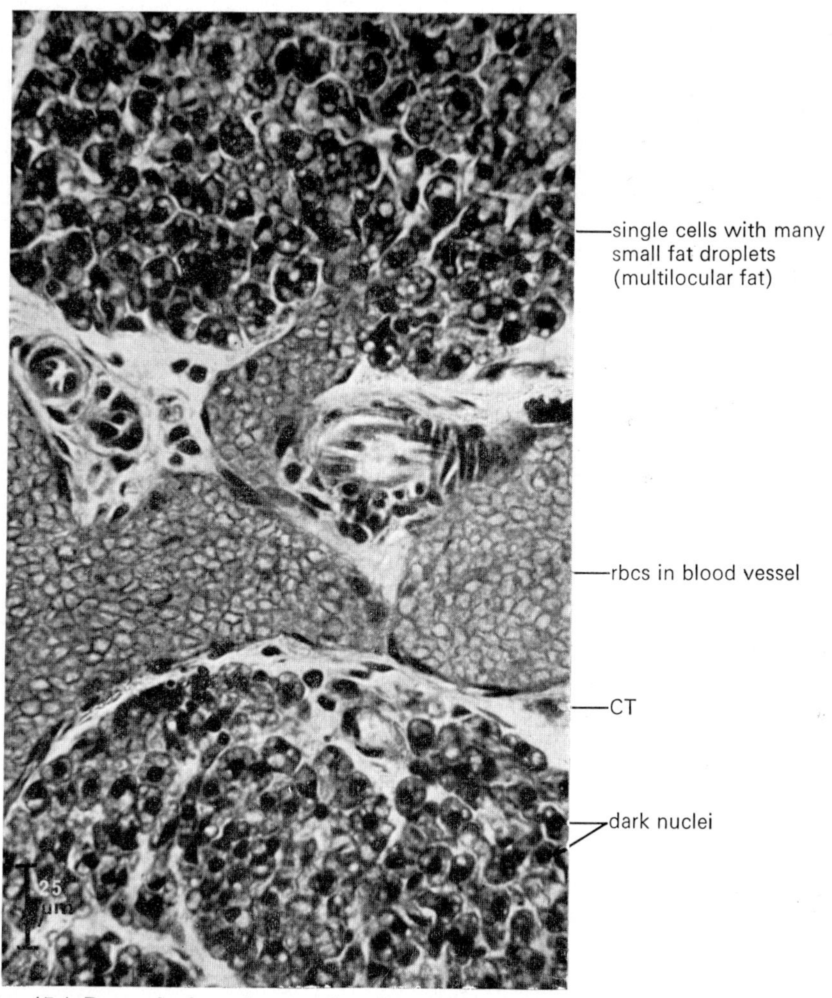

FIG. 17.4. Brown fat from the ventral neck region of a new-born mouse. Note the multilocular fat globules and the extensive blood supply.

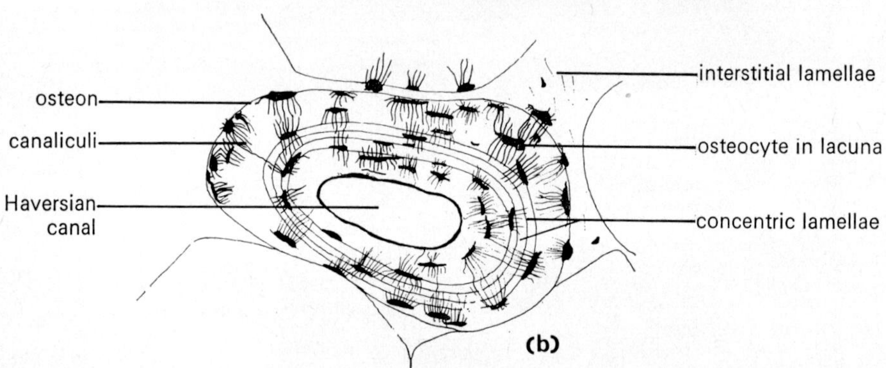

FIG. 17.5 (a) Section through hard lamellar bone from an adult mouse; (b) drawing of one osteon seen in (a).

calcified matrix. Between these lamellae the mature osteocytes are situated in lacunae which communicate by minute canaliculi. The Haversian canals carry blood-vessels and are linked to one another and to the central marrow cavity by canals of Volkmann.

The vertebral column is very flexible, and is characterized by the centra, which can be either gastrocoelous or acoelous (flat) and which possess thin discs of bone (epiphyses) on both faces. Inter-vertebral fibrocartilage discs contain the remnants of the notochord. Some of the more posterior vertebrae are fused to form a sacrum, used for supporting the hind limbs on land. Ribs, which articulate with the vertebrae, are confined to the thoracic region. Most of them fuse with the sternum to produce a rigid thoracic cage which forms a support for the forelimbs and a protective covering for the heart and lungs.

Nervous system

The central nervous system of mammals is capable of great plasticity of response, and of intelligent thought and learning. Associated with these features the cerebral hemispheres are large and have an extensive cortex of layers of neurones (Fig. 17.6): this is the chief co-ordination centre. The mouse forebrain is not particularly well developed, but in some species the cortex becomes extremely convoluted and covers most of the brain. There is also an increase in the number of cross-connections between the two sides of the brain, including the corpus callosum or great cerebral commissure (see Fig. 17.1). The diencephalon is reduced in size and bears a small glandular pineal, but there is no parapineal. The optic tectum, instead of appearing as paired optic lobes as in all other vertebrates except snakes, is composed of four small collections of neurones (corpora quadrigemina) concerned specifically with sight and hearing. Below the convoluted cerebellum a thick mass of fibres (known as the pons or pontine bridge) links the cerebellum with the anterior centres. As in other amniotes, the medulla oblongata (Fig. 17.7) consists of numerous "nuclei" of nerve cell bodies and acts as the main relay centre of the autonomic nervous system. The ANS is clearly divided in mammals into thoraco-lumbar (sympathetic) and cranio-sacral (parasympathetic) portions, which have reciprocal functions. The mammalian CNS is surrounded by three meninges; the pia mater and arachnoid membrane, constituting the inner leptomeninx, and an outer dura mater.

Sense organs

The primary senses are particularly well developed in mammals, although their relative importance varies from one animal to another according to its mode of life.

The eye [Fig. 17.8(a)] is round and often large, especially in nocturnal species. Stereoscopic vision is a characteristic of mammals, and is achieved by frontal positioning of the eyes and by incomplete decussation of the

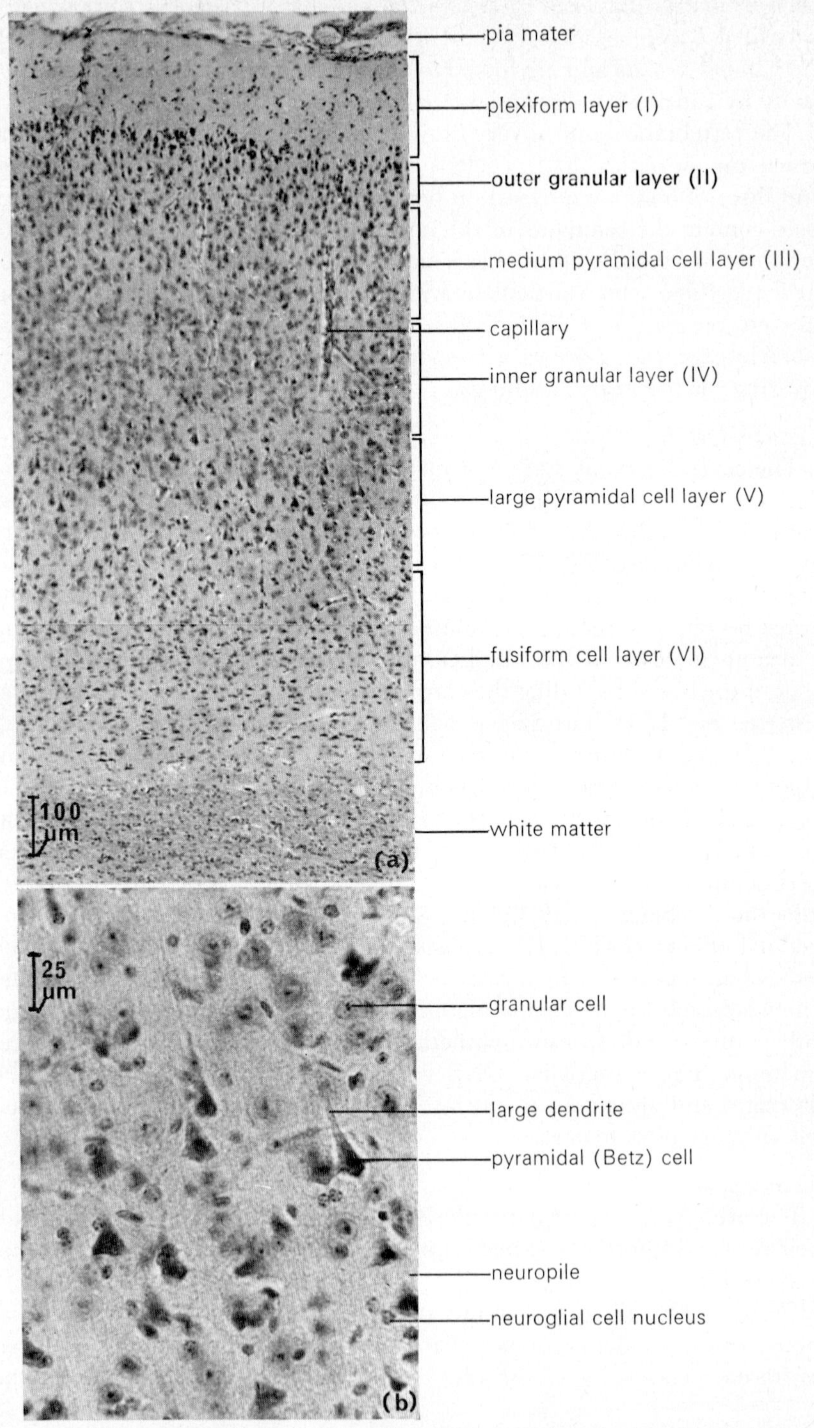

FIG. 17.6 (a) Section through a mammalian cerebral cortex, showing the six layers of neurones; (b) detail of the large pyramidal cell layer (V) in (a).

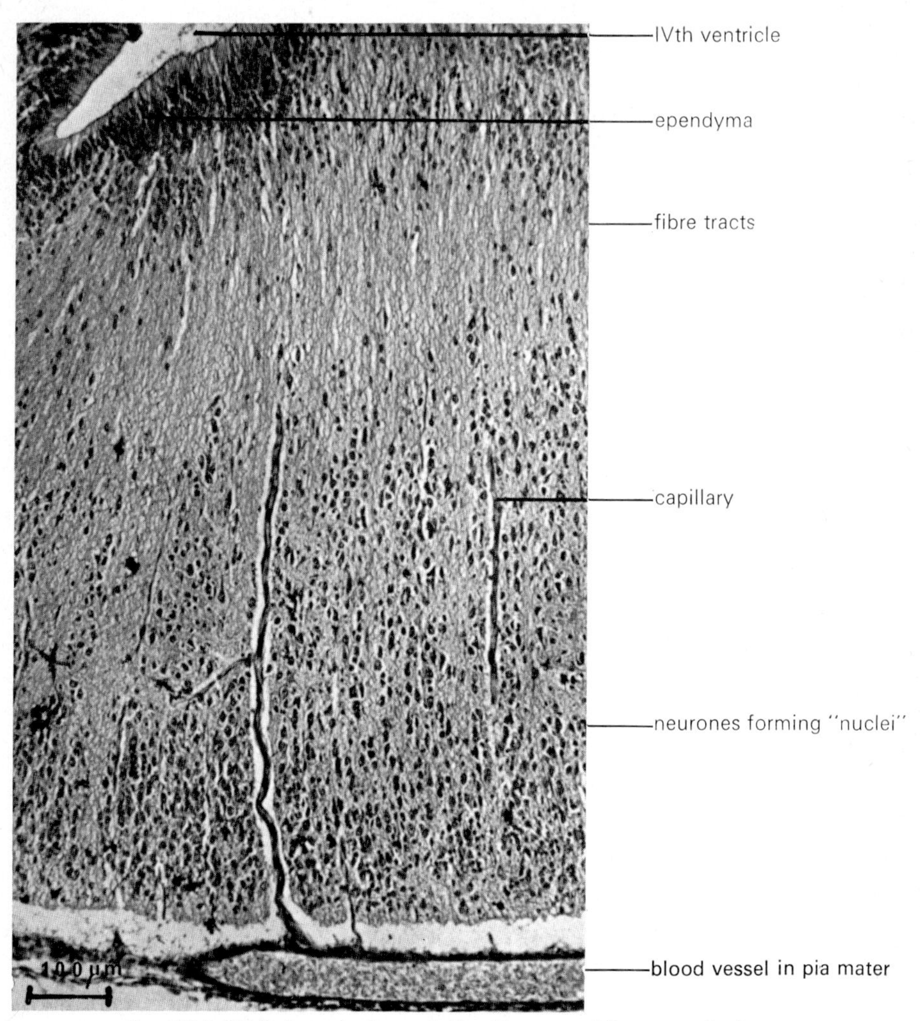

FIG. 17.7. VLS through the medulla oblongata of the mouse brain.

optic tracts (i.e. tracts from the outside of each retina run ipsilaterally while those from the inner edge run contralaterally) [Fig. 17.8(b)]. The eye is fringed with two lids operated by facial muscles: the nictitating membrane is normally vestigial. The scleroid coat is tough and fibrous but has no strengthening ossicles. The choroid is thick and heavily vascularized but bears no inward projection such as the pecten of birds, although it does form a reflective *tapetum lucidum* in many nocturnal species. The lens is spherical in rodents, although in other mammals it is a biconvex disc with the greater curvature on the posterior face. Accommodation is achieved by altering the shape of the lens, but this is brought about in a different way to that found in the birds and reptiles. In mammals the lens is suspended from the ciliary body by fine, tough fibres which hold the elastic lens in a stretched position [they are difficult to see on photographs as they are transparent, but their position is marked in Fig. 17.8(a)]. Contraction of the ciliary muscles moves the region of attachment of the fibres in towards the centre of the eye, thus releasing tension on the lens and allowing it to assume a more rounded shape (for near vision). The retinal composition depends on way of life, rods predominating in the retinas of nocturnal mammals and cones in diurnal retinas. True lacrymal glands are found in the orbital region of mammals; they secrete a watery fluid which lubricates the eye surface.

The sense of hearing is better developed in mammals as a whole than in any other class. The ear has the same basic plan as in other amniotes, but it is modified in certain major respects. The external ear is complex, consisting of a deep channel (external auditory meatus) protected by cartilage or bone and fringed by a pinna composed of elastic cartilage. The pinna is used for catching sound, and is very large in the mouse. The middle ear is enclosed in a bony bulla, and the columella (found in birds) is replaced by three small bony ossicles: the stapes, derived from the columella, and the incus and malleus derived from jaw articulation bones. In the inner ear the cochlea is long and coiled (for compactness), and possesses the same components as that of the bird, although it is a little more complex (Fig. 17.9). The mammalian ear is an engineering masterpiece, and can pick up very faint sounds. Air-borne vibrations are amplified by resonance as they pass down the external auditory meatus, and set up vibrations in the taut tympanic membrane whence they are conveyed to the chain of ossicles. Due to the arrangement and articulation of the three bones, large movements of the tympanic membrane are converted into smaller movements of the stapes. Since the tympanic membrane has a large area compared with the small oval window on which the stapes rests, the concentration of force that ensues will amplify the sound still further. Many mammals including mice can hear sounds of very high frequency (ultrasound), while some (e.g. bats) use echolocation of ultrasound extensively in navigation.

In a few mammals (e.g. bats and whales), the sense of smell is reduced,

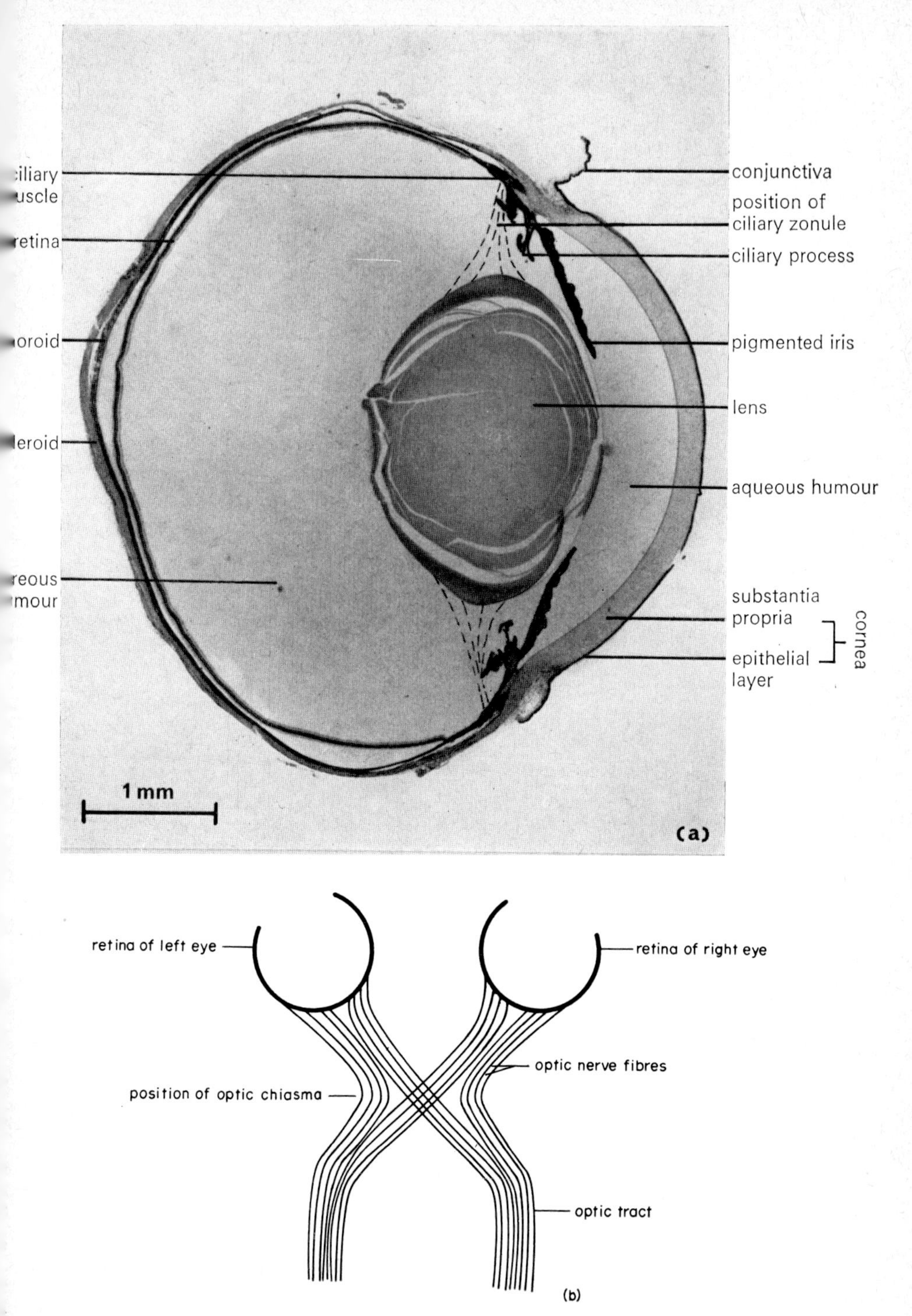

Fig. 17.8 (a) Section through a mammal eye. The positions of the lens suspensory fibres are marked by dotted lines; (b) diagram of the mammalian optic chiasma, showing the incomplete decussation of the optic nerve fibres from the right and left eyes.

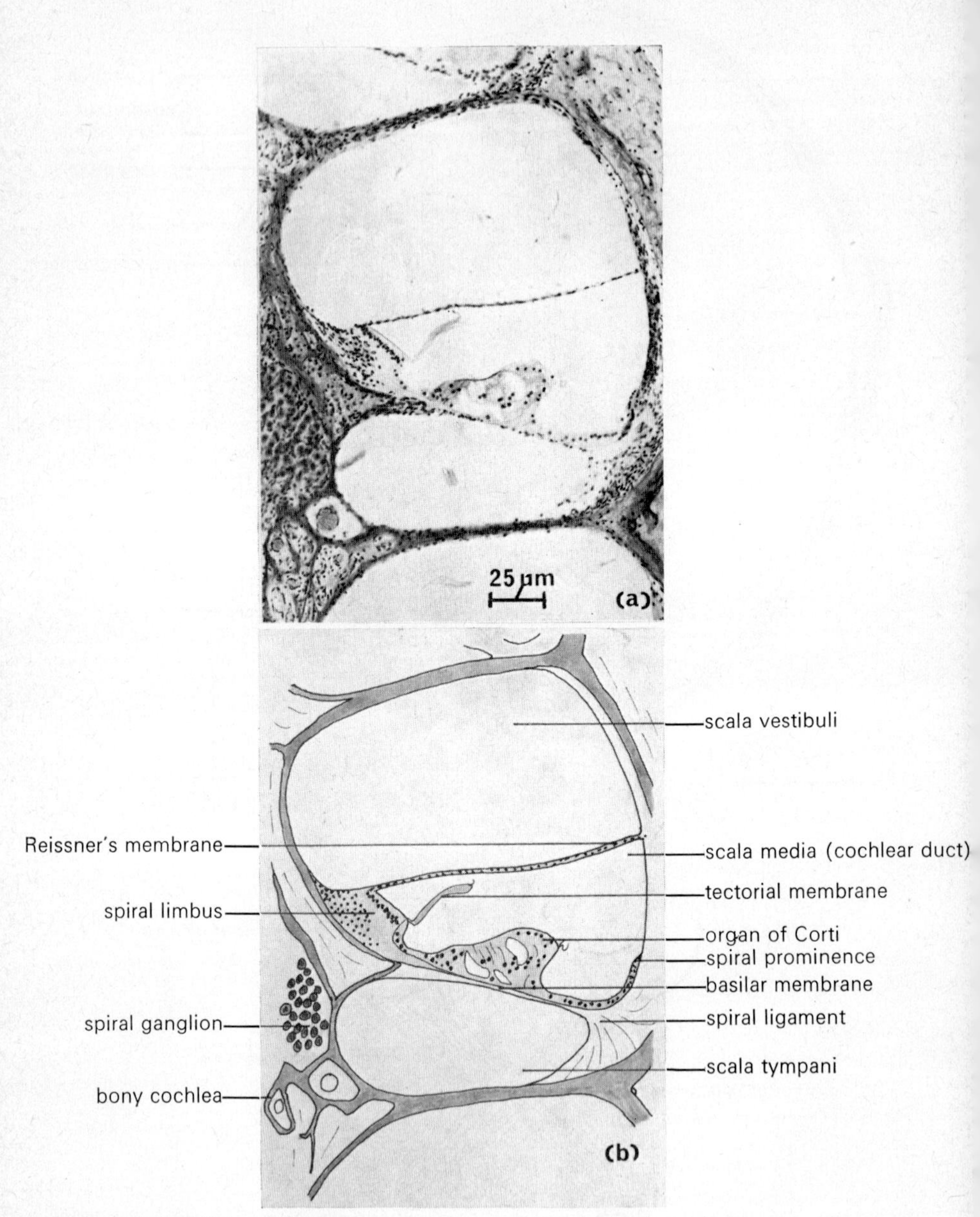

FIG. 17.9 (a) TS and (b) drawing of one turn of the cochlea from the inner ear of a mouse.

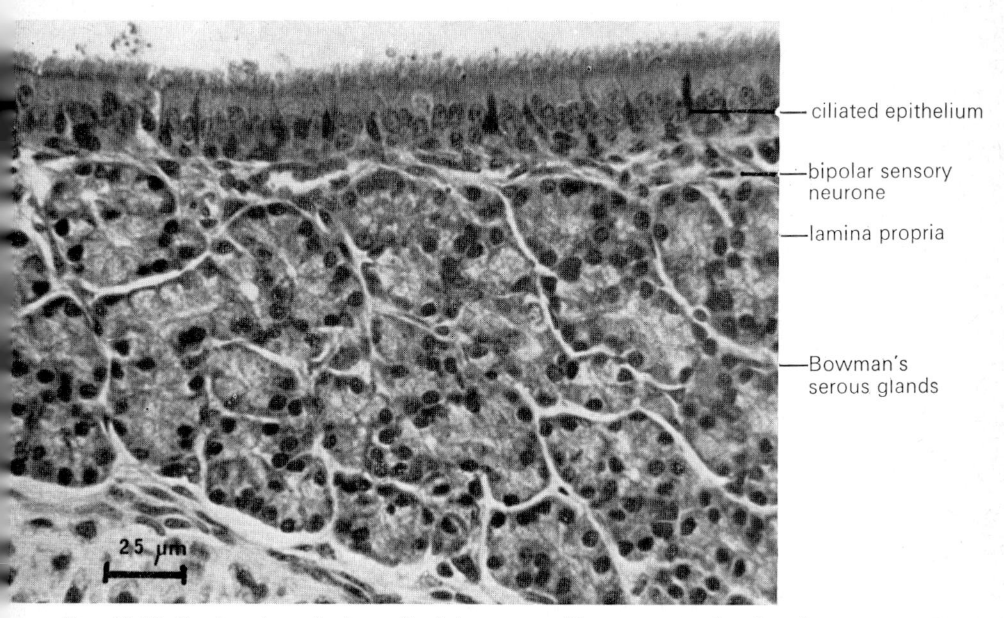

FIG. 17.10. Section through the wall of the mouse olfactory organ showing the sensory epithelium and the extensive serous Bowman's glands.

but in most species it is of particular importance. The olfactory organ has a large vestibule merging with the main nasal cavity which is sub-divided by conchae (turbinal bones) into a series of pockets and sinuses. The epithelial area is restricted dorsally. It has the typical structure of vertebrate olfactory epithelia, but is distinguished from that of other classes by the thick layer of Bowman's glands in the lamina propria (Fig 17.10). These are tubulo-acinar serous glands which secrete a watery fluid that moistens the olfactory epithelium and acts as a solvent for scent particles. Ducts from the orbital lacrymal glands, and from the small Jacobson's organ lying along the nasal septum, join the long naso-pharyngeal duct before it enters the buccal cavity.

Endocrine organs

The mammalian pituitary [Fig. 17.11(a, b)] has an extensive anterior lobe derived from Rathke's pouch, and a large posterior lobe of nervous origin. The anterior lobe is divided into a pars glandularis [Fig. 17.11(c)] concerned with the secretion of trophic hormones, an apparently non-secretory pars tuberalis surrounding the pituitary stalk [Fig. 17.11(d)], and a secretory pars intermedia [Fig. 17.11(e)]. The function of the pars tuberalis is so far unknown, while the pars intermedia, although implicated in colour change in amphibians, as yet plays no clear role in mammals. The pars nervosa [Fig. 17.11(f)] is linked by the infundibular stalk to the median eminence of the hypothalamus, and releases neurosecretions produced by various hypothalamic nuclei. Small quantities of stored neurosecretory material appear as dark Herring bodies in light microscope sections: the Herring bodies probably represent swollen nerve endings. Hormones released by the pars nervosa are involved in control of water balance by the kidney, and in control of smooth muscle (of blood-vessels, uterus and mammary glands).

The thyroid is a bilobed encapsulated organ lying at the base of the neck. Some of its follicles are of ultimobranchial origin, as these organs are present in the embryo and are later absorbed by the thyroid. Their function of controlling calcium and phosphate balance is taken over by four small parathyroids which are closely applied to the thyroid.

Inter-renal and chromaffin tissues are intimately associated in mammals, and form the paired adrenal glands which are sited just anterior to the kidneys. The two tissues remain separate with the inter-renal tissue forming a distinct cortex round the medulla of chromaffin tissue [Fig. 17.12 (a)]. Experimental evidence suggests that the different zones found in the mammalian adrenal cortex [Fig. 17.12(b–d)] secrete different hormones. Thus it is thought that the zona glomerulosa produces mineralocorticoids which control water and electrolyte balance; that the zona fasciculata secretes glucocorticoids involved in the regulation of metabolism; and that

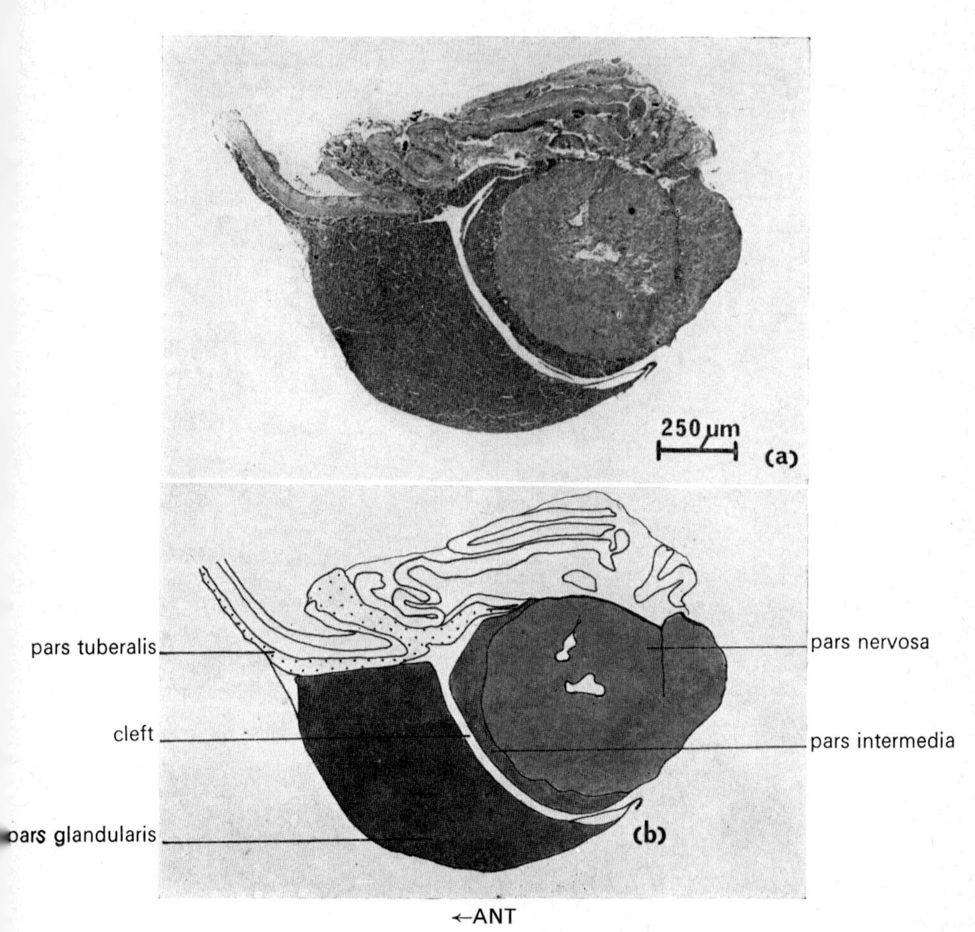

FIG. 17.11 (a) and (b)—see p. 691 for legend

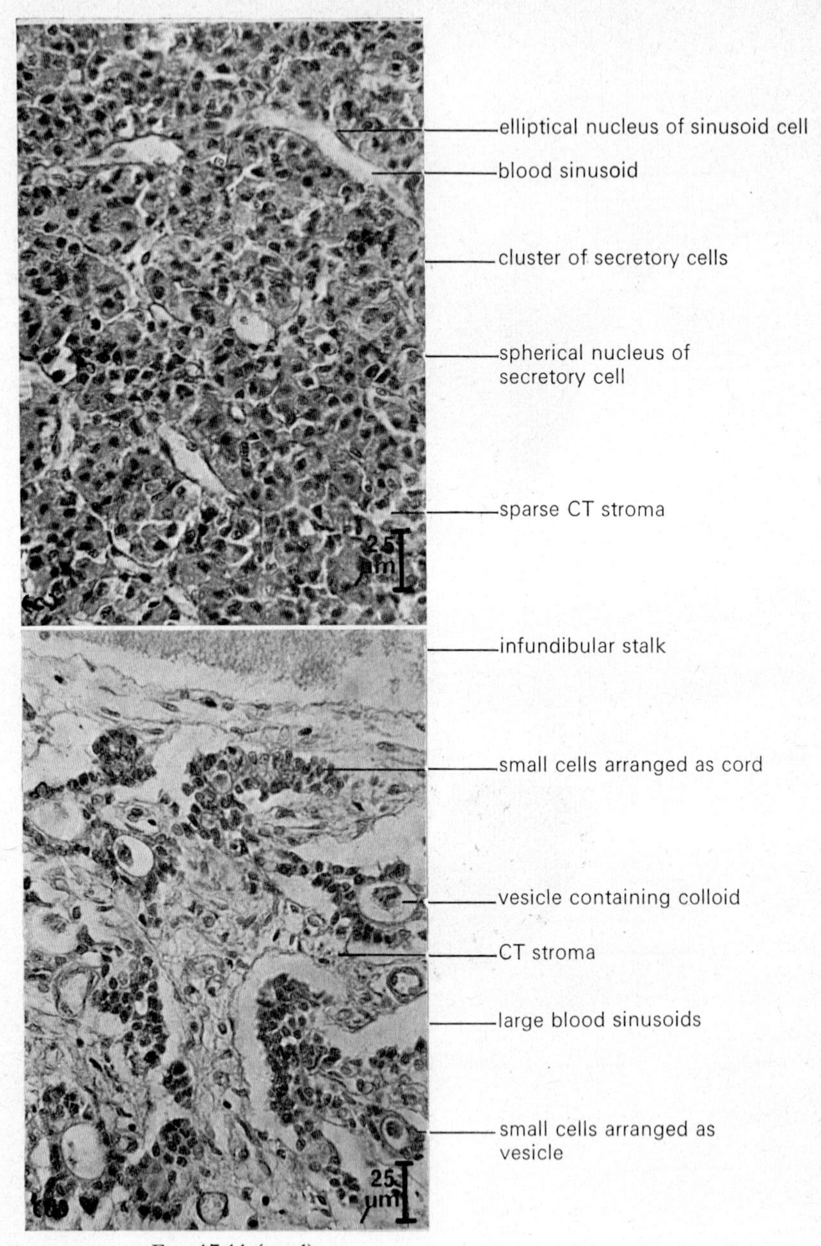
elliptical nucleus of sinusoid cell
blood sinusoid
cluster of secretory cells
spherical nucleus of secretory cell
sparse CT stroma
25 µm
infundibular stalk
small cells arranged as cord
vesicle containing colloid
CT stroma
large blood sinusoids
small cells arranged as vesicle
25 µm

Fig. 17.11 (c—d)

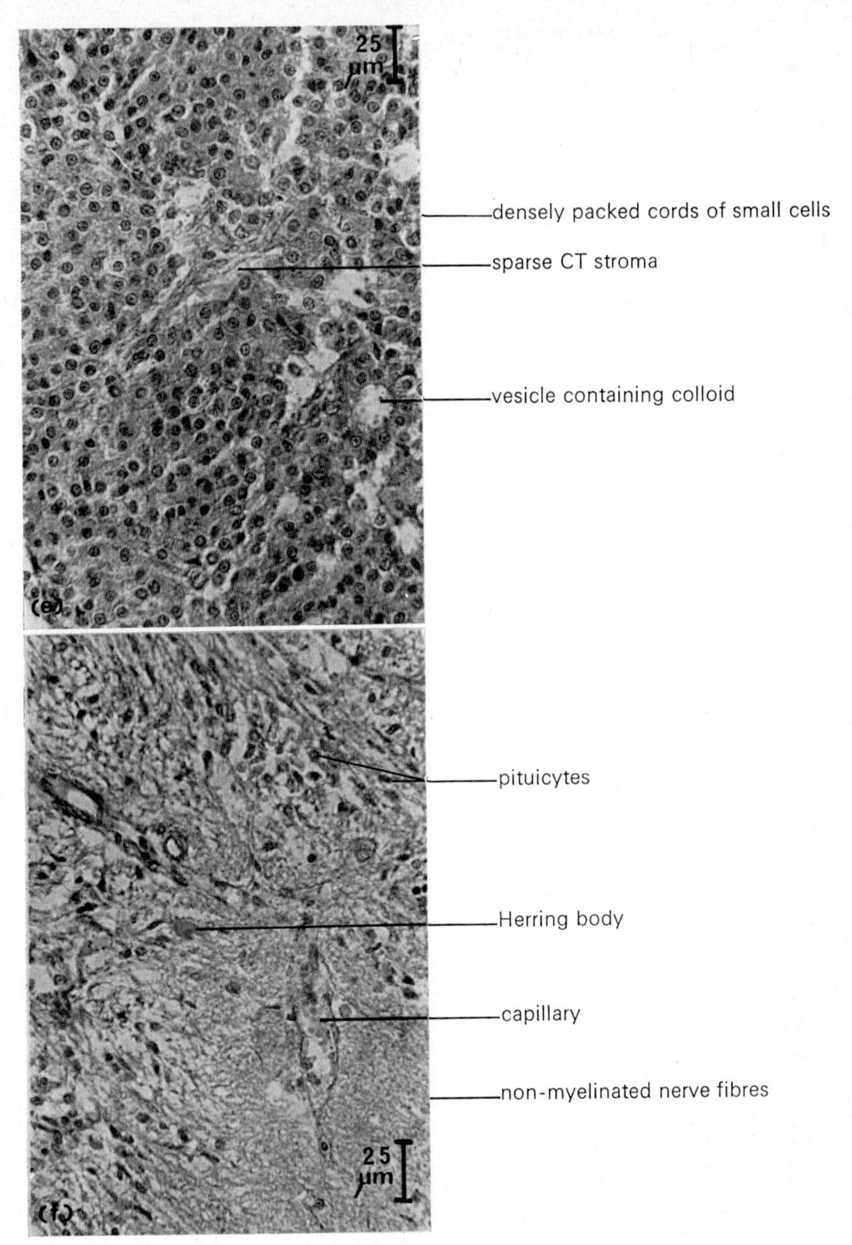

FIG. 17.11 (a) VLS and (b) drawing of a mammalian pituitary to show the glandular and nervous portions. Details of (c) pars glandularis, (d) pars tuberalis, (e) pars intermedia and (f) pars nervosa of the mammalian pituitary.

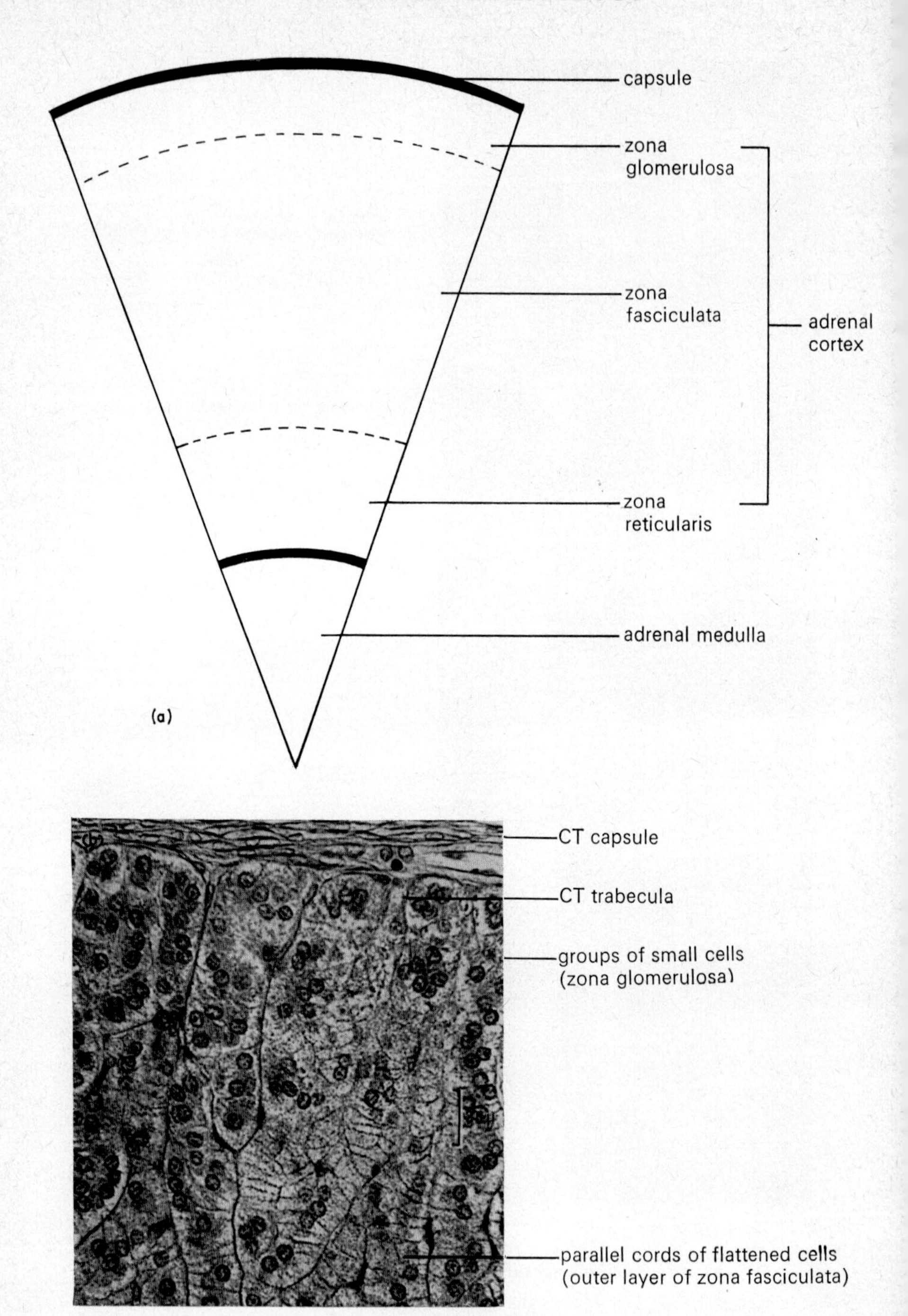

capsule
zona glomerulosa
zona fasciculata
adrenal cortex
zona reticularis
adrenal medulla
(a)
CT capsule
CT trabecula
groups of small cells (zona glomerulosa)
parallel cords of flattened cells (outer layer of zona fasciculata)

FIG. 17.12 (a—b)

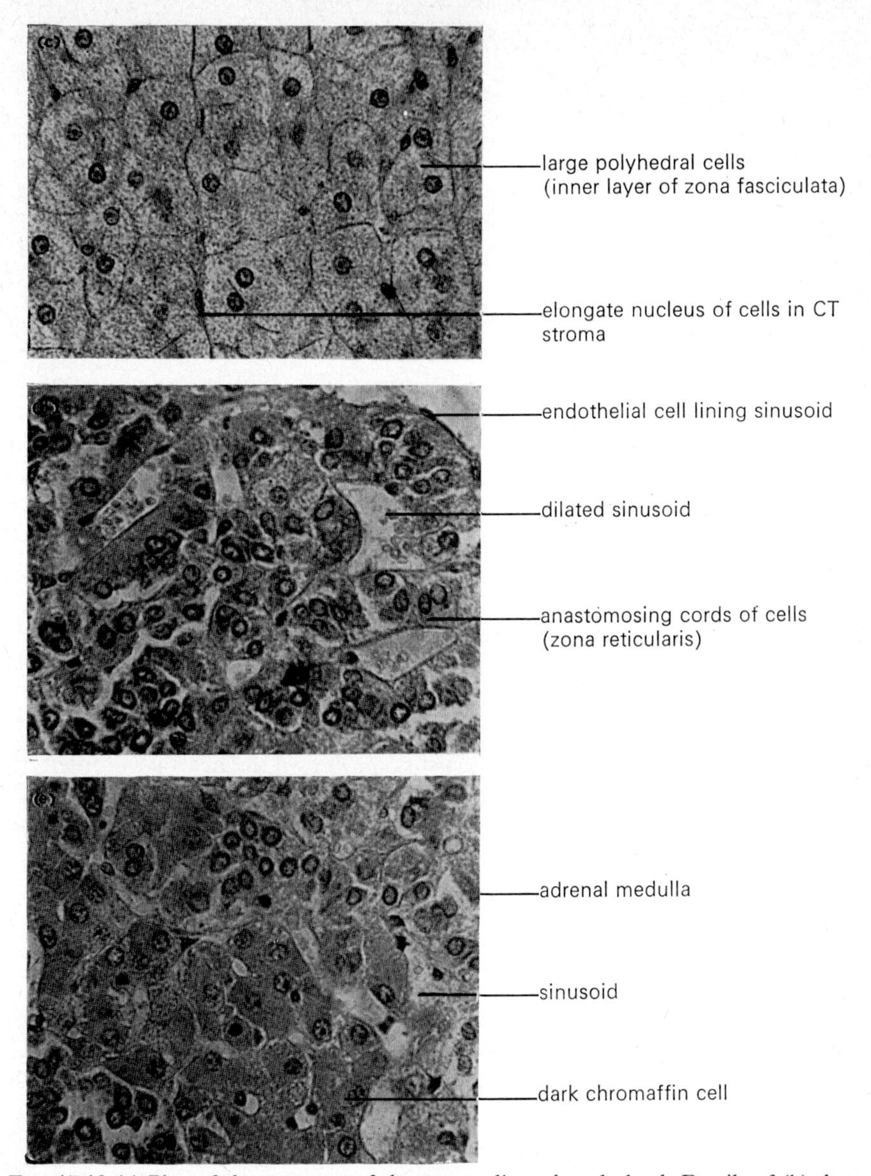

FIG. 17.12 (a) Plan of the structure of the mammalian adrenal gland. Details of (b) the outer zone of the adrenal cortex, (c) the middle zone of the adrenal cortex; (d) the inner zone of the adrenal cortex, and (e) the adrenal medulla.

sex hormones may be secreted by the inner zona reticularis. The adrenal medulla [Fig. 17.12(e)], being a modified sympathetic nerve ending, secretes catecholamines involved in metabolism and stress responses.

Respiration

All mammals breathe air using paired lungs, even aquatic mammals which have to surface to do so. The lungs are usually divided into lobes and have a large surface area divided up into numerous saccular alveoli (Fig. 17.13). The alveoli have very thin walls and are surrounded by capillary networks. The trachea and bronchi outside the lung boundary are strengthened by incomplete cartilaginous rings or plates. At the entrance of the trachea is the larynx, a cartilaginous box containing taut ridges of elastic tissue which form the vocal cords and are responsible for vocal sounds in mammals. The lungs are situated in pleural sacs in the thoracic cavity, and are ventilated by action of the muscular diaphragm and the intercostal muscles (between the ribs). Since the lungs form a blind-ended sac and air flow past the respiratory surface must, therefore, be tidal (rather than continuous as in birds), there is inevitably some dead-space in the lungs. The powerful suction pump formed by the enclosed thoracic cavity and its respiratory muscles compensates partially for this reduced efficiency of oxygen exchange.

Circulation

The circulatory system of mammals is double and the heart is completely divided, by a muscular septum, into two atria and two ventricles. The atria and ventricles are separated by a fibrous ring, and the only muscular connection between them is a bundle of modified muscle fibres (Purkinje fibres) which conduct impulses between the two sets of chambers. The sinus venosus is only represented in the adult by two small muscular nodes (SA and AV nodes) in the wall of the right atrium which initiate the heart beat and conduct it to the Purkinje fibres. In the mammal heart each atrium bears a small saccular chamber (auricle) of unknown function, and all the heart valves are fibrous (the right atrio-ventricular valve having three flaps while the left one retains two). The rest of the circulatory system is similar to that of birds except that in mammals only the left aortic arch remains (Fig. 17.14), and all traces of a renal-portal system have disappeared.

Mammal erythrocytes are produced in the bone marrow and destroyed in the spleen. In the embryo they are large and nucleated, while in the adult they are small (7–10 μ), enucleate biconcave discs [see Fig. 17.28 (b, c)], except in the camel where they are ovoid. Mammal blood also contains small non-nucleated platelets. The lymphoid thymus is very large in the embryo (see Fig. 17.1) but regresses in the adult. Its structure is similar to that of birds. Several lymphoid tonsils are found in the mouth and neck region.

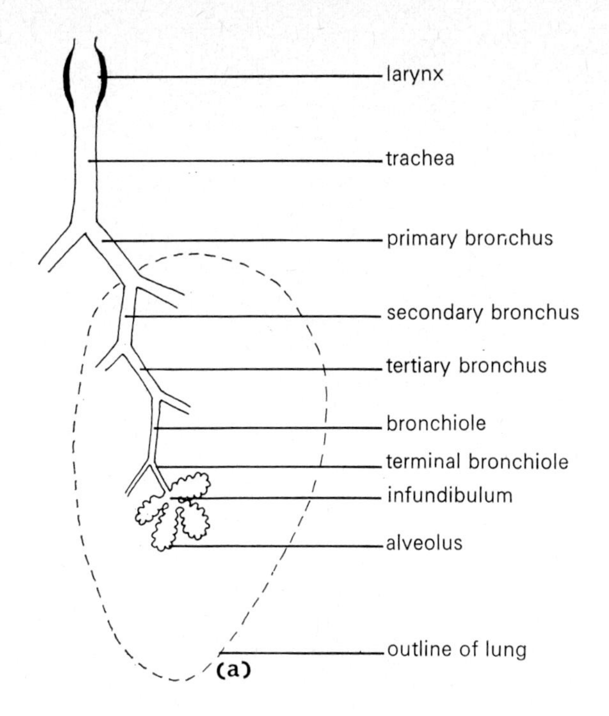

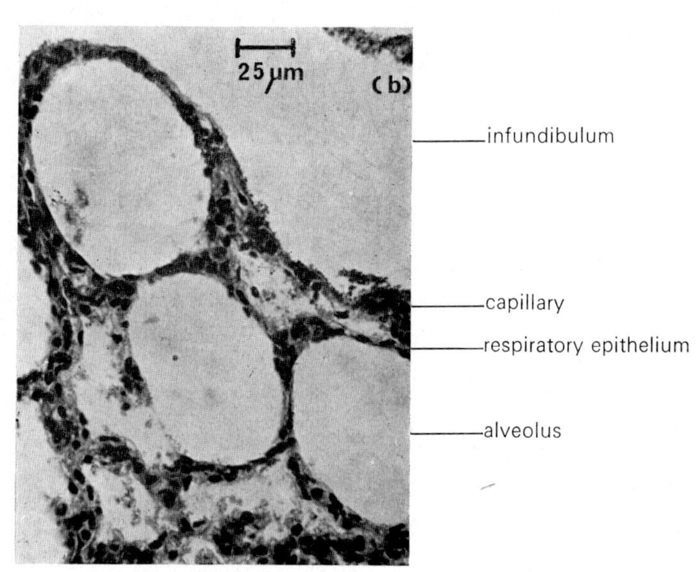

FIG. 17.13 (a) Plan of the mammalian respiratory system; (b) detail of lung structure of the mouse.

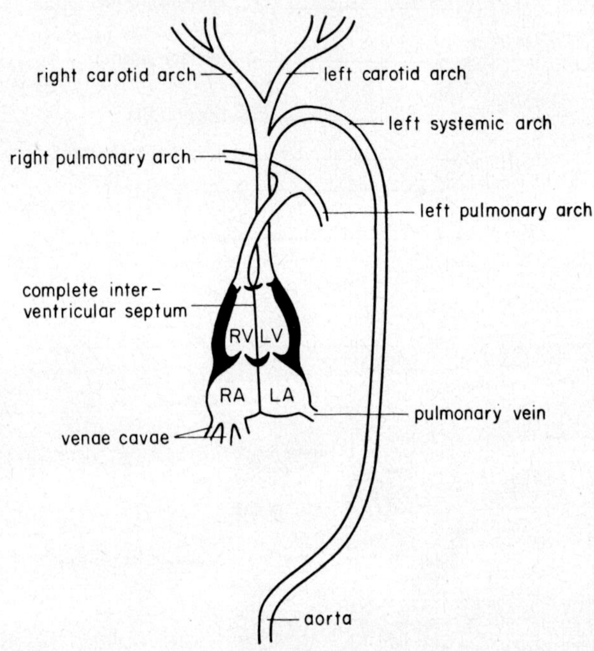

FIG. 17.14. Plan of the heart and aortic arches of a mammal. (LA = left atrium, RA = right atrium, LV = left ventricle, RV = right ventricle).

Digestion

Mice are omnivorous, and their digestive tracts are therefore representative of generalized mammalian guts. Those species with restricted diets show specializations of various regions of the digestive tract.

In mice the mouth is provided with lips, teeth, and mobile cheeks and tongue, all of which assist feeding. The tongue [Fig. 17.15(a)] is composed of complex muscle layers and is divided down the midline by a vertical lingual septum of fibrous elastic tissue containing fragments of cartilage. It is lined with stratified epithelium thrown up into papillae of various shapes which possess taste buds or serous glands, or are cornified for rasping [Fig. 17.15(b)]. The posterior part of the tongue contains lymphoid masses and mucous glands. At the base of the tongue and in other regions surrounding the buccal cavity are situated branched alveolar salivary glands (Fig. 17.16) which secrete mucus, water, calcium and enzymes. The teeth are embedded in holes in the jaw and are held in place by cement. Figure 17.17 shows a developing tooth from a newborn mouse. A layer of odontoblasts round the dermal papilla produce a hard material (dentine) which is harder even than compact bone and forms the bulk of the adult tooth. The dermal papilla becomes invaded by nerves and blood-vessels and forms the pulp. At a later stage the surrounding enamel epithelial cells (amyeloblasts) deposit an extremely hard calcareous enamel on top of the dentine.

A short pharynx lined with simple tubular mucous glands leads into an oesophagus which traverses the diaphragm before joining the stomach. The oesophagus (Fig. 17.18) is lined with a layer of stratified squamous epithelium, which is thick and cornified in herbivorous and gramnivorous mammals but fairly thin in carnivores. The oesophagus possesses two distinct external muscle layers, but these are not arranged in the regular circular and longitudinal fashion such as can be seen lower down the gut. Both muscle layers run obliquely, and in the top portion of the oesophagus they are striated (as shown in Fig. 17.18) and are responsible for swallowing food. The muscular U-shaped stomach is often modified according to diet, and in mice, which eat hard food, it has a horny lining at the cardiac (proximal) end. The fundic (middle) region is well-supplied with gastric glands lined with two different kinds of cells secreting hydrochloric acid and digestive enzymes (Fig. 17.19) and it is here that the preliminary stages of digestion occur. The stomach of some herbivorous mammals (ruminants) is enlarged and modified as a fermentation chamber.

All mammals show a well-developed "enzyme chain" throughout the gut for the sequential breakdown of food; the sites of production of these enzymes being the salivary glands, stomach, pancreas, duodenum and the proximal end of the small intestine. The distal end of the small intestine and the large intestine are concerned with absorption of breakdown products and water. The duodenum (Fig. 17.20) has a folded lining, and is

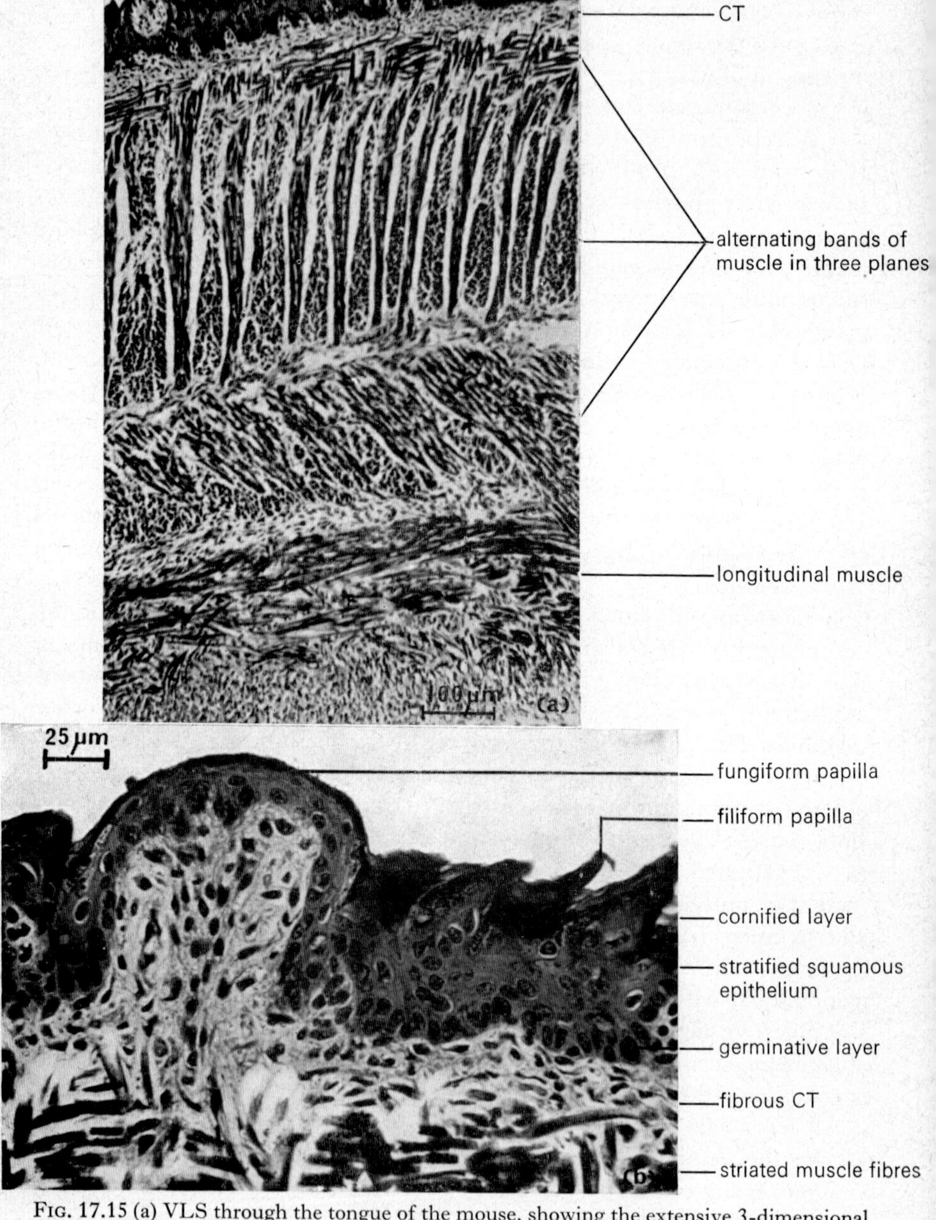

FIG. 17.15 (a) VLS through the tongue of the mouse, showing the extensive 3-dimensional muscle network; (b) detail of the epithelial lining of the mouse tongue.

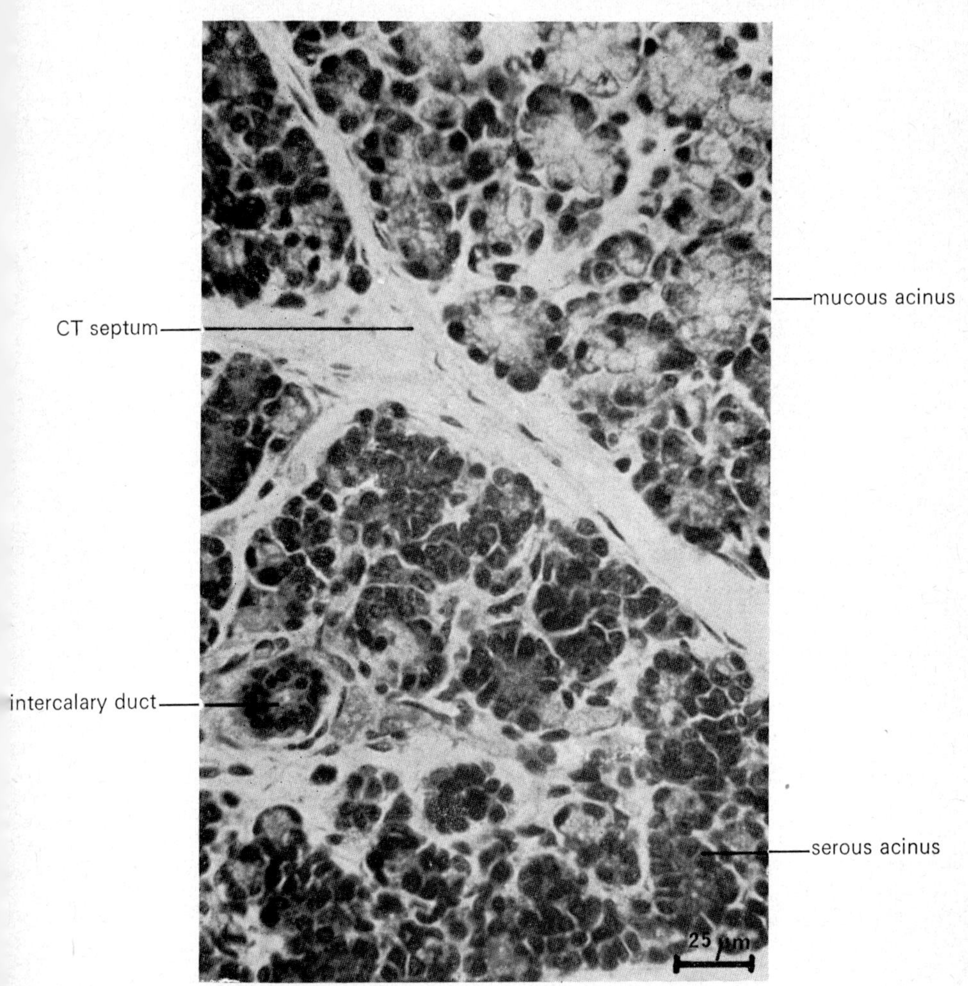

Fig. 17.16. Section through a mixed (mucous and serous) submaxillary salivary gland in the ventral neck region of the mouse.

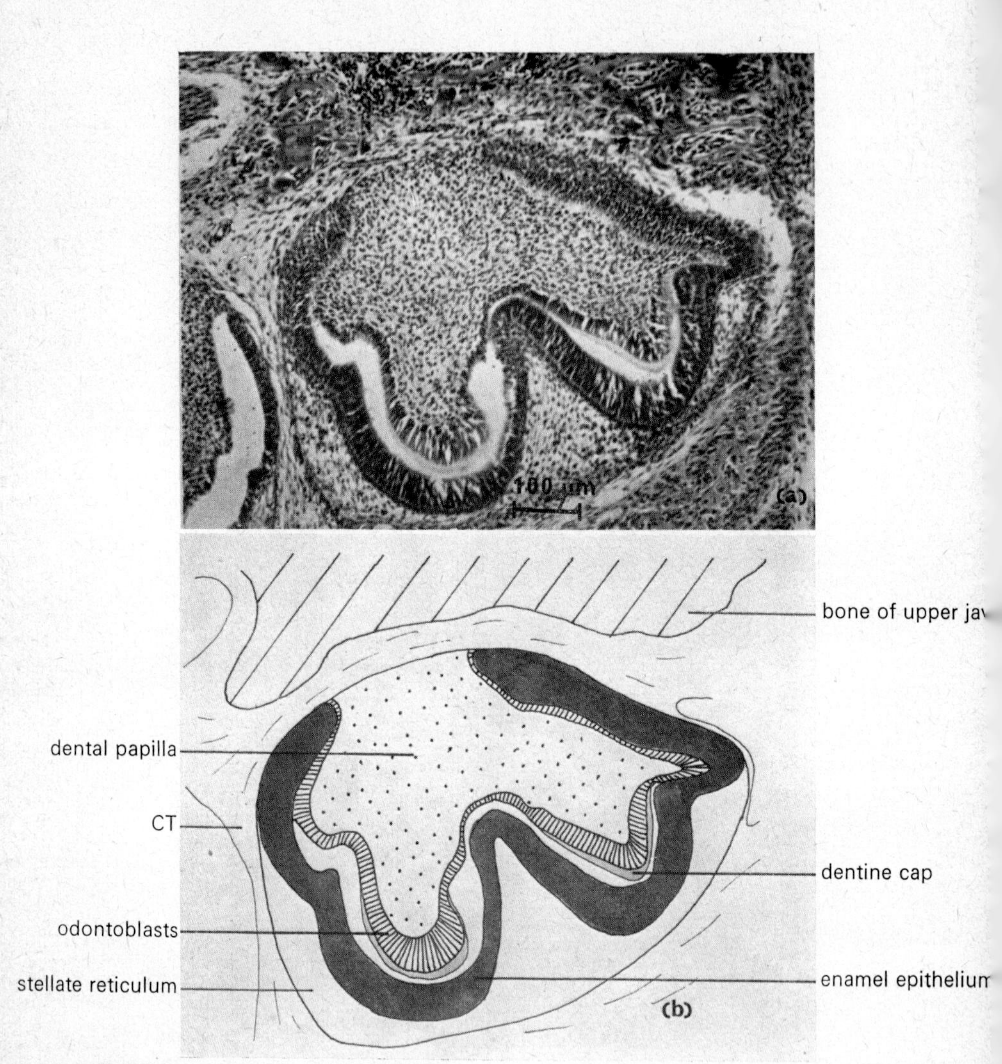

FIG. 17.17 (a) VLS and (b) drawing of part of the upper jaw of a new-born mouse, showing a developing tooth.

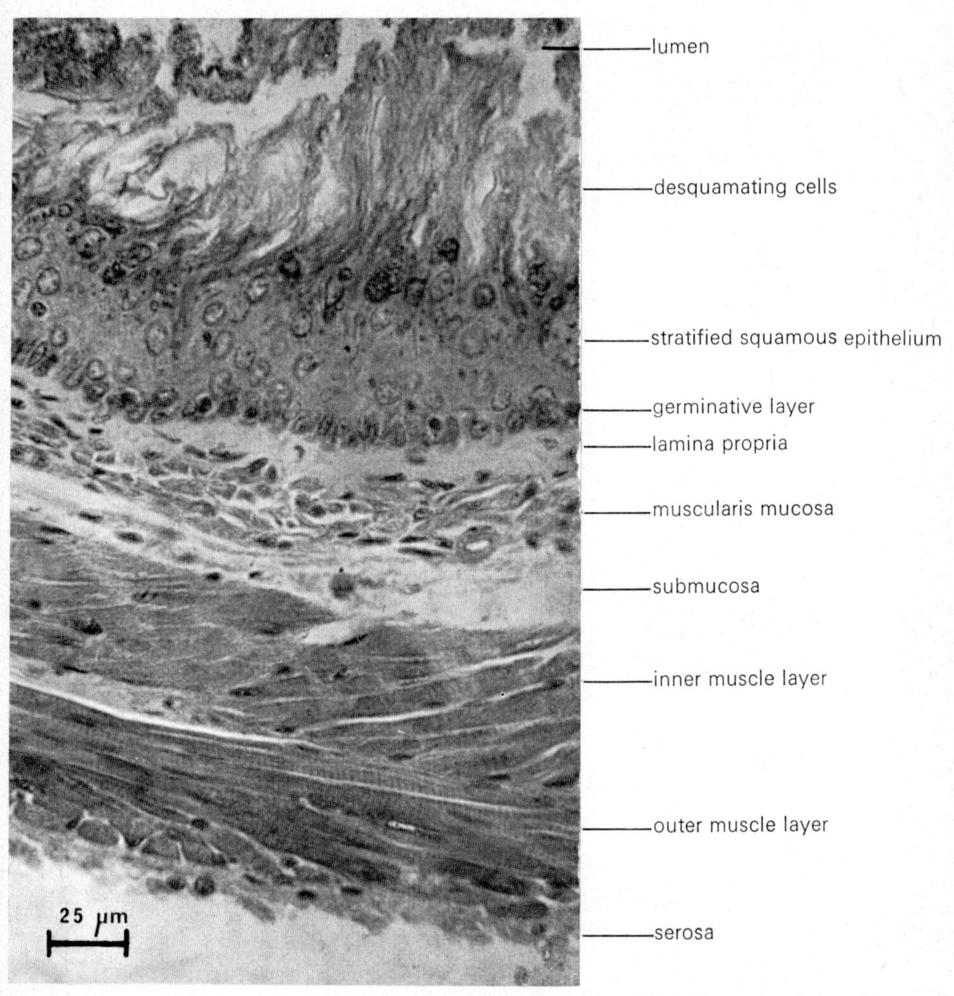

FIG. 17.18. Oblique section through the mouse oesophagus. Note the striated muscle fibres.

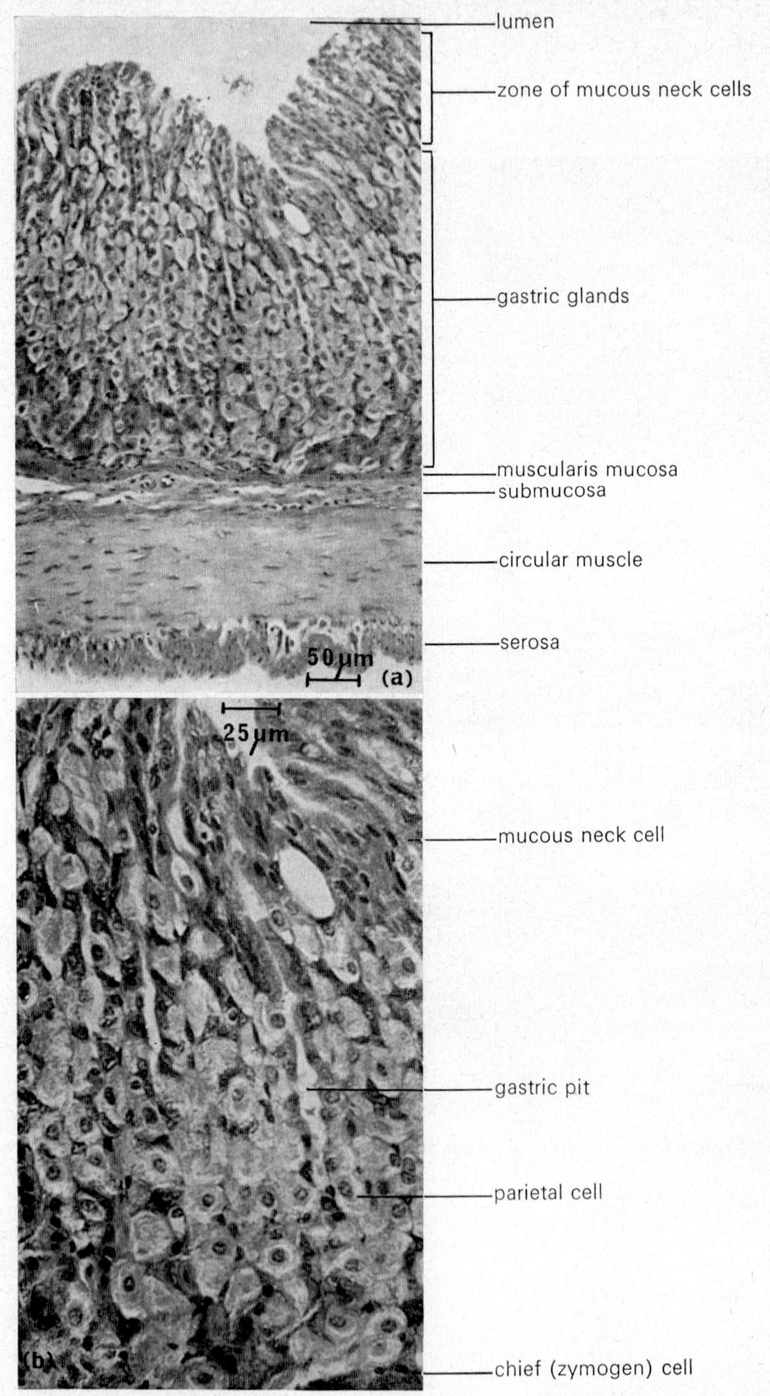

FIG. 17.19 (a) TS through the mouse stomach wall in the fundic region; (b) detail of the gastric glands in the fundic stomach.

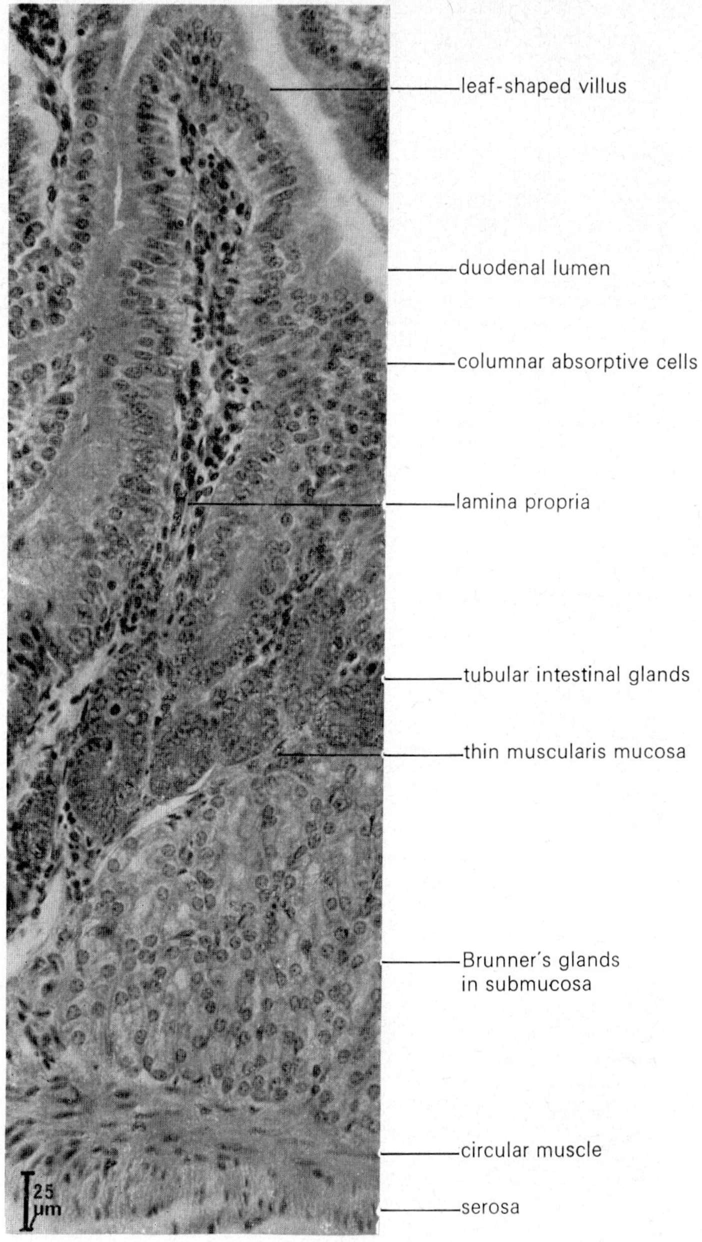

FIG. 17.20. TS through the wall of the mouse duodenum, showing the submucosal Brunner's glands.

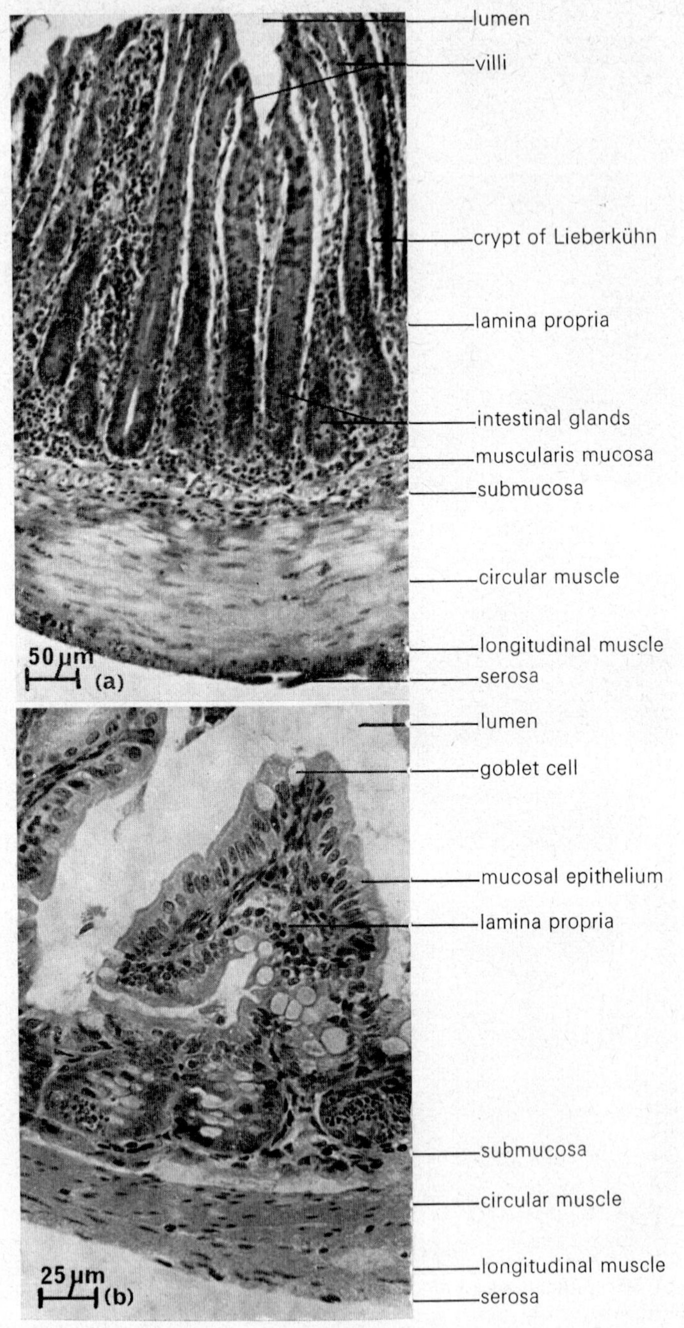

FIG. 17.21 (a) TS through the wall of the small intestine (jejunum) of the mouse; **(b)** TS through the large intestine (colon) wall of the mouse.

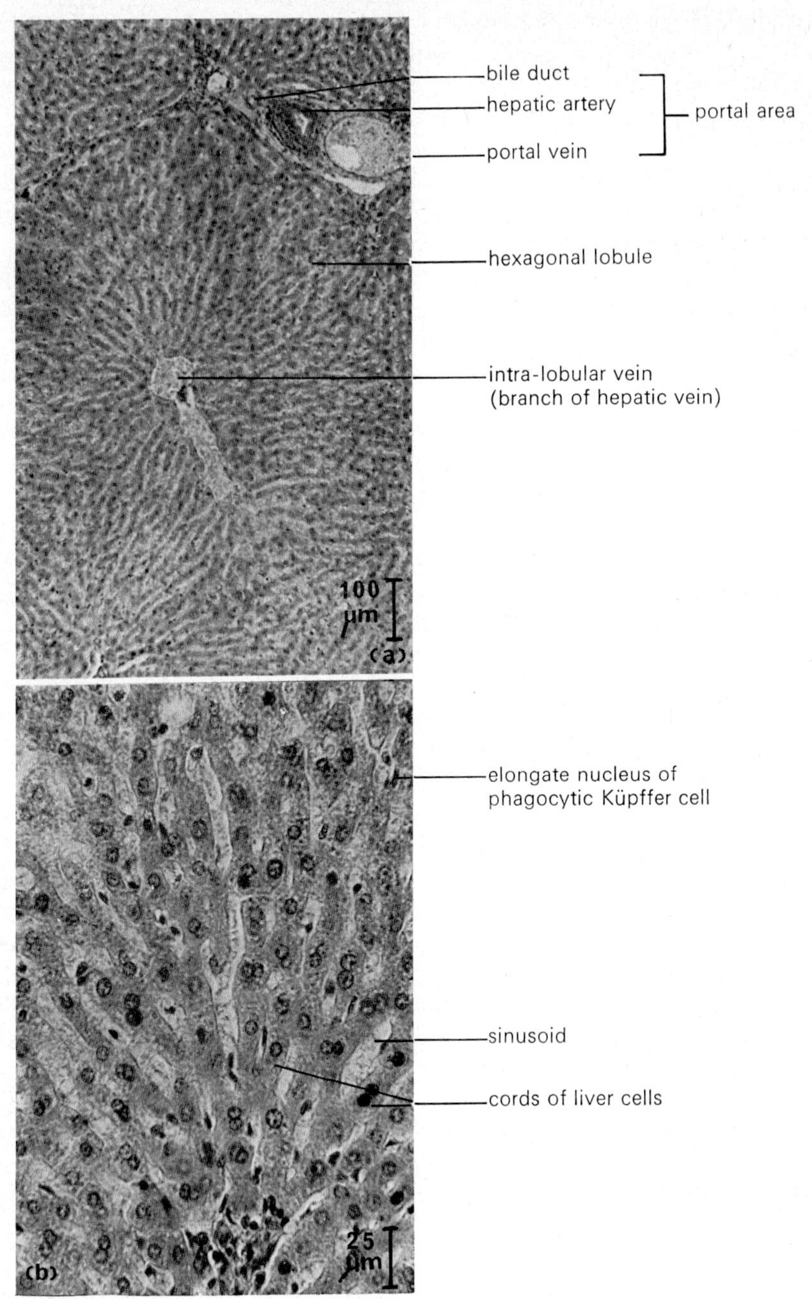

FIG. 17.22 (a) Section through the liver showing the hexagonal lobes characteristic of mammals; (b) detail of liver structure.

generally the only part of the gut to bear submucosal tubular (Brunner's) glands. The small intestine consists of the jejunum, bearing long finger-like villi and many deep crypts lined with simple tubular glands [Fig. 17.21(a)]: and an ileum with a less-folded mucosal lining and numerous lymph nodes (Peyer's patches). The large intestine is differentiated into a wide colon [Fig. 17.21(b)] provided with many mucous cells; a large anterior caecum, also bearing lymph nodes, which acts as a fermentation chamber; and a rectum which opens directly to the outside by an anus guarded by a muscular sphincter. There is no cloaca.

The typical mammalian liver (Fig. 17.22) is large, bilobed, and divided into hexagonal lobules surrounding a central hepatic vein. Portal areas between adjacent lobules contain other blood-vessels, lymphatics and branches of the bile ducts. The liver cells are arranged in characteristic laminae of one cell thickness, and are separated by blood sinusoids. Bile canaliculi form a laminar network of spaces between the liver cells and drain into the bile ducts. Bile is stored in a thick-walled gall bladder tucked beneath the right liver lobe. The duct from the gall bladder enters the duodenum, together with the pancreatic duct, through a raised ampulla of Vater.

The pancreas is an irregular-shaped organ consisting of masses of acinar cells secreting digestive enzymes with large groups of islet cells scattered among them. At least two metabolic hormones (insulin and glucagon) are secreted by the islet cells.

Excretion

The paired metanephric kidneys are the main excretory organs in mammals, although small amounts of waste material are lost across the skin and lungs. Mammalian kidneys are compact, bean-shaped organs clearly divided into a cortex and medulla [Fig. 17.23(a)]. The cortex [Fig.17.23(b)] contains numerous (about 20,000 in the mouse) Bowman's capsules and their associated glomeruli, proximal and distal portions of the nephric tubles, and the juxtaglomerular apparatus consisting of myoepithelial cells which are probably concerned with secretion of the hormone renin, a pressor substance that plays a part in regulating renal blood flow. The medulla is heavily vascularized and is packed with parallel tubules of the collecting ducts and the ascending and descending loops of Henle [Fig. 23(c)]. The loop of Henle is much better developed than in birds, and is particularly long in desert mammals (such as the gerbil shown in the photograph) which can reabsorb almost all the water filtered off from the blood by the kidneys. The collecting tubules drain into a renal pelvis and from thence into the ureter. Urine passes from the paired ureters into a distensible allantoic bladder, and to the outside through the urethra.

As in cyclostomes, but in no other vertebrates, mammalian kidneys are supplied only by the arterial system, all traces of a renal-portal system having disappeared.

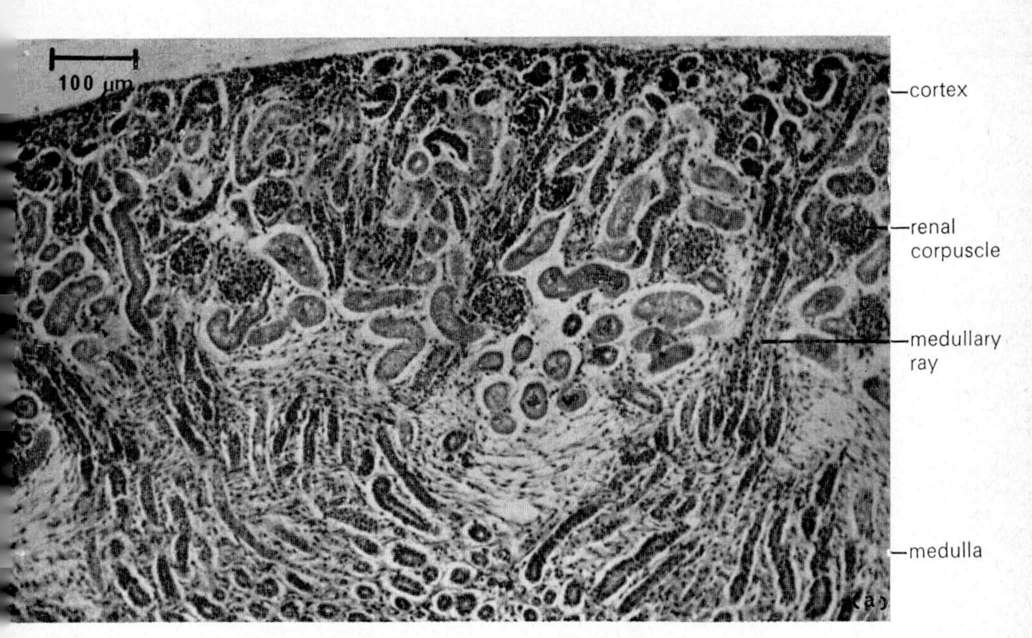

FIG. 17.23 (a)—see p. 709 for legend.

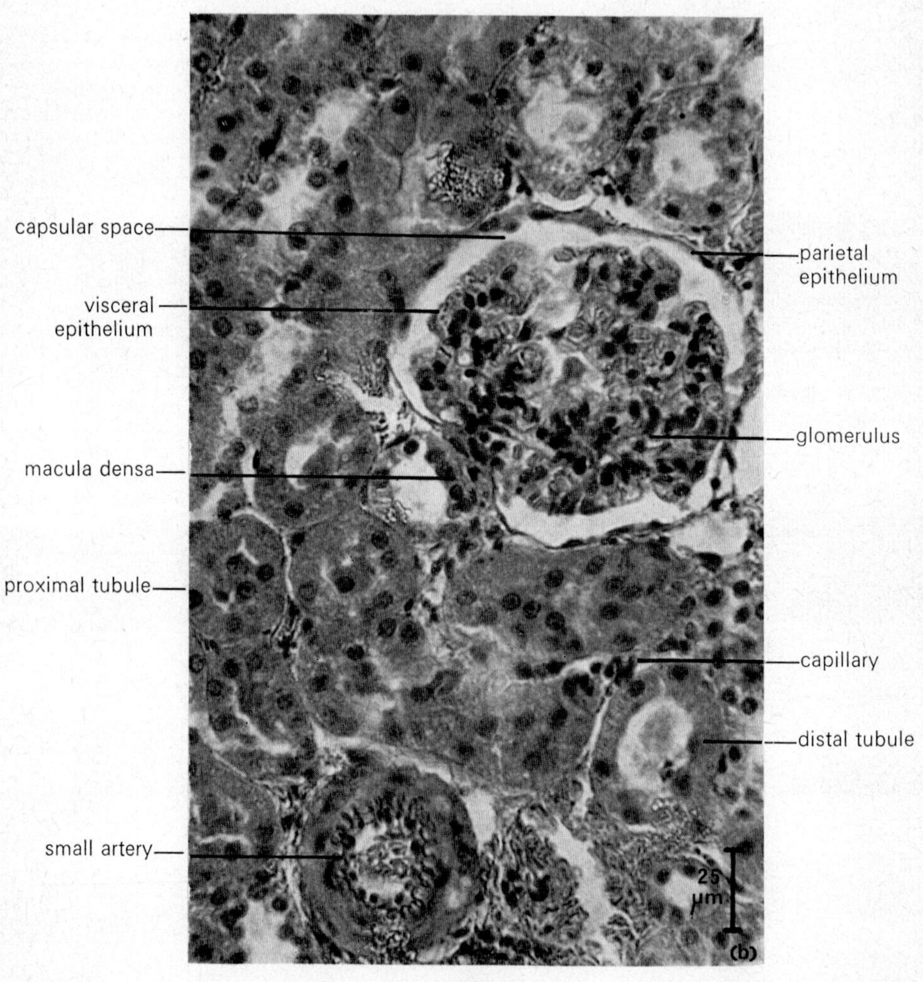
capsular space
visceral epithelium
macula densa
proximal tubule
small artery
parietal epithelium
glomerulus
capillary
distal tubule
25 μm
(b)

FIG. 17.23 (b)

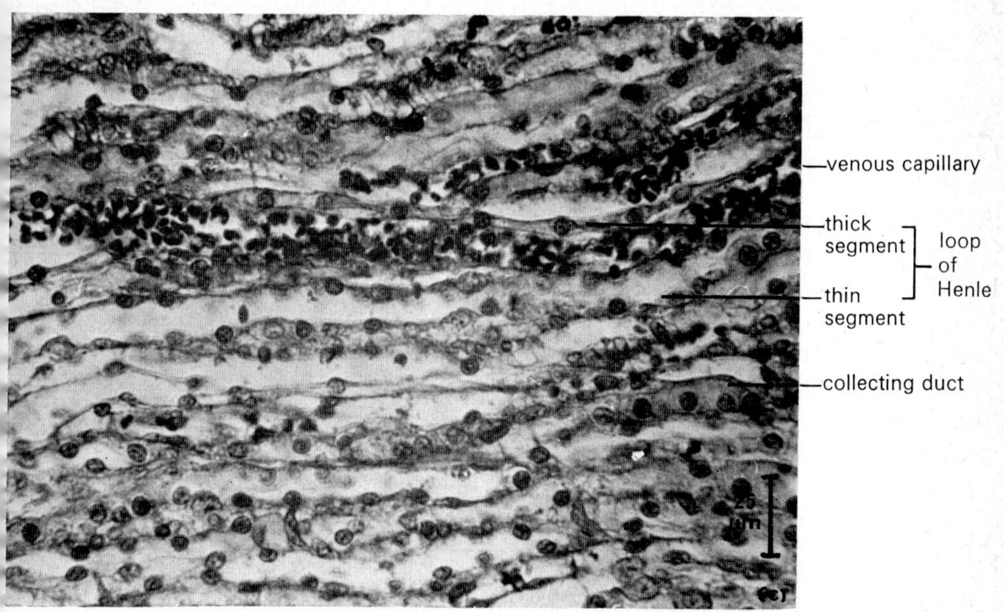

FIG. 17.23 (a) Section through the mouse kidney to show the cortex, medulla and medullary rays; (b) detail of the cortical region of the mouse kidney, showing a renal corpuscle, parts of the proximal and distal tubules and a small artery. Note the *macula densa*, which is a small group of tall narrow cells in the wall of the distal tubule. The *macula densa* lies in contact with the juxtaglomerular apparatus (not visible) and may also play a part in renal blood flow; (c) detail of the medulla from the kidney of a gerbil (a small desert rodent) showing the close association of the loops of Henle, venous capillaries and collecting ducts in this region.

Reproduction

Mammals are dioecious, and both sexes have paired gonads and genital ducts. The ducts are of mesonephric origin, but, whereas they remain separate along nearly their whole length in the male, in the female the ducts are often joined for a considerable part of their length. The gonads of both sexes secrete reproductive hormones.

In male rodents the testes are normally housed within the body cavity, whence they descend into scrotal sacs hanging outside the abdominal cavity during the breeding season. In most other male eutherians the testes descend into the scrotum at an early age, and remain there. Each round testis is subdivided by connective tissue septa into lobules, each composed of numerous looped seminiferous tubules. In section it can be seen that waves of spermatogenesis pass down the tubules. Thus all tubules are either immature (containing only spermatogonia) or maturing, in which case they show all stages of sperm development (Fig. 17.24). Ripe sperm are shed into the tubule lumina, and pass through the epididymes and paired vasa deferentia into a tubule unique to mammals, the urethra. The urethra is homologous with the urino-genital sinus of other vertebrates, and runs through a muscular penis. Opening into the urethra are mucous cells, tubular mucous glands of Littré and two sets of branched tubulo-alveolar glands (prostate and Cowper's glands) producing secretions which assist the processes of copulation and fertilization.

Mammalian ovaries consist of a large, highly vascularized medulla (stroma) surrounded by a thick cortex filled with developing ova, and covered by a layer of germinal epithelium. As the germ cells ripen they move inwards and embed deeply in the stroma, where they are nourished and grow (Fig. 17.25). Ripe ova are released by rupture of the follicles, are captured by ostia with frilled edges, and pass down the paired ciliated oviducts. In many mammals the oviducts join at their posterior ends to form a single uterus, but in the mouse the duplex condition is found where there are two separate uteri. It is characteristic of rodents that many embryos develop simultaneously *in utero*. There is a single thin-walled, distensible vagina lined with stratified epithelium (Fig. 17.26). Ruptured ovarian follicles become corpora lutea and release hormones involved in pregnancy.

The uteri have thick walls composed of an external muscular myometrium, and a vascularized glandular endometrium which is capable of forming a decidual chorio-allantoic placenta to nourish the fertilized egg and developing embryo. The mouse placenta (Fig. 17.27) is disc-shaped (*placenta discoidalis*). During the early stages of development, embryonic trophoblast tissue is still present (as in the photograph) and the placenta is of the haemochorial type. Trophoblast tissue can be seen in Fig. 17.28 (a–d) which show details of various regions of the placenta. Just before term in mice, the trophoblast tissue disappears leaving only the foetal

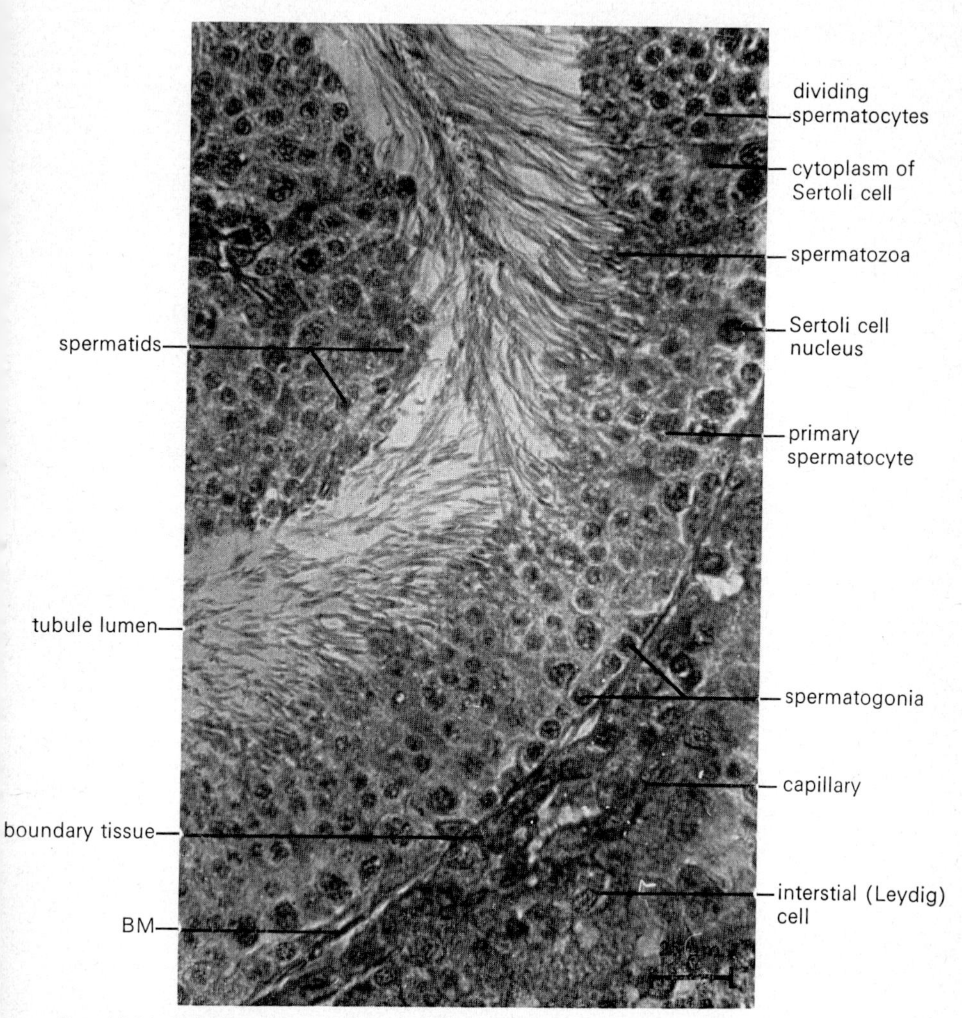

FIG. 17.24. LS through part of a seminiferous tubule from the testis of an adult male mouse. Note the asynchronous sperm development.

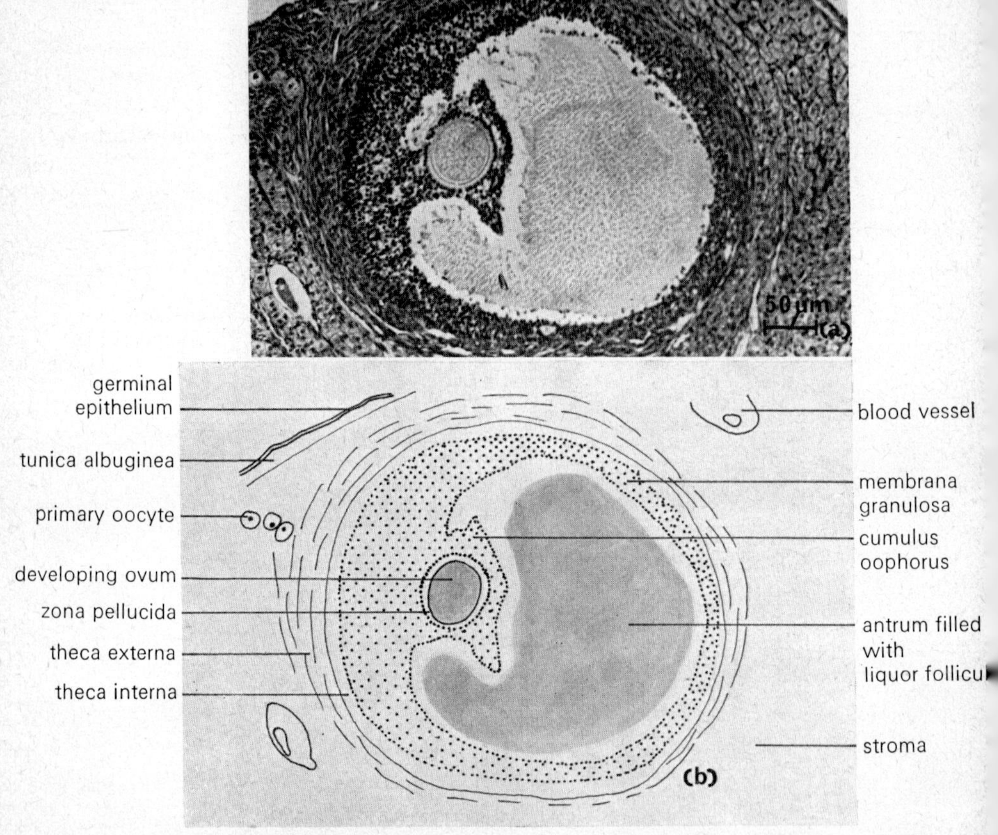

Fig. 17.25 (a) Section through part of a mammalian ovary, and (b) interpretive drawing.

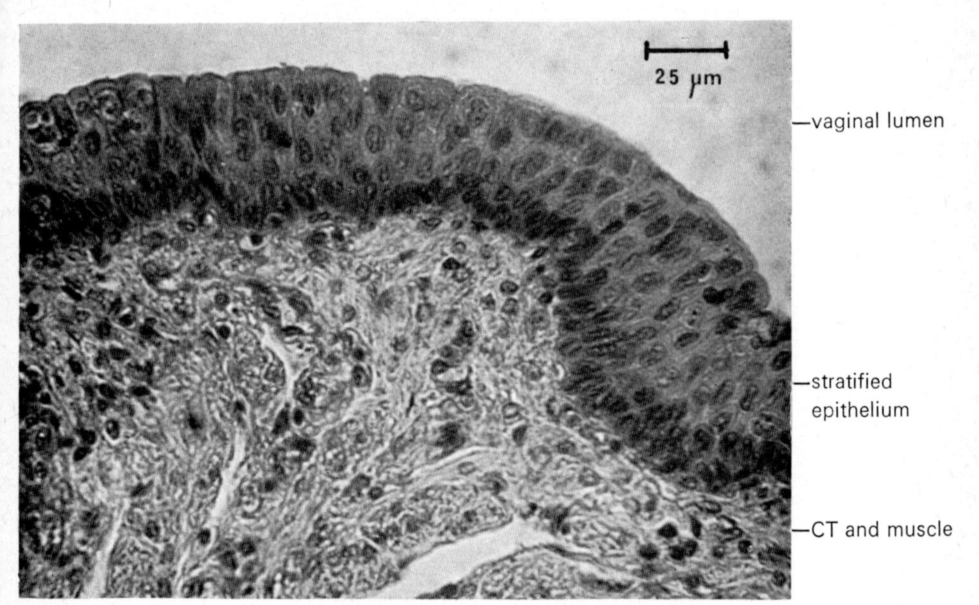

FIG. 17.26. Section through the vagina wall of an adult female mouse.

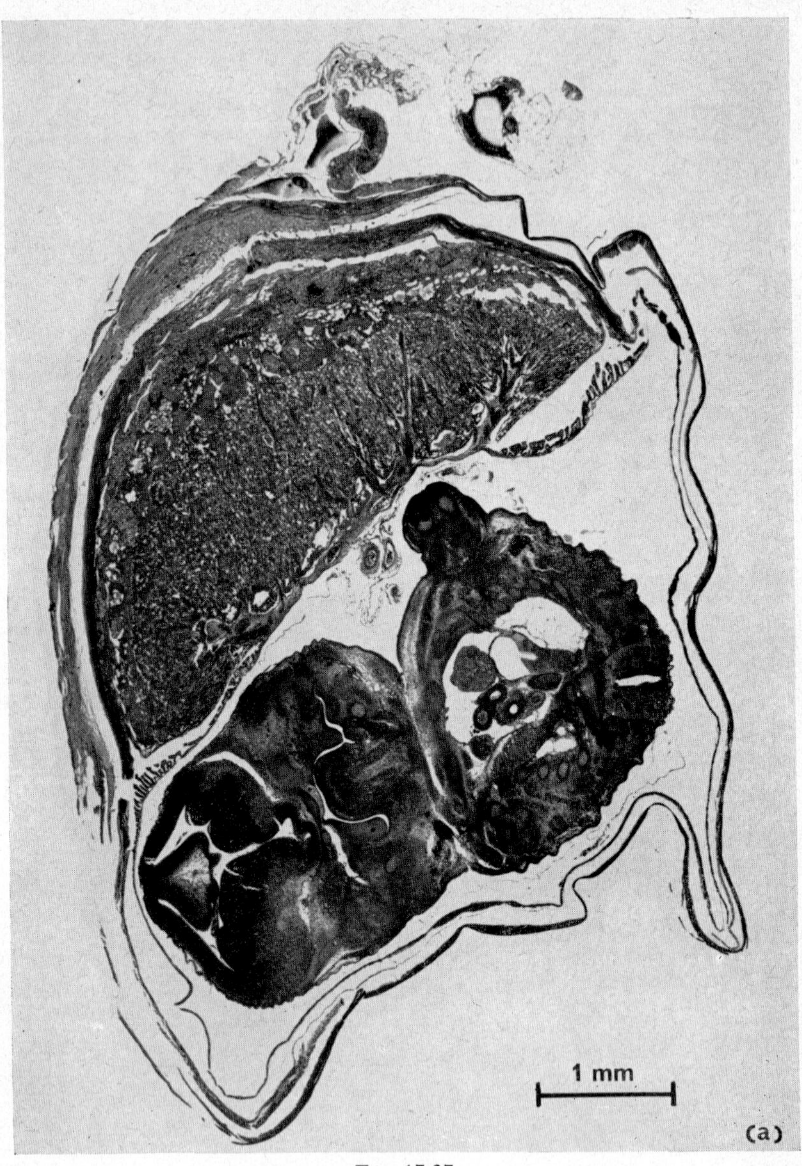

FIG. 17.27

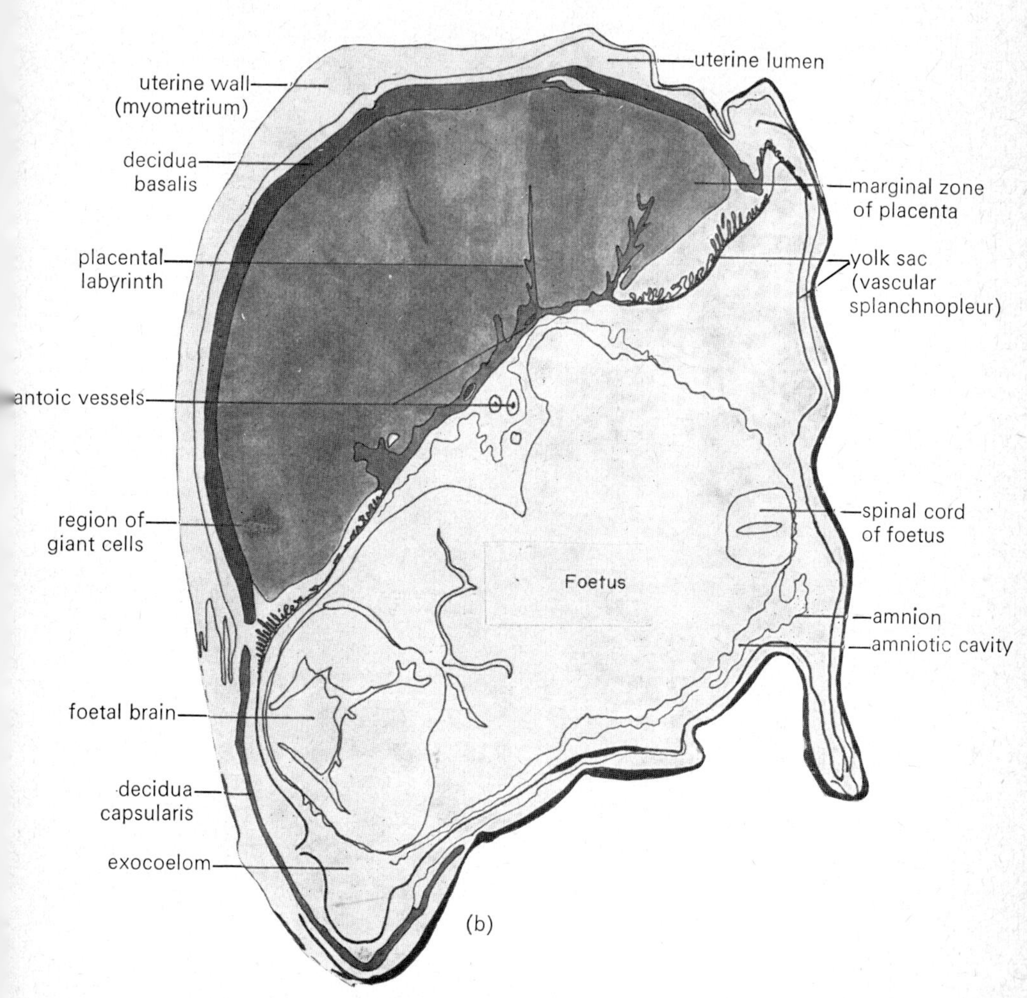

Fig. 17.27 (a) TS and (b) drawing of the uterus of a pregnant mouse, showing the discoidal placenta and a 15-day embryo *in utero*.

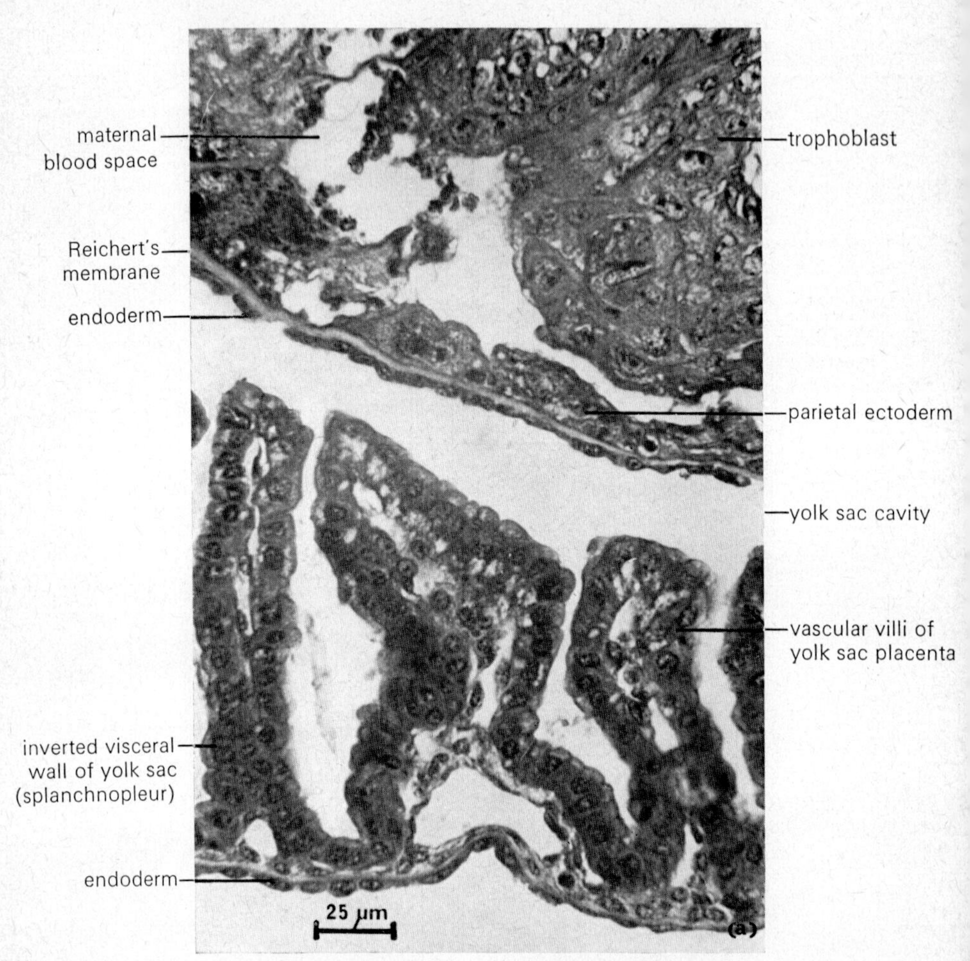

FIG. 17.28 (a)—see p. 718 for legend.

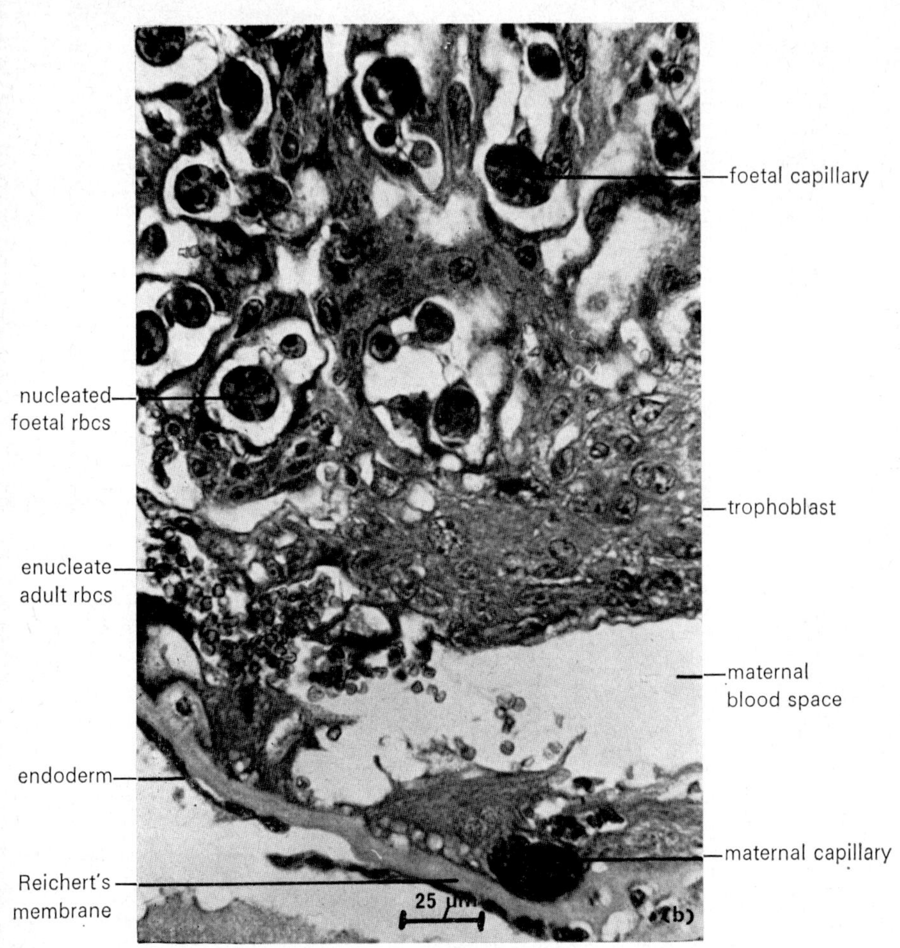

FIG. 17.28 (b)—see p. 718 for legend.

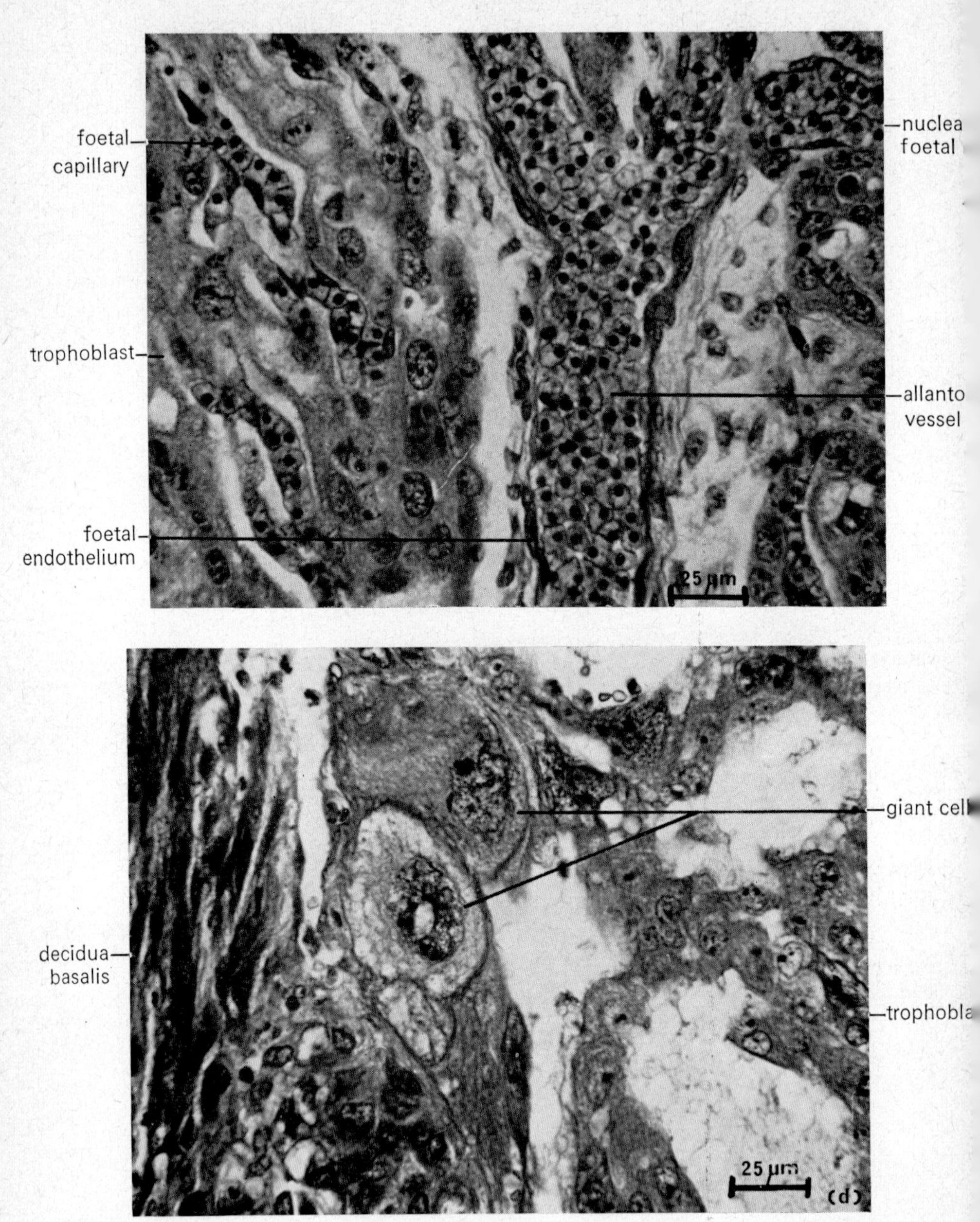

FIG. 17.28. Details of the structure of the pregnant mouse placenta seen in Fig. 17.27; (a) detail of the yolk-sac wall and part of the marginal zone of the placenta; (b) detail showing the close association of foetal and maternal blood vessels in the marginal zone; (c) allantoic capillaries in the labyrinthine region of the placenta; (d) giant cells in the marginal zone of the placenta. Note that the decidua basalis is part of the uterus wall while the trophoblast tissue is foetal in origin.

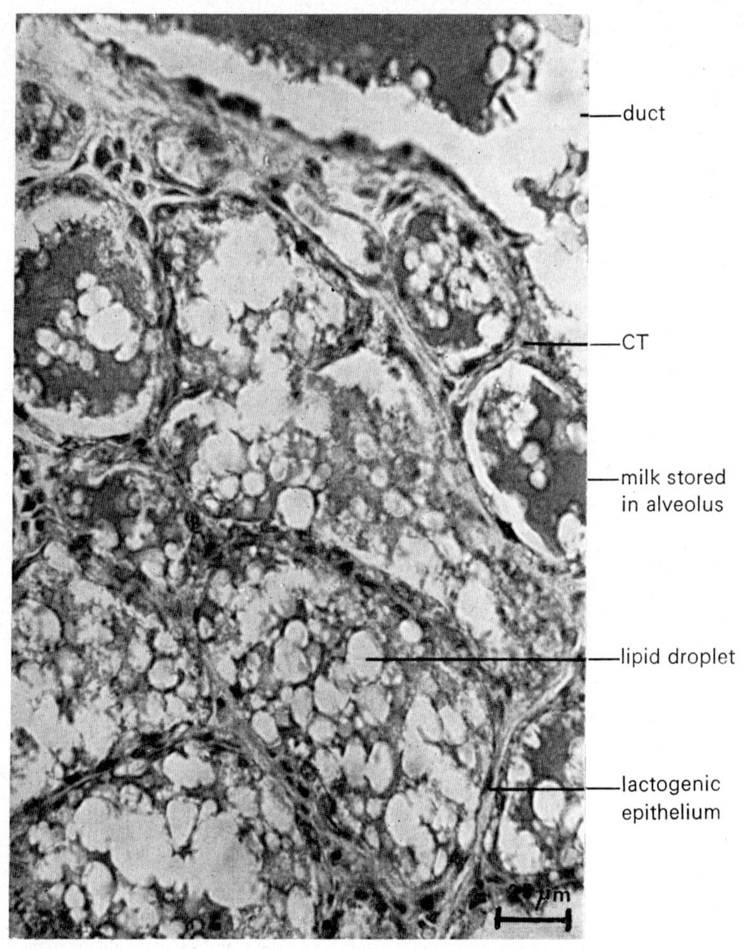

FIG. 17.29. Section through the mammary gland of a lactating mouse.

epithelium separating foetal blood from the maternal blood: at this stage the placenta is haemoendothelial. Rodent placentas are characterized by the presence of phagocytic giant cells in the marginal region [Fig. 17.28 (d)]. These are of foetal origin and destroy decidual cells. Part of the trophoblast tissue is a site for the production of gonadotrophic hormones which play an important part in maintaining the placenta.

Developing mammals are nourished to a relatively advanced stage by the placenta, which eventually breaks down under hormonal influence and is expelled with, or soon after, the foetus. Thereafter the young mammal is nourished by the mother's mammary glands (Fig. 17.29) which are modified sebaceous glands on the ventral surface. For the first few days after birth they secrete a clear fluid (colostrum) with a high protein content, and later they produce milk. It is a particular feature of mammals that they care for their young for a long period of time after birth.

Bibliography

General Books on Form and Function

Alexander, R. McN. (1975). *The Chordates*. Cambridge: Cambridge Univ. Press.

Barnes, R. D. (1968). *Invertebrate Zoology*, 2nd ed. Philadelphia: W. B. Saunders Co.

Barrington, E. J. W. (1967). *Invertebrate Form and Function*. London: Nelson.

Borrodaile, L. A., Potts, F. A., Eastham L. E. S. and Saunders, J. T. (Revised Kerkut, G. A.) (1967). *The Invertebrata*, 4th ed. London: Cambridge Univ. Press.

Bullock, T. H. and Horridge, G. A. (1965). *Structure and Function in the Nervous Systems of Invertebrates*, Vols. I and II. London: W. H. Freeman & Co.

Bullough, W. S. (1962). *Practical Invertebrate Anatomy*. London: McMillan & Co. Ltd.

Cox, F. E. G., Dales, R. Phillips, Green, J., Morton, J. E., Nichols, D. and Wakelin, D. (1969). *Practical Invertebrate Zoology*. London: Sidgwick & Jackson.

Florey, E. (1966). *An Introduction to General and Comparative Physiology*. Philadelphia: W. B. Saunders Co.

Freeman, W. H. and Bracegirdle, B. (1971). *An Atlas of Invertebrate Structure*. London: Heinemann.

Gardiner, M. S. (1972). *The Biology of Invertebrates*. New York: McGraw-Hill.

Goodrich, E. S. (1958). *Studies on the Structure and Development of Vertebrates*, Vols. I and II. London: Dover Publications Inc.

Gordon, M. S. (1972). *Animal Physiology: Principles and Adaptations*, 2nd ed. London: Collier-McMillan Ltd.

Grassé, P. P. (Ed.) (1948). *Traité de Zoologie*. Paris: Masson et Cie.

Hoar, W. S. (1966). *General and Comparative Physiology*. New York: Prentice-Hall Inc.

Jollie, M. (1962). *Chordate Morphology*. New York: Reinhold Publ. Co.

Kent, G. C. (Jr) (1969). *Comparative Anatomy of the Vertebrates*, 2nd ed. St. Louis: C. V. Mosby Co.

Laverack, M. S. and Dando, J. (1974). *Lecture Notes on Invertebrate Zoology*. Oxford: Blackwell Scientific Publications.

McCauley, W. J. (1971). *Vertebrate Physiology*. Philadelphia: W. B. Saunders Co.

Meglitsch, P. A. (1967). *Invertebrate Zoology*, 2nd ed. London: Oxford Univ. Press.

Prosser, C. L. and Brown, F. A. (Jr) (1965). *Comparative Animal Physiology*, 2nd ed. Philadelphia: W. B. Saunders Co.

Romer, A. S. (1970). *The Vertebrate Body*, 4th ed. Philadelphia: W. B. Saunders Co.

Russell-Hunter, W. D. (1968). *A Biology of Lower and Higher Invertebrates*, Vols. I and II. London: Collier-McMillan Ltd.

Saunders, J. T. and Manton, S. M. (1967). *A Manual of Practical Vertebrate Morphology*, 3rd ed. London: Oxford Univ. Press.

Wilson, J. A. (1972). *Principles of Animal Physiology*. London: Collier-McMillan Ltd.

Young, J. Z. (1966). *The Life of Vertebrates*, 2nd ed. London: Oxford Univ. Press.

Histology

Andrew, W. (1959). *Textbook of Comparative Histology*. London: Oxford Univ. Press.

Bloom, W. and Fawcett, D. W. (1968). *A Textbook of Histology*, 9th ed. Philadelphia: W. B. Saunders Co.

Freeman, W. H. and Bracegirdle, B. (1968). *An Atlas of Histology*, 2nd ed. London: Heinemann.

Leeson, T. S. and Leeson, C. R. (1970). *Histology*. Philadelphia: W. B. Saunders Co.

Le Gros Clark, W. (1971). *The Tissues of the Body*, 6th ed. Oxford: Clarendon Press.

Patt, D. I. and Patt, G. R. (1969). *Comparative Vertebrate Histology*. New York: Harper and Row.

Reith, E. J. and Ross, M. H. (1965). *Atlas of Descriptive Histology*. New York: Hoeber.

Chapter 1 (Platyhelminths)

Hyman, L. H. (1951). *The Invertebrates*, Vol. II—Platyhelminthes and Rhynchocoela. New York: McGraw-Hill.

Smyth, J. D. (1962). *Introduction to Animal Parasitology*. London: English Univ. Press. Ltd.

Smyth, J. D. (1969). *The Physiology of Cestodes*. London: Oliver & Boyd.

Smyth, J. D. (1966). *The Physiology of Trematodes*. London: Oliver & Boyd.

Chapters 2–3 (Acanthocephala and Nematodes)

Crofton, H. D. (1966). *Nematodes*. London: Hutchinson Univ. Library.

Hyman, L. H. (1951). *The Invertebrates*, Vol. III—Acanthocephala, Aschelminthes and Entoprocta. New York: McGraw-Hill.

Lee, D. L. (1965). *The Physiology of Nematodes*. London: Oliver and Boyd.

Nicholas, W. L. (1967). The Biology of the Acanthocephala: In—*Advances in Parasitology*, Vol. 5 (Ed. Dawes, B.) London & New York: Academic Press.

Smyth, J. D. (1962). *Introduction to Animal Parasitology*. London: English Univ. Press.

Chapter 4 (Annelids)

Dales, R. Phillips (1967). *Annelids*, 2nd ed. London: Hutchinson Univ. Library.

Laverack, M. S. (1963). *The Physiology of Earthworms*. Oxford: Pergamon Press.

Mann, K. H. (1961). *Leeches (Hirudinea): Their Structure, Physiology, Ecology and Embryology*. Oxford: Pergamon Press.

Chapters 5–6 (Onychophora and Arthropods)

Chapman, R. F. (1969). *The Insects: Structure and Function*. London: English Univ. Press Ltd.

Clarke, K. U. (1973). *The Biology of the Arthropoda*. London: Edward Arnold (Publishers) Ltd.

Cornwell, P. B. (1968). *The Cockroach*, Vol. I (The Rentokil Library). London: Hutchinson.

Guthrie, D. M. and Tindall, A. R. (1968). *The Biology of the Cockroach*. London: Edward Arnold (Publishers) Ltd.

Rockstein, M. (Ed.) (1964). *The Physiology of the Insecta*, Vols I–III. London & New York: Academic Press.

Savory, T. (1964). *Arachnida*. London & New York: Academic Press.

Waterman, T. H. (Ed.) (1960–1961). *The Physiology of Crustacea*, Vols 1–2. London & New York: Academic Press.

Wigglesworth, V. B. (1972). *The Principles of Insect Physiology*, 7th ed. London: Chapman & Hall.

Chapter 7 (Molluscs)

Hyman, L. H. (1967). *The Invertebrates*, Vol. VI—Mollusca 1. New York: McGraw-Hill.

Morton, J. E. (1967). *Molluscs*, 4th ed. London: Hutchinson Univ. Library.

Purchon, R. D. (1968). *The Biology of the Mollusca*. Oxford: Pergamon Press.

Thompsett, D. H. (1939). L.M.B.C. Memoirs No. XXXII – *Sepia*. Liverpool: University Press of Liverpool.

Wilbur, K. M. and Yonge, C. M. (Eds.) (1964–1966). *Physiology of Mollusca*, Vols. I & II. London & New York: Academic Press.

Chapter 8 (Echinoderms)

Binyon, J. (1972). *Physiology of Echinoderms*. Oxford: Pergamon Press.

Boolootian, R. A. (Ed.) (1966). *Physiology of Echinodermata*. New York: John Wiley & Sons.

Chadwick, H. C. (1939). L.M.B.C. Memoirs No. XXV – *Asterias*, No. III – *Echinus*. Liverpool: University Press of Liverpool.

Hyman, L. H. (1955). *The Invertebrates*, Vol. IV – Echinodermata. New York: McGraw-Hill.

Nichols, D. (1966). *Echinoderms*, 2nd ed. London: Hutchinson Univ. Library.

Chapters 9–10 (Hemichordates and Protochordates)

Barrington, E. J. W. (1965). *The Biology of the Hemichordata and Protochordata.* London: Oliver and Boyd.

Colman, J. S. (1953). L.M.B.C. Memoirs No. XXXV – *Ciona.* Liverpool: University Press of Liverpool.

Hyman, L. H. (1959). *The Invertebrates*, Vol. V – Smaller Coelomate Groups. New York: McGraw-Hill.

Chapters 11–13 (Fishes)

Hardisty, M. W. and Potter, I. C. (Eds.) (1971–1972). *The Biology of Lampreys*, Vols I and II. London & New York: Academic Press.

Hoar, W. S. and Randall, D. J. (Eds.) (1969–1971). *Fish Physiology*, Vols I–V. London & New York: Academic Press.

Chapters 14–15 (Amphibia and Reptiles)

Bellairs, A. d'A. (1968). *Reptiles.* London: Hutchinson Univ. Library.

Bellairs, A. d'A. (1969). *The Life of Reptiles*, Vols. I and II. London: Weidenfeld and Nicolson.

Gans, C. and Parsons, T. S. (Eds.) (1970). *Biology of the Reptilia*, Vols. II and III – Morphology B and C. London & New York: Academic Press.

Moore, J. A. (Ed.) (1970). *Physiology of the Amphibia.* London & New York: Academic Press.

Smith, M. (1969). *The British Amphibians and Reptiles*, 4th ed. London: Collins (New Naturalist).

Chapter 16 (Birds)

Marshall, A. J. (Ed.) (1966). *Biology and Comparative Physiology of Birds*, Vols. I and II. London & New York: Academic Press.

Yapp, W. B. (1970). *The Life and Organization of Birds.* London: Edward Arnold (Publishers) Ltd.

Chapter 17 (Mammals)

Cook, M. J. (1965). *The Anatomy of the Laboratory Mouse.* London & New York: Academic Press.

Harrison Matthews, L. (1968). *British Mammals.* London: Collins (New Naturalist).

Southern, H. N. (1965). *The Handbook of British Mammals.* Oxford: Blackwell Scientific Publications.

Young, J. Z. (1970). *The Life of Mammals.* Oxford: Clarendon Press.

Index of Animals

Subject Index

Italic numbers indicate pages on which figures appear

R